W9-BMW-352

STATISTICAL CONCEPTS

With Applications to Business and Economics

STATISTICAL CONCEPTS

With Applications to Business and Economics

Richard W. Madsen
Melvin L. Moeschberger

University of Missouri–Columbia

PRENTICE-HALL, INC., ENGLEWOOD CLIFFS, NEW JERSEY 07632

LIBRARY OF CONGRESS CATALOGING IN PUBLICATION DATA

Madsen, Richard W (date)
Statistical concepts.

Includes index.
1. Statistics. 2. Social sciences—Statistical methods. I. Moeschberger, M. L., joint author. II. Title.
HA29.M233 519.5 79-22947
ISBN 0-13-844878-7

Printed in the United States of America

10 9 8 7 6 5 4 3

Cover design by Wanda Lubelska
Manufacturing buyer: R. Anthony Caruso

Prentice-Hall International, Inc., London
Prentice-Hall of Australia Pty. Limited, Sydney
Prentice-Hall of Canada, Ltd., Toronto
Prentice-Hall of India Private Limited, New Delhi
Prentice-Hall of Japan, Inc., Tokyo
Prentice-Hall of Southeast Asia Pte. Ltd., Singapore
Whitehall Books Limited, Wellington, New Zealand

To **Carole, Eric, and Peter**
Sandra, Shelley, Suzy, and Scott

CONTENTS

PREFACE

Probability and statistics are a part of daily life for most people. In order for probability and statistics to be a useful and helpful part of professional life as well as daily life, it is necessary to devote time and effort to their study. Many individuals, including those in various business areas, feel that quantitative areas deserve more emphasis in the curriculum of undergraduates. Consequently, more courses in mathematical areas are now being required of business undergraduates. This textbook was written with these ideas in mind. Specifically, it was written as a text that could be (and has been) used for a two-semester course in probability and statistics for business undergraduates having had a mathematical background stronger than high school algebra. We consider the mathematical level to be "intermediate," meaning that a background of college algebra is required. Students having had an introductory course in calculus (so that they can differentiate and integrate simple polynomials) should be able to follow all the material in the book. Students without such background can skip over the portions using calculus. These portions have been placed at the end of the sections where used and clearly set off by a series of bullets (•). Exercises requiring calculus have also been marked by bullets. The book can also be used as a text for a one-semester course by covering only Chapters 1 through 8 or 9 and by deleting material not strictly necessary for only these chapters. An outline is given at the end of this prcfacc.

Although the primary audience for this textbook is undergraduates in the areas of business and economics, we have included examples and applications to several other disciplines. Some examples and exercises are based on real data. These examples and exercises are denoted by a star (*). Other examples and exercises, although involving artificial data, are based on studies that could have, realistically, been carried out. We have also emphasized uses and abuses of statistics in daily life. By giving attention to these various facets of statistics it is hoped that the material will be fresh, meaningful, and helpful to a future worker who needs a quantitative background on his or her job.

We would like to address our next remarks directly to students using this book for a text. First, to the 12% of college students who enjoy mathematics: We believe that you will find this material no more difficult than other math courses you have taken. We hope that you may find it more enjoyable than more theoretical courses in view of the applicability of statistics to real-world problems. Next, to the 83% of college students who find mathematics to be a chore: We have written this book in a somewhat informal style and have included some attempts at humor in an effort to make reading this book more enjoyable for you. Realistically, however, the burden of the work must fall on you. We believe that you will learn this material more easily and get a better grade (which in itself will make you feel more positive toward the area of statistics in particular and mathematics in general) if you do the following: (1) Casually read through the material *before* your instructor lectures on it. (2) Carefully read through the material *after* your instructor has lectured on it, this time with pencil, paper, and a calculator at hand. When reading the examples, work through the calculations yourself and check to see that you agree with the answers given. Do this before attempting to work any exercises. (3) Work as many exercises as you can. Answers to the odd-numbered exercises are given in Appendix A4. (You may have noticed that we have only addressed 95% of the college students. The other 5% won't read the preface anyway, so there is nothing we can say to them. These percentages are based on the author's experience.)

We gratefully acknowledge the fact that many other people besides ourselves have contributed to the existence of this book. Most of the typing of the original manuscript was ably done by Glenda Perry and Carole Madsen. We had useful discussions with several of our colleagues at the University of Missouri–Columbia and received some very helpful suggestions from the reviewers. The reviewers were Professor Austin Bonis (Rochester Institute of Technology), Professor Richard Haase (Drexel University), Professor Art Kraft (University of Nebraska–Lincoln), Professor Terry Seaks (University of North Carolina, Greensboro), Professor Paul Shaman (Wharton School, University of Pennsylvania), and Dr. David Pack (Oak Ridge National Laboratory). Our thanks go to Joan Resler, Ron Ledwith, Barbara Zeiders, and the other people associated with Prentice-Hall for the part they had in the production of this book.

OUTLINE FOR ONE-SEMESTER COURSE

The following outline simply suggests sections to be covered in a one-semester course. Sections could be added or deleted at the instructor's discretion. A more detailed outline is given in the Teacher's Manual. It is assumed that all portions of sections involving calculus will be omitted.

Chapter 1. Sections 1.1–1.6, 1.8–1.10
Chapter 2. Sections 2.1–2.8
Chapter 3. Sections 3.1–3.5
Chapter 4. Sections 4.1, 4.2, 4.8, 4.10
Chapter 5. Sections 5.1–5.4, 5.6–5.8
Chapter 6. Sections 6.1–6.6
Chapter 7. Sections 7.1–7.7
Chapter 8. Sections 8.1–8.6
Chapter 9. Sections 9.1, 9.2, 9.4

R. W. MADSEN
M. L. MOESCHBERGER

STATISTICAL CONCEPTS

With Applications to Business and Economics

1

PROBABILITY

1.1 INTRODUCTION

A saying familiar to us all is that there is nothing certain in this life except death and taxes. This, of course, is a cynical way of conveying the idea that life is full of uncertainty and variability. It also illustrates the way people make inferences about the future by considering their past experience. "Uncertainty and variability" and "inference" are two concepts with which we will be concerned throughout this text. *Probability* is a mathematical means of studying uncertainty and variability and *statistics* is primarily concerned with making inferences based on partial information and past experience.

Those readers familiar with the Peanuts comic strip are aware of Lucy's habit of pulling away the football when Charlie Brown is about to kick it (see the next page for a graphic example of this). This has happened so often that when you see the first panel you make an inference about what the final outcome will be; that is, you make an inference about the future based on past experience. (We could speculate about what would happen if Lucy actually let him kick the football, but we will leave such speculations to you, the reader.)

As an illustration of uncertainty, variability, and inference we might consider a typical intercollegiate football game between, say, the University of Missouri and Kansas University to be played at Columbia, Missouri. On the day before the game, could you give precise answers to any of the following questions? What will the paid

attendance at the game be? Will the toss of the coin at the beginning of the game be heads or tails? How many field goals will be attempted? How many hot dogs will be sold? What will the final score be? Which team will win? At what time will the game be over? How many car accidents will there be within the city limits in the hour after the game? Of course, you cannot give precise answers to these questions because of the uncertainty involved. Furthermore, although the answers to some of these questions are of interest only to a sports fan, the answers to other questions are important to various officials: the stadium official who must assign ticket takers to various gates, the concessionaire who must order the hot dogs, the police chief who assigns traffic control officers, and so on. How do these officials make their decisions? They must make inferences based on partial information and past experience. You will see as you read this text how probability and statistics can be helpful in answering questions such as these. You will also see from the examples used throughout the text that probability and statistics have application to fields such as agronomy, forestry, genetics, medical technology, political science, quality control, social science, and zoology, as well as business and economics. We are tempted to say "everything from A to Z," but honesty compels us to admit that we have not included examples from fields beginning with the letters J, T, and X! While we are on the subject of honesty we should say that there are some people who try to use statistics dishonestly. This is no surprise to you, of course. You have often heard such phrases as "figures don't lie but liars figure," and such phrases have arisen because of the way people have misused statistics. We will point out some ways in which people misuse and abuse statistics, not so that you will do the same, but so that you can recognize such abuses and not be misled by them.

1.2 SAMPLE SPACES

What do you think of when someone mentions the word "experiment"? For some people the word conjures up visions of a mad scientist and his helper Igor working in a dimly lit lab surrounded by bubbly test tubes. For others it brings memories of a high school or college physics or chemistry lab. In any case we might think of an *experiment* as being a well-defined act or procedure leading to some outcome. Frequently, experiments, under very well controlled conditions, have the property of *reproducibility*; that is, the very same sequence of steps done at different points in time lead to the exact same outcome.

However, sometimes experimental outcomes differ because of changes in hard-to-control conditions (e.g., gas experiments are sensitive to changes in pressure and temperature); for other experiments the outcomes differ because of an inherent randomness in the experiment (e.g., counting the number of cosmic rays that strike a satellite in a 1-minute time interval). We will be primarily interested in experiments whose outcomes differ from one experimental trial to the next due to an inherent randomness. We will refer to such experiments as *random experiments*.

Since we cannot know beforehand exactly what the outcome of a random experiment will be, it is important to know what the possible outcomes are. This leads us to the following definition.

Definition 1.1 The *sample space*, S, of a random experiment is the set of all possible outcomes of that experiment.

Example 1.1 A referee flips a coin at the start of a football game. The two possible outcomes are "heads" (H) or "tails" (T), so $S = \{\text{H, T}\}$.

Example 1.2 Old Mother Hubbard has six children: Alan, Betty, Carl, Diane, Eric, and Frieda, ages 1 though 6, respectively. She wants to choose one of the children to feed the dog and decides to do it by rolling a die and choosing the child whose age appears on the die. In this case

$$S = \{\text{Alan, Betty, Carl, Diane, Eric, Frieda}\}.$$

Example 1.3 To entertain fans during the half-time of a basketball game, a local merchant sponsors a free-throw-shooting contest. A rather portly fan, J. Stout, was chosen randomly from the fans in attendance and, since he had not touched a basketball in many years, was allowed to shoot until he made a basket. The outcome of interest is the number of shots Mr. Stout would take before making a basket. Here we have $S = \{1, 2, 3, \ldots\}$.

In the first two examples we see that the sample space is finite; that is, there is a natural number that corresponds to the number of points or elements of S. In the first example there are two points in S, in the second example there are six. However, in the third example we see that, theoretically, any number of shots might be required for Mr. Stout to make a basket, so S must contain all the integers. If a sample space is such that all elements could be written down in a list but the list could never be finished because it is infinitely long, the number of elements in S is said to be *countably infinite*.

Definition 1.2 If a sample space contains a finite or countably infinite number of elements, it is said to be *countable*. Such sample spaces are also called *discrete*.

Example 1.4 The kick-off time for a football game is 1:30 P.M. The outcome of interest is the time of the final gun. Since each quarter of a football game lasts 15 minutes and since the half-time intermission is 20 minutes, the earliest the game could be over (the time of the final gun) is at 2:50 P.M. Of course, the time is generally later than that because of time-outs, out-of-bounds plays, incompleted passes, and so on (all of which cause the "official" clock to be stopped). Since the sample space is the set of all possible outcomes, we would have $S = \{t : t \geq 2{:}50 \text{ P.M.}\}$.

If we were to make the further assumption that all games would be automatically ended at 6:00 P.M. due to darkness, we would have $S = \{t : 2{:}50 \text{ P.M.} \leq t \leq 6{:}00 \text{ P.M.}\}$.

Whenever we attempt to measure time, we are limited by our measuring devices. In Example 1.4 we might expect the time of the final gun to be reported to the nearest minute or second, in which case we would have to say that S is discrete. However, ignoring our inability to measure time to any number of decimal places, time is actually a continuous quantity, and consequently S consists of all numbers in some interval of the real line.

Definition 1.3 If a sample space contains the set of all numbers in some interval of the real line, that sample space is said to be *continuous*.

1.3 EVENTS

Let us imagine that in Example 1.2, Mrs. Hubbard's dog, a chihuahua named Fang, has noticed that when one of the girls is chosen to feed him he gets more food than when a boy is chosen. Since Fang enjoys eating, he has more than an academic interest in the outcome of Mrs. Hubbard's dice-rolling experiment. In fact, he is hoping that one of the three girls is chosen and not one of the three boys. In other words, if we let $G = \{\text{Betty, Diane, Frieda}\}$ and $B = \{\text{Alan, Carl, Eric}\}$, Fang would hope that one of the elements of G rather than of B would be selected.

This is an example of a situation where a person (using the term loosely) is interested in knowing whether a member of a certain subset of S was chosen without being interested in which particular member of that subset was chosen.

Definition 1.4 An *event* is a subset of a sample space S.

Events are generally denoted by capital letters near the beginning of the alphabet. B and G as defined earlier are examples of events. Events may be defined by a verbal description or by listing the elements of S which satisfy that description. For example,

$$A = \{\text{child chosen is at least four years old}\} = \{\text{Diane, Eric, Frieda}\}.$$

There are many other events that could be defined using this sample space (see Exercise 8). In fact, since there are six elements of S, there are $2^6 = 64$ different events that can be defined. (We will show in Section 1.4 that if a sample space contains n distinct points, the number of events that can be defined using that sample space is 2^n.)

Example 1.5 The sample space found in Example 1.3 was

$$S = \{1, 2, 3, \ldots\}.$$

In that case the outcome of the experiment was the number of shots required for Mr. Stout to make a basket. Some events that could be defined are:

$$A = \{\text{an even number of shots are taken}\}$$
$$B = \{\text{an odd number of shots are taken}\}$$
$$C = \{\text{more than 4 shots are taken}\}$$
$$D = \{\text{2 or fewer shots are taken}\}.$$

Example 1.6 In Example 1.4 we saw that if a football game was terminated at 6:00 P.M. due to darkness, the sample space giving the possible times of the final gun would be

$$S = \{t : 2{:}50 \text{ P.M.} \leq t \leq 6{:}00 \text{ P.M.}\}.$$

Here we could define:

$$A = \{\text{game is over after 4:00}\}$$
$$B = \{\text{game is over no later than 4:00}\}$$
$$C = \{\text{game is over between 3:00 and 4:00}\}$$
$$D = \{\text{game is over by 2:30}\}.$$

You will recall from your study of sets that a set which has no members (e.g., the set of four-footed ducks) is said to be empty. We will denote the empty set by the symbol $\varnothing$. The event D defined in Example 1.6 is actually the empty set. Why? Since S consists of times between 2:50 and 6:00, it is impossible for the game to be over by 2:30; hence D contains no points of S and consequently must be empty.

In Examples 1.5 and 1.6 you can see that it is impossible for events A and B to occur simultaneously. In set terminology the simultaneous occurrence of two sets is called the *intersection* of those sets. Since events are merely subsets of some sample space, it makes sense to talk about the intersection of two events, denoted by $\cap$. In set terminology if two sets A and B have an empty intersection, they are said to be *disjoint*. When the sets happen to be events, a different term is used to convey the same notion.

Definition 1.5 Two events, A and B, are said to be *mutually exclusive* if $A \cap B = \varnothing$.

You must be careful here not to be confused by the notation. As you can see, the letters A and B denote quite different events in the different examples. In both Examples 1.5 and 1.6, the events A and B are mutually exclusive. However, in the example of Mrs. Hubbard's children, $A \cap B = \{\text{Eric}\} \neq \varnothing$, so here A and B are *not* mutually exclusive.

We also wish to emphasize that the concept of "mutually exclusive" relates to pairs of events. It is easy to define three different events, say A, B, and C, such that $A \cap B \cap C = \varnothing$ while $A \cap B \neq \varnothing$, $A \cap C \neq \varnothing$, and $B \cap C \neq \varnothing$. It would not be correct to say that A, B, and C are mutually exclusive. If we say that events $A_1, A_2, \ldots, A_n$ are *pairwise mutually exclusive*, we will mean that $A_i \cap A_j = \varnothing$ for all $i \neq j$.

Since events are sets, we can consider other set operations in addition to intersections. For example, we can consider unions and complements of sets. In this book we will use a "bar" to denote the complement of a set (e.g., $\bar{A}$). (Other books may use A', $\tilde{A}$, A^c, etc., to denote complements.)

Example 1.7 A gas station operator, Tex Echo, has eight gas pumps at his station. Tex periodically counts the number of cars using the gas pumps. For this experiment, $S = \{0, 1, 2, \ldots, 8\}$. Defining

$$A = \{\text{an odd number of pumps are in use}\} = \{1, 3, 5, 7\}$$
$$B = \{\text{at least half of the pumps are in use}\} = \{4, 5, 6, 7, 8\}$$
$$C = \{\text{at most half of the pumps are in use}\} = \{0, 1, 2, 3, 4\},$$

find the events $\bar{A}$, $B \cup C$, and $B \cap C$.

We have

$$\bar{A} = \{0, 2, 4, 6, 8\}$$
$$B \cup C = \{0, 1, 2, \ldots, 8\} = S$$
$$B \cap C = \{4\}.$$

In some situations it is helpful to have a diagram to aid in finding the elements of a set formed by unions, intersections, or complements. One such diagram is called a *Venn diagram*. In these diagrams the sample space S is represented by a rectangle and events are represented by portions of S, usually, but not necessarily, circles. If

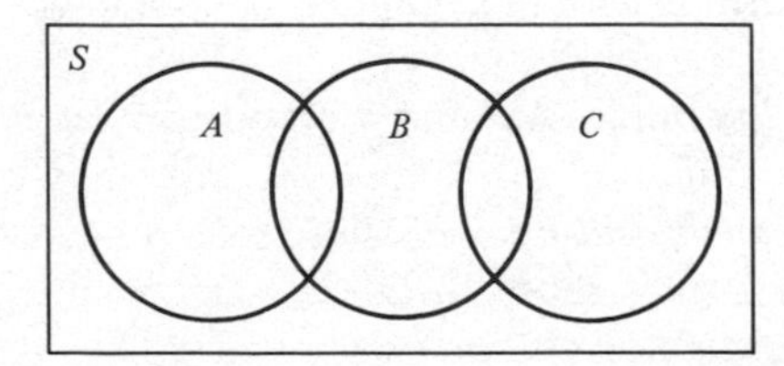

FIGURE 1.1
Venn Diagram

events are known to be mutually exclusive, the corresponding circles are shown as nonoverlapping. Figure 1.1 shows a Venn diagram in which A and C are mutually exclusive but A and B are not mutually exclusive, nor are B and C. In Figure 1.2 the shaded regions correspond to $\bar{A}$, $A \cap B$, and $A \cup B$, respectively.

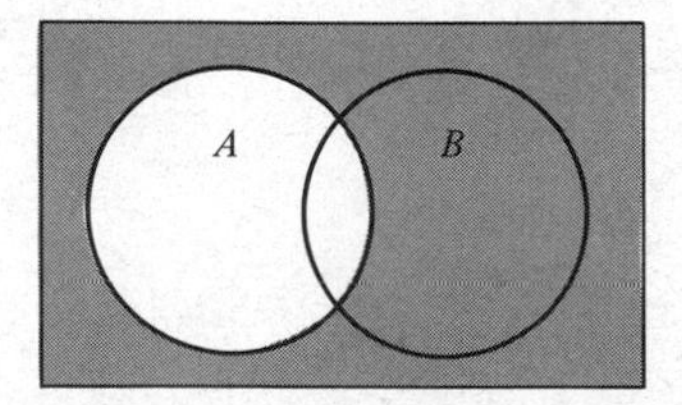

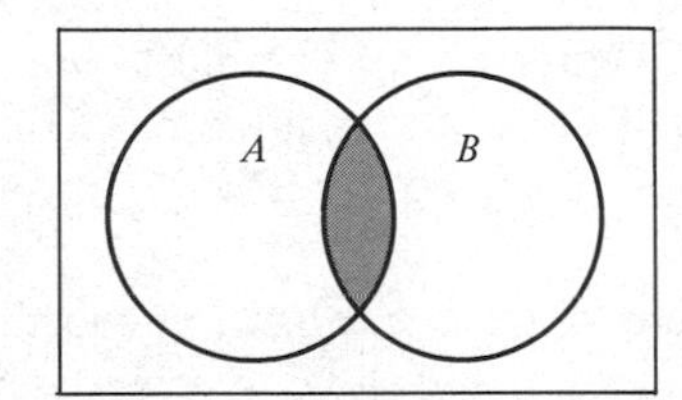

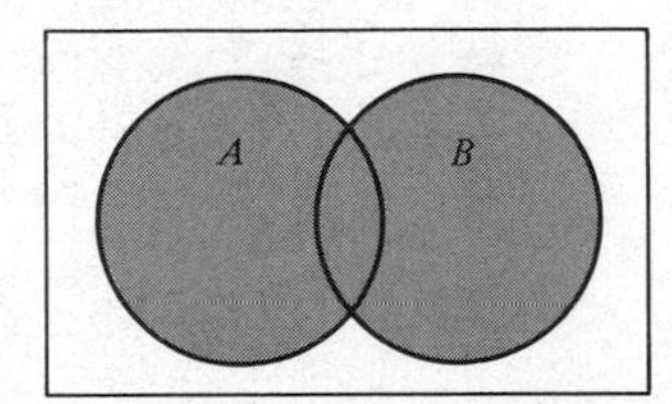

FIGURE 1.2
Venn Diagrams Showing Complements, Intersections, and Unions

1.4 METHODS OF COUNTING

We will see in the next section that in certain situations it is necessary to know the total number of points in a given sample space as well as the number of points in events defined on that sample space. In some cases it is very easy to determine the total number of points (elements) in a sample space. For example, in Example 1.1, S contains two points; in Example 1.2, S contains six points. However, in other cases it is not so easy.

Example 1.8 A monkey is placed in front of a typewriter keyboard containing only the letters of the alphabet and is allowed to type three letters. The sample space consists of all possible three-letter "words" (i.e., sets of three letters). How many points are there in S?

You can see that it is not as easy to find the number of points here as it was in previous examples. We can find the answer to the question by considering the "experiment" to be a sequence of three simpler experiments (namely, the monkey's choice of first letter, of second letter, and of third letter) and then using the following counting rule.

Counting Rule Consider a sequence of k experiments. If the first experiment has n_1 different possible outcomes and, for each of those outcomes, the second experiment has n_2 different possible outcomes and, for each of those outcomes, the third experiment has n_3 different possible outcomes, and so on, then the sequence of k experiments has $n_1 \cdot n_2 \cdot n_3 \cdot \ldots \cdot n_k$ different possible outcomes.

Example 1.8 *(continued)* Since there are 26 letters from which to choose the first letter, $n_1 = 26$. Similarly, $n_2 = n_3 = 26$, so the total number of different outcomes is $n_1 \cdot n_2 \cdot n_3 = 26 \cdot 26 \cdot 26 = 17{,}576$.

Example 1.9 At the Pizza Palace pizzas are made with two different types of crust, thin or thick, and with any of four meat toppings and five vegetable toppings available. If they will make pizzas with any combination of ingredients on request, how many different types of pizzas can they advertise as being available?

We can think of a pizza being specified by a sequence of questions. First, what kind of crust? Second, should the first kind of meat be included? Third, should the second kind of meat be included? . . . Tenth, should the fifth kind of vegetable be included? Since there are two answers or outcomes to each of these questions, $n_1 = n_2 = \ldots = n_{10} = 2$, so $n_1 \cdot n_2 \cdot \ldots \cdot n_{10} = 2^{10} = 1024$.

Example 1.10 Miss Natalie Atyred has a color-coordinated set of clothes consisting of 3 slacks, 2 skirts, 4 sweaters, and 2 pairs of shoes. For how many different days can she wear a different outfit?

You must be a bit careful here. Assuming that she chooses either slacks or skirt (and not both), she has 5 choices, so $n_1 = 5$. If the second and third choices are sweater and shoes, $n_2 = 4$ and $n_3 = 2$. Consequently, she could wear a different outfit for $n_1 \cdot n_2 \cdot n_3 = 5 \cdot 4 \cdot 2 = 40$ different days.

Example 1.11 If a sample space consists of n distinct points, how many different subsets (events) can be defined?

We can approach this problem in a manner similar to the method we used to "construct" a pizza in Example 1.9. In particular, we can look at each point in S and ask whether or not it should be included in the subset. Since there are two choices for each point (to be or not to be included) and since there are n different points, the total number of subsets is 2^n.

If you read over the counting rule, you will see the phrase "for each of those outcomes" repeatedly. This is an important phrase to keep in mind when applying the counting rule. For example, say that the license plates in the country of Lower Peskarania consist of two letters, but with the condition that repeated letters (e.g., AA) are not allowed. How many different license plates can be issued? It is true that there are 26 choices for the first position and 26 for the second position, however it is *not* true that for *each* choice in the first position there are 26 for the second. In fact, for each choice for the first position there are 25 choices for the second position. Consequently, the total number of license plates that can be issued is $n_1 \cdot n_2 = 26 \cdot 25 = 650$.

There are many situations where we are called upon to rank various items. In advertisements we see people asked to choose the best cleanser, detergent, coffee, and so on. We see lists of the 10 most admired persons, best dressed, worst dressed, and so on. These are examples of rankings. We will see how ranks can be used in statistics in Chapter 11 when we study nonparametric statistics. For now we will be content with finding the number of different ways items can be ranked.

Example 1.12 Before deciding where to order pizzas for their pizza party a fraternity chooses one of its members, Pepper Oni, to sample pizzas from four establishments:

Antony's, Bruno's, Clyde's, and Dominic's. Pepper is asked to rank the pizzas on overall quality. In how many different ways could these be ranked?

If we think of this as a sequence of experiments where the first experiment is to choose the best of the four, the second to choose the second best from the remaining three, and so on, we see that the rankings can be done in $n_1 \cdot n_2 \cdot n_3 \cdot n_4 = 4 \cdot 3 \cdot 2 \cdot 1 = 24$ ways.

The same reasoning would lead us to conclude that if there were 10 pizza establishments in town, there would be

$$10 \cdot 9 \cdot 8 \cdot 7 \cdot 6 \cdot 5 \cdot 4 \cdot 3 \cdot 2 \cdot 1 = 3{,}628{,}800$$

different ways in which Pepper could order them. You can see that there would be an advantage to a shorthand notation for expressing products like the previous one, so we will introduce the "factorial" notation.

Definition 1.6 By $n!$ (read "n factorial") we mean

$$n \cdot (n-1) \cdot (n-2) \cdot \ldots \cdot 2 \cdot 1.$$

If Pepper were asked to rank the two best out of the four, in how many ways could this be done? In this case he would only perform the first two "experiments": choosing the best of the four and then choosing the second best of the remaining three. The number of ways is

$$n_1 \cdot n_2 = 4 \cdot 3 = 12.$$

Since the number of possibilities is not too large, we could list them as follows:

AB	BA	CA	DA
AC	BC	CB	DB
AD	BD	CD	DC.

If he were asked to rank the top 4 out of 10, the number of ways would be

$$10 \cdot 9 \cdot 8 \cdot 7 = 5040.$$

Here we can use the factorial notation to advantage by writing

$$10 \cdot 9 \cdot 8 \cdot 7 = \frac{10 \cdot 9 \cdot 8 \cdot 7 \cdot 6 \cdot 5 \cdot 4 \cdot 3 \cdot 2 \cdot 1}{6 \cdot 5 \cdot 4 \cdot 3 \cdot 2 \cdot 1} = \frac{10!}{6!} = \frac{10!}{(10-4)!}.$$

In general, if we wish to rank k items out of n, the number of ways in which this can be done is

$$\frac{n!}{(n-k)!}.$$

Such a ranking is frequently referred to as a *permutation* of k of n distinguishable objects and can be denoted by ${}_nP_k$:

$$ {}_nP_k = \frac{n!}{(n-k)!}. \tag{1.1}$$

Example 1.13 There are eight runners in the finals of the men's 200-meter dash at the Olympics. In how many different ways might the gold, silver, and bronze medals (i.e., first, second, and third places) be awarded?

We simply apply equation (1.1) with $n = 8$ and $k = 3$ to get

$${}_8P_3 = \frac{8!}{(8-3)!} = 8 \cdot 7 \cdot 6 = 336.$$

Example 1.14 There are eight runners in the semifinals of the men's 200-meter dash at the Olympics. In how many ways might four runners be chosen to advance to the finals?

In this case all four of the top runners are treated equally in the sense that they all go to the finals regardless of the order of finish. Since we are not interested in *ranking* the top four finishers, equation (1.1) does not apply.

There are many situations like this where we are interested in choosing k objects out of n but where the order of choosing has no relevance. We will denote the number of ways of choosing k objects out of n by ${}_nC_k$ and will determine, indirectly, how to actually evaluate that quantity. In particular, we will consider the problem of *ranking* k objects out of n as a two-stage experiment. The first stage is to choose the k objects that will be ranked, the second is to rank those k chosen. We already know that the number of ways of ranking k objects out of n is

$$\frac{n!}{(n-k)!}$$

and we must get this same numerical answer whether we think of the ranking procedure as a one-stage or a two-stage procedure. Treating it like a two-stage procedure, we can use the counting rule to determine the total number of outcomes:

Experiment 1: Choose k of n; number of ways $= n_1 = {}_nC_k$
Experiment 2: Rank k of k; number of ways $= n_2 = k!$

Total number of ways $= n_1 \cdot n_2 = {}_nC_k \cdot k!$

Since the one-stage and two-stage procedures lead to the same numerical answer, we get the equation

$${}_nC_k \cdot k! = \frac{n!}{(n-k)!},$$

which can be solved for ${}_nC_k$ to give

$$ {}_nC_k = \frac{n!}{k!\,(n-k)!} \tag{1.2}$$

An alternative notation for this quantity is

$${}_nC_k = \binom{n}{k}.$$

Example 1.14 (*continued*) To find the number of ways four runners could be chosen from eight to advance to the finals, we must evaluate (1.2) with $n = 8$ and $k = 4$. We get

$$ {}_8C_4 = \binom{8}{4} = \frac{8!}{4!\,(8-4)!} = \frac{8 \cdot 7 \cdot 6 \cdot 5 \cdot 4 \cdot 3 \cdot 2 \cdot 1}{4 \cdot 3 \cdot 2 \cdot 1 \cdot 4 \cdot 3 \cdot 2 \cdot 1} = 70. \tag{1.3}$$

The number of ways of choosing k objects out of n is also referred to as the number of *combinations* of k objects out of n. By way of warning, we would like to tell you that in finding the numerical answer to a combinations problem such as (1.3), you should *not* multiply out all the terms of the numerator and denominator and then divide to get your answer (even if you have a hand calculator). There will always be cancellations you can (and should) make before carrying out any further operations. This will save time and may lessen the chance of error. If your calculator happens to have a factorial key, you may wish to investigate the effects of roundoff error.

Example 1.12 (*continued*) Let us assume that Pepper was asked to choose two of the four pizza establishments for further study by a three-man committee chosen from the fraternity members. In a sense Pepper's responsibility is to choose the "finalists," who will be further investigated. In how many ways can these be chosen?

Since ranking is not important, we use (1.2) with $n = 4$ and $k = 2$ to get

$$\binom{4}{2} = \frac{4!}{2!\,2!} = \frac{4 \cdot 3 \cdot 2!}{2 \cdot 1 \cdot 2!} = \frac{4 \cdot 3}{2} = 6.$$

The six pairs are AB, AC, AD, BC, BD, and CD.

One way that Pepper might communicate his results to the committee would be to put a sign in the window of each establishment, say an F for the finalists and an A for the also-rans. These could be put in six different ways, as in Figure 1.3. From this you can see that the number of ways of choosing two objects from four objects and the number of ways of putting two A labels and two F labels on four objects and the number of ways of arranging two A's and two F's in a row are all identical. The following theorem generalizes this idea.

Theorem 1.1 The number of ways of arranging n_1 items of one kind, n_2 items of a second kind, . . . , and n_k items of a kth kind in a row is

$$\frac{n!}{n_1!\,n_2!\ldots n_k!}, \tag{1.4}$$

where $n = n_1 + n_2 + \ldots + n_k$.

This theorem can be proven by using repeated applications of the counting rule. We will not give the details of the proof.

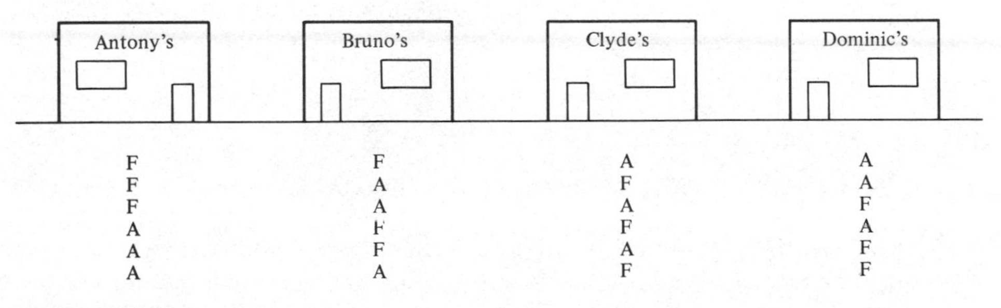

FIGURE 1.3

Example 1.15 One of Peter's chores is to put away the canned goods that his mother brings home from the store. If among the canned goods are 3 cans of tomato soup, 4 cans of vegetable soup, 2 cans of noodle soup, and 1 can of pea soup, in how many different ways can Peter line up these cans of soup on a shelf?

Using (1.4) with $n_1 = 3$, $n_2 = 4$, $n_3 = 2$, $n_4 = 1$, and $n = 10$, we get

$$\frac{10!}{3!\,4!\,2!\,1!} = \frac{10 \cdot 9 \cdot 8 \cdot 7 \cdot 6 \cdot 5 \cdot 4!}{3 \cdot 2 \cdot 2 \cdot 4!} = 12{,}600.$$

To be helpful, Peter decides to remove all the labels from the soup cans. Much to his chagrin, however, he finds that his mother does not consider this to be very helpful, and he decides that he should replace the labels. If all the cans are the same size, in how many ways can he put the labels onto the cans? In view of the discussion preceding Theorem 1.1, the answer to this is exactly the same as the answer to the first part of this example: namely, 12,600.

Example 1.16 How many different "words" can be made by different arrangements of the letters of the word "statistics"?

If n_1 = number of s's, n_2 = number of t's, n_3 = number of a's, n_4 = number of i's, and n_5 = number of c's, we can use (1.4) to get

$$\frac{10!}{3!\,3!\,1!\,2!\,1!} = \frac{10 \cdot 9 \cdot 8 \cdot 7 \cdot 6 \cdot 5 \cdot 4 \cdot 3!}{3!\,3 \cdot 2 \cdot 2} = 50{,}400.$$

In the examples we have given in this section, the solutions have been found by a straightforward application of the counting rule or equations (1.1), (1.2), or (1.4). Many interesting counting problems cannot be solved by a simple application of one of these equations, but rather require a combination of these equations. You should keep this in mind as you try to work the exercises.

Example 1.17 Three couples, the Tees, the Dees, and the Ems, ordered season basketball tickets together and requested that they be next to each other in the same row. This request was granted, but unfortunately seats 1, 2, 3, and 4 were separated from seats 5 and 6 by an aisle. (This actually happened to one of the authors!) In how many different ways could the couples seat themselves if couples sit next to each other?

This can be solved by viewing the problem as a sequence of decisions: first, decide which couple will be assigned to which pair of seats; second, decide how the Tees will sit in the seats assigned to them; third, decide how the Dees will sit in the seats assigned to them; fourth, decide how the Ems will sit in the seats assigned to them. We have $n_1 = 3! = 6$, $n_2 = n_3 = n_4 = 2! = 2$, so the total number of ways is

$$n_1 \cdot n_2 \cdot n_3 \cdot n_4 = 6 \cdot 2 \cdot 2 \cdot 2 = 48.$$

1.5 ASSIGNMENT OF PROBABILITIES—EQUALLY LIKELY OUTCOMES

The weatherman (your friendly local meteorologist) is an example of a person who deals with variable quantities. We are all familiar with the limitations of the weatherman when it comes to predictions, but we have come to accept the fact that the predictions cannot be perfect and that imperfect predictions are usually better than no predictions at all. Take a few minutes to consider how a weatherman deals with the problems of variability. In predicting tomorrow's high (or low) temperature, instead of giving a single number he will frequently give a range of values. ("Tomorrow's high will be between 20 and 25 degrees," rather than "Tomorrow's high will be 22 degrees." This type of prediction will be discussed in Chapter 7.) In predicting precipitation he will frequently give precipitation probabilities. ("The probability of measurable precipitation is 80% this afternoon, decreasing to 60% tonight.")

How do you interpret precipitation probabilities? If you had the option of having a picnic tomorrow or next week, what would you do if the weather report said "The probability of measurable precipitation tomorrow is 80%"? What if it said "10%"? Why?

Without going into mathematical detail at this point, we can say that the assignment of a probability to an event (such as "measurable precipitation during the next 12 hours") should be some number between 0 and 1 (equivalently 0% and 100%) and that the more certain the event is to happen, the closer the probability assigned is to 1. We assign the number 1 only if the event is absolutely sure to happen and the number 0 if the event cannot possibly happen.

Our objective in the next several sections will be to devise methods for assigning probabilities to events defined on various sample spaces. We pointed out in the last section that if a sample space contains n points, there are 2^n different events which can be defined on that sample space. For example, if there are $n = 10$ points, then $2^{10} = 1024$ different events can be defined. Would it be easier to assign probabilities to each of 10 points or to each of 1024 events? Clearly, the former! Consequently, we will first discuss how to assign probabilities to individual points of a sample space and then how to use those probabilities to assign probabilities to events.

In many situations when you consider all the elements of a sample space, there is no reason to expect any particular outcome to be more likely to occur than any other. For example, when the referee of Example 1.1 flips a coin, there is no reason to expect "heads" to be more likely to occur than "tails." When Mrs. Hubbard of Example 1.2 chooses a child by rolling a die, there is no reason to expect any particular child to be chosen over any other one. In such cases we assign the same probability to each point in the sample space. In particular,

> if a sample space S contains n distinct points and if each of these points is equally likely to occur, then to each point in S we assign probability $1/n$.

Notation. The individual points or elements of S will be denoted by e or e_i. Since a subset of S containing just one point, e_i, is still an event, we will refer to such events as *elementary events*. We will denote the probability of an event A by $P(A)$.

Example 1.18 In Example 1.1 the sample space consisted of two points, H and T. Since it is reasonable to assume that these two outcomes are equally likely, we have $P(\text{head}) = P(\text{tail}) = \frac{1}{2}$.

Example 1.19 Assuming Mrs. Hubbard's die of Example 1.2 is "fair," since S

contains six points, we have

$$P(\text{Alan}) = P(\text{Betty}) = P(\text{Carl}) = P(\text{Diane}) = P(\text{Eric}) = P(\text{Frieda}) = \tfrac{1}{6}.$$

Example 1.20 The nurse in the delivery room of a large hospital notes the time of birth of babies to the nearest 5 minutes. What is the probability that the time of the first birth tomorrow morning will be noted as 15 minutes after the hour?

There are 12 possible outcomes and it is reasonable to consider each one as being equally likely, so we would assign a probability of $\frac{1}{12}$.

There are times when we simplify a problem somewhat by saying that an item is chosen *at random* from a group of items. When we use this term, we generally mean that each item is equally likely to be chosen.

If outcomes in a sample space are known or assumed to be equally likely, we assign probabilities to events as follows:

if a sample space S contains n equally likely points and

if an event A contains n_A points, then

$$P(A) = \frac{n_A}{n}. \tag{1.5}$$

Example 1.21 Again referring to Mrs. Hubbard's children, the event B was defined to be "a boy is chosen." Since S contains 6 points and B contains 3 points, it follows from (1.5) that

$$P(B) = \tfrac{3}{6} = .50.$$

Example 1.22 Frieda came home from school yesterday and told her mother how happy she was to have been chosen by Miss Prim, her first-grade teacher, to be on the Halloween party committee. Further inquiry revealed that (1) although Frieda is on the committee, Mrs. Hubbard will have to do the work; (2) the committee consists of 5 girls; and (3) Miss Prim claims to have chosen the committee at random from the class, which consists of 10 girls and 10 boys. Does Mrs. Hubbard have reason to doubt Miss Prim's claim of random selection?

The sample space of interest in this problem is the set of possible committees of size 5 that could be chosen from a class of 20 students. The total number of ways of choosing 5 from 20 is $\binom{20}{5} = 15{,}504$, so this is the number of elements in S. Under the assumption of random selection, each of these outcomes is equally likely. If we assume that Mrs. Hubbard is wondering about the fact that the committee consists of all girls, we can define the event A to be {committee consists of all girls}. To find $P(A)$, we will need to find n_A, the number of different committees made up of 5 girls, or the number of ways of choosing 5 girls from the 10 girls in the class. This is $\binom{10}{5} = 252$. Hence we get $P(A) = 252/15{,}504 = .0163$. If you think back to the discussion earlier in this section, you will recall that probabilities of events near zero are rather unlikely to happen. Mrs. Hubbard might reason that while a truly random selection *could* lead to a committee of all girls, such an event is rare enough to warrant some doubt as to the randomness of the selection.

1.6 ASSIGNMENT OF PROBABILITIES—RELATIVE FREQUENCIES

Mealtime at any household having small children can be an interesting experience. This is always true at the Hubbard household, with its six children, but breakfast time is especially interesting. In fact, an observer with an eye out for quantities that may vary from one day to the next can have a real field day. One might watch for (1) the number of times a glass of milk is spilled, (2) the number of ounces of orange juice that Diane tries to pour into a 4-ounce glass, (3) the number of squabbles over who gets the chintzy toy from the box of Munchi-Crunchies, (4) the number of pieces of jelly bread that are dropped onto the floor, and so on. If a piece of jelly bread is dropped onto the floor, we might consider this an "experiment" and ask about the possible outcomes. There are two possible outcomes in this case: bread lands jelly-side up or bread lands jelly-side down. We might like to assign probabilities to these two elementary events and so we ask whether these two outcomes are equally likely. What do you think? Mrs. Hubbard says she does not like to think of this as an "experiment" but that she knows from the number of times she has had to clean up the floor that the bread is more likely to land jelly-side down!

It is not hard to think of instances such as this where the different possible outcomes are *not* equally likely. How can probabilities be assigned in situations such as these? If the experiment is one that can be repeated under virtually identical conditions a large number of times, then, although probabilities cannot be found exactly, they can be approximated (estimated) by using relative frequencies.

Definition 1.7 Let e be some elementary outcome for an experiment that can be repeated an arbitrarily large number of times under virtually identical conditions. If f is the number of times that e is observed to occur in n repetitions of the experiment, the *relative frequency* of occurrence of e is f/n.

It can be shown mathematically that as n gets larger and larger, there is a better and better chance that f/n will be very close to $P(e)$, the actual probability of e. Because of this, for large n, we can use f/n as an estimate of $P(e)$. (This will be discussed in more detail in Chapter 7.)

Example 1.23 An ordinary thumbtack, when dropped onto a hard surface, will land either point up or point down. As a class experiment, 26 students each dropped

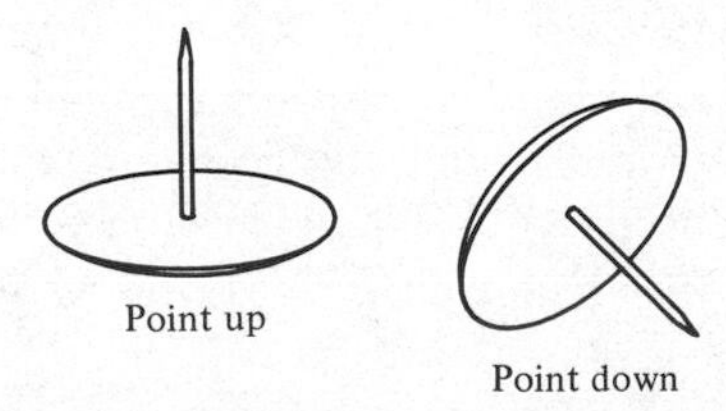

a tack (under conditions agreed upon in advance) 50 times and recorded the number of times it landed point up. The total number of times it landed point up was 684. The relative frequency of occurrence for $26 \times 50 = 1300$ repetitions was

$$\frac{f}{n} = \frac{684}{1300} = .526.$$

Example 1.24 According to an article in *U.S. News and World Report* (April 12, 1976) concerning persons indicted for criminal tax fraud, ". . . 1,470 such cases were disposed of in the federal courts. Of these, 1,046 pleaded guilty or no contest, and 173 were convicted after a trial." Based on this information, what is the approximate probability that a criminal tax-fraud indictment will result in a plea of guilty, no contest, or a conviction?

The number of repetitions of this "experiment" was $n = 1470$ and the number of times the event in question occurred was $f = 1046 + 173 = 1219$; hence the relative frequency of occurrence was

$$\frac{f}{n} = \frac{1219}{1470} = .829.$$

1.7 ASSIGNMENT OF PROBABILITIES—SUBJECTIVE PROBABILITIES

In the previous section we discussed how probabilities could be assigned to elementary events when the outcomes are not all equally likely but where the experiment of interest could be repeated a large number of times under similar conditions. It is not hard to think of "experiments" that either cannot be repeated or cannot be repeated under similar conditions. For example, say that we were interested in:

1. The probability that the Democratic candidate will win the next presidential election.
2. The probability that UCLA will win the national championship in basketball next year.
3. The probability that a particular coffee substitute will gain 10% of the coffee market the first year it is introduced.
4. The probability that you (your cousin, brother, sister, best friend—pick one) will get married this year.
5. The probability that Podunk U., perennial underdogs, will beat Great State, perennial overdogs, in the Big Game next week (you should fill in names of teams with which you are familiar).

In some of these cases you might have some information on past records that you could use to get a relative frequency (e.g., in case 1, 2, or 5). However, relative frequencies are only valid if conditions under which observations have been made remain relatively constant. For example, were conditions the same in 1932 (Hoover vs. Roosevelt) as in 1964 (Johnson vs. Goldwater) or as in 1968 (Humphrey vs. Nixon) or as in 1976 (Ford vs. Carter)? Does UCLA have a dominant player such as Lew Alcindor (Jabbar) or Bill Walton? Is Podunk U. having its first undefeated season in 42 years while Great State is having a losing season? In these situations you might wish to make use of past history to make assignments of probability, but you would not want to use only past history.

In situations such as 3 and 4, you certainly cannot use a relative-frequency approach to the problem of assigning probabilities. However, in many situations like these you do have some intuitive feeling about the likelihood of the event in question actually occurring. It may not be entirely reasonable to simply act as if no information whatsoever is available. One solution to this dilemma is to use subjective probabilities.

Definition 1.8 *Subjective probability* assigned to an event is a probability that reflects the degree of belief of an individual as to the likelihood of the occurrence of that event.

Since such probabilities are, by their very nature, subjective, the probability assigned to the same event can be different for different individuals. For example, Mr. Don Khee, a dyed-in-the-wool Democrat, and Miss Ellie Fant, a diehard Republican, might assign probabilities of .80 and .40, respectively, to the event "Democratic candidate will win the next presidential election." If the numbers .80 and .40 really do reflect the degree of belief of these individuals, they should be considered to be correct. This contrasts with assigning probabilities to events where all outcomes are equally likely and where there is precisely one correct answer. It also contrasts with assigning probabilities to events by using relative frequencies. With relative frequencies, if two individuals each perform a particular experiment, say 100 times, the relative frequency of occurrence will most likely differ by some amount. However, if they were patient enough to repeat the experiment 10,000 or 100,000 times each, the relative frequency of occurrence would be practically the same for the two individuals.

Since we have discussed three different methods of assigning probabilities to events, you might reasonably ask: Which is best? This is difficult to answer, since the methods tend to overlap somewhat. For example, if you are given a coin, say a Peskaranian penny, and asked about the probability that it will come up heads when flipped, you might examine it, decide that it is relatively symmetric, and conclude that "heads" and "tails" are equally likely, so you say $P(\text{heads}) = \frac{1}{2}$. In a sense you have assigned the probability on the basis of equally likely outcomes, but in another sense you have used your subjective reasoning (by examining the coin and deciding that it is symmetric), so you might say that you have assigned a subjective probability. These differences are somewhat subtle and really will not play much of a role at this introductory level. In most cases the probabilities needed to work exercises will be given explicitly or will be able to be deduced by straightforward methods.

We would like to make one further point, which will be developed more fully in Chapter 12. We will assume that people studying this book are willing to use logical and reasonable approaches to the study of uncertainty and variability. With the possible exception of rabid fans and wild-eyed fanatics, we assume that people would be willing to modify their subjective probability of an event in light of empirical evidence. For example, if Mr. Khee read a Gallup poll one week before the election which showed the Republican candidate leading by five percentage points, he might want to change the probability he assigned to a Democratic victory. Similarly, if you were to flip the Peskaranian penny 20 times and observe 19 heads, you might want to increase the probability that you assigned to the event "heads."

1.8 AXIOMATIC PROBABILITY

What would you think if you turned the radio on one morning and heard the local weatherman, Jack Frost, say "The probability of measurable precipitation today is 90%," and then 2 minutes later heard him say "The probability of *no* measurable precipitation today is 80%"? You might wonder if he had stayed out in the cold too long because logic tells you that the probabilities of precipitation and *no* precipitation should not add up to 170%! The fact is that to make logical and consistent assignments of probabilities to events requires a systematic approach. We will state three axioms that must be satisfied when assigning probabilities to events in order to have a consistent system and then deduce from those axioms certain relationships that must hold among probabilities assigned to events.

Let S denote the entire sample space and let $A_1, A_2, \ldots$ denote events. The following axioms must be satisfied for a mathematically correct assignment of probabilities:

Axiom 1 $P(A_i) \geq 0$ for all events A_i.

Axiom 2 $P(S) = 1$.

Axiom 3 $P(A_i \cup A_j) = P(A_i) + P(A_j)$ if $A_i \cap A_j = \varnothing$.

The first axiom simply says that probabilities assigned to events must be nonnegative. This fact gives you a crude but effective way of checking your answers to probability problems. If you are trying to determine the probability of an event and you obtain a negative answer, you *know* that the answer you have is wrong.

The second axiom says that you must assign probability one to the event S, the entire sample space. This is reasonable since if S consists of *all possible* outcomes of an experiment, then one of those outcomes, and consequently S, is sure to occur.

The third axiom says that if two events are mutually exclusive, the probability that one or the other occurs (they cannot both occur simultaneously, by the very definition of mutually exclusive) is equal to the sum of the probabilities that each occurs. This can be illustrated by use of a special Venn diagram (Figure 1.4). Say

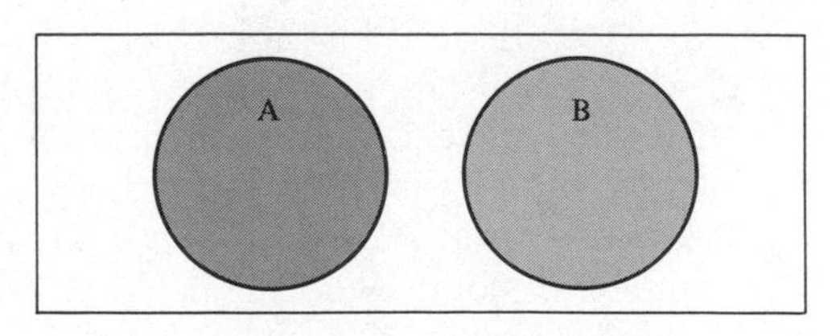

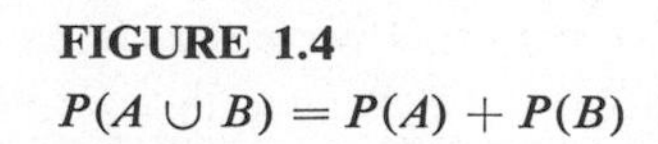

FIGURE 1.4
$P(A \cup B) = P(A) + P(B)$

that we agree to draw a rectangle representing S in such a way that the total area is 1 square unit. Then, from Axiom 2, it would be reasonable to equate area with probability. We could then let the area inside a circle representing a particular event be equal to the probability of that event. Then since the area associated with the event $A \cup B$ is the sum of the areas of A and of B (assuming that $A \cap B = \varnothing$), the probability of $A \cup B$ would be equal to $P(A)$ plus $P(B)$.

Theorem 1.2 $P(\varnothing) = 0$.

Proof: Since $S \cap \varnothing = \varnothing$, we can use Axiom 3 to say

$$P(S \cup \varnothing) = P(S) + P(\varnothing).$$

However, we know that $S \cup \varnothing = S$, so we get

$$P(S) = P(S \cup \varnothing) = P(S) + P(\varnothing). \tag{1.6}$$

Subtracting $P(S)$ from both sides of (1.6) gives $P(\varnothing) = 0$.

We could use the relationship between area and probability to give a less mathematical proof of Theorem 1.2. How much area is associated with the empty set $\varnothing$? Since $\varnothing$ contains no points, it contains no area, and consequently is assigned zero probability.

Theorem 1.3 $P(\bar{A}) = 1 - P(A)$.

Proof: Since $A \cap \bar{A} = \varnothing$, Axiom 3 is applicable. Also, $A \cup \bar{A} = S$, so we get

$$P(S) = P(A \cup \bar{A}) = P(A) + P(\bar{A}). \tag{1.7}$$

Since we know from Axiom 2 that $P(S) = 1$, we replace the left-hand side of (1.7) with a 1 and equate it to the right-hand side to get

$$1 = P(A) + P(\bar{A}), \tag{1.8}$$

which can be solved for $P(\bar{A})$ to yield the desired result.

Note that by rearranging equation (1.8) we also get $P(A) = 1 - P(\bar{A})$.

Example 1.25 In Example 1.3 we presented a situation where Mr. Stout was at the free-throw line trying to make a basket. Let us assume that on his first shot the probability of making a basket is .10. We will not make any further assumptions about his probability of making the second or any subsequent shots. However, we can still find the probability of

$$\text{event } A = \{\text{Mr. Stout requires two or more shots to make a basket}\}.$$

How? We note that the complement of A, $\bar{A}$, is simply the event that he makes a basket on his first shot, so

$$P(A) = 1 - P(\bar{A}) = 1 - P(\text{makes basket on first shot}) = 1 - .10 = .90.$$

Theorem 1.4 For any event A, $P(A) \leq 1$.

Proof: Since $\bar{A}$ is an event, we know that $P(\bar{A}) \geq 0$. Consequently,

$$1 - P(\bar{A}) \leq 1,$$

so

$$P(A) = 1 - P(\bar{A}) \leq 1.$$

Theorem 1.5 For any two events A and B,

$$P(A \cup B) = P(A) + P(B) - P(A \cap B).$$

Proof: In Figure 1.5 you can see that $A \cup B$ can be written as the union of three sets, in such a way that each distinct pair are mutually exclusive events. We can

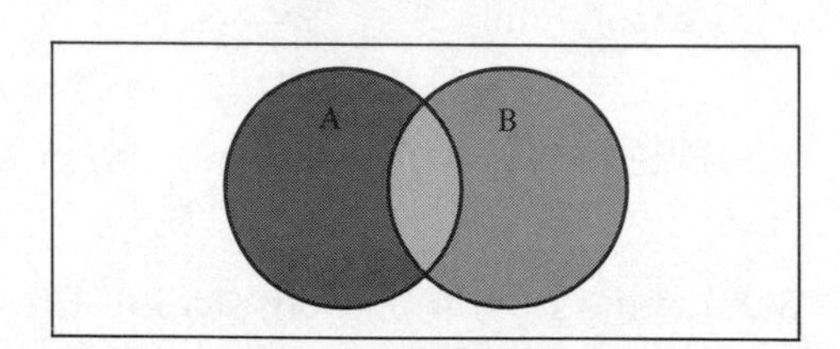

FIGURE 1.5
Decomposition of $A \cup B$

think of this as a decomposition of $A \cup B$. One set is $A \cap \bar{B}$, the second is $A \cap B$, and the third is $\bar{A} \cap B$. Since $(A \cap \bar{B})$ and $(A \cap B)$ are disjoint sets whose union is the set A, we can write

$$P(A) = P((A \cap \bar{B}) \cup (A \cap B)) = P(A \cap \bar{B}) + P(A \cap B). \tag{1.9}$$

Similarly, $A \cup B$ can be written as the union of the disjoint sets $A \cap \bar{B}$ and B, so

$$P(A \cup B) = P((A \cap \bar{B}) \cup B) = P(A \cap \bar{B}) + P(B). \tag{1.10}$$

Solving for $P(A \cap \bar{B})$ in (1.9) gives

$$P(A \cap \bar{B}) = P(A) - P(A \cap B),$$

and substituting this into (1.10) gives

$$\begin{aligned} P(A \cup B) &= P(A) - P(A \cap B) + P(B) \\ &= P(A) + P(B) - P(A \cap B). \end{aligned}$$

If we were to prove this theorem using the relationship between area and probability, we would note that the area corresponding to $A \cup B$ is the sum of the areas of $A \cap \bar{B}$, $A \cap B$, and $\bar{A} \cap B$, while if we add together the areas of A and B, we would have counted the area of $A \cap B$ twice. Hence the area of $A \cup B$ is equal to the area of A plus the area of B minus the area of $A \cap B$. Equating area and probability, the result follows.

Example 1.26 In the game of Parcheesi, a player rolls two dice and may move a token on the board forward by an amount equal to the number showing on the first die, the number on the second die, or the sum of the numbers on the two dice. (For example, if the dice show a 1 and a 3, a token may be moved 1, 3, or 4 spaces forward.) Find the probability that at least one of the two dice rolled shows a five.

The sample space for this experiment consists of 36 pairs of values, where the first member of the pair represents the number showing on the first die and the second

1, 1	1, 2	1, 3	1, 4	1, 5	1, 6
2, 1	2, 2	2, 3	2, 4	2, 5	2, 6
3, 1	3, 2	3, 3	3, 4	3, 5	3, 6
4, 1	4, 2	4, 3	4, 4	4, 5	4, 6
5, 1	5, 2	5, 3	5, 4	5, 5	5, 6
6, 1	6, 2	6, 3	6, 4	6, 5	6, 6

FIGURE 1.6
Possible Outcomes When Rolling Two Dice

member represents the number showing on the second die (see Figure 1.6). We assume that if the dice are fair, all 36 outcomes are equally likely. Now define event A to be "5 shows on the first die" and event B to be "5 shows on the second die." Then the event that we are interested in is $A \cup B$. From Theorem 1.5 we see that to find $P(A \cup B)$ we will need to find $P(A)$, $P(B)$, and $P(A \cap B)$. Since the outcomes in S are equally likely, we use Theorem 1.5 to get

$$P(A) = P(B) = \tfrac{6}{36} = \tfrac{1}{6}$$

and

$$P(A \cap B) = P(5 \text{ shows on both dice}) = \tfrac{1}{36},$$

so

$$P(A \cup B) = \tfrac{1}{6} + \tfrac{1}{6} - \tfrac{1}{36} = \tfrac{11}{36}.$$

Example 1.27 In the Parcheesi game, say you are interested in the probability that either a 5 shows on at least one of the dice or the sum of the numbers on the two dice equals 5. (This allows you to move a token out of home base.) Let us define the event C to be "5 shows on at least one of the dice" and the event D to be "sum of numbers on the two dice equals 5." In Example 1.26 we found $P(C)$ to be $\frac{11}{36}$. From Figure 1.6 you can see that $P(D) = \frac{4}{36}$. In this case $C \cap D = \varnothing$ (which you can easily see since if one of the dice is a 5, the sum must be greater than 5, while if the sum is 5, each face individually must be less than 5), so $P(C \cap D) = P(\varnothing) = 0$, which gives

$$P(C \cup D) = P(C) + P(D) - P(C \cap D) = \tfrac{11}{36} + \tfrac{4}{36} - 0 = \tfrac{15}{36}.$$

This last example illustrates a situation where Theorem 1.5 is applied to two events that are mutually exclusive. Note that if A and B are two mutually exclusive events, then $P(A \cap B) = P(\varnothing) = 0$, so $P(A \cup B) = P(A) + P(B)$, which is exactly the statement of Axiom 3.

1.9 CONDITIONAL PROBABILITY

A headline on the sports page of the *Columbia Daily Tribune*, under an AP byline, read "Do Dolphins Have A Bookie?" The article went on to relate how an alleged bookie had access to the Dolphins' locker room and consequently might have had information about the team personnel not available to other members of his profession. The article also stated that no club members were suspected of illegal activities. Why was this circumstance newsworthy? Have you heard of other similar situations? The reason for concern is that odds (which are directly related to probabilities) set for athletic competition are subjective and can be changed as conditions change. Any information gained by an individual and not available to the general public can increase the chances of that individual making money at the expense of the general public. (There are obvious legal and moral ramifications of such behavior which we will not pursue at this point.) As an example, consider a basketball game between two teams, Podunk U. and Cow College, which, based on performances mid-way through the season, is rated a toss-up. If you found that P.U.'s star center, Stretch Owt, and high scoring guard, Mark Oneup, were just declared academically ineligible, would it affect who you would pick to win the game? Certainly.

The notion of modifying the probability assigned to an event based on additional information is not restricted to subjective probabilities. Consider, for example, the sample space in Example 1.26 corresponding to the outcomes when two dice are thrown. If I tell you that at least one of the dice is a four (but do not tell you anything more), what is the probability that the sum of the numbers on the two dice is equal to five? In Example 1.27 we found that

$$P(D) = P(\text{sum of numbers on two dice equals 5}) = \tfrac{4}{36},$$

but in that case we were not given the additional information that one of the dice is a four (call this event F). If, in fact, we know that event F has occurred, we know that one of the outcomes

4,1 4,2 4,3 4,4 4,5 4,6 1,4 2,4 3,4 5,4 6,4

has occurred. Hence for all practical purposes we can consider these 11 points to be the "reduced sample space" ("reduced" because it contains fewer points than the original sample space). Since we assumed that the outcomes in the original sample space were equally likely, it is reasonable to assume that the outcomes in the reduced sample space are also equally likely. Exactly two of these outcomes yield a sum of 5, so the probability of observing a sum of 5 given the additional information that one of the dice is a four is $\tfrac{2}{11}$. The notation used for this is

$$P(D \mid F) = \tfrac{2}{11},$$

where $P(D \mid F)$ is read "the probability of event D given that event F has occurred" or more simply "the probability of D given F." We can get a visual picture of this situation by considering Figure 1.7, where we represent F by a rectangle rather than the usual circle. We will not label the points, but you can see that F contains 11 points and D contains 4 points. Furthermore, the intersection of F and D contains 2 points,

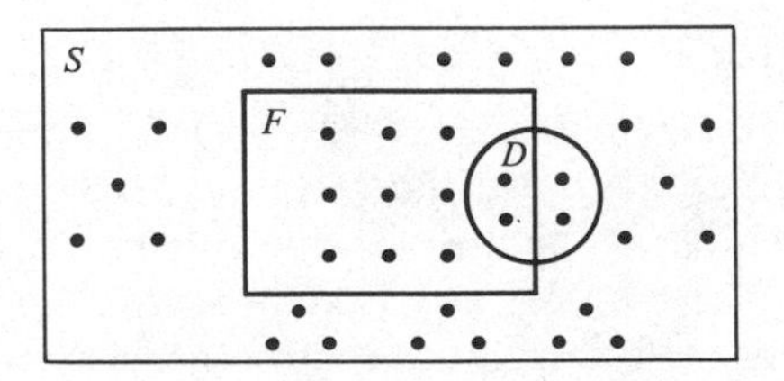

FIGURE 1.7

namely (1, 4) and (4, 1). By reviewing the previous discussion, you can see that

$$P(D \mid F) = \frac{n_{D \cap F}}{n_F}; \tag{1.11}$$

that is, the conditional probability of D given F is found by taking the ratio of the number of points in the intersection of D and F to the number of points in F, the reduced sample space. If we divide the numerator and denominator of the right-hand side of equation (1.11) by n_S, the number of points in the sample space, we get

$$P(D \mid F) = \frac{n_{D \cap F}/n_S}{n_F/n_S} = \frac{P(D \cap F)}{P(F)}. \tag{1.12}$$

Although the argument we gave to derive (1.12) was based on the assumption that all points in S are equally likely, the same result would hold if outcomes were not equally likely. In view of this discussion we give the following definition.

Definition 1.9 If A and B are two events such that $P(B) > 0$, then

$$P(A \mid B) = \frac{P(A \cap B)}{P(B)}. \tag{1.13}$$

[It should be clear that we require $P(B)$ to be strictly positive, since mathematically we cannot allow division by zero.]

Example 1.28 Referring to the Hubbard children of Example 1.2, what is the probability that the child chosen is at least four years old (A) given that a boy is chosen (B)? Since $P(B) = \frac{3}{6}$ and $P(A \cap B) = \frac{1}{6}$, we get

$$P(A \mid B) = \frac{\frac{1}{6}}{\frac{3}{6}} = \frac{1}{3}.$$

Remember that there is no great significance attached to the letters associated with events. If the information given corresponds to an event A, we can write

$$P(B \mid A) = \frac{P(B \cap A)}{P(A)}. \tag{1.14}$$

In other words, you need not relabel the events so that event B is always the event that is given to have occurred.

Example 1.29 If two dice are rolled and events D and F are defined by

$$D = \{\text{sum of numbers on two dice equals 5}\}$$
$$F = \{\text{one of the dice shows a 4}\},$$

what is $P(F \mid D)$?

$$P(F \mid D) = \frac{P(F \cap D)}{P(D)} = \frac{\frac{2}{36}}{\frac{4}{36}} = \frac{1}{2}.$$

We can see from this example that $P(F \mid D)$ need not equal $P(D \mid F)$. In general, given any two events A and B, $P(A \mid B)$ may or may not equal $P(B \mid A)$. Considering further properties of conditional probabilities, we might wonder if any order relationship exists between conditional and unconditional probabilities for a given event. That is, which, if any, of the following are always true: $P(A) > P(A|B)$, $P(A) < P(A|B)$, $P(A) = P(A|B)$? The answer is that none of these is always true (see Exercise 46).

It is true, however, that, provided we always condition on the same event, the properties of probability discussed in Section 1.8 still hold. For example, if we assign probabilities to events given that event C, say, has occurred, each of the following holds:

$$P(A|C) \geq 0 \qquad P(S|C) = 1$$

(1.15)
$$P(A \cup B|C) = P(A|C) + P(B|C) \qquad \text{if } A \cap B = \varnothing$$
$$P(\varnothing|C) = 0$$
$$P(\bar{A}|C) = 1 - P(A|C)$$
$$P(A \cup B|C) = P(A|C) + P(B|C) - P(A \cap B|C).$$

Example 1.30 In Example 1.28 we found that the probability that the child chosen is at least four years old given that a boy is chosen is $\frac{1}{3}$. If we want to find the probability that the child chosen is at most three years old given that a boy is chosen, we simply note that "at least four years old" and "at most three years old" are complementary, so using (1.15) we get

$$P(\bar{A}|B) = 1 - P(A|B) = 1 - \tfrac{1}{3} = \tfrac{2}{3}.$$

In the examples given so far, we have used equation (1.13) to find conditional probabilities by using the probability of the intersection. This, of course, assumes that it is relatively easy to find the probability of the intersection. There are many situations where it is relatively easy to find the conditional probability and the conditional probability can then be used to find the probability of the intersection. In particular, solving (1.13) for $P(A \cap B)$, we get

(1.16)
$$P(A \cap B) = P(B) \cdot P(A|B).$$

An alternative way of finding $P(A \cap B)$ is to solve (1.14) to get

$$P(A \cap B) = P(A) \cdot P(B|A).$$

Example 1.31 Felix is the owner of a small pet store, Something Fishy. One of the tanks of tropical fish in his store contains 6 tetras, 4 zebras, and 2 blue moons. Felix randomly nets two fish to put into a smaller tank. Assuming that the fish are all equally likely to be caught, what is the probability that Felix nets two tetras? [Note that the assumption of fish being equally likely to be caught may not really be true, but it allows us to illustrate an application of (1.16).]

Let A denote the event that the first fish caught is a tetra, and let B denote the event that the second fish caught is a tetra. We are interested in finding $P(A \cap B)$. It is easy to find $P(A)$ using equally likely outcomes, namely, $P(A) = \frac{6}{12}$. Furthermore, if we consider event A as given (i.e., if we know that a tetra has been removed from the tank), we know that there are 5 tetras remaining among 11 fish. Consequently, the probability that the second fish caught is a tetra given that the first caught was a tetra is $P(B|A) = \frac{5}{11}$. Combining these results, we get

$$P(A \cap B) = P(A) \cdot P(B|A) = \tfrac{6}{12} \cdot \tfrac{5}{11} = \tfrac{5}{22} = .227.$$

Example 1.32 After Felix has transferred the fish to the smaller tank, you observe that the smaller tank contains a tetra and a zebra. What is the probability that this would happen?

From just observing the smaller tank we cannot tell the order in which these fish were caught. However, this is not relevant, since we are only interested in the fact that one tetra and one zebra were caught. Obviously, this could happen by catching a tetra and then a zebra (event TZ), or by catching a zebra and then a tetra (event ZT). Clearly, these events are mutually exclusive (since if the first fish was a tetra, it is impossible for the first fish to be a zebra), so the probability of the union of these two events will simply be the sum of the respective probabilities:

$$\begin{aligned} P(\text{a tetra and zebra are caught}) &= P(TZ \cup ZT) = P(TZ) + P(ZT) \\ &= \tfrac{6}{12} \cdot \tfrac{4}{11} + \tfrac{4}{12} \cdot \tfrac{6}{11} = \tfrac{2}{11} + \tfrac{2}{11} = .364. \end{aligned}$$

Using conditional probabilities to find the probability of an intersection can be extended to intersections of three or more events. For example,

$$P(A \cap B \cap C) = P(A) \cdot P(B|A) \cdot P(C|A \cap B),$$

$$P(A \cap B \cap C \cap D) = P(A) \cdot P(B|A) \cdot P(C|A \cap B) \cdot P(D|A \cap B \cap C),$$

and so on.

Example 1.33 Four cards are dealt from a well-shuffled deck. What is the probability that all four are aces?

Let A_i denote the event that the ith card dealt is an ace. Then $P(A_1) = \frac{4}{52}$, $P(A_2|A_1) = \frac{3}{51}$, $P(A_3|A_1 \cap A_2) = \frac{2}{50}$, and $P(A_4|A_1 \cap A_2 \cap A_3) = \frac{1}{49}$, so

$$\begin{aligned} P(\text{all four are aces}) &= P(A_1 \cap A_2 \cap A_3 \cap A_4) \\ &= \frac{4}{52} \cdot \frac{3}{51} \cdot \frac{2}{50} \cdot \frac{1}{49} \\ &= \frac{1}{270{,}725} = .0000037. \end{aligned}$$

1.10 INDEPENDENT EVENTS

A likable panhandler, Ben Broke, offers you some information about tomorrow's soccer match between the Denver Donkeys and the Boston Bulls in exchange for 50 cents for coffee and a donut. Past history indicates that Ben's information is not always the best, but you agree to the deal anyway. Ben then tells you that he has information relating to the weather conditions and that in particular he received a call from his sister in Denver saying that it had rained there yesterday. It happens that the game is to be played in Boston, and you now wonder whether Ben's "information" is really worth anything. Was it? Not likely, since Denver and Boston are so far apart. However, such information might be of interest for cities closer together, such as Pittsburgh and Philadelphia, Kansas City and St. Louis, and so on.

In the previous illustration it seems clear that the information given was not such as to make you want to change any probabilities you might have assigned to events occurring in Boston. For example,

$$P(\text{rain tomorrow in Boston})$$

and

$$P(\text{rain tomorrow in Boston} \mid \text{rain yesterday in Denver})$$

would likely be assigned the same probabilities. Intuitively, you would probably feel that the weather in Boston is independent of the weather in Denver.

Definition 1.10 Events A and B are said to be *independent* if $P(A \mid B) = P(A)$.

Example 1.34 A red die and a green die are tossed. You are interested in the event $A = \{\text{red die shows a six}\}$ and are told that event $B = \{\text{doubles appear}\}$ has occurred. Are A and B independent?

$$P(A \mid B) = \frac{P(A \cap B)}{P(B)} = \frac{\frac{1}{36}}{\frac{1}{6}} = \frac{1}{6}$$

and

$$P(A) = \tfrac{6}{36} = \tfrac{1}{6},$$

so A and B are independent.

If it is known that A and B are independent, the probability of $A \cap B$ is simply the product of the probability of A and the probability of B.

Theorem 1.6 Events A and B are independent if and only if

$$P(A \cap B) = P(A) \cdot P(B). \tag{1.17}$$

Proof: If A and B are independent, then since $P(A \mid B) = P(A)$, we have, from (1.16),

$$P(A \cap B) = P(B) \cdot P(A \mid B) = P(B) \cdot P(A).$$

Similarly, if $P(A \cap B) = P(A) \cdot P(B)$, then

$$P(A \mid B) = \frac{P(A \cap B)}{P(B)} = \frac{P(A) \cdot P(B)}{P(B)} = P(A),$$

so by Definition 1.10, A and B are independent.

Corollary 1.1 If A and B are independent, then

$$P(B \mid A) = P(B).$$

Proof: From (1.14) and (1.17), we have

$$P(B \mid A) = \frac{P(A \cap B)}{P(A)} = \frac{P(A) \cdot P(B)}{P(A)} = P(B).$$

Example 1.34 *(continued)* It is easy to see from the definition of events A and B that

$$\begin{aligned} P(B \mid A) &= P(\text{doubles appear} \mid \text{red die shows a six}) \\ &= P(\text{green die shows a six} \mid \text{red die shows a six}) \\ &= \tfrac{1}{6} = P(B), \end{aligned}$$

so A and B are independent.

To make use of Theorem 1.6 to calculate the probability of an intersection, you must know (or assume) that two events are independent. In some cases it is quite reasonable to assume independence, while in other cases it may be quite unreasonable. For example, when we consider weather conditions in Denver on one day and in Boston on the next day, it is reasonable to assume independence. If we consider the height of a father and the height of his son, we might well expect these to not be independent.

Example 1.35 What is the probability that Podunk U. wins the flip of the coin at the start of two consecutive home football games?

Let A_1 denote the event that P.U. wins the flip at its first home game, A_2 at its second. It is certainly true that $P(A_i) = \frac{1}{2}$ for $i = 1, 2$, and it is logical to assume independence. (Can the coin being flipped really "remember" what happened the previous week?). Consequently,

$$P(A_1 \cap A_2) = P(A_1) \cdot P(A_2) = \tfrac{1}{2} \cdot \tfrac{1}{2} = \tfrac{1}{4}.$$

Theorem 1.6 can be generalized to more than two events. In particular, we will say that events $A_1, A_2, \ldots, A_n$ are totally independent if and only if for any collection of two, three, . . . , or all n events, the probability of the intersection of those events is equal to the product of the probabilities of each of the events. For example,

$$\begin{aligned} P(A_i \cap A_j) &= P(A_i) \cdot P(A_j) && i \neq j \\ P(A_i \cap A_j \cap A_k) &= P(A_i) \cdot P(A_j) \cdot P(A_k) && i \neq j \neq k \\ P(A_i \cap A_j \cap A_k \cap A_l) &= P(A_i) \cdot P(A_j) \cdot P(A_k) \cdot P(A_l) && i \neq j \neq k \neq l \end{aligned} \tag{1.18}$$

and so on.

Example 1.36 What is the probability that P.U. wins the flip of the coin at the start of all five home football games next season?

Using the same notation and reasoning as in Example 1.35, we get

$$\begin{aligned} P(A_1 \cap A_2 \cap A_3 \cap A_4 \cap A_5) &= P(A_1) \cdot P(A_2) \cdot P(A_3) \cdot P(A_4) \cdot P(A_5) \\ &= (\tfrac{1}{2})^5 = \tfrac{1}{32}. \end{aligned}$$

Example 1.37 The proportion of people having blood types A, B, AB, and O are approximately .12, .08, .03, and .77, respectively. If two students from P.U. are selected at random from the students currently enrolled, what is the probability that both have the same blood type? What would this probability be if two brothers were chosen from the set of all pairs of brothers enrolled at P.U.?

In this first case, it is reasonable to assume independence between individuals and consequently between events defined in terms of those individuals. Let A_1 (A_2) denote the event that the first (second) individual chosen has type A blood, and define B_1, B_2, AB_1, AB_2, O_1, and O_2 similarly. Then

$$\begin{aligned}
P(\text{both have same blood type}) \\
&= P((A_1 \cap A_2) \cup (B_1 \cap B_2) \cup (AB_1 \cap AB_2) \cup (O_1 \cap O_2)) \\
&= P(A_1 \cap A_2) + P(B_1 \cap B_2) + P(AB_1 \cap AB_2) + P(O_1 \cap O_2) \\
&= P(A_1)P(A_2) + P(B_1)P(B_2) + P(AB_1) \cdot P(AB_2) + P(O_1) \cdot P(O_2) \\
&= (.12)(.12) + (.08)(.08) + (.03)(.03) + (.77)(.77) \\
&= .6146.
\end{aligned}$$

When we consider the problem when the individuals chosen are brothers, it is not clear that the assumption of independence between individuals is valid. When we put this question to a physician, we were told that there would not be independence. As one example, if both parents are of blood type O, all children will have blood type O. By making certain assumptions it is possible to calculate the probability that brothers (or more generally, siblings) have the same blood type, but the calculations are more involved than we care to do here.

We will conclude this section with a warning to you not to confuse the concepts of mutually exclusive and independence. When thinking of the concept of independence of events A and B, it is easy to think that "the occurrence of A has nothing to do with the occurrence of B." In a sense this is right, but in thinking of a Venn diagram showing events A and B, the phrase "has nothing to do with" becomes interpreted as "nonoverlapping" and consequently gets confused with mutually exclusive events. The two concepts are quite different, as we will see. To avoid the trivial case, assume that A and B are two events with positive probability, and to be explicit say that $P(A) = .4$ and $P(B) = .2$. If A and B are independent, then

$$P(A \cap B) = P(A) \cdot P(B) = (.4) \cdot (.2) = .08$$

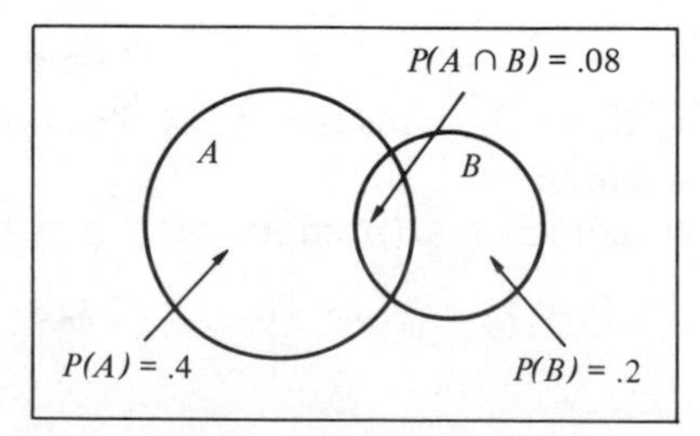

Independent Events

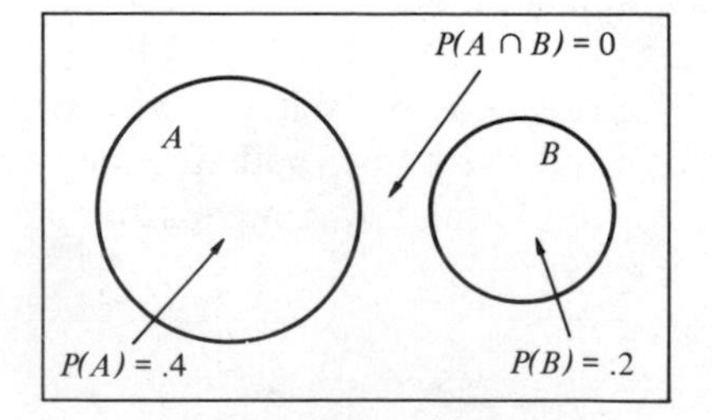

Mutually Exclusive Events

FIGURE 1.8

Venn Diagrams Showing Independent and Mutually Exclusive Events while if they are mutually exclusive,

$$P(A \cap B) = P(\varnothing) = 0.$$

Figure 1.8 shows Venn diagrams for these two cases.

Example 1.38 A red die and a green die are tossed together. Define events A, B, and C as follows: $A = \{\text{red die shows a 2}\}$, $B = \{\text{red die shows an odd number}\}$, $C = \{\text{green die shows an even number}\}$. Which pairs of events, if any, are mutually exclusive? Independent?

If we approach the problem logically, we see that events A and B cannot both happen at the same time, so they are mutually exclusive. Similarly, since the green die certainly can't "know" what happened to the red die, the two events logically must be independent.

Approaching the problem mathematically, we can see (with the help of Figure 1.6) that

$$P(A) = \tfrac{6}{36} = \tfrac{1}{6}, \qquad P(B) = \tfrac{18}{36} = \tfrac{1}{2}, \qquad P(C) = \tfrac{18}{36} = \tfrac{1}{2},$$
$$P(A \cap B) = 0, \qquad P(A \cap C) = \tfrac{3}{36} = \tfrac{1}{12}, \qquad P(B \cap C) = \tfrac{9}{36} = \tfrac{1}{4},$$

so

$$P(A \cap B) = 0$$
$$P(A \cap C) = P(A) \cdot P(C)$$
$$P(B \cap C) = P(B) \cdot P(C),$$

which means that A and B are mutually exclusive while events A and C and events B and C are independent. These relationships are shown in Figure 1.9.

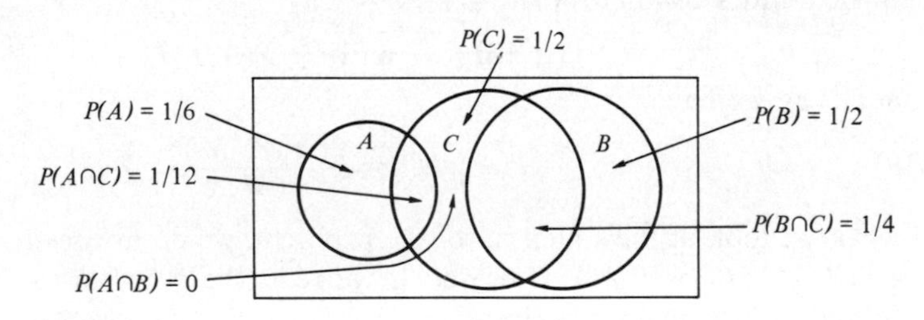

FIGURE 1.9
Venn Diagram for Example 1.38

Of course, it is possible for pairs of events to be neither mutually exclusive nor independent, as you can tell by reviewing Examples 1.31 to 1.33. However, it should be clear from this discussion that if two events A and B each have positive probability, they cannot be both mutually exclusive and independent at the same time.

1.11 BAYES' THEOREM

In Section 1.9 we discussed conditional probability and stated that there are situations where additional information allows us to change the probability assigned to an event. In the simplest situations this is done through a direct application of equation (1.13). However, there are situations where (1.13) can be used, but must be used indirectly.

Example 1.39 Two secretaries, Alice and Ann, do office work for Mr. Dithers, their main job being the filing of papers. Of all the papers that come into the office, Alice files 60% and Ann files the rest. Of course, each secretary occasionally misfiles a paper, and in fact Alice misfiles 5% of the papers she files and Ann misfiles 7% of those she files. Mr. Dithers has been looking for the Dingby contract and has found that it has been misfiled. He is very anxious to berate one of the secretaries, but does not wish to scold the wrong one. What is the probability that the Dingby contract was filed by Alice? By Ann?

This is a problem in conditional probability. Given no other information, we would assign a probability of .6 to the event that Alice filed the Dingby contract and a probability of .4 to the event that Ann filed it. However, we are given the additional information that this contract has been misfiled, so we will want to use conditional probabilities. We begin by defining some relevant events.

$$A_1 = \text{paper was filed by Alice}$$
$$A_2 = \text{paper was filed by Ann}$$
$$M = \text{paper was misfiled.}$$

The probabilities of interest here are

$$P(A_1 \mid M) \qquad \text{and} \qquad P(A_2 \mid M).$$

From (1.13), we have

$$P(A_1 \mid M) = \frac{P(A_1 \cap M)}{P(M)}, \tag{1.19}$$

but when we look at the statement of the problem, we do not see (directly) the quantities $P(A_1 \cap M)$ or $P(M)$. However, if we read the problem carefully, we see that the following information is given:

$$P(A_1) = .60, \qquad P(A_2) = .40$$
$$P(M \mid A_1) = .05, \qquad P(M \mid A_2) = .07.$$

In trying to solve (1.19), we see that we can get $P(A_1 \cap M)$ by

$$P(A_1 \cap M) = P(A_1) \cdot P(M \mid A_1)$$
$$= (.60)(.05) = .030.$$

To find $P(M)$, we note that a paper can be misfiled in two distinct ways, by Alice or by Ann. In set notation this means that we can write the event M as the union of two mutually exclusive events, $A_1 \cap M$ and $A_2 \cap M$:

$$M = (A_1 \cap M) \cup (A_2 \cap M),$$

so

$$\begin{aligned} P(M) &= P(A_1 \cap M) + P(A_2 \cap M) \\ &= P(A_1) \cdot P(M \mid A_1) + P(A_2) \cdot P(M \mid A_2) \\ &= (.60)(.05) + (.40)(.07) = .058. \end{aligned}$$

Substituting into (1.19) gives

$$P(A_1 \mid M) = \frac{.030}{.058} = .517$$

and similar calculations show that $P(A_2 \mid M) = .483$. Consequently, Mr. Dithers is more likely to be right if he chooses to scold Alice. Although Alice does a better job of filing papers than Ann (misfiling 5% as opposed to 7%), her larger work load (60% to 40%) comes into play, making it more likely that she is to blame. The Venn diagram in Figure 1.10 illustrates the preceding situation.

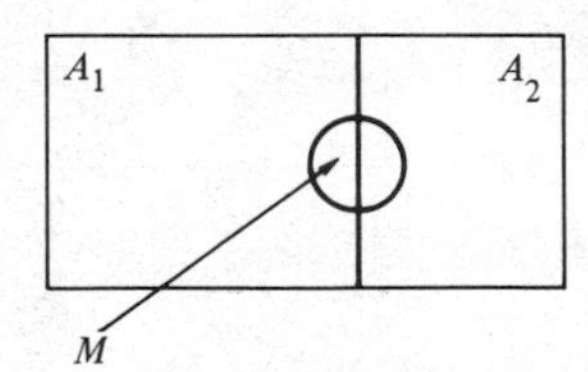

FIGURE 1.10
Venn Diagram for Example 1.39

We could easily generalize Example 1.39 to include any number of secretaries. However, to be specific and to illustrate a slightly more general situation, let us assume that three secretaries, Alice, Ann, and Arlene, file 50, 30, and 20% of the papers, respectively, and that they misfile 5, 7, and 10% of the papers they file. What is the conditional probability that a misfiled paper was filed by Alice, Ann, or Arlene? Letting A_3 denote the event that paper was filed by Arlene, we note that

$$M = (M \cap A_1) \cup (M \cap A_2) \cup (M \cap A_3) \tag{1.20}$$

and that the events on the right-hand side of (1.20) are mutually disjoint, so that

$$P(M) = P(M \cap A_1) + P(M \cap A_2) + P(M \cap A_3).$$

Using

$$P(A_1) = .50, \qquad P(A_2) = .30, \qquad P(A_3) = .20$$
$$P(M \mid A_1) = .05, \qquad P(M \mid A_2) = .07, \qquad P(M \mid A_3) = .10,$$

we get

$$\begin{aligned} P(A_1 \mid M) &= \frac{P(A_1 \cap M)}{P(M)} \\ &= \frac{P(A_1) \cdot P(M \mid A_1)}{P(A_1) \cdot P(M \mid A_1) + P(A_2) \cdot P(M \mid A_2) + P(A_3) \cdot P(M \mid A_3)} \end{aligned}$$

$$= \frac{(.50)(.05)}{(.50)(.05) + (.30)(.07) + (.20)(.10)}$$

$$= \frac{.025}{.025 + .021 + .020} = .379.$$

Similarly, we find that

$$P(A_2 \mid M) = \frac{.021}{.066} = .318.$$

We could find $P(A_3 \mid M)$ by the same approach or by using (1.15) to get

$$P(A_3 \mid M) = 1 - (.379 + .318)$$
$$= .303.$$

The Venn diagram corresponding to this situation is given in Figure 1.11.

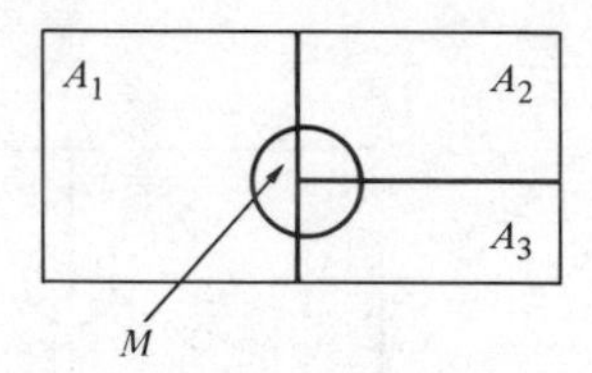

FIGURE 1.11
Venn Diagram for Example 1.39 with Three Secretaries

Note that the sets A_1, A_2 (or A_1, A_2, and A_3) considered are (pairwise) disjoint sets and that in both examples the union of these sets was the entire sample space. The following theorem generalizes the technique used in Example 1.39.

Theorem 1.7 (*Bayes' Theorem*) If $A_1, A_2, \ldots, A_k$ are (pairwise) mutually exclusive events whose union is S, and if B is some event such that $P(B) > 0$, then

$$(1.21) \qquad P(A_i \mid B) = \frac{P(A_i) \cdot P(B \mid A_i)}{P(A_1) \cdot P(B \mid A_1) + P(A_2) \cdot P(B \mid A_2) + \cdots + P(A_k) \cdot P(B \mid A_k)}$$

for $i = 1, 2, \ldots, k$.

Proof: Event B may or may not intersect with each of the events A_i, but in any case we can write B as the union of k (pairwise) mutually exclusive events (see Figure 1.12). That is,

$$B = (B \cap A_1) \cup (B \cap A_2) \cup \ldots \cup (B \cap A_k).$$

Using this fact, we get

$$P(A_i \mid B) = \frac{P(A_i \cap B)}{P(B)}$$

$$= \frac{P(A_i \cap B)}{P(A_1 \cap B) + P(A_2 \cap B) + \cdots + P(A_k \cap B)}$$

$$= \frac{P(A_i) \cdot P(B \mid A_i)}{P(A_1) \cdot P(B \mid A_1) + P(A_2) \cdot P(B \mid A_2) + \cdots + P(A_k) \cdot P(B \mid A_k)}.$$

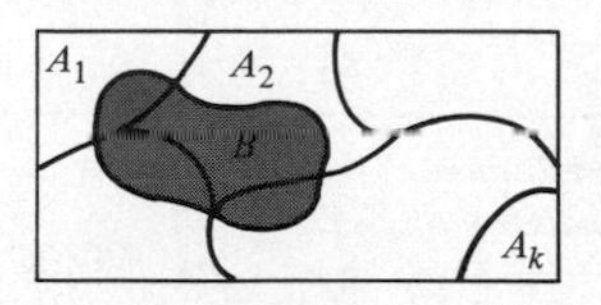

FIGURE 1.12
Venn Diagram for Bayes' Theorem

Example 1.40 Mr. Dee Syder is an inspector at the Watzit Manufacturing Company. Dee inspects watzits coming out of a watzit maker. As in all manufacturing processes, some of the watzits are defective. Dee has noticed that on some days 1% of the watzits produced are defective, while on other days 5 or 10% are defective. More explicitly, 60% of the time 1% of the production is bad, while 5 and 10% are bad 30 and 10% of the time, respectively. If the first item Dee inspected today was defective, what is the probability that the watzit maker is currently producing 1% defectives? 5%? 10%?

To solve this problem, it might be helpful to visualize a watzit maker (as in Figure 1.13), which has an internal dial that can be set to determine the rate of defectives. We might also think of this dial as being relatively inaccessible to the machine operator and set to the various positions by gremlins before the day's work is begun. Define events A_1, A_2, A_3, and B as follows:

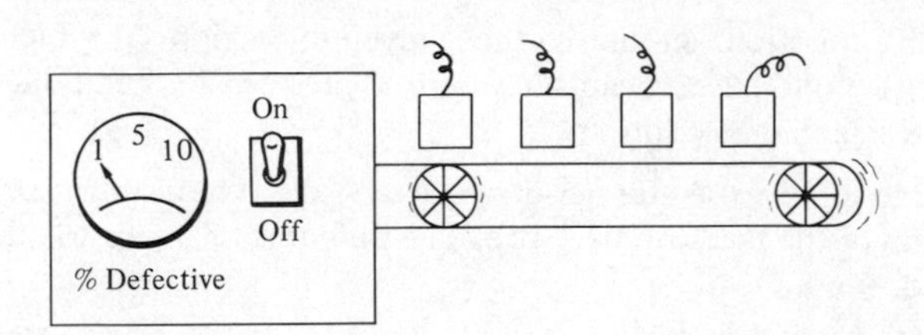

FIGURE 1.13
Watzit Maker Set at 1% Defective Rate

$$A_1 = \text{dial is set at } 1\%$$
$$A_2 = \text{dial is set at } 5\%$$
$$A_3 = \text{dial is set at } 10\%$$
$$B = \text{inspected watzit is defective.}$$

From the statement of the problem, we get

$$P(A_1) = .60, \quad P(A_2) = .30, \quad P(A_3) = .10$$
$$P(B \mid A_1) = .01, \quad P(B \mid A_2) = .05, \quad P(B \mid A_3) = .10.$$

Using equation (1.21) with $k = 3$ and $i = 1$, we get

$$P(A_1 \mid B) = \frac{.006}{.006 + .015 + .010} = .1935.$$

and similarly, $P(A_2 \mid B) = .4839$ and $P(A_3 \mid B) = .3226$.

It is often helpful to put all this information in a table, as is done in Table 1.1. Since the second and last columns correspond to probabilities assigned to the events

TABLE 1.1

Setting, A_i	Initial Probability, $P(A_i)$	Conditional Probability of Observed Event, $P(B \mid A_i)$	Product, $P(A_i) \cdot P(B \mid A_i)$	Revised Probability, $P(A_i) \cdot P(B \mid A_i)/\sum = P(A_i \mid B)$
A_1	.60	.01	.006	.1935
A_2	.30	.05	.015	.4839
A_3	.10	.10	.010	.3226
	1.00		$\sum = .031$	1.0000

A_i, events that are pairwise mutually exclusive and whose union is S, it follows that these columns must sum to 1. Also the sum of the fourth column, denoted by $\sum$, is equal to $P(B)$, the probability that an inspected watzit is defective.

EXERCISES

1. Make a list of five questions similar to those given in Section 1.1 which might be asked about (a) a political convention where a presidential candidate is to be chosen; (b) a presidential election.

2. In the game of Monopoly, the number of spaces a token is moved forward is determined by the sum of the faces on two dice. The outcome of interest is the number of spaces moved. Find S.

3. Mr. Hugh Dini, a magician, asks a member of his audience to choose a card at random from a well-shuffled deck. Find S.

4. Mr. I. M. Syk has a 9:00 A.M. appointment with his doctor. (a) If he is interested in the amount of time he will have to wait beyond 9 o'clock before being called into the examining room, what is S? (b) Assume that Mr. Syk is also interested in counting the number of emergency patients the doctor must treat before he gets his chance to be examined. What is S in this case?

5. The manager of the Gro-Sirs Supermarket has a special this week on frozen orange juice, with a limit of 10 cans. As an experiment the assistant manager will count the number of cans of this orange juice bought by a customer. Find S.

6. Determine whether S is finite, countably infinite, or continuous in (a) Exercise 2; (b) Exercise 4(a); (c) Exercise 4(b).

7. Determine whether S is finite, countably infinite, or continuous in (a) Exercise 3; (b) Exercise 5.

8. The sample space from Example 1.2 of Section 1.2 consisted of the six Hubbard children: Alan, Betty, Carl, Diane, Eric, and Frieda, ages 1 to 6, respectively. List the elements of S satisfying the following conditions: (a) $C = \{$child chosen is younger than 4$\}$; (b) $D = \{$child chosen is a girl younger than four$\}$; (c) $E = \{$child chosen is a boy who is at least four years old$\}$; (d) $F = \{$child chosen is a four-year-old boy$\}$.

9. Using the sample space described in Exercise 8, define three events A_1, A_2, and A_3

in such a way that $A_1 \cap A_2 \cap A_3 = \varnothing$ while $A_1 \cap A_2 \neq \varnothing$, $A_1 \cap A_3 \neq \varnothing$, and $A_2 \cap A_3 \neq \varnothing$.

10. Using the events C, D, E, and F as defined in Exercise 8, determine which (if any) pairs of events are mutually exclusive. (*Hint:* There are six distinct pairs for you to consider.)

11. Using the events C, D, E, and F as defined in Exercise 8, list the elements of the following:

(a) $C \cap D$; (b) $C \cup E$;
(c) $\bar{F}$; (d) $C \cup D \cup E$.

12. Using events C, D, and E as defined in Exercise 8, carefully construct a Venn diagram which shows the relationships that exist among these sets.

13. A card is chosen from a well-shuffled deck of ordinary playing cards. Consider the following events: $A = \{\text{ace is chosen}\}$, $B = \{\text{black card is chosen}\}$, $C = \{\text{club is chosen}\}$, $D = \{\text{diamond is chosen}\}$, $E = \{\text{even-numbered card (2, 4, 6, 8, or 10) is chosen}\}$, $F = \{\text{face card chosen}\}$. Determine which (if any) pairs of events are mutually exclusive. (*Hint:* There are 15 distinct pairs for you to consider.)

14. Referring to Exercise 13, find the number of elements of S that are elements of the following events: (a) $A \cap B$; (b) $B \cup \bar{B}$; (c) $C \cap D$; (d) $\bar{D}$; (e) $E \cup F$; (f) $F \cup D$.

15. At the Pizza Palace (Example 1.9) pizzas can be made with a choice of thick or thin crust and any of four meat toppings and five vegetable toppings. If pizzas can be made with at most one meat topping and/or at most one vegetable topping, how many different types of pizza can be made?

16. A men's clothing store offers leisure suits consisting of two pants, three jackets, and two vests. Assuming that a vest need not be worn, how many different outfits are possible?

17. License plates in a certain state are of the form

letter–number–letter–number–number–number

(e.g., A1A120). (a) Assuming that any letter and any number can be used any number of times, how many different license plates could be issued? (b) How many of these would not contain the letter A? (c) How many would contain the letter A?

18. In Exercise 17 we assumed that the first letter of the license plate could be any letter. In fact, the first letter corresponds to the month in which the plates are issued. Consequently, it is necessary to choose 12 letters to correspond to the months (e.g., A $\equiv$ January, S $\equiv$ September, Z $\equiv$ December). Assume that these 12 letters have been chosen. How many different license plates can be issued?

19. A grievance committee of three is to be chosen from 40 employees in a small company. How many different committees could be chosen?

20. How many different "words" can be made by rearranging all the letters of the word "Mississippi"?

21. During the last football season, a certain college team won the pregame toss of the coin only 3 times in the 11 games. The actual sequence of wins and losses of the coin toss was L L L W L L W L L W L. How many different sequences of 3 W's and 8 L's are there?

22. The football team mentioned in Exercise 21 had a 3–6–2 won–loss–tied record. The actual sequence of wins, losses, and ties was L L T W T L W L L W L. In how many different ways could their final record have come out 3–6–2?

23. Six people are lined up waiting to go into a bank. In how many ways might they be lined up (a) if there are no restrictions?; (b) if two individuals, Amos Hatfield and Zeke McCoy, refuse to stand next to each other?

24. In how many ways could the six persons be seated in the situation as described in Example 1.17 of Section 1.4 if the three wives insist on sitting together?

25. Show that $\binom{n}{k} = \binom{n}{n-k}$. Discuss why this result should logically hold when viewed from the point of view of "choosing k objects out of n."

26. (a) Write out the terms of $(x + y)^2$, $(x + y)^3$, and $(x + y)^4$. (b) Find $\binom{2}{0}$, $\binom{2}{1}$, $\binom{2}{2}$; $\binom{3}{0}$, $\binom{3}{1}$, $\binom{3}{2}$, $\binom{3}{3}$; $\binom{4}{0}$, $\binom{4}{1}$, $\binom{4}{2}$, $\binom{4}{3}$, $\binom{4}{4}$. Compare the coefficients of the terms of (a) with the corresponding parts of (b). It can be shown in general that the coefficient of $x^k y^{n-k}$ in the expansion of $(x + y)^n$ is $\binom{n}{k}$.

27. "Three-way" light switches can be wired into the circuitry of a house so that a light can be controlled by either of two switches. Each switch has three terminals, to which three wires are to be connected. In how many different ways can the wires be connected? Assuming that only one way is correct, what is the probability that an amateur electrician, connecting the wires at random, connects them correctly on the first try?

28. Give two ways (other than those given in Section 1.5) in which a weather report deals with the problems of variability.

29. A single fair die is thrown. Find the probability assigned to each of the events A, B, C, and D defined by $A = \{\text{odd number appears}\}$, $B = \{\text{even number appears}\}$, $C = \{\text{number less than 4 appears}\}$, $D = \{\text{number greater than 2 appears}\}$.

30. Using the events A, B, C, and D defined in Exercise 29, find (a) $P(A \cap B)$; (b) $P(A \cup B)$; (c) $P(\bar{C})$; (d) $P(C \cap D)$; (e) $P(C \cup D)$.

31. Elinor, a realtor, has a list of 20 rental units: 7 single-family dwellings, 5 duplex units, and 8 apartments. Five of the single-family dwellings have more than two bedrooms, while three of the duplexes and two of the apartments have more than two bedrooms. In her haste to meet a family looking for a rental unit, she has only brought a key for one of these units with her. Find the probabilities of the following events: (a) $A = \{\text{key is for a duplex}\}$; (b) $B = \{\text{key is for a unit having more than two bedrooms}\}$; (c) $\bar{A}$; (d) $(A \cap B)$; (e) $A \cup B$.

32. Assume that the six people mentioned in Exercise 23 are lined up at random. What is the probability that in this random ordering Amos and Zeke are standing next to each other?

33. If a "word" is made by choosing at random one of the permutations of the word "STOP," what is the probability that it will actually be an English word?

34. If a "word" is made by randomly arranging the letters of the word "LOOP," what is the probability that it will actually be an English word?

35. Consider the sample space defined in Exercise 2. Are each of these points equally likely? Explain.

36. Drop an ordinary thumbtack onto a hard surface 100 times and record the number of times it lands point up. Find the relative frequency of occurrence. (If this experiment is done by several individuals with the intent of combining the results, make sure that the conditions of the experiment are as uniform as possible.)

37. Take a coin, hold it on edge with one finger, and spin it by flicking it with your free hand. Be sure to do this on a hard surface and be sure that it gets a good spin. Observe whether it lands heads up or tails up. Repeat this experiment 50 times. Find the relative frequency of the number of heads. Using the same coin, flip it 50 times and find the relative frequency of the number of heads.

38. To find the relative number of cars made by different manufacturers, it is proposed to observe cars traveling on a certain main highway and classify them as manufactured by Chrysler, Ford, General Motors, or "Other." If this is done, the sample space contains exactly four points. Do you believe that these are equally likely? If so, what probability would you assign to each elementary event in S? If not, based on your experience and intuition, what probability would you personally estimate for each elementary event in S?

39. To check the estimates made in Exercise 38, actually perform the experiment indicated there by observing 100 cars. Describe the location you used to make your observations. Explain how a poor choice of location could influence your results.

40. Give an argument based on Venn diagrams and the relationship of area and probability to justify Theorem 1.3, which states that $P(\bar{A}) = 1 - P(A)$.

41. In the game of Parcheesi, two dice are rolled and a player may move a token forward by a number of spaces equal to the number showing on the first die, the number showing on the second die, or the sum of the numbers on the two dice (see Example 1.26 of Section 1.8). Margie is playing Parcheesi. Find the probability that on her next turn she will have the opportunity to move her token (a) exactly four spaces (event A); (b) exactly five spaces (event B); (c) either four or five spaces (event $A \cup B$).

42. In the game of Monopoly, if a player is sent to jail, he may get out of jail free if on his next turn he throws "doubles" (i.e., if the same number faces up on both dice). What is the probability of him throwing doubles on his next turn?

43. In a game of Monopoly, assume that your token is on the Shortline Railroad, two spaces from Park Place and four spaces from the Boardwalk, both of which are owned by your opponents. Assume that you will go bankrupt if you land on one of these, but not otherwise. What is the probability that you will go bankrupt on your next turn?

44. Using Axiom 3, prove that if A_1, A_2, and A_3 satisfy $A_i \cap A_j = \varnothing$ for $i \neq j$, then $P(A_1 \cup A_2 \cup A_3) = P(A_1) + P(A_2) + P(A_3)$.

45. Assume that A_1, A_2, and A_3 are events such that each pair is mutually exclusive. Determine which assignments of probability are consistent with the axioms and/or theorems given in Section 1.8. For those which are not consistent, state which axiom or theorem is violated.

(a) $P(A_1) = .2$, $P(A_2) = .3$, $P(A_3) = .5$;
(b) $P(A_1) = .5$, $P(A_2) = -.1$, $P(A_3) = .6$;

(c) $P(A_1) = .3$, $P(A_2) = .4$, $P(A_3) = .5$;
(d) $P(A_1) = .2$, $P(A_2) = .3$, $P(A_3) = .4$.

46. Determine which of the following assignments of probabilities are consistent with the axioms and/or theorems of Section 1.8. For those which are not consistent, state which axiom or theorem is violated.
(a) $P(A) = .6$, $P(B) = .6$, $P(A \cap B) = .1$;
(b) $P(A \cap B) = 0$, $P(A) = .1$, $P(A \cup B) = .15$, $P(B) = .1$;
(c) $P(A) = .1$, $P(B) = .2$, $P(A \cup B) = .4$;
(d) $P(A) = .2$, $P(B) = .5$, $P(A \cap B) = .1$.

47. A card is chosen from a well-shuffled deck of playing cards. Define events A_1, A_2, A_3, and B as follows: A_1 = ace is chosen, A_2 = club is chosen, A_3 = diamond is chosen, B = black card is chosen. Determine whether $P(A_i)$ is greater than, less than, or equal to $P(A_i \mid B)$ for $i = 1$, 2, and 3.

48. A tank of tropical fish contains 6 tetras, 4 zebras, and 2 blue moons (see Example 1.31, Section 1.9). If Felix nets two fish at random, what is the probability that he nets two different species of fish?

49. If two dice are rolled, find the probability that at least one shows a 4 given that the sum is greater than or equal to 8.

50. Ray Nershine is a sales representative for a company that sells home solar heating equipment. The company has put an ad in the local newspaper and Ray will follow up on leads obtained from response to this ad. Of the 80 people responding to the ad, 10 are actually interested in equipment for a new home, while 20 are interested in equipment to add to an already built home; the remainder of the people are merely curious. If Ray is given three leads at random from the 80, what is the probability that (a) all three are interested in equipment for a new home?; (b) at least one is interested in equipment to add to an already built home?

51. A track official has bought a box of blanks to be used with a starting gun. The box contains 5 bad shells among 50. (a) If the official chooses shells from this box, what is the probability that three bad shells in a row will be chosen? (b) If 5 races need to be started with these shells, what is the probability that at least one race will not be started properly because of a bad shell?

52. A dish of M & M's contains 10 green, 20 brown, and 20 yellow candies. Scott is allowed to take three candies. Assuming that he chooses them at random, what is the probability that he gets at least two different colored candies among his three pieces? (*Hint:* You may wish to consider the complement of this event and use Theorem 1.3.)

53. In playing the game of Scrabble, you see that, using some letters already on the board, you will be able to put the word "quixotic" over a "triple word score" space if you can draw a "u" and an "i" to replace two letters you just used. If there are two u's and three i's among the 20 remaining tiles, what is the probability that you will draw the letters you need?

54. The weather in Camelot, as in many places, is unpredictable. However, Merlin has noted that if it rains one day, the probability of rain the next day is 40%, while if there is no rain on a given day, the probability of rain the next day is only 20%. Arthur can get his scheduled tournament completed if there is no rain on at least 2 of the next 3 days. Given that it is sunny today and that the tournament is scheduled to start tomorrow, what is the probability that Arthur's tournament will be completed?

55. To show that n events $A_1, A_2, \ldots, A_n$ are totally independent, it must be shown that several equations of the form of (1.18) hold. Show that in fact $2^n - n - 1$ such equations must be verified.

56. A fair coin is flipped five times in succession. What is the probability of having a "streak" of 3 or more heads in a row in these five flips?

57. Assume that the probability of a white Christmas in Kansas City is .6. What is the probability of at least 1 white Christmas there in the next 4 years?

58. Would you expect the events of "a white Christmas in K.C." and a "white New Year's Day in K.C." to be independent? Explain.

59. The Bankamericharge Credit Company offers credit limits of \$200, \$400, . . . , \$1000 to card holders, depending on their credit rating. Assume that the probability is .20 that a credit-card holder selected at random owes half or more of his limit and that 10% of the credit card holders have a zero balance. (a) What is the probability that a payment received in this afternoon's mail will be from an account owing less than half its credit limit? (b) What is the probability that the first three payments received in this afternoon's mail will all be from accounts owing less than half of their credit limit?

60. Let A and B be two events such that $P(A) = .4$ and $P(B) = .6$. Find $P(A \cup B)$ if A and B are (a) mutually exclusive; (b) independent. Construct Venn diagrams depicting (a) and (b).

61. Let A and B be two events such that $P(A) = .4$ and $P(A \cup B) = .6$. Find $P(B)$ if A and B are (a) mutually exclusive; (b) independent.

62. A red die and a green die are tossed together. Define events A, B, and D as follows: $A = \{\text{red die shows a 3}\}$, $B = \{\text{red die shows an even number}\}$, and $D = \{\text{sum of both dice is} \leq 4\}$. Which pairs of events, if any, are (a) mutually exclusive?; (b) independent?

63. Mrs. Hubbard has a yellow cookie jar containing 20 chocolate chip and 10 peanut butter cookies and a green cookie jar containing 20 chocolate chip and 30 peanut butter cookies. Carl randomly picked a cookie jar to snitch from and took a cookie from the jar at random. Mrs. Hubbard found him eating a chocolate chip cookie. What is the probability that he snitched from (a) the yellow cookie jar?; (b) the green cookie jar?

64. Consider the same situation as in Exercise 63 but assume that the snitcher this time is Diane. Since Diane's favorite color is green, she is twice as likely to snitch from the green jar as from the yellow jar. If Diane is found eating a chocolate chip cookie, what is the probability that she snitched from (a) the yellow cookie jar?; (b) the green cookie jar?

65. Three identical wallets each contain two bills. One contains two \$5 bills, another contains two \$10 bills, and the third contains one \$5 and one \$10 bill. One of these three wallets is randomly selected and a randomly chosen bill is removed. If the bill is found to be a \$5, what is the probability that the remaining bill is also a \$5?

66. Jack Pyne runs a mail-order lumber and supply business located in Seattle. He receives orders from four areas of the country as follows:

Area	*Percent of Orders*
East	10
Midwest	30
South	20
West	40

Forty percent of the orders which come from the East, 20% from the Midwest, 25% from the South, and 10% from the West are for at least \$25. What is the probability that (a) an order is not from the West?; (b) an order from the South is for less than \$25?; (c) an order is from the West given that it is for at least \$25?; (d) an order is for at least \$25?

67. Let A_1, A_2, A_3, and B be events such that $P(A_1) = .2$, $P(A_2) = .3$, $P(A_3) = .5$, $P(B|A_1) = .10$, $P(B|A_2) = .20$, and $P(B|A_3) = .40$. Assume that A_1, A_2, and A_3 are pairwise mutually exclusive. Find (a) $P(A_1|B)$; (b) $P(B)$.

2

RANDOM VARIABLES—PROBABILITY DISTRIBUTIONS AND EXPECTATIONS

2.1 RANDOM VARIABLES

In the examples and exercises of Chapter 1 you saw several examples of sample spaces. Some of those sample spaces consisted of elements that were real numbers (Example 1.3), some had elements that would be easy to associate with real numbers (Example 1.2), and others had no obvious relationship to real numbers (Example 1.1).

Let us begin by considering some examples where there is a "natural" association of real numbers to points in a sample space. In Example 1.2 it seems rather natural to associate the age (a real number) of each Hubbard child with that child. Another fairly natural association of real numbers to points in this sample space would be the weight of the child chosen. Figure 2.1 shows these associations. It would not be hard to find other natural associations using this particular sample space.

In mathematical terms, a rule for associating a real number with each element in a sample space is called a function. In the study of probability and statistics, these functions are called random variables.

Definition 2.1 A *random variable* is a function that assigns a real number to each point in a sample space.

In mathematics, functions are generally denoted by f, and if several different functions are being considered they might be distinguished by using f, g, and h or

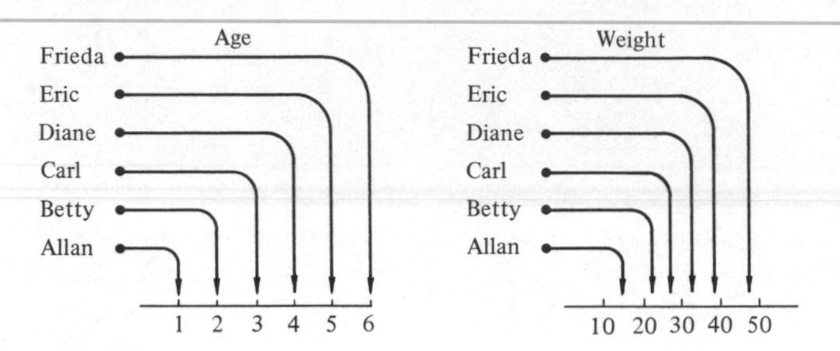

FIGURE 2.1

Associating Real Numbers with Points in a Sample Space

f_1, f_2, f_3, and so on. In probability and statistics, random variables are generally denoted by X, and if several different random variables are being considered they might be distinguished by using X, Y, and Z or X_1, X_2, X_3, and so on.

Example 2.1 Consider the sample space of the Hubbard children (Example 1.2) and define

$$X = \text{age of child chosen}$$

$$Y = \text{weight of child chosen.}$$

Here X and Y are random variables.

You might wonder why the particular term "random variable" was adopted as a name for a function. The "variable" part of the name relates to the fact that any of several values might be assigned to the function (e.g., in Example 2.1 the value of X might be 1, 2, 3, 4, 5, or 6). The "random" part relates to the fact that the particular value that will be assigned depends on the outcome of a random experiment.

Example 2.2 In Example 1.1 we mentioned the experiment of flipping a coin. In that case the sample space consists of two points, H and T. Although no "natural" association of real numbers with these points comes immediately to mind, we could arbitrarily define a random variable X by

$$X = \begin{cases} 10 & \text{if H occurs} \\ -5 & \text{if T occurs.} \end{cases}$$

While there may be no logical reason to assign these particular values, X is nevertheless a perfectly valid random variable. On the other hand, if a person agreed to pay you \$10 if a coin came up heads when flipped, if you paid him \$5 if it came up tails, X would represent your gain from one trial of this experiment.

Using this same sample space, we could define Y to be the number of heads that appear on a single coin flip. In this case we would have

$$Y = \begin{cases} 1 & \text{if H occurs} \\ 0 & \text{if T occurs.} \end{cases}$$

Example 2.3 At the Fizzy Cola Bottling Company, Fizzy Cola is put into "16-ounce" bottles. However, the actual amount put into a bottle is not always 16 ounces. (You can verify this yourself. The next time you go to a grocery store, notice the different amounts of cola in the bottles in a carton! If you cannot find Fizzy Cola, check some other brand.) Mr. Peeper's job is to see that the right amount of cola goes into the bottles, and to this end he measures the amount of cola in every 1000th bottle. The sample space for this experiment is continuous and, for the sake of illustration, consists of values between 15 and 17 ounces, say. We could define the random variable X to be the amount of cola measured (i.e., if 16.2 ounces is measured, $X = 16.2$, etc.) We could also define

$$Y = \begin{cases} -1 & \text{if amount of cola is} < 15.75 \text{ oz} \\ 0 & \text{if amount of cola is between 15.75 and 16.25 oz} \\ 1 & \text{if amount of cola is} > 16.25 \text{ oz.} \end{cases}$$

Example 2.4 In Appendix A3 you will find some data that were gathered from 200 students enrolled in a statistics course at the University of Missouri–Columbia. The data themselves are as given to us by the students. If a student is chosen at random from this group, we could define

$$X_1 = \text{age of student chosen}$$

$$X_2 = \text{height of student chosen}$$

$$X_3 = \begin{cases} 1 & \text{if student chosen is a female} \\ 0 & \text{if student chosen is a male,} \end{cases}$$

and so on.

It is often convenient to write a general expression for the value assumed by a random variable, rather than a specific number. We will let lowercase letters represent values assumed by a random variable. For instance, in Example 2.3 we might describe the values that the random variables X and Y might assume by

$$X = x \qquad 15 \leq x \leq 17$$

$$Y = y \qquad y = -1, 0, +1.$$

2.2 PROBABILITY DISTRIBUTIONS FOR DISCRETE RANDOM VARIABLES

The distinction between discrete and continuous random variables is much like the distinction between discrete and continuous sample spaces (see Section 1.2). More explicitly, a discrete random variable is one that can assume either a finite or a countably infinite number of different values. A continuous random variable is one that assumes values over a continuous range of values. In this section we will concentrate our attention on discrete random variables.

There is a direct relationship between the values assumed by a random variable and the points in a sample space. This relationship allows us to define events by using random variables. In Example 2.1 we could express the event $A =$ {child chosen is at most three years old} by $\{X \leq 3\}$. You can see that a certain economy of notation is attained by this kind of expression.

Example 2.5 From Figure 2.1 and the definition of the random variable Y in Example 2.1, we can write the following alternative expressions for the following events:

$$\begin{aligned} A &= \{\text{child chosen weighs less than 20 lb}\} = \{Y < 20\} \\ B &= \{\text{child chosen weighs between 20 and 30 lb (inclusive)}\} \\ &= \{20 \leq Y \leq 30\} \\ C &= \{\text{child chosen weighs at least 30 lb}\} = \{Y \geq 30\}. \end{aligned} \tag{2.1}$$

If probabilities have been assigned to points in a sample space, it is possible to assign probabilities to any events defined on that sample space, and in particular it is possible to assign probabilities to events defined in terms of a random variable.

Example 2.6 Each of the children referred to in Example 2.1 is equally likely to be chosen, so a probability of $\frac{1}{6}$ is assigned to each point of S. Since there are three children who are "at most three years old," it follows that

$$P[X \leq 3] = \tfrac{3}{6} = \tfrac{1}{2}.$$

Also, it is easy to see from Figure 2.1 that

$$\begin{aligned} P[Y \leq 20] &= \tfrac{1}{6} \\ P[20 \leq Y \leq 30] &= \tfrac{2}{6} \\ P[Y \geq 30] &= \tfrac{3}{6}. \end{aligned}$$

Given a particular random variable, it is convenient to be able to assign probabilities to events like those in (2.1) without always referring back to the sample space itself. This can be done by describing all possible values of the random variable and the associated probabilities. For example, using X as defined in Example 2.1, we have

x:	1	2	3	4	5	6
$P[X = x]$:	$\frac{1}{6}$	$\frac{1}{6}$	$\frac{1}{6}$	$\frac{1}{6}$	$\frac{1}{6}$	$\frac{1}{6}$.

Definition 2.2 The *probability distribution of a discrete random variable* X is the set of possible values of X and the corresponding probabilities.

A probability distribution can be expressed by a table listing the values that X can assume, together with the corresponding probabilities or by a mathematical function. Referring again to the random variable X of Example 2.1, we could write

$$P[X = x] = \begin{cases} \frac{1}{6} & \text{if } x = 1, 2, \ldots, 6 \\ 0 & \text{otherwise.} \end{cases}$$

Example 2.7 Two-year-old Betty is stringing beads. She is choosing beads at random from a box containing 5 red beads and 10 blue beads. Let X be the number of red beads among the first two Betty strings. What is the probability distribution of X?

The possible values of X are 0, 1, and 2. By defining R_i to be the event that the ith bead chosen is red and using conditional probability, we see that

$$\begin{aligned}
P[X=0] &= P[\bar{R}_1 \cap \bar{R}_2] = P[\bar{R}_1] \cdot P[\bar{R}_2 \mid \bar{R}_1] \\
&= \tfrac{10}{15} \cdot \tfrac{9}{14} = \tfrac{9}{21} \\
P[X=1] &= P[(R_1 \cap \bar{R}_2) \cup (\bar{R}_1 \cap R_2)] \\
&= P[R_1] \cdot P[\bar{R}_2 \mid R_1] + P[\bar{R}_1] \cdot P[R_2 \mid \bar{R}_1] \\
&= \tfrac{5}{15} \cdot \tfrac{10}{14} + \tfrac{10}{15} \cdot \tfrac{5}{14} = \tfrac{10}{21} \\
P[X=2] &= P[R_1 \cap R_2] = P[R_1] \cdot P[R_2 \mid R_1] \\
&= \tfrac{5}{15} \cdot \tfrac{4}{14} = \tfrac{2}{21},
\end{aligned}$$

so the probability distribution is given by

x:	0	1	2
$P[X=x]$:	$\frac{9}{21}$	$\frac{10}{21}$	$\frac{2}{21}$.

The same result can be expressed in functional form using combinations by

$$P[X=x] = \begin{cases} \binom{5}{x}\binom{10}{2-x} \Big/ \binom{15}{2} & x = 0, 1, 2 \\ 0 & \text{otherwise.} \end{cases} \tag{2.2}$$

[Note that the denominator of (2.2) is the total number of ways of choosing 2 beads from 15 while the numerator is the total number of ways of choosing x red beads from the 5 red available times the total number of ways of choosing the remaining $2 - x$ blue beads from the 10 blue available.]

The probability distribution for any discrete random variable must satisfy the following two conditions:

$$P[X=x] \geq 0 \qquad \text{for all } x \tag{2.3}$$

$$\sum_{\text{all } x} P[X=x] = 1. \tag{2.4}$$

These two conditions follow directly from Axioms 1 and 2 of Section 1.8. You can easily verify that these conditions are satisfied for the examples of probability distributions given so far in this section. (If you need to review the $\sum$ notation, read Appendix A1.)

The probability distribution of a random variable can be used to find the probability that a random variable assumes values in some interval of values. For example, using the probability distribution given in (2.2), we can find

$$P[1 \leq X \leq 2] = \tfrac{12}{21}$$

by simply adding the probabilities for those values x satisfying $1 \leq x \leq 2$ (i.e., those values between 1 and 2 inclusive). If we change the interval slightly by not including one (or both) of the end points, we may also change the probability associated with that interval. In particular,

$$P[1 < X \leq 2] = \tfrac{2}{21}$$

(2.5)
$$P[1 \leq X < 2] = \tfrac{10}{21}$$

$$P[1 < X < 2] = 0.$$

From this you can see that in calculating the probability that a discrete random variable lies between two values, a and b say, you must be careful to specify which end points are to be included. We will see in the next section that this is not as important for continuous random variables. It should be noted that the probability distribution for a random variable can be used to find the probability that a random variable assumes a value in a set A other than an interval. This is done as follows:

(2.6)
$$P[X \in A] = \sum_{x \in A} P[X = x].$$

For example, using the probability distribution

x:	0	1	2
$P[X = x]$:	$\frac{9}{21}$	$\frac{10}{21}$	$\frac{2}{21}$

(see Example 2.7), we find that

$$P[|X - 1| \geq 1] = P[X \geq 2 \text{ or } X \leq 0] = P[X \in A]$$
$$= \sum_{x \in A} P[X = x] = \tfrac{9}{21} + \tfrac{2}{21} = \tfrac{11}{21},$$

where $A = \{x : x \leq 0 \text{ or } x \geq 2\}$.

We have said that the probability distribution for a discrete random variable can be expressed by means of a table or by means of a function. In either case it might be desirable to get a visual picture of the probability distribution by graphical means. One way of doing this is by means of a probability histogram.

Definition 2.3 If a discrete random variable X assumes only integer values, then a *probability histogram* for the probability distribution of X can be constructed. A probability histogram is a bar graph with (a) bars centered at x having (b) unit width and (c) height equal to $P[X = x]$.

The probability histogram for the probability distribution given in (2.2) is given in Figure 2.2. Notice that it follows from (a), (b), and (c) of Definition 2.3 that the *area* of the bar centered at x is equal to the *probability* that X equals x.

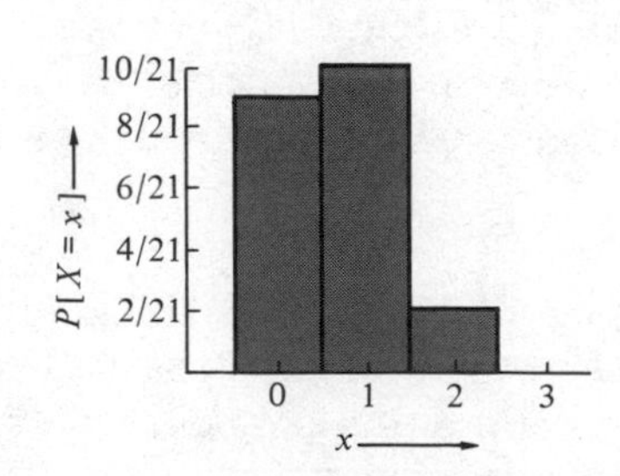

FIGURE 2.2
Probability Histogram

If a discrete random variable assumes values other than integer values, it may not be as easy to draw a bar graph having the area of the bar equal to the probability. In such a case it is possible to draw a graph having a vertical line at the value x with the height of the line equal to $P[X = x]$. See, for example, Figure 2.13 in Section 2.6.

We shall see that there are many random variables that arise from very different experiments but have the same probability distribution. In particular, consider the following situations:

1. A fair coin is flipped twice, and each time it is noted whether a head or a tail occurs. The sample space consists of four points. Define X to be the number of heads that appear.
2. A fair die is rolled twice and the number facing up each time is noted. The sample space consists of 36 points. Define Y to be the number of times an odd number faces up.
3. A pediatrician checks his files for families consisting of exactly two children, noting the sex of each child in the family. The sample space consists of four points. Define W to be the number of girls in the family.

In each of these situations it is easy to see that the random variable defined can assume the values 0, 1, and 2 and it is not too hard to see that the corresponding probabilities are $\frac{1}{4}$, $\frac{1}{2}$, and $\frac{1}{4}$. (This assumes that the probability of a child being a girl is $\frac{1}{2}$ and that the sex of children within the same family will be independent.) In view of the fact that such different situations can lead to random variables having the same probability distribution, we will often consider only the probability distribution itself, without reference to the particular sample space that gave rise to the random variable.

2.3 PROBABILITY DISTRIBUTIONS FOR CONTINUOUS RANDOM VARIABLES

In Example 2.3 we defined a random variable X to be the amount of Fizzy Cola put into a 16-ounce bottle and pointed out that X would be a continuous random variable. We assumed, for the sake of illustration, that the amount put into the bottles was in the range 15 to 17 ounces. As in the case of discrete random variables, we would like to describe the probability distribution of this continuous random variable so that we could assign probabilities to events such as $P[X \geq 16]$ and $P[15.8 \leq X \leq 16.2]$, for example. As we consider the problem of assigning probabilities to events such as these we will find some very marked differences from the case of discrete random variables.

Let us imagine for now that the machine that fills the bottles is presently in very good working order. Using your intuition, which of the following events would you think should be assigned a larger probability:

$$A = [15.0 \leq X \leq 15.5] \quad \text{or} \quad B = [15.75 \leq X \leq 16.25]?$$

If the machine is working well, you would have to agree that event B is more likely, since the machine is in fact aiming at getting 16 ounces into each bottle. Consequently, we would think that to adequately describe this real-world situation we would have to assign more probability to B than to A. Can we assign probabilities to events such as A and B by assigning probabilities to all possible values that the random variable can assume, as in the case of discrete random variables? The answer is no, as we shall see.

Say we define C and D by

$$C = [15.9 \leq X \leq 16.1]$$
$$D = [15.99 \leq X \leq 16.01].$$

Again using your intuition, which event would you consider more likely, B or C? C or D? It seems apparent that as we "shrink" the interval around the value 16, we would assign less and less probability to that interval. Pursuing this idea to its logical conclusion, we would have to agree that as the size of the interval decreased to zero, the amount of probability assigned would have to decrease to zero as well. However, when the size of the interval is zero, we have the event

$$E = [16 \leq X \leq 16]$$

or $X = 16$. Consequently, we would have to say $P[X = 16] = 0$. In fact, the same logic holds for any value, not just 16, so we must say that

$$P[X = x] = 0 \qquad \text{for all } x \tag{2.7}$$

for a continuous random variable.

In spite of the logic of the previous argument, many people feel ill at ease with assigning zero probability to each point. The feeling that one gets from the study of discrete random variables and from the common everyday usage of probability is that events which are assigned zero probability are impossible (and, in fact, this is correct for discrete random variables). "Aha," you say. "But if I fill a bottle, it must have *some* amount in it, so that particular value should have been assigned a positive probability!" However, there are some mathematical reasons for *not* assigning positive probability and furthermore, for continuous random variables, a probability of zero assigned to an event does not mean that the event is impossible in the usual sense. On the other hand, if you view the problem in the following light, you will see that the event may be impossible in a practical sense. Assume that Mr. Peepers, the quality control engineer at the Fizzy Cola plant, can measure the contents of a bottle of cola to any number of decimal places of accuracy. If he asks you to fill a bottle to *exactly* 16 ounces and then checks on your attempt, do you think you'll be able to do it? No! You're sure to be a little high or a little low, no matter how carefully you do it. Of course, all of this is only of theoretical or philosophical importance. Practically, Mr. Peepers can only measure to a certain degree of accuracy and practically he would be satisfied if you filled the bottle to within some small tolerance of 16 ounces.

If we are going to assign positive probabilities to events like A, B, C, and D but zero probability to events like E, how can we do it? We can do it by equating area and probability. More precisely, we will use a probability density function (pdf) to find the area (and hence the probability) associated with an event.

Definition 2.4 The *probability density function* (pdf) of a continuous random variable X is a function, denoted by $f(x)$, with the property that the area bounded by $f(x)$, the horizontal axis, and the values a and b is equal to $P[a \leq X \leq b]$.

Example 2.8 Assume that the probability density function for the continuous random variable X, the amount of Fizzy Cola put into a 16-ounce bottle, is given by

$$f(x) = \begin{cases} \frac{1}{2} & 15 \leq x \leq 17 \\ 0 & \text{otherwise.} \end{cases}$$

Graph the pdf and find $P[15.0 \leq X \leq 15.5]$ and $P[15.75 \leq X \leq 16.25]$.

The graph of the pdf is shown in Figure 2.3 and the areas corresponding to the

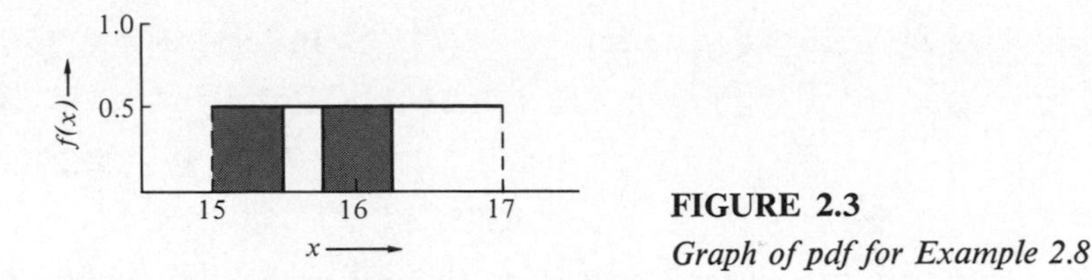

FIGURE 2.3
Graph of pdf for Example 2.8

events $[15.0 \leq X \leq 15.5]$ and $[15.75 \leq X \leq 16.25]$ are shaded. Since the regions whose area must be found are simply rectangles, we get

$$\text{area} = \text{base} \cdot \text{height} = (.5)(.5) = .25$$

in both cases; consequently, the probabilities in both cases are .25.

The probability density function of this example is easy to work with, but it does not adequately describe the real-world situation of the variable nature of filling soda bottles. As indicated in our previous discussion, we would expect to have more probability "concentrated" around the value 16 with a more gradual tapering off on either side than we see here. The pdf of the next example seems more realistic in this regard.

Example 2.9 Assume that the probability density function for X, the amount of Fizzy Cola put into a 16-ounce bottle, is given by

$$f(x) = \begin{cases} x - 15 & 15 \leq x \leq 16 \\ 17 - x & 16 < x \leq 17 \\ 0 & \text{otherwise.} \end{cases}$$

Graph the pdf and find $P[15.0 \leq X \leq 15.5]$ and $P[15.75 \leq X \leq 16.25]$.

The graph of the pdf is shown in Figure 2.4, and the areas corresponding to the events of interest are shaded in. In the first case, the region is a triangle with base equal to $\frac{1}{2}$ and height equal to $f(15.5) = \frac{1}{2}$, so

$$P[15.0 \leq X \leq 15.5] = \tfrac{1}{2}bh = \tfrac{1}{2} \cdot \tfrac{1}{2} \cdot \tfrac{1}{2} = \tfrac{1}{8}.$$

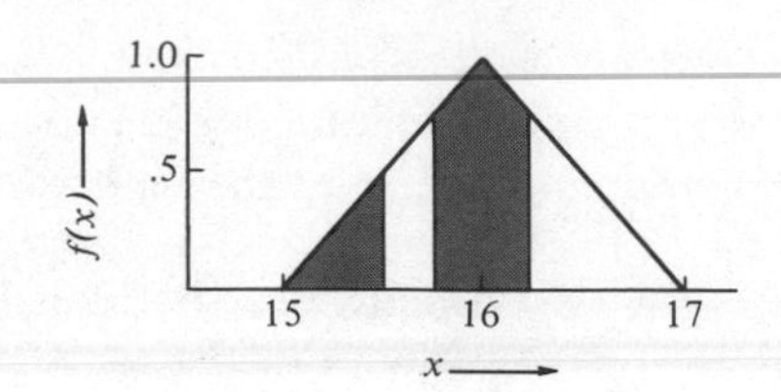

FIGURE 2.4
Graph of pdf for Example 2.9

In the second case, the region is not of such a simple shape and we will have to use some ingenuity to find the area. One approach is to make use of the fact that the pdf is symmetric about the value $x = 16$ and that the total area under $f(x)$ must be equal to 1 (since the total probability must be equal to 1). The regions corresponding to the events $[X > 16.25]$ and $[X < 15.75]$ are both triangular and of the same area, so we get

$$\begin{aligned} P[15.75 \leq X \leq 16.25] &= 1 - (P[X > 16.25] + P[X < 15.75]) \\ &= 1 - 2P[X > 16.25] \\ &= 1 - 2(\tfrac{1}{2})(.75)(.75) = 1 - .5625 \\ &= .4375. \end{aligned}$$

There are two general properties that all probability density functions must satisfy:

(2.8) $\quad f(x) \geq 0 \quad$ for all x

(2.9) $\quad$ total area bounded by $f(x)$ and the horizontal axis $= 1$.

These results are analogous to (2.3) and (2.4) for the discrete case.

Note that by equating area and probability we satisfy equation (2.7), that $P[X = x] = 0$ for all x. In particular, to find $P[X = 16] = P[16 \leq X \leq 16]$ in Example 2.9, we would find the area bounded by $f(x)$, the horizontal axis, and $a = b = 16$. Hence we are finding the area under a single point, and the area under a point is zero. Furthermore, since the probability associated with individual points is zero, the following probabilities are all equal:

$$\begin{aligned} P[15.75 \leq X \leq 16.25] &= P[15.75 < X \leq 16.25] \\ &= P[15.75 \leq X < 16.25] = P[15.75 < X < 16.25]. \end{aligned}$$

Of course, the same fact would be true for end points other than 15.75 and 16.25. In fact, we can say that if X is a continuous random variable, then for any numbers a and b,

(2.10) $$\begin{aligned} P[a \leq X \leq b] &= P[a < X \leq b] = P[a \leq X < b] \\ &= P[a < X < b]. \end{aligned}$$

Note how this differs from the case where X is a discrete random variable [see (2.5)].

In most instances in this book you will not be expected to find a pdf for a given random variable. In general, the pdf will simply be given to you and you will be asked to find the probabilities (areas) needed to solve a particular problem.

• • • • •

In the examples given so far in this section, we have been able to find areas by using simple geometric shapes. In many situations the density functions are not simply straight lines, and finding the area under a curve is a bit more complicated. However, the concept of the integral of calculus comes to our aid in this situation. Since the definite integral of a function gives exactly the area bounded by the function, the horizontal axis, and the ordinates a and b, it follows that if $f(x)$ is the probability density function for a random variable X, then

$$P[a \leq X \leq b] = \int_a^b f(x)\,dx. \tag{2.11}$$

Example 2.10 The probability density function for a continuous random variable Y is given by

$$f(y) = \begin{cases} \frac{3}{8}y^2 & 0 \leq y \leq 2 \\ 0 & \text{otherwise.} \end{cases}$$

Graph $f(y)$ and find $P[0 \leq Y \leq 1.0]$ and $P[1.5 \leq Y \leq 2.0]$.

The graph of the pdf is shown in Figure 2.5 and the areas corresponding to the

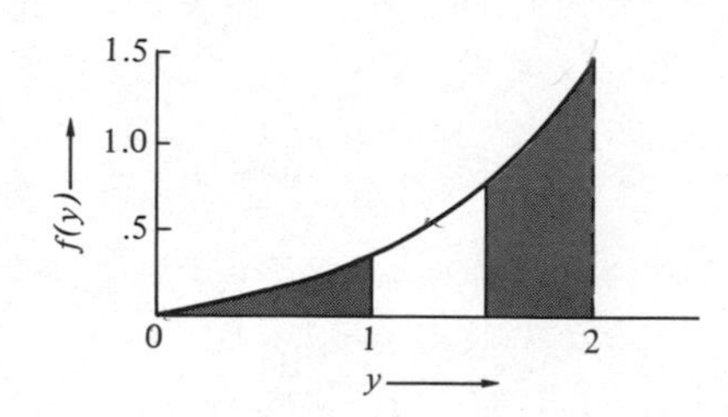

FIGURE 2.5
Graph of pdf for Example 2.10

events of interest are shaded in. We find the probabilities of interest by integration. In the first case we take $a = 0$ and $b = 1$, to get

$$P[0 \leq Y \leq 1] = \int_0^1 \frac{3}{8}y^2\,dy = \left[\frac{3}{8} \cdot \frac{y^3}{3}\right]_0^1 = \frac{1}{8}.$$

In the second case we have $a = 1.5$ and $b = 2$, so

$$P[1.5 \leq Y \leq 2.0] = \int_{1.5}^2 \frac{3}{8}y^2\,dy = \left[\frac{3}{8} \cdot \frac{y^3}{3}\right]_{1.5}^2$$

$$= \frac{1}{8}(8 - 3.375) = .578$$

(to three decimal places).

We could also calculate probabilities for intervals that extend beyond the range of values for which the pdf is positive. The main thing to remember in these situations is that if over some region the density function is always zero, then the area (and hence the probability associated with that area) between the density function and the axis is zero. Furthermore, if an interval includes subintervals where the density function is zero and other subintervals where it is positive, the integral can be divided into separate integrals corresponding to these subintervals. For example, if we want to find $P[-2 \leq Y \leq 1]$, we use the fact that $f(y)$ is defined to be zero for y between -2 and 0 and to be $3y^2/8$ for y between 0 and 1, to get

$$P[-2 \leq Y \leq 1] = \int_{-2}^{1} f(y)\,dy = \int_{-2}^{0} 0\,dy + \int_{0}^{1} \left(\frac{3y^2}{8}\right) dy$$
$$= 0 + \left[\frac{y^3}{8}\right]_0^1 = \frac{1}{8}.$$

Similarly, to find $P[Y \leq 1]$, we note that no finite lower limit has been set for Y, so we can write

$$P[Y \leq 1] = P[-\infty < Y \leq 1] = \int_{-\infty}^{1} f(y)\,dy$$
$$= \int_{-\infty}^{0} 0\,dy + \int_{0}^{1} \left(\frac{3y^2}{8}\right) dy = 0 + \left[\frac{y^3}{8}\right]_0^1 = \frac{1}{8}.$$

Example 2.11 Using the probability density function given in Example 2.8,

$$f(x) = \begin{cases} \frac{1}{2} & 15 \leq x \leq 17 \\ 0 & \text{otherwise.} \end{cases}$$

use integration to find (a) $P[15.75 \leq X \leq 16.25]$; (b) $P[X \leq 16]$; (c) $P[X \leq t]$ for t between 15 and 17.

Using (2.11), we get

(a) $$P[15.75 \leq X \leq 16.25] = \int_{15.75}^{16.25} (\tfrac{1}{2})\,dx = [\tfrac{1}{2}x]_{15.75}^{16.25}$$
$$= \tfrac{1}{2}(16.25 - 15.75) = .25.$$

(b) $$P[X \leq 16] = P[-\infty < X \leq 16] = \int_{-\infty}^{16} f(x)\,dx$$
$$= \int_{-\infty}^{15} 0\,dx + \int_{15}^{16} \tfrac{1}{2}\,dx$$
$$= 0 + [\tfrac{1}{2}x]_{15}^{16} = \tfrac{1}{2}(16 - 15) = \tfrac{1}{2}.$$

(c) $$P[X \leq t] = P[-\infty < X \leq t] = \int_{-\infty}^{t} f(x)\,dx$$
$$= \int_{-\infty}^{15} 0\,dx + \int_{15}^{t} \tfrac{1}{2}\,dx$$
$$= 0 + [\tfrac{1}{2}x]_{15}^{t} = \frac{t - 15}{2}.$$

If we had solved part (c) of Example 2.11 before (b), we could have used that result to get the answer to (b). More explicitly, we see that (b) is a special case of (c)—the case where $t = 16$. Given that $P[X \leq t] = (t - 15)/2$ for $15 \leq t \leq 17$, we can immediately write

$$P[X \leq 16] = \frac{16 - 15}{2} = \frac{1}{2}.$$

2.4 CUMULATIVE DISTRIBUTION FUNCTIONS

In the previous two sections we calculated probabilities associated with various intervals and in particular considered probabilities of intervals of the form $[X \leq b]$ for some value b. We will see that the entire probability distribution of a random variable, discrete or continuous, is completely determined by assigning probabilities to events like $[X \leq b]$ for all possible values of b. In many cases this is the most efficient way of expressing a probability distribution for a random variable. In fact, many of the important tables found in the Appendix and needed later in this book are expressed this way.

Example 2.12 Using the probability distribution found in Example 2.7, find $P[X \leq b]$ for $b = 0, .2, .5$, and 1.

For convenience we rewrite that probability distribution here:

x:	0	1	2
$P[X = x]$:	$\frac{9}{21}$	$\frac{10}{21}$	$\frac{2}{21}$.

Since $x = 0$ is the only value of x assigned positive probability which also satisfies $x \leq 0$, it follows that

$$P[X \leq 0] = \tfrac{9}{21}.$$

Similarly, $x = 0$ is the only value of x assigned positive probability which also satisfies $x \leq .2$, so

$$P[X \leq .2] = \tfrac{9}{21}.$$

Likewise,

$$P[X \leq .5] = \tfrac{9}{21}.$$

However, $x = 0$ and $x = 1$ both satisfy $x \leq 1$, so it follows from (2.6) that

$$P[X \leq 1] = \tfrac{9}{21} + \tfrac{10}{21} = \tfrac{19}{21}.$$

It should be clear that the answer to $P[X \leq b]$ will depend on the value of b chosen and consequently $P[X \leq b]$ is actually a function of b. In mathematics we are more likely to see the quantity x used rather than b as the argument of a function. Following that convention and using F to denote the function, we give the following definition.

Definition 2.5 The *cumulative distribution function* (CDF) of a random variable X is defined to be

$$F(x) = P[X \leq x]$$

for all real values of x.

Returning to Example 2.12, it is easy to see that for any number x strictly between 0 and 1 we will have

$$P[X \leq x] = \tfrac{9}{21}.$$

Also, if we take x to be any number satisfying $1 \leq x < 2$, then

$$P[X \leq x] = \tfrac{19}{21}.$$

Furthermore, since it is impossible for X to assume values less than 0 in this situation, we have, for $x < 0$,

$$P[X < x] = 0,$$

and since X *must* assume a value less than or equal to 2, we have

$$P[X \leq x] = 1$$

for $x \geq 2$. All of this can be summarized by the cumulative distribution function for this random variable.

$$F(x) = \begin{cases} 0 & \text{if } x < 0 \\ \frac{9}{21} & \text{if } 0 \leq x < 1 \\ \frac{19}{21} & \text{if } 1 \leq x < 2 \\ 1 & \text{if } x \geq 2. \end{cases}$$

(Note that the CDF is completely specified only when the values are given for all real x.)

The graph of x vs. $F(x)$ is given in Figure 2.6. (We have placed the vertical axis to the left of zero so that the behavior at zero can be more easily seen. The dashed

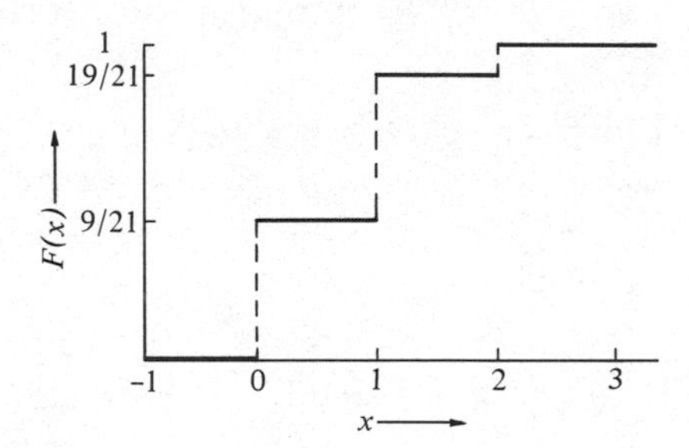

FIGURE 2.6
The Cumulative Distribution Function for Example 2.12

vertical lines are also a convenience to help you visualize the jumps that occur in the graph.) This graph is typical of the graphs of all CDFs in that it is nondecreasing going from left to right. It is also typical of the graphs of CDFs for discrete random variables in that it is a *step function* (although in this case it appears as if the steps have been constructed by a singularly inept carpenter!). Earlier we said that the cumulative distribution function of a random variable completely determines the probability distribution of a random variable. Closer examination of the graph of this particular CDF shows how the graph can be used to find the probability distribution. In particular, notice that:

1. Jumps occur at values of x having positive probability.
2. The size of the jump at x is equal to $P[X = x]$.

[Note that (2) is true for all x, since if there is no jump at x, then the size of the jump and hence the probability assigned to x is zero.]

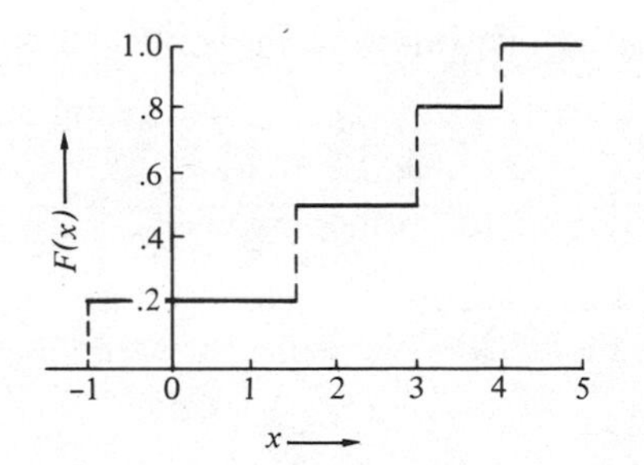

FIGURE 2.7
CDF for Example 2.13

Example 2.13 Find the probability distribution for the random variable whose cumulative distribution function is given in Figure 2.7.

By examining the graph of the CDF, we see that jumps occur at $x = -1, 1.5$, 3, and 4 and the size of the jumps are .2, .3, .3, and .2, respectively; hence the probability distribution for this random variable is given by

x:	-1	1.5	3	4
$P[X = x]$:	.2	.3	.3	.2.

By the very definition of the cumulative distribution function, it can be used to find probabilities of intervals such as $[X \leq b]$, since $P[X \leq b]$ is simply $F(b)$. It can also be used to find probabilities of intervals such as $[a < X \leq b]$, since

$$P[a < X \leq b] = F(b) - F(a). \tag{2.12}$$

To see that this is true, note that the event $[X \leq b]$ can be written as the union of two mutually exclusive events as follows:

$$[X \leq b] = [X \leq a] \cup [a < X \leq b].$$

Because the events on the right-hand side are mutually exclusive, Axiom 3 of Section 1.8 applies, so we can write

$$P[X \leq b] = P[X \leq a] + P[a < X \leq b]$$

or

$$F(b) = F(a) + P[a < X \leq b].$$

Equation (2.12) follows from this last equation by subtracting $F(a)$ from both sides. As an illustration of this result, by referring to Figure 2.7 we can find

$$P[1 < X \leq 2] = F(2) - F(1) = .5 - .2 = .3,$$

$$P[0 < X \leq 3] = F(3) - F(0) = .8 - .2 = .6,$$

and

$$P[-1 < X \leq 5] = F(5) - F(-1) = 1 - .2 = .8.$$

If you look back at the definition of the cumulative distribution function for a random variable (Definition 2.5), you will see that we did not specify whether the definition pertained to discrete or continuous random variables. This was not an oversight on our part, since in fact the very same definition applies to both discrete and continuous random variables.

Example 2.14 Find the cumulative distribution function for the continuous random variable defined in Example 2.8:

$$f(x) = \begin{cases} \frac{1}{2} & 15 \leq x \leq 17 \\ 0 & \text{otherwise.} \end{cases}$$

Since it is impossible, in this example, for the random variable X to assume a value less than 15, it follows that $P[X \leq b] = 0$ if $b < 15$:

$$F(b) = 0 \qquad \text{if } b < 15.$$

Similarly, X *must* assume a value less than 17, so $P[X \leq b] = 1$ if $b > 17$:

$$F(b) = 1 \qquad \text{if } b > 17.$$

The "interesting" case is for those values of x between 15 and 17. Remembering that the area under the probability density function for a certain region corresponds to the probability assigned to that region, we find that $P[X \leq b] = (b - 15)/2$ for $15 \leq b \leq 17$ (see Figure 2.8). This is true since the area of the shaded rectangle is found by multiplying the base $(b - 15)$ times the height $(\frac{1}{2})$ to get $(b - 15)/2$. (As in our earlier discussion, we have used b to denote the argument of F to avoid possible confusion with the x used to denote the scale on the horizontal axis in Figure 2.8.) We can summarize these results [writing $F(x)$ instead of $F(b)$] as follows:

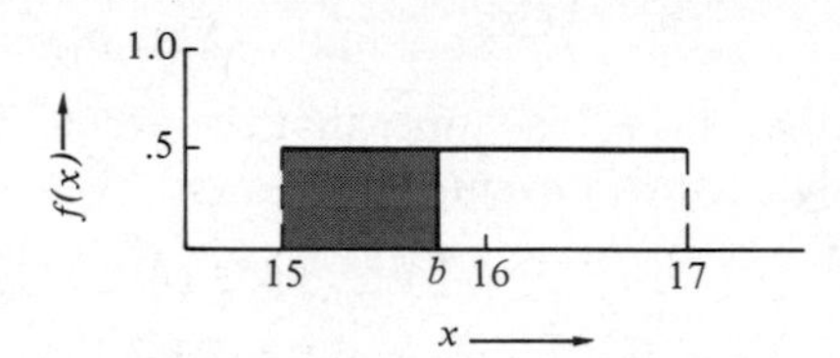

FIGURE 2.8

$$F(x) = \begin{cases} 0 & \text{if } x < 15 \\ \dfrac{x - 15}{2} & \text{if } 15 \leq x \leq 17 \\ 1 & \text{if } x > 17. \end{cases} \tag{2.13}$$

The graph of x vs. $F(x)$ is given in Figure 2.9. This graph is typical of graphs of CDFs for continuous random variables. You should compare this graph with the one in Figure 2.6, which corresponds to the CDF for a discrete random variable.

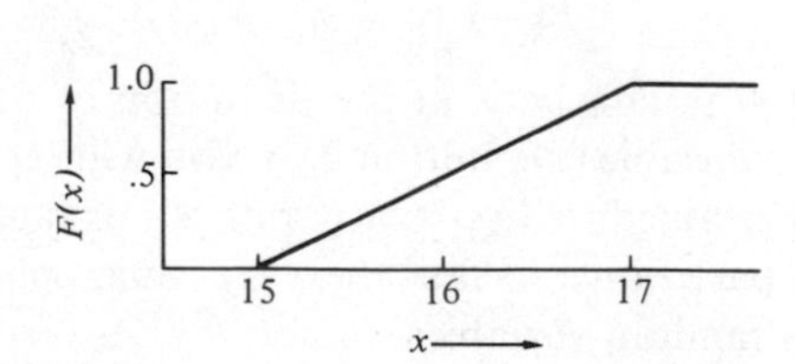

FIGURE 2.9

Since equation (2.12) is valid for all random variables, discrete or continuous, we can use the CDF found in Example 2.14 to find

$$P[15.0 < X \leq 15.5] = F(15.5) - F(15.0)$$
$$= \frac{15.5 - 15}{2} - \frac{15.0 - 15}{2}$$
$$= .25$$

and

$$P[15.75 < X \leq 16.25] = F(16.25) - F(15.75)$$
$$= \frac{16.25 - 15}{2} - \frac{15.75 - 15}{2}$$
$$= .25.$$

How do these examples compare with the events and corresponding probabilities found in Example 2.8? The difference is subtle and perhaps easy to overlook, but if you look carefully you will see in the one case the left inequality is strict ($<$) and in the other it is not strict ($\leq$), but the corresponding probabilities are the same. The reason the probabilities are the same is that X in this case is a continuous random variable [see (2.10)]. The answers could be quite different if X were discrete.

• • • • •

In finding the CDF for the continuous random variable in Example 2.14 we found the area corresponding to the event $[X \leq b]$ by finding the area of a rectangle. We could have used integration to obtain the same result. In particular, it follows from (2.11) that

$$P[X \leq b] = \int_{-\infty}^{b} f(x)\, dx.$$

Remember that if $f(x)$ is defined differently in different regions, we might have to write the integral as a sum of integrals each taken over a particular region. This was done in Example 2.11. In fact, in that example we essentially found the CDF of Example 2.14 without actually calling it that! We have, if $b < 15$,

$$F(b) = P[X \leq b] = \int_{-\infty}^{b} f(x)\, dx = \int_{-\infty}^{b} 0\, dx = 0;$$

if $b \in [15, 17]$,

$$F(b) = P[X \leq b] = \int_{-\infty}^{b} f(x)\, dx = \int_{-\infty}^{15} 0\, dx + \int_{15}^{b} \tfrac{1}{2}\, dx$$
$$= 0 + [\tfrac{1}{2}x]_{15}^{b} = \frac{b - 15}{2}$$

and if $b > 17$,

$$F(b) = P[X \leq b] = \int_{-\infty}^{b} f(x)\, dx$$
$$= \int_{-\infty}^{15} 0\, dx + \int_{15}^{17} \tfrac{1}{2}\, dx + \int_{17}^{b} 0\, dx$$
$$= 0 + [\tfrac{1}{2}x]_{15}^{17} + 0 = 1.$$

If we were to write $F(x)$ instead of $F(b)$, we would get (2.13).

Example 2.15 Find the cumulative distribution function for the random variable Y given in Example 2.10. Graph this CDF.

The probability density function for this example is

$$f(y) = \begin{cases} \frac{3}{8}y^2 & 0 \leq y \leq 2 \\ 0 & \text{otherwise.} \end{cases}$$

It should be clear to you from the previous examples that

$$F(y) = 0 \qquad \text{if } y < 0$$

and

$$F(y) = 1 \qquad \text{if } y > 2.$$

Now if b is some number between 0 and 2,

$$F(b) = F[Y \leq b] = \int_{-\infty}^{b} f(y)\,dy = \int_{-\infty}^{0} 0\,dy + \int_{0}^{b} \frac{3}{8}y^2\,dy$$

$$= 0 + \left[\frac{1}{8}y^3\right]_0^b = \frac{b^3}{8},$$

so

$$F(y) = \frac{y^3}{8} \qquad \text{if } 0 \leq y \leq 2.$$

Summarizing, we have

$$F(y) = \begin{cases} 0 & \text{if } y < 0 \\ \frac{y^3}{8} & \text{if } 0 \leq y \leq 2 \\ 1 & \text{if } y > 2. \end{cases}$$

The graph of y vs. $F(y)$ is given in Figure 2.10.

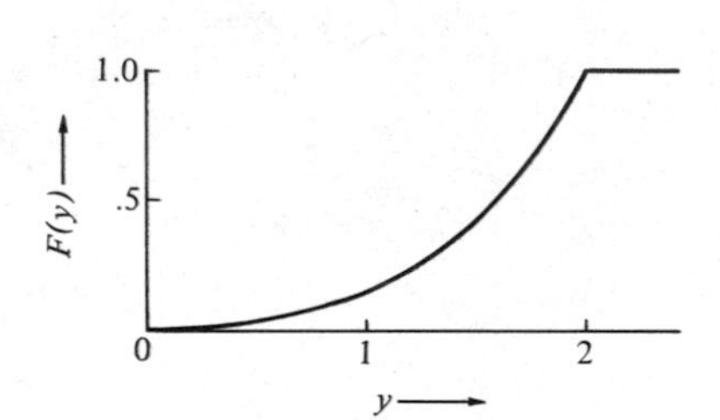

FIGURE 2.10

If in a particular discussion we restrict our attention to a single random variable, there should be no difficulty with the notation of F to represent the CDF of that random variable. However, if there are two or more random variables being considered in a discussion, the notation F is no longer unambiguous. To avoid ambiguity we can use subscripts. For example, if continuous random variables X and Y are under consideration, we could denote their respective probability density functions by $f_X(x)$ and $f_Y(y)$ and the corresponding CDFs by $F_X(x)$ and $F_Y(y)$.

We saw earlier in this section that for discrete random variables, the probability distribution could be used to find the CDF and the CDF could be used to find the probability distribution. A similar result is true when the random variable being considered is continuous. The CDF can be found from the probability density function by integration and the probability density function can be found from the CDF by the "inverse" operation of integration, differentiation. Remembering that the derivative of a constant is zero, we see that if we differentiate $F(x)$ as given in (2.13), we get

$$\frac{dF(x)}{dx} = \frac{d}{dx}(0) = 0 \qquad \text{if } x < 15$$

$$\frac{dF(x)}{dx} = \frac{d}{dx}\left(\frac{x-15}{2}\right)$$

$$= \frac{d}{dx}(\tfrac{1}{2}x) - \frac{d}{dx}(\tfrac{15}{2}) = \tfrac{1}{2} \qquad \text{if } 15 \le x \le 17$$

$$\frac{dF(x)}{dx} = \frac{d}{dx}(1) = 0 \qquad \text{if } x > 17.$$

Since $f(x) = dF(x)/dx$, we get

$$f(x) = \begin{cases} \frac{1}{2} & 15 \le x \le 17 \\ 0 & \text{otherwise,} \end{cases}$$

which, of course, is exactly the probability density function we started with in Example 2.14.

Example 2.16 The CDF found in Example 2.15 was

$$F(y) = \begin{cases} 0 & y < 0 \\ \dfrac{y^3}{8} & 0 \le y \le 2 \\ 1 & y > 2. \end{cases}$$

Differentiate this function to obtain the probability density function.

Since $F(y)$ is defined differently in different regions, we must consider each case separately.

$$\frac{dF(y)}{dy} = \frac{d}{dy}(0) = 0 \qquad \text{if } y < 0$$

$$\frac{dF(y)}{dy} = \frac{d}{dy}\left(\frac{y^3}{8}\right) = \frac{3y^2}{8} \qquad \text{if } 0 \le y \le 2$$

$$\frac{dF(y)}{dy} = \frac{d}{dy}(1) = 0 \qquad \text{if } y > 2,$$

so

$$f(y) = \begin{cases} \dfrac{3y^2}{8} & \text{if } 0 \le y \le 2 \\ 0 & \text{otherwise.} \end{cases}$$

2.5 EXPECTED VALUE

Have you or members of your family ever had the opportunity to enter a sweepstakes? When you receive in the mail an "Opportunity of a Lifetime" or "Six Chances to Win," all with "No Purchase Necessary," knowing that "You May Already Be a Winner," you certainly have to stop and think about the fact that "This May Be Your Lucky Day!" Clichés aside, when confronted with an opportunity to enter such a contest, you must make a decision as to enter or not. It is obvious that thousands of people do decide to enter and that other thousands decide not to enter. In most cases, we would conjecture, the decision is not really made on a rational, clearly thought out, objective basis. We would like to discuss one procedure (certainly not the only one) that would allow you to make such a decision more objectively.

In addition to deciding *whether* to enter a particular contest, you may be confronted with an additional decision of *which* contest to enter if you happen to receive two such "Once in a Lifetime" opportunities within a few days! Fortunately, when this happened to us, we were able to determine from the literature (propaganda?) that came with the entry blanks, the dollar value of the various prizes and the corresponding probabilities of winning those prizes. (We did have to make some simplifying assumptions, which we won't discuss here.) For obvious reasons, the literature did not specify the probability of getting nothing, but this probability is easy to find [using (2.4)] and has been included in Table 2.1.

We can use the information in Table 2.1 to help us decide which, if any, of these sweepstakes we should enter. What we see here, however, is a special case of the more general problem of comparing probability distributions. If X represents your winnings when entering Sweepstakes R and Y represents your winnings when entering

TABLE 2.1
Sweepstakes Prizes and Probabilities

Sweepstakes R		*Sweepstakes C*	
Prize	*Probability*	*Prize*	*Probability*
0	268/269 = .996282	0	329/330 = .996970
5	1/325 = .003077	1.5	1/394 = .002538
10	1/1,566 = .000639	10	1/3,300 = .000303
1,000	1/469,700 = .000002129	25	1/6,600 = .000152
3,000	1/1,565,667 = .000000639	50	1/33,000 = .000030
5,000	1/2,348,500 = .000000426	100	1/165,000 = .000006
10,000	1/4,697,000 = .000000213	250	1/660,000 = .000001515
50,000	1/14,630,000 = .000000068	500	1/1,650,000 = .000000606
		1,500	1/3,300,000 = .000000303
		5,000	1/5,500,000 = .000000182
		10,000	1/16,500,000 = .000000061
		24,000	1/16,500,000 = .000000061
		125,000	1/16,500,000 = .000000061

Sweepstakes C, then Table 2.1 acutally gives the probability distributions of X and Y. It is certainly difficult to make a direct comparison between these distributions, since there is quite a difference between the possible values (i.e., the prize values) that can be assumed by the two random variables. We would like to find a single number for each distribution which in some sense characterizes the distribution and then use those numbers as a basis for comparing the distributions.

To see how such a number might be obtained, let us consider a much simpler game than the sweepstakes. In particular, assume that you are given the opportunity to flip a fair coin three times and that you will be paid \$1 for each head that comes up. How much is such an opportunity worth to you? Think about it. If you play this game just one time, you could get as little as \$0 or as much as \$3. If X is the amount (in dollars) that you win in a play of this game, it is not too hard to see that the probability distribution of X is

(2.14)
$$\begin{array}{rcccc} x: & 0 & 1 & 2 & 3 \\ P[X=x]: & \frac{1}{8} & \frac{3}{8} & \frac{3}{8} & \frac{1}{8}, \end{array}$$

so you can actually assign a probability to the various amounts you might win. However, on a single play of this game you just don't know what will happen. Another way to view this problem, then, is not to concern yourself with a single trial but to ask: "Over an extremely large number of trials, how much can I expect to get on the average?" This is a question that we can answer.

Let us assume that you are allowed to play the aforementioned game 1000 times. We would like to find what you might expect to get on the average for these thousand games, so we begin by asking how many of those times you might expect to win \$0. To determine this, we consider again the concept of relative frequency as discussed in Section 1.6. In that section we said that for a large number of trials we would expect the relative frequency of occurrence of some event $A, f/n$, to be close to the actual probability of A's occurrence, $P(A)$. Now we wish to consider the same idea from a different point of view. The event A we are interested in is the event of winning \$0, or equivalently the event $X = 0$. However, in this case we know that $P(A)$ [i.e., $P(X = 0)$] is equal to $\frac{1}{8}$, and we also know that $n = 1000$; we want to make a guess at f, the number of times you win \$0 in 1000 tries. If

$$\frac{f}{n} \approx P(A),$$

it is not unreasonable to guess that

$$f = nP(A) = 1000(\tfrac{1}{8}) = 125.$$

The same kind of reasoning leads us to say that we would expect to win \$1 about $(1000)(\frac{3}{8}) = 375$ times, to win \$2 about 375 times, and to win \$3 about 125 times. What, then, is the total amount you might expect to win? The total is

$$0(125) + 1(375) + 2(375) + 3(125) = 1500,$$

while the average amount per trial is found by dividing both sides of this equation by 1000 to get

(2.15)
$$0(.125) + 1(.375) + 2(.375) + 3(.125) = 1.50.$$

Hence the average amount you might expect to win per trial, when averaged over a large number of trials, is \$1.50. If you examine (2.15) more closely, you will see that the left-hand side is found by taking the sum of products of the value of the random variable times the probability that the random variable assumes that value:

$$0 \cdot P[X = 0] + 1 \cdot P[X = 1] + 2 \cdot P[X = 2] + 3 \cdot P[X = 3].$$

This leads us to the following definition.

Definition 2.6 The *expected value of a discrete random variable X*, denoted by $E(X)$, is defined as

$$E(X) = \sum_{\text{all } x} x \cdot P[X = x]. \tag{2.16}$$

Notation: $E(X)$ is also called the mean of X and is denoted by the Greek lowercase letter mu (μ).

We should emphasize that the expected value of a random variable is the *long-run average value* of a random variable. If you were to play the game of tossing three coins (as defined earlier) many times, the average amount you would gain per try would be \$1.50. If you could find a benefactor who would allow you to play this game and who would charge you less than \$1.50 to play, it would certainly be to your advantage to do so. A more likely situation would be for you to find someone who would like to charge you more than \$1.50 for the privilege of playing this game. (This, by the way, is the way it is in gambling casinos. We would hope that the study of expectations would point out to you that there is a mathematical as well as a moral basis for saying that you "can't win" by gambling!) We would also like to point out that the expected value is not the value you "expect" a random variable to assume on a given observation. This should be clear from the coin example, since on a single observation the observed value will be 0, 1, 2, or 3 while the expected value is 1.50.

Example 2.17 Find the expected values for X and Y, the winnings in Sweepstakes R and C, respectively.

Using (2.16), we get

$$\begin{aligned} E(X) &= 0 \cdot P[X = 0] + 5 \cdot P[X = 5] + \ldots + 50{,}000 \cdot P[X = 50{,}000] \\ &= 0\left(\frac{268}{269}\right) + 5\left(\frac{1}{325}\right) + \ldots + 50{,}000\left(\frac{1}{14{,}630{,}000}\right) \\ &= 0 + .0154 + \ldots + .0034 = .0335 \end{aligned}$$

and

$$\begin{aligned} E(Y) &= 0 \cdot P[Y = 0] + 1.50 \cdot P[Y = 1.50] + \ldots + 125{,}000 \cdot P[Y = 125{,}000] \\ &= 0\left(\frac{329}{330}\right) + 1.50\left(\frac{1}{394}\right) + \ldots + 125{,}000\left(\frac{1}{16{,}500{,}000}\right) \\ &= 0 + .0038 + \ldots + .0076 = .0244. \end{aligned}$$

Hence the expected value for Sweepstakes R is approximately 3 cents, and the expected value for Sweepstakes C is approximately 2 cents.

Based on the results of this example, which sweepstakes is better to enter? Using expected value as a criterion, Sweepstakes R is better. Should you enter either one of these sweepstakes? What do you have to lose? Well, if it were free to enter, you would have nothing to lose; however, generally you have to pay the postage, so you stand to lose the cost of a stamp! When you next receive a sweepstakes offer, you will of course make your own decision as to enter or not. We would recommend, however, that if you are out of money and down to your last postage stamp, you use the stamp to send a letter to your folks requesting assistance rather than to enter the sweepstakes. Your chances of getting \$5,000 or \$10,000 back are probably about the same, while your chances of getting \$5 or \$10 are probably increased considerably!

We have indicated in our discussion earlier in this section that expected values could be used to compare probability distributions. We will see in later sections that expected values of functions of random variables in addition to the expected values of the random variables themselves can be useful in comparing probability distributions. To this end we give the following definition.

Definition 2.7 If X is a discrete random variable and if $g(X)$ is some function of X, the *expected value of* $g(X)$ is defined by

$$E(g(X)) = \sum_{\text{all } x} g(x) \cdot P[X = x]. \tag{2.17}$$

Note the similarity between (2.16) and (2.17). In each case a sum of products of a particular value times the probability of that value occurring is found. The function $g(X)$ could be any mathematical function of X (e.g., sin X, log X, etc.), but we generally will consider only simple functions such as polynomials.

Example 2.18 The probability distribution for X given in (2.14) was

x:	0	1	2	3
$P[X = x]$:	$\frac{1}{8}$	$\frac{3}{8}$	$\frac{3}{8}$	$\frac{1}{8}$.

Use this probability distribution to find $E(2X)$, $E(X + 2)$, and $E(X^2)$.

Using (2.17), we get

$$\begin{aligned} E(2X) &= \sum_{\text{all } x} (2X) \cdot P[X = x] \\ &= 2(0)(\tfrac{1}{8}) + 2(1)(\tfrac{3}{8}) + 2(2)(\tfrac{3}{8}) + 2(3)(\tfrac{1}{8}) \\ &= 3.0, \\ E(X + 2) &= \sum_{\text{all } x} (x + 2) \cdot P[X = x] \\ &= (0 + 2)(\tfrac{1}{8}) + (1 + 2)(\tfrac{3}{8}) + (2 + 2)(\tfrac{3}{8}) + (3 + 2)(\tfrac{1}{8}) \\ &= 3.5, \end{aligned}$$

and

$$\begin{aligned} E(X^2) &= \sum_{\text{all } x} (x^2) \cdot P[X = x] \\ &= 0^2(\tfrac{1}{8}) + 1^2(\tfrac{3}{8}) + 2^2(\tfrac{3}{8}) + 3^2(\tfrac{1}{8}) = 3.0. \end{aligned}$$

Remembering that for this example $E(X) = 1.5$, do you see any relationship between $E(2X)$ and $E(X)$ or between $E(X + 2)$ and $E(X)$? Since it is hard to see a relationship when so little information is available, you might take a moment to verify that $E(4X) = 6.0$ and $E(X + 5) = 6.5$ and then see if a relationship appears to exist. Such a relationship does in fact exist, as we shall see in the following theorems. The proofs of these theorems are given for the case where X is a discrete random variable, but the theorems are true for all random variables, discrete or continuous.

Theorem 2.1 If X is a random variable and if b is a constant, then

$$E(bX) = bE(X). \tag{2.18}$$

Proof: Using (2.17) with $g(X) = bX$ and using properties of summations (Appendix A1), we get

$$E(bX) = \sum_{\text{all } x} (bx)P[X = x] = b \sum_{\text{all } x} xP[X = x] = bE(X).$$

Theorem 2.2 If X is a random variable and if c is a constant, then

$$E(X + c) = E(X) + c. \tag{2.19}$$

Proof: Using (2.17) with $g(X) = (X + c)$, we get

$$\begin{aligned} E(X + c) &= \sum_{\text{all } x} (x + c) \cdot P[X = x] = \sum_{\text{all } x} (x \cdot P[X = x] + c \cdot P[X = x]) \\ &= \sum_{\text{all } x} x \cdot P[X = x] + \sum_{\text{all } x} c \cdot P[X = x] \\ &= \sum_{\text{all } x} x \cdot P[X = x] + c \sum_{\text{all } x} P[X = x] \\ &= E(X) + c. \end{aligned}$$

[In this proof we used (2.4) of Section 2.2 to say that $c \sum P[X = x] = c$.]

Theorem 2.3 If X is a random variable and if c is a constant, then

$$E(c) = c. \tag{2.20}$$

Proof: We will let you do this proof yourself. [*Hint:* Let $g(X) = c$.]

Theorem 2.4 If X is a random variable and if $g(X)$ and $h(X)$ are two functions of X, then

$$E(g(X) + h(X)) = E(g(X)) + E(h(X)). \tag{2.21}$$

Proof: Once again use (2.17), to get

$$\begin{aligned} E(g(X) + h(X)) &= \sum_{\text{all } x} (g(x) + h(x)) \cdot P[X = x] \\ &= \sum_{\text{all } x} (g(x) \cdot P[X = x] + h(x) \cdot P[X = x]) \\ &= \sum_{\text{all } x} g(x) \cdot P[X = x] + \sum_{\text{all } x} h(x) \cdot P[X = x] \\ &= E(g(X)) + E(h(X)). \end{aligned}$$

Example 2.19 For a random variable X it was found that $E(X) = 5$ and $E(X^2) = 50$. Using Theorems 2.1 to 2.4, find $E(4X)$, $E(X + 5)$, and $E(2X + X^2)$.

From Theorem 2.1,

$$E(4X) = 4E(X) = 4(5) = 20;$$

from Theorem 2.2,

$$E(X + 5) = E(X) + 5 = 5 + 5 = 10;$$

and from Theorems 2.4 and 2.1,

$$\begin{aligned} E(2X + X^2) &= E(2X) + E(X^2) \\ &= 2E(X) + E(X^2) = 2(5) + 50 = 60. \end{aligned}$$

Example 2.20 Show that for any random variable X,

$$E(X - \mu) = 0. \tag{2.22}$$

To show this we need to note that although X is a random variable, its expected value, denoted by either μ or $E(X)$, is a constant. From (2.19) we have

$$E(X - \mu) = E(X) - \mu,$$

but since $E(X)$ and μ are simply different symbols for the same quantity, it must be that $E(X) - \mu = 0$.

A note of warning here: although you may get that impression from the previous theorems, expected values of functions of X cannot always be expressed as functions of $E(X)$. For example, it is *not* generally true that $E(X^2) = [E(X)]^2$ or that $E(\sqrt{X}) = \sqrt{E(X)}$.

● ● ● ● ●

The mean or expected value of a continuous random variable can be found by an equation analogous to (2.16). The summation symbol is replaced by an integral and the discrete probability $P[X = x]$ is replaced by the probability density function $f(x)$.

Definition 2.8 The *expected value of a continuous random variable* X, denoted by $E(X)$, is defined as

$$E(X) = \int_{-\infty}^{\infty} x \cdot f(x)\, dx. \tag{2.23}$$

Example 2.21 Find the expected value of X for the density function given in Example 2.8,

$$f(x) = \begin{cases} \frac{1}{2} & 15 \le x \le 17 \\ 0 & \text{otherwise.} \end{cases}$$

Using (2.23), we get

$$\begin{aligned} E(X) &= \int_{-\infty}^{\infty} xf(x)\, dx = \int_{-\infty}^{15} (x \cdot 0)\, dx + \int_{15}^{17} x \cdot \frac{1}{2}\, dx + \int_{17}^{\infty} (x \cdot 0)\, dx \\ &= 0 + \frac{1}{2}\left[\frac{x^2}{2}\right]_{15}^{17} + 0 = \frac{17^2 - 15^2}{4} = \frac{289 - 225}{4} = 16. \end{aligned}$$

If we give an interpretation to $E(X)$ in the context of Example 2.8, we would say that $E(X) = 16$ means that the average amount of cola put into a bottle, when averaged over a large number of bottles, is 16 ounces.

You can also see from this example that when integrating, the only region that will lead to a nonzero result is that region where $f(x)$ is positive. We use this fact in the next example.

Example 2.22 Find the expected value of Y for the density function given in Example 2.10,

$$f(y) = \begin{cases} \frac{3}{8}y^2 & 0 \le y \le 2 \\ 0 & \text{otherwise.} \end{cases}$$

Using (2.23), we get

$$E(Y) = \int_0^2 y\left(\frac{3}{8}\right)y^2\,dy = \frac{3}{8}\int_0^2 y^3\,dy$$

$$= \frac{3}{8}\left[\frac{y^4}{4}\right]_0^2 = \frac{3}{8}\left(\frac{2^4 - 0}{4}\right) = \frac{3}{2} = 1.5.$$

To find the expected value of a function of a continuous random variable, we use a continuous analog of (2.17).

Definition 2.9 If X is a continuous random variable and if $g(X)$ is some function of X, then the *expected value of* $g(X)$ is defined by

$$E(g(X)) = \int_{-\infty}^{\infty} g(x)f(x)\,dx. \tag{2.24}$$

Example 2.23 A continuous random variable X has a density function given by

$$f(x) = \begin{cases} 6(x - x^2) & 0 \le x \le 1 \\ 0 & \text{otherwise.} \end{cases}$$

Find $E(X^2)$.

Using (2.24), we have

$$E(X^2) = \int_{-\infty}^{\infty} x^2 \cdot f(x)\,dx$$

$$= \int_0^1 x^2[6(x - x^2)\,dx] = \int_0^1 6(x^3 - x^4)\,dx$$

$$= 6\left[\frac{x^4}{4} - \frac{x^5}{5}\right]_0^1 = 6\left(\frac{1}{4} - \frac{1}{5}\right) = .3.$$

Example 2.24 Using the density function of Example 2.22, find $E(5Y^2 - 2Y + 4)$.

We could use (2.24) directly and evaluate

$$\int_0^2 (5y^2 - 2y + 4)(\tfrac{3}{8})y^2\,dy$$

or we could use Theorems 2.1 to 2.4 to write

$$E(5Y^2 - 2Y + 4) = 5E(Y^2) - 2E(Y) + 4. \tag{2.25}$$

Since we already found $E(Y)$ in Example 2.22, we can evaluate (2.25) as soon as we find $E(Y^2)$. We will use this approach.

$$E(Y^2) = \int_0^2 y^2\left(\frac{3}{8}\right)y^2\,dy = \frac{3}{8}\int_0^2 y^4\,dy = \frac{3}{8}\left[\frac{y^5}{5}\right]_0^2$$

$$= \frac{3}{8}\left(\frac{2^5 - 0}{5}\right) = \frac{12}{5},$$

so

$$E(5Y^2 - 2Y + 4) = 5(\tfrac{12}{5}) - 2(\tfrac{3}{2}) + 4 = 13.$$

2.6 MEASURES OF CENTRAL TENDENCY

In giving an interpretation of the expected value of a random variable, we have referred to it as a long-run average value of the random variable. We have also seen that the expected value of a random variable is a middle value in the sense that if X can assume values between two numbers a and b, then $E(X)$ will be some value between a and b. For example, if *all* bottles of Fizzy Cola are filled with an amount of soda between 15 and 17 ounces, then certainly the average (i.e., expected) amount per bottle must also be between 15 and 17 ounces. In some sense, then, $E(X)$, the mean, gives a measure of the center of a probability distribution. In this section we will see that there are several different ways of measuring the "center" of a probability distribution (i.e., several measures of "central tendency").

Consider an imaginary situation where an eccentric but wealthy gentleman, Phil Anthropist, offers to give you \$25 if you can guess the number that will come up when a die is tossed. If the die is fair, the probability distribution for X, the number that comes up, is given by

(2.26)

x:	1	2	3	4	5	6
$P[X = x]$:	$\frac{1}{6}$	$\frac{1}{6}$	$\frac{1}{6}$	$\frac{1}{6}$	$\frac{1}{6}$	$\frac{1}{6}$.

Assume further that if you guess wrong, you will get nothing. What guess would you make? Of course in this case one guess is as good as another, because all the outcomes are equally likely. What would your guess be if Phil were to make the same offer when an unbalanced die is tossed? In particular, say that the probability distribution for Y, the number that comes up on the unbalanced die, is given by

(2.27)

y:	1	2	3	4	5	6
$P[Y = y]$:	.1	.3	.2	.2	.1	.1.

What guess would you make in this case? After a bit of reflection, you would likely decide to guess the number 2, since in this situation you will have the highest probability of your guess being correct. With this in mind, we give the following definition.

Definition 2.10 The *mode* of the probability distribution for a *discrete random variable* X is that value x for which $P[X = x]$ is largest.

Example 2.25 Find the mode for the probability distribution given in (2.27).

The mode is 2, since $P[Y = 2] > P[Y = y]$ for all $y \neq 2$.

Example 2.26 In Example 2.7 of Section 2.2, two-year-old Betty was stringing beads chosen at random from a box containing 5 red beads and 10 blue beads. The random variable X was defined to be the number of red beads among the first two beads strung. What number of red beads are most likely to be strung: 0, 1, or 2?

To answer this question, we would find the mode of the probability distribution of X. This distribution was found to be

x:	0	1	2
$P[X = x]$:	$\frac{9}{21}$	$\frac{10}{21}$	$\frac{2}{21}$.

Since $P[X = 1] > P[X = x]$ for $x = 0$ or 2, the mode is equal to 1. Consequently, it will most likely be the case that one red bead will have been strung among the first two.

It would certainly not be hard to define probability distributions where two (or more) different values are tied for having the largest probability. [For example, the distribution corresponding to the number coming up on a fair die, (2.26).] We could talk about bimodal, trimodal, and so on, distributions, or we could simply say that a mode does not exist. Since this is not a very important issue, we do not mind compromising. Let us agree that if two values of the random variable are tied for having the largest probability, we will call both of the values modes and say that the distribution is *bimodal.* If three or more values are tied, we will say that a mode does not exist.

The definition given earlier for the mode of a discrete probability distribution cannot be used in the case of a continuous probability distribution. The reason for this is that $P[X = x] = 0$ for all x [see (2.7)]. Hence we give a special definition of the mode for continuous random variables.

Definition 2.11 The *mode* of the probability distribution for a *continuous random variable* X is that value x for which $f(x)$ is largest.

The mode of the probability distribution for a continuous random variable can be easily determined by looking at the graph of the probability density function. For instance, you can see from Figure 2.4 of Section 2.3 that the mode for that distribution occurs at $x = 16$.

Example 2.27 In a Mexican restaurant, La Casa de Fuego, bottles hold about 100 grams of hot sauce. These bottles are filled at the beginning of the day. At the end of the day, X, the amount remaining in a randomly chosen bottle has a distribution given by

$$f(x) = \begin{cases} \dfrac{x}{5000} & 0 \leq x \leq 100 \\ 0 & \text{otherwise.} \end{cases}$$

Find the mode of this probability distribution.

A graph of this function is given in Figure 2.11. You can see that $f(x)$ is largest when $x = 100$, so 100 is the mode.

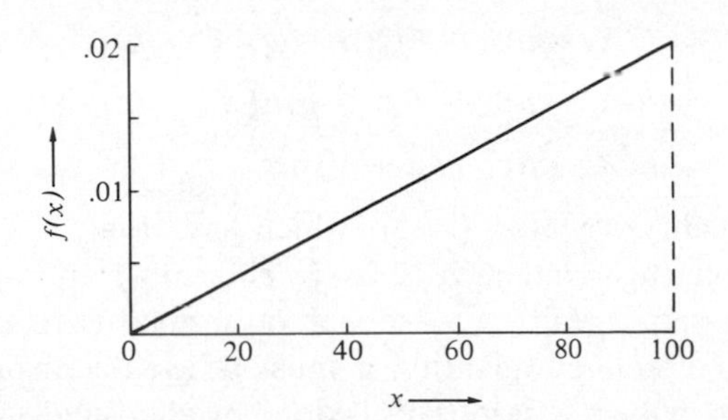

FIGURE 2.11

In many real-world problems, the mode of a probability distribution will be somewhere in the "middle" of the probability distribution (i.e., if a and b are the smallest and largest values that the random variable can assume, the mode will lie between a and b). Consequently, it is not unreasonable to discuss the concept of a mode in this section on "measures of central tendency." However, there are situations where the mode is not between a and b—but at a and/or b—so that the mode is at an end point. The probability distribution of Example 2.27 illustrates this situation.

Let us return again to Phil's offer relative to guessing at the outcome of a fair die. Assume that Phil modifies his offer from an "all-or-nothing" situation (call it offer A) to the following (offer B):

> If you guess correctly, you get $25.
>
> If you guess g and the outcome is x, you get $25 - (x - g)^2$ dollars.

In this case there is a penalty for wrong guesses, but you do not necessarily lose everything. Now remember that you are free to choose g, your guess, but the value x is determined by the outcome on the die. If we define Y to be your random payoff, then

$$Y = 25 - (X - g)^2.$$

Assuming that your objective is to maximize your payoff, what would you choose for g? Would you choose an extreme value such as 1 or 6, or would you choose a middle value such as 3 or 4? The logical course of action here would be for you to choose a middle value. By using expected values, we can determine objectively which middle value to choose.

We know, of course, that the actual payoff, Y, is a random variable no matter what guess, g, is made. We might choose g in such a way as to maximize the expected payoff. However, since

$$E(Y) = E[25 - (X - g)^2] = 25 - E(X - g)^2,$$

maximizing $E(Y)$ is equivalent to minimizing $E(X - g)^2$. To find the value of g that minimizes $E(X - g)^2$, we will use a mathematical trick and use certain properties of expectations: namely, (2.18), (2.20), and (2.21). The mathematical "trick" involves adding and subtracting μ to $(X - g)$. We then group certain terms together, expand the square, and use the properties of expectation to get

$$
\begin{aligned}
(2.28) \qquad E[(X-g)^2] &= E[(X-\mu+\mu-g)^2] = E[((X-\mu)+(\mu-g))^2] \\
&= E[(X-\mu)^2 + 2(\mu-g)(X-\mu) + (\mu-g)^2] \\
&= E(X-\mu)^2 + E[2(\mu-g)(X-\mu)] + E(\mu-g)^2 \\
&= E(X-\mu)^2 + 2(\mu-g)E(X-\mu) + (\mu-g)^2 \\
&= E(X-\mu)^2 + (\mu-g)^2.
\end{aligned}
$$

[To get the last equality we used (2.22), which says that $E(X-\mu) = 0$.] Now, remember that we are trying to minimize (2.28) by choosing g appropriately. The quantity $E(X-\mu)^2$ does not depend on g, so our only hope is to minimize $(\mu-g)^2$. Since this quantity is a squared quantity, it must be greater than or equal to zero, so the smallest we can make it is zero. This is done by choosing g to be μ. Recall that we said that our guess would be a middle value, so the previous analysis which led us to guess μ reinforces the idea that μ is, in fact, a measure of the middle of a probability distribution.

Example 2.28 To maximize the expected payoff under offer B, what should your guess be for (a) the fair die, (b) the unbalanced die with probability distribution given by (2.27)?

For a fair die $\mu = 3.5$, so this is the best guess. For the probability distribution (2.27), we have

$$\mu_Y = \sum_{y=1}^{6} y \cdot P[Y = y] = 3.2,$$

so this would be the best guess for the unbalanced die.

You may have noticed that in these two cases, by guessing at the mean value, which is not one of the values that could appear on the die, you give up the opportunity of getting the full \$25. However, in terms of expected value (i.e., your average gain in the long run), guessing μ is your wisest course of action.

We have seen that μ is a measure of the center of a probability distribution in the sense that it is the value of g that minimizes $E(X-g)^2$. It is also the center in another sense, which may be easier to understand intuitively: μ is the "center of gravity" of a probability distribution. Mathematically, the center of gravity would be the value c that satisfies $E(X-c) = 0$. Of course, we already know from (2.22) that $c = \mu$ will satisfy this condition. "Wait a minute," you say. "That's not very intuitive unless you happen to be a math or physics major!" You are right, of course, so let us view this a bit differently. Let us say that the Hubbard children go to a park and that Frieda (48 pounds) and Diane (32 pounds) wish to ride a seesaw. You are no doubt familiar with what happens if they both sit on the ends with the fulcrum in the center. However, by adjusting their positions or by moving the fulcrum, it is possible to get the seesaw to balance. When the children are on the ends, the point along the board at which they balance is the center of gravity (see Figure 2.12). If you now imagine a certain amount of probability being placed on each end of a board and if you think of probability as "weight," you can see that a balance point can be found in

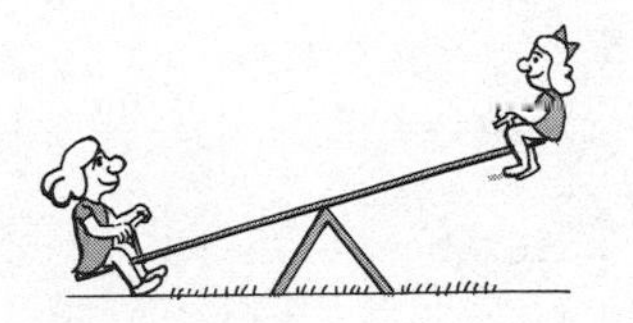

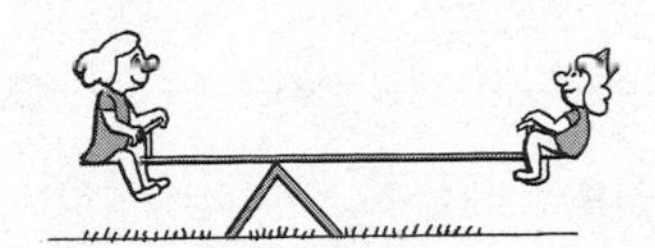

FIGURE 2.12

Finding the Center of Gravity

this case as well. The balance point is the mean. It is also possible for all the Hubbard children to sit at various places on the seesaw at the same time and to still make it balance. Similarly, a random variable can have several different values (which relate to the position of different children on the seesaw) and corresponding probabilities (which relate to the weight of the corresponding children), and the mean will still be the balance point or center of gravity (see Figure 2.13). The same kind of result would hold for continuous probability distributions. You can imagine a solid sheet of material cut out in the shape of a particular probability density function. The balance point of this cutout would be the mean. Of course, if a distribution is symmetric, it is easy to find the mean, since it would correspond to the value about which the distribution is symmetric [see Figure 2.14(a)].

Let us return once again to Phil's offers relative to guessing at the outcome when a fair die is rolled. We have already considered two offers, and now we will consider a third offer, which, at the risk of being called unimaginative, we will call offer C:

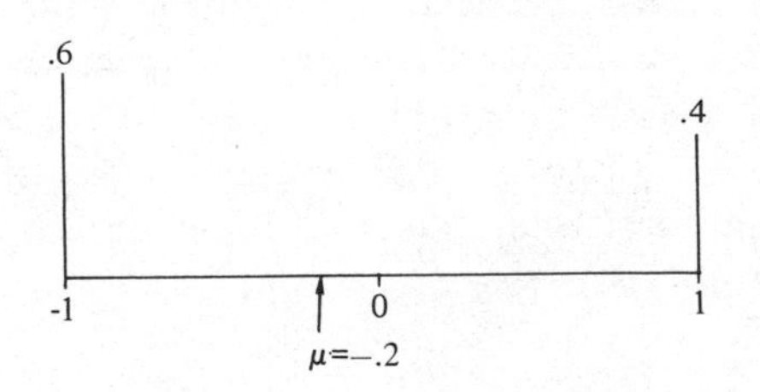

FIGURE 2.13

The Mean as the Center of Gravity

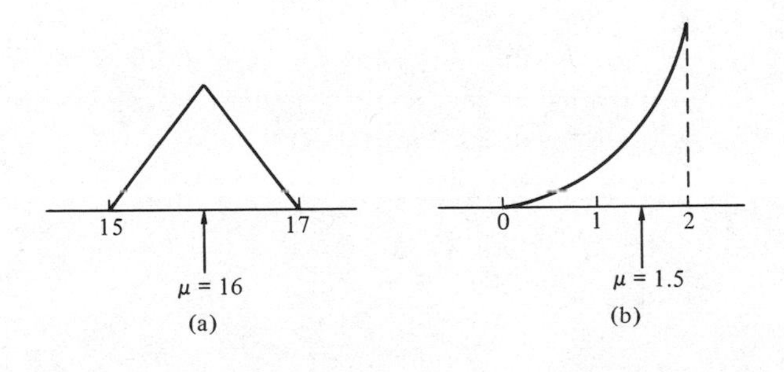

FIGURE 2.14

The Mean for Continuous Distributions

If you guess correctly, you get \$25.

If you guess g and the outcome is x, you get $25 - |x - g|$ dollars.

(Remember that the absolute value of a number is the magnitude of that number without regard to sign.) What would you choose for g in this case, an extreme value or a central value? You would probably choose a central or middle value. If we define

$$Z = 25 - |X - g|$$

to be the payoff under offer C and if we choose g to maximize the expected payoff, we will find that it is maximized if we choose g to be the median of the probability distribution.

Definition 2.12 A *median* of the probability distribution of a random variable X is a value $\tilde{\mu}$ such that

$$P[X \leq \tilde{\mu}] \geq .5 \quad \text{and} \quad P[X \geq \tilde{\mu}] \geq .5. \tag{2.29}$$

Example 2.29 Find the median of the probability distribution corresponding to the unbalanced die (2.27).

Since

$$P[Y \leq 3] = .6 \geq .5 \quad \text{and} \quad P[Y \geq 3] = .6 \geq .5$$

we see that $\tilde{\mu}_Y = 3$. (This means that if the unbalanced die is to be rolled and you are to guess its outcome under the conditions of offer C, your best strategy would be to guess that a 3 will come up.)

You may have noticed that in Definition 2.12 we used the indefinite article "a" instead of the definite article "the" before the word "median." The reason for this is that for some discrete probability distributions, there is not a unique value satisfying (2.29). However, there will be a smallest value, say c, and a largest value, say d, satisfying (2.29). If you would like to have a unique value to call the median, you could take $(c + d)/2$ as the median. We will follow this convention.

Example 2.30 Find the median of the probability distribution corresponding to the outcome when a fair die is rolled (2.26).

We have $P[X \leq 3] = \frac{3}{6} \geq .5$ and $P[X \geq 3] = \frac{4}{6} \geq .5$, while $P[X \leq 4] = \frac{4}{6} \geq .5$ and $P[X \geq 4] = \frac{3}{6} \geq .5$. Furthermore, any number smaller than 3 or greater than 4 will not satisfy (2.29), so we take $\tilde{\mu} = (3 + 4)/2 = 3.5$.

For continuous probability distributions the process of finding the median is somewhat simplified, since individual points are not assigned positive probability. Consequently, we need only find the value $\tilde{\mu}$ that satisfies

$$P[X \leq \tilde{\mu}] = .5$$

or, equivalently,

$$F(\tilde{\mu}) = .5. \tag{2.30}$$

If the CDF of a probability distribution is known, finding the median simply involves setting the CDF equal to .5 and solving.

Example 2.31 Find the median of the continuous probability distribution given in Example 2.8 of Section 2.3.

In Example 2.14 we found $F(x)$ to be $(x - 15)/2$ for $15 \leq x \leq 17$. Setting $F(\tilde{\mu}) = .5$ and solving, we get

$$F(\tilde{\mu}) = \frac{\tilde{\mu} - 15}{2} = .5$$

$$\tilde{\mu} - 15 = 2(.5)$$

$$\tilde{\mu} = 16.$$

Referring to Figure 2.3 of Section 2.3, you can see that a vertical line drawn at the median $\tilde{\mu} = 16$ divides the probability (area) into two equal parts. Hence the median is a measure of the center of a probability distribution in the sense that at least half of the probability lies at or above the median and at least half of the probability lies at or below the median. As is the case with the mean, if a distribution is symmetric, the median will be equal to the value about which the distribution is symmetric. If you look at Figure 2.4 of Section 2.3, corresponding to the distribution of Example 2.9, you can see that the distribution is symmetric about the value 16, so $\tilde{\mu} = 16$.

Example 2.32 The CDF for the probability distribution of X, the amount of hot sauce remaining in bottles in La Casa de Fuego at the end of the day (Example 2.27), is given by

$$F(x) = \begin{cases} 0 & x < 0 \\ \dfrac{x^2}{10{,}000} & 0 \leq x \leq 100 \\ 1 & x > 100. \end{cases}$$

Find the median for this distribution.

Solving $F(\tilde{\mu}) = .5$ for $\tilde{\mu}$, we find

$$F(\tilde{\mu}) = \frac{\tilde{\mu}^2}{10{,}000} = .5$$

$$\tilde{\mu}^2 = 5000$$

$$\tilde{\mu} = \sqrt{5000} = 70.71 \text{ grams.}$$

(This means that half of the bottles would have 70.71 grams or less of hot sauce left, while the rest have more than this amount.)

If the CDF is a linear or quadratic function, it will not be too hard to solve equation (2.30). However, for more complicated functions, solving that equation may be more difficult. In situations like this, it is possible to get an estimate of the median by using a graph of the CDF. In particular, locate the value .5 on the vertical axis, draw a horizontal line from that point to the curve $F(x)$, and drop a vertical line from that point to the x axis. The corresponding x value will be the median. Figure 2.15 illustrates this method for the CDF corresponding to the probability density function given in Example 2.10. From the graph the value of the median would appear to be approximately 1.6.

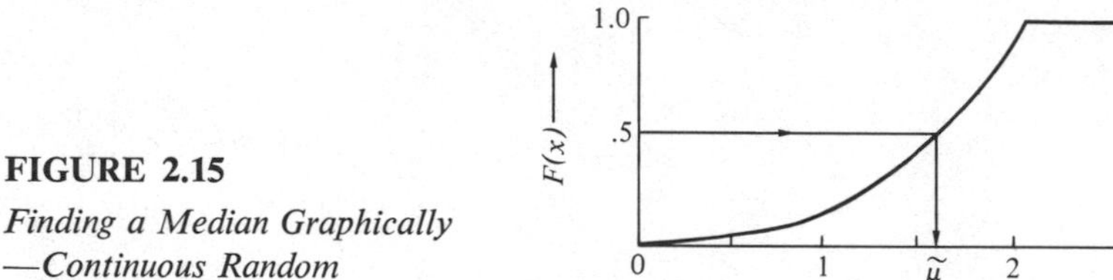

FIGURE 2.15
Finding a Median Graphically —Continuous Random Variables

A similar graphical technique can be used for finding the median of a discrete distribution. Recall that the CDF for a discrete random variable is a step function. Using the terminology of carpentry, the vertical portion of the step is called the "riser" and the horizontal portion is called the "tread." The median can be estimated as follows. Locate the value .5 on the vertical axis, draw a horizontal line until it hits a riser, and drop a vertical line to the x axis. The corresponding x value will be the median. If the horizontal line happens to hit a tread rather than a riser, drop a vertical line from halfway across the tread to the x axis to find the median. Both of these situations are shown in Figure 2.16 for the distributions corresponding to Examples 2.29 and 2.30.

Although this is somewhat of a digression from the general theme of measures of central tendency, we will devote some space here to the topic of fractiles in general. You are no doubt familiar with the concept of percentiles from your experiences in taking standardized examinations. The median is the fiftieth percentile, *deciles*

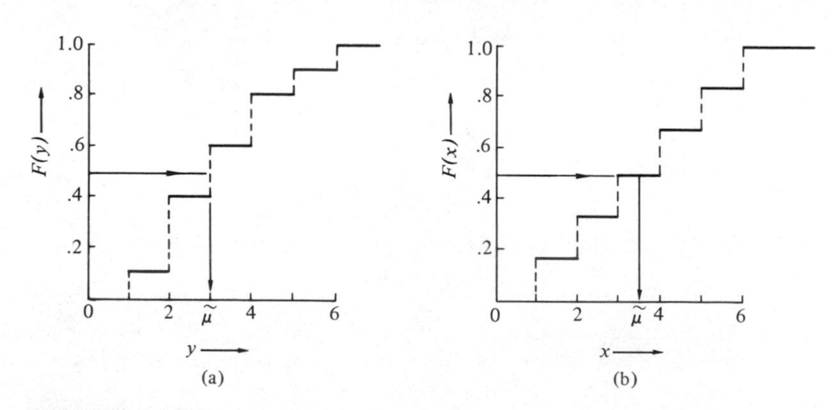

FIGURE 2.16
Finding a Median Graphically—Discrete Random Variables

are the tenth, twentieth, . . . , ninetieth percentiles, and *quartiles* are the twenty-fifth, fiftieth, and seventy-fifth percentiles. All of these taken together are called *fractiles.*

Definition 2.13 The number x_p is the $100 \cdot p$ percentile of a probability distribution if

$$P[X \leq x_p] \geq p \quad \text{and} \quad P[X \geq x_p] \geq 1 - p, \tag{2.31}$$

for $0 \leq p \leq 1$.

Once a graph of the CDF for a random variable X is drawn, it is easiest to find percentiles using a graphical approach, as described earlier. Locate the value p on the vertical axis, draw a horizontal line to intersect with the CDF, drop a vertical line to the x axis, and read the x value as the percentile. For example, from Figure 2.16(a) you can read the 25th and 75th percentiles as 2 and 4, respectively; from Figure 2.16(b) you can see that the 25th and 75th percentiles are 2 and 5.

We have defined three measures of central tendency—the mode, the mean, and the median—each of which would correspond to the best guess as to the observed value of a random variable under offers A, B, and C, respectively. How do these measures compare with one another? We have already said that if a distribution is symmetric, the mean and median will be equal. What about the mode? It may or may not be equal to this common value, as you can see from Figure 2.17. If a distribution is not symmetric but has a longer "tail" to one side than the other, it is said to be *skewed.* If the longer tail is to the right (left), the distribution is said to be skewed to the right (left). For unimodal skewed distributions such as those in Figure 2.18, the mean will be farthest out toward the long tail and the median will be between the mean and mode.

Measures of central tendency are frequently used to define a "typical" or "representative" value of a random variable. If you were asked to give a measure of central tendency that was "representative" of a particular probability distribution, which

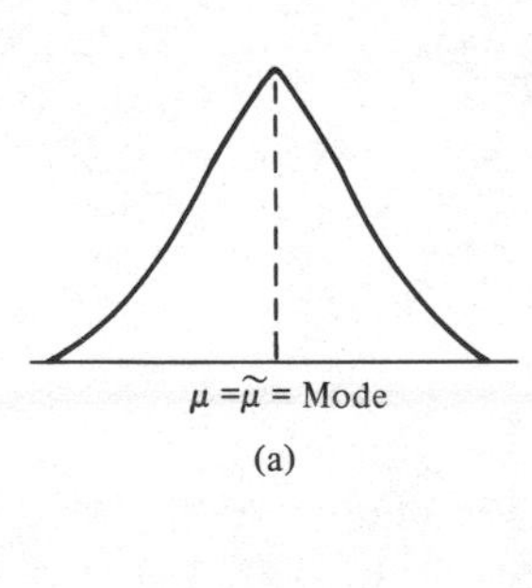

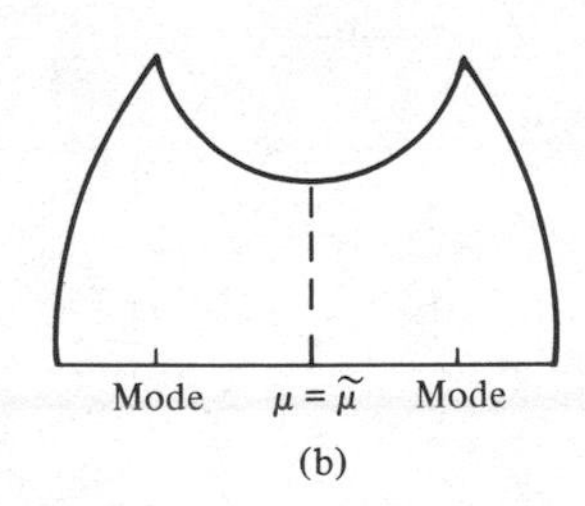

FIGURE 2.17
Examples of Symmetric Distributions

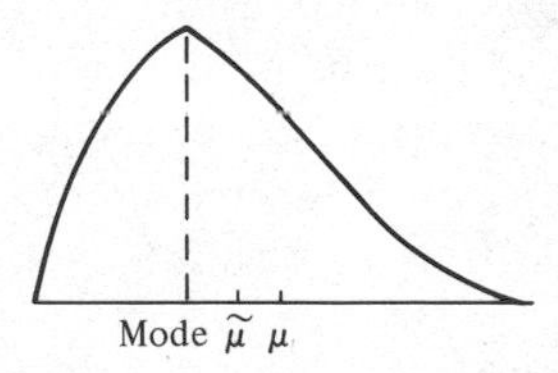

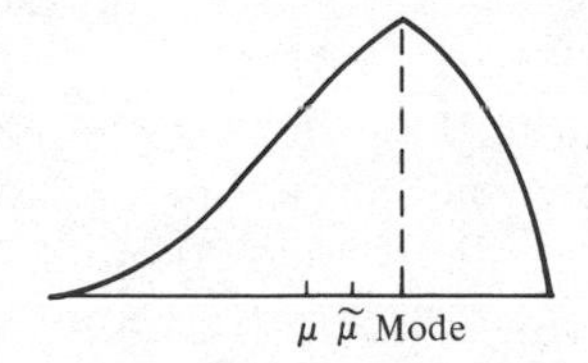

FIGURE 2.18
Examples of Skewed Distributions

would you give? If the distribution in question were like the one in Figure 2.17(a), it would not matter, since all three values are the same. On the other hand, for a highly skewed distribution, the values of the mean and median, for example, can be quite widely separated. One example of a highly skewed distribution would be the distribution of incomes of families in the United States. As you probably know, most incomes are in the \$5,000 to \$25,000 range, but the distribution is highly skewed to the right because of a relatively few individuals earning extremely large amounts of money. In this situation the median is considered more "representative" than the mean.

The mean and median, as well as the mode, are commonly referred to as *averages*. Since these values can be quite different for skewed distributions, a well-written news article (for example) will clearly state whether a stated "average" is actually a mean or median value. A worthwhile exercise is to observe how the news media handle this particular situation.

2.7 MEASURES OF VARIABILITY

In the previous section we saw that under Phil's offer B, getting $25 - (X - g)^2$ dollars for guessing g when X is observed on the roll of a die, the best guess we could make would be μ, the mean of the probability distribution of X. Let us assume now that Phil is going to allow you to play this guessing game and is also going to allow you to choose which of three unbalanced dice you will throw. The probability distributions for the number appearing up on the first, second, and third dice, respectively (denoted by X, Y, and Z), are given in Table 2.2. It is not hard to see that the mean

TABLE 2.2
Probability Distributions for Three Unbalanced Dice

	Number Up					
	1	*2*	*3*	*4*	*5*	*6*
$P[X = x]$	.30	.10	.20	.20	.10	.10
$P[Y = y]$	.20	.20	.30	.10	.10	.10
$P[Z = z]$	.05	.35	.30	.20	.05	.05

in each case is equal to 3, so no matter which die you choose your guess will be the same. Which die would you choose to use? You might eliminate the first die on the grounds that the probability of the mean actually occurring is less. Even more important than that, however, is that with the second die, the outcome is more likely to be at *or near* the mean than with the first die. Similarly, in comparing the second and third distributions, you see that the probabilities are clustered more around the mean for the third than the second. Another way of saying this is that there is less variability, relative to the mean, in the third distribution than in the first or second. Most people, when given a choice in a situation like this, would generally choose the third die. We can approach this problem more objectively by comparing the expected payoffs in each case. We find

$$E[25 - (X - \mu_X)^2] = 25 - E(X - \mu_X)^2 = 25 - 2.80 = 22.20$$
$$E[25 - (Y - \mu_Y)^2] = 25 - E(Y - \mu_Y)^2 = 25 - 2.40 = 22.60$$
$$E[25 - (Z - \mu_Z)^2] = 25 - E(Z - \mu_Z)^2 = 25 - 1.40 = 23.60,$$

so in fact our choice of the third die proves to be correct in terms of maximizing the expected payoff.

You will notice that the quantity which distinguishes the expected payoffs in each case is $E(X - \mu)^2$. This, as you can see from the previous discussion, can be used as a measure of the variability of the probability distribution of a random variable.

Definition 2.14 The *variance* of the probability distribution of a random variable X is defined by

$$V(X) = E(X - \mu_X)^2. \tag{2.32}$$

An alternative notation for the variance of a random variable is

$$\sigma_X^2.$$

It is important to note that $V(X)$ and σ_X^2 are identical quantities and that the difference is one of notation only.

Example 2.33 Show that the variance of the distribution of X given in Table 2.2 is 2.80.

From equation (2.17), we see that we find $E(X - \mu)^2$ by

$$\begin{aligned} E(X - \mu)^2 &= \sum_{\text{all } x} (x - \mu)^2 \cdot P[X = x] = \sum_{x=1}^{6} (x - 3)^2 \cdot P[X = x] \\ &= (1 - 3)^2(.30) + (2 - 3)^2(.10) + (3 - 3)^2(.20) + (4 - 3)^2(.20) \\ &\quad + (5 - 3)^2(.10) + (6 - 3)^2(.10) \\ &= 4(.30) + 1(.10) + 0(.20) + 1(.20) + 4(.10) + 9(.10) \\ &= 2.80. \end{aligned}$$

Since it is necessary to find $\mu = E(X)$ before calculating the variance, it is sometimes convenient to use the following mathematically equivalent formula for calculating the variance:

$$V(X) = E(X^2) - [E(X)]^2. \tag{2.33}$$

We can prove that (2.32) and (2.33) are equivalent by using the properties of expectations found in Section 2.5: namely, (2.18), (2.20), and (2.21). Recalling that μ and $E(X)$ are different symbols for the same quantity and that μ is in fact a constant, we proceed as follows:

$$\begin{aligned} V(X) &= E[(X-\mu)^2] = E(X^2 - 2\mu X + \mu^2) \\ &= E(X^2) + E(-2\mu X) + E(\mu^2) \\ &= E(X^2) - 2\mu E(X) + \mu^2 \\ &= E(X^2) - 2\mu \cdot \mu + \mu^2 \\ &= E(X^2) - \mu^2 = E(X^2) - [E(X)]^2. \end{aligned}$$

Example 2.34 Calculate the variance of the distribution of X given in Table 2.2 using equation (2.33).

From previous discussion we have that $E(X) = 3$. We get

$$\begin{aligned} E(X^2) &= \sum_{\text{all } x} x^2 \cdot P[X = x] = \sum_{x=1}^{6} x^2 \cdot P[X = x] \\ &= (1)^2(.3) + (2)^2(.1) + (3)^2(.2) + (4)^2(.2) + (5)^2(.1) + (6)^2(.1) \\ &= 1(.3) + 4(.1) + 9(.2) + 16(.2) + 25(.1) + 36(.1) \\ &= 11.8. \end{aligned}$$

Using (2.33), we get $V(X) = 11.8 - (3)^2 = 2.8$, which of course agrees with the results of Example 2.33.

When we speak of the number appearing up when a die is tossed, we are considering a number, pure and simple. In many situations, however, the random variable under consideration has some units of measurement naturally associated with it. For instance, in Example 2.4 of Section 2.1 we defined random variables on a sample space related to students chosen at random from a large statistics class. The random variable X_1, the age of the student chosen, would have units of "years" associated with it, while the random variable X_2, the height of the student chosen, would have units of "inches" associated with it. When we calculate the expected value of X_1, for example, we would also associate the units of years with the expected value. [Note that since probabilities are pure numbers (i.e., unitless), when we calculate $E(X_1) = \sum (x \text{ years}) \cdot P[X_1 = x]$ we end up with units of years for $E(X_1)$.] Because the units on X_1 and $E(X_1)$ are the same, it makes sense to consider a quantity such as $X_1 - E(X_1)$. However, when we calculate the variance of X_1, we get

$$V(X_1) = \sum_x [x \text{ years} - E(X_1) \text{ years}]^2 P[X_1 = x],$$

and consequently the variance of X_1 would have the units "years-squared" associated with it. Just as it does not make sense to "compare apples with oranges," it would not make sense to consider a quantity such as $X_1 - V(X_1)$, because it wouldn't make sense to subtract "years-squared" from "years." Besides what are "years-squared" anyway? To avoid embarrassing questions such as these, we find it convenient to introduce a measure of variability that will have associated with it the same units as are associated with the random variable in question. The simplest way to do this is to take the square root of the variance.

Definition 2.15 The *standard deviation* of the probability distribution of a random variable X is defined by

$$\sigma_X = \sqrt{V(X)}.$$

[Now perhaps you can see that there was some logic in using the symbol σ_X^2 as an alternative notation for $V(X)$. The Greek letter corresponding to "s," the first letter of the word "standard deviation," is sigma (σ).]

Example 2.35 Find the standard deviation for the probability distributions of X, Y, and Z given in Table 2.2.

Since we know that

$$V(X) = 2.80, \qquad V(Y) = 2.40, \qquad V(Z) = 1.40,$$

we simply take the square roots of each of these quantities to get (to two decimal places)

$$\sigma_X = 1.67, \qquad \sigma_Y = 1.55, \qquad \sigma_Z = 1.18.$$

There are many situations where a random variable is modified by means of an additive or multiplicative constant. In these situations the variance of the original random variable can be used to find the variance of the modified random variable. By continuing with our hypothetical problem of guessing at the outcomes on various dice, we can at least get an intuitive idea as to the relationship between the variances of the original and modified random variables. For example, if the die with probability distribution denoted by X were modified by adding two dots to each face, the probability distribution would be

(2.34)

Value:	3	4	5	6	7	8
Probability:	.30	.10	.20	.20	.10	.10.

Remembering that the payoff is to be $25 - (X - g)^2$ dollars, would there be any advantage in using the modified die instead of the original? No! While you would use a different value for your guess in the modified situation, there is no change in the variability. This result is explained by the following theorem.

Theorem 2.5 For any random variable X,

$$V(X + c) = V(X).$$

Proof: To prove this, we need to use equation (2.19), which states that

$$E(X + c) = E(X) + c.$$

Using (2.32), we have

$$\begin{aligned} V(X + c) &= E[(X + c) - \mu_{X+c}]^2 \\ &= E[(X + c) - (E(X) + c)]^2 = E[X + c - E(X) - c]^2 \\ &= E[X - E(X)]^2 = E(X - \mu_X)^2 = V(X). \end{aligned}$$

As another example, consider the probability distribution denoted by X in Table 2.2, modified by doubling the number of dots on each face (i.e., modifying by use of a multiplicative constant). In this case the probability distribution would be

(2.35)

Value:	2	4	6	8	10	12
Probability:	.30	.10	.20	.20	.10	.10.

In this situation the mean, and hence your guess, would be 6. However, you can see that there is more variability. In the original case, if you guessed the mean of 3, the worst that could happen would be for a 6 to appear, leaving you with $25 - (6 - 3)^2 = 16$ dollars. However, now the worst would be for a 12 to appear, leaving you with $25 - (12 - 6)^2 = -11$ dollars. A "can't-miss" proposition has turned into one that could cost you $11! The variance has no doubt increased and the following theorem can be used to determine just what the increase was.

Theorem 2.6 For any random variable X,

$$V(cX) = c^2 V(X).$$

Proof: To prove this theorem we use equation (2.18), which states that

$$E(cX) = cE(X).$$

Again using (2.32), we have

$$\begin{aligned} V(cX) &= E[cX - \mu_{cX}]^2 = E[cX - E(cX)]^2 \\ &= E[cX - cE(X)]^2 = E[c(X - E(X))]^2 \\ &= E[c^2(X - E(X))^2] \\ &= c^2 E[X - E(X)]^2 = c^2 V(X). \end{aligned}$$

It follows from the relationship between standard deviation and variance that

$$\sigma_{X+c} = \sigma_X \quad \text{and} \quad \sigma_{cX} = \sqrt{c^2 \sigma_X^2}.$$

A third result about variances that we can give is

$$V(c) = 0.$$

That is, if a random variable always assumes the same constant value, its variance will be zero. (Obviously, this is a very special case, since such a "random variable" is not very random nor is it variable!)

Example 2.36 Find the variances and standard deviations for the probability distributions given in (2.34) and (2.35).

Since the distributions in (2.34) and (2.35) correspond to $X + 2$ and $2X$, respectively (for X in Table 2.2), we have

$$V(X + 2) = V(X) = 2.80, \qquad \sigma_{X+2} = \sigma_X = 1.67$$

and

$$V(2X) = (2)^2 V(X) = 4V(X) = 11.20, \qquad \sigma_{2X} = 2\sigma_X = 3.34.$$

It is relatively easy to get an intuitive feeling for the concept of the mean of a probability distribution because most of us are rather familiar with the concept of an average. By using the concept of center of gravity we can also get a visual feeling for the location of the mean on a graph of a probability distribution. Unfortunately, we are not as familiar with the concept of variability, so many people who can mechanically compute a variance (or standard deviation) ask "What have I got?" or "How do I know if this quantity I found is large or small?" As to questions concerning the magnitude of the variance or standard deviation, we can only answer that that will depend on the context of the problem and/or the units used. [For example, if X = actual diameter (measured in inches) of a "quarter-inch-diameter" bolt, then $V(X) = .001$ would be considered very "large"; whereas if $Y =$ actual diameter (measured in micrometers) of 2-millimeter-diameter wire, then $V(Y) = 100$ would be considered relatively "small."] In Section 2.9 we will see that Chebyshev's inequality can give some more insight into what the standard deviation can tell you.

The variance and standard deviation are the most important measures of variability in a probability distribution. This importance is related to the fact that they measure the variability in terms of deviations from the mean, and the mean, because of certain mathematical properties, is the most important measure of central tendency. There are other less important measures of variability which also relate to deviations about the mean which we will not discuss here. We will briefly mention a measure of variability that can be used when the median is used as a measure of central tendency rather than the mean. This is the *expected absolute deviation about the median*, which, since there is no standard notation for this quantity, we will call EAD($\tilde{\mu}$) and define by

$$\text{EAD}(\tilde{\mu}) = \text{expected absolute deviation about } \tilde{\mu} = E(|X - \tilde{\mu}_X|).$$

Example 2.37 A discrete random variable X has a probability distribution given by

x:	0	1	2	3	4
$P[X = x]$:	.10	.15	.20	.25	.30.

Find the median and the expected absolute deviation about the median.

The median is 3, since

$$P[X \leq 3] = .70 \geq .50 \qquad \text{and} \qquad P[X \geq 3] = .55 \geq .50.$$

Since $|X - \tilde{\mu}|$ is a function of X, we find its expected value by using equation (2.17) of Section 2.5.

$$\begin{aligned} E(|X - \tilde{\mu}|) &= \sum_{\text{all } x} |x - \tilde{\mu}| \cdot P[X = x] \\ &= |0 - 3|(.10) + |1 - 3|(.15) + |2 - 3|(.20) \\ &\quad + |3 - 3|(.25) + |4 - 3|(.3) \\ &= 3(.10) + 2(.15) + 1(.20) + 0(.25) + 1(.3) \\ &= 1.1. \end{aligned}$$

2.8 STANDARDIZED RANDOM VARIABLES

Two farmers were sitting in front of the general store having a discussion as to who was the better farmer. Zeke, who raises soybeans, claimed that he was the better farmer, since he had gotten a yield of 40 bushels/acre of soybeans, whereas Zack, a wheat farmer, had gotten a yield of only 36 bushels/acre. Zack maintains that it is not right to compare yields in these terms, since it's like comparing apples and oranges. "No," says Zeke, "We're comparing soybeans and wheat!" At this point Zephaniah, the owner of the general store, decided to make use of his college education (he had a joint major in business, psychology, and statistics) to settle the argument. How do you think he would proceed?

Zeph suggested that they compare themselves not with each other but with other farmers raising the same crops. If, in fact, the mean yield for soybeans is 34 bushels/acre and for wheat it is 30 bushels/acre (when all U.S. production is considered), how would Zeke and Zack compare? If we let x represent Zeke's yield and μ_X the mean soybean production and let y represent Zack's yield with μ_Y the mean wheat production, we see that

$$x - \mu_X = 40 - 34 = 6 \quad \text{and} \quad y - \mu_Y = 36 - 30 = 6.$$

Hence in this case the deviation from the mean is the same for both farmers and the argument is not yet settled. What other factor should be considered for comparing two different variables? The variability as measured by the standard deviation could be considered. Zeph suggests using the standard deviation as a divisor of the deviation from the mean. He reasons that being 6 units from the mean is not too unexpected if there is a lot of variability but would be more unexpected if there is a relatively small amount of variability. Since $\sigma_X = 6$ and $\sigma_Y = 4$, Zeph's calculations show that

$$\frac{x - \mu_X}{\sigma_X} = \frac{40 - 34}{6} = 1.0$$

while

$$\frac{y - \mu_Y}{\sigma_Y} = \frac{36 - 30}{4} = 1.5.$$

In these terms, Zeph concludes, Zack is the better farmer.

The procedure used in this illustration is known as *standardizing*. In general, if X is any random variable having mean μ_X and standard deviation σ_X, then Z defined by

$$Z = \frac{X - \mu_X}{\sigma_X} \tag{2.36}$$

is said to be standardized. Theorem 2.7 shows that any standardized random variable has a mean equal to zero and a variance equal to 1.

Theorem 2.7 If X is a random variable having mean μ_X and standard deviation σ_X, and if Z is defined by (2.36), then

$$E(Z) = 0 \quad \text{and} \quad V(Z) = 1.$$

Proof: To prove this theorem we simply use properties of expectation and variance derived in Sections 2.5 and 2.7. In particular, using (2.18) and (2.22) we see that

$$E(Z) = E\left(\frac{X - \mu_X}{\sigma_X}\right) = \frac{1}{\sigma_X}E(X - \mu_X) = \frac{1}{\sigma_X}(0) = 0.$$

Similarly, using Theorems 2.5 and 2.6 we see that

$$V(Z) = V\left(\frac{X - \mu_X}{\sigma_X}\right) = \frac{1}{\sigma_X^2} V(X - \mu_X)$$

$$= \frac{1}{\sigma_X^2} V(X) = \frac{\sigma_X^2}{\sigma_X^2} = 1.$$

The observed value of z obtained by substituting an observed value of x into (2.36) is referred to as a *standard score*. Standard scores for random variables having different distributions are more comparable than the values of the random variables themselves. This is the reason that Zeph suggested using standard scores for comparing Zeke and Zack's yields. Figure 2.19 shows what the distributions of soybean and wheat

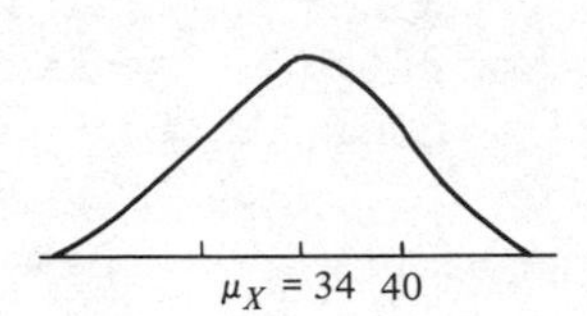

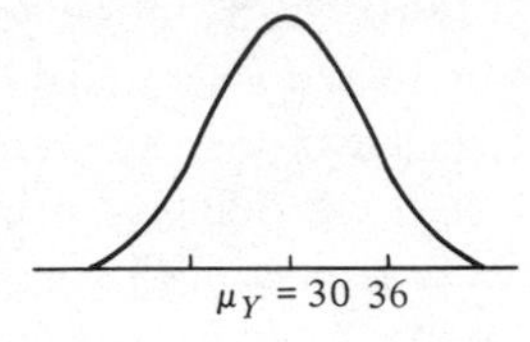

FIGURE 2.19
Possible Distributions for Soybean and Wheat Production (bushels/acre)

production *might* look like. Note that because of the greater variability in X, soybean yields, being 6 units above the mean is not as impressive a feat as being 6 units above the mean for Y, the wheat yields. In particular, the proportion of soybean farmers whose yield exceeds 40 bushels/acre is greater than the proportion of wheat farmers whose yield exceeds 36 bushels/acre.

Example 2.38 The average score on an accounting exam was 65, with a standard deviation of 10. Find the standard scores corresponding to scores of 60 and 80. Also find the score corresponding to standard scores of -1 and $+2$.

To find the standard scores use (2.36) with $\mu = 65$ and $\sigma = 10$. This gives

$$z = \frac{60 - 65}{10} = -.5$$

and

$$z = \frac{80 - 65}{10} = +1.5$$

as the standard scores corresponding to 60 and 80. To find the score corresponding to a particular standard score, simply solve (2.36) for X as follows:

$$Z = \frac{X - \mu}{\sigma}$$

$$Z\sigma = X - \mu;$$

therefore,

$$X = \mu + Z\sigma. \tag{2.37}$$

Substituting into (2.37), we get

$$x = 65 + (-1)(10) = 55$$

and

$$x = 65 + (2)(10) = 85.$$

2.9 CHEBYSHEV'S INEQUALITY

In Figure 2.19 we showed probability distributions that could describe the yield (in bushels/acre) of soybeans and wheat for farmers in the United States. It is rather easy to see visually that the amount of area to the right of 40 in the first (X) distribution is greater than the area to the right of 36 in the second (Y) distribution and, consequently, in view of the relationship between area and probability, that $P[X \geq 40] > P[Y \geq 36]$. Now consider only the X distribution, soybean yields. We have assumed that $\mu_X = 34$ and $\sigma_X = 6$. Is the distribution given for X in Figure 2.19 the *only* distribution having a mean of 36 and a standard deviation of 6? No! There are many different distributions having that mean and standard deviation. Furthermore, these different distributions may assign a different probability to the event $[X \geq 40]$. We cannot, in fact, assign a definite answer to $P[X \geq 40]$ unless we know exactly the form of the probability density function $f_X(x)$.

You might ask, then, whether it does any good to know the mean and standard deviation of a random variable if you don't know the exact form of the probability density function (or, for discrete random variables, the probability distribution). Can we make any statements about probabilities assigned to events? The answer to this question is a qualified "yes." We cannot assign exact probabilities to all events, but we can say that for events of the form

$$[|X - \mu| \geq k\sigma] \tag{2.38}$$

the probability assigned, *no matter what the exact pdf or probability distribution*, must be less than or equal to $1/k^2$.

Example 2.39 A random variable X has a mean $\mu = 20$ and a standard deviation $\sigma = 2$. What can be said about the probability of $[|X - 20| \geq 6]$?

By comparing this event with (2.38), we see that $k\sigma = 6$. Since we know that $\sigma = 2$, it follows that

$$k = \frac{6}{\sigma} = \frac{6}{2} = 3,$$

so the probability that $|X - 20| \geq 6$ must be less than or equal to $1/k^2 = 1/3^2 = \frac{1}{9}$:

$$P[|X - 20| \geq 6] = P[|X - \mu| \geq 3\sigma] \leq \frac{1}{3^2} = \frac{1}{9}.$$

We will formally state this result as a theorem and give a proof of it for the case where X is a discrete random variable.

Theorem 2.8 (*Chebyshev's Inequality*) If X is a random variable having mean μ and variance σ^2, then for any $k > 0$,

$$P[|X - \mu| \geq k\sigma] \leq \frac{1}{k^2}.$$

Proof: If X is a discrete random variable, then σ^2 is found from

$$\sigma^2 = E(X - \mu)^2 = \sum_{\text{all } x} (x - \mu)^2 \cdot P[X = x]. \tag{2.39}$$

The first "trick" in this proof is to partition the values of x into two sets: the set $A_1 = \{x : |x - \mu| \geq k\sigma\}$ and the set $A_2 = \{x : |x - \mu| < k\sigma\}$. Then instead of writing a single sum over "all x" values, we can write the right-hand side of (2.39) using two sums, one over the x values in A_1 and the other over the x values in A_2:

$$\sum_{\text{all } x} (x - \mu)^2 \cdot P[X = x] = \sum_{x \in A_1} (x - \mu)^2 P[X = x] + \sum_{x \in A_2} (x - \mu)^2 P[X = x]. \tag{2.40}$$

Next notice that both summands on the right-hand side of (2.40) are nonnegative, since for each value of x, the quantities $(x - \mu)^2$ and $P[X = x]$ are nonnegative. If we consider just the first summand of the right-hand side of (2.40), that must be no larger than the sum of both summands:

$$\sum_{x \in A_1} (x - \mu)^2 P[X = x] + \sum_{x \in A_2} (x - \mu)^2 P[X = x] \geq \sum_{x \in A_1} (x - \mu)^2 P[X = x]. \tag{2.41}$$

Next we see that for all $x \in A_1$, $|x - \mu| \geq k\sigma$, so it must be that

$$(x - \mu)^2 = |x - \mu|^2 \geq k^2\sigma^2.$$

Hence

$$\sum_{x \in A_1} (x - \mu)^2 P[X = x] \geq \sum_{x \in A_1} k^2\sigma^2 P[X = x]. \tag{2.42}$$

Since the term $k^2\sigma^2$ does not involve x, the quantity $k^2\sigma^2$ can be factored from the right-hand side of (2.42) to give

$$\sum_{x \in A_1} k^2\sigma^2 P[X = x] = k^2\sigma^2 \sum_{x \in A_1} P[X = x]. \tag{2.43}$$

Finally, it follows from equation (2.6) that

$$k^2\sigma^2 \sum_{x \in A_1} P[X = x] = k^2\sigma^2 P[A_1] = k^2\sigma^2 P[|X - \mu| \geq k\sigma]. \tag{2.44}$$

If we were to now write out equations (2.39) to (2.44) in a line, we would get

$$\sigma^2 \geq k^2\sigma^2 P[|X - \mu| \geq k\sigma]. \tag{2.45}$$

Solving (2.45) for $P[|X - \mu| \geq k\sigma]$, we get

$$P[|X - \mu| \geq k\sigma] \leq \frac{\sigma^2}{k^2\sigma^2} = \frac{1}{k^2}.$$

(Note that although Chebyshev's inequality is actually true for all $k > 0$, it is only of practical interest for $k > 1$. Why is it only of interest when $k > 1$?)

Example 2.40 In Example 2.38 we noted that the average score on an accounting exam was 65, with a standard deviation of 10. If a test paper is chosen at random from this set of examinations, what is the probability that the score is 85 or above or 45 or below?

We cannot answer this question precisely without knowing the individual scores on the exams (or at least knowing considerably more information than we are given). We can, however, give an upper bound on the answer using Chebyshev's inequality. That is, we can find a value p such that the precise probability is less than or equal to p. The main thing we need to do is to rewrite the event of interest into the form $|X - \mu| \geq k\sigma$. If X represents the score on the paper chosen, we can write

$$[X \geq 85 \text{ or } X \leq 45] = [X - 65 \geq 20 \text{ or } X - 65 \leq -20]$$
$$= [|X - 65| \geq 20] = [|X - 65| \geq 2(10)].$$

Since $65 = \mu$ and $10 = \sigma$, we find, using Chebyshev's inequality, that

$$P[X \geq 85 \text{ or } X \leq 45] = P[|X - \mu| \geq 20] \leq \frac{1}{2^2} = \frac{1}{4}.$$

Hence the actual probability must be less than or equal to $\frac{1}{4}$.

Here we found an upper bound on the probability that a score would deviate from the mean of 65 by more than 20 points (i.e., by more than 2 standard deviations). If we were to ask for an upper bound on the probability that a score would differ from the mean by more than 30 points (3 standard deviations), the answer would be $1/3^2$, or $\frac{1}{9}$. In general, for any probability distribution, the probability that a random variable differs from its mean by more than k standard deviations is less than or equal to $1/k^2$.

We have said that Chebyshev's inequality can be used when the mean and variance of a random variable are known but the exact probability distribution is not known. If the probability distribution *is* known, Chebyshev's inequality could still be used, but since exact probability statements can be made, Chebyshev's inequality is generally not used. However, for the sake of illustration, we will give an example where the probability distribution is known and where we use this distribution to calculate exact probabilities and compare the results with the bounds given by Chebyshev's inequality.

Example 2.41 A discrete random variable X has a probability distribution given by

x:	1	2	3	4	5	6	7
$P[X = x]$:	.03	.04	.07	.72	.07	.04	.03.

Using Chebyshev's inequality, find an upper bound on the probabilities of the events $|X - \mu| \geq k\sigma$ for $k = 1, 2, 3$, and 4. Also find the exact probabilities for these events.

It is not hard to see that, because of the symmetry of the distribution, $\mu = 4$. We also see that

$$\sigma^2 = E(X - \mu)^2 = \sum_{x=1}^{7} (x - 4)^2 P[X = x]$$
$$= (-3)^2(.03) + (-2)^2(.04) + \ldots + (+3)^2(.03) = 1,$$

so $\sigma = 1$. Using Chebyshev's inequality, we find

$$P[|X - \mu| \geq 1\sigma] \leq \frac{1}{1^2} = 1 = 1.0000$$

$$P[|X - \mu| \geq 2\sigma] \leq \frac{1}{2^2} = \frac{1}{4} = .2500$$

$$P[|X - \mu| \geq 3\sigma] \leq \frac{1}{3^2} = \frac{1}{9} = .1111$$

$$P[|X - \mu| \geq 4\sigma] \leq \frac{1}{4^2} = \frac{1}{16} = .0625.$$

The exact probability for $|X - \mu| \geq 1\sigma$ is simply

$$\begin{aligned} P[|X - 4| \geq 1 \cdot 1] &= P[|X - 4| \geq 1] \\ &= P[(X - 4) \geq +1] + P[(X - 4) \leq -1] \\ &= P[X \geq 5] + P[X \leq 3] = .14 + .14 = .28. \end{aligned}$$

Similarly,

$$\begin{aligned} P[|X - \mu| \geq 2\sigma] &= .14 \\ P[|X - \mu| \geq 3\sigma] &= .06 \\ P[|X - \mu| \geq 4\sigma] &= 0. \end{aligned}$$

You can see in this example that the exact probabilities are considerably less than the upper bounds given by Chebyshev's inequality. Remember that if the exact probability distribution is known, it is better to calculate probabilities exactly. Chebyshev's inequality is generally used only if the exact probabilities are *not* known.

EXERCISES

1. In Example 2.4 of Section 2.1 three random variables were defined using the information given in Appendix A3. Define three other random variables on this sample space.

2. In Example 2.3 of Section 2.1 two random variables were defined based on the measured amount of cola in selected bottles. Determine whether or not the following are random variables. (a) Z = amount over or under 16 ounces; (b) U = "good" if contents are within $\frac{1}{4}$ ounce of correct, U = "bad" if not; (c) $W = 0$ if contents are within $\frac{1}{4}$ ounce of correct, $W = +1$ if not.

3. A Peskaranian penny has probability $p = .6$ of coming up heads when flipped. Let X be the number of heads observed in two flips of a Peskaranian penny. (a) Find the probability distribution for X. (b) Graph the corresponding probability histogram.

4. Let Y be the number of heads observed in three flips of the Peskaranian penny described in Exercise 3. (a) Find the probability distribution for Y. (b) Graph the corresponding probability histogram.

5. Determine which, if any, of the following are legitimate probability distributions. For those which are not, explain why they are not.

(a)	x:	0	1	2	3
	$P[X = x]$:	$\frac{1}{2}$	0	$\frac{1}{3}$	$\frac{1}{6}$
(b)	y:	10	100	1000	
	$P[Y = y]$:	.5	.6	−.1	
(c)	z:	−1	0	+1	
	$P[Z = z]$:	.2	.3	.4	
(d)	w:	−2	−1	0	
	$P[W = w]$:	.3	.4	.5	

6. Thrifty-Nifty, a discount store, offers a coupon special on peanuts with a limit of three jars per coupon at the special price. The number of jars, X, bought by a customer making a purchase at the store was found to have the following probability distribution:

x:	0	1	2	3
$P[X = x]$:	.45	.25	.10	.20.

(a) Graph the corresponding probability histogram.
(b) Find $P[1 \leq X \leq 2]$, $P[0 \leq X \leq 1]$, $P[X \leq 1]$, and $P[X > 2]$.

7. Use the random variable X as defined in Exercise 6 to express the following events:
(a) A customer buys some peanuts.
(b) A customer buys at most one jar of peanuts.
(c) A customer buys at least two jars of peanuts.
(d) Assign probabilities to the events defined in parts (a), (b), and (c).

8. In Example 1.20 reference was made to the time of birth of babies. Assume that the exact time of the birth of babies could be recorded. If $X =$ time of birth (in minutes, after the hour), the probability density function for the continuous random variable X is given by

$$f(x) = \begin{cases} \frac{1}{60} & 0 \leq x < 60 \\ 0 & \text{otherwise.} \end{cases}$$

(a) Graph this density function.
(b) Find $P[10 \leq X \leq 20]$, $P[0 \leq X \leq 15]$, $P[X \leq 15]$, and $P[X > 30]$.

9. The probability density function for a continous random variable Y is given by

$$f(y) = \begin{cases} \frac{y}{2} & 0 \leq y \leq 2 \\ 0 & \text{otherwise.} \end{cases}$$

(a) Graph this density function.
(b) Find $P[Y \leq 1]$, $P[Y \geq 1]$, $P[0 \leq Y \leq 3]$, and $P[Y > 2]$.

10. A probability density function for X, the amount of cola in a "16-ounce" bottle of Fizzy Cola, could be the following:

$$f(x) = \begin{cases} \frac{x - 15}{2} & 15 \leq x \leq 17 \\ 0 & \text{elsewhere.} \end{cases}$$

(a) Graph this density function.
(b) Find $P[15 \leq X \leq 15.5]$, $P[X > 16]$, $P[X \leq 16.5]$, and $P[15.5 \leq X \leq 16.5]$.

• ***11.*** Dr. Pepper, a statistician for the Fizzy Cola company, wishes to find a probability density function for X that is quadratic in nature. He determines that it should be of the form

$$f(x) = \begin{cases} k(x^2 - 32x + 255) & 15 \leq x \leq 17 \\ 0 & \text{elsewhere.} \end{cases}$$

(a) Show that k must be equal to $-\frac{3}{4}$ in order for $f(x)$ to be a probability density function.
(b) Graph $f(x)$.
(c) Find $P[15 \leq X \leq 16]$. (*Hint:* This can be done in either of two ways.)
(d) Find $P[X \leq 15.5]$, $P[X \leq 16.5]$, and $P[15.5 \leq X \leq 16.5]$.

12. The cumulative distribution function for a discrete random variable X is given by

$$F(x) = \begin{cases} 0 & x < -1 \\ .2 & -1 \leq x < 1 \\ .5 & 1 \leq x < 2 \\ 1 & x \geq 2. \end{cases}$$

Find the probability distribution for X.

13. The Gro-Sirs supermarket has four checkout lanes. The manager has determined that the probability distribution of the number of lanes in use at 9:00 A.M. is given by

x:	0	1	2	3	4
$P[X = x]$:	.10	.15	.15	.20	.40.

Find the cumulative distribution function for this distribution. Graph this CDF.

14. A fair coin is flipped three times. Let Y denote the number of heads. (a) Find the probability distribution for Y. (b) Find the cumulative distribution function $F(y)$. (c) Graph this CDF.

15. Shown is a graph of the CDF for a discrete random variable Y.

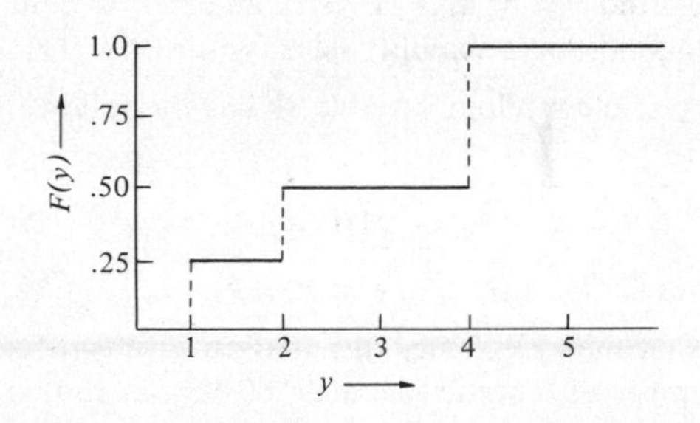

(a) Using the graph, find $P[Y \leq 0]$, $P[Y \leq 1]$, $P[Y \leq 1.5]$, and $P[Y \leq 5]$.
(b) Find the corresponding discrete probability distribution.

16. Pat works at the Pizza Palace and has determined that the number of toppings X ordered for a pizza has the following probability distribution:

x:	0	1	2	3	4
$P[X = x]$:	.10	.55	.20	.10	.05.

(a) Find $F(x)$. (b) Express $P[X \leq 1]$, $P[X > 2]$, $P[0 < X \leq 2]$ in terms of $F(x)$ [e.g., $P[X \leq 3] = F(3)$]. (c) Evaluate the expressions in part (b).

• *17.* The probability density function for a continuous random variable Y is given by

$$f(y) = \begin{cases} \frac{y}{2} & 0 \leq y \leq 2 \\ 0 & \text{otherwise.} \end{cases}$$

(Cf. Exercise 9.) (a) Find $F(y)$. (b) Express $P[Y \leq 1]$, $P[Y \geq 1]$, $P[0 \leq Y \leq 3]$, and $P[Y > 2]$ in terms of $F(y)$ [e.g., $P[Y \leq .5] = F(.5)$]. (c) Evaluate the expressions in part (b).

18. The CDF for the density function given in Exercise 8 is

$$F(x) = \begin{cases} 0 & x < 0 \\ \frac{x}{60} & 0 \leq x < 60 \\ 1 & x \geq 60. \end{cases}$$

(a) Graph this CDF.
(b) Use this CDF to evaluate the probabilities of Exercise 8(b).

• *19.* (a) Find the CDF for the density function given in Exercise 11.
(b) Use this CDF to evaluate the probabilities of Exercise 11(d).

• *20.* (a) Find the CDF for the density function given in Exercise 10.
(b) Use this CDF to evaluate the probabilities of Exercise 10(b).

21. The CDF for a continuous random variable Y is given by

$$F(y) = \begin{cases} 0 & y < 1 \\ \frac{(y-1)^3}{8} & 1 \leq y \leq 3 \\ 1 & y > 3. \end{cases}$$

(a) Find $P[Y \leq 4]$, $P[Y \leq 2]$, and $P[Y > 1.5]$.
• (b) Find the probability density function $f(y)$.

22. A discrete random variable W has the following probability distribution:

w:	0	1	4	9
$P[W = w]$:	.4	.3	.2	.1.

Find $E(W)$, $E(\sqrt{W})$, and $E(W^2)$.

23. Using the probability distribution given in Exercise 13, find $E(X)$, the expected number of lanes in use at 9:00 A.M. at the Gro-Sirs supermarket.

24. Tricky Ricky, who has a reputation not above reproach, will pay you \$1 for each tail you flip on a coin that he pulls from his pocket. He asks, however, that you pay him \$1.75 for the privilege of flipping this coin 4 times. Is this a reasonable request (from your point of view) if (a) the coin is fair?; (b) the coin is a Peskaranian penny with probability $p = .6$ of coming up heads?

25. The Flash Photo store offers outdated film at a reduced rate but with no guarantee. (The usual guarantee is that you are refunded the cost of each picture which does not come out.) The usual cost for a roll of 6 pictures is \$3. The reduced rate is \$2.

The probability distribution for the number of pictures that come out is given by the following function:

$$P[X = x] = \begin{cases} \binom{6}{x}(\frac{1}{64}) & x = 0, 1, 2, \ldots, 6 \\ 0 & \text{otherwise.} \end{cases}$$

(a) Is this a good buy? Why? (b) Would it be a good buy at \$1 per roll? Why? (c) At \$2/roll, what is the probability that you will get your money's worth? (*Hint:* Consider the number of pictures per dollar for the film with the guarantee and the expected number of pictures per dollar for the film with no guarantee.)

26. A discrete random variable U has the following probability distribution:

u:	0	1	4	9	16
$P[U = u]$:	.64	.25	.09	.01	.01.

(a) Find $E(U)$, $E(U^2)$, and $E(\sqrt{U})$.

(b) Using the results of part (a), find $E(U - 2\sqrt{U})$, $[E(U^2 - 4U)]^{1/2}$, and $E[U^2 - 2U + 3\sqrt{U} - 4]$.

27. Let X and Y be two random variables with $E(X) = 4$ and $E(Y) = -2$. Find (a) $E(4X)$; (b) $E(X + 4)$; (c) $E(2Y + 4)$; (d) $E(3Y - 14)$.

28. If $a = 2$, $b = 3$, and $c = -4$ and if $E(X) = 5$ and $E(Y) = 7$, find (a) $E(aX + b)$; (b) $E(bY - c)$; (c) $E(cX - a)$; (d) $E(abY)$.

29. (a) Find the expected number of jars of peanuts bought by a customer making a purchase at Thrifty-Nifty (Exercise 6). (b) If Thrifty-Nifty makes 12 cents per jar profit, what is their expected profit per customer?

30. Prove Theorem 2.3 of Section 2.5.

• ***31.*** Using the density function given in Exercise 9, find $E(Y)$ and $E(Y^2)$.

• ***32.*** Using the density function given in Exercise 8 for the time of birth of babies, find $E(X)$ and $E(X^2)$.

• ***33.*** Using the density function given in Exercise 11, find the expected amount of cola in a 16-ounce bottle of Fizzy Cola.

• ***34.*** Using the density function given in Exercise 10, find the expected amount of cola in a 16-ounce bottle of Fizzy Cola.

35. Find the mean, median, and mode for a discrete random variable X having the following probability distribution:

x:	—1	0	1	2
$P[X = x]$:	.35	.30	.20	.15.

36. Find the mean, median, and mode for the probability distribution of the discrete random variable U given in Exercise 26.

37. Find the mean, median, and mode for the number of toppings ordered for a pizza made at the Pizza Palace, using the probability distribution given in Exercise 16.

• ***38.*** On the basis of the graph made in Exercise 10(a), is the density function of that exercise symmetric or skewed? On the basis of this, what can you say about the relative magnitudes of the mean, median, and mode? Find the mean, median, and mode for this distribution. (*Hint:* The results of Exercises 20 and 34 may be helpful.)

• *39.* Find the mean, median, and mode for the probability distribution of Exercise 11. [*Hint:* The graph from Exercise 11(b) may be helpful.]

40. Using the CDF given in Exercise 12, find the 10th, 25th, 50th, and 90th percentiles. (*Hint:* It may be helpful to first graph the CDF.)

41. Using the graph of the CDF constructed in Exercise 13, find the 10th, 25th, 50th, and 90th percentiles.

42. Using the CDF given in Exercise 18, find (a) the median; (b) the 25th percentile.

43. Using the graph of the CDF given in Exercise 15, find the median. Also find the 10th, 75th, and 90th percentiles.

44. Using the CDF given in Exercise 21, find (a) the median; (b) the 12.5th percentile; (c) a general expression for the $(100 \cdot p)$ percentile $(0 \leq p \leq 1)$.

45. An ad for a well-known income tax return service read "Congratulations. You just saved \$5 by doing your own income tax. And all it cost you was three long, sleepless nights." (a) If you saw this ad, would you assume that \$5 was meant to convey a "typical" or a "representative" cost?

Reading the ad further, it said "Charges start at \$5 and the average cost was under \$12.50 for the 7 million families we served last year." (b) Could \$5 be the mean value? (c) Could it be the median value? (d) Can you tell from the wording of the ad what type of "average" is being used?

46. Find $E(25 - |X - g|)$ using the probability distribution for the unbalanced die (2.27) of Section 2.6 when (a) $g = 2$; (b) $g = 3$; (c) $g = 4$.

47. Obtain a clipping or a copy of an article in a newspaper or magazine that uses the term "average." Was the average used a mean, median, or mode? Was this made clear from the exposition given in the article?

48. Find the variance and standard deviation of the number of checkout lanes in use at the Gro-Sirs Supermarket at 9:00 A.M. using the probability distribution given in Exercise 13.

49. Find the variance and standard deviation of the number of toppings ordered for a pizza made at the Pizza Palace using the probability distribution given in Exercise 16.

50. A random variable X has mean $E(X) = 4$ and variance $V(X) = 16$. Find (a) $V(X + 12)$; (b) $V(4X)$; (c) $V(4X + 12)$; (d) $E(X^2)$; (e) the standard deviation of $9X + 25$.

51. The selling price X of gasoline in cents at El Shiek Service Station varies from day to day. The probability distribution of X is given by

x:	180	184	188	192	196
$P[X = x]$:	.1	.2	.4	.2	.1.

(a) Find the variance of X.
(b) What would be the variance of Y, the selling price of gas expressed in dollars rather than cents?

(c) What would be the variance of the selling price of gas (in cents) if all prices are raised 20 cents?

• **52.** Using the density function given in Exercise 10, find the variance and standard deviation of the amount of cola in a 16-ounce bottle of Fizzy Cola. (*Hint:* If you have worked Exercise 28, you already have the expected value.)

• **53.** Using the density function

$$f(y) = \begin{cases} \frac{y}{2} & 0 \leq y \leq 2 \\ 0 & \text{otherwise,} \end{cases}$$

find $E(Y)$, $E(Y^2)$, and $V(Y)$ (cf. Exercise 9).

• **54.** Find the mean and variance of the amount of hot sauce remaining in bottles in La Casa de Fuego restaurant using the density function given in Example 2.27:

$$f(x) = \begin{cases} \frac{x}{5000} & 0 \leq x \leq 100 \\ 0 & \text{otherwise.} \end{cases}$$

55. Seth Kimo can rent an ice cream pushcart for \$8/day. Dry ice for the freezer costs an additional \$6/day. He makes 15 cents on each ice cream sale, but the number sold on a given day is a random variable having a mean of 300/day and a standard deviation of 120/day. Find (a) his expected daily profit; (b) the variance of his daily profit; (c) the standard deviation of his daily profit. [*To think about:* Which, if any, of the answers to (a), (b), and (c) would change if the rental cost went up to \$10/day?]

56. Beth, Seth's sister, works selling ice cream in the same area as he, and her daily sales follow the same pattern as his, a mean of 300/day and a standard deviaton of 120/day. However, for her pay she receives \$8/day from her employer plus 7 cents commission on each ice cream sale. Find (a) her expected daily profit; (b) the variance of her daily profit; (c) the standard deviation of her daily profit. (*To think about:* How do these values compare with Seth's? Whose income is more variable? Would you rather get paid like Seth or like Beth? Why?)

57. Dow Jonesetti, a Peskaranian stockbroker, offers two different stock portfolios, each requiring an investment of 1000∂ [an emid (∂) is Peskaranian currency]. The Peskaranian economy being as simple as it is, you are able to determine the following possible values and corresponding probabilities for each portfolio for 1 year after your investment.

	Value			
Portfolio	*800∂*	*1000∂*	*1200∂*	*1400∂*
I	.05	.50	.35	.10
II	.15	.25	.05	.55

If you had 1000∂ to invest, which portfolio would you choose? Explain why you would make that choice. (*To think about:* Do you consider yourself a "risk taker" or a "risk avoider"?)

58. If the current exchange rate is 5∂ for \$1, express the standard deviation of the values of the portfolios in Exercise 57 in terms of dollars.

59. Prove that if a discrete random variable assumes only one value, say $X = c$, with $P[X = c] = 1$, then $V(X) = 0$.

60. The scores on a college entrance examination are known to have a mean of 500 and a standard deviation of 100. (a) Find the standard scores corresponding to scores of 450, 500, 625, and 700. (b) Find the raw (unstandardized) score corresponding to standardized scores of -1.5, 0, .75, and 1.25.

61. The Owl and the Pussycat were having a discussion as to who was the better mouser. The owl claimed that he was better since he caught 6 mice per night while the cat only caught 4 mice per night. Who should the argument be settled in favor of if the mean number of mice per night caught by owls and cats in general is 5 and 2.8, respectively, and the standard deviation for owls and cats is .5 and .4, respectively? (*To think about:* Is it reasonable to settle the argument using standard scores?)

62. The probability distribution for a discrete random variable W is given by

w:	-2	-1	0	1	2
$P[W = w]$:	.08	.18	.48	.18	.08.

(a) Find μ_W and σ_W. (b) Find $P[|W - \mu_W| \geq 1.5\sigma_W]$ by using probability distribution of W. (c) Find the bound on this probability using Chebyshev's inequality.

• *63.* A continuous random variable X has a probability density function $f_X(x)$ defined by

$$f_X(x) = \begin{cases} \dfrac{3x^2}{8} & 0 \leq x \leq 2 \\ 0 & \text{otherwise.} \end{cases}$$

(a) Find $E(X)$ and σ_X.
(b) Find $P[|X - 1.5| \geq .775]$ by using the density function.
(c) Find an upper bound for $P[|X - 1.5| \geq .775]$ using Chebyshev's inequality.

64. On their first visit to the planet Vulcan, scientists estimated that the mass of an edible fruit called an elppa had a mean mass of 1.5 kilograms (kg) with a variance of .09 $(\text{kg})^2$. The scientists did not, unfortunately, have time to determine the exact distribution of the mass. A later group of scientists picked an elppa at random from a tree and determined its mass to be 3 kg. Do these results appear to be consistent? Explain.

3

TWO RANDOM VARIABLES

3.1 JOINT PROBABILITY DISTRIBUTIONS

In Example 2.4 of Section 2.1 we pointed out that it is possible to define many different random variables on the same sample space. In that example we defined

$$X_1 = \text{age of student chosen}$$
$$X_2 = \text{height of student chosen}$$

and we could also define, for example,

$$X_4 = \text{weight of student chosen}$$
$$X_5 = \text{grade-point average of student chosen.}$$

Up to now we have considered random variables singly. However, in many situations some relationship exists between random variables (e.g., a height–weight relationship), which leads us to study random variables jointly (i.e., in pairs). (Actually, later we will consider more than two variables at a time, but it is convenient to begin by studying them in pairs.) By studying the joint probability distribution of two random variables, we can determine whether or not a relationship exists between two random variables. If a relationship does exist, information about one random variable can be used to make predictions about the value of the related variable. For example, a person's height might be useful in predicting the person's weight, the amount of rainfall might be useful in predicting crop yields, the size of a sales force might be useful in predicting total annual sales, and so on.

Definition 3.1 The *joint probability distribution for two discrete random variables X* and Y is the set of all pairs of values (x, y) along with the corresponding probabilities, $P[X = x \cap Y = y]$.

(For notational convenience it is common to write $P[X = x \cap Y = y]$ as $P[X = x, Y = y]$. We will follow this convention.) In the following example we will obtain the joint distribution for two pairs of random variables by using the probabilities assigned to points in the sample space on which these random variables are defined.

Example 3.1 Three fair coins, a penny, a nickel, and a dime, are tossed and it is noted whether they come up heads or tails. Let $X =$ the number of heads on the first two coins, $Y =$ the number of heads on the last two coins, and $Z =$ the number of heads on the last coin. Find the joint probability distributions of X and Y and of X and Z.

To solve this problem we first write out the sample space and then, for each point in S, write the corresponding values of the random variables. This is done in Table 3.1. For example, for the outcome H–H–T, we see that $X = 2$, $Y = 1$, and

TABLE 3.1
Values of X, Y, and Z for Example 3.1

S			*X*	*Y*	*Z*
Outcome on:			*Number of Heads on Penny and Nickel*	*Number of Heads on Nickel and Dime*	*Number of Heads on Dime*
Penny	*Nickel*	*Dime*			
H	H	H	2	2	1
H	H	T	2	1	0
H	T	H	1	1	1
H	T	T	1	0	0
T	H	H	1	2	1
T	H	T	1	1	0
T	T	H	0	1	1
T	T	T	0	0	0

$Z = 0$. The most convenient way of expressing the joint probability distributions here is by means of a table. On the left side and top of this table we write the values of the random variables, and in the table we give the corresponding probability. For example, in Table 3.2 we see that the entry in the upper left-hand corner of the X, Y table is $\frac{1}{8}$. This means $P[X = 0, Y = 0] = \frac{1}{8}$. We determine this by examining Table 3.1 and noting that $X = 0$ and $Y = 0$ (together) only when the outcome is T–T–T. Using independence among coins we see that $P(\text{T–T–T}) = P(\text{T}) \cdot P(\text{T}) \cdot P(\text{T}) = (\frac{1}{2})^3 = \frac{1}{8}$. (In fact, each point in S has probability $\frac{1}{8}$.) Consequently, $P[X = 0, Y = 0] = P[\text{T–T–T}] = \frac{1}{8}$. The other values in the table are obtained similarly.

TABLE 3.2

Joint Probability Distributions for X, Y and X, Z

		Y					Z	
		0	*1*	*2*			*0*	*1*
	0	$\frac{1}{8}$	$\frac{1}{8}$	0		*0*	$\frac{1}{8}$	$\frac{1}{8}$
X	*1*	$\frac{1}{8}$	$\frac{2}{8}$	$\frac{1}{8}$	X	*1*	$\frac{2}{8}$	$\frac{2}{8}$
	2	0	$\frac{1}{8}$	$\frac{1}{8}$		*2*	$\frac{1}{8}$	$\frac{1}{8}$

The joint probability distribution can be used to find probabilities for events defined in terms of both random variables. For example, to find the probability that X and Y assume values in some two-dimensional set A, we use an equation similar to equation (2.6) of Section 2.2: namely,

$$P[(X, Y) \in A] = \sum_{(x,y) \in A} P[X = x, Y = y]. \tag{3.1}$$

For example, using the joint probability distribution for X and Y defined in Example 3.1, we find that

$$\begin{aligned} P[0 \leq X \leq 1, 0 \leq Y \leq 1] &= \sum_{(x,y) \in A} P[X = x, Y = y] \\ &= \tfrac{1}{8} + \tfrac{1}{8} + \tfrac{1}{8} + \tfrac{2}{8} \\ &= \tfrac{5}{8}, \end{aligned}$$

where $A = \{(x, y)\colon 0 \leq x \leq 1 \text{ and } 0 \leq y \leq 1\}$. Once you understand that it is necessary to add all probabilities corresponding to values (x, y) satisfying a particular condition, it is not necessary to explicitly write out a definition of the set A. To find $P[X = Y]$, we simply add $P[X = 0, Y = 0]$, $P[X = 1, Y = 1]$, and $P[X = 2, Y = 2]$ to get

$$P[X = Y] = \tfrac{1}{8} + \tfrac{2}{8} + \tfrac{1}{8} = \tfrac{1}{2}.$$

You can verify that the joint probability distributions found in Example 3.1 satisfy the following two conditions, which are analogous to conditions (2.3) and (2.4):

$$P[X = x, Y = y] \geq 0 \qquad \text{for all } (x, y) \tag{3.2}$$

$$\sum_{\text{all }(x,y)} P[X = x, Y = y] = 1. \tag{3.3}$$

In fact, (3.2) and (3.3) must hold for all joint probability distributions for discrete random variables.

Example 3.2 Mrs. Mumm has noticed that her four-year-old twins, Maxie and Minnie, frequently spill their milk at mealtimes. After a child spills milk twice at a given meal, he gets no more milk for that meal. Mrs. Mumm has determined that the joint distribution for the number of glasses spilled at a meal by Maxie (X) and by Minnie (Y) is as follows:

		Y		
		0	1	2
	0	.10	.05	.05
X	1	.25	.20	.05
	2	.15	.05	.10

Verify that this is a valid joint probability distribution and find the probability that Maxie spills more glasses of milk than Minnie.

It is easy to see that (3.2) is satisfied, since all entries in the table are positive. Since

$$.10 + .05 + .05 + .25 + .20 + .05 + .15 + .05 + .10 = 1,$$

(3.3) is also satisfied. The event "Maxie spills more glasses of milk than Minnie" can be expressed in terms of X and Y by $X > Y$. We find that

$$\begin{aligned} P[X > Y] &= \sum_{(x,y):x>y} P[X = x,\ Y = y] \\ &= P[X = 1,\ Y = 0] + P[X = 2,\ Y = 0] + P[X = 2, Y = 1] \\ &= .25 + .15 + .05 = .45. \end{aligned}$$

• • • • •

When we consider the joint probability distribution for two continuous random variables, we must find a two-dimensional analog to the probability density function. Such a function, which we will call a *joint probability density function* (joint pdf), allows us to calculate probabilities by integration.

Definition 3.2 The joint probability density function for two continuous random variables X and Y is a function $f(x, y)$ such that for all constants a, b, c, and d,

$$P[a \leq X \leq b, c \leq Y \leq d] = \int_c^d \int_a^b f(x, y)\, dx\, dy.$$

A brief digression is perhaps in order here, since many persons having an introductory course in calculus have not been introduced to double integrals. By adhering to the following rules, you should be able to work all the problems of this type in this book.

1. Evaluate the integral iteratively as if it were parenthesized.

$$\text{(3.4)} \qquad \int_c^d \int_a^b f(x, y)\, dx\, dy = \int_c^d \left(\int_a^b f(x, y)\, dx \right) dy.$$

2. Be sure that the limits of integration agree with the variable of integration. For example, in the inner integral of (3.4), since the integration is with respect to x, the limits are those on X. Note that the *order* of integration is not important as long as the limits are assigned appropriately:

$$\int_c^d \left(\int_a^b f(x, y)\, dx \right) dy = \int_a^b \left(\int_c^d f(x, y)\, dy \right) dx.$$

3. When evaluating $\int_a^b f(x, y)\, dx$, treat y *exactly* as you would treat a constant,

and when evaluating $\int_c^d f(x, y)\,dy$, treat x *exactly* as you would treat a constant.

4. Use the results of the first (inner) integration as the integrand for the second (outer) integral.

Example 3.3 Given that

$$f(x, y) = \begin{cases} \dfrac{3xy^2}{4} & 0 \le x \le 1, 0 \le y \le 2 \\ 0 & \text{otherwise} \end{cases}$$

is the joint probability density function for two continuous random variables X and Y, find $P[0 \le X \le 1/2, 0 \le Y \le 1]$.

We will solve this problem two ways to emphasize step (2).

$$\begin{aligned} P[0 \le X \le 1/2, 0 \le Y \le 1] &= \int_0^1 \int_0^{1/2} \frac{3xy^2}{4}\,dx\,dy \\ &= \int_0^1 \left(\int_0^{1/2} \frac{3xy^2}{4}\,dx \right) dy = \int_0^1 \left(\frac{3x^2y^2}{8} \Big|_0^{1/2} \right) dy \\ &= \int_0^1 \frac{3y^2}{32}\,dy = \frac{y^3}{32}\Big|_0^1 = \frac{1}{32}. \end{aligned}$$

Alternatively,

$$\begin{aligned} P[0 \le X \le 1/2, 0 \le Y \le 1] &= \int_0^{1/2} \int_0^1 \frac{3xy^2}{4}\,dy\,dx \\ &= \int_0^{1/2} \left(\int_0^1 \frac{3xy^2}{4}\,dy \right) dx = \int_0^{1/2} \left(\frac{xy^3}{4} \Big|_0^1 \right) dx \\ &= \int_0^{1/2} \frac{x}{4}\,dx = \frac{x^2}{8}\Big|_0^{1/2} = \frac{1}{32}. \end{aligned}$$

A joint density function for two continuous random variables must always satisfy the following conditions, which are analogous to (3.2) and (3.3):

(3.5) $$f(x, y) \ge 0 \qquad \text{for all } x, y.$$

(3.6) $$\int_{-\infty}^{\infty} \int_{-\infty}^{\infty} f(x, y)\,dx\,dy = 1.$$

(We emphasized in Section 2.3 that, for a single random variable X, probability is associated with *area* under a curve. When considering the joint distribution for two random variables, X and Y, probability is associated with *volume* under a surface. Since drawing good graphs in three dimensions is generally difficult, we will not pursue this generalization further.)

Example 3.4 Robin D. Hood, a well-known archer, is shooting arrows at a square target measuring 2 feet on a side. Robin has determined that the horizontal distance

(X) and the vertical distance (Y) that an arrow hits from the center of the target can be described by the joint probability density

$$f(x, y) = \begin{cases} \frac{3}{16}(2 - x^2 - y^2) & -1 \leq x \leq 1, -1 \leq y \leq 1 \\ 0 & \text{otherwise.} \end{cases}$$

Find the probability that the next arrow he shoots will land in a 1-foot square centered on the larger target, as in Figure 3.1.

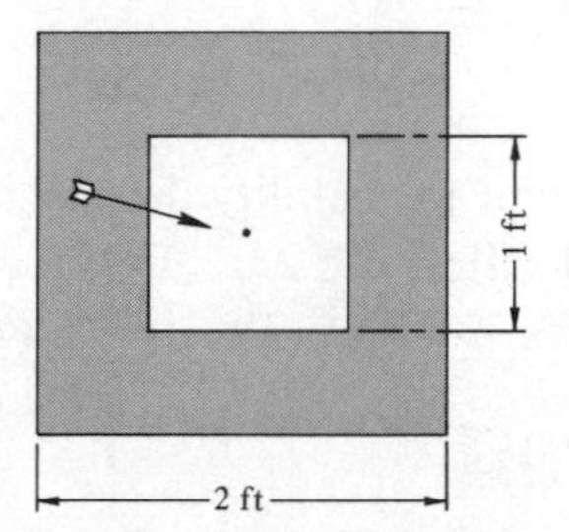

FIGURE 3.1
Robin's Target

In terms of distance from the center of the target, we need to find

$$P[-1/2 \leq X \leq 1/2, -1/2 \leq Y \leq 1/2] = \int_{-1/2}^{1/2} \int_{-1/2}^{1/2} f(x, y)\, dx\, dy.$$

Following the rules stated earlier in this section, we get

$$\begin{aligned}
\int_{-1/2}^{1/2} \left(\int_{-1/2}^{1/2} \frac{3}{16}(2 - x^2 - y^2)\, dx \right) dy &= \int_{-1/2}^{1/2} \left(\frac{3}{16}\left(2x - \frac{x^3}{3} - xy^2\right)\Big|_{-1/2}^{1/2} \right) dy \\
&= \int_{-1/2}^{1/2} \frac{3}{16}\left(\left[2\left(\frac{1}{2}\right) - \frac{1}{3}\left(\frac{1}{2}\right)^3 - \left(\frac{1}{2}\right)y^2\right] \right. \\
&\qquad \left. - \left[2\left(-\frac{1}{2}\right) - \frac{1}{3}\left(-\frac{1}{2}\right)^3 - \left(-\frac{1}{2}\right)y^2\right] \right) dy \\
&= \int_{-1/2}^{1/2} \frac{3}{16}\left(\frac{23}{12} - y^2\right) dy \\
&= \frac{3}{16}\left(\frac{23}{12}y - \frac{y^3}{3}\right)\Big|_{-1/2}^{1/2} \\
&= \frac{3}{16}\left(\left[\frac{23}{12}\left(\frac{1}{2}\right) - \frac{1}{3}\left(\frac{1}{2}\right)^3\right] \right. \\
&\qquad \left. - \left[\frac{23}{12}\left(-\frac{1}{2}\right) - \frac{1}{3}\left(-\frac{1}{2}\right)^3\right] \right) \\
&= \frac{3}{16}\left(\frac{22}{12}\right) = \frac{11}{32} = .34375.
\end{aligned}$$

3.2 MARGINAL AND CONDITIONAL DISTRIBUTIONS

It is possible to use the joint distribution of the random variables X and Y to find the probability distribution of X (or Y) alone. For example, if Mrs. Mumm was only interested in the probability distribution of the amount of milk spilled by Maxie (X), she could determine that from the joint distribution of X and Y. In this case the values that X can assume are 0, 1, and 2, and we can reason that if we are only concerned with the event that $X = 0$, we can let Y assume any value. That is,

$$\begin{aligned} P[X = 0] &= P[X = 0 \text{ and } Y = \text{anything}] \\ &= P[X = 0 \text{ and } Y = 0, 1, \text{ or } 2] \\ &= P[(X = 0, Y = 0) \text{ or } (X = 0, Y = 1) \text{ or } (X = 0, Y = 2)] \\ &= P[X = 0, Y = 0] + P[X = 0, Y = 1] + P[X = 0, Y = 2] \\ &= .10 + .05 + .05 = .20. \end{aligned}$$

Similarly, we can find

$$\begin{aligned} P[X = 1] &= P[X = 1, Y = 0] + P[X = 1, Y = 1] + P[X = 1, Y = 2] \\ &= .25 + .20 + .05 = .50 \end{aligned}$$

and

$$\begin{aligned} P[X = 2] &= P[X = 2, Y = 0] + P[X = 2, Y = 1] + P[X = 2, Y = 2] \\ &= .15 + .05 + .10 = .30. \end{aligned}$$

[Note that we can add the probabilities because the events ($X = 0, Y = 0$), ($X = 0, Y = 1$), ($X = 0, Y = 2$), etc., are mutually exclusive.] Consequently, the probability distribution for X is given by

x:	0	1	2
$P[X = x]$:	.20	.50	.30.

Using summation notation, we could express these probabilities by

$$P[X = 0] = \sum_y P[X = 0, Y = y],$$

$$P[X = 1] = \sum_y P[X = 1, Y = y],$$

$$P[X = 2] = \sum_y P[X = 2, Y = y],$$

or, more generally, by

$$P[X = x] = \sum_y P[X = x, Y = y]. \tag{3.7}$$

By arguments identical to those leading to (3.7), we can see that the probability distribution for Y can be obtained from

$$P[Y = y] = \sum_x P[X = x, Y = y].$$

Continuing with the example of the Mumm twins, the probability distribution for Y can be found from

$$\begin{aligned}P[Y=0] &= P[X=0,\ Y=0] + P[X=1,\ Y=0] + P[X=2,\ Y=0]\\ &= .10 + .25 + .15 = .50\\ P[Y=1] &= P[X=0,\ Y=1] + P[X=1,\ Y=1] + P[X=2,\ Y=1]\\ &= .05 + .20 + .05 = .30\\ P[Y=2] &= P[X=0,\ Y=2] + P[X=1,\ Y=2] + P[X=2,\ Y=2]\\ &= .05 + .05 + .10 = .20.\end{aligned}$$

Since the probability distribution of X or Y can be obtained by adding the entries in a particular row or column of the table giving the joint distribution, it is convenient and conventional to write those sums in the margins of the table opposite the corresponding x or y value as in Table 3.3; probabilities found in this way are referred to as *marginal probabilities*.

TABLE 3.3

		Y 0	1	2	
X	0	.10	.05	.05	.20
	1	.25	.20	.05	.50
	2	.15	.05	.10	.30
		.50	.30	.20	

Example 3.5 Find the marginal probabilities for X and Y and for X and Z using the joint probability distributions of Example 3.1.

In Table 3.4 we have rewritten Table 3.2 but have added the row and column

TABLE 3.4
Marginal Totals

		Y 0	1	2	
X	0	$\frac{1}{8}$	$\frac{1}{8}$	0	$\frac{2}{8}$
	1	$\frac{1}{8}$	$\frac{2}{8}$	$\frac{1}{8}$	$\frac{4}{8}$
	2	0	$\frac{1}{8}$	$\frac{1}{8}$	$\frac{2}{8}$
		$\frac{2}{8}$	$\frac{4}{8}$	$\frac{2}{8}$	

		Z 0	1	
X	0	$\frac{1}{8}$	$\frac{1}{8}$	$\frac{2}{8}$
	1	$\frac{2}{8}$	$\frac{2}{8}$	$\frac{4}{8}$
	2	$\frac{1}{8}$	$\frac{1}{8}$	$\frac{2}{8}$
		$\frac{4}{8}$	$\frac{4}{8}$	

totals in the margins. From these totals we can easily write down the marginal distributions for X, Y, and Z.

$$\begin{array}{rccc} x: & 0 & 1 & 2 \\ P[X=x]: & \frac{1}{4} & \frac{1}{2} & \frac{1}{4}. \\ y: & 0 & 1 & 2 \\ P[Y=y]: & \frac{1}{4} & \frac{1}{2} & \frac{1}{4}. \\ z: & 0 & 1 & \\ P[Z=z]: & \frac{1}{2} & \frac{1}{2}. & \end{array}$$

You can see from Table 3.4 that the probability distribution for X, the number of heads observed on the first two coins, is exactly the same whether it is obtained from the joint distribution of X and Y or from the joint distribution of X and Z. This is, of course, as it should be, since in both cases, X is defined in exactly the same way. You can also see that although X and Y are different random variables, they have exactly the same probability distribution.

We said at the beginning of Section 3.1 that by studying the joint distribution of two random variables, we could determine whether or not a relationship exists between two random variables. By studying conditional probability distributions, we can see if such a relationship exists. To find the conditional probability distribution for discrete random variables, we will use conditional probabilities as defined in Section 1.8. Recall that for any two events A and B,

$$P[A \mid B] = \frac{P[A \cap B]}{P(B)}. \tag{3.8}$$

Now consider the random variables X and Y defined in Example 3.2, referring to the number of glasses of milk spilled by each of the twins. In the beginning of this section we found the marginal distribution of X, the number of glasses of milk spilled by Maxie. In particular, we found that $P[X = 1] = .50$. If we had the additional information that Minnie spilled one glass of milk (i.e., that $Y = 1$), what probability would we assign to the event that Maxie spilled one glass (i.e., $X = 1$)? Since $(X = 1)$ and $(Y = 1)$ are simply events, we can use (3.8) to get

$$\begin{aligned} P[X = 1 \mid Y = 1] &= \frac{P[X = 1, Y = 1]}{P[Y = 1]} \\ &= \frac{.20}{.30} = \frac{2}{3}. \end{aligned} \tag{3.9}$$

(We found both $P[X = 1, Y = 1]$ and $P[Y = 1]$ from Table 3.3.) You see that when we had no information about Y, we assigned a probability of .5 to the event $(X = 1)$, but when we knew that Y was equal to 1 [i.e., given $(Y = 1)$], we assigned a probability of $\frac{2}{3}$ to that event. You can see from this (as Mrs. Mumm knew all along) that there is a relationship between the random variables X and Y.

Continuing with this illustration we could also find the probability that $X = 0$

or that $X = 2$ given that $Y = 1$. Using equations similar to (3.9) and finding the requisite probabilities from Table 3.3, we get

$$P[X = 0 \mid Y = 1] = \frac{P[X = 0, Y = 1]}{P[Y = 1]}$$

$$= \frac{.05}{.30} = \frac{1}{6}$$

$$P[X = 2 \mid Y = 1] = \frac{P[X = 2, Y = 1]}{P[Y = 1]}$$

$$= \frac{.05}{.30} = \frac{1}{6}.$$

You can see that these conditional probabilities can be summarized as follows:

x:	0	1	2
$P[X = x \mid Y = 1]$:	$\frac{1}{6}$	$\frac{2}{3}$	$\frac{1}{6}$.

If you examine these probabilites more closely, you will see that they are nonnegative and sum to 1. That is, the values taken together form a probability distribution for X given that $Y = 1$. It should be clear that there is nothing very special about the value of 1 for Y, and in fact any value of y could be used.

Definition 3.3 The conditional probability distribution for the discrete random variable X given that $Y = y$ is found from

$$P[X = x \mid Y = y] = \frac{P[X = x, Y = y]}{P[Y = y]}. \tag{3.10}$$

Similarly, the conditional probability distribution for the discrete random variable Y given that $X = x$ is found from

$$P[Y = y \mid X = x] = \frac{P[X = x, Y = y]}{P[X = x]}. \tag{3.11}$$

Example 3.6 The joint probability distribution for X, the number of children in a family, and Y, the number of dogs and/or cats belonging to a family randomly chosen from families living in Petsville, is given in Table 3.5. Find the conditional probability distribution of Y given that $X = 2$.

TABLE 3.5

			Y		
		0	*1*	*2*	
	0	.03	.07	0	.10
X	*1*	.12	.12	.06	.30
	2	.12	.18	.10	.40
	3	0	.15	.05	.20

Since we need to use $P[X = 2]$ in this problem, we have written the marginal distribution of X in the margin of Table 3.5. We find the conditional probability distribution of Y given $X = 2$ by using (3.11):

$$P[Y = 0 \mid X = 2] = \frac{P[X = 2, Y = 0]}{P[X - 2]} = \frac{.12}{.40} = .30$$

$$P[Y = 1 \mid X = 2] = \frac{P[X = 2, Y = 1]}{P[X = 2]} = \frac{.18}{.40} = .45$$

$$P[Y = 2 \mid X = 2] = \frac{P[X = 2, Y = 2]}{P[X = 2]} = \frac{.10}{.40} = .25$$

or

y:	0	1	2
$P[Y = y \mid X = 2]$:	.30	.45	.25.

Example 3.7 Using the joint probability distributions for X and Y and X and Z for the three coins (Table 3.4), find the conditional distribution for X given that Y equals 0. Also, find the conditional distribution for X given that Z equals 0.

Using (3.10), we get

$$P[X = 0 \mid Y = 0] = \frac{P[X = 0, Y = 0]}{P[Y = 0]} = \frac{\frac{1}{8}}{\frac{2}{8}} = \frac{1}{2}$$

$$P[X = 1 \mid Y = 0] = \frac{P[X = 1, Y = 0]}{P[Y = 0]} = \frac{\frac{1}{8}}{\frac{2}{8}} = \frac{1}{2}$$

$$P[X = 2 \mid Y = 0] = \frac{P[X = 2, Y = 0]}{P[Y = 0]} = \frac{0}{\frac{2}{8}} = 0$$

or

x:	0	1	2
$P[X = x \mid Y = 0]$:	$\frac{1}{2}$	$\frac{1}{2}$	0.

Using an analogous equation for Z instead of Y, we get

$$P[X = 0 \mid Z = 0] = \frac{P[X = 0, Z = 0]}{P[Z = 0]} = \frac{\frac{1}{8}}{\frac{4}{8}} = \frac{1}{4}$$

$$P[X = 1 \mid Z = 0] = \frac{P[X = 1, Z = 0]}{P[Z = 0]} = \frac{\frac{2}{8}}{\frac{4}{8}} = \frac{1}{2}$$

$$P[X = 2 \mid Z = 0] = \frac{P[X = 2, Z = 0]}{P[Z = 0]} = \frac{\frac{1}{8}}{\frac{4}{8}} = \frac{1}{4}$$

or

x:	0	1	2
$P[X = x \mid Z = 0]$:	$\frac{1}{4}$	$\frac{1}{2}$	$\frac{1}{4}$.

Were you surprised in this example to find a conditional probability of zero? If you remember that $Y = 0$ means tails on both the nickel and the dime and that $X =$ number of heads on the penny and the nickel, you will see that if $Y = 0$, then X cannot possibly be 2.

Since conditional probability distributions are perfectly good probability distributions, they can be used to find expected values. However, these expected values are conditional on the value of another random variable. For example, we can use the conditional distribution of Y given $X = 2$ as found in Example 3.6 to find the expected number of pets in a family given that the family has two children. We use

$$E(Y \mid X = 2) = \sum y \cdot P[Y = y \mid X = 2]$$
$$= 0(.30) + 1(.45) + 2(.25) = .95.$$

We will leave it to you to verify that

$$E(Y \mid X = 0) = 0(.30) + 1(.70) + 2(0) = .70,$$
$$E(Y \mid X = 1) = 0(.40) + 1(.40) + 2(.20) = .80,$$

and

$$E(Y \mid X = 3) = 0(0) + 1(.75) + 2(.25) = 1.25.$$

The idea of the conditional expectation of Y given X will play an important role in the study of regression in Chapter 9.

• • • • •

Marginal and conditional probability distributions can be defined for continuous random variables by analogy with the definitions given for discrete random variables. Recalling that integrals are continuous analogs of sums, we would find the marginal probability density function for X and Y by analogy with (3.7):

(3.12) $$f_X(x) = \int_{-\infty}^{\infty} f(x, y)\, dy,$$

and similarly we would find the marginal density for Y from

(3.13) $$f_Y(y) = \int_{-\infty}^{\infty} f(x, y)\, dx.$$

(Since in problems of this type there are two marginal densities, one for each variable, we distinguish them by using subscripts on the function f.) To evaluate (3.12) or (3.13) you will follow rule (3) given in Section 3.1 for evaluating double integrals.

Example 3.8 Using the joint probability density function

$$f(x, y) = \begin{cases} \frac{3}{16}(2 - x^2 - y^2) & -1 \le x \le 1, \quad -1 \le y \le 1 \\ 0 & \text{otherwise} \end{cases}$$

(cf. Example 3.4), find the marginal density functions for X and Y.

To find $f_X(x)$, we need to evaluate

$$f_X(x) = \int_{y=-1}^{1} \tfrac{3}{16}(2 - x^2 - y^2)\, dy$$
$$= \tfrac{3}{16}[2y - x^2y - y^3/3]_{y=-1}^{1}$$
$$= \tfrac{3}{16}[(2 - x^2 - \tfrac{1}{3}) - (-2 + x^2 + \tfrac{1}{3})]$$
$$= \tfrac{3}{8}[(\tfrac{5}{3} - x^2)] \qquad -1 \le x \le 1.$$

To find $f_Y(y)$, we need to evaluate

(3.14) $$f_Y(y) = \int_{x=-1}^{1} \tfrac{3}{16}(2 - x^2 - y^2)\,dx.$$

However, because $f(x, y)$ is symmetric in x and y, we know that the result will be

$$f_Y(y) = \tfrac{3}{8}(\tfrac{5}{3} - y^2) \qquad -1 \le y \le 1.$$

[If you have any doubts about this, you should evaluate (3.14) to verify this for yourself.]

Example 3.9 Using the joint probability density function

$$f(x, y) = \begin{cases} \dfrac{x + y}{3} & 0 \le x \le 1, 0 \le y \le 2 \\ 0 & \text{otherwise,} \end{cases}$$

find the marginal density functions for X and Y.

To find $f_X(x)$, we need to integrate $f(x, y)$ with respect to y. Since the limits on Y are from 0 to 2, we get

$$f_X(x) = \int_{y=0}^{2} f(x, y)\,dy = \int_0^2 \frac{x + y}{3}\,dy$$

$$= \frac{1}{3}\left[xy + \frac{y^2}{2}\right]_0^2 = \frac{1}{3}[(2x + 2) - 0] = \frac{2(x + 1)}{3} \qquad 0 \le x \le 1.$$

To find $f_Y(y)$ we use (3.13), noting that the limits on X are from 0 to 1.

$$f_Y(y) = \int_{x=0}^{1} f(x, y)\,dx = \int_0^1 \frac{x + y}{3}\,dx$$

$$= \frac{1}{3}\left[\frac{x^2}{2} + xy\right]_0^1 = \frac{1}{3}\left[\left(\frac{1}{2} + y\right) - 0\right] = \frac{(2y + 1)}{6} \qquad 0 \le y \le 2.$$

To find the conditional density of X given that $Y = y$ or the conditional density of Y given that $X = x$, we again proceed by analogy with the discrete case. By analogy with (3.10) and (3.11), we have

(3.15) $$f(x \mid y) = \frac{f(x, y)}{f_Y(y)}$$

and

(3.16) $$f(y \mid x) = \frac{f(x, y)}{f_X(x)}$$

Example 3.10 In Example 3.8 we found the marginal distributions for X and Y for the joint probability distribution given in Example 3.4. Recall that X and Y were defined to be the horizontal and vertical distances that an arrow hits from the center of the target. Using these results, find the conditional distribution of the horizontal distance from the center of the target (X) given that the vertical distance from the center of the target (Y) is zero.

Using (3.15) and the marginal density for Y found in Example 3.8, we get as a general solution

$$f(x \mid y) = \frac{\tfrac{3}{16}(2 - x^2 - y^2)}{\tfrac{3}{8}(\tfrac{5}{3} - y^2)}.$$

(By a general solution we mean that this is a general form for the conditional distribution for X given any value of y.) The specific solution when $y = 0$ is

$$f(x \mid y = 0) = \frac{\frac{3}{16}(2 - x^2 - 0^2)}{\frac{3}{8}(\frac{5}{3} - 0^2)}$$

$$= \tfrac{3}{10}(2 - x^2) \qquad -1 \leq x \leq 1.$$

Just as we were able to find expected values using discrete conditional probability distributions, we can also find them using continuous conditional probability densities. We simply use (2.23) with the conditional density replacing the unconditional density. For example, to find $E(X \mid Y = 0)$ for the conditional density just found, we use

$$E(X \mid Y = 0) = \int_{-1}^{1} x \cdot f(x \mid y = 0)\, dx = \int_{-1}^{1} x\left(\frac{3}{10}(2 - x^2)\right) dx$$

$$= \frac{3}{10}\left[\frac{2x^2}{2} - \frac{x^4}{4}\right]_{-1}^{1} = \frac{3}{10}\left[\left(1 - \frac{1}{4}\right) - \left(1 - \frac{1}{4}\right)\right] = 0.$$

In the context of Example 3.4, we would say that given that Robin's arrow hit the center of the target in the vertical direction (i.e., given that $Y = 0$), his expected distance from the center of the target in the horizontal direction is zero.

3.3 INDEPENDENT RANDOM VARIABLES

In finding the conditional probability distribution for a discrete random variable X, say, given a particular value of Y or Z, we get some idea as to the interrelationship between the random variables. Consider again Example 3.1, where a penny, nickel, and d me were flipped and where X was defined to be the number of heads observed on the penny and nickel, Y the number of heads on the nickel and dime, and Z the number of heads on the dime. By comparing the marginal distribution of X (found in Example 3.5) and the conditional distribution of X given that $Y = 0$ (found in Example 3.7), we can see that these distributions differ. In particular, we see that

$$P[X = 0] = \tfrac{1}{4} \qquad \text{while} \qquad P[X = 0 \mid Y = 0] = \tfrac{1}{2}.$$

In this case it is relatively easy to see that the random variables X and Y are in fact related, since the value of each depends to a large degree on the number of heads showing on the nickel.

On the other hand, if we look at the marginal distribution of X and compare it with the conditional distribution of X given $Z = 0$ (Examples 3.5 and 3.7), we see that these distributions are identical. In fact, if we were to calculate $P[X = x \mid Z = 1]$, we would find that it too is the same as the marginal distribution of X. That is,

(3.17) $$P[X = x] = P[X = x \mid Z = z] \qquad \text{for all } z.$$

If two random variables X and Z satisfy (3.17), then we will say that they are *independent*. In this situation we relate independence to marginal and conditional distributions. By using a formula analogous to (3.10), we can also relate independence to marginal and joint probability distributions. In particular, it follows that if X and Z are independent, we can write

(3.18) $$P[X = x \mid Z = z] = P[X = x]$$

and, by analogy with (3.10), we can write

(3.19) $$P[X = x \mid Z = z] = \frac{P[X = x, Z = z]}{P[Z = z]}.$$

By equating the right-hand sides of (3.18) and (3.19), we get

$$P[X = x] = \frac{P[X = x, Z = z]}{P[Z = z]}$$

or

(3.20) $$P[X = x, Z = z] = P[X = x] \cdot P[Z = z].$$

Equation (3.20) must hold true for *all* values of x and z. If it fails to hold for *any* values of x and z, these random variables will be dependent.

Example 3.11 Verify that the random variables X and Z as defined in Example 3.1 are independent.

By examining Table 3.4, where the marginal totals are given along with the joint distribution of X and Z, you can see that (3.20) holds for all x and z.

Example 3.12 Verify that the random variables X and Y as defined in Example 3.1 are *not* independent.

Once again we refer to Table 3.4. In this case you can see that

$$P[X = 1, Y = 0] = \tfrac{1}{8} = \tfrac{1}{2} \cdot \tfrac{1}{4} = P[X = 1] \cdot P[Y = 0],$$

so you may be tempted to conclude that X and Y are independent. However, if you look further you will see that

$$P[X = 0, Y = 0] = \tfrac{1}{8} \neq \tfrac{1}{4} \cdot \tfrac{1}{4} = P[X = 0] \cdot P[Y = 0].$$

Since it is *not* true that $P[X = x, Y = y] = P[X = x] \cdot P[Y = y]$ for *all* x and y values, we conclude that X and Y are not independent.

In these examples we determined whether or not two random variables were independent by using the joint distribution. However, we can reverse the situation and find the joint distribution of two random variables if we know in advance that they are independent. We discussed a similar situation in Section 1.9 in dealing with independent events. As in that case, you can generally assume that random variables will be independent if they are "logically" independent. For instance, since $X =$ number of heads on a penny and nickel and $Z =$ number of heads on a dime (Example 3.1) are logically independent, we correctly conclude that they are independent random variables. This idea can be extended to three or more random variables.

Theorem 3.1 If $X_1, X_2, \ldots, X_n$ are independent discrete random variables, the joint probability distribution of these random variables can be found by taking the product of the marginal distributions. That is,

$$P[X_1 = x_1, X_2 = x_2, \ldots, X_n = x_n] = P[X_1 = x_1] \cdot P[X_2 = x_2] \cdot \ldots \cdot P[X_n = x_n].$$

It is not difficult to extend the notion of independence to continuous random variables. We can express independence in terms of either marginal or joint probability distributions by analogy with equation (3.17) or (3.20).

If X and Y are continuous random variables such that

$$f_X(x) = f(x \mid y) \qquad \text{for all } x \text{ and } y \tag{3.21}$$

or

$$f(x, y) = f_X(x) \cdot f_Y(y) \qquad \text{for all } x \text{ and } y \tag{3.22}$$

then X and Y are independent random variables. If (3.21) or (3.22) fail to hold for any x or y, the random variables X and Y are dependent.

Example 3.13 The joint density function for two continuous random variables is

$$f(x, y) = \begin{cases} \dfrac{3x^2y}{2} & 0 \leq x \leq 1, 0 \leq y \leq 2 \\ 0 & \text{otherwise.} \end{cases}$$

The marginal densities were found to be

$$f_X(x) = \begin{cases} 3x^2 & 0 \leq x \leq 1 \\ 0 & \text{otherwise} \end{cases} \qquad f_Y(y) = \begin{cases} \dfrac{y}{2} & 0 \leq y \leq 2 \\ 0 & \text{otherwise.} \end{cases}$$

Verify that X and Y are independent random variables.

It is easy to see that

$$f_X(x) \cdot f_Y(y) = (3x^2)\left(\frac{y}{2}\right) = \frac{3x^2y}{2} = f(x, y)$$

for all x and y, so it follows from (3.22) that X and Y are independent.

Example 3.14 Using the joint density function from Example 3.4 and the corresponding marginal densities for X and Y found in Example 3.8, determine whether or not X and Y are independent random variables.

The joint density function is

$$f(x, y) = \begin{cases} \frac{3}{16}(2 - x^2 - y^2) & -1 \leq x \leq 1, -1 \leq y \leq 1 \\ 0 & \text{otherwise,} \end{cases}$$

and the marginal densities are

$$f_X(x) = \begin{cases} \frac{3}{8}(\frac{5}{3} - x^2) & -1 \leq x \leq 1 \\ 0 & \text{otherwise} \end{cases}$$

$$f_Y(y) = \begin{cases} \frac{3}{8}(\frac{5}{3} - y^2) & -1 \leq y \leq 1 \\ 0 & \text{otherwise.} \end{cases}$$

Multiplying $f_X(x)$ and $f_Y(y)$, we get

$$f_X(x) \cdot f_Y(y) = \begin{cases} \frac{9}{64}(\frac{25}{9} - \frac{5}{3}(x^2 + y^2) + x^2y^2) & -1 \leq x \leq 1, -1 \leq y \leq 1 \\ 0 & \text{otherwise,} \end{cases}$$

which is certainly not the same as $f(x, y)$. Consequently, we conclude that X and Y are dependent random variables.

3.4 EXPECTATIONS FOR FUNCTIONS OF TWO RANDOM VARIABLES

In Section 2.5 we presented methods for finding the expected value for a random variable X and for finding the expected value for functions of X. When we are given two random variables, we may be interested in finding the expected value of some function of those random variables. As a simple example of this, we might consider the two random variables defined in Example 3.2: namely, the number of glasses of milk spilled at a meal by each of the Mumm twins. It might be of interest to know the total number of glasses spilled, which would be $X + Y$, and the expected total number of glasses spilled, $E(X + Y)$. Here the function of X and Y we are interested in is

$$g(X, Y) = X + Y.$$

We will give a formula which can be used to find the expected value of any function of two discrete random variables and then illustrate the use of the formula by applying it to the specific example given earlier.

Definition 3.4 If X and Y are two discrete random variables, and if $g(X, Y)$ is some function of X and Y, then

$$E(g(X, Y)) = \sum_{\text{all } x,y} g(x, y) \cdot P[X = x, Y = y]. \tag{3.23}$$

Note that the procedure used here is very much like the case for a single random variable as described after Definition 2.7 of Section 2.5. The expected value is found by taking a sum of products of a particular value times the probability of that value occurring.

Example 3.15 Using the joint probability distribution given in Example 3.2, find $E(X + Y)$, the expected number of glasses of milk spilled by the Mumm twins.

For easy reference we will rewrite the joint distribution of X and Y in Table 3.5. Using this distribution and (3.23), we find the expected value of $X + Y$ from

$$\begin{aligned}
E(X + Y) &= \sum_{\text{all } x,y} (x + y) \cdot P[X = x, Y = y] \\
&= (0 + 0) \cdot P[X = 0, Y = 0] + (0 + 1) \cdot P[X = 0, Y = 1] \\
&\quad + (0 + 2) \cdot P[X = 0, Y = 2] \\
&\quad + (1 + 0) \cdot P[X = 1, Y = 0] + (1 + 1) \cdot P[X = 1, Y = 1] \\
&\quad + (1 + 2) \cdot P[X = 1, Y = 2] \\
&\quad + (2 + 0) \cdot P[X = 2, Y = 0] + (2 + 1) \cdot P[X = 2, Y = 1] \\
&\quad + (2 + 2) \cdot P[X = 2, Y = 2] \\
&= 0(.10) + 1(.05) + 2(.05) + 1(.25) + 2(.20) \\
&\quad + 3(.05) + 2(.15) + 3(.05) + 4(.10) \\
&= 0 + .05 + .10 + .25 + .40 + .15 + .30 + .15 + .40 \\
&= 1.80.
\end{aligned}$$

TABLE 3.5

		Y		
		0	1	2
	0	.10	.05	.05
X	1	.25	.20	.05
	2	.15	.05	.10

In the context of the previous example you can see that it might be of interest to calculate the expected value of a sum of two random variables. If X represented the (random) length of a rectangle and Y represented the (random) width, then $E(2X + 2Y)$ would give the expected value of the perimeter of the rectangle and $E(X \cdot Y)$ would give the expected value of the area of the rectangle.

Example 3.16 In order to earn credit toward a Junior Woodchuck Woodworking Merit Badge, Chip selects two boards at random from a pile of boards of various lengths. He then cuts each of these in half and nails them together to make a rectangular box. Let X and Y denote the length and width (in inches) of the box, respectively. Using the joint probability distribution given in Table 3.6, find $E(2X + 2Y)$ and $E(XY)$.

TABLE 3.6

		Y		
		5	10	15
	5	$\frac{2}{90}$	$\frac{10}{90}$	$\frac{6}{90}$
X	10	$\frac{10}{90}$	$\frac{20}{90}$	$\frac{15}{90}$
	15	$\frac{6}{90}$	$\frac{15}{90}$	$\frac{6}{90}$

We have

$$
\begin{aligned}
E(2X + 2Y) &= \sum_{\text{all } x,y} (2x + 2y)P[X = x, Y = y] \\
&= [2(5) + 2(5)](\tfrac{2}{90}) + [2(5) + 2(10)](\tfrac{10}{90}) + \dots + [2(15) + 2(15)](\tfrac{6}{90}) \\
&= \tfrac{40}{90} + \tfrac{300}{90} + \dots + \tfrac{360}{90} = 42.
\end{aligned}
$$

Using (3.23) with $g(X, Y) = XY$, we get

$$
\begin{aligned}
E(XY) &= \sum_{\text{all } x,y} xyP[X = x, Y = y] \\
&= (5)(5)(\tfrac{2}{90}) + (5)(10)(\tfrac{10}{90}) + (5)(15)(\tfrac{6}{90}) \\
&\quad + (10)(5)(\tfrac{10}{90}) + (10)(10)(\tfrac{20}{90}) + (10)(15)(\tfrac{15}{90}) \\
&\quad + (15)(5)(\tfrac{6}{90}) + (15)(10)(\tfrac{15}{90}) + (15)(15)(\tfrac{6}{90}) \\
&= \tfrac{9800}{90} = 108.\bar{8}.
\end{aligned}
$$

Although most practical people would prefer to calculate expected values for functions that have some real-world interpretation (e.g., perimeter or area), it is not strictly necessary to be able to interpret a function in real-world terms. Some good, believe it or not, can come from considering some expectations on a theoretical level. For example, although we may not be able to *interpret* $E(XY)$ for the Mumm twins, we can still calculate it.

Example 3.17 Using the joint probability distribution given in Examples 3.2 and 3.15 (Mumm twins), find $E(XY)$.

Using equation (3.23) with $g(X, Y) = XY$, we get

$$\begin{aligned} E(XY) &= \sum_{\text{all } x,y} xyP[X = x, Y = y] \\ &= (0)(0)(.10) + (0)(1)(.05) + (0)(2)(.05) \\ &\quad + (1)(0)(.25) + (1)(1)(.20) + (1)(2)(.05) \\ &\quad + (2)(0)(.15) + (2)(1)(.05) + (2)(2)(.10) \\ &= 0 + 0 + 0 + 0 + .20 + .10 + 0 + .10 + .40 \\ &= .80. \end{aligned}$$

You might wonder if any of these expected values could be found by some indirect method. For example, can $E(X)$ and $E(Y)$ be used to find $E(X + Y)$ or $E(XY)$? The following two theorems answer these questions. To prove these theorems, however, you will need to understand the concept of a "double summation."

In calculating the expected value of a function of two random variables, we have used the notation

$$\sum_{\text{all } x,y}$$

to represent the sum of certain products taken over all possible x and y values. If you reread Examples 3.15 to 3.17, you will see that we went through all possible x and y values in a systematic way. For example, in Example 3.15 we started with $x = 0$ and considered all possible y values, $y = 0$, 1, and 2. Then we took $x = 1$ and considered $y = 0$, 1, and 2, and so on. In other words, for each fixed x we summed over all possible y values. We could write this as a double sum as follows:

$$\sum_{\text{all } x,y} \quad \text{or} \quad \sum_{x=0}^{2} \sum_{y=0}^{2} \quad \text{or} \quad \sum_{x} \sum_{y}. \tag{3.24}$$

Of course, we could also have systematically started with y and considered all possible x values. In this case the double sum would be

$$\sum_{y=0}^{2} \sum_{x=0}^{2} \quad \text{or} \quad \sum_{y} \sum_{x}. \tag{3.25}$$

With this notation in mind, we can prove the following theorems.

Theorem 3.2 For any random variables X and Y,

$$E(X + Y) = E(X) + E(Y). \tag{3.26}$$

Proof: This theorem is true for either continuous or discrete random variables, but the proof we give is for discrete random variables. The proof uses the relationships given in (3.23) to (3.25) and equations like (3.7) to get marginal probability distributions from the joint probability distribution.

$$\begin{aligned} E(X+Y) &= \sum_{\text{all } x,y} (x+y) \cdot P[X=x, Y=y] \\ &= \sum_{\text{all } x,y} xP[X=x, Y=y] + \sum_{\text{all } x,y} yP[X=x, Y=y] \\ &= \sum_x \sum_y xP[X=x, Y=y] + \sum_y \sum_x yP[X=x, Y=y] \\ &= \sum_x x(\sum_y P[X=x, Y=y]) + \sum_y y(\sum_x P[X=x, Y=y]) \\ &= \sum_x xP[X=x] + \sum_y yP[Y=y] = E(X) + E(Y). \end{aligned}$$

The same kind of argument can be used to show that if $g(X)$ and $h(Y)$ are functions of X and Y, then

$$E(g(X) + h(Y)) = E(g(X)) + E(h(Y)).$$

The following corollary generalizes Theorem 3.2 to consideration of more than two random variables.

Corollary 3.1 For any sequence of random variables $X_1, X_2, \ldots, X_n$,

$$E(X_1 + X_2 + \ldots + X_n) = E(X_1) + E(X_2) + \ldots + E(X_n).$$

Example 3.18 Find $E(X+Y)$, the expected total number of glasses of milk spilled by the Mumm twins, using equation (3.26).

In Section 3.2 we determined the marginal distributions of X and Y to be

x:	0	1	2	y:	0	1	2
$P[X=x]$:	.2	.5	.3	$P[Y=y]$:	.5	.3	.2,

from which we get

$$E(X) = 0(.2) + 1(.5) + 2(.3) = 1.10$$

and

$$E(Y) = 0(.5) + 1(.3) + 2(.2) = .70.$$

It follows that

$$E(X+Y) = E(X) + E(Y) = 1.10 + .70 = 1.80.$$

Theorem 3.3 If X and Y are *independent* random variables, then

$$E(XY) = E(X) \cdot E(Y). \tag{3.27}$$

Proof: Remember that if X and Y are independent random variables, the joint probability distribution can be written as the product of the marginal probability distributions of X and Y. Using this fact, we get

$$\begin{aligned} E(XY) &= \sum_{\text{all } x,y} xyP[X=x, Y=y] \\ &= \sum_{\text{all } x,y} xyP[X=x] \cdot P[Y=y] \end{aligned}$$

$$= \sum_x \sum_y xyP[X = x] \cdot P[Y = y]$$

$$= \sum_x (xP[X = x](\sum_y yP[Y = y]))$$

$$= \sum_x (xP[X = x] \cdot E(Y)) = (\sum_x xP[X = x])E(Y) = E(X)E(Y).$$

As in the case of Theorem 3.2, we can generalize this argument to show that if $g(X)$ and $h(Y)$ are functions of X and Y and *if X and Y are independent*, then

$$E(g(X) \cdot h(Y)) = E(g(X)) \cdot E(h(Y)).$$

It must be emphasized that while Theorem 3.2 holds for *any* random variables, we must know that the random variables are *independent* in order to *guarantee* that (3.27) will hold. It is also important to know that if (3.27) *fails* to hold for two random variables, those two random variables must not be independent.

Example 3.19 Compare $E(XY)$ and $E(X) \cdot E(Y)$ for the Mumm twins example. What can you conclude?

Since $E(XY) = .80$ and $E(X) \cdot E(Y) = (1.1)(.7) = .77$ are not equal, we can conclude that X and Y are not independent.

Example 3.20 Two fair dice are tossed, one red and one green. Let X and Y denote the number facing up on the red and green dice, respectively. Find $E(XY)$.

Since X and Y are independent, it follows from Theorem 3.3 that $E(XY)$ will equal $E(X)$ times $E(Y)$. Since each face is equally likely to turn up, it follows that

$$E(X) = E(Y) = 1(\tfrac{1}{6}) + 2(\tfrac{1}{6}) + \ldots + 6(\tfrac{1}{6}) = 3.5,$$

so $E(XY) = (3.5)(3.5) = 12.25$.

If X and Y are random variables, then of course $Z = g(X, Y)$ is also a random variable. We have discussed how to find the expected value of Z and would like to point out that the variance of Z can also be found. By using the definition of the variance of a random variable (Definition 2.14), we see that

$$V(Z) = E(Z - \mu_Z)^2 = E[g(X, Y) - E(g(X, Y))]^2,$$

so to get the variance of $g(X, Y)$ we need to find the expected value of another function of X and Y:

$$h(X, Y) = [g(X, Y) - E(g(X, Y))]^2.$$

The procedure used is simply to apply (3.23) to $h(X, Y)$. Alternatively, we could find $V(Z)$ by using $E(Z^2) - [E(Z)]^2$. We will consider this problem further in the next section.

• • • • •

Expected values for functions of two random variables can be determined for continuous as well as discrete random variables. If X and Y are continuous random variables having joint probability density function $f(x, y)$, then

$$E(g(X, Y)) = \int_{-\infty}^{\infty} \int_{-\infty}^{\infty} g(x, y)f(x, y)\, dy\, dx. \tag{3.28}$$

Example 3.21 Using the joint density

$$f(x, y) = \begin{cases} \frac{3}{16}(2 - x^2 - y^2) & -1 \le x \le 1, -1 \le y \le 1 \\ 0 & \text{otherwise} \end{cases}$$

(see Example 3.4), find $E(XY)$.

Using equation (3.28) with $g(X, Y) = XY$, we get

$$\begin{aligned} E(XY) &= \int_{-\infty}^{\infty} \int_{-\infty}^{\infty} xyf(x, y)\,dy\,dx \\ &= \int_{-1}^{1} \int_{-1}^{1} xy\left(\frac{3}{16}\right)(2 - x^2 - y^2)\,dy\,dx \\ &= \frac{3}{16} \int_{-1}^{1} \int_{-1}^{1} (2xy - x^3y - xy^3)\,dy\,dx \\ &= \frac{3}{16} \int_{-1}^{1} \left[\frac{2xy^2}{2} - \frac{x^3y^2}{2} - \frac{xy^4}{4}\right]_{-1}^{1} dx \\ &= \frac{3}{16} \int_{-1}^{1} 0\,dx = 0. \end{aligned}$$

If we were to pursue this example further we would find (using the marginal densities found in Example 3.8) that

$$\begin{aligned} E(X) &= \int_{-1}^{1} x\left(\frac{3}{8}\right)\left(\frac{5}{3} - x^2\right) dx \\ &= \frac{3}{8}\left[\frac{5}{3}\left(\frac{x^2}{2}\right) - \left(\frac{x^4}{4}\right)\right]_{-1}^{1} = 0 \end{aligned}$$

and that $E(Y) = 0$. Hence in this case $E(XY) = 0 = E(X)E(Y)$. Can we conclude from this, in view of Theorem 3.3, that X and Y are independent? No! In fact, we *know* from Example 3.14 that these random variables are *dependent*. This illustrates the fact that while independence of X and Y implies that $E(XY)$ will equal $E(X) \cdot E(Y)$, it is *not* true that $E(XY) = E(X) \cdot E(Y)$ implies that X and Y are independent.

Example 3.22 Using the joint probability density function given in Example 3.13 for the independent random variables X and Y, find $E(XY)$. We find

$$E(X) = \int_{0}^{1} x(3x^2)\,dx = \left[\frac{3x^4}{4}\right]_{0}^{1} = \frac{3}{4}$$

and

$$E(Y) = \int_{0}^{2} y\left(\frac{y}{2}\right) dy = \left[\frac{y^3}{6}\right]_{0}^{2} = \frac{4}{3},$$

so it follows that $E(XY) = E(X)E(Y) = (\frac{3}{4})(\frac{4}{3}) = 1$.

Example 3.23 For the joint distributions given in the two previous examples, find $E(X + Y)$.

Using the joint distribution referred to in Example 3.21, we get

$$E(X+Y) = E(X) + E(Y) = 0 + 0 = 0,$$

while for the joint distribution of Example 3.22 we have

$$E(X+Y) = E(X) + E(Y) = \tfrac{3}{4} + \tfrac{4}{3} = \tfrac{25}{12}.$$

Note that we simply used the results of Theorem 3.2 to obtain the answers and that the independence of the random variables is not an issue.

3.5 COVARIANCE AND CORRELATION

In many situations it is of interest to consider the sum of two (or more) random variables. One example we used in the previous section was $X + Y$, the total number of glasses of milk spilled at a meal by the Mumm twins. We have seen that calculating the expected value of a sum of random variables is relatively easy (using Theorem 3.2) and would now like to consider the problem of finding an expression for the variance of a sum of two random variables. The derivation we give follows from the properties of expected values found in Section 2.5, from the definition of variance given in Definition 2.14, and from Theorem 3.2.

$$\begin{aligned}
(3.29)\quad V(X+Y) &= E[(X+Y) - E(X+Y)]^2 \\
&= E[(X+Y) - (E(X) + E(Y))]^2 \\
&= E[X + Y - E(X) - E(Y)]^2 = E[(X - E(X)) + (Y - E(Y))]^2 \\
&= E[(X - E(X))^2 + 2(X - E(X))(Y - E(Y)) + (Y - E(Y))^2] \\
&= E(X - E(X))^2 + E[2(X - E(X))(Y - E(Y))] + E[Y - E(Y)]^2 \\
&= V(X) + 2E[(X - E(X))(Y - E(Y))] + V(Y) \\
&= V(X) + V(Y) + 2E[(X - E(X))(Y - E(Y))].
\end{aligned}$$

From (3.29) we see that the variance of the sum of X and Y is the sum of the variance of X plus the variance of Y *plus* an additional term.

Definition 3.5 The *covariance* of two random variables X and Y is defined to be

$$\text{Cov}(X, Y) = E[(X - E(X))(Y - E(Y))].$$

Using this definition, we can write

$$V(X+Y) = V(X) + V(Y) + 2\,\text{Cov}(X, Y).$$

Using the fact that $E(X - E(X)) = 0$ [equation (2.22)], we can show that

$$(3.30)\qquad E[(X - E(X))(Y - E(Y))] = E(XY) - E(X) \cdot E(Y).$$

We show this as follows: rewriting the left-hand side of (3.30), we get

$$\begin{aligned}
E[(X - E(X))(Y - E(Y))] &= E[(X - E(X)) \cdot Y - (X - E(X)) \cdot E(Y)] \\
&= E[(X - E(X)) \cdot Y] - E[(X - E(X)) \cdot E(Y)] \\
&= E[XY - E(X) \cdot Y] - E(Y)E[X - E(X)] \\
&= E(XY) - E[E(X) \cdot Y] - E(Y) \cdot 0 \\
&= E(XY) - E(X) \cdot E(Y).
\end{aligned}$$

Using (3.30) and Theorem 3.3, you can see that if X and Y are independent, then Cov $(X, Y) = 0$. (However, the converse is not true.) This is so since if X and Y are independent, then

$$E(XY) = E(X) \cdot E(Y),$$

so

$$E(XY) - E(X)E(Y) = 0.$$

In a crude sense, then, this indicates that we might be able to use Cov (X, Y) as a measure of dependence between two random variables. We have already seen that if X and Y are independent, then Cov (X, Y) will be zero. If X and Y are related in such a way that large values of X tend to go with large values of Y while small values of X tend to go with small values of Y, then the covariance of X and Y will tend to be positive. On the other hand, if large values of X tend to go with small values of Y and vice versa, the covariance of X and Y will tend to be negative.

Example 3.24 Let X and Y be as defined in Example 3.1: $X =$ number of heads appearing on a penny and nickel and $Y =$ number of heads appearing on a nickel and a dime. Find Cov (X, Y)

First try to use your intuition and guess as to whether the covariance will be positive or negative. Does it appear that "large" X values will tend to go with "large" Y values? That is, if X is equal to 1 or 2, is it likely that Y will also be equal to 1 or 2? If so, then you would have to guess that the covariance will be positive.

Using Table 3.2, we can find

$$\begin{aligned} E(XY) &= (0)(0)(\tfrac{1}{8}) + 0(1)(\tfrac{1}{8}) + \ldots + (2)(2)(\tfrac{1}{8}) \\ &= 0 + 0 + 0 + 0 + \tfrac{2}{8} + \tfrac{2}{8} + 0 + \tfrac{2}{8} + \tfrac{4}{8} \\ &= \tfrac{10}{8}. \end{aligned}$$

Furthermore, $E(X) = E(Y) = 0(\frac{1}{4}) + 1(\frac{1}{2}) + 2(\frac{1}{4}) = 1$, so

$$\text{Cov}\,(X, Y) = \tfrac{10}{8} - (1)(1) = \tfrac{2}{8},$$

which is positive.

Example 3.25 Find the variance for the total number of glasses of milk spilled by the Mumm twins.

In previous examples we have calculated $E(X)$, $E(Y)$, and $E(XY)$, so we can see that

$$\begin{aligned} \text{Cov}\,(X, Y) &= E[(X - E(X))(Y - E(Y))] = E(XY) - E(X)E(Y) \\ &= .80 - (1.1)(.7) = .80 - .77 = .03. \end{aligned}$$

[Note that this result comes as no surprise to Mrs. Mumm. She has told the children many times that when they fool around at the table, they are both likely to spill their milk. Consequently, she expects large values of X to go with large values of Y (i.e., she expects the covariance to be positive!).]

In order to find $V(X + Y)$, we also need to find $V(X)$ and $V(Y)$. We will leave it to you to verify that $V(X) = .49$ and $V(Y) = .61$. Using these results in (3.29), we get

$$V(X+Y) = V(X) + V(Y) + 2\,\text{Cov}\,(X, Y)$$
$$= .49 + .61 + 2(.03) = 1.16.$$

You should be able to see that if X and Y are independent random variables, then $\text{Cov}\,(X, Y) = 0$, so (3.29) will simplify.

Theorem 3.4 If X and Y are independent random variables, then $V(X+Y) = V(X) + V(Y)$.

Corollary 3.2 If $X_1, X_2, \ldots, X_n$ are independent random variables, then

$$V(X_1 + X_2 + \ldots + X_n) = V(X_1) + V(X_2) + \ldots + V(X_n).$$

In Example 3.16 we found the expected value of the perimeter of a box made by Chip for his woodworking merit badge. The perimeter is found from $P = 2X + 2Y$. In this case we are not simply interested in a sum of random variables but in a linear combination of random variables. To find the variance of a linear combination of two random variables, we use

$$V(aX + bY) = a^2 V(X) + b^2 V(Y) + 2ab\,\text{Cov}\,(X, Y). \tag{3.31}$$

We will not bother proving equation (3.31), since the proof is similar to the proof of (3.29), but we will illustrate its use by an example.

Example 3.26 Using the joint probability distribution given in Example 3.16, find $V(P) = V(2X + 2Y)$.

From equation (3.31) with $a = b = 2$, we see that

$$V(P) = V(2X + 2Y) = 4V(X) + 4V(Y) + 8\,\text{Cov}\,(X, Y).$$

You can verify that

$$E(X) = E(Y) = 10.5$$
$$E(X^2) = E(Y^2) = 122.50,$$

so

$$V(X) = V(Y) = 122.50 - (10.5)^2 = 122.50 - 110.25 = 12.25.$$

In Example 3.16 we found that $E(XY) = 108.\bar{8}$, so we get

$$\text{Cov}\,(X, Y) = E(XY) - E(X)E(Y)$$
$$= 108.8\bar{8} - 110.25 = -1.36\bar{1}.$$

Hence

$$V(2X + 2Y) = 4(12.25) + 4(12.25) + 8(-1.36\bar{1})$$
$$= 87.\bar{1}.$$

In our earlier discussion about covariance, we indicated that the *sign* of the covariance could give some indication about the relationship between two random variables X and Y. It is reasonable to ask if the *magnitude* of the covariance gives any indication as to the relationship between random variables. We will see that this is not the case. In Example 3.16 Chip measured the length and width of the box he was making in *inches*. Had he measured the same boards using the metric system, he would have found the lengths to be 12.7, 25.4, and 38.1 centimeters. Using the same probability distribution as in Example 3.16 with the measurements in centimeters rather than inches, we would find the covariance to be -8.781 instead of -1.361. We have only changed the units of measurement and not the relationship between length and width. This indicates that the magnitude of the covariance does not provide useful information about the relationship between two random variables. We did, however, face a problem similar to this before in resolving the discussion between Zack and Zeke about grain production. In that case we standardized the random variables and we can do a similar thing here.

Definition 3.6 The *correlation* between two random variables X and Y is defined to be

$$\rho_{X,Y} = \frac{\text{Cov}(X, Y)}{\sigma_X \sigma_Y}. \tag{3.32}$$

By writing out the covariance term, we see that (3.32) can be rewritten as

$$\begin{aligned}\rho_{X,Y} &= \frac{E[(X - E(X))(Y - E(Y))]}{\sigma_X \sigma_Y} \\ &= E\left[\left(\frac{X - E(X)}{\sigma_X}\right)\left(\frac{Y - E(Y)}{\sigma_Y}\right)\right]\end{aligned}$$

so that $\rho_{X,Y}$ is a standardized version of the covariance.

Example 3.27 Find the correlation between the length X and the width Y of the box being made by Chip.

All the necessary quantities have already been calculated in Example 3.26. We have

$$\begin{aligned}\rho_{X,Y} &= \frac{\text{Cov}(X, Y)}{\sigma_X \sigma_Y} = \frac{-1.361}{\sqrt{(12.25)(12.25)}} \\ &= -0.111.\end{aligned}$$

[Had we expressed the measurements in centimeters rather than inches, we would have found the variance of X (and of Y) to be $(12.25)(2.54)^2 = 79.032$. The covariance would have been -8.781, but ρ would have come out the same.]

It follows from our earlier discussions that if X and Y are independent random variables, then

$$\rho_{X,Y} = \frac{\text{Cov}(X, Y)}{\sigma_X \sigma_Y} = \frac{0}{\sigma_X \sigma_Y} = 0. \tag{3.33}$$

It can be shown mathematically that for any random variables X and Y,

$$-1 \leq \rho_{X,Y} \leq +1.$$

Although we will not dwell on the fact here, we should point out that $\rho_{X,Y}$ only measures the strength of the *linear relationship* between random variables. If $|\rho_{X,Y}|$ is close to 1, there is a strong linear relationship between X and Y. However, it may be that $|\rho_{X,Y}|$ is close to zero but that there is a *nonlinear* relationship between X and Y. These ideas will be discussed in greater detail in Chapter 9.

EXERCISES

1. The joint probability distribution for two random variables X and Y is given by the following table:

			Y	
		-2	0	2
	-1	.06	.18	.06
X	0	.08	.24	.08
	1	.06	.18	.06

(a) Find $P[X \leq 0,\ Y \leq 0]$.
(b) Find $P[X \leq Y]$.

2. The American Bureau of Credit has made a study of the number of credit cards held by and the number of credit purchases (in one week) made by persons living in the community of Frugalville. ABC has determined that for a person chosen at random from Frugalville, the joint probability distribution of X and Y, the number of credit cards held by and the number of credit purchases made by the person chosen, respectively, is as follows:

				Y		
		0	1	2	3	4
	0	.10	0	0	0	0
X	1	.15	.11	.10	.02	.02
	2	.09	.05	.08	.05	.03
	3	.06	.04	.02	.03	.05

(a) Find $P[2 \leq X \leq 3,\ 2 \leq Y \leq 4]$.
(b) Find $P[Y \leq X]$.

3. A survey in the city of Wheeling showed that X, the number of cars, and Y, the number of bicycles, owned by members of a randomly chosen family has the following joint probability distribution.

				Y		
		0	1	2	3	4
	0	0	.03	.05	.02	0
X	1	.05	.08	.12	.10	.05
	2	.07	.09	.16	.08	0
	3	.03	.05	.02	0	0

(a) Find $P[X > Y]$.
(b) Find the probability that the family has two or more cars.
(c) Find $P[X \geq 2,\ Y \geq 2]$.

• ***4.*** Given that

$$f(x, y) = \begin{cases} \dfrac{3x(x+y)}{5} & 0 \leq x \leq 1, 0 \leq y \leq 2 \\ 0 & \text{otherwise,} \end{cases}$$

verify that

(a) $\int_0^2 \int_0^1 f(x, y)\,dx\,dy = 1$;

(b) $f_X(x) = \int_0^2 f(x, y)\,dy = 6(x^2 + x)/5, \quad 0 \le x \le 1$;

(c) $f_Y(y) = \int_0^1 f(x, y)\,dx = (2 + 3y)/10, \quad 0 \le y \le 2$;

(d) $P[0 \le X \le \frac{1}{2}, 0 \le Y \le 1] = \frac{1}{16}$.

• **5.** Given that

$$f(x, y) = \begin{cases} \dfrac{x + xy}{2} & 0 \le x \le 1, 0 \le y \le 2 \\ 0 & \text{otherwise,} \end{cases}$$

verify that

(a) $\int_0^2 \int_0^1 f(x, y)\,dx\,dy = 1$;

(b) $f_X(x) = \int_0^2 f(x, y)\,dy = 2x, \quad 0 \le x \le 1$;

(c) $f_Y(y) = \int_0^1 f(x, y)\,dx = (y + 1)/4, \quad 0 \le y \le 2$;

(d) $P[0 \le X \le \frac{1}{2}, 0 \le Y \le 1] = \frac{3}{32}$.

• **6.** A joint density function for two continuous random variables is

$$f(x, y) = \begin{cases} x^2 & 0 \le x \le 1, 0 \le y \le 3 \\ 0 & \text{otherwise.} \end{cases}$$

(a) Find $P[0 \le X \le \frac{1}{2}, 0 \le Y \le 1]$.
(b) Find the marginal densities for X and Y.

• **7.** Given that the joint density function of X and Y is

$$f(x, y) = \begin{cases} \dfrac{3x^2 y}{2} & 0 \le x \le 1, 0 \le y \le 2 \\ 0 & \text{otherwise,} \end{cases}$$

show that the marginal densities are

$$f_X(x) = \begin{cases} 3x^2 & 0 \le x \le 1 \\ 0 & \text{otherwise} \end{cases} \quad \text{and} \quad f_Y(y) = \begin{cases} \dfrac{y}{2} & 0 \le y \le 2 \\ 0 & \text{otherwise} \end{cases}$$

(cf. Example 3.13).

8. Use the joint probability distribution for X and Y given in Exercise 1 to find
(a) the marginal distribution for X;
(b) the marginal distribution for Y;
(c) the conditional probability distribution for X given that $Y = 0$;
(d) $E(X \mid Y = 0)$.

9. Using the joint probability distribution given in Exercise 2 (credit cards held and number of credit purchases), find
(a) the marginal probability distribution of X;
(b) the marginal probability distribution of Y;

(c) the conditional probability distributions for Y given $X = x$, for $x = 0, 1, 2,$ and 3;

(d) $E(Y|X = x)$ for $x = 0, 1, 2, 3$. (*To think about:* How could Frugalvillian parents use these results in a discussion with their child, who asked for a credit card of his own?)

10. Using the joint distribution given in Exercise 3, find (a) the expected number of bicycles owned by a family known to have two cars; (b) the expected number of cars owned by a family known to have two bicycles.

• ***11.*** (a) Using the information given in Exercise 4, find the conditional distribution of Y given that $X = 1$. (b) Use the results of part (a) to find $E(Y|X = 1)$.

• ***12.*** (a) Using the information given in Exercise 5, find the conditional distribution of X given that $Y = 1$. (b) Use the results of part (a) to find $E(X|Y = 1)$.

13. Determine whether or not the random variables having the joint probability distribution defined in (a) Exercise 4, (b) Exercise 5, (c) Exercise 7 are independent. Justify your answer.

14. Determine whether or not the random variables having the joint probability distribution defined in (a) Exercise 1, (b) Exercise 2, (c) Exercise 3 are independent. Justify your answer. (*To think about:* In the context of Exercises 2 and 3, would you expect these random variables to be independent? Why or why not?)

15. Determine whether or not the random variables having the joint probability distributions defined in (a) and (b) are independent.

(a)

		Y: −1	0	1
X	−2	.10	.20	.10
	0	.05	.10	.05
	2	.20	0	.20

(b)

		Y: −1	0	1
X	−2	.12	.15	.13
	0	.05	.10	.05
	2	.13	.15	.12

16. Using the joint distribution given in Exercise 15(b), find (a) $E(X)$; (b) $E(Y)$; (c) $E(XY)$; (d) $E(X^2Y)$.

17. Referring to Exercise 2, find (a) $E(X)$; (b) $E(Y)$; (c) $E(XY)$. If there is a 50-cent monthly charge for having a credit card and a 10-cent charge for each credit purchase, then the total monthly dollar cost of having credit cards is $.50X + .10Y$. (d) Find $E(.50X + .10Y)$.

18. In the city of Wheeling the city council has passed a "wheel tax." Each family must pay 50 cents for each wheel owned (not counting spare tires, etc.) Find the expected tax for a randomly chosen family living in Wheeling. (See Exercise 3.)

• ***19.*** Referring to Exercise 4, find (a) $E(X)$; (b) $E(Y)$; (c) $E(XY)$.

• ***20.*** Referring to Exercise 5, find (a) $E(2X)$; (b) $E(3Y)$; (c) $E(4XY)$.

21. Let X and Y be two random variables with $E(X) = 0$, $E(Y) = 2$, $V(X) = 1$, $V(Y) = 16$, and $E(XY) = 2$. Find (a) Cov (X, Y); (b) $\rho_{X,Y}$; (c) $V(X + Y)$.

22. Let U and W be two random variables with $E(U) = 0$, $E(W) = 2$, $E(U^2) = 1$, $E(W^2) = 20$, and $E(UW) = -2$. Find (a) Cov (U, W); (b) $\rho_{U,W}$; (c) $V(U - W)$; (d) $V(2U + W - 3)$.

23. Find Cov (X, Y) using the joint distributions given in Exercise 15(a) and (b). (*To think about:* Are your answers here consistent with the answers you obtained in Exercise 15?)

24. Find $V(.50X + .10Y)$, the variance of the monthly cost of having credit cards for a Frugalvillian family (cf. Exercises 2 and 17).

25. Find the variance and standard deviation of the amount of wheel tax paid by a randomly chosen family from the city of Wheeling (cf. Exercises 3 and 18).

26. The Silver City Bank has two types of accounts, checking and savings. Let X denote the number of checking accounts and Y the number of savings accounts held at this bank by a randomly chosen family from Silver City. The joint distribution of X and Y is given by

		Y			
		0	1	2	3
	0	.30	.10	.03	.02
X	1	.08	.15	.10	.02
	2	.02	.05	.07	.06

(a) What proportion of Silver City families have either a checking or savings account at the Silver City Bank?
(b) Find the expected number of accounts held in this bank by a randomly chosen Silver City family.
(c) Find the standard deviation of the number of accounts held in this bank by a randomly chosen Silver City family.
(d) Find $\rho_{X,Y}$.

27. Using the joint density function

$$f(x, y) = \begin{cases} \dfrac{3x^2y}{2} & 0 \leq x \leq 1, 0 \leq y \leq 2 \\ 0 & \text{otherwise,} \end{cases}$$

find Cov (X, Y). (*Hint:* Look at Exercises 7 and 13.)

• **28.** The joint density function for two continuous random variables X and Y is given by

$$f(x, y) = \begin{cases} \dfrac{x + 2y}{12} & 0 \leq x \leq 2, 0 \leq y \leq 2 \\ 0 & \text{otherwise.} \end{cases}$$

The marginal densities for X and Y are

$$f_X(x) = \begin{cases} \dfrac{2x + 4}{12} & 0 \leq x \leq 2 \\ 0 & \text{otherwise} \end{cases} \qquad f_Y(y) = \begin{cases} \dfrac{4y + 2}{12} & 0 \leq y \leq 2 \\ 0 & \text{otherwise.} \end{cases}$$

Find Cov (X, Y) and $\rho_{X,Y}$.

29. Prove that $V(aX + bY) = a^2V(X) + b^2V(Y) + 2ab \operatorname{Cov}(X, Y)$. [*Hint:* Follow the proof of (3.29).]

4

SPECIAL PROBABILITY DISTRIBUTIONS

4.1 INTRODUCTION

In Chapter 2 we discussed random variables and their probability distributions in general terms. We also found formulas whereby the expected values and variances could be calculated from the probability distributions. In this chapter we will describe specific families of probability distributions that occur over and over again in real-world problems. We will give some formulas that will allow the means and variances for these special distributions to be easily found.

At the end of Section 2.2 we gave examples of three random variables, each arising from a different physical situation, having the exact same probability distribution. Although the physical situations are really quite different, some aspects of these situations are quite similar. For example, in each case the experiment could be classified as having one of two events happening: head or tail, odd or even, or boy or girl. We will see that by concentrating on certain characteristics (such as the occurrence of exactly one of two events), we can define families or classes of random variables. These families have names that you will come to recognize—names such as Binomial, Poisson, Normal, and others.

Members of human families are all unique but at the same time have certain family traits or characteristics in common (prominent nose, brown eyes, high forehead, etc.) (Figure 4.1). We will see that families of probability distributions also have certain "family traits." Furthermore, we will see that different members of the same family can be distinguished by different parameter values. (In fact, we could define a

FIGURE 4.1
Family Characteristics

parameter as a quantity that characterizes a probability distribution.) Finally, we will see that some of the families are related to each other.

For each of the special families of probability distributions that we study in this chapter, we will (1) give conditions under which a random variable can be expected to have a particular probability distribution, (2) find a mathematical expression for the probability distribution, and (3) find a mathematical expression for the mean and variance of a random variable having that distribution.

4.2 THE BINOMIAL DISTRIBUTION

One of the conditions that one must look for to know that a random variable has a binomial distribution is a sequence of experiments, each of which can be classified in exactly one of two ways. We have described several situations like this, one example in particular being the machine producing watzits (Example 1.40). In that example each watzit produced is either defective or nondefective (good). Let us assume for the moment that the dial which controls the proportion of defectives is set at 10%. Further assume that each item is produced independently of the other items. Under these conditions, what is the probability that the first two items produced on a particular day are defective and the next three items are not defective? To solve this problem it will be helpful to define some events. Let

$$D_1 = \{\text{first item produced is defective}\}$$
$$D_2 = \{\text{second item produced is defective}\}$$
$$G_3 = \{\text{third item produced is good (nondefective)}\}$$
$$G_4 = \{\text{fourth item produced is good}\}$$
$$G_5 = \{\text{fifth item produced is good}\}.$$

The event we are interested in can be expressed as the intersection of these five events:

$$D_1 \cap D_2 \cap G_3 \cap G_4 \cap G_5.$$

By the assumption of independence, we can write [in view of equation (1.18)]

$$P(D_1 \cap D_2 \cap G_3 \cap G_4 \cap G_5) = P(D_1) \cdot P(D_2) \cdot P(G_3) \cdot P(G_4) \cdot P(G_5).$$

Now if the dial is set to produce 10% defectives, it must be true that

$$P(D_i) = .10 \qquad \text{for } i = 1, 2$$

and

$$P(G_i) = .90 \qquad \text{for } i = 3, 4, 5.$$

Consequently,

$$P(D_1 \cap D_2 \cap G_3 \cap G_4 \cap G_5) = (.1)(.1)(.9)(.9)(.9) = (.1)^2(.9)^3 = .00729.$$

Now let us ask a similar, but slightly different question: What is the probability that the first three items produced are nondefective and the next two items are defective? Using a notation similar to the one we just used, we can write this event as

$$G_1 \cap G_2 \cap G_3 \cap D_4 \cap D_5.$$

Here, too, we can find the probability of this event by using the assumption of independence to get

$$\begin{aligned} P(G_1 \cap G_2 \cap G_3 \cap D_4 \cap D_5) &= P(G_1) \cdot P(G_2) \cdot P(G_3) \cdot P(D_4) \cdot P(D_5) \\ &= (.9)(.9)(.9)(.1)(.1) = (.1)^2(.9)^3 = .00729. \end{aligned}$$

It should now come as no surprise if we say that the probability of observing two defective and three nondefective items *in some particular order* will be .00729.

Next consider the following question: What is the probability of observing two defective and three nondefective items in the next five items produced? Note that we have not specified any particular order in this case. Therefore, the event $E =$ {two defective and three nondefective items} will occur if

$$\begin{array}{lr} D_1 \cap D_2 \cap G_3 \cap G_4 \cap G_5 & \textit{or} \\ D_1 \cap G_2 \cap D_3 \cap G_4 \cap G_5 & \textit{or} \\ D_1 \cap G_2 \cap G_3 \cap D_4 \cap G_5 & \textit{or} \\ \vdots & \vdots \\ G_1 \cap G_2 \cap G_3 \cap D_4 \cap D_5 & \end{array} \tag{4.1}$$

occurs. That is, the event E can be written as the *union* of the events listed in (4.1). Finding the probability of a union of several events can be easy *if* the events are mutually exclusive. Are the events listed in (4.1) mutually exclusive? Can any two of them occur simultaneously? No they cannot. For example, in comparing the first two sequences of (4.1) you see that they cannot occur at the same time, since if the second item is defective (D_2) it cannot also be good (G_2). Since they are mutually exclusive, it follows from Axiom 3 of Section 1.8 that

$$\begin{aligned} P(E) = P(D_1 \cap D_2 \cap G_3 \cap G_4 \cap G_5) + P(D_1 \cap G_2 \cap D_3 \cap G_4 \cap G_5) \\ + \ldots + P(G_1 \cap G_2 \cap G_3 \cap D_4 \cap D_5). \end{aligned} \tag{4.2}$$

From our earlier remarks we know that each of the summands on the right-hand side of (4.2) is equal to .00729. If we knew how many summands there were, we could find the answer we are after. As we shall see, there are 10 summands, so

$$P(E) = \underbrace{.00729 + .00729 + \ldots + .00729}_{10 \text{ terms}} = 10(.00729) = .0729.$$

How did we know that there are 10 summands? By looking at the list given in (4.1), you can see that each event is distinguishable from the others in that the D's and G's are in different orders. In how many different ways can 2 D's and 3 G's be arranged in a row? The answer is given by Theorem 1.1 of Section 1.4. Since there are a total of five items, we have

$$\frac{5!}{2!\,3!} = \binom{5}{2} = \frac{5 \cdot 4 \cdot 3 \cdot 2 \cdot 1}{2 \cdot 1 \cdot 3 \cdot 2 \cdot 1} = 10.$$

The previous discussion can be summarized as follows:

$$P[2 \text{ defectives in 5 items}] = \binom{5}{2}(.1)^2(.9)^3. \tag{4.3}$$

What we have presented here has been a very specific example. The important characteristics are the following: (1) each "experiment," in this case the production of a watzit, can be classified as having exactly one of two possible outcomes; (2) the experiments are independent; and (3) the probability that any item would be classified as defective remained constant throughout all experiments.

If we were to be more general, we could denote the probability of a defective item by p, where p could assume the values .01, .05, or .10 for the watzit-making machine or, in other examples, p could assume any value between 0 and 1. In this case (4.3) could be written as

$$P[2 \text{ defectives in 5 items}] = \binom{5}{2}p^2(1-p)^3. \tag{4.4}$$

Instead of considering 5 items, a more general approach would be to consider n items. If 2 items were defective, it follows that $n - 2$ items would be nondefective; consequently, (4.4) could be rewritten as

$$P[2 \text{ defectives in } n \text{ items}] = \binom{n}{2}p^2(1-p)^{n-2}. \tag{4.5}$$

Finally, instead of specifying two defectives, we might specify x defectives (and, consequently, $n - x$ nondefectives) to get, instead of (4.5),

$$P[x \text{ defectives in } n \text{ items}] = \binom{n}{x}p^x(1-p)^{n-x}. \tag{4.6}$$

Example 4.1 If the watzit machine is producing 5% defectives, what is the probability that exactly 1 of the next 4 watzits produced is defective?

To solve this, use equation (4.6) with $n = 4$, $x = 1$, and $p = .05$ to get

$$P[1 \text{ defective in 4 items}] = \binom{4}{1}(.05)^1(.95)^3 = .1715.$$

Any experiment whose outcome can be classified in exactly one of two ways can be said to be *dichotomous*. It is customary to refer to one of these ways as a "success" and the other way as a "failure." You must remember that a "success" does not necessarily denote something desirable and a "failure" does not necessarily denote something undesirable. When discussing the birth of babies, labeling the birth of a boy as a "success" is simply a matter of convenience and does not imply that the birth of a girl is something undesirable! It is also customary to denote the probability of a "success" by p and the probability of a "failure" by q, where $q = 1 - p$.

With this background in mind, we can now define conditions under which a random variable can be expected to have a binomial distribution.

Given a sequence of n independent experiments where each experiment can be classified in exactly one of two ways, say success or failure, and where p, the probability that an experiment results in a success, remains constant from one experiment to the next, the random variable X defined to be the number of successes in these n trials will have a *binomial* distribution.

It should be clear from our earlier discussion that the probability distribution of X will depend on the two quantities n and p. These two quantities distinguish different members of the family of binomial distributions, and consequently n and p are the parameters of the binomial distribution. By modifying (4.6) slightly to indicate the dependence of the probability on n and p, we can write the probability distribution of X as

$$P[X = x \mid n, p] = \binom{n}{x} p^x q^{n-x} \qquad x = 0, 1, \ldots, n. \tag{4.7}$$

All members of the binomial family of distributions have the general form of (4.7) for their probability distribution. The parameters n and p distinguish different members of this family.

Example 4.2 What is the probability that your football team will win the toss of the coin before 4 of the 5 home games next season?

We would like to use the binomial distribution to find this probability, so we must first verify that the necessary conditions are satisfied. The experiment in question involves the flip of a coin. Are there exactly two outcomes of interest? Yes. Are the experiments independent? Yes. Does the probability of a success (i.e., winning the coin flip) remain constant from one experiment to the next? Yes. Since these conditions are satisfied, we know that X, defined to be the number of successes or number of times the coin flip is won, has a binomial distribution. It remains to identifiy the parameters. You should be able to see that $n = 5$ and $p = \frac{1}{2}$. Using (4.7), we get

$$P[X = 4 \mid n = 5, p = \tfrac{1}{2}] = \binom{5}{4}(.5)^4(.5)^1 = .15625.$$

Example 4.3 A fair die is rolled 5 times. What is the probability that three 6's appear?

In this situation, when we ask whether there are exactly two outcomes of interest we might be tempted to say "no." Why? Because there are six possible outcomes when a fair die is rolled. However, we are really only interested in whether the die turns up a 6 (a success) or some number other than 6 (a failure). As far as this particular problem is concerned, we are only interested in two outcomes; hence it is not unreasonable to try to use a binomial distribution. You should verify for yourself that the other conditions necessary for X, the number of 6's that appear in 5 rolls of a fair die, to have a binomial distribution are satisfied. Since the parameters are $n = 5$ and $p = \frac{1}{6}$, we get

$$P[X = 3 \mid n = 5, p = \tfrac{1}{6}] = \binom{5}{3}(\tfrac{1}{6})^3(\tfrac{5}{6})^2 = .0321.$$

In these examples the size of n has been small enough to allow us to do these calculations by using a hand calculator. You can no doubt imagine that these calculations become more troublesome for larger values of n. For this reason we have included tables in the Appendix that give probabilities obtained from (4.7) for $n = 1, 2, \ldots, 20$ and for selected values of p. We have also given tables of the cumulative distribution function for these values of n and p. You should become familiar with both of these tables, as they can save you time (and avoid potential error) over hand calculations.

Example 4.4 The random variable X has a binomial distribution with parameters $n = 15$ and $p = .3$. (a) Find $P[5 \leq X \leq 7]$. (b) What would this probability be if p were equal to .6?

(a) Using Table A1, we find

$$\begin{aligned} P[5 \leq X \leq 7] &= P[X = 5 \mid n = 15, p = .3] + P[X = 6 \mid n = 15, p = .3] \\ &\quad + P[X = 7 \mid n = 15, p = .3] \\ &= .2061 + .1472 + .0811 \\ &= .4344. \end{aligned}$$

The cumulative distribution function can also be used to solve this problem. Since

$$P[5 \leq X \leq 7] = P[4 < X \leq 7] = P[X \leq 7] - P[X \leq 4],$$

we find from the cumulative binomial tables (Table A2),

$$P[X \leq 7] - P[X \leq 4] = .9500 - .5155 = .4345.$$

(The difference of .0001 between these answers is due to roundoff errors in the tables.)

(b) When p is greater than .5, the binomial probabilities cannot be found directly from the tables. However, they can be found indirectly by considering the number of failures instead of the number of successes. If we let X^* denote the number of failures and p^* the probability of a failure, it is not hard to see, for example, that

$$P[X = 5 \text{ "successes"} \mid n = 15, p = .6] = P[X^* = 10 \text{ "failures"} \mid n = 15, p^* = .4].$$

This is true since

$$\binom{15}{5}(.6)^5(.4)^{15-5} = \binom{15}{10}(.4)^{10}(.6)^{15-10}.$$

From Table A1 we find $P[X^* = 10 \mid n = 15, p^* = .4] = .0245$ and by adding terms we find

$$P[X = 5, 6, \text{ or } 7 \mid n = 15, p = .6] = P[X^* = 10, 9, \text{ or } 8 \mid n = 15, p^* = .4] = .2038.$$

By examining Table A1, you can see that the binomial probability distribution is symmetric only when $p = .5$. For a given n, the degree of skewness becomes greater as p gets close to zero or 1. (See Exercises 11 and 12.)

Now that we know how to find the probability distribution for a random variable having a binomial distribution, we can try to use this distribution to find the expected value and variance of that random variable. To find the expected value using equation (2.16), we would have to evaluate

$$\sum_{x=0}^{n} x\binom{n}{x}p^x q^{n-x}.$$

While this expression can be evaluated, the algebra involved is a bit unpleasant, so we will use an indirect approach to find $E(X)$.

Example 4.5 In years gone by, mattresses at the Featherbed Mattress Company were tested for comfort by having an inspector sleep on each mattress. If the inspector could sleep through his entire 8-hour shift, the mattress was deemed acceptable. Because of this procedure, an inspector could only test one mattress per shift. Times have changed and an inspector can now test a mattress in a much shorter time. However, union rules still require that an inspector test only one mattress per shift. In order to get more mattresses tested, the company has hired three inspectors to work each shift. Tom, Dick, and Harry work the first shift. If the probability that a mattress is judged acceptable is denoted by p and if this probability is the same for all inspectors and if the inspectors work independently, what is the expected value and variance of X, where $X =$ total number of mattresses judged acceptable? What would the answer be if n inspectors worked this shift? (Whew! Would you like to try reading that through again?)

You should be able to see from the conditions stated in the problem that X has a binomial distribution with parameters 3 and p in the first case and parameters n and p in the second case. Considering the first case, if we define

$$X_1 = \text{number of mattresses judged acceptable by Tom}$$
$$X_2 = \text{number of mattresses judged acceptable by Dick}$$
$$X_3 = \text{number of mattresses judged acceptable by Harry},$$

then the total number of mattresses judged acceptable must be

$$X = X_1 + X_2 + X_3.$$

If we restrict our attention to just X_1, we see that X_1 is a binomial random variable where the number of trials is equal to 1. In this case it is very easy to write down the probability distribution of X_1 and to then use that probability distribution to find $E(X_1)$ and $V(X_1)$. In particular,

$$P[X_1 = x] = \binom{1}{x} p^x q^{1-x} \qquad x = 0, 1,$$

so

$$P[X_1 = 0] = \binom{1}{0} p^0 q^{1-0} = q$$

and

$$P[X_1 = 1] = \binom{1}{1} p^1 q^{1-1} = p.$$

From this we get

$$E(X_1) = 0 \cdot P[X_1 = 0] + 1 \cdot P[X_1 = 1] = 0 \cdot q + 1 \cdot p = p$$

and

$$E(X_1^2) = 0^2 \cdot P[X_1 = 0] + 1^2 \cdot P[X_1 = 1] = 0 \cdot q + 1 \cdot p = p.$$

Consequently,

$$V(X_1) = E(X_1^2) - (E(X_1))^2 = p - p^2 = p(1 - p) = pq,$$

so we find that $E(X_1) = p$ and $V(X_1) = pq$. Now what would happen if we considered just X_2? X_2 is also a binomial random variable where the number of trials is equal to 1. Since we have assumed that each inspector has the same probability p of judging a mattress acceptable, it follows that X_1 and X_2, although different random variables will have the same probability distribution and consequently the same mean and variance. That is, $E(X_2) = p$ and $V(X_2) = pq$. Clearly, the same reasoning would lead us to conclude that $E(X_3) = p$ and $V(X_3) = pq$. Now we are in a position to find $E(X)$ and $V(X)$. It follows from Corollary 3.1 that

$$E(X) = E(X_1 + X_2 + X_3) = E(X_1) + E(X_2) + E(X_3) = p + p + p = 3p.$$

Furthermore, since we have assumed independence, it follows from Corollary 3.2 that

$$V(X) = V(X_1 + X_2 + X_3) = V(X_1) + V(X_2) + V(X_3) = pq + pq + pq = 3pq.$$

It should not be hard for you to see that if n inspectors work on the shift, then

$$\begin{aligned} E(X) &= E(X_1 + X_2 + \ldots + X_n) \\ &= E(X_1) + E(X_2) + \ldots + E(X_n) = p + p + \ldots + p = np \end{aligned}$$

and

$$\begin{aligned} V(X) &= V(X_1 + X_2 + \ldots + X_n) \\ &= V(X_1) + V(X_2) + \ldots + V(X_n) = pq + pq + \ldots + pq = npq. \end{aligned}$$

It should be clear that in this example there is nothing very special about the fact that the binomial random variable came about in the context of testing mattresses. The same results are true for any binomial random variable.

Theorem 4.1 If X is a random variable having a binomial distribution with parameters n and p, then

$$E(X) = np \quad \text{and} \quad V(X) = npq. \tag{4.8}$$

Example 4.6 Find the expected value and variance of the number of times your football team will win the toss before the home football games next season (Example 4.2).

Since $n = 5$ and $p = .5$, it follows from (4.8) that

$$E(X) = np = 5(.5) = 2.5$$

and

$$V(X) = npq = 5(.5)(.5) = 1.25.$$

Example 4.7 A fair die is rolled 5 times. Find the expected value and standard deviation of the number of 6's that appear.

Since $n = 5$ and $p = \frac{1}{6}$, we have

$$E(X) = np = 5(\tfrac{1}{6}) = \tfrac{5}{6} = .83\bar{3}$$

and

$$\sigma_x = \sqrt{V(X)} = \sqrt{npq} = \sqrt{5(\tfrac{1}{6})(\tfrac{5}{6})} = \tfrac{5}{6} = .83\bar{3}.$$

4.3 THE MULTINOMIAL DISTRIBUTION

One frequently sees tires on sale in the automotive section of stores. Many times when you read the fine print at the bottom of a sales banner, it will say "factory seconds" or "slight blemishes." Frequently, these tires will carry the same guarantee as their unblemished cousins. What does this tell you about the inspection process at the manufacturing plant? It tells you that the tires are not simply classified into two categories as "defective" or "nondefective," but rather are classified into three (or possibly more) categories, perhaps "nondefective," "minor defect," and "serious defect."

You can see from this example that there are situations which are similar to a binomial but where the experiment has three or more outcomes of interest. By generalizing the conditions of a binomial slightly, we can get conditions necessary for random variables to have a multinomial distribution.

> Consider a sequence of n independent experiments where each experiment can be classified as having occurred in exactly one of k ways, say $C_1, C_2, \ldots, C_k$. Let p_i denote the probability that an experiment will result in outcome C_i and assume that p_i remains constant from one experiment to the next. If X_i is defined to be the number of times that outcome C_i occurs in the n experiments, the joint distribution of $X_1, X_2, \ldots, X_k$ is a *multinomial* distribution.

The joint probability distribution of $X_1, \ldots, X_k$ depends on the parameters $n, p_1, p_2, \ldots,$ and p_k. The joint distribution is given by extension of equation (4.7):

$$(4.9) \qquad P[X_1 = x_1, X_2 = x_2, \ldots, X_k = x_k \mid n, p_1, p_2, \ldots, p_k]$$

$$= \frac{n!}{x_1!\,x_2!\ldots x_k!} p_1^{x_1} p_2^{x_2} \ldots p_k^{x_k},$$

where each p_i is nonnegative, $\sum_{i=1}^{k} p_i = 1$, and each x_i is a nonnegative integer with $\sum_{i=1}^{k} x_i = n$.

Example 4.8 When tires are manufactured by the Goodstone Tire Company, 80% of the tires inspected are classified as good, 15% have minor defects, and the remainder have major defects. Assuming independence among tires manufactured, what is the probability that of the next 10 tires produced, 8 are good and one has a minor defect?

You can see that the conditions for a multinomial distribution are satisfied. In this case $k = 3$ and the outcomes of the experiment are: C_1 = tire is classified as good, C_2 = tire is classified as having a minor defect, C_3 = tire is classified as having a major defect. We also have $n = 10$ and X_1 = number out of 10 classified as good, X_2 = number out of 10 classified as having a minor defect, and X_3 = number out of 10 classified as having a major defect. It remains to identify $p_1, p_2,$ and p_3 and x_1, x_2, x_3. From reading the problem you can see that $p_1 = .80, p_2 = .15, x_1 = 8,$ and $x_2 = 1$. What about p_3 and x_3? These were not explicitly stated in the problem but they can be deduced from the conditions $\sum p_i = 1$ and $\sum x_i = n$. You can see that

p_3 must be .05 and x_3 must be 1. From (4.9), we get

$$P[X_1 = 8, X_2 = 1, X_3 = 1 \mid n = 10, p_1 = .80, p_2 = .15, p_3 = .05]$$
$$= \frac{10!}{8!\,1!\,1!}(.80)^8(.15)^1(.05)^1 = .1132.$$

This example illustrates the fact that only $k - 1$ of the values $x_1, \ldots, x_k$ and $k - 1$ of the values $p_1, \ldots, p_k$ need to be specified. Once these values are specified, the remaining x value and p value are uniquely determined. You can also see that when $k = 2$, the multinomial problem really becomes a binomial problem. Hence these two families of distributions are intimately related.

When discussing the binomial distribution, we only explicitly defined one random variable, the number of "successes," and one value of p, the probability of a success. Of course, in this situation there is another random variable, the number of "failures." Clearly, the number of successes totally determines the number of failures, just as the probability of a success, p, totally determines the probability of a failure. Since it suffices to know the number of successes and p, we chose to not explicitly talk about the number of failures.

Example 4.9 Six fair dice are tossed at once. What is the probability that one each of the faces 1 through 6 are facing up?

In this example we are interested in all six different outcomes (as opposed to a similar problem in Example 4.3, where there were essentially two outcomes of interest). Since the conditions of the multinomial distribution are satisfied and since we have $k = 6, n = 6, p_1 = p_2 = \ldots = p_6 = \frac{1}{6}$ and $x_1 = x_2 = \ldots = x_6 = 1$, it follows from (4.9) that

$$P[X_1 = 1, X_2 = 1, \ldots, X_6 = 1 \mid n = 6, p_1 = \tfrac{1}{6}, \ldots, p_6 = \tfrac{1}{6}]$$
$$= \frac{6!}{1!\,1!\ldots 1!}(\tfrac{1}{6})^1(\tfrac{1}{6})^1 \ldots (\tfrac{1}{6})^1 = .01543210.$$

Since the multinomial distribution is really a *joint* probability distribution for the random variables $X_1, X_2, \ldots, X_k$, you might wonder whether or not these random variables are independent and what the marginal distributions are like. (Of course, you might not wonder about these things, but we'll tell you about them anyway!) The question of independence can be answered rather easily on intuitive grounds. Consider the situation of Example 4.9, where six dice are tossed. If we were to give you information about the value of X_1, would that indirectly give you information about the values of $X_2, \ldots, X_6$? In particular, if you were told that X_1 equals 5 (i.e., five of the six dice turn up 1's), would this give you any information about $X_2, \ldots, X_6$? It sure would! In particular, it would tell you that exactly one of the random variables $X_2, \ldots, X_6$ will assume the value 1 and the rest will assume the value 0. This, of course, implies that the random variables $X_1, \ldots, X_6$ are not independent.

Finding the marginal distribution of X_1, say, is also a relatively easy matter. If we are only interested in X_1, the number of times outcome C_1 occurs, then we can classify getting outcome C_1 as a success and classify all other outcomes as "failures."

Note that this is exactly what we did in Example 4.3. Consequently, the probability distribution of X_1, when considered by itself, is a binomial distribution with parameters n and p_1. We can use this fact to find expressions for the mean and variance of each X_i.

Theorem 4.2 If $X_1, X_2, \ldots, X_k$ have a multinomial distribution with parameters $n, p_1, p_2, \ldots, p_k$, then

$$E(X_i) = np_i \quad \text{and} \quad V(X_i) = np_i(1 - p_i) \tag{4.10}$$

for $i = 1, 2, \ldots, k$.

Example 4.10 In Example 4.8 we determined that X_1, X_2, and X_3, the number of Goodstone tires classified as being good, having minor defects, and having major defects, respectively, would have a multinomial distribution with parameters $n = 10$, $p_1 = .80$, $p_2 = .15$, and $p_3 = .05$. Find the expected value and variance for each of these random variables.

Substituting the parameter values into equation (4.10), we find that

$$\begin{aligned} E(X_1) &= np_1 = 10(.80) = 8.0 & V(X_1) &= np_1(1 - p_1) = 1.600 \\ E(X_2) &= np_2 = 10(.15) = 1.5 & V(X_2) &= np_2(1 - p_2) = 1.275 \\ E(X_3) &= np_3 = 10(.05) = 0.5 & V(X_3) &= np_3(1 - p_3) = 0.475. \end{aligned}$$

4.4 THE NEGATIVE BINOMIAL DISTRIBUTION

Going to a carnival can be an enjoyable experience for children who like cotton candy, popcorn, and other such "food." It can be a painful experience for the parents who pay inflated prices for rides that were much better and longer just a few years back when they were kids. It can also be a learning experience for a student of probability and of human nature. It is clear to objective observers that even the games of "skill" are more like games of chance and that the odds strongly favor the operators and not the participant. One common game of skill is a "ring-toss" game (Figure 4.2), where you "toss 'til you win." Of course, the sticks that are easiest to get the ring on are the ones with the smallest prize (a plastic comb, a Superman ring, etc.) A student of probability might watch this game being played with an eye toward the number of rings a player tosses before getting a ring on a stick. Will a player always

FIGURE 4.2
Winner (?) at Ring Toss

get a ring on the first toss? On the second toss? Can you be sure? No. The number of tosses required to get a ring on a stick is a random variable.

The random variable described in the previous illustration has some characteristics in common with a random variable having a binomial distribution. In particular, we consider a sequence of independent experiments where each experiment results in a "success" or a "failure." In this illustration, however, the number of trials or experiments is not fixed, but is a random variable. You keep getting more tries until you get a success. You might think of a slightly modified version of watching this game. For example, if a person paid his money twice, he would get to toss until he obtained his second "success." Similarly, if he paid r times, he would get to toss until he obtained r successes.

Consider a sequence of independent experiments where each experiment can be classified in exactly one of two ways, success or failure, and where p, the probability that an experiment results in a success, remains constant from one experiment to the next. The random variable X defined to be the number of trials necessary to attain r successes will have a *negative binomial* distribution. (Some authors refer to this distribution as a "Pascal" distribution.)

The probability distribution of X will depend on the two parameters r and p. If you compare the conditions necessary for a random variable to have a binomial distribution with those necessary for a random variable to have a negative binomial distribution, you will see three identical conditions: (1) independent experiments, (2) dichotomous outcomes, and (3) constant probability of a success. The principal differences are that with the binomial distribution, n, the number of trials, is fixed and *number of successes* observed during those n trials *is a random variable*; with the negative binomial distribution, r, the number of successes, is fixed and the *number of trials* necessary to obtain those r successes *is a random variable*. [In keeping with our custom of using X to represent a random variable, we have used the same symbol to denote a binomial random variable and a negative binomial random variable (as well as several other random variables in Chapter 2). If in a particular problem we wish to consider *both* types of random variables, binomial and negative binomial, we will distinguish them by subscripts: X_B will denote a binomial random variable and X_{NB} will denote a negative binomial random variable.]

In order to find the probability distribution for a negative binomial random variable, we will start with a specific example.

Example 4.11 When Lucky Lucy plays ring toss she has a probability $p = .4$ of getting a ring on a stick. If she plays the game 3 times, what is the probability that she will require exactly 6 tosses to get her three prizes? Assume independence between tosses.

We have a negative binomial distribution with $r = 3$ and $p = .4$ and we are interested in $P[X = 6]$. Let us consider the event $E = \{X = 6\}$. If you are told that E has occurred, what do you know about what transpired on the sixth toss? Since Lucy stops throwing rings when she obtains her third success, it must be that she obtained a success (i.e., got a ring on) on the sixth toss. What can be said about the number of successes in the first five tries? There must have been *exactly* two successes in the first five. In view of these facts, we can write the event E as the intersection of two events,

$$E = \{X = 6\} = \{2 \text{ successes in first 5 tries}\} \cap \{\text{success on sixth try}\}.$$

Since, by assumption, there is independence between tosses, we can write

$$\begin{aligned} P[E] &= P[X = 6] \\ &= P[2 \text{ successes in first 5 tries}] \cdot P[\text{success on sixth try}]. \end{aligned} \tag{4.11}$$

To evaluate (4.11), we note that when we consider *the particular event* $\{X = 6\}$, we need to find P[2 successes in 5 tries]. The "5" tries is a fixed number of trials, so the probability of 2 successes in 5 tries can be evaluated by using the binomial distribution. Evaluating the probability of a success on the sixth trial is easy, since the probability of a success on any single trial is $p = .4$. This allows us to evaluate (4.11) as

$$\begin{aligned} P[X = 6] &= P[2 \text{ successes in first 5 tries}] \cdot P[\text{success on sixth try}] \\ &= \left[\binom{5}{2} p^2(1-p)^3\right] \cdot p \\ &= \left[\binom{5}{2}(.4)^2(.6)^3\right] \cdot (.4) = (.3456)(.4) = .13824. \end{aligned}$$

With this example in mind, we can now find a general expression for the probability distribution of a negative binomial random variable X with parameters r and p. To find $P[X = x \mid r, p]$, remember that x is a fixed value that the random variable X can assume. It should be clear that to get r successes you will need at least r tries, so the values X can assume are $r, r + 1, r + 2, \ldots$. The event $\{X = x\}$ is the event that "the rth success occurs on trial number x." Of course, this means that on trial number x, a success is observed. What does it mean about the number of successes that occurred on the first $(x - 1)$ trials? It means that there must have been $(r - 1)$ successes in the first $(x - 1)$ trials. Since all trials are independent, it follows that we can write

$$\begin{aligned} P[X = x \mid r, p] \\ &= P[r - 1 \text{ successes in first } x - 1 \text{ tries}] \cdot P[\text{success on } x\text{th try}]. \end{aligned} \tag{4.12}$$

Just as in the example with Lucky Lucy, we evaluate (4.12) by noting that the first factor on the right-hand side of (4.12) can be treated as a binomial probability and that the second factor is simply p, the probability of a success on any try. Since $x - 1$ represents the (fixed) number of trials and $r - 1$ represents the number of successes, it follows from (4.7) that

$$\begin{aligned} P[r - 1 \text{ successes in first } x - 1 \text{ tries}] &= \binom{x-1}{r-1} p^{r-1} q^{(x-1)-(r-1)} \\ &= \binom{x-1}{r-1} p^{r-1} q^{x-r}. \end{aligned}$$

Hence we can write (4.12) as

$$P[X = x \mid r, p] = \left[\binom{x-1}{r-1} p^{r-1} q^{x-r}\right] \cdot p \tag{4.13}$$

$$= \binom{x-1}{r-1} p^r q^{x-r} \qquad x = r, r + 1, r + 2, \ldots. \tag{4.14}$$

Example 4.12 Elmer Phudd is going duck hunting. Whenever he shoots there is a .05 probability that he will down a duck. If there is a limit of two ducks/day, what is the probability that he will use exactly 3 shots to get his limit? What is the probability that he will need more than 3 shots to get his limit?

Since Elmer will keep trying until he has gotten two successes, it should be clear that X, the number of shots necessary to get two successes, has a negative binomial distribution. What are the parameters in this problem? The parameters are $r = 2$ and $p = .05$. To solve the first part of the problem, we use (4.13) with $x = 3$. [Since we plan to evaluate $P[X = 3]$ using binomial tables to get the first factor, we use (4.13). If we were to evaluate the probability directly, we would use (4.14).] This gives

$$P[X = 3 \mid r = 2, p = .05] = \left[\binom{3-1}{2-1}(.05)^1(.95)^1\right](.05)$$

$$= \left[\binom{2}{1}(.05)^1(.95)^1\right](.05) = (.0950)(.05) = .00475.$$

To solve the second part of the problem, we need to find

$$P[X > 3] = P[X \geq 4] = P[X = 4] + P[X = 5] + P[X = 6] + \dots.$$

(Here, for convenience, we have not rewritten the parameters r and p each time.) We see that we must evaluate an infinitely long series. (Actually, we would only evaluate terms until they had zeros in the first four decimal places, for example.) Can you think of a better way to solve this problem? We could solve this problem by finding the probability of the complement of this event and using Theorem 1.3. Doing this, we get

$$P[X > 3] = 1 - P[X \leq 3] = 1 - \{P[X = 2] + P[X = 3]\}. \tag{4.15}$$

Since we have already found $P[X = 3]$, we need only find $P[X = 2]$. Using (4.13), we get

$$P[X = 2] = \left[\binom{2-1}{2-1}(.05)^1(.95)^0\right](.05)$$

$$= \left[\binom{1}{1}(.05)^1(.95)^0\right](.05) = (.05)(.05) = .0025.$$

Using these results in (4.15), we get

$$P[X > 3] = 1 - (.00250 + .00475) = .99275.$$

The algebra involved in finding the mean and variance of a random variable having a negative binomial distribution is somewhat complicated, so we will merely state the expressions for the mean and variance without proof.

Theorem 4.3 If X is a random variable having a negative binomial distribution with parameters r and p, then

$$E(X) = \frac{r}{p} \quad \text{and} \quad V(X) = \frac{rq}{p^2}. \tag{4.16}$$

Example 4.13 What are the mean and variance of the number of tosses that Lucy will need to make to win three prizes?

From Example 4.11 we see that $r = 3$ and $p = .4$, so

$$E(X) = \frac{r}{p} = \frac{3}{.4} = 7.5$$

and

$$V(X) = \frac{rq}{p^2} = \frac{3(.6)}{(.4)^2} = 11.25.$$

Example 4.14 What are the mean and standard deviation of the number of shots that Elmer will need to take to get his limit?

From Example 4.12 we see that $r = 2$ and $p = .05$, so

$$E(X) = \frac{r}{p} = \frac{2}{.05} = 40$$

and

$$\sigma_X = \sqrt{\frac{rq}{p^2}} = \sqrt{\frac{2(.95)}{(.05)^2}}$$

$$= \sqrt{760} = 27.6.$$

Example 4.15 To ascertain the quality of a large shipment of potatoes, an inspector for the Pot-O-Potatoes Co., C. M. Good, cuts open potatoes one at a time until a bad potato is found. If there is no limit on the number of potatoes C. M. inspects and if the probability of finding a bad potato is .10, how many potatoes can we expect C. M. to inspect from each shipment? What is the probability that 4 or fewer will be inspected? What are the answers to these questions if the probability of finding a bad potato is .50?

If we assume that the probability of getting a bad potato remains constant throughout the shipment, then X, the number of potatoes that must be cut open to find a bad potato, will have a negative binomial distribution with parameters $r = 1$ and $p = .10$. It follows from (4.16) that

$$E(X) = \frac{r}{p} = \frac{1}{.10} = 10.$$

To find the probability that 4 or fewer will be inspected, we will need to take a sum of terms. Using (4.13) and the binomial tables, we get

$$P[X \leq 4 \mid r = 1, p = .10] = \sum_{x=1}^{4} \left[\binom{x-1}{r-1} p^{r-1} q^{x-r}\right] \cdot p$$

$$= \left[\sum_{x=1}^{4} \binom{x-1}{0} (.1)^0 (.9)^{x-1}\right](.1)$$

$$= (1 + .900 + .810 + .729)(.1) = .3439.$$

If the value of p is .50, we get

$$E(X) = \frac{r}{p} = \frac{1}{.50} = 2$$

and

$$P[X \leq 4 \mid r = 1, p = .5] = \left[\sum_{x=1}^{4} \binom{x-1}{0} (.5)^0 (.5)^{x-1}\right](.5)$$

$$= (1 + .500 + .250 + .125)(.5) = .9375.$$

4.5 THE HYPERGEOMETRIC DISTRIBUTION

In Example 4.15 we assumed that each potato had a given probability of actually being bad. This is similar to the situation with the watzit maker, where each watzit produced had a given probability of coming out defective. Prior to examination of a particular shipment of potatoes (or watzits), the actual number of bad potatoes (or defective watzits) is unknown. There are situations, however where the number of items having various attributes are known in advance of examination.

Example 4.16 Pat boiled a dozen eggs to take on a picnic. Jody, one of Pat's children, decided to play a practical joke and replaced 3 of the hard-boiled eggs with uncooked eggs. Pat is about to try to peel 4 eggs chosen from the dozen. What is the probability that all 4 eggs are actually hard-boiled?

In this situation we (and Jody, too, but not Pat) know that this particular dozen eggs contains 9 eggs having one attribute (hard-boiled) and 3 eggs having a second attribute (uncooked). (Note how this differs from a situation where you take 12 watzits from the assembly line but don't know *before inspection* how many are good and how many are defective.) If we define a random variable X to be the number of raw eggs among the 4 eggs chosen by Pat, we are interested in the probability of the event $\{X = 0\}$. To find this probability, we make the assumption that the eggs are chosen randomly from the carton so that all subsets of 4 eggs from 12 are equally likely to be chosen. By using the "equally likely outcomes" approach of Section 1.5, we see that there are

$$\binom{12}{4} = \frac{12!}{4!\,8!} = 495$$

ways that 4 eggs can be chosen from 12,

$$\binom{9}{4} = \frac{9!}{4!\,5!} = 126$$

ways that 4 hard-boiled eggs could be chosen from 9 hard-boiled eggs, and

$$\binom{3}{0} = \frac{3!}{0!\,3!} = 1$$

way of choosing no uncooked eggs from the 3 uncooked eggs, so

$$P[X = 0] = \frac{\binom{3}{0}\binom{9}{4}}{\binom{12}{4}} = \frac{(1)(126)}{495} = .2545.$$

In this example you see that the problem involves taking a sample of objects from a group of objects without replacement. The objects can be classified as having (or not having) a particular attribute. The random variable of interest is the number of objects *in the sample* having a particular attribute.

Consider a collection of N objects where A of these objects have a certain attribute and the remaining $N - A$ objects do not have this attribute. If a sample of n objects are chosen at random and without replacement from this collection of objects, then X, the number of objects in the sample having the attribute, is a random variable having a *hypergeometric distribution.*

To find the probability distribution for X, we use arguments similar to those advanced in Example 1.22 of Section 1.5 and Example 4.16. Since the n objects are chosen randomly from the N objects available, there are $\binom{N}{n}$ different possible subsets of n objects which could be chosen. To find $P[X = x]$, we need to know the number of these subsets that have x objects having the attribute (and $n - x$ objects not having the attribute). There are $\binom{A}{x}$ ways of choosing x objects from the A having the attribute and $\binom{N-A}{n-x}$ ways of choosing $n - x$ objects from the $N - A$ not having the attribute. The quantities n, N, and A are parameters of this distribution, as indicated by the following notation:

$$P[X = x \mid n, N, A] = \frac{\binom{A}{x}\binom{N-A}{n-x}}{\binom{N}{n}}. \tag{4.17}$$

The expression given in (4.17) is well defined only if each of the combinations symbols on the right-hand side of (4.17) is well defined. This means that x must satisfy the following two relationships:

$$0 \leq x \leq A \quad \text{and} \quad 0 \leq n - x \leq N - A.$$

Example 4.17 A dish of M & M candies has been sitting on the table for several hours and the children have been taking and eating their favorite colored candies. The dish now contains 25 candies, of which 15 are brown and 10 are green. If Chris takes three pieces of candy at random, what is the probability that exactly 2 are brown?

In this case we are considering a collection of 25 pieces of candy, 15 of which have the attribute of being brown and the remaining 10 candies have the attribute of being green, or, more to the point, do not have the attribute of being brown. Since we are assuming that Chris takes the pieces of candy at random and since in most cases candy is chosen without replacement, it is reasonable to assume that X, the number of brown M & M's, will have a hypergeometric distribution. Using (4.17), we get

$$P[X = 2 \mid n = 3, N = 25, A = 15] = \frac{\binom{15}{2}\binom{10}{1}}{\binom{25}{3}} = \frac{\frac{15 \cdot 14}{2 \cdot 1} \cdot \frac{10}{1}}{\frac{25 \cdot 24 \cdot 23}{3 \cdot 2 \cdot 1}}$$

$$= \frac{1050}{2300} = .4565.$$

Example 4.18 A 40-gram bag of M & M's was found to contain 20 brown, 9 yellow, 5 tan, 10 orange, and 5 green M & M's. If Chris chooses 3 at random from this bag, what is the probability that exactly 2 are brown? What is the probability that none are orange?

Here we see that the candies can be classified according to color into five categories. However, we can still treat this problem as a hypergeometric distribution by considering candies to be brown or not brown in the first part of the problem and orange or not orange in the second part. For the first part we get

$$P[X = 2 \mid n = 3, N = 49, A = 20] = \frac{\binom{20}{2}\binom{29}{1}}{\binom{49}{3}} = \frac{(190)(29)}{18424} = .2991,$$

while in the second part we get

$$P[X = 0 \mid n = 3, N = 49, A = 10] = \frac{\binom{10}{0}\binom{39}{3}}{\binom{49}{3}} = \frac{(1)(9139)}{18424} = .4960.$$

By using a method similar to that used in Section 4.2 to find the expected value of a random variable having a binomial distribution, it is possible to find the expected value of a random variable having a hypergeometric distribution. Finding the variance of a hypergeometric random variable is a bit more complicated, so we will not present any proofs, but merely state these results in a theorem.

Theorem 4.4 If X is a random variable having a hypergeometric distribution with parameters n, N, and A, then

$$(4.18) \qquad E(X) = n\left(\frac{A}{N}\right) \quad \text{and} \quad V(X) = n\left(\frac{A}{N}\right)\left(1 - \frac{A}{N}\right)\left(\frac{N-n}{N-1}\right).$$

Example 4.19 Pat chooses 4 eggs from a carton containing 9 hard-boiled eggs and 3 uncooked eggs. Find the expected value and variance of the number of uncooked eggs among the 4 chosen by Pat (cf. Example 4.16).

In this case we know that X, the number of uncooked eggs in the sample of 4, has a hypergeometric distribution with parameters $n = 4$, $N = 12$, and $A = 3$. It follows from (4.18) that

$$E(X) = n\left(\frac{A}{N}\right) = 4\left(\frac{3}{12}\right) = 1$$

and

$$V(X) = n\left(\frac{A}{N}\right)\left(1 - \frac{A}{N}\right)\left(\frac{N-n}{N-1}\right) = 4\left(\frac{3}{12}\right)\left(1 - \frac{3}{12}\right)\left(\frac{12-4}{12-1}\right)$$
$$= 4\left(\frac{3}{12}\right)\left(\frac{9}{12}\right)\left(\frac{8}{11}\right) = \frac{6}{11}.$$

Example 4.20 Sweet Georgia Brown is selling boxes of candy, of which, according to the distributor from whom she buys the candy, at least half are chocolate-covered cherries. She chooses, at random, 10 pieces of candy from a box containing 40 pieces and finds that only 1 of the pieces is a cherry. If the box actually contained half cherries, how many cherries should she have expected to get? What is the probability that she would get 1 cherry or less? What does this suggest about the distributor's claim?

Since the number of pieces of candy that are cherries in the 10 chosen, X, is a random variable having a hypergeometric distribution, we need to identify the parameters of this distribution. We have $n = 10$, $N = 40$, and $A = 20$. From (4.18) we see that the expected value is

$$E(X) = n\left(\frac{A}{N}\right) = 10(\tfrac{20}{40}) = 5.$$

The probability of getting 1 cherry or less is

$$P[X \leq 1] = \sum_{x=0}^{1} \frac{\binom{A}{x}\binom{N-A}{n-x}}{\binom{N}{n}} = \sum_{x=0}^{1} \frac{\binom{20}{x}\binom{20}{10-x}}{\binom{40}{10}}$$

$$= \frac{\binom{20}{0}\binom{20}{10}}{\binom{40}{10}} + \frac{\binom{20}{1}\binom{20}{9}}{\binom{40}{10}} = \frac{1{,}771{,}978}{847{,}660{,}528} = .00209.$$

Since this probability is so small, we can see that it is rather rare to get 1 or fewer cherries if, in fact, the box contains half cherries. We might conclude that the distributor was misrepresenting the number of cherries in these boxes of candy.

In situations where candy is being sampled, it is certainly not unreasonable to assume that sampling is done without replacement. However, there are situations when sampling is done *with replacement*. As a specific example, we might assume that Chris randomly chooses an M & M from a dish containing 20 brown candies and 30 candies that are not brown, and that the sampling is done with replacement. In this case the probability of getting a brown M & M is $p = .4$, and this probability remains constant from one trial to the next. In this case the trials are independent and it follows that X, the number of brown candies in a sample of size n, will have a binomial distribution. Note that if sampling were done *without replacement*, the trials would not be independent and X would have a hypergeometric distribution. It is useful to compare these two situations, so let us consider two random variables: X_B, the number of brown candies in a sample of size n when sampling *with* replacement and X_H, the number of brown candies in a sample of size n when sampling *without* replacement. Note that in sampling without replacement, the largest sample we can choose is equal to N. Also note that p, the parameter for X_B, will be equal to A/N. Comparing means and variances we see that

$$E(X_B) = np = n\left(\frac{A}{N}\right), \qquad E(X_H) = n\left(\frac{A}{N}\right)$$

and

$$V(X_B) = npq = n\left(\frac{A}{N}\right)\left(1 - \frac{A}{N}\right), \qquad V(X_H) = n\left(\frac{A}{N}\right)\left(1 - \frac{A}{N}\right)\left[\left(\frac{N-n}{N-1}\right)\right].$$

Consequently, the expected values are the same for both cases, but the variances differ by a factor of $[(N - n)/(N - 1)]$. You can see that if n is small relative to the magnitude of N, then $(N - n)/(N - 1)$ will be relatively close to 1, so the variances of X_B and X_H will be almost equal. [For example, if $n \leq .10N$, then $(N - n)/(N - 1)$ will be greater than or equal to .90.] Since the means and variances are close and since the random variables are measuring the same quantity (in this example, the number of brown candies), we might expect the probability distributions to be close. This is true, as the following example indicates.

Example 4.21 Assume that Chris' dish of M & M's contains 100 candies, of which 40 are brown. If a sample of 10 candies is chosen, compare $P[X_B = x]$ with $P[X_H = x]$ for $x = 1, 4,$ and 7.

We have $n = 10$, $A = 40$, $N = 100$, and $p = 40/100 = .4$. The probabilities for X_B are easily found from the tables to be

$$P[X_B = 1] = .0403, \qquad P[X_B = 4] = .2508, \qquad P[X_B = 7] = .0425.$$

Using (4.17), we find

$$P[X_H = 1] = \frac{\binom{40}{1}\binom{60}{9}}{\binom{100}{10}} = .0342, \qquad P[X_H = 4] = \frac{\binom{40}{4}\binom{60}{6}}{\binom{100}{10}} = .2643,$$

$$P[X_H = 7] = \frac{\binom{40}{7}\binom{60}{3}}{\binom{100}{10}} = .0369.$$

Comparing these results, we can see that the values of the probabilities are relatively close.

If you were to actually work out some of the foregoing hypergeometric probabilities, you would see that the calculations involve a number of multiplications and divisions and take a bit of time to do. As a time-saving device, there is a definite advantage to using binomial probabilities as an *approximation* to hypergeometric probabilities. The approximation is satisfactory for $n \leq .10N$ (assuming that A is neither very close to zero nor very close to N) and the approximation is excellent if $n \leq .05N$.

Example 4.22 Seven hundred out of 1000 registered voters in Kleentown favor an antilitter proposition to be voted on in the next election. If 20 voters are selected at random and interviewed, what is the probability that 15 will favor the proposition?

Although it is not explicitly stated, it is reasonable to assume that sampling will be done without replacement. (It might prove embarrassing to ask the same persons twice whether they favor the proposition. Besides, it is intuitively clear that you gain a bit more information in the long run by sampling without replacement.) Since the sampling is done without replacement, we know that the hypergeometric distribution should be used to calculate the exact probability of interest. However, evaluating

$$\binom{700}{15}\binom{300}{5}\bigg/\binom{1000}{20}$$

is no simple matter, so we might consider using a binomial approximation. Since $n = 20$ is less than $.05N = 50$, we can expect the approximation to be very good. Since $n = 20$, $A = 700$, and $N = 1000$, we take $p = (A/N) = .7$ and get

$$P[X_H = 15 \mid n = 20, A = 700, N = 1000] \approx P[X_B = 15 \mid n = 20, p = .7] = .1789.$$

[Long and tedious calculations show the exact answer (to four decimal places) to be .1801.]

You will recall that the multinomial distribution is a generalization of the binomial distribution which is appropriate when the experiment under consideration has more than two possible outcomes. In a similar way, there is a distribution that can be used to generalize the hypergeometric distribution when the objects in a collection can be classified as having exactly one of k different attributes rather than one of two attributes.

Consider a collection of N objects where A_1 of these have one attribute, A_2 have a second attribute, . . . , and the remaining A_k have a kth attribute. If a sample of n objects is chosen at random and without replacement from this collection of objects and if

$$X_1 = \text{number of objects in the sample having attribute } 1,$$
$$X_2 = \text{number of objects in the sample having attribute } 2,$$
$$\vdots$$
$$X_k = \text{number of objects in the sample having attribute } k,$$

then the joint distribution of these random variables will be an *extended hypergeometric distribution* (or a multivariate hypergeometric distribution).

The joint probability distribution for the random variables $X_1, \ldots, X_k$ can be given by generalizing (4.17) as follows:

$$(4.19) \quad P[X_1 = x_1, X_2 = x_2, \ldots, X_k = x_k \mid n, N, A_1, A_2, \ldots, A_k] = \frac{\binom{A_1}{x_1}\binom{A_2}{x_2}\cdots\binom{A_k}{x_k}}{\binom{N}{n}},$$

where $0 \leq x_i \leq A_i$ for $i = 1, 2, \ldots, k$, $\sum_{i=1}^{k} A_i = N$, and $\sum_{i=1}^{k} x_i = n$.

Example 4.23 In Example 4.18 we described the contents of a 40-gram bag of M & M's in terms of the number of each of five different colors. If Chris chooses 3 candies at random from this bag, what is the probability of getting 1 brown, 1 yellow, and 1 orange?

In this case the different colors will be the different attributes in question, so $k = 5$. In the order listed in Example 4.18, we have

$$A_1 = 20, \qquad A_2 = 9, \qquad A_3 = 5, \qquad A_4 = 10, \qquad A_5 = 5.$$

It follows from the statement of the problem that $x_1 = x_2 = x_4 = 1$ and $x_3 = x_5 = 0$, so using (4.19), we get

$$P[X_1 = 1, X_2 = 1, X_3 = 0, X_4 = 1, X_5 = 0 \mid n = 3, N = 49, A_1 = 20, A_2 = 9, A_3 = 5, A_4 = 10, A_5 = 5]$$

$$= \frac{\binom{20}{1}\binom{9}{1}\binom{5}{0}\binom{10}{1}\binom{5}{0}}{\binom{49}{3}} = \frac{(20)(9)(1)(10)(1)}{18{,}424} = .0977.$$

4.6 THE POISSON DISTRIBUTION

Georgia Brown (Example 4.20) is contemplating the possibility of putting up a small candy stand near one end of a small shopping mall. She is concerned about the number of customers who might come to her stand in a 5-minute time period. She feels that if the number is too small, she will not be able to make enough profit, and if the number is too large, she may lose impatient customers or have to hire someone to help her wait on customers. You might try to think of ways that you could get information that would help Georgia make her decision as to whether or not to put up a candy stand. In the meantime, imagine that the candy stand has been put up. Is it possible to predict *exactly* how many customers will come to the stand in a certain time period? If you think like we do, you will conclude that the answer is no. If this is so, the number of customers arriving in a certain time period must be a random variable. One question we might ask about this random variable is: What are the possible values that this random variable can assume? The values could be 0, 1, 2, and so on. (Practically speaking, of course, there would be some upper limit to the possible values, but from a theoretical point of view, any positive integer could be allowed.)

If we define X to be the number of customers coming to Georgia's stand in a certain time period, we would like to be able to find the probability distribution of X. How could we assign a value to $P[X = 1]$, for example? On what factors would this probability depend? Two factors come to mind: the average *rate* of arrivals of customers and the *length* of the time period. It seems reasonable to assume that the average rate of arrivals (measured in terms of customers/minute or perhaps customers/hour) will differ for different hours of the day. The rate might be higher in the afternoon than in the morning, for instance. (In the following discussions we will assume that the *rate remains approximately constant* over the whole interval being considered.) Also, a longer time period should increase the probability of having more customers. We will see that the probability distribution of X will depend directly on these two factors.

Consider a process where some well-defined events occur at a certain fixed rate, λ. Assume that these events occur singly (i.e., one at a time) and are occurring during a certain time interval or over a certain space interval, t. If the random variable X is defined to be the number of these events occurring over a specified interval, the distribution of X will be a Poisson distribution with parameters λ and t.

Actually, we have simplified the statement of the conditions under which a random variable will have a Poisson distribution. Certain mathematical conditions must be satisfied before we can say that a random variable has a Poisson distribution. We will mention these conditions briefly in Section 4.10, but the aforementioned conditions should be satisfactory for our purposes. The probability distribution for X is given by

$$(4.20) \qquad P[X = x \mid \lambda, t] = \frac{e^{-\lambda t}(\lambda t)^x}{x!} \qquad x = 0, 1, 2, \ldots.$$

These probabilities can be found in Table A3.

Example 4.24 Assume that customers arrive at Georgia's stand at the rate of 30/hour. What is the probability that exactly 3 customers will arrive at her stand in the next 5 minutes?

If we define X to be the number of arrivals in the next 5 minutes, then X will have a Poisson distribution. The thing to be most careful about here is to be sure that λ and t are expressed in the same kind of units. Since λ is given as 30 per *hour* and t is given as 5 *minutes*, it is necessary to change the units on one of these two quantities. We could write λ as $\frac{1}{2}$ per minute to get

$$\lambda t = (\tfrac{1}{2}/\text{minute}) \cdot (5 \text{ minutes}) = \tfrac{5}{2} = 2.5,$$

or we could write t as $\frac{1}{12}$ of an hour to get

$$\lambda t = (30/\text{hour}) \cdot (\tfrac{1}{12} \text{ hour}) = \tfrac{30}{12} = 2.5.$$

As you can see, the answer is the same in either case. Reading from the tables we find that

$$P[X = 3 \mid \lambda t = 2.5] = .2138.$$

Earlier we stated that *under certain conditions* a random variable X will have a Poisson distribution. If these conditions are met perfectly, the Poisson probabilities will perfectly describe the probability distribution of X. If the conditions are only approximately satisfied, then the Poisson probabilities will only approximately, although for practical purposes adequately, describe the probability distribution of X. The following are situations where the Poisson distribution may adequately describe the random variable of interest:

The number of cosmic rays striking a satellite in a 1-minute interval.
The number of cars leaving an exit ramp of an interstate highway in 2 minutes.
The number of pickup trucks passing an underpass in a 5-minute interval.
The number of hurricanes striking the Gulf Coast in a season.
The number of bacteria on a 1-mm² portion of a slide.
The number of flaws in a 4-ft by 8-ft sheet of plywood.
The number of creosote bushes in a 10-m² portion of a desert.
The number of chocolate chips in a 2-inch-diameter cookie.

This, of course, is just a small portion of the list of possibilities. Note that the first four are examples of events occurring over *time* intervals, while the last four are examples of events occurring over *space* intervals.

To find expressions for the mean and variance of a random variable having a Poisson distribution, it is necessary to use a couple of algebraic tricks. We will give these expressions in the next theorem, but we will not go through all the proof. We will just show how the expected value can be found.

Theorem 4.5 If X is a random variable having a Poisson distribution with parameters λ and t, then

$$E(X) = \lambda t \quad \text{and} \quad V(X) = \lambda t. \tag{4.21}$$

Proof: Since for any probability distribution $\sum P[X = x] = 1$, we know that

$$\sum_{x=0}^{\infty} P[X = x \mid \lambda t] = \frac{e^{-\lambda t}(\lambda t)^0}{0!} + \frac{e^{-\lambda t}(\lambda t)^1}{1!} + \frac{e^{-\lambda t}(\lambda t)^2}{2!} + \ldots = 1. \tag{4.22}$$

We will use this result in the following. Using the definition of $E(X)$, Definition 2.6, we get

$$\begin{aligned} E(X) &= \sum_{x=0}^{\infty} x \cdot P[X = x \mid \lambda t] \\ &= 0 \cdot P[X = 0 \mid \lambda t] + 1 \cdot P[X = 1 \mid \lambda t] + 2 \cdot P[X = 2 \mid \lambda t] + \ldots \\ &= 0 + 1 \cdot P[X = 1 \mid \lambda t] + 2 \cdot P[X = 2 \mid \lambda t] + 3 \cdot P[X = 3 \mid \lambda t] + \ldots \\ &= 1 \cdot \frac{e^{-\lambda t}(\lambda t)^1}{1!} + 2 \cdot \frac{e^{-\lambda t}(\lambda t)^2}{2!} + 3 \cdot \frac{e^{-\lambda t}(\lambda t)^3}{3!} + \ldots . \end{aligned} \tag{4.23}$$

At this point when you examine (4.23) you see that a factor λt can be factored from each term and that the coefficient in each term is the same as the factorial in the denominator of each term. Factoring out λt and rewriting the factorial term in the denominator, we have that (4.23) is equal to

$$\begin{aligned} &\lambda t \left\{ \frac{1 \cdot e^{-\lambda t}(\lambda t)^0}{1!} + \frac{2 \cdot e^{-\lambda t}(\lambda t)^1}{2 \cdot 1!} + \frac{3 \cdot e^{-\lambda t}(\lambda t)^2}{3 \cdot 2!} + \ldots \right\} \\ &\qquad = \lambda t \left\{ \frac{e^{-\lambda t}(\lambda t)^0}{0!} + \frac{e^{-\lambda t}(\lambda t)^1}{1!} + \frac{e^{-\lambda t}(\lambda t)^2}{2!} + \ldots \right\} \\ &\qquad = \lambda t. \end{aligned}$$

The last equality follows from the fact that, according to (4.22), the expression in braces is equal to 1.

Example 4.25 Find the mean and variance of the number of customers arriving at Georgia's stand in a 5-minute period.

Since we found in Example 4.24 that $\lambda t = 2.5$, it follows from the equalities given in (4.21) that

$$E(X) = 2.5 \quad \text{and} \quad V(X) = 2.5.$$

★***Example 4.26*** Twenty-one raisin cookies taken from a larger batch of cookies were given to students in a statistics class. Each student counted the raisins in his cookie and a total of 98 raisins were found. Find the approximate probability that a cookie from this batch will contain 2 or fewer raisins.

If the cookies are of approximately the same size and if the raisins were well mixed throughout the dough, the number of raisins per cookie, X, should follow a Poisson distribution approximately. To find probabilities for X, we need to know the value of λt. Since λt is equal to $E(X)$, the *long-run average* number of raisins per cookie, it is not unreasonable to estimate the value of λt by using the "short-run" average number of raisins per cookie in the cookies examined. Since the average is $\frac{98}{21} = 4.6\bar{6}$, we will take λt to be equal to 4.7, the value in the tables closest to $4.6\bar{6}$. Hence the approximate probability that a cookie will contain 2 or fewer raisins is

$$
\begin{aligned}
P[X \leq 2 \mid \lambda t = 4.7] &= \sum_{x=0}^{2} P[X = x \mid \lambda t = 4.7] \\
&= .0091 + .0427 + .1005 = .1523.
\end{aligned}
$$

(The numbers given in this example were actually obtained in a class experiment we did. As you can see, real data do not always yield neat solutions as do artificial data. C'est la vie.)

4.7 THE UNIFORM DISTRIBUTION

The special probability distributions we have discussed up to this point have all corresponded to discrete random variables. In this section, and in the following two sections, we will describe some special continuous probability distributions.

Consider a simple experiment where you stand in front of a store and record the time, measured in seconds after the minute that customers enter the store. If you measured the time only to the nearest second, the possible values you could observe would be 1, 2, 3, . . . , 60. Are any of these times more likely to occur than any other? No. Since it is reasonable to assume that these times are all equally likely, we would assign a probability of $\frac{1}{60}$ to each. If you were able to measure time to any degree of accuracy, the time you could observe, X, would be any value in the interval from 0 to 60 seconds. Since time is continuous, we would like to find a continuous probability density function which would reflect the fact that all these possibilities are "equally likely." This can be done by defining

$$
f(x) = \begin{cases} \frac{1}{60} & 0 \leq x \leq 60 \\ 0 & \text{otherwise.} \end{cases}
$$

The graph of $f(x)$ is given in Figure 4.3. The "equally likely" nature of this random variable can be seen by considering the probability that X will fall in some interval of length l. As long as the interval is kept entirely between 0 and 60, the probability that X will fall in the interval will always be $l/60$, no matter where the interval is located. Furthermore, this will be true no matter what the length of the interval.

If X is a random variable that can assume values only between the numbers a and b and if the probability that X assumes a value in any interval of length l is the same no

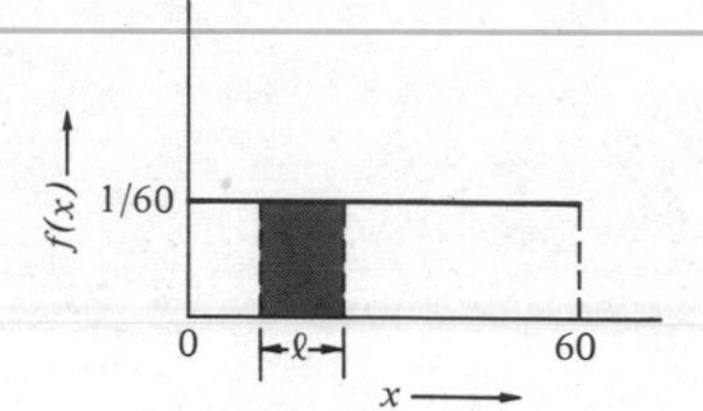

FIGURE 4.3
Example of a pdf for a Uniform Distribution

matter where the interval is located (between a and b), then X is said to have a *uniform distribution* over the interval (a, b).

The numbers a and b are parameters of this distribution and the probability density function of a random variable having this distribution is

$$f(x) = \begin{cases} \dfrac{1}{b-a} & a \leq x \leq b \\ 0 & \text{otherwise.} \end{cases}$$

The graph of this general pdf is given in Figure 4.4. In the situation described at the beginning of this section the parameters would be $a = 0$ and $b = 60$. In Example 2.8 of Section 2.3 we gave a simple pdf which was used in an effort to describe the amount of cola put into a 16-ounce bottle. In that example the pdf was for a uniform random variable having $a = 15$ and $b = 17$ (cf. Figure 2.3).

To find the mean of a random variable having a uniform distribution, we need only note that the pdf is symmetric about the midpoint between a and b, so the mean is equal to $(a + b)/2$. (This idea was discussed in Section 2.6.) To find the variance, we can use the fact that

$$E(X^2) = \frac{b^3 - a^3}{3(b-a)}$$

and then use equation (2.33) to get

$$\begin{aligned} V(X) &= E(X^2) - [E(X)]^2 \\ &= \frac{b^3 - a^3}{3(b-a)} - \left(\frac{a+b}{2}\right)^2 = \frac{(b-a)^2}{12}. \end{aligned}$$

These results are summarized in the next theorem.

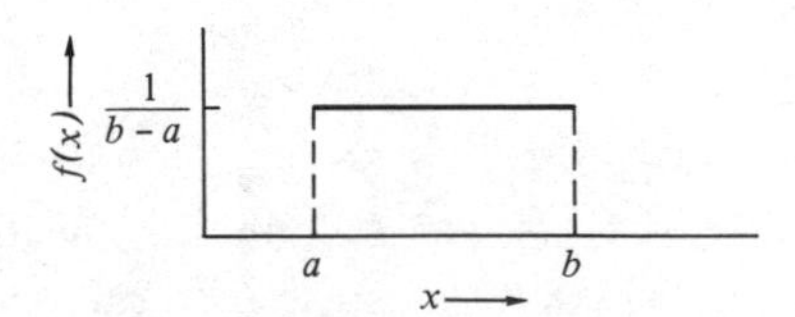

FIGURE 4.4
Graph of the pdf of a Random Variable Having a Uniform Distribution

Theorem 4.6 If X is a random variable having a uniform distribution with parameters a and b, then

(4.24) $$E(X) = \frac{a+b}{2} \quad \text{and} \quad V(X) = \frac{(b-a)^2}{12}.$$

Example 4.27 The Robinson family always uses 4-ounce tubes of Grit toothpaste. Let X denote the number of ounces remaining in the tube currently being used by the family. Find the mean and standard deviation of X.

In this situation it is not unreasonable to assume that X has a uniform distribution over the interval (0, 4), so $a = 0$ and $b = 4$. Using (4.24), we find that

$$E(X) = \frac{a+b}{2} = \frac{0+4}{2} = 2$$

and

$$\sigma_X = \sqrt{V(X)} = \sqrt{\frac{(b-a)^2}{12}} = \sqrt{\frac{(4-0)^2}{12}} = 1.155.$$

Example 4.28 Mr. Earl E. Rizer is planning to get an early start for the family camping trip tomorrow. He hopes to get up at 5:00 A.M. Unfortunately, his four-year-old son Joe set the family alarm clock when he went to bed at 8:00 P.M. The unfortunate part is that Joe does not know how to set alarm clocks. Assuming that the time for the alarm to go off was set randomly, what is the expected time for the alarm to ring? What is the probability that the alarm will ring later than Earl wants (i.e., after 5 o'clock)?

If we assume that the time of the alarm was chosen at random, a uniform distribution is appropriate. The problem that we encounter in this example is how to distinguish 8 P.M. from 8 A.M., since these are the times between which the alarm will ring. We could identify 8 P.M. with -4 and 8 A.M. with $+8$, or we could identify 8 P.M. with $+8$ and 8 A.M. with $+20$. Taking the former approach, we would set $a = -4$ and $b = +8$ (Figure 4.5). If X represents the time at which the alarm will ring, then $E(X) = (a + b)/2 = [8 + (-4)]/2 = 2$, so the expected value of the time at which the alarm will ring is 2 A.M. To find $P[X > 5]$, we need only find the area shaded in on the graph in Figure 4.5. Since the height of this rectangular is $1/(b - a) = \frac{1}{12}$ and the width is $8 - 5 = 3$, the area, and hence the probability, is $\frac{3}{12}$:

$$P[X > 5] = \tfrac{3}{12} = .25.$$

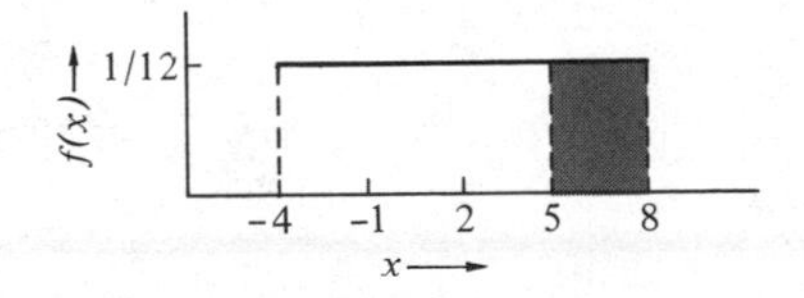

FIGURE 4.5
The pdf for Example 4.28

4.8 THE NORMAL DISTRIBUTION

In Section 2.3 we defined a random variable X to be the amount of soda put into a 16-ounce bottle of Fizzy Cola. In Example 2.8 we considered the possibility of X having a uniform distribution over the interval from 15 to 17, but we decided that that distribution was rather unrealistic in describing the filling of soda bottles. We concluded that the triangular-shaped distribution given in Example 2.9 could more realistically describe the way we would expect a reasonably well adjusted machine to fill soda bottles. Although the triangular distribution is a much better model than the uniform distribution, it still has some shortcomings. Go back and study Figure 2.4 and try to find some shortcomings. One that comes to mind is that the triangular distribution given there has a positive density function only for values of x between 15 and 17. Consequently, this implicitly assumes that no bottle ever contains less than 15 ounces, for example. A probability density function that allows for X to be less than 15 (but assigns small probability to that event) but still retains some of the positive properties of the triangular distribution (such as having the probability "concentrated" around the value 16 with a gradual tapering off on either side) is given in Figure 4.6.

The shape of the curve given in Figure 4.6 resembles a bell and, not surprisingly, is sometimes referred to as a "bell-shaped" curve. You are no doubt familiar with other situations where the distribution of a random variable can be described by a curve similar to this one. For example, if you think about the heights of college-age women whom you know, you will see that most have heights in the vicinity of 63 to 66 inches, with a smaller percent having heights less than 63 inches or greater than 66 inches. A similar thing is true for college-age men, except that most have heights in the vicinity of 68 to 71 inches. Approximate distributions for X, the heights of college-age men, and for Y, the heights of college-age women, are given in Figure 4.7. Actually, if we were to consider the heights of children of a given sex and age, we would find that their heights would have a distribution of approximately this same shape. If you were to consider the size of a particular variety of fruit, for example, a similar kind of distribution might be found. There are, in fact, a great number of situations where a particular measurement might have a distribution with this characteristic symmetric bell shape.

A word of warning is in order here, however. It is certainly *not true* that all measurements have such distributions. For example, if you were to measure the heights

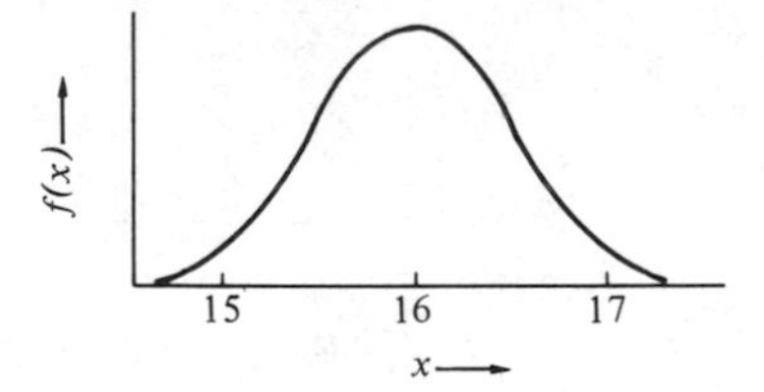

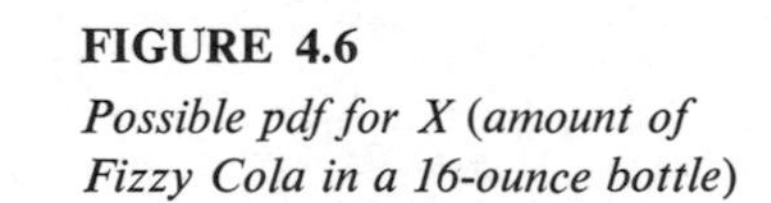

FIGURE 4.6
Possible pdf for X (amount of Fizzy Cola in a 16-ounce bottle)

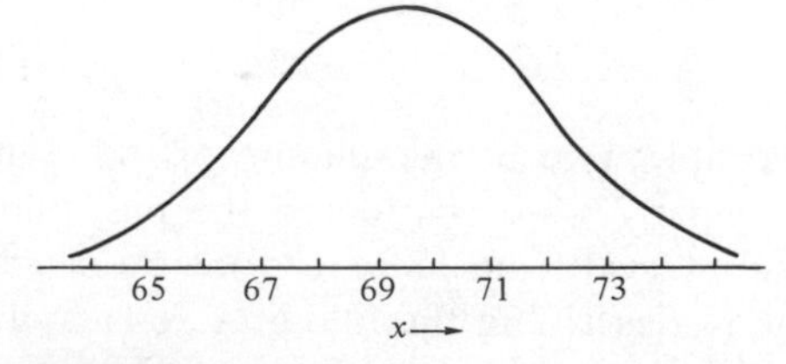

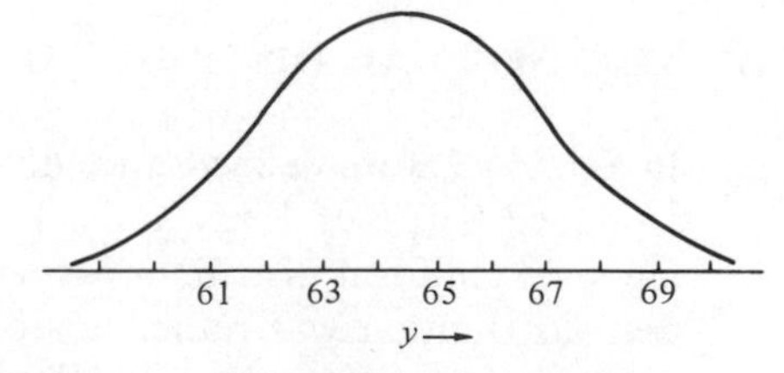

FIGURE 4.7
Approximate Distribution for the Heights of College-Age Males (X) and Females (Y)

of all children of elementary school age (grades K–6) in a large city, the distribution might be better approximated by a uniform distribution. The reason for this is because of the mixture of ages of the children in these grades. Another example would be the weights of individuals. Considering the weights of college-age men, you would find that the distribution is skewed to the right. You might think about this statement relative to your own experience to see if it is reasonable. You might also use the data on weights of statistics students given in Appendix A3 to see that the distribution of weights really is skewed. (If you are still not convinced, perhaps we should tell you that this statement is supported by information obtained from the National Center for Health Statistics.) Of course, a similar result holds for the weights of college-age women as well.

Although there are several symmetric distributions having a graph with a "bell shape" (some of which we will discuss in Chapter 7), the most important is the *normal distribution*. Unfortunately, it is very difficult to state general conditions under which a random variable (usually a measurement) will have a normal distribution. In most problems in this book you will simply be told to "assume that X has a normal distribution . . . ." How will this help you in a real-world problem if there is no one around to tell you to make this assumption? To be honest, it won't help you! However, in Chapter 11 we will discuss some methods that you can use to check on the reasonableness of this assumption. We will also see in Chapter 5 that certain special random variables will have an approximately normal distribution.

If you now consider the graphs given in Figures 4.6 and 4.7, you will see that there is not just one normal distribution, but rather there is a family of normal distributions. The "family trait" they have in common is the bell shape. What is it, then, that distinguishes the different members of the family? There are two distinguishing features: the location of the center of the distribution and the amount of variability. The two graphs in Figure 4.7 (corresponding to heights of men and women) have approximately the same amount of variability, but the location of the center (i.e., the mean) differs. In each case about 90% of the observations will fall within about 4 units of the mean. However, in Figure 4.6 (corresponding to the amount of cola in a bottle), about 90% of the observations fall within 1 unit of the mean. This reflects differences in variability.

As you might expect in view of the two distinguishing features, the normal family of distributions has two parameters. One of these parameters gives the value of x where the curve is the highest. Since this is a symmetric unimodal distribution, it follows from the discussion given in Section 2.6 that this value will be the mean. Consequently, this parameter is denoted by the symbol μ. The second parameter describes the variability and in fact is equal to the variance for this distribution, so it is logical to denote this parameter by σ^2. The form of the probability density function for a random variable having parameters μ and σ^2 is given by

$$(4.25) \qquad f(x \mid \mu, \sigma^2) = \frac{1}{\sqrt{2\pi\sigma^2}} e^{-(1/2)[(x-\mu)^2/\sigma^2]} \qquad -\infty < x < \infty.$$

(The parameter μ can assume any real value, and of course σ^2 must be positive.) To those of you not familiar with exponential notation, this equation may look somewhat imposing, but it is really a relatively simple mathematical function.

In many problems dealing with normal random variables it will be helpful to you to make a sketch of the density function. By following the guidelines we will give, you will be able to make a reasonably good sketch without explicitly evaluating the density function. The necessary information is contained in Figure 4.8. First, draw a horizontal line and label the values μ, $\mu - \sigma$, $\mu + \sigma$, $\mu - 2\sigma$, and $\mu + 2\sigma$ (where, of course, $\sigma = \sqrt{\sigma^2}$) to establish the scale. Second, plot a point at any height you wish above μ and call this height h. [That is, plot the point (μ, h). If you are writing in your notes, you might choose h equal to 1 or 2 inches, or if writing on a blackboard, you might choose h equal to 1 foot. The actual size is not important.] Above $\mu + \sigma$ on the horizontal scale plot a point about $\frac{6}{10}$ as high as the point at μ [i.e., plot the point $(\mu + \sigma, .60h)$]. Since the distribution is symmetric about the value μ, also plot the point $(\mu - \sigma, .60h)$. Now plot points $(\mu + 2\sigma, .15h)$ and $(\mu - 2\sigma, .15h)$. (A very precise graph would use the values $.6067h$ and $.1354h$ instead of $.60h$ and $.15h$, but these latter figures will do nicely for our purposes. Finally, connect the points by a smooth curve. The curve between $\mu - \sigma$ and $\mu + \sigma$ should be facing "downward" (like an inverted bowl) and the remaining part of the curve should be facing "upward."

To illustrate this technique, we have drawn in Figure 4.9 three different normal curves on the same axis. Keep in mind that the total area under each of these curves must equal 1, so we must be somewhat careful in choosing h when plotting different normal curves on the same axis. Since σ and h are actually inversely proportional, if $\sigma_1 = \frac{1}{2}\sigma_2$, then $h_1 = 2h_2$; if $\sigma_1 = \frac{1}{3}\sigma_2$, then $h_1 = 3h_2$; and so on. The first curve has parameters $\mu_1 = 3$, $\sigma_1^2 = 4$; the second, $\mu_2 = 8$, $\sigma_2^2 = 4$; and the third, $\mu_3 = 8$ and $\sigma_3^2 = 1$. By comparing curves 1 and 2, you can see that if μ is different for two normal distributions while σ^2 is the same, then the two curves are identical in shape but are centered at different locations. By comparing curves 2 and 3, you can see that if μ is the same for two normal distributions while σ^2 is different, then the centers

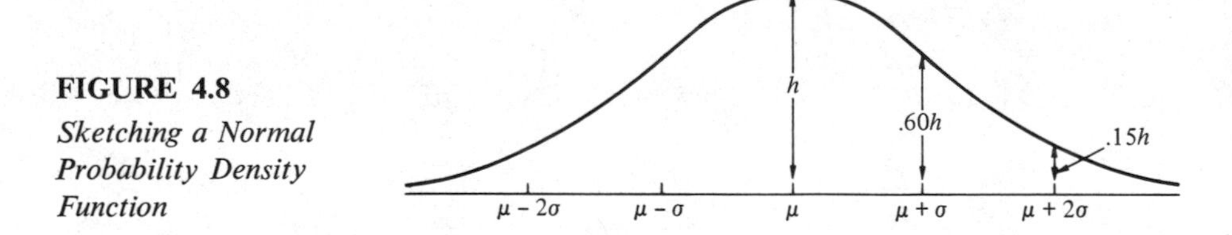

FIGURE 4.8
Sketching a Normal Probability Density Function

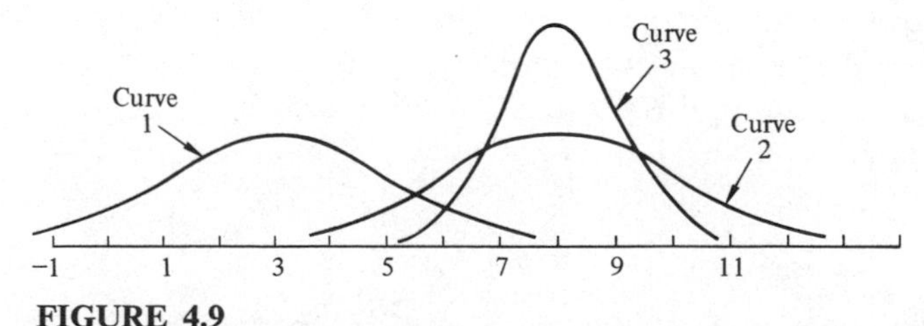

FIGURE 4.9
Normal Curves for Different Parameter Values

of the distributions are the same but the shapes are different. The distribution with the smaller variance will be more peaked.

If you are told that a random variable X has a normal distribution with known parameter values, say $\mu = 3$ and $\sigma^2 = 4$, no further calculations need be made to find the expected value and variance, since these are simply the parameter values. For instance, in this case

$$E(X) = \mu = 3 \qquad \text{and} \qquad V(X) = \sigma^2 = 4.$$

The notation that we will use to indicate that a random variable X has a normal distribution with parameters μ and σ^2 is

$$X \sim N(\mu, \sigma^2).$$

The symbol "$\sim$" means "is distributed as." If we were to write

$$Z \sim N(0, 1) \qquad \text{or} \qquad Z \sim N(\mu = 0, \sigma^2 = 1),$$

we would mean that Z is a random variable which has a normal distribution with a mean of zero and a variance of 1. Such a random variable is called a *standard normal random variable.*

In order to find probabilities for events defined in terms of a normal random variable, it is necessary to use tables. [The density function given in (4.25) is not integrable, so even readers familiar with calculus could not use integration to find probabilities! The probabilities must be found by using numerical integration.] If Z is a standard normal random variable, the cumulative distribution function

$$F_Z(z) = P[Z \leq z]$$

can be found for values of z between 0 and 3.99 from Table A4. By using complements it is also possible to find

$$P[Z > z] = 1 - F_Z(z)$$

and by using (2.12) of Section 2.4 it is possible to find

$$P[a < Z \leq b] = F_Z(b) - F_Z(a).$$

Since the standard normal distribution is symmetric about zero, if $z < 0$,

$$F_Z(z) = 1 - F_Z(-z).$$

(If there is little chance for confusion, the Z subscript on F will be omitted.) We should point out that some authors denote the CDF of a standard normal random variable by the symbol Φ rather than F.

Example 4.29 If Z is a standard normal random variable, find $P[Z \leq 1]$, $P[Z > 2]$, $P[0 < Z \leq 1.5]$, and $P[Z \leq -2]$.

Using Table A4 and the relationships just given we find

$$\begin{aligned} P[Z \leq 1] &= F(1) = .8413, \\ P[Z > 2] &= 1 - F(2) = 1 - .9772 = .0228, \\ P[0 < Z \leq 1.5] &= F(1.5) - F(0) = .9332 - .5000 = .4332, \\ P[Z \leq -2] &= F(-2) = 1 - F(+2) = .0228. \end{aligned}$$

Note that since Z is a continuous random variable, the probability associated with any individual point is zero. As in equation (2.10),

$$P[a \leq Z \leq b] = P[a < Z \leq b] = P[a \leq Z < b]$$
$$= P[a < Z < b].$$

In Section 2.6 we discussed how the cumulative distribution function can be used to find the median and other percentiles of a particular distribution. Since the values of the cumulative distribution function for the standard normal distribution are tabulated, it should be possible to use these tables to find percentiles. (Actually, finding the median is not much of a challenge, since the distribution is symmetric. If it does strike you as a challenge, refer back to Figure 2.17.) To find the 99th percentile, for example, we need to find the constant a such that

$$P[Z \leq a] = F(a) = .9900.$$

Looking in the tables we find

$$F(2.32) = .9898$$

and

$$F(2.33) = .9901.$$

Since the latter value is slightly closer, we could say that the 99th percentile is approximately 2.33. (Using more detailed tables, we would get a value of 2.326.)

Example 4.30 If Z is a standard normal random variable, find the 10th, 90th, and 95th percentiles, $z_{.10}$, $z_{.90}$, and $z_{.95}$.

Using the tables and the symmetry of the normal distribution, we find

$$F(-1.28) = .1003 \qquad \text{and} \qquad F(-1.29) = .0985,$$

so we would take -1.28 to be the 10th percentile:

$$z_{.10} = -1.28.$$

$$F(+1.28) = .8997 \qquad \text{and} \qquad F(+1.29) = .9015,$$

so we would take $+1.28$ to be the 90th percentile:

$$F(1.64) = .9495 \qquad \text{and} \qquad F(1.65) = .9505.$$

In this case the interpolation is so easy that we would take 1.645 to be the 95th percentile.

Notice that the 10th and 90th percentiles are negatives of each other. This is true since, as we indicated earlier, the standard normal distribution is symmetric about zero. With this in mind, could you find 5th percentile from the information given in Example 4.30? You should be able to do so.

Example 4.31 If Z is a standard normal random variable, find two values a and b such that $P[a \leq Z \leq b] = .95$.

If a and b are chosen to be symmetrically located with respect to zero, the approximate location should be as shown in Figure 4.10. If 95% of the area lies between a and b, then 2.5% of the area must lie in each tail (i.e., below a and above b).

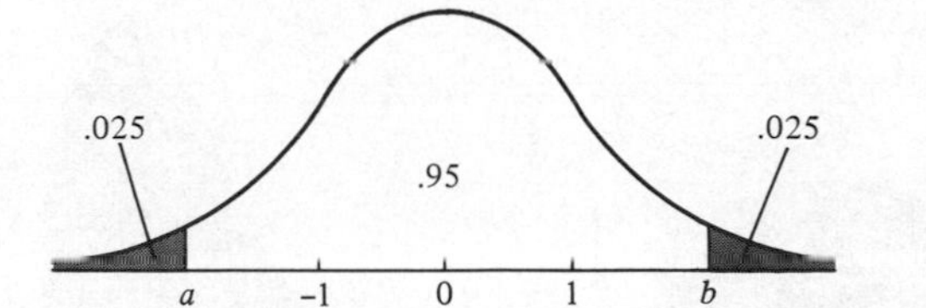

FIGURE 4.10
Graph for Example 4.31

If 2.5% lies *above b*, 97.5% must lie *below b*. Consequently, b must be the 97.5 percentile and, in view of our previous discussions, a must be the negative of b. Since $F(1.96) = .9750$, we have $a = z_{.025} = -1.96$ and $b = z_{.975} = +1.96$.

We have now discussed how to find probabilities and percentiles for standard normal random variables. It may appear as if we have only scratched the surface, since there are an infinite number of choices for μ and σ^2 for nonstandard normal random variables and we would like to be able to find probabilities and percentiles for these as well. Fortunately, by using the technique of standardizing as discussed in Section 2.8, we can use standard normal tables to find probabilities and percentiles for nonstandard normal variables.

Theorem 4.7 If X is a normal random variable having mean μ and variance σ^2, then

$$Z = \frac{X - \mu}{\sigma}$$

will be a normal random variable having mean zero and variance 1 (i.e., Z will be standard normal).

Say that $X \sim N(\mu = 3, \sigma^2 = 4)$. What is the probability that X will be less than or equal to 5? Remembering that algebraically we must perform the same operations on both sides of an inequality to maintain that inequality, we get

$$(4.26) \qquad [X \leq 5] = [X - \mu \leq 5 - \mu] = \left[\frac{X - \mu}{\sigma} \leq \frac{5 - \mu}{\sigma}\right].$$

It follows from Theorem 4.7 that we can replace $(X - \mu)/\sigma$ on the left-hand side of (4.26) with Z, where $Z \sim N(0, 1)$, and we can replace $(5 - \mu)/\sigma$ with

$$\frac{5 - \mu}{\sigma} = \frac{5 - 3}{2} = 1$$

on the right-hand side. From this we conclude that the event $X \leq 5$ is equivalent to the event $Z \leq 1$. Consequently,

$$P[X \leq 5] = P\left[\frac{X \quad \mu}{\sigma} \leq \frac{5 \quad \mu}{\sigma}\right] = P[Z \leq 1] = .8413.$$

In Figure 4.11 we show the graphs of X and Z using the same horizontal and vertical scales and we have shaded in the areas corresponding to the events $[X \leq 5]$ and

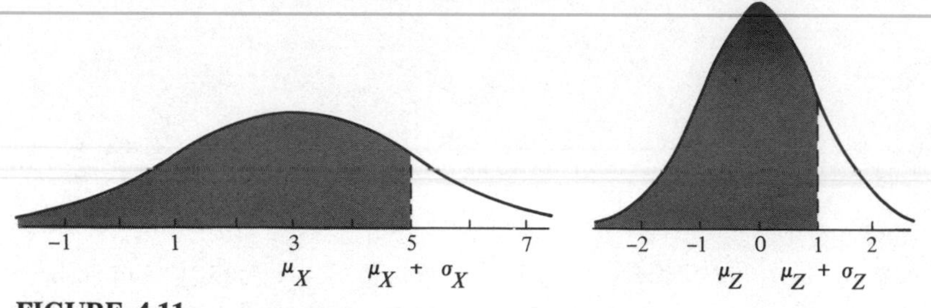

FIGURE 4.11

Untransformed and Transformed Normal Random Variables

$[Z \leq 1]$. In each case the shaded region gives the area below the line corresponding to 1 standard deviation above the mean. (The value 5 is equal to $\mu_X + \sigma_X = 3 + 2$, while the value 1 is equal to $\mu_Z + \sigma_Z = 0 + 1$.)

Example 4.32 The amount of coffee in an "8-ounce" jar of Sancafé instant coffee is actually a random variable, call it X. If the amount of coffee put into a jar is a normal random variable with $\mu = 8.00$ ounces and $\sigma^2 = .04$ oz^2, find the probability that a randomly chosen jar will contain less than 7.80 ounces.

By assumption X is a normal random variable and it is certainly not standard normal. To standardize X we will need to know the value of σ. In this case σ^2 is given, so $\sigma = \sqrt{\sigma^2} = \sqrt{.04 \text{ oz}^2} = .20$ ounce. Using this, we get

$$P[X < 7.80] = P\left[\frac{X - \mu}{\sigma} < \frac{7.80 - 8.00}{.20}\right]$$

$$= P\left[Z < \frac{7.80 - 8.00}{.20} = -1\right] = .1587.$$

In general, if X is a normal random variable with parameters μ and σ^2, then by standardizing we can find probabilities using the standard normal tables. In particular,

$$P[a \leq X \leq b] = P\left[\frac{a - \mu}{\sigma} \leq Z \leq \frac{b - \mu}{\sigma}\right].$$

Finding percentiles for nonstandard normal random variables can be done through the same kind of transformation as that used for finding probabilities. For example, if $X \sim N(\mu = 3, \sigma^2 = 4)$ and if we wish to find the constant a such that

$$P[X \leq a] = F_X(a) = .9900$$

(i.e., the 99th percentile for X), we write

$$P[X \leq a] = P\left[\frac{X - \mu}{\sigma} \leq \frac{a - \mu}{\sigma}\right]$$

$$= P\left[Z \leq \frac{a - 3}{2}\right] = .9900.$$

Remember that the unknown quantity here is a. Since we know from the discussion preceding Example 4.30 that $z_{.99} = 2.326$, that is, that

$$P[Z \leq 2.326] = .9900,$$

it must be true that

$$\frac{a-3}{2} = 2.326.$$

Solving for a, we get

$$a = 3 + 2(2.326) \doteq 7.65.$$

Example 4.33 If $X \sim N(\mu = 3, \sigma^2 = 4)$, find the 95th percentile. Also find two values a and b such that $P[a \leq X \leq b] = .95$.

Suppose that c represents the 95th percentile of X. In Example 4.30 we found that $P[Z \leq 1.645] = .95$, so

$$P[X \leq c] = P\left[\frac{X-\mu}{\sigma} \leq \frac{c-\mu}{\sigma}\right]$$

$$= P\left[Z \leq \frac{c-3}{2} = 1.645\right] = .95,$$

so

$$\frac{c-3}{2} = 1.645$$

or

$$c = 3 + 2(1.645) = 6.29.$$

Using the results of Example 4.31, we find that

$$P[a \leq X \leq b] = P\left[\frac{a-\mu}{\sigma} \leq \frac{X-\mu}{\sigma} \leq \frac{b-\mu}{\sigma}\right]$$

$$= P\left[\frac{a-3}{2} \leq Z \leq \frac{b-3}{2}\right]$$

$$= P\left[-1.96 = \frac{a-3}{2} \leq Z \leq \frac{b-3}{2} = 1.96\right] = .95.$$

Solving for a and b gives

$$a = 3 + 2(-1.96) = -.92 \qquad \text{and} \qquad b = 3 + 2(1.96) = 6.92.$$

Example 4.34 Ninety percent of all "8-ounce" jars of Sancafé Coffee (as described in Example 4.32) contain at least b ounces. Find b.

We can write this as

$$P[X \geq b] = .90$$

or

$$P[X \leq b] = .10.$$

In Figure 4.12 we have sketched the distribution for X so that we can get a rough idea as to the answer to this problem. From the graph we would guess that b would be some value between 7.6 and 7.8. Making a sketch and a guess as to the answer in

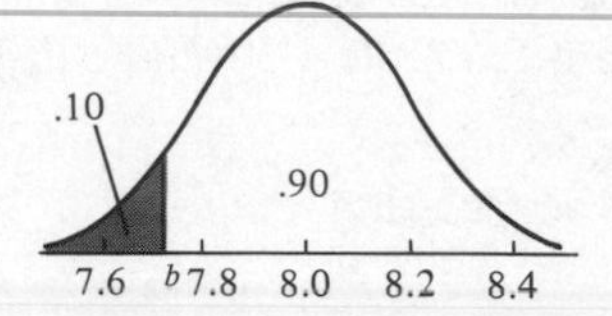

FIGURE 4.12
Distribution for Random Variable of Example 4.34

problems like these can help you to avoid making gross errors. We now find

$$P[X \leq b] = P\left[\frac{X-\mu}{\sigma} \leq \frac{b-\mu}{\sigma}\right]$$
$$= P\left[Z \leq \frac{b-8.0}{.20} = -1.28\right],$$

so

$$b = 8.0 + (.20)(-1.28) = 7.744.$$

This answer is consistent with our rough guess.

4.9 THE EXPONENTIAL DISTRIBUTION

In Section 4.6 we discussed Georgia's candy stand, paying particular attention to the number of customers coming to her stand in a given time interval. We could view this problem somewhat differently and ask about the amount of time she would have to wait after she opened her stand for her first customer to arrive. It should not be hard for you to see that the waiting time will be a random variable, say T. If, in fact, "experience is the best teacher," then by performing a simple experiment yourself you will become convinced that T really is a random variable. Observe some process, such as cars entering a parking lot or customers entering a store. Start timing when the secondhand on your watch is on "12" and record the amount of time until the next arrival. If you repeat this experiment a few times, you will become convinced that T is a random variable.

Consider a process where some well-defined events occur over time. If the number of events occurring in a specified time interval follows a Poisson distribution with parameter λ, then the waiting time until the first occurrence, T, will follow an *exponential distribution* with parameter λ.

We will prove later in this section that a random variable T having an exponential distribution will have a cumulative distribution function and probability density function given by

$$F(t) = \begin{cases} 1 - e^{-\lambda t} & t \geq 0 \\ 0 & t < 0 \end{cases} \tag{4.27}$$

and

$$f(t) = \begin{cases} \lambda e^{-\lambda t} & t \geq 0 \\ 0 & t < 0. \end{cases} \tag{4.28}$$

Keeping in mind that $e^0 = 1$, we see that

$$f(0) = \lambda e^{-\lambda(0)} = \lambda,$$

so sketching the density function becomes a relatively easy matter. In Figure 4.13 we have sketched the graphs of pdf's for T_1 and T_2, exponential random variables having parameters $\lambda_1 - 1$ and $\lambda_2 - 2$, respectively.

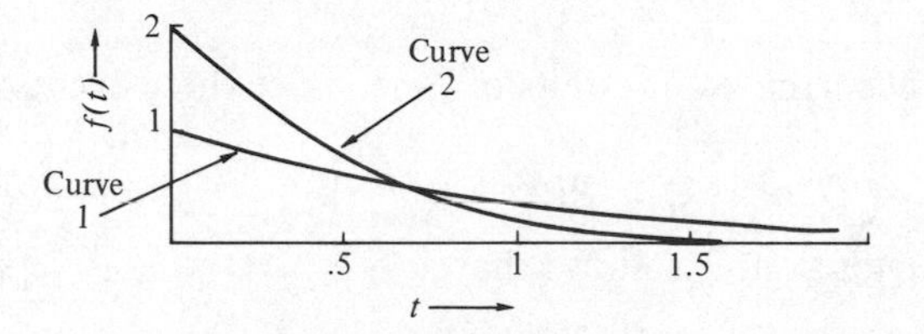

FIGURE 4.13
Probability Density Functions for Exponential Random Variables

If you remember that λ is the *rate of arrivals*, you should be able to make an intuitive guess as to which of the random variables, T_1 or T_2, will have the longer expected waiting time. Since $\lambda_2 = 2$ is greater than $\lambda_1 = 1$, events are occurring at a greater rate for the second situation. In waiting for the occurrence of an event, in which case would you expect to wait longer? You'd expect to wait longer when the rate of arrivals is less. Similar reasoning might also lead you to expect more variability when the rate of arrivals is less. That these facts are true can be seen from the following theorem.

Theorem 4.8 If T is a random variable having an exponential distribution with parameter λ, then

(4.29) $$E(T) = \frac{1}{\lambda} \quad \text{and} \quad V(T) = \frac{1}{\lambda^2}.$$

Example 4.35 Compare the means and variances of the random variables T_1 with parameter $\lambda_1 = 1$ and T_2 with parameter $\lambda_2 = 2$ (see Figure 4.13).

Since for T_1 we have $\lambda_1 = 1$, it follows from (4.29) that

$$E(T_1) = \frac{1}{\lambda_1} = 1 \quad \text{and} \quad V(T_1) = \frac{1}{\lambda_1^2} = 1.$$

Similarly, since $\lambda_2 = 2$, we get

$$E(T_2) = \frac{1}{\lambda_2} = \frac{1}{2} \quad \text{and} \quad V(T_2) = \frac{1}{\lambda_2^2} = \frac{1}{4}.$$

We should now address ourselves to the problem of finding probabilities for events defined in terms of T. The easiest way to do this is to make use of the relationship between the Poisson distribution and the exponential distribution. Since T is the waiting time until the first occurrence of an event where the total number of occurrences in a given time interval follows a Poisson distribution, it should not be surprising that a relationship does exist. In particular, if we consider a fixed amount of time t, then the event

$$T > t$$

is simply the event that the waiting time until the first occurrence is longer than t units of time. If you were told that the *waiting time* for the first occurrence is longer than t (say $t = 5$ minutes, to be explicit), what would you know about the *number of occurrences* in those t units (5 minutes)? There must be no occurrences; otherwise, the waiting time until the first occurrence would be less than t (5 minutes)! Consequently, the event $T > t$ is equivalent to the event

$$X = 0,$$

where X is the number of occurrences in t units of time. Since these events are equivalent, it follows that

$$P[T > t \mid \lambda] = P[X = 0 \mid \lambda, t].$$

Since X has a Poisson distribution, it follows that

$$P[X = 0 \mid \lambda, t] = \frac{e^{-\lambda t}(\lambda t)^0}{0!} = e^{-\lambda t}. \tag{4.30}$$

To find the cumulative distribution function for T, we use

$$\begin{aligned} F_T(t) = P[T \leq t \mid \lambda] &= 1 - P[T > t \mid \lambda] \\ &= 1 - P[X = 0 \mid \lambda, t] \\ &= 1 - e^{-\lambda t}. \end{aligned}$$

[Compare this result with (4.27).]

Example 4.36 We assumed in Example 4.24 that customers arrive at Georgia's stand at the rate of 30/hour (or .5/minute). Assuming that there are no customers waiting when she opens the stand, what is the expected waiting time until the first customer arrives? What is the probability that the first customer will arrive between 3 and 5 minutes after she opens the stand?

To find the expected waiting time, use Theorem 4.8:

$$E(T) = \frac{1}{\lambda} = \frac{1}{.5/\text{minute}} = 2 \text{ minutes.}$$

To find the requisite probability, we use

$$\begin{aligned} P[3 \leq T \leq 5] &= F(5) - F(3) \\ &= (1 - e^{-\lambda(5)}) - (1 - e^{-\lambda(3)}) \\ &= e^{-\lambda(3)} - e^{-\lambda(5)} \\ &= e^{-1.5} - e^{-2.5}. \end{aligned}$$

To evaluate $e^{-1.5}$ and $e^{-2.5}$, we use the results of (4.30) and the Poisson tables.

$$P[X = 0 \mid \lambda = .5, t = 3] = e^{-1.5} = .2231$$

and

$$P[X = 0 \mid \lambda = .5, t = 5] = e^{-2.5} = .0821,$$

so

$$P[3 \leq T \leq 5] = .2231 - .0821 = .1410.$$

The exponential distribution plays an important role in various areas of applied probability, one of those areas being "queuing theory." A queue is simply a waiting line. Anyone who has ever gone to a grocery store, department store, or bank, or who has ever gone through a college registration line, is familiar with queues! The average length of queues is determined by factors such as the rate of arrivals, the service rate (e.g., the amount of time required for a customer to have his groceries checked), and the number of servers. You can imagine that considerations such as these should play a role in the planning of any facility offering services of the type suggested. (A nonmathematical "theorem" that has been suggested as describing the behavior of waiting lines is: "the shortest line always moves the slowest!" Is that consistent with your experience?)

• • • • •

By using differentiation, it is possible to find the probability density function for an exponential random variable. In Section 2.4 we saw that the pdf can be obtained from the CDF by differentiating. Differentiating the CDF given in (4.27), we get

$$\frac{dF(t)}{dt} = \frac{d}{dt}(1 - e^{-\lambda t}) = (-\lambda)(-e^{-\lambda t})$$

$$= \lambda e^{-\lambda t} \qquad t \geq 0$$

and

$$\frac{dF(t)}{dt} = \frac{d}{dt}(0) = 0 \qquad t < 0.$$

Combining these results, we get

$$f_T(t) = \begin{cases} \lambda e^{-\lambda t} & t \geq 0 \\ 0 & t < 0 \end{cases}$$

as in (4.28).

4.10 THE NORMAL APPROXIMATION TO THE BINOMIAL AND OTHER APPROXIMATIONS

In attempting to evaluate probabilities associated with random variables having a binomial distribution, we have used tables whenever possible rather than performing direct calculations. Of course, Tables A1 and A2 are very limited and, because of space and cost constraints, must remain so. As we will see, for $n > 20$ we will be able to approximate binomial probabilities by using probabilities calculated from other distributions.

Let us begin by considering a random variable X_B having a binomial distribution with parameters $n = 16$ and $p = .5$. A probability histogram (see Definition 2.3) for this random variable is given in Figure 4.14. You will recall that since $p = .5$, the distribution is symmetric. As you look at Figure 4.14, are you reminded of some other distribution? You should be reminded of a normal distribution, as the graph in Figure 4.14 is approximately bell-shaped. As you know, there is a family of normal distributions and we would like to find the particular member of the family whose graph is closest to the probability histogram of X_B. Since X_B has an expected value of

$$np = 16(.5) = 8,$$

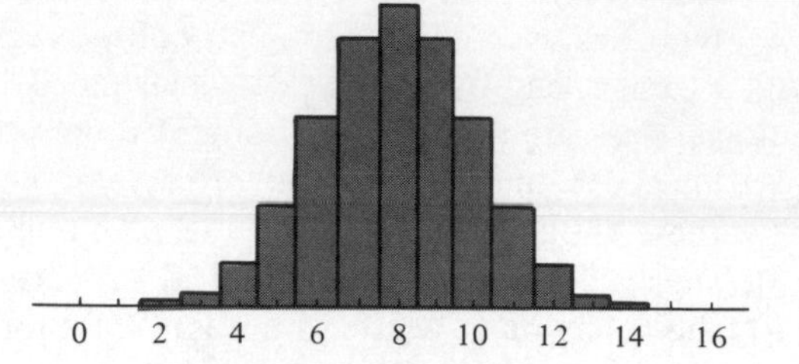

FIGURE 4.14
Probability Histogram for a Binomial Random Variable

it seems reasonable to choose the parameter μ for the normal random variable, which we will call X_N, to be the same value. That is, choose $\mu = np = 8$. Similarly, since

$$V(X_B) = npq = 16(.5)(.5) = 4,$$

we can choose $\sigma^2 = V(X_N)$ to be 4 as well. By choosing the member of the normal family that has the same mean and variance as the member of the binomial family we are considering, we should get a graph that is very close to the probability histogram of X_B. In Figure 4.15 we have given a graph of the probability density function for X_N, which is $N(\mu = 8, \sigma^2 = 4)$.

It seems reasonable to assume that since the graphs in Figures 4.14 and 4.15 are similar, probabilities associated with specific *events* might also be similar. We will see that rather than events, however, we should give attention to similar *regions*. For example, if we consider events $X_B \leq 5$ and $8 \leq X_B \leq 10$, we find from the tables that

$$P[X_B \leq 5] = .0667 + .0278 + .0085 + .0018 + .0002 = .1050,$$

$$P[8 \leq X_B \leq 10] = .1964 + .1746 + .1222 = .4932.$$

To find probabilities for these same events for the normal random variable, we get

$$P[X_N \leq 5] = P\left[\frac{X_N - \mu}{\sigma} \leq \frac{5-8}{2}\right]$$
$$= P[Z \leq -1.5] = .0668$$

and

$$P[8 \leq X_N \leq 10] = P\left[\frac{8-8}{2} \leq \frac{X_N - \mu}{\sigma} \leq \frac{10-8}{2}\right]$$
$$= P[0 \leq Z \leq 1] = .8413 - .5000 = .3413.$$

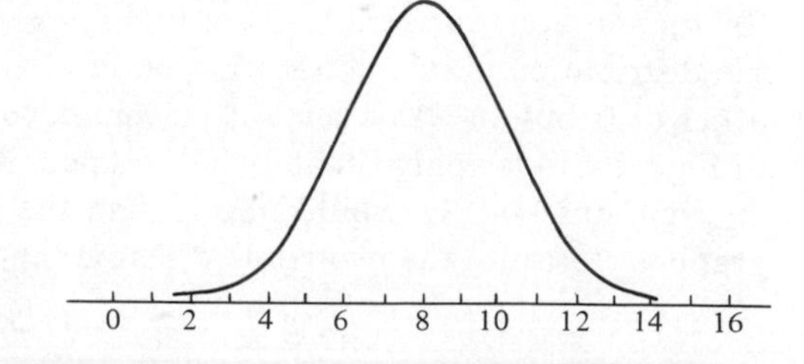

FIGURE 4.15
Pdf for a Normal Random Variable

Unfortunately, when we compare the probabilities for these *events*, we see that $P[X_B \leq 5]$ and $P[X_N \leq 5]$, for example, are not too close! Similarly, $P[8 \leq X_B \leq 10]$ and $P[8 \leq X_N \leq 10]$ are not very close.

What is the reason for the discrepancy, when in fact the graphs appear to be so similar? We can get some explanation by considering Figure 4.16, where we have placed the probability histogram for X_B and the pdf for X_N on the same axes. We have also shaded in the regions on the probability histogram corresponding to the events $[X_B \leq 5]$ and $[8 \leq X_B \leq 10]$. Recall from our discussion in Section 2.2 that the area of the rectangle centered at x gives the probability of the event $X = x$. Consequently, the probability that X_B will be less than or equal to 5 is found by adding together the areas of the rectangles centered at integer values less than or equal to 5. Since the rectangle corresponding to 5 actually extends to 5.5, the corresponding region under the normal curve would be all values to the left of 5.5. The probability is

$$P[X_N \leq 5.5] = P\left[\frac{X_N - \mu}{\sigma} \leq \frac{5.5 - 8}{2}\right]$$
$$= P[Z \leq -1.25] = .1056.$$

As you can see, this value is much closer to the probability of the event $[X_B \leq 5]$. Likewise, the rectangles whose areas must be added together to get $P[8 \leq X_B \leq 10]$ begin at 7.5 and extend to 10.5. The area under the normal curve between 7.5 and 10.5 is

$$P[7.5 \leq X_N \leq 10.5] = P\left[\frac{7.5 - 8}{2} \leq \frac{X_N - \mu}{\sigma} \leq \frac{10.5 - 8}{2}\right]$$
$$= P[-.25 \leq Z \leq 1.25] = .8944 - .4013 = .4931.$$

Here, too, you can see that the probability is very close to $P[8 \leq X_B \leq 10]$.

We have tried to show by the previous illustration that the normal distribution can be used to find approximate probabilities for binomial random variables. The approximations can be quite good provided that consideration is made for the fact that the rectangles whose areas must be added together extend .5 above and below their nominal values. This difference is referred to as the *continuity correction factor*. Such a correction is necessary when approximating a discrete distribution (such as the binomial) by a continuous distribution (such as the normal).

Is it true that the probability of any event defined in terms of a binomial random variable can be well approximated by the probability of an event defined in terms of a normal random variable, provided that a continuity correction is used? You might

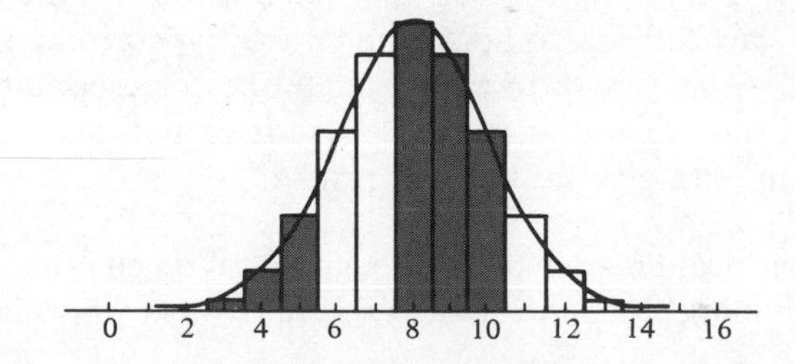

FIGURE 4.16
Probability Histogram for X_B and pdf for X_N

get that impression from the illustration given, but it is a false impression. The main reason that the approximation was good was that we chose $p = .5$ so that the distribution of X_B would be symmetric. If you have worked Exercise 11 of this chapter you will have seen that for $p \neq .5$, the binomial distributions are *not* symmetric. Although this is true, the lack of symmetry becomes less noticeable, for any value of p, as n gets increasingly large. As a general rule of thumb, we will say that if $np \geq 5$ and $nq \geq 5$, the binomial distribution is symmetric enough for the normal approximation to be used. Of course, if $n \leq 20$, no approximation is necessary, since exact probabilities can be found from the tables or by hand calculations. We can summarize this discussion in a theorem.

Theorem 4.9 If X_B is a binomial random variable having parameters n and p satisfying (i) $n > 20$ and (ii) $np \geq 5$ and $nq \geq 5$, and if X_N is a normal random variable with mean $\mu = np$ and variance $\sigma^2 = npq$, then for any integers i and j,

$$(4.31) \qquad P[i \leq X_B \leq j] \approx P[i - .5 \leq X_N \leq j + .5]$$
$$= P\left[\frac{(i - .5) - np}{\sqrt{npq}} \leq \frac{X_N - np}{\sqrt{npq}} \leq \frac{(j + .5) - np}{\sqrt{npq}}\right]$$
$$= P\left[\frac{(i - .5) - np}{\sqrt{npq}} \leq Z \leq \frac{(j + .5) - np}{\sqrt{npq}}\right].$$

Example 4.37 During the course of a game of Monopoly, a pair of dice was thrown 180 times. Find the approximate probability that doubles (i.e., both faces being the same) occurred between 24 and 32 times inclusive.

Here we have $n = 180$ and $p = P[\text{doubles appear}] = \frac{1}{6}$, so

$$np = 30 \qquad \text{and} \qquad nq = 150.$$

Since the conditions of Theorem 4.9 are satisfied, we can use (4.31) with $i = 24$ and $j = 32$. This gives

$$P[24 \leq X_B \leq 32] \approx P\left[\frac{23.5 - 30}{\sqrt{25}} \leq Z \leq \frac{32.5 - 30}{\sqrt{25}}\right]$$
$$= P\left[\frac{-6.5}{5} \leq Z \leq \frac{2.5}{5}\right]$$
$$= P[-1.3 \leq Z \leq .5] = .6915 - .0968 = .5947.$$

In Section 4.5 we discussed the fact that in sampling from a finite population *without replacement* we could expect to observe a random variable having a hypergeometric distribution. We also said that if the sample size is less than 5% of the population size, the binomial distribution can be used to approximate the hypergeometric distribution. If n is greater than 20, we may wish to approximate the corresponding binomial distribution by using a normal approximation. Believe it or not, such an "approximation to an approximation" can give satisfactory results.

Example 4.38 The president of the local chapter of the League of Women Voters, Polly Tishen, is interested in determining the number of registered voters who plan to vote on a special tax issue next Tuesday. If 300 out of 1000 registered voters do in fact plan to vote, what is the probability that in a random sample of 50 registered voters, taken without replacement, 12 or fewer will say they plan to vote?

Since sampling is being done without replacement, the number of persons in the sample who will say they plan to vote, say X_H, will follow a hypergeometric distribution with parameters $n = 50$, $A = 300$, and $N = 1000$. To approximate X_H we would use a binomial random variable X_B with parameters $n = 50$ and $p = .30$, since 30% of the population of interest are assumed to plan to vote. Using a one-sided version of (4.31), we find that

$$P[X_H \leq 12] \approx P[X_B \leq 12] \approx P\left[\frac{X_N - np}{\sqrt{npq}} \leq \frac{12.5 - 15}{\sqrt{10.5}}\right]$$

$$= P\left[Z \leq \frac{-2.5}{3.24} = -.77\right] = .2206.$$

We have already mentioned the fact that the normal distribution is a good approximation to the binomial distribution provided that the binomial distribution is symmetric enough. When, because of a small value of p (or q), the distribution is too skewed to be approximated by a normal distribution, the binomial distribution can be well approximated by an appropriate Poisson distribution. Let us say that X_B is a binomial random variable with parameters n and p, with p being a small value. The expected value of X_B is

$$E(X_B) = np.$$

It would seem reasonable to choose a Poisson random variable X_P having the same expected value in order to obtain a good approximation. Since

$$E(X_P) = \lambda t,$$

we choose λt to be equal to np. (Remember that in finding probabilities for Poisson random variables, you need to know the value of the product $\lambda \cdot t$. It is not necessary to know the values of λ and t themselves.) You will recall that in choosing a member of the normal family to approximate a binomial distribution, we choose μ and σ^2 so that they matched the mean and variance of the binomial distribution being approximated. If we compare the variances of the binomial and Poisson distributions being considered here, we find that

$$V(X_B) = npq \qquad V(X_P) = \lambda t = np.$$

Since we assumed that p was a small value (i.e., close to zero), it follows that q will be close to 1. Consequently,

$$V(X_B) = npq \approx np = \lambda t = V(X_P).$$

You can see from this that the closer to zero p is, the better the approximation will be.

Example 4.39 Using the tables, compare the probability distributions of X_B and X_P when $n = 20$, $p = .10$, and $\lambda t = 2$. Also, compare these distributions when $n = 20$, $p = .05$, and $\lambda t = 1$.

Note that in each case we are taking $\lambda t = np$ so that the distributions have the same means. You can verify that in the first case the tabulated values are

x:	0	1	2	3	4	5	6	7
$P[X_B = x]$:	.1216	.2702	.2852	.1901	.0898	.0319	.0089	.0020
$P[X_P = x]$:	.1353	.2707	.2707	.1804	.0902	.0361	.0120	.0034

and in the second case the values are

x:	0	1	2	3	4	5	6
$P[X_B = x]$:	.3585	.3774	.1887	.0596	.0133	.0022	.0003
$P[X_P = x]$:	.3679	.3679	.1839	.0613	.0153	.0031	.0005.

You can see from this example that the smaller p gives somewhat better agreement between the two distributions. Remember also that by interchanging the roles of "success" and "failure" in the definition of a binomial random variable, you interchange the roles of p and q. Consequently, if p is close to 1, you can redefine a success and make p close to zero. The following theorem is stated under the assumption that a "success" is defined so that p is small.

Theorem 4.10 If X_B is a binomial random variable having parameters n and p satisfying (i) $n > 20$ and (ii) $np < 5$, and if X_P is a Poisson random variable such that $\lambda t = np$, then

$$P[X_B = x \mid n, p] \approx P[X_P = x \mid \lambda t = np].$$

Example 4.40 In a *Time* magazine article (April 11, 1977) it was stated that in flying on a major U.S. airline, the probability of reaching your destination safely is 99.999%. Al Timiter, who sells aircraft equipment, does a lot of flying on business trips. If he makes one round trip (two flights) a day, 5 days a week, 50 weeks a year for the next 20 years, what is the approximate probability that he will reach his destination safely every time?

In a very real sense Al would tend to call the event of reaching his destination safely a success. In this case p, the probability of a success, would be .99999. To satisfy our agreed-upon convention of having p small, however, we will have to consider a success to be *not* reaching safely so that p will be small. Since with this definition p equals .00001 and since $n = 2 \cdot 5 \cdot 50 \cdot 20 = 10{,}000$, we have $np = .10$. It follows from Theorem 4.10 that since we want the probability of no successes, we should find

$$P[X_B = 0 \mid n = 10{,}000, p = .0001] \approx P[X_P = 0 \mid \lambda t = .1] = .9048.$$

In Section 4.6 when we first introduced you to the Poisson distribution, we said that there were certain mathematical conditions which, if satisfied, would guarantee that a random variable would have a Poisson distribution. These same conditions give a mathematical explanation for the fact that the binomial and Poisson distributions are similar for small values of p. Furthermore, even a somewhat superficial understanding of these conditions might help you decide whether or not it is reasonable to assume that a particular physical situation could lead to a Poisson random variable. For these reasons we will state these conditions and indicate how they can be used to explain the Poisson approximation to the binomial. For convenience, these conditions are stated in terms of events occurring in time, although they also apply to events occurring in space.

Condition 1 Events defined in terms of nonoverlapping intervals are independent.

Condition 2 In a very small interval of time, the probability of two or more occurrences is very small.

Condition 3 The probability of a single event occurring in an interval of time Δt remains the same no matter where Δt is located.

Condition 4 The probability of a single event occurring in an interval of time Δt is approximately proportional to Δt.

Condition 4 has been included for the purpose of being mathematically precise. Actually, conditions 1 to 3 are sufficient for our purposes.

Consider an experiment in which we count radioactive particles being detected by a counting device over some time interval t. Let λ denote the rate of occurrences. Now consider the time interval t to be divided into n equal-size intervals of size $\Delta t = t/n$ (see Figure 4.17). Then the total number of particles detected in time t will be equal to the sum of the number of particles detected in each small interval Δt. (Note that this is very similar to the approach used in Example 4.5.)

We can now consider this to be a sequence of n experiments: count the number of particles detected in the first interval, the number detected in the second interval, . . . , the number detected in the nth interval. According to condition 1, each of these experiments will be independent. *If n is large enough*, then Δt will be small and according to condition 2, the probability of two or more occurrences during Δt is very small. Consequently, there will be essentially only two outcomes: either one particle will be detected or no particles will be detected (i.e., we are considering a sequence of independent dichotomous trials). Furthermore, condition 3 implies that the probability of a single event occurring, a "success," remains constant no matter where Δt is located. These are exactly the conditions needed for a random variable defined as the total number of successes in n trials to have a binomial distribution. (Note, however, that the condition of having only two possible outcomes is only approximately satisfied.) Consequently,

$$X_P = \text{total number of particles detected in time } t$$

and

$$X_B = \text{total number of "successes" in } n \text{ intervals of length } \Delta t$$

should have approximately the same distribution *if n is large enough*. It remains only to find the value of p for the binomial distribution. If we reason that X_P and X_B should

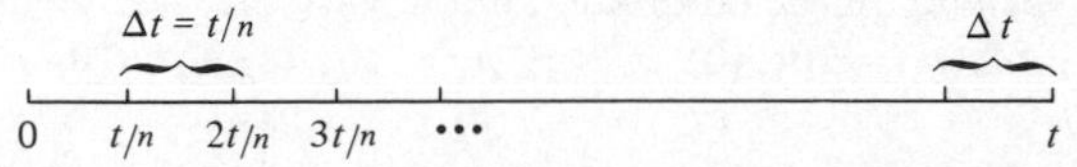

$\Delta t = t/n$ $\quad$ Δt

0 $\quad t/n \quad 2t/n \quad 3t/n \quad \cdots \quad t$

FIGURE 4.17
Counting Particles Detected in Time

have the same expected values, then

$$E(X_P) = \lambda t = E(X_B) = np.$$

From this we deduce that p must be $\lambda t/n$. Note that the approximation is good only when n is large, in which case

$$p = \frac{\lambda t}{n}$$

will be small. Summarizing, we would say that for large n and small p, the distribution of X_P (Poisson with expected value λt) will be approximately the same as the distribution of X_B (binomial with parameters n and $p = \lambda t/n$), having expected value np.

EXERCISES

1. Which of the following experiments can be classified as being dichotomous?
(a) Observing whether a coin falls heads or tails.
(b) Observing the outcome of a football game.
(c) Observing whether an arrow hits a bull's-eye.
(d) Observing the outcome of a roll of a fair die.

2. If X is a binomial random variable with $n = 3$ and $p = .12$, find (a) $P[X = 2]$; (b) $P[X = 0]$; (c) $P[X \geq 1]$.

3. Using the binomial table A1, find the following probabilities:
(a) $P[X = 0 \mid n = 5, p = .10]$; (b) $P[X \geq 1 \mid n = 8, p = .20]$;
(c) $P[X = 4 \mid n = 10, p = .70]$; (d) $P[X \geq 10 \mid n = 12, p = .80]$.

4. Using the cumulative binomial table A2, find the following probabilities:
(a) $P[X \leq 4 \mid n = 5, p = .15]$; (b) $P[X \geq 4 \mid n = 8, p = .20]$;
(c) $P[4 \leq X \leq 7 \mid n = 10, p = .30]$; (d) $P[5 \leq X \leq 9 \mid n = 12, p = .80]$.

5. Sixty percent of the customers entering the Golden Bank make a deposit into either their checking or savings account. Of the next 10 customers, what is the probability that (a) exactly 6 will make a deposit?; (b) at least 8 will make a deposit?

★*6.* Over a certain time is was noted that 12 girls in a row were born to families of staff members in the statistics department of a Midwestern university. (a) Find the probability that in the next 12 births to families of this group, all babies will be girls. (b) Find the probability that in the next 12 births to families of this group, all children will be the same sex. (*To think about:* Is the assumption of independence reasonable in this situation?)

★*7.* During a televised basketball game between the University of Missouri and Kansas University, the commentator noted that a player about to shoot a free throw was making 90% of his free throws. The player then missed his next three free throws. Assuming a binomial distribution, find the probability that a player making 90% of his free-throw shots would miss all of his next three shots. Is the assumption of independence reasonable in this situation? Explain.

8. Find the mean and variance of the binomial random variables having the following parameters: (a) $n = 5$, $p = .10$; (b) $n = 5$, $p = .90$; (c) $n = 10$, $p = .05$; (d) $n = 10$, $p = .95$.

9. Find the mean and standard deviation of the binomial random variables having the following parameters: (a) $n = 25$, $p = .2$; (b) $n = 25$, $p = .8$; (c) $n = 100$, $p = .10$; (d) $n = 100$, $p = .90$.

10. Using Table A1 to get the probability distribution for a binomial random variable X with $n = 5$ and $p = .30$, find $E(X)$ using the fomula $\sum_{x=0}^{5} x \cdot P[X = x]$. Compare your answer with the result obtained using the formula $E(X) = np$. Explain any discrepancy.

11. Using the tables for the binomial distribution, graph probability histograms for the following binomial distributions: (a) $n = 10$, $p = .5$; (b) $n = 10$, $p = .25$; (c) $n = 10$, $p = .85$.

12. Find the mean, median, and mode for each distribution given in Exercise 11.

13. In many states high school seniors must pass a state examination to show that they have attained a minimum proficiency in reading and mathematics before receiving a high school diploma. In past years 90% of the seniors at the Bay Area High School have passed this examination. (a) Estimate the probability that at least 16 of the 20 students in Mr. Ebenezar's class will pass this examination. (What assumptions have you made?) (b) How many of 310 seniors at B.A.H.S. would you expect to pass the examination this year?

14. The manager of the A & B Supermarket has found that the four types of cake mixes sold in his store, A & B, Betty Cookers, Duncan-Fines, and Billsberry sell at different rates. In fact, the four brands have 15, 30, 20, and 35%, respectively, of the cake-mix market in his store. (a) Find the probability that the next 8 sales of cake mixes will include 2 sales of each brand. (b) Find the expected number of Billsberry sales out of the next 20 sales. (c) The manager offers a special on the store brand (A & B) of cake mixes. Out of the next 20 sales he notices that 8 were A & B. Assuming that the special actually had no effect on sales, find the probability of 8 or more A & B sales. (d) In view of the answer to part (c), what might you say about the effect of the special on sales?

15. The random variables X_1, X_2, and X_3 have a multinomial distribution with $p_1 = .1$, $p_2 = .3$, and $p_3 = .6$. If $n = 10$, find

(a) $P[X_1 = 1, X_2 = 3, X_3 = 6]$; (b) $P[X_1 = 5, X_2 = 3, X_3 = 2]$;
(c) $P[X_3 \leq 3]$; (d) $E(X_1)$ and $V(X_1)$.

16. A large flock of geese consists mainly of Canada geese, with about 5% each being blue geese and snow geese. In a banding project, conservation workers used a special netting system to trap geese. Assuming that 10 geese were caught at once, find the probability that there would be 8 Canadas, 1 blue goose, and 1 snow goose. What assumption did you make?

17. Referring to Exercise 16, assuming the geese freely intermingle, what is the probability of netting at most 12 Canadas out of 20? What might you conclude if workers netted 20 and found only 12 Canadas?

18. The Detroit Dragons play a 10-game football schedule, 5 games at home and 5 games away. At home, the probability of their winning, losing, or tying is .8, .1, and .1, respectively, while away the probabilities are .7, .2, and .1. Assuming independence among games, find the probability that they will end the season with a 9–0–1 record.

19. If $X_1, X_2, \ldots, X_k$ have a multinomial distribution, the marginal distribution of X_i is binomial with parameters n and p_i. Use this fact to show that the random variables $X_1, X_2, \ldots, X_k$ are not independent.

20. Assume that J. Stout (see Example 1.3) has probability .15 of making a basket from the free-throw line on any given shot. If he is allowed to shoot until he makes a basket, find (a) the probability that he requires exactly 4 shots; (b) the expected number of shots he will have to take.

21. Let X be a random variable having a negative binomial (Pascal) distribution with parameters $p = .10$ and $r = 3$. Find (a) $P[X = 5]$; (b) $P[X \geq 4]$; (c) $E(X)$; (d) $V(X)$.

22. (a) Assuming that the geese referred to in Exercise 16 freely intermingle, find the probability that at least one snow goose is netted among 10 captured. (b) Assuming that 10 geese are captured each time the net is thrown and that workers will keep trying until they capture at least one snow goose, find the probability that they will need 3 or more throws of the net. (c) Find the expected number of throws of the net necessary to catch at least one snow goose.

★23. Before the start of the 1974 baseball season, Henry Aaron had hit 713 home runs in 12,725 times at bat (this number of times at bat includes walks and sacrifice flies). He needed 2 more home runs to total 715 and break Babe Ruth's record. Assuming that he would hit home runs at the same rate as in the past and assuming independence, and assuming four turns at bat in a game, find the probability that he would hit home run 715 before the end of the second game of the season. Find the expected number of turns at bat needed to hit home run 715.

24. A proposed amendment to the Constitution of the United States must be ratified by at least three-fourths of the states in order to be approved. (a) If the probability is .80 that any given state will ratify the proposed amendment, and if each state legislature considers the proposed amendment only once, find the expected number of state legislatures that will need to consider this amendment before it is approved. (b) Write out an expression for the probability that 45 legislatures will need to consider the amendment before it is approved.

25. Repeat Exercise 24(a) under the assumption that the probability is .60 that any given state will ratify the proposed amendment. What does your answer suggest concerning the prospect of this amendment being approved?

26. Given that a population consisting of 30 items contains 10 items of type A and 20 of type B and that X_1 represents the number of items of type A in a sample of 6 items, find (a) $P[X_1 = 2]$; (b) $P[X_1 \geq 1]$; (c) the expected value of X_1; (d) the variance of X_1.

27. An auditor, Mr. I. Checkum, wishes to check the savings accounts in a small bank owned by Mr. Rip M. Auf. The bank has 1000 accounts, and if there is reason to believe that 5% of the accounts are in error, the auditor will do a complete audit. Mr. Checkum decides to sample 20 accounts and do a complete audit if 1 or more accounts are found to be in error. Find the probability that Mr. Checkum will do a complete audit if (a) 10, (b) 50, (c) 100 of the accounts are actually in error.

28. A lot of 20 watzits contains 3 items with minor defects and 2 with major defects. An inspector chooses 5 items from this lot for inspection and will accept the entire lot if he finds at most one item with nothing worse than a minor defect. (a) Find the probability that he will accept this lot. (b) Find the probability that he will accept the lot if it contains 6 items with minor defects and 4 with major defects.

29. Mr. Ken D. Barr owns a small candy store. He sells individually wrapped chocolate candies for 2 cents each. Of 200 candies, 150 are wrapped in silver foil and the rest are wrapped in red foil. If Billy buys 5 candies, find the probability that he gets exactly 2 red (a) using the hypergeometric distribution; (b) using the binomial approximation to the hypergeometric. (Assume random selection.)

30. In the lower house of the parliament of the country of Upper Midovia, there are 250 representatives. Although they have not made a public declaration, 150 of these representatives actually favor a controversial tax bill. A reporter interviews seven representatives at random. (a) What is the exact probability that more than 5 of these representatives will be in favor of this bill? (b) Use the binomial distribution to approximate this answer. Under the assumption that only 100 representatives favor the bill, (c) repeat part (a); and (d) repeat part (b). [*To think about:* Intuitively, would you expect the probability found in part (a) to be larger or smaller than the one found in part (c)? Is it?]

31. If X is a Poisson random variable with parameter $\lambda t = 2$, find (a) $P[X = 4]$; (b) $P[X \geq 2]$; (c) $E(X)$; (d) $V(X)$.

32. The number of accidents per week at the intersection of Broadway and Main can be assumed to follow a Poisson distribution with parameter $\lambda = .5$/week. Find the probability of (a) at least 2 accidents at this intersection next week; (b) no accidents in the next 2 weeks; (c) the expected number of accidents next year.

★33. During the years 1950–1973 (inclusive), 18 tornadoes were reported to have touched down in Story County, Iowa. (a) If the number of tornadoes touching down in a year follows a Poisson distribution, find (approximately) the expected number of tornadoes touching down next year. Find the probability of (b) at least one tornado; (c) at most two tornadoes touching down next year.

★34. The following data give the number of deaths from the kick of a horse per corps per year for 10 Prussian Army Corps for 20 years (a total of 200 observations).

Number of deaths/corps/year:	0	1	2	3	4
Observed frequency:	109	65	22	3	1.

(a) Find the mean number of occurrences, that is, the total number of deaths divided by the total number of observations. (b) Assuming that the number of deaths/corps/year follows a Poisson distribution with λt equal to the mean value found in part (a), find the probabilities corresponding to 0, 1, . . . , 4 occurrences. (c) The expected number of times that there would be zero occurrences would be $n \cdot p_0 = 200\,(P[X = 0])$. Calculate this expected value as well as the expected number of times that there would be 1, 2, 3, or 4 occurrences. (Now compare the expected frequencies with the observed frequencies. Does there appear to be good agreement?)

35. Iris is the owner of the Stork Shop, a store specializing in maternity needs. She has found that female shoppers enter her store at the rate of 30/hour, while male shoppers have a rate of only 5/hour. Find the probability that exactly 2 customers (without regard to their sex) will enter the Stork Shop in the next 6 minutes. (Assume that customers enter one at a time.)

36. Think of a counting experiment that *you* can perform which might adequately be described by a Poisson random variable. *Carefully* describe the experiment. This should be an experiment that involves counting the number of occurrences of some event over a time or space interval. It should be an experiment that can be repeated 50 times at minimal cost and should take no more than 1 hour.

37. Perform the experiment described in Exercise 36 fifty times. (a) List the 50 counts. (b) Make a frequency distribution as in Exercise 34. (c) Find the mean number of occurrences. [You may wish to round this number when using the tables in part (d).] (d) Assuming a Poisson distribution, find the probability corresponding to 0, 1, 2, . . . occurrences. (e) Find the expected number of times there will be 0, 1, 2, . . . occurrences. Does there appear to be good agreement? (Cf. Exercise 34.)

38. If X has a uniform distribution over the interval 1 to 5, find (a) $E(X)$; (b) $V(X)$.

39. A random variable Y has a uniform distribution over the interval 5 to 10. (a) Sketch the density function. Find (b) $P[4 \leq X \leq 8]$; (c) $E(X)$; (d) $V(X)$.

40. The exact number of minutes after the hour that a baby is born, X, can be considered a uniform random variable over the interval 0 to 60. Find (a) $P[0 \leq X \leq 30]$; (b) $E(X)$; (c) σ_X.

41. Given that X is a random variable having a uniform distribution over the interval (a, b) and that $E(X) = 10$ and $V(X) = 27$, find a and b.

• **42.** Show that if X is a random variable having a uniform distribution over the interval (a, b), then $E(X^2) = (b^3 - a^3)/[3(b - a)]$.

43. Let Z be a standard normal random variable. Find

(a) $P[Z \leq 1]$; (b) $P[Z \leq .5]$; (c) $P[Z \geq .82]$;
(d) $P[Z \leq -1.5]$; (e) $P[Z \geq -1.96]$.

44. Let Z be a standard normal random variable. Find the constant b such that (a) $P[Z \leq b] = .50$; (b) $P[Z \leq b] = .70$; (c) $P[Z \geq b] = .30$. Also find (d) $z_{.40}$; (e) $z_{.80}$.

45. Let X be a normal random variable with mean $\mu = 2$ and variance $\sigma^2 = 9$. Find (a) $P[X \leq 5]$; (b) $P[X \leq 4]$; (c) $P[X \geq 6]$; (d) $P[X \leq 1]$; (e) $P[X \geq 0]$.

46. Let X be a normal random variable with mean $\mu = -10$ and variance $\sigma^2 = 400$. Find (a) $P[X \leq 0]$; (b) $P[X \geq 10]$; (c) $P[X \leq -20]$; (d) $P[X \geq -30]$.

47. Let X be a normal random variable with mean $\mu = 2$ and variance $\sigma^2 = 9$. Find the constant b such that (a) $P[X \leq b] = .50$; (b) $P[X \leq b] = .60$; (c) $P[X \geq b] = .25$; (d) $P[X \geq b] = .95$.

48. Let X be a normal random variable with mean $\mu = -10$ and variance $\sigma^2 = 400$. Find the constant b such that (a) $P[X \leq b] = .50$; (b) $P[X \leq b] = .75$; (c) $P[X \geq b] = .35$; (d) $P[X \geq b] = .99$.

★**49.** According to figures provided by the National Center for Health Statistics (NCHS), the heights of seven-year-old girls are approximately normally distributed, with mean $\mu = 47.5$ inches and standard deviation 2.1 inches. Find the probability that a randomly selected seven-year-old girl will be between (a) 46.0 and 49.0 inches tall; (b) 45.0 and 48.0 inches tall.

50. Using the information given in Exercise 49, it can be said that 90% of the seven-year-old girls are taller than b inches. Find b. Twenty-five percent of the seven-year-old girls are taller than c inches. Find c. Aunt Minnie has come to visit and remarks that seven-year-old Susie seems tall for her age. Susie is actually 49 inches tall. Comment on Aunt Minnie's remark.

51. The amount of cola dispensed by a bottling machine into a "16-ounce" bottle of Fizzy Cola is a normal random variable with a mean of 16.10 ounces and a standard deviation of 0.50 ounce. (a) Find the probability that a bottle will contain at least the advertised amount. (b) If the bottles will hold exactly 17 ounces, what proportion of bottles will be filled to overflowing?

52. (a) Referring to Exercise 51, find the constant c such that 70% of the bottles filled contain at least c ounces. (b) Find the probability that at least 6 of 8 bottles in a carton of Fizzy Cola contain at least c ounces.

53. A package of four 60-watt General Electric standard light bulbs is labeled "Avg. life 1000 hours." Assume that the light bulbs have a lifetime that is normally distributed with a standard deviation of 200 hours. (a) What is the probability that a light bulb of this type will last for at least 1200 hours? (b) What is the probability that all four bulbs in this package will last 1000 hours or more?

54. Mr. Eric Peter Arbuckle has kept careful records of the gas mileage obtained by his Mini-Mac, a compact pickup. EPA estimates that his average mileage in city driving is 26.4 mpg with a standard deviation of 2.0 mpg. (a) Assuming that the mileage on a tank of gas is normally distributed, find the probability that he will average better than 30.0 mpg on a tank. (b) Eighty percent of the time he averages at least d mpg. Find d. (*To think about:* Might you expect gas mileages to be normally distributed, or are they more likely to be skewed?)

55. If T is a continuous random variable having an exponential distribution with parameter $\lambda = 2$, find (a) $P[T > 3]$; (b) $E(T)$; (c) $V(T)$.

56. Let T be a continuous random variable having an exponential distribution with parameter $\lambda = 4$. Find (a) $P[1 \leq T \leq 2]$; (b) $E(T)$; (c) σ_T.

57. Mr. Ken Rite, a reporter for the Daily Exposé, has written an article about the dangerous Broadway and Main intersection (see Exercise 32). He plans to publish the article after the next accident at that intersection. (a) What is the expected amount of time he will have to wait before publishing the article? (b) What is the probability that he will have to wait at least 4 weeks?

58. The amount of time required for a customer to be checked out at the B-Q (Be Quick) supermarket is exponentially distributed with a mean time of 2 minutes per customer. (a) Find λ. (b) If there is one customer ahead of you, what is the probability that it will take longer than 3 minutes for his checkout? (c) If there are two customers ahead of you, what is the probability that at least one of them will take longer than 3 minutes to check out?

59. Let X_B be a binomial random variable with $n = 400$ and $p = .50$. Use the normal approximation to estimate (a) $P[X_B \leq 210]$; (b) $P[X_B \geq 185]$.

60. A binomial random variable Y_B has parameters $n = 100$ and $p = .10$. Use the normal approximation to estimate $P[8 \leq Y_B \leq 15]$.

61. A coin that is thought to be fair was flipped 900 times and 488 heads were observed. (a) Estimate the probability of observing 488 or more heads if the coin is, in fact, fair. (b) What might you conclude about the fairness of this coin?

62. A box contains 36 packages of four 60-watt bulbs (a total of 144 bulbs, as described in Exercise 53). Find the probability that 80 or more of these bulbs will last at least 1000 hours.

63. Mr. Philip D. Tangk operates a small gas station. Phil estimates that 20% of all cars entering his station are in need of an oil change. What is the probability that of the next 100 cars entering his station (a) at least 15 need an oil change?; (b) at most 25 need an oil change?; (c) exactly 20 need an oil change?

64. A large tank of tropical fish contains 100 guppies. The probability that a guppy will contract the dreaded fishus-stinkus disease is .10. (a) Assuming independence, what is the probability that at least 5 guppies will contract this disease? (b) Answer the same question if the probability of contracting the disease is .01. (*To think about:* Is the assumption of independence reasonable?)

65. A binomial random variable X_B has parameters $n = 100$ and $p = .02$. Use the Poisson approximation to estimate (a) $P[X_B \leq 1]$; (b) $P[X_B \geq 3]$.

66. A binomial random variable Y_B has parameters $n = 500$ and $p = .01$. Estimate $P[3 \leq Y_B \leq 5]$ and $P[6 \leq Y_B \leq 8]$ using (a) the normal approximation to the binomial; (b) the Poisson approximation to the binomial.

67. Assume that the probability that a mouse subjected to a low dosage of a suspected carcinogenic chemical will develop a tumor is .01. If 50 mice are subjected to this dosage, what is the probability that (a) exactly one mouse, (b) two or more mice will develop tumors.

68. Assume that the probability of a triple play occurring in a baseball game is .001. If you plan to keep going to baseball games until you see a triple play, how many games should you expect to attend? Estimate the probability of seeing a triple play for the first time in the 1001st game you attend.

5

SAMPLING THEORY AND SAMPLING DISTRIBUTIONS

5.1 INTRODUCTION

Did you ever come home from school and find your mother baking cookies? If your response to such an occurrence was typical, you probably asked if you could have one (assuring mother that such an indulgence would have minimal effect on your appetite). If given approval (and perhaps even if not), you would choose a cookie from those available. If you have ever done this, you have engaged in the statistical practice of sampling.

Now let us state some definitions. A collection of objects having a certain well-defined set of characteristics will be called a *population*. Any subset of a population will be called a *sample* from that population. In many situations, although we are interested in some aspect of a particular population, we cannot examine the entire population because of cost and/or time considerations. In situations such as these we generally settle for examining a portion of the population (i.e., a sample), in the hopes that the sample will "mirror" the population or be "representative" of the population.

Example 5.1 When George Gallup, Lou Harris, and other well-known pollsters are interested in the proportion of Americans favoring a political candidate or issue, they would, theoretically, like to poll all Americans. However, they do not have the resources to interview the entire population. Instead, they must resort to selecting a portion of the population in the hopes that the sample they choose will accurately reflect the population at large.

Other situations where one might use sampling instead of examining an entire population would be an auditor who checks a portion of the accounts held by a certain business, a personnel manager who examines the records of a few employees for absenteeism, or a health inspector who visits a few of the restaurants in a large city. In still other situations a person in upper management might interview a sample of employees to find their attitudes toward their immediate supervisor, or a medical doctor may treat a sample of animals or patients (depending on the stage of experimentation) to determine the efficacy of a particular drug and/or radiation treatment in combating a particular type of cancer. No doubt you can think of dozens of other situations where sampling is done.

Let us return again to the situation described earlier where your mother had given you permission to take a cookie. What did you do after receiving permission? You carefully selected a cookie looking for a large one that looked perfectly done. You then proceeded to eat the cookie and then made a qualitative judgment about its goodness. (If it was a really great cookie, you might have asked for another, while if it was really bad, you might have given it to the baby for use as a teething biscuit.) There are two points we want to make here. First, the cookie you chose was not randomly selected. Although your cookie did in fact comprise a sample, it did not comprise a "random sample." Second, you were not interested in the cookie per se, but you were interested in making a "measurement" on the cookie you chose. If the measurement were a judgment about the overall quality, you might simply rate the cookie as "great," "good," "fair," "poor," or "terrible." You, of course, do this by tasting the cookie in hopes that it will receive a top rating. On the other hand, there are other more objective measurements that you might make; for example, you might count the number of raisins (or nuts or chocolate chips) in the cookie. We must emphasize that most sampling involves two stages: first, obtaining a portion of the objects from the entire population, and second, making some appropriate measurement on each object. The purpose of sampling is to allow us to use information gained from the sample to make statements about the entire population from which the sample was taken.

5.2 RANDOM SAMPLING

What is wrong with carefully selecting the biggest and best-looking cookie from those available instead of randomly selecting one? From your point of view, of course, nothing is wrong. However, if your sister were to see your cookie and not see the entire batch (population) of cookies, she could certainly get the wrong idea as to the quality of the remaining cookies. Your sample does not reflect the population as a whole. If your cookie were chosen at random, it would more likely be typical of the population.

As you might imagine, much grief can be caused by not choosing a sample properly. Think about this for a while. If, for example, you decided to gather opinions about a political candidate by using a telephone survey and placed your calls to households during the hours of 9 A.M. to 5 P.M., who would tend to answer the phone? Who would not be there to answer the phone? Would the persons answering be a fair reflection of the entire population in your area?

One way to choose a sample of objects from a population of objects so that the sample is likely to be representative is to choose a simple random sample.

Definition 5.1 A *simple random sample of n objects* chosen from a population of objects is one chosen in such a way that all samples of size n are equally likely to be chosen.

A random sample of objects can be chosen either with or without replacement. Clearly, there are situations where sampling with replacement would not be reasonable. A child sampling cookies will not sample with replacement; an inspector testing flashbulbs will not put back a used flashbulb; a pollster will not want to allow the possibility of interviewing the same person more than once. However, we will see later that the mathematics involved in analysis is somewhat simpler if sampling is done with replacement. Furthermore, if the population size is large relative to the sample size, then the analysis used when sampling with replacement gives a good approximation to the results that would have been obtained for the analysis used when sampling without replacement (cf. binomial–hypergeometric comparisons in Section 4.10). For these reasons we will assume that sampling is done with replacement unless stated otherwise.

Choosing a random sample of objects is not as simple as it might appear. If you want to randomly select a cookie, you might reach into the cookie jar, thoroughly mix the cookies, and take one out without looking. This could lead to a random selection, but you might imagine some difficulties with such a scheme. Furthermore, such a procedure would not be workable if you were trying to take a sample of students from a class or rattlesnakes from a cage! An alternative procedure would be to assign a number to each member of a population, select a number at random, and then choose the corresponding member of the population to be in your sample. You can probably think of various ways that a number can be selected at random. One way is to use a list of random digits as given in Table A5, as illustrated in the next example.

Example 5.2 For simplicity we will take our population to be the first 100 students listed in Appendix A3. Using the random number tables, choose a random sample of five students from this population with replacement.

We should choose a "random" starting point in Table A5, but we will start in the upper left-hand corner of the table for the convenience of this example. Since our population consists of 100 members, we will look at two-digit numbers. Reading from left to right we find

10 48 01 50 11,

so the corresponding students would comprise our sample. (Notice, by the way, that since no number was repeated in this sample of numbers, we would have gotten the very same sample if we had been sampling without replacement.)

At the end of Section 5.1 we said that most sampling involves two stages: obtaining the sample of objects and making the appropriate measurement on each object. So far in this section we have only considered the first stage. We now wish to turn our attention to the second stage. Let us begin by considering a sample of size 1 being taken from a population. The actual sampling process can be considered an experiment, so the set of all possible outcomes of the sampling procedure can be considered to be the sample space S. S will simply be the entire population of objects. Now if the measurement made on the object selected by the sampling procedure is a real number, we are assigning a real number to each element of S, and hence are defining a random variable. If, for example, we were interested in the age of the student sampled, the "measurement" we make would be to find the age, and we could define a random variable X by

$$X = \text{age of student chosen.}$$

Similarly, we might measure the height or weight or grade-point average of the student chosen, in which case we could define corresponding random variables Y, W, and Z, say. (We will discuss different types of measurements that might be made in Section 6.7.)

Continuing with the situation described in Example 5.2, where the first 100 students listed in Appendix A3 comprised our population, we find that 47 students are 19 years old while 40, 11, and 2 are 20, 21, and 22 years old, respectively. Hence, if a student is chosen at random from among these 100 students, then the probability that the student will be 19 years old will be 47/100. Using this reasoning it follows that the probability distribution for X, the age of the student chosen, will be

(5.1)

x:	19	20	21	22
$P[X = x]$:	.47	.40	.11	.02.

If we were to choose five students, say, with replacement from this population and were to "measure" the age of each one, we would define five random variables as follows:

$$X_1 = \text{age of first student chosen,}$$
$$X_2 = \text{age of second student chosen,}$$
$$\vdots$$
$$X_5 = \text{age of the fifth student chosen.}$$

Each of these random variables will have a probability distribution associated with it. It should not be hard to see that the distribution of X_1 will be exactly the same as the distribution of X given in (5.1), since both random variables relate to the age of a randomly chosen student. That is, the distribution of X_1 is given by

$$\begin{array}{lcccc} x_1: & 19 & 20 & 21 & 22 \\ P[X_1 = x_1]: & .47 & .40 & .11 & .02. \end{array}$$

If you remember that the sampling is being done with replacement, it will not be hard to see that X_2 will also have exactly the same distribution as that of X (or X_1) and that X_1 and X_2 will be independent. Similarly, the distributions of X_3, X_4, and X_5 will have the same distribution as that of X and, in fact, the sequence of random variables $X_1, X_2, \ldots, X_5$ will be independent. From this we see that if we start with a random sample of objects from a population of objects and if we consider measurements made on each of those objects, we are led to a sequence of independent random variables all having the same probability distribution. In statistical terminology this sequence of random variables is said to constitute a random sample from the probability distribution of the random variable X.

Definition 5.2 A sequence of random variables $X_1, X_2, \ldots, X_n$ is said to be a *random sample from the probability distribution of a random variable* X if they are independent and if each has the same probability distribution as X.

Example 5.3 The random variables $X_1, X_2, \ldots, X_5$ as defined earlier are a random sample from the probability distribution of the random variable X defined in (5.1). For the particular sample found in Example 5.2, the observed sample values are

$$20, \quad 21, \quad 19, \quad 20, \quad 19.$$

Note that if a sample of objects is chosen *without* replacement, the corresponding sequence of random variables, based on appropriate measurements, will *not* be independent, although they will all have the same probability distribution. It is this lack of independence that complicates the mathematical analysis and leads us to generally consider sampling with replacement.

It is possible to obtain a random sample from a probability distribution in the sense of Definition 5.2 without taking measurements on a set of objects obtained by sampling in the sense of Definition 5.1. If, for example, you were to roll an ordinary die n times and defined

$$\begin{aligned} X_1 &= \text{number appearing on first toss,} \\ X_2 &= \text{number appearing on second toss,} \\ &\vdots \\ X_n &= \text{number appearing on } n\text{th toss,} \end{aligned}$$

these would constitute a sequence of independent random variables all having the same probability distribution. In this case there is no population of objects that were sampled to obtain these values. Nevertheless, we can think of these values as a sample in the sense that they constitute just a portion of all possible observations that might be made.

We will see in Chapter 7 that there are types of sampling other than simple random sampling. These other types of sampling generally are used in an effort to ensure that a sample will be more representative of the population. The fact that they are more representative can lead to increased precision and/or can be more cost-effective. Unfortunately, the gain in assuring a more representative sample is obtained at the cost of more complicated mathematical analysis.

5.3 NONSAMPLING ERRORS

As you continue reading this chapter, you must keep in mind the reason for sampling. An individual would like information about a certain population but is unable to examine the entire population. Consequently, a portion of the population (i.e., a sample) must be examined. Frequently, a simple random sample is taken in the hopes that the sample so obtained will be representative of the population at large. Of course, it is perfectly possible for a simple random sample to be "unrepresentative." For example, a sample of students could result in finding all students of ages 21 or 22, even though these ages represent the "extreme" of the distribution. Similarly, a set of 1000 accounts may contain only 10 accounts with errors. However, an auditor taking a random sample of 10 accounts *may* find that all accounts in the sample contain errors. Events such as these have small probability of occurring, but they can happen. This type of "error" gives a distorted view of the population, but it is accounted for in the statistical analysis. Errors of this type, due to chance variations, are called *sampling errors*. The error is not a mistake or blunder, but simply a result of the fact that only a portion of the entire population is being examined.

Now consider the situation where you have been hired by a newly formed polling company, the Ask-All-Tell-All Company. It is your job to interview randomly chosen individuals at their homes. (Or you may wish to put the shoe on the other foot and imagine yourself to be the interviewee rather than the interviewer.) What kind of answers might you get to the following questions:

"How old are you?"

"How much money did you earn last year?"

"Did you cheat on your income tax?"

"Do you spend more money on alcoholic beverages than you give to charity?"

"How often do you beat your spouse?"

Do you think you will always get honest, accurate answers to questions such as these? Do you think the answers you would get would be the same as those which would be given to someone of the opposite sex? You should be able to see from this that answers given (if any) will depend on the sensitivity of the question and the degree to which a respondent feels threatened by the question or by the interviewer.

What kind of results do you think you would get if you analyzed the data obtained in response to questions such as these? Certainly, the results will be of very little value, no matter how careful the statistical analysis, if the data obtained are inaccurate (i.e., the results will be in error). These errors are not due to the fact that a sample is used instead of the entire population, but are due to factors other than sampling variation. Consequently, they are called *nonsampling errors*.

Another kind of nonsampling error that can be made is an error in recording or transcribing information. A carelessly written "7" might later be read as a "1", or vice versa. In punching data onto computer cards, it is very easy to punch the wrong number. Errors of this kind can completely destroy the validity of a carefully designed and analyzed statistical study.

Another kind of nonsampling error can come about due to nonrespondents. A nonrespondent is a person selected to be in a sample from whom data cannot be collected either because he cannot be contacted (e.g., his telephone is out of order) or because he chooses not to answer questions put to him either in person or on a questionnaire. This is often a serious problem with using questionnaires as a method of obtaining data. The proportion of questionnaires returned is often quite low, frequently in the range 20 to 30%. The reason this causes difficulty is that the respondents (from whom you collect data) may be quite different from the nonrespondents (from whom you collect no data). For example, if the Ask-All-Tell-All Company sent to randomly selected voters a letter for the local Republican committee soliciting donations and also asking questions about certain issues and candidates, could they expect the respondents and nonrespondents to have the same attitudes toward these issues? Would you expect the responses to be typical of all voters? You should be able to see that nonresponse can cause a severe problem of biased results.

In most of our work from this point on we will assume that the data we have are free of nonsampling error. However, if in the future you are ever called upon to collect or analyze data in a real-world situation, you should remember the importance of controlling nonsampling errors.

5.4 STATISTICS

If you were given a list of the names and telephone numbers of 100 students at your school and were given 20 minutes to come up with an estimate of the mean number of credit cards held per student, what would you do? Think about it for a minute. You certainly would not have time to call everyone, so you might just start calling students, recording the number of credit cards for each one who answered, and continue doing this until your time was up. Then you would find the "average" number of cards held by those who answered and use this average as an estimate of the mean number of credit cards held by all people on the list. The usual method for finding the "average" is to add up the numbers and divide by the number of observations. In statistical terminology this quantity is referred to as the sample mean.

Definition 5.3 Given a sequence of random variables $X_1, X_2, \ldots, X_n$, a random sample from a probability distribution, the *sample mean* is defined by

(5.2)
$$\bar{X} = \frac{X_1 + X_2 + \ldots + X_n}{n}$$

or

$$\bar{X} = \frac{\sum_{i=1}^{n} X_i}{n}.$$

Since the quantity $\bar{X}$ will depend on the random variables $X_1, X_2, \ldots, X_n$, it, too, is a random variable. The sample mean is an example of a "statistic."

Definition 5.4 A *statistic* is any quantity calculated from sample observations.

Statistics are often sample analogs of parameters. Recall from Section 4.1 that parameters are quantities that characterize the probability distribution of a random variable. The mean and median, for example, characterize the center of the probability distribution while the variance and standard deviation characterize the variability of the probability distribution. In a similar way, statistics are quantities that are characteristic of a sample. Generally, parameters are constant values, while statistics are random variables.

Example 5.4 Find the observed value of $\bar{X}$ if in the 20 allotted minutes you were able to call five students and found that they had 3, 0, 4, 0, and 1 credit cards, respectively.

Using (5.2), we find that

$$\bar{x} = \frac{3 + 0 + 4 + 0 + 1}{5} = 1.6.$$

(Note that we have followed our usual convention of letting uppercase letters denote random variables and lowercase letters denote values assumed by random variables.)

Definition 5.5 Given a random sample of observations, $X_1, X_2, \ldots, X_n$, the *sample median*, $\tilde{X}$, is defined to be the middle value when the values are arranged in ascending order. If n is even, the sample median is taken to be the mean of the middle two observations.

Example 5.5 Find the observed value of $\tilde{X}$ using the information from Example 5.4. What would the median be if a sixth student was found to have two credit cards?

The first step in finding the sample median is to arrange the observed values in increasing order to get

$$0, \quad 0, \quad 1, \quad 3, \quad 4.$$

Since there are five values, the third value is the middle one, hence

$$\tilde{x} = 1.$$

(Note that had you taken the middle value of the observations in their original order, you would have wrongly determined the median to be 4.)

Had a sixth student reported having two credit cards, the observed values in increasing order would be

$$0, \quad 0, \quad 1, \quad 2, \quad 3, \quad 4.$$

Since there are six values, the third and fourth are the middle two values, so

$$\tilde{x} = \frac{1 + 2}{2} = 1.5.$$

We could define the sample analog of the mode of a probability distribution to be the sample mode, the observed value that occurs most frequently. This quantity is of very little practical use; hence, we will not dwell on it here.

The sample mean and median (as well as the mode) are considered to be measures of the center of the sample values. We are often interested in the variability in a set of data and hence define the sample variance as follows:

Definition 5.6 Given a random sample of observations, $X_1, X_2, \ldots, X_n$ the *sample variance* S^2 is defined by

(5.3)
$$S^2 = \sum_{i=1}^{n} \frac{(X_i - \bar{X})^2}{n - 1}.$$

Note the close analogy here to the variance of a probability distribution, σ^2. In Section 2.7 we defined σ^2 by $E(X - \mu)^2$, or in words, the long-run average value of the squared deviation of X about its mean μ. In Definition 5.6 we defined the sample variance to be a sample average value of the squared deviation of the sample observations X_i about their sample mean $\bar{X}$. Of course, we generally think of a sample average as determined by dividing a sum by the number of observations, which in this case would be n. Our reason for dividing by $n - 1$ will be explained in Chapter 7. Note that S^2 will be undefined when $n = 1$, since the divisor in (5.3) would be zero. This does not really present a problem, since if there is only one observation in a sample, there is no basis for determining variability anyway.

Example 5.6 A bank supervisor recorded the number of errors made by five bank tellers on a Friday. The numbers were 2, 0, 0, 1, and 2. Find the sample variance.

The first step in finding the sample variance is to find the sample mean. We get

$$\bar{x} = \frac{\sum_{i=1}^{5} x_i}{5} = \frac{2 + 0 + 0 + 1 + 2}{5} = 1.$$

Using this in (5.3), we find that

$$\begin{aligned} s^2 &= \frac{\sum_{i=1}^{5} (x_i - \bar{x})^2}{5 - 1} \\ &= \frac{(2 - 1)^2 + (0 - 1)^2 + (0 - 1)^2 + (1 - 1)^2 + (2 - 1)^2}{4} \\ &= \frac{1 + 1 + 1 + 0 + 1}{4} = 1. \end{aligned}$$

Using the fact that $\sum X_i = n\bar{X}$, we can derive a formula for finding S^2 which is algebraically equivalent to (5.3), but which is sometimes easier to use in making calculations. By expanding the term $(X_i - \bar{X})^2$ and using properties of the summation operation, we obtain the following:

$$\begin{aligned}\sum (X_i - \bar{X})^2 &= \sum (X_i^2 - 2X_i\bar{X} + \bar{X}^2)\\ &= \sum X_i^2 + \sum(-2X_i\bar{X}) + \sum \bar{X}^2\\ &= \sum X_i^2 - 2\bar{X}\sum X_i + n\bar{X}^2\\ &= \sum X_i^2 - 2\bar{X}(n\bar{X}) + n\bar{X}^2 = \sum X_i^2 - n\bar{X}^2.\end{aligned}$$

Using this, we find that

$$S^2 = \frac{\sum X_i^2 - n\bar{X}^2}{n-1}. \tag{5.4}$$

Example 5.7 The number of errors recorded by a bank supervisor for five tellers was 2, 0, 0, 1, and 2 (see Example 5.6). Find the sample variance S^2 using (5.4).

We already found that $\bar{x} = 1$. We next find that

$$\begin{aligned}\sum x_i^2 &= 2^2 + 0^2 + 0^2 + 1^2 + 2^2\\ &= 4 + 0 + 0 + 1 + 4 = 9.\end{aligned}$$

From (5.4), we find that

$$s^2 = \frac{9 - 5(1^2)}{5-1} = \frac{4}{4} = 1.$$

Just as the standard deviation σ is the square root of the variance σ^2, the sample standard deviation, which we will denote by S, is the square root of the sample variance S^2. There are other measures of variability in a sample, although these are not nearly as important as the sample variance or standard deviation. We will just briefly mention that the sample range is defined to be the difference between the largest and smallest values in a sample. In the data of Example 5.6, and 5.7, for example, the range is $2 - 0 = 2$. Another measure of variation is the sample analog of the expected absolute deviation about the median defined in Section 2.7. This would be defined by

$$\frac{\sum_{i=1}^{n} |X_i - \tilde{X}|}{n}$$

and might be called the sample mean deviation about the median. Many authors define the sample mean deviation about the mean by

$$\frac{\sum_{i=1}^{n} |X_i - \bar{X}|}{n}.$$

Neither of these measures of variations is used very much because of the mathematical difficulties involved using absolute values.

5.5 EXACT PROBABILITY DISTRIBUTION FOR SOME STATISTICS

At the beginning of the last section we pointed out that the value of the sample mean $\bar{X}$ could be used as an estimate of a population mean μ. In Chapter 7 we will discuss the problem of estimation of parameters in more detail. In order to have such discussions, it will first be necessary to learn more about characteristics of the statistic $\bar{X}$ and of other statistics.

When we defined the sample mean in the previous section we pointed out the fact that the sample mean, and, in fact, all statistics, are random variables. All random variables have a probability distribution, so it follows that the sample mean, sample median, sample variance, and so on, will each have a probability distribution. If we can find the probability distribution for a given statistic, that will allow us to find probabilities of certain events and to find expected values. We will see that finding the exact probability distribution for sample statistics based on even small samples is rather tedious, although not very difficult in theory.

We will consider the relatively simple situation of obtaining the distribution of $\bar{X}$ and S^2 when samples of size 2 are taken from the probability distribution (5.1) (i.e., the distribution of ages of a randomly selected student from the list of 100 students). For convenience we will rewrite that probability distribution here and then use that distribution to find the mean and variance of X. The probability distribution is

(5.5)

x:	19	20	21	22
$P[X = x]$:	.47	.40	.11	.02,

from which we find that

$$E(X) = \mu_X = \sum x \cdot P[X = x] = 19(.47) + 20(.40) + 21(.11) + 22(.02)$$
$$= 19.68$$

and

$$V(X) = \sigma_X^2 = \sum (X - \mu_X)^2 P[X = x]$$
$$= (19 - 19.68)^2(.47) + (20 - 19.68)^2(.40) + (21 - 19.68)^2(.11)$$
$$+ (22 - 19.68)^2(.02)$$
$$= .5576.$$

(Note that if you were to add up all the ages in the population of 100 students and then divide by the number in the population, 100, you would obtain the value 19.68. For this reason, μ_X is often referred to as the *population mean*. Similarly, σ_X^2 is referred to as the *population variance*.)

We now wish to consider random samples of size 2 taken from this probability distribution with replacement. Denote the observations by X_1 and X_2:

$$X_1 = \text{age of first student chosen}$$
$$X_2 = \text{age of second student chosen.}$$

To find the probability distribution of $\bar{X}$, remember that the probability distribution of any random variable is specified by (1) listing the possible values the random variable can assume, and (2) listing the corresponding probabilities for each of these values. What is the smallest value that $\bar{X}$, the sample mean, can assume? The smallest value is 19. What values in the sample will lead to this value? $\bar{X}$ will equal 19 if and only if $X_1 = 19$ and $X_2 = 19$. This information allows us to find the probability corresponding to the event $\bar{X} = 19$. We have

$$P[\bar{X} = 19] = P[X_1 = 19 \text{ and } X_2 = 19]. \tag{5.6}$$

Now finding the probability on the right-hand side of (5.6) can be greatly simplified when we remember that X_1 and X_2 are independent and that X_1 and X_2 both have the same probability distribution as that of X given in (5.5) (see Definition 5.2). With this in mind, we have

$$\begin{aligned} P[X_1 = 19 \text{ and } X_2 = 19] &= P[X_1 = 19] \cdot P[X_2 = 19] \\ &= (.47)(.47) = .2209. \end{aligned}$$

The next larger value that $\bar{X}$ can assume is 19.5. This value can occur in two mutually exclusive ways: ($X_1 = 19$ and $X_2 = 20$) or ($X_1 = 20$ and $X_2 = 19$). Hence we find that

$$\begin{aligned} P[\bar{X} = 19.5] &= P[(X_1 = 19, X_2 = 20) \text{ or } (X_1 = 20, X_2 = 19)] \\ &= P[X_1 = 19, X_2 = 20] + P[X_1 = 20, X_2 = 19] \\ &= P[X_1 = 19] \cdot P[X_2 = 20] + P[X_1 = 20] \cdot P[X_2 = 19] \\ &= (.47)(.40) + (.40)(.47) \\ &= .1880 + .1880 = .3760. \end{aligned}$$

TABLE 5.1

Value of X_1 (*Age of First Student*)	*Value of* X_2 (*Age of Second Student*)	$P[X_1 = x_1, X_2 = x_2]$		$\bar{X}$	S^2
19	19	.2209		19.0	0.00
19	20	.1880		19.5	0.50
19	21	.0517	←	20.0	2.00
19	22	.0094		20.5	4.50
20	19	.1880		19.5	0.50
20	20	.1600	←	20.0	0.00
20	21	.0440		20.5	0.50
20	22	.0080		21.0	2.00
21	19	.0517	←	20.0	2.00
21	20	.0440		20.5	0.50
21	21	.0121		21.0	0.00
21	22	.0022		21.5	0.50
22	19	.0094		20.5	4.50
22	20	.0080		21.0	2.00
22	21	.0022		21.5	0.50
22	22	.0004		22.0	0.00

We could continue in this way of finding the values that $\bar{X}$ can assume and the corresponding probabilities. It might be easier if we approach the problem in a slightly different, but systematic, way. In Table 5.1 we have listed all possible samples of size 2, that is, all possible X_1, X_2 pairs, the probability corresponding to each pair, the value of $\bar{X}$ corresponding to each pair, and, for future reference, the value of S^2 corresponding to each pair. The table can be used to find the probability distribution of $\bar{X}$ as follows. To find the probability that $\bar{X}$ will equal a specific value, say 20, find all values "20" in the $\bar{X}$ column, find the corresponding probabilities (marked by arrows in the table), and add these probabilities. This gives

$$P[\bar{X} = 20] = .0517 + .1600 + .0517 = .2634.$$

(Note that finding $P[\bar{X} = 19.5]$ by this method yields the same result as obtained earlier, as it should.) Continuing these calculations allows us to write out the entire probability distribution for $\bar{X}$:

(5.7)

$\bar{x}$:	19.0	19.5	20.0	20.5	21.0	21.5	22.0
$P[\bar{X} = \bar{x}]$:	.2209	.3760	.2634	.1068	.0281	.0044	.0004.

The same approach can be used to find the exact probability distribution of S^2, the sample variance. The last column of Table 5.1 gives the values of S^2 for each sample. For example, if $X_1 = 19$ and $X_2 = 20$ (second row of the table), then $\bar{X} = 19.5$, so

$$s^2 = \frac{(19 - 19.5)^2 + (20 - 19.5)^2}{2 - 1} = \frac{(-.5)^2 + (.5)^2}{1} = .50.$$

To find the probability that S^2 will equal .50, find all values ".50" in the S^2 column, find the corresponding probabilities, and add these probabilities.

$$P[S^2 = .50] = .1880 + .1880 + .0440 + .0440 + .0022 + .0022$$
$$= .4684.$$

You should verify for yourself that the probability distribution for S^2 is given by

(5.8)

s^2:	0	.50	2	4.50
$P[S^2 = s^2]$:	.3934	.4684	.1194	.0188.

As we indicated earlier, once these probability distributions are obtained, it is possible to calculate probabilities of certain events as well as to find expected values.

Example 5.8 If a random sample of size 2 is taken from the distribution of ages, find (a) the probability that $\bar{X}$ will be less than or equal to 20, (b) $E(\bar{X})$, and (c) $V(\bar{X})$.

Using the probability distribution for $\bar{X}$ in (5.7), we find that

$$P[\bar{X} \leq 20] = P[\bar{X} = 19] + P[\bar{X} = 19.5] + P[\bar{X} = 20]$$
$$= .8603.$$

Using the definition of the expected value of a random variable,

$$E(\bar{X}) = \sum \bar{x} \cdot P[\bar{X} = \bar{x}] = 19(.2209) + 19.5(.3760) + \ldots + 22(.0004)$$
$$= 19.68$$

and

$$\begin{aligned} V(\bar{X}) &= E(\bar{X} - \mu_{\bar{X}})^2 = \sum (\bar{x} - 19.68)^2 \cdot P[\bar{X} = \bar{x}] \\ &= (19 - 19.68)^2(.2209) + (19.5 - 19.68)^2(.3760) \\ &\quad + \ldots + (22 - 19.68)^2(.0004) \\ &= .2788. \end{aligned}$$

Example 5.9 If a random sample of size 2 is taken from the distribution of ages, find (a) the probability that S^2 will be greater than or equal to 2, (b) $E(S^2)$, and (c) $V(S^2)$.

Using the probability distribution for S^2 in (5.8), we find that

$$\begin{aligned} P[S^2 \geq 2] &= P[S^2 = 2] + P[S^2 = 4.5] \\ &= .1382. \end{aligned}$$

To find the expected value, we take

$$\begin{aligned} E(S^2) &= \sum s^2 \cdot P[S^2 = s^2] \\ &= (0)(.3934) + (.5)(.4684) + (2)(.1194) + (4.5)(.0188) \\ &= .5576, \end{aligned}$$

and to find the variance

$$\begin{aligned} V(S^2) &= E(S^2 - \mu_{S^2})^2 = \sum (s^2 - .5576)^2 \cdot P[S^2 = s^2] \\ &= (0 - .5576)^2(.3934) + (.5 - .5576)^2(.4684) \\ &\quad + \ldots + (4.5 - .5576)^2(.0188) \\ &= .6645. \end{aligned}$$

By considering this example it should not be too hard for you to see that for even moderately small sample sizes, it is a very tedious task to find the exact probability distribution of sample statistics. Once the distribution is found, expectations and variances can be calculated, although this task becomes rather tiresome if there are a large number of possible values that the statistic can assume. We will see in the next sections that there are some mathematical results that allow us to side-step these problems.

[In Section 5.2 we said that although in practice sampling is usually done without replacement, the analysis is somewhat simpler if sampling is done with replacement. Furthermore, if the population size is large relative to the sample size, then the analysis based on sampling with replacement gives a good approximation to the results that would have been obtained using the analysis based on sampling without replacement. To illustrate this fact, we give here the exact probability distribution of $\bar{X}$ based on samples of size 2 taken without replacement.

$\bar{x}$:	19	19.5	20	20.5	21	21.5	22
$P[\bar{X} = \bar{x}]$:	.2184	.3798	.2620	.1079	.0273	.0044	.0002.

Compare this result with the distribution given in (5.7).]

5.6 THE MEAN AND VARIANCE OF SOME STATISTICS

In the previous section we found the mean and variance of the random variable X to be

$$\mu_X = E(X) = 19.68 \qquad \text{and} \qquad \sigma_X^2 = V(X) = .5576.$$

We also found that if samples of size 2 are taken from this distribution, then

$$\mu_{\bar{X}} = E(\bar{X}) = 19.68 \qquad \text{and} \qquad \sigma_{\bar{X}}^2 = V(\bar{X}) = .2788$$

while

$$\mu_{S^2} = E(S^2) = .5776.$$

Comparing these results, do you see any relationship between them? It should be fairly obvious that $E(X)$ and $E(\bar{X})$ are equal, as are $V(X)$ and $E(S^2)$. It is perhaps less obvious that $V(\bar{X})$ is exactly $V(X)$ divided by 2, and 2 happens to be the sample size. Although this could be a coincidence, in fact, it is not, as we shall see.

Theorem 5.1 If $X_1, X_2, \ldots, X_n$ is a random sample of size n taken from the probability distribution of a random variable X having mean $E(X) = \mu$ and variance $V(X) = \sigma^2$, then the sample mean $\bar{X}$ will have expectation equal to μ and variance equal to σ^2 divided by n:

(5.9)
$$E(\bar{X}) = \mu \qquad \text{and} \qquad V(\bar{X}) = \frac{\sigma^2}{n}.$$

The proof of this theorem depends on Theorems 3.2 and 3.4. You should take a few moments to review those theorems. Another fact crucial to the proof of this theorem follows from Definition 5.2. Since $X_1, X_2, \ldots,$ and X_n all have the same probability distribution as that of X, it follows that they will all have the same mean and variance as that of X:

$$E(X_1) = E(X_2) = \ldots = E(X_n) = \mu$$

and

$$V(X_1) = V(X_2) = \ldots = V(X_n) = \sigma^2.$$

Combining these facts along with Theorem 2.1, we get

$$E(\bar{X}) = E\left(\frac{X_1 + X_2 + \ldots + X_n}{n}\right) = \frac{1}{n}[E(X_1 + X_2 + \ldots + X_n)]$$

$$= \frac{1}{n}[E(X_1) + E(X_2) + \ldots + E(X_n)]$$

$$= \frac{1}{n}(\mu + \mu + \ldots + \mu) = \frac{n\mu}{n} = \mu.$$

Similarly, using Theorem 2.6, we get

$$V(\bar{X}) = V\left(\frac{X_1 + X_2 + \ldots + X_n}{n}\right) = \frac{1}{n^2}V(X_1 + X_2 + \ldots + X_n)$$

$$= \frac{1}{n^2}[V(X_1) + V(X_2) + \ldots + V(X_n)]$$

$$= \frac{1}{n^2}(\sigma^2 + \sigma^2 + \ldots + \sigma^2) = \frac{n\sigma^2}{n^2} = \frac{\sigma^2}{n}.$$

The importance of Theorem 5.1 is that it allows us to find the mean and variance of the sample mean $\bar{X}$ without going to the trouble of finding the exact probability distribution of $\bar{X}$.

Example 5.10 The daily receipts at the Eager Weavers, a small upholstery shop, have a mean of \$150 and a standard deviation of \$42. If the shop is open 5 days a week, what are the mean and variance of the mean daily income for a week?

Since we have $\mu = 150$ and $\sigma^2 = (42)^2$ with $n = 5$, it follows that $\bar{X}$, the mean daily income for a week, will have an expectation given by

$$E(\bar{X}) = \mu = 150$$

and a variance given by

$$V(\bar{X}) = \frac{\sigma^2}{n} = \frac{(42)^2}{5} = 352.8.$$

In working problems like these, you must be very careful to distinguish between the variance and standard deviation. If Example 5.10 had asked for the standard deviation of $\bar{X}$ instead of the variance, the answer would have been

$$\sigma_{\bar{X}} = \sqrt{V(\bar{X})} = \frac{\sigma}{\sqrt{n}} = \frac{42}{\sqrt{5}} = 18.78.$$

At the beginning of Sections 5.4 and 5.5 we mentioned that the value of $\bar{X}$ could be used as an estimate of an unknown mean μ. By considering (5.9) we can get some idea as to how good that estimate can be. Since $E(\bar{X}) = \mu$, we know that "on the average" $\bar{X}$ will be equal to μ. Furthermore, since $V(\bar{X}) = \sigma^2/n$, we see that as n gets large, the variability about the value μ will tend to get small. Consequently, when n is large, there is a high probability that $\bar{X}$ will be close to μ. This notion can be quantified after you finish studying Section 5.8 (see Exercise 24).

We pointed out at the beginning of this section that for a previously worked example, the expected value of the sample variance was equal to the population variance [i.e., $E(S^2) = V(X) = \sigma^2$]. The next theorem tells us that this is always the case.

Theorem 5.2 If $X_1, X_2, \ldots, X_n$ is a random sample of size n taken from the probability distribution of a random variable X having variance $V(X) = \sigma^2$, the sample variance S^2 will have expectation equal to σ^2:

$$E(S^2) = \sigma^2. \tag{5.10}$$

The proof of this theorem, although not very difficult, is a bit more involved than we would like, so we will not go through it here. An outline of the proof is given in Exercise 16 at the end of this chapter.

Example 5.11 Find the expected value of the sample variance S^2 when S^2 is found using the daily receipts for 5 days from the Eager Weaver shop (see Example 5.10).

Since $\sigma = 42$, it follows from Theorem 5.2 that

$$E(S^2) = \sigma^2 = (42)^2 = 1764.$$

The results given in Theorem 5.1 apply when $X_1, X_2, \ldots, X_n$ are independent random variables. If sampling is done from a finite population without replacement, the resulting random variables found will not be independent. It will still be true that

$$E(\bar{X}) = \mu,$$

but the variance of $\bar{X}$ will be given by

$$V(\bar{X}) = \frac{\sigma^2}{n}\left(\frac{N-n}{N-1}\right).$$

5.7 THE DISTRIBUTION OF A LINEAR COMBINATION OF NORMAL RANDOM VARIABLES

We saw in Section 5.5 that it can be very difficult to find the exact probability distribution of the sample mean $\bar{X}$. Fortunately, there are some exceptions to this situation. In particular, if a random sample is taken from a normal distribution, it is very easy to find the exact probability distribution of $\bar{X}$. More generally, we will see that it is easy to find the exact probability distribution of any linear combination of normal random variables.

Definition 5.7 If $X_1, X_2, \ldots, X_n$ is a sequence of random variables and if $a_1, a_2, \ldots, a_n$ is a sequence of constants, the random variable Y defined by

$$Y = a_1X_1 + a_2X_2 + \ldots + a_nX_n$$

is said to be a *linear combination* of the random variables $X_1, X_2, \ldots, X_n$.

Example 5.12 An instructor in Marketing 234 gives two examinations, one a midterm and the other a final exam. The final grade in the course is determined by weighting the midterm $\frac{1}{3}$ and the final exam $\frac{2}{3}$. Express the final grade as a linear combination of the two exam scores.

If we let X_1 denote the score of the midterm exam and X_2 the score of the final exam, the final grade, say Y, will be denoted by

$$Y = \tfrac{1}{3}X_1 + \tfrac{2}{3}X_2;$$

that is, we take $a_1 = \frac{1}{3}$ and $a_2 = \frac{2}{3}$.

Example 5.13 The manager of the Swedish Stockholm Smorgasbord Restaurant, Sven Svenson, used to charge a fixed price to each customer. Each customer was allowed to eat as much as he desired. Most managers of such establishments charge one price for children under twelve and another price for everyone else twelve and over. Sven, however, noticed that males tend to eat more than females, whether adult or child, and that heavier people (e.g., 220-pound males) tend to eat more than lighter people (e.g., 150-pound males). Consequently, Sven has decided to charge people according to the format shown in Figure 5.1. Express the cost to a family of four with one boy and one girl under twelve. Find the exact cost to the Dees if Sven's scale shows Dick, Pat, Mike, and Suzy to weigh 205, 110, 70, and 60 pounds, respectively.

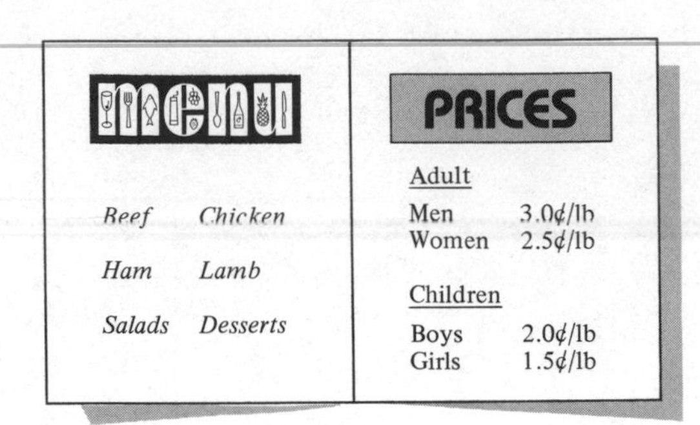

FIGURE 5.1
Sven's Menu

If we let X_1, X_2, X_3, and X_4 denote the weights of the husband, wife, boy, and girl, respectively, the cost (in cents) to a family of four (with children under twelve) would be given by

$$Y = 3X_1 + 2.5X_2 + 2X_3 + 1.5X_4.$$

The exact cost to the Dee family would be

$$\begin{aligned} y &= 3(205) + 2.5(110) + 2(70) + 1.5(60) \\ &= 1120 \text{ cents} \end{aligned}$$

(or $11.20).

Once a linear combination of random variables is defined, it is often of interest to find the mean and/or variance of that linear combination. The necessary formulas for doing this are given in the next two theorems.

Theorem 5.3 If Y is a linear combination of the random variables $X_1, X_2, \ldots, X_n$ having means $E(X_1) = \mu_1, E(X_2) = \mu_2, \ldots, E(X_n) = \mu_n$, then

$$\mu_Y = E(Y) = a_1\mu_1 + a_2\mu_2 + \ldots + a_n\mu_n. \tag{5.11}$$

Theorem 5.4 If Y is a linear combination of the *independent* random variables $X_1, X_2, \ldots, X_n$, having variances $V(X_1) = \sigma_1^2, V(X_2) = \sigma_2^2, \ldots, V(X_n) = \sigma_n^2$, then

$$\sigma_Y^2 = V(Y) = a_1^2\sigma_1^2 + a_2^2\sigma_2^2 + \ldots + a_n^2\sigma_n^2. \tag{5.12}$$

The proofs of these theorems are left to the exercises because they are similar to the proof of Theorem 5.1. We would emphasize, however, that the assumption of independence is necessary in Theorem 5.4; otherwise, equation (5.12) would have to include a number of covariance terms.

Example 5.14 If the mean weights of adult men and women and of boys and girls under twelve are 170, 140, 58, and 60 pounds, respectively, what is the expected cost at the Swedish Stockholm Smorgasbord for a family of four as described in Example 5.13?

Using (5.11), we find that

$$\begin{aligned} E(Y) &= 3\mu_1 + 2.5\mu_2 + 2\mu_3 + 1.5\mu_4 \\ &= 3(170) + 2.5(140) + 2(58) + 1.5(60) \\ &= 1066 \text{ cents.} \end{aligned}$$

Example 5.15 Assume that the variances of the weights of adult men and women and of boys and girls under twelve are 750, 575, 300, and 400 (lb^2), respectively. Spence and Judy belong to the Big Brothers and Big Sisters organizations and plan to take their "adopted" little brother and sister to the Smorgasbord. What is the standard deviation of the cost to a group of four such as this one?

Since we are interested in the standard deviation, we would first find the variance, and then take the square root. Using (5.12), we get

$$\begin{aligned} V(Y) &= (3)^2\sigma_1^2 + (2.5)^2\sigma_2^2 + (2)^2\sigma_3^2 + (1.5)^2\sigma_4^2 \\ &= 9(750) + 6.25(575) + 4(300) + 2.25(400) \\ &= 12{,}443.75, \end{aligned}$$

so

$$\sigma_Y = \sqrt{V(Y)} \approx 111 \text{ cents} = \$1.11.$$

(You might give some thought here as to why we did not attempt to find the standard deviation of the cost to a family of four as described in Example 5.13.)

As you know, having a knowledge of the mean and variance of a random variable is not as helpful as knowing the actual probability distribution of that random variable. The following theorem allows us to calculate the probability distribution of a linear combination of random variables in a special case.

Theorem 5.5 If Y is a linear combination of independent random variables $X_1, X_2, \ldots, X_n$ each having a normal distribution, then Y will have a normal distribution.

The proof of this theorem requires much more mathematics than we are assuming in this text, so we will not offer a proof.

One very important linear combination of random variables is the sample mean. If we let $a_1 = a_2 = \ldots = a_n = 1/n$, then

$$\begin{aligned} &a_1X_1 + a_2X_2 + \ldots + a_nX_n \\ &\qquad = \frac{1}{n}X_1 + \frac{1}{n}X_2 + \ldots + \frac{1}{n}X_n \\ &\qquad = \frac{1}{n}(X_1 + X_2 + \ldots + X_n) = \bar{X}, \end{aligned}$$

so $\bar{X}$ is actually a linear combination of the random variables in a random sample. This fact, in conjunction with Theorem 5.1, justifies the statement of the following theorem.

Theorem 5.6 If $X_1, X_2, \ldots, X_n$ is a random sample of size n taken from a normal distribution having mean μ and variance σ^2, then $\bar{X}$ will be normally distributed with $\mu_{\bar{X}} = \mu$ and $\sigma_{\bar{X}}^2 = \sigma^2/n$. Consequently,

$$\frac{\bar{X} - \mu}{\sigma/\sqrt{n}}$$

will have a standard normal distribution.

Example 5.16 In preliminary tests, Gooey Glue has shown a mean holding strength of 3000 pounds per square inch (psi) with a standard deviation of 400 psi. The variation is due to the incomplete mixing of the glue. (The glue is hard to mix because it keeps sticking to the stirring rods.) Assume that the holding strength for different tubes follows a normal distribution. Find the probability that the sample mean holding strength for a sample of 25 tubes will be less than or equal to 2900 psi.

Using Theorem 5.6 with $\mu = 3000$, $\sigma = 400$, and $n = 25$, we find that

$$P[\bar{X} \leq 2900] = P\left[\frac{\bar{X} - \mu}{\sigma/\sqrt{n}} \leq \frac{2900 - \mu}{\sigma/\sqrt{n}}\right]$$

$$= P\left[Z \leq \frac{2900 - 3000}{400/\sqrt{25}}\right]$$

$$= P[Z \leq -1.25] = .1056.$$

[Note that since $(\bar{X} - \mu)/(\sigma/\sqrt{n})$ is a standard normal random variable in this case, we have replaced it by Z. The probability obtained here was found from the standard normal tables.]

You might give some thought as to whether or not it would be appropriate to apply Theorem 5.5 to the problem of finding probabilities associated with the random variables defined in Examples 5.12 and 5.13. In either case, is it reasonable to assume that the random variables X_i are normally distributed? Is it reasonable to assume that they are independent? If they are not independent, would it be easy to find σ_Y? If σ_Y is not known, is it possible to standardize Y?

5.8 THE CENTRAL LIMIT THEOREM

We have now seen that finding the exact probability distribution of the sample mean $\bar{X}$ can be a very simple matter (if the sample is taken from a normal distribution) or a very complicated matter (if the sample is taken from a nonnormal distribution such as the ages of statistics students, as found in Section 5.4). It is unfortunate but true that by far most random variables do not follow a normal distribution. Consequently, in most situations, finding the *exact* probability distribution of the sample mean is a very complicated matter. As we shall see, however, it is often a simple matter to find an approximate probability distribution for the sample mean.

In Section 5.5 we found the exact probability distribution for $\bar{X}$ based on random samples of size 2 taken from a certain distribution. A graph of this probability dis-

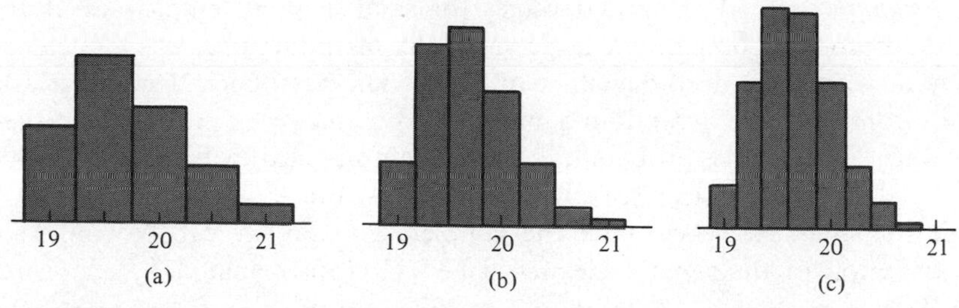

FIGURE 5.2

tribution is given in Figure 5.2(a). In Figure 5.2(b) and (c) you will find graphs showing the exact probability distribution of $\bar{X}$ based on random samples of size 3 and 4, respectively. (Our primary interest among these graphs is the shape. To make them more easily comparable, the vertical scales on each are different, but the total area enclosed in each is the same.) What differences do you see among these graphs? In particular, does there appear to be any "trend" as n gets larger? You may notice that the first graph is skewed sharply to the right, but that the skewness becomes less as the sample size increases. As n increases, the distribution becomes more symmetric. You may also notice that the distribution is starting to resemble a normal distribution in shape as n gets larger. If we were to graph the distribution for $\bar{X}$ for samples of size $n = 25$, say, the distribution would look very much like a normal distribution. A most remarkable fact is that this tendency toward normality would hold true for essentially any distribution from which we might sample. A mathematical statement of this fact is given in the following theorem.

Theorem 5.7 (*Central Limit Theorem*) If $X_1, X_2, \ldots, X_n$ is a random sample of size n taken from a distribution having a mean μ and finite variance σ^2, then for large n, $\bar{X}$ will have an *approximately* normal distribution with $\mu_{\bar{X}} = \mu$ and $\sigma_{\bar{X}}^2 = \sigma^2/n$, that is, the quantity

$$\frac{\bar{X} - \mu}{\sigma/\sqrt{n}}$$

will have a standard normal distribution (approximately).

A proof of this theorem must be left to advanced theoretical courses in mathematical statistics. The importance of this theorem to further developments in this text can hardly be overemphasized. The Central Limit Theorem is of practical importance for two reasons. First, the condition that the distribution from which the sample is taken have a finite variance is satisfied by almost every distribution encountered in applied problems. Hence, the Central Limit Theorem is almost always applicable. Second, the sample size n necessary for the normal approximation to be good is not all that large. As a general rule of thumb, if $n \geq 30$, the distribution of $\bar{X}$ will be well approximated by a normal distribution. In fact, if the sample is taken from a continuous distribution that is "normalish" in shape (say unimodal and not too skewed), $\bar{X}$ may be well approximated by a normal distribution for samples as small as 5 or 10.

Example 5.17 The annual wages (or salary) of all employees at the Leprechaun Automobile Company, a large small-car manufacturer, has a mean of $18,400 per year with a standard deviation of $3200. The personnel manager, Kathleen O'Leary, is going to take a random sample of 36 employees and calculate the sample mean wage. What is the probability that the sample mean will exceed $19,000?

Typically, wages of all employees at any large facility will not be normally distributed. However, since the sample size is 36, we can be confident that the distribution of the sample mean will be well approximated by a normal distribution. With $n = 36$, $\mu = 18{,}400$, and $\sigma = 3200$, we get

$$\begin{aligned} P[\bar{X} > 19{,}000] &= P\left[\frac{\bar{X}-\mu}{\sigma/\sqrt{n}} > \frac{19{,}000-\mu}{\sigma/\sqrt{n}}\right] \\ &= P\left[\frac{\bar{X}-\mu}{\sigma/\sqrt{n}} > \frac{19{,}000-18{,}400}{3200/\sqrt{36}} = 1.125\right] \\ &\approx P[Z > 1.125] = .1303. \end{aligned}$$

Example 5.18 The company doctor, Pat O'Reilly, is examining the records of employees who are over thirty years of age. Dr. O'Reilly is interested in the weight of these employees and plans to take a random sample of 16 employees, find their weights, and calculate the sample mean. If the true mean is 161 pounds, with a standard deviation of 25 pounds, what is the probability that the sample mean will be no more than 165 pounds?

We would not expect the weights of the employees to be normally distributed, but to be skewed to the right. However, since the skewness is not too extreme, a sample of size 16 should be large enough for the approximation to be good. Using the Central Limit Theorem with $n = 16$, $\mu = 161$, and $\sigma = 25$, we find that

$$\begin{aligned} P[\bar{X} \le 165] &= P\left[\frac{\bar{X}-\mu}{\sigma/\sqrt{n}} \le \frac{165-\mu}{\sigma/\sqrt{n}}\right] \\ &= P\left[\frac{\bar{X}-\mu}{\sigma/\sqrt{n}} \le \frac{165-161}{25/4} = .64\right] \\ &\approx P[Z \le .64] = .7389. \end{aligned}$$

As we indicated earlier in this chapter, we have assumed that sampling is being done with replacement or, equivalently, that the random variables $X_1, X_2, \ldots, X_n$ are independent with the same probability distribution. If the sampling is done without replacement, as indicated at the end of Section 5.6, the variance of the random variable $\bar{X}$ is given by

$$V(\bar{X}) = \frac{\sigma^2}{n}\left(\frac{N-n}{N-1}\right). \tag{5.13}$$

The only difference between the variance of $\bar{X}$ in sampling without replacement instead of with replacement is the factor $[(N-n)/(N-1)]$ [cf. equation (5.9)]. The distribution of $\bar{X}$ will be approximately normal as long as n is large enough while not being too large. As a rule of thumb, we can say that if $n \ge 30$ while $n < N/2$, then if $X_1, X_2, \ldots, X_n$ is a sequence of dependent random variables obtained from a random sample taken without replacement from a distribution having mean μ and variance σ^2,

$$\frac{\bar{X} - \mu}{\frac{\sigma}{\sqrt{n}}\sqrt{\frac{N-n}{N-1}}}$$

will be distributed (approximately) as a standard normal random variable. This rule of thumb assumes that the distribution of X is not too skewed.

EXERCISES

1. Make a list of five different situations where sampling is done. (See Example 5.1 and subsequent discussions.) In each case, carefully describe the population being sampled.

2. Using the list of 200 students in Appendix A3 as a population, say you were interested in taking a sample of 20 students. The sampling scheme you decide to use is a systematic scheme where you choose at random an integer from 0 to 9 and choose every tenth student from that point on. For example, if you start with the number 3, your sample consists of students 3, 13, 23, . . . , 193; if you start with the number 0, your sample consists of students 10, 20, 30, . . . , 200, and so on. (a) Are the samples obtained this way random samples? Why or why not? (b) Are they representative samples? Why or why not?

3. Referring to Exercise 2, say you have determined that there are 61 females and 139 males in the population. To take a sample of 20, you decide to take a random sample of six females from the 61 females in the population and random sample of 14 males from the 139 in the population. (a) Are the samples obtained in this way random samples? Why or why not? (b) Are they representative samples? Why or why not?

4. Referring to Exercise 2, say you decide to take the sample by choosing one student at random from the first 10 students, one at random from the second 10 students, and so on, to obtain a sample of 20. (a) Are the samples obtained in this way random samples? Why or why not? (b) Are they representative samples? Why or why not?

5. Say you have determined that 90% of the people in your town have telephones and that 90% of the people having telephones have their telephone numbers listed in the directory. Since you realize that it is almost impossible to get a complete list of all people living in your town, you decide to take a survey using the phone directory. Would you expect the sample to be "representative" or "biased" with respect to answers to the following questions: (a) "Are the New York Yankees your favorite baseball team?" (b) "Do you own a Lincoln or a Cadillac?" Why?

6. Using the random number tables, choose a random starting point and choose a random sample of 25 students from the first 100 in the list *with* replacement. List the numbers chosen and the corresponding ages. Do the same for the corresponding weights. (These data can be used in Exercise 3 of Chapter 6.)

7. Using the random number tables, choose a random starting point (different from the one used in Exercise 6) and choose a random sample of 25 students from the first 100 in the list *without* replacement. List the numbers chosen and the corresponding heights. Do the same for the corresponding grade-point averages. (These data can be used in Exercise 4 of Chapter 6.)

8. Five observed values of a random variable X were found to be

$$x_1 = 10, \quad x_2 = 14, \quad x_3 = 21, \quad x_4 = 17, \quad x_5 = 13.$$

(a) Find $\bar{x}$. (b) Find the deviations from $\bar{x}$. That is, find $(x_1 - \bar{x})$, $(x_2 - \bar{x})$, . . . , $(x_5 - \bar{x})$. (c) Calculate $\sum_{i=1}^{5} (x_i - \bar{x})$ for these numbers. (d) Show that in general, $\sum_{i=1}^{n} (x_i - \bar{x}) = 0$.

9. The following are daily high temperatures (in degrees Fahrenheit) recorded at the airport of a Midwestern city for 10 randomly selected days of the last year:

$$10, \quad 82, \quad 28, \quad 57, \quad 82, \quad 75, \quad 98, \quad 17, \quad 68, \quad 93.$$

(a) Find the sample mean, $\bar{x}$. (b) Find the sample median, $\tilde{x}$. (c) Find the sample standard deviation, s. (*To think about:* If you were thinking of moving to this city and were curious as to the climate, would knowing $\bar{x}$ be helpful to you? Why?)

10. The amounts of money spent by six customers at the Havitall Hardware Store were

$$\$2.44, \quad 1.97, \quad 24.88, \quad 3.69, \quad 5.43, \quad 2.09.$$

(a) Find the sample mean, $\bar{x}$. (b) Find the sample median, $\tilde{x}$. (c) Find the sample variance, s^2. (*To think about:* If Mr. Havitall claimed that a "typical" customer spent \$6 or \$7 at his store, would you agree? Why?)

11. When sampling with replacement from first- and second-grade children at T. R. Roosevelt Elementary School, you find the probability distribution of the ages to be

x (age):	6	7	8
$P[X = x]$:	.3	.4	.3.

Assume that samples of size 3 are to be chosen. Find the probability distribution of (a) the sample mean, $\bar{X}$; (b) the sample median, $\tilde{X}$. (*Hint:* Make a table showing all possible samples and the corresponding probabilities. For each possible sample, calculate the corresponding value of $\bar{x}$ and $\tilde{x}$.)

12. (a) Using the probability distribution found in Exercise 11(a), find $E(\bar{X})$. (b) Using the probability distribution found in Exercise 11(b), find $E(\tilde{X})$. (c) Using the original distribution of ages given in Exercise 11, find $E(X) = \mu_X$.

13. (a) Using the probability distribution found in Exercise 11(a), find $V(\bar{X})$. (b) Using the probability distribution found in Exercise 11(b), find $V(\tilde{X})$. (*To think about:* Would $\bar{X}$ or $\tilde{X}$ be better to use if you wanted an estimate of μ_X, assuming it to be unknown? Why?)

★14. The weights of a group of college-age males are *not* normally distributed. The mean weight μ is 165 pounds and the standard deviation is 20.6 pounds. Find the mean and variance of $\bar{X}$, the sample mean weight, if $\bar{X}$ is based on samples of (a) size 5; (b) size 10; (c) size 20.

★15. The weights of a group of college-age females are *not* normally distributed. The mean weight μ is 123 pounds and the standard deviation is 15.5 pounds. Find the mean and standard deviation of $\bar{Y}$, the sample mean weight, if $\bar{Y}$ is based on samples of (a) size 4; (b) size 25; (c) size 100.

16. Prove Theorem 5.2. [*Hint:* Write $E(S^2)$ as $E[\sum (X_i - \bar{X})^2/(n-1)]$. Expand $(X_i - \bar{X})^2$ and use properties of expectations and summations to simplify the resulting expression. To evaluate $E(X_i\bar{X})$, write this as $E[X_i(X_i + \sum_{j \neq i}^{n} X_j)/n]$. Use the assumption of independence to evaluate $E(X_iX_j)$. Also use the fact that $E(X^2) = \sigma^2 + \mu^2$.]

17. Prove Theorem 5.3. (*Hint:* Use Theorem 2.1 and Corollary 3.1.)

18. Prove Theorem 5.4. (*Hint:* Use Theorem 2.6 and Corollary 3.2.)

19. Let W, X, and Y be independent normal random variables with parameters $\mu_W = 10$, $\mu_X = 30$, $\mu_Y = 25$ and $\sigma_W^2 = 9$, $\sigma_X^2 = 16$, $\sigma_Y^2 = 50$. What is the distribution of the random variable (a) $T = W + X + 2Y$?; (b) $U = 2W + X - 2Y$? Find (c) $P[T < 100]$; (d) $P[U > 20]$.

20. On a box containing a replacement headlight for a car is a note saying "Install in pairs. . . . When one headlamp burns out, chances are the other may burn out at any time." With this in mind, let X denote the lifetime of the left headlight in a car and let Y denote the lifetime of the right headlight. Assume that X and Y are independently distributed, both having a normal distribution with mean $\mu = 720$ hours and $\sigma = 50$ hours. (a) Find the probability that the headlights will burn out within 50 hours of each other, assuming that they were installed at the same time. (*Hint:* Consider the distribution of $X - Y$.) (b) What would the answer to part (a) be if μ were 1000 hours? (*To think about:* Is the statement on headlight box essentially accurate? How do you reconcile this with the assumption that X and Y are independent?)

21. Assume that the heights of men in $\mu\epsilon\nu$ fraternity are approximately normally distributed with a mean of 70.8 inches and a standard deviation of 2.7 inches, while the heights of women in $\Gamma\alpha\lambda$ sorority are approximately normal with mean and standard deviation 64.6 and 2.5 inches, respectively. A man from the fraternity is randomly chosen to have a blind date with a randomly chosen woman from the sorority. What is the probability that the man will be taller than the woman?

★22. The heights of a group of college-age males are approximately normally distributed with mean $\mu = 70.8$ inches and standard deviation $\sigma = 2.7$ inches. If a random sample of size 4 is taken and if $\bar{X}$ is the sample mean, find (a) $P[\bar{X} > 68]$; (b) $P[\bar{X} \leq 71]$.

★23. The heights of a group of college-age females are approximately normally distributed with a standard deviation of $\sigma = 2.5$ inches. If the mean μ is actually unknown and is estimated by the mean of a random sample, $\bar{Y}$, find the probability that $\bar{Y}$ differs (in absolute value) from μ by 1 inch or less if the sample is based on (a) 4; (b) 25; (c) 64 observations.

24. Let $X_1, X_2, \ldots, X_n$ be a random sample from a $N(\mu, \sigma^2 = 100)$ population with μ unknown. Find the following:
(a) $P[|\bar{X} - \mu| < .1]$ when $n = 9$; (b) $P[|\bar{X} - \mu| < .1]$ when $n = 36$;
(c) $P[|\bar{X} - \mu| < .1]$ when $n = 144$.

25. The population of times measured by "3-minute" egg timers is approximately normally distributed, with $\mu = 3$ minutes and $\sigma = .2$ minute. If samples of 25 timers are tested, find the time that would be exceeded by 95% of the sample means.

26. In the city of Greenville the mean family income is \$14,600 per year. The standard deviation of these incomes is \$2400. If $\bar{X}$ is the mean of the income of 36 families chosen at random from this city, estimate (a) $P[14{,}000 \leq \bar{X} \leq 15{,}000]$; (b) $P[\bar{X} \leq 15{,}300]$; (c) $P[\bar{X} \geq 13{,}800]$. (*To think about:* Would you expect the incomes to be normally distributed? Did you base your solution to this problem on Theorem 5.6 or Theorem 5.7?)

27. Crates to be delivered by the Carrihall Freight Company have a mean weight of of 240 pounds and a standard deviation of 66 pounds. What is the probability that the total weight of 36 crates taken at random and loaded onto a truck will exceed the specified capacity of the truck, which is 10,000 pounds?

6

DESCRIPTIVE STATISTICS

6.1 ORGANIZING AND SUMMARIZING DATA

In Chapter 5 we gave as a primary reason for sampling an interest in a particular population but an inability to investigate the entire population. We said that sampling is frequently a two-stage operation, first obtaining objects from the population of objects and then making a measurement on the objects selected. We generally refer to these measurements as sample data. Generally, the sample data are used to make an *inference* about the population sampled: that is, to draw some conclusion about the population based on observations made on a sample. Statistical inference will be the subject of the remaining chapters in this book. In this chapter, however, we will be primarily concerned with *descriptive statistics*, that is, with methods of describing and summarizing the data collected itself, without making any inference about the population from which it came.

Data can be collected as a result of sampling as previously discussed, but it can also be collected as an ongoing process in time. For example, you might randomly select 20 (working) days from the last 5 years and find the closing prices of the Dow Jones industrial average on each of those days. On the other hand, you may read the newspaper today and for the next 19 working days and record the Dow Jones averages for each of these days. In either case you will end up with a total of 20 observations (although in the latter case the observations may not be independent). Another situation where data are collected as an ongoing process is during the course of an athletic contest. During the course of a football game, for instance, a "statistician" records as much information as possible about each play. (By the way, many people think that this is the primary role of a statistician—being a data collector. We hope that when you finish studying this book you will have a more accurate picture of the role of a professional statistician!) If you happened to miss a particular game but were given the complete set of data as collected by the sports statistician, you would be able to re-create just about the entire game, play by play. For many of us, however, it would be difficult to make a lot of sense out of the information available if we only had a small amount of time to devote to the task. However, when these same data are organized appropriately and summarized or condensed properly, we can learn the more important results (e.g., Denver beat Houston 20–17, Walter Payton carried 22 times for 147 yards, Thidwick Fumblemeyer carried 6 times for minus 14 yards, etc.). Of course, you must be careful not to make the condensation of information too extreme. Not many sports fans would be satisfied reading on the sports page on the day after the game *only* the information that "Denver beat Houston for the American Football Conference title!" This is an important rule to keep in mind: It is important to organize and summarize data into an easily assimilated form, but not at the expense of the loss of too much information.

6.2 FREQUENCY DISTRIBUTIONS

Data can be summarized in various ways, two of the most common ways being tables and graphs. Newspapers and magazines, especially news magazines such as *Time*, *Newsweek*, and *U.S. News & World Report*, contain large numbers of these tables and graphs, especially in sections dealing with business and economics. In this section we will consider ways of summarizing large amounts of data in tabular form.

Let us consider the set of data on statistics students given in Appendix A3. There are various measurements that we might consider, but we will start with one of the simpler ones, namely, the year in school of the individuals. Since we use the designation "freshman" $\equiv 1$, "sophomore" $\equiv 2$, and so on, only small integer values can occur in the data. In fact, since there are no graduate students among these 200 students, only the values 1 through 4 do occur. We construct a *frequency table* by listing the possible values and the corresponding frequency of occurrence of each. For large data sets, this type of construction would generally be done by the computer, but for smaller data sets, say 20 to 100 observations, this construction can be done by hand. In constructing a table by hand, first list the possible values; second, go through the data set making a tally mark next to the appropriate value; and third, find the frequency for each possible value by totaling the tally marks for each. This procedure is illustrated in Table 6.1 for the first 20 students in the list. In Table 6.2 we give the frequency table for the year in school data for all 200 students. Since the computer did the tabulation in the second case, there is no tally column given in Table 6.2.

We could make a frequency distribution for the ages of these students as well as the year in school. There is one slight difference between the measurement of age and the measurement of the year in school. It is customary to report one's age as of the last birthday, so students reporting their age as "19" may have just had their nineteenth

TABLE 6.1
Frequency Table—20 Observations

Year in School	*Tally*	*Frequency*
1		0
2	~~||||~~ ~~||||~~ ~~||||~~ ||	17
3	|||	3
4		0

TABLE 6.2
Frequency Table—200 Observations

Year in School	*Frequency*
1	3
2	146
3	42
4	9

birthday or they may be just a few days short of their twentieth birthday. The ages reported as "19" actually, then, represent a range of ages that might be thought of as "at least 19 but less than 20." On the other hand, we generally think of a person as a sophomore through his entire second year and don't think of the year in school as "sophomore and one-third," and so on. In the case of ages, then, we actually have a range of values grouped together into what we will call a *class*. The frequency table for ages is given in Table 6.3.

The idea of grouping values together into a class is useful from another point of view. If instead of data on 200 college students, you had data on 1000 people living in a small town, how would you present a frequency table of the ages of these people? You could, of course, list all the individual ages (as of the last birthday) as was done in Table 6.3. This new table would have to include values from 0 to 80 or 90. Such a table would be so long that our purpose of providing an easily understood summary of the data would not be accomplished. One solution to this problem is to group several ages together into an artificial class. We say "artificial" because there is really no

TABLE 6.3
Frequency Table for Ages

Age (At Last Birthday)	*Frequency*
19	95
20	74
21	26
22	5

natural grouping that comes to mind. We could group together 5 or 10 or 15 ages, for example. Which of these would be preferred? That is really a difficult question to answer and might depend on the purpose of gathering information. The local school board might answer the question differently than the Golden Age Society. As a rough rule of thumb, let us agree to use between 5 and 20 classes, using more when a large amount of data is to be summarized. The number of classes used determines the number of ages grouped together. You are no doubt familiar with other examples of grouping values together. The weight classes used in boxing or wrestling and weight classes or horsepower classes for automobiles are some examples.

Example 6.1 Make a frequency distribution for the weights of male statistics students given in Appendix A3. Use an appropriate number of classes.

This is obviously a situation where we will want to use classes to group together several weights, since the range of weights for these 139 students is from 130 to 240 pounds. We could use anywhere from 8 to 12 classes, depending on the amount of detail we wish to retain. (More classes implies more detail retained.) We will use nine classes. Since these classes must include 111 different values (from 130 to 240 *inclusive*), we can group together 13 weights and thus cover the range of values necessary. The frequency distribution as found by the computer is given in Table 6.4. The first class was taken to start at 128 pounds.

TABLE 6.4

Frequency Table for Weights of Males

Weight Class	*Frequency*
128–140	12
141–153	27
154–166	46
167–179	27
180–192	16
193–205	3
206–218	4
219–231	2
232–244	2

Example 6.2 Make a frequency distribution for the weights of female statistics students given in Appendix A3. Use an appropriate number of classes.

The weights of the female students are listed here for convenience.

127	135	140	115	110	100	130	125	145	115	115
120	155	115	135	107	115	120	103	112	160	125
105	130	150	125	120	123	125	140	100	118	100
115	115	135	130	128	145	115	122	120	155	88
114	110	135	115	112	105	125	127	160	118	115
140	107	140	115	115	134					

In looking over this list you see that the smallest value is 88 and the largest value is 160. Since we have 61 students, we can use from seven to nine classes. Let us choose seven classes. Each class should include 11 weights. We can start the first class at 86 pounds. With this number of observations it is not too hard to construct the frequency table by hand. The results are given in Table 6.5.

TABLE 6.5

Frequency Table for Weights of Females

Weight Class	*Tally*	*Frequency*
86–96	|	1
97–107	~~||||~~ |||	8
108–118	~~||||~~ ~~||||~~ ~~||||~~ ||||	19
119–129	~~||||~~ ~~||||~~ ||||	14
130–140	~~||||~~ ~~||||~~ ||	12
141–151	|||	3
152–162	||||	4

We would now like to give several definitions related to the grouping of observations into classes. First, note that in the data all weights are recorded in whole pounds. Hence the weight classes can be defined by listing whole pound values that *might be* included in the class. Such values will be referred to as class limits.

Definition 6.1 The *class limits* for a given class are the smallest and largest values that might occur in the data and will be put into that class.

The class limits for the first class given in Table 6.5 are 86 and 96. Of course, since weight is a continuous quantity, it is possible for a person to weigh 96.3 or 96.7 pounds, say. The former value would be rounded to 96 and counted in the first class while the latter value would be rounded to 97 and counted in the second class. What is the dividing line between these classes, then? It is 96.5. An alternative way of defining the classes would be 85.5 to 96.5, 96.5 to 107.5, and so on. You might be concerned about the fact that the value 96.5 occurs in both classes. Where would you put an observation whose value was 96.5, in the first or second class? This is not a cause for concern for two reasons. First, you will not be given a value such as 96.5 to classify since all the data are expressed in terms of whole pounds. Second, since weight is a continuous quantity, it is virtually impossible to observe a value of 96.500 . . . . Another way of handling the problem is to make a convention whereby any value falling on the boundary will be put into the higher (larger values) class.

Definition 6.2 The *class boundaries* for a given class are halfway between the class limits of that class and the next larger (or smaller) class.

The class boundaries for the first class defined in Table 6.5 are 85.5 and 96.5. The first value is referred to as the lower class boundary, the second as the upper class boundary. How big is the first class? Its size can be determined by the difference in the class boundaries, namely

$$96.5 - 85.5 = 11.$$

Definition 6.3 The *class size* for a given class is the difference between the upper class boundary and the lower class boundary.

The last two definitions are virtually self-explanatory. We are including them for future reference.

Definition 6.4 The *class midpoint* for a given class is halfway between the upper and lower class boundaries.

Definition 6.5 The *class frequency* for a given class is equal to the number of observations in the data set that fall within the class boundaries.

In later sections we will use the class midpoint as a single value to represent all the observations in the class. For this reason it is desirable, although certainly not necessary, for the class midpoint to be a value that could occur in the data. For example, in the classes defined in Table 6.5 the class midpoints are 91, 102, 113, . . . , 157.

6.3 FREQUENCY HISTOGRAMS AND FREQUENCY POLYGONS

Summarizing data into a frequency table is certainly one way of presenting data in a form in which it can be more easily assimilated. If, however, "a picture is worth two thousand words" (inflation is even affecting clichés these days), we might like to present data in a graphical form. We will see that the graphs discussed in this and the next sections can be considered as sample analogs of quantities discussed earlier for random variables.

You are all no doubt familiar with bar graphs. The information presented in any frequency table can also be presented in graphical form using a bar graph. The bars are generally labeled by their class and the height of the bar is equal to the class frequency. Such bar graphs, called *frequency histograms*, are similar to a probability histogram for a discrete random variable. In Figure 6.1 we give a frequency histogram for the analysis of reasons for nonresponse given by interviewers who were unable to make contact with interviewees. In this example the classes simply represent various reasons for nonresponse and the location of these along the horizontal axis really is a matter of individual choice. The vertical bars in a situation such as this need not be touching. If, however, the classes are made up of measurements such as height or weight, then the vertical bars are drawn touching. More precisely, the bars are drawn so that the edges are at the class boundaries and consequently so that they are centered at the class midpoints. Figure 6.2, showing a frequency histogram for weights of male statistics students, is an example of this type of histogram.

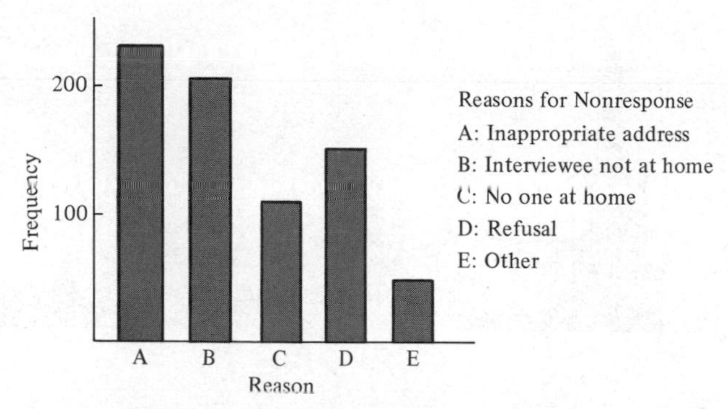

FIGURE 6.1
Frequency Histogram

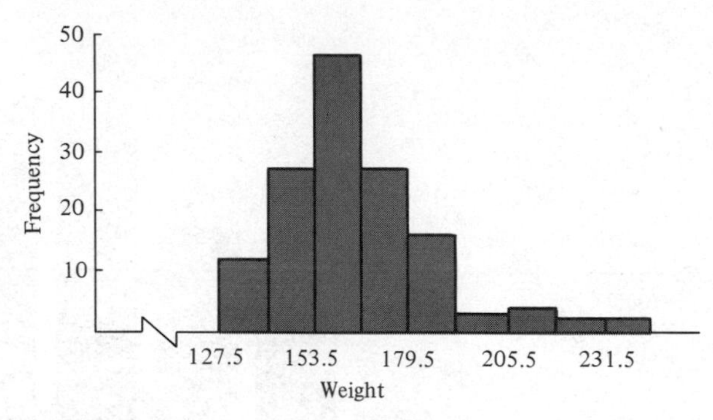

FIGURE 6.2

One feature of this histogram that we would like to point out is the relative size of the scales used on the horizontal and vertical axes. As you can see, the height corresponding to the highest frequency is approximately three-fourths of the total width allotted to the range of weight values (127.5 to 244.5). This kind of proportion results in a graph that is aesthetically pleasing. There is nothing very special about the ratio of 3 to 4, but if the ratio is too much different than this, a somewhat distorted picture is obtained. Compare the graph given in Figure 6.2 with the graphs of Figure 6.3, which show the same information but using different scales.

One reason for trying to construct a histogram which is aesthetically pleasing is to avoid the apparent distortion that results from disproportionate scales on the axes, that is, to try to present an honest picture. This is also the reason for the presence of the scale break (——⁄\/—) on the horizontal axis. Most of us are accustomed to having a set of axes start at zero and the scale break draws attention to the fact that there is a break in the scale between 0 and 127.5. There are times when the vertical axis does not start at zero, and in these cases a scale break should be shown on the vertical axis as well. If such a break is not shown, a distorted impression can be made. Unfortunately, not all such distortions are inadvertent. Figure 6.4 shows a graph that appeared in an advertisement placed in a newspaper by a bank. While we will

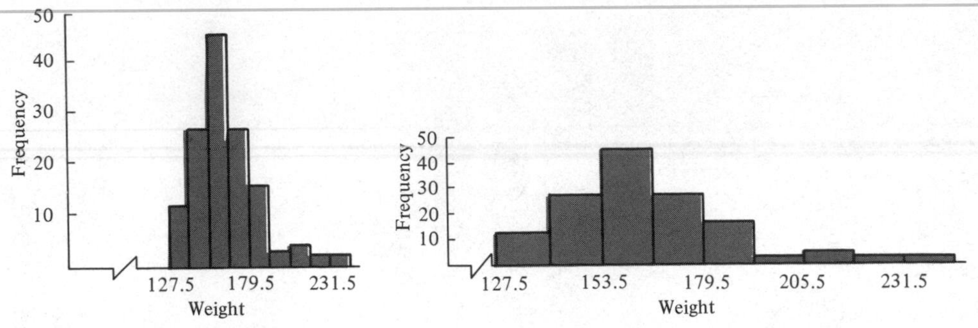

FIGURE 6.3

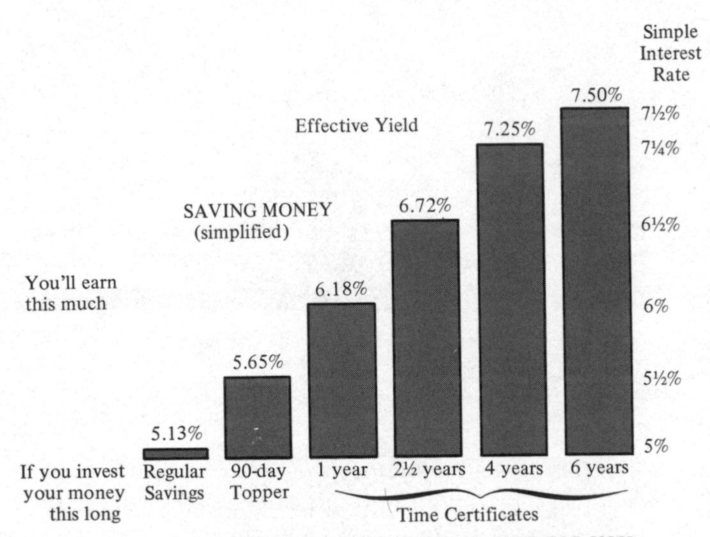

FIGURE 6.4

not accuse the designers of this ad of purposely distorting this picture (and indeed the vertical axis is labeled properly), we believe that the end result is the same. What impression do you get from the ad about the relative magnitude of the interest rate for regular savings and time certificates? What do you think the purpose of the ad is? How does the impression you get differ from the one you get from Figure 6.5, where the vertical scale starts at zero instead of 5%? We hope that this example will cause you to look more carefully at graphs you see in the media!

In Section 6.2 we mentioned that the choice of the number of classes used in summarizing data is a subjective matter and that too few classes could lead to excessive loss of detail while too many classes would not sufficiently condense the data. This point can be illustrated by considering the histograms given in Figure 6.6. In Figure 6.6(a) we give a histogram for the weight data treated in Example 6.1 but where only

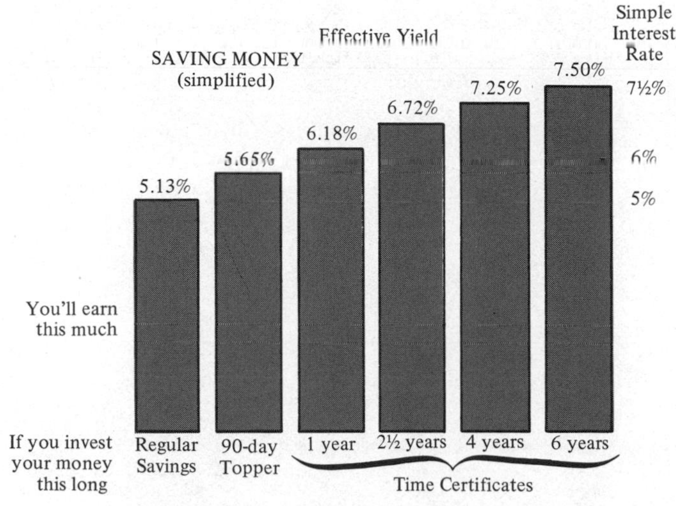

FIGURE 6.5

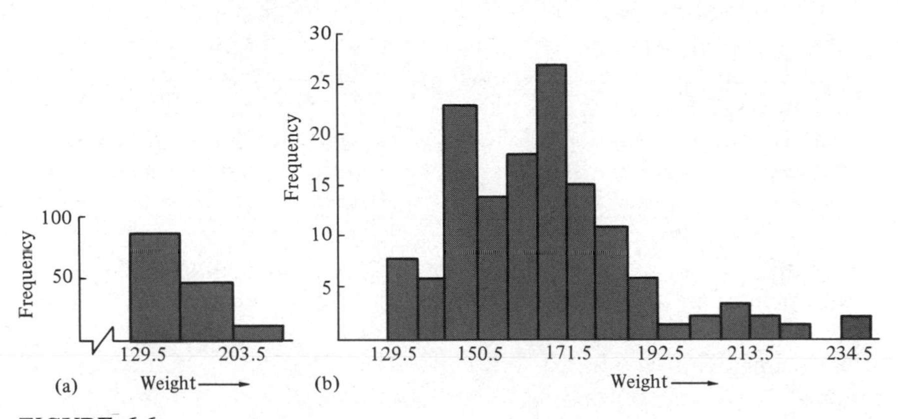

FIGURE 6.6

three classes are used; in Figure 6.6(b), 16 classes are used. Compare these histograms with the one given in Figure 6.2.

If the measurements given in the data are continuous, it might be desirable to have a means of providing a graphical summary of the set of data. A frequency polygon can serve such a purpose. A *frequency polygon* is constructed by joining the points (x_i, y_i) with straight-line segments, where x_i is the midpoint of the ith class and y_i is the frequency of the ith class. A frequency polygon is "tied down" to the horizontal axis by including one class smaller and one larger than those given in a frequency table. These classes will have zero frequency.

Example 6.3 Construct a frequency polygon for the weights of male statistics students given in Appendix A3.

Using the classes and frequencies given in Table 6.4 and noting that the classes 115 to 127 and 245 to 257 have zero frequency, we can construct the frequency polygon given in Figure 6.7.

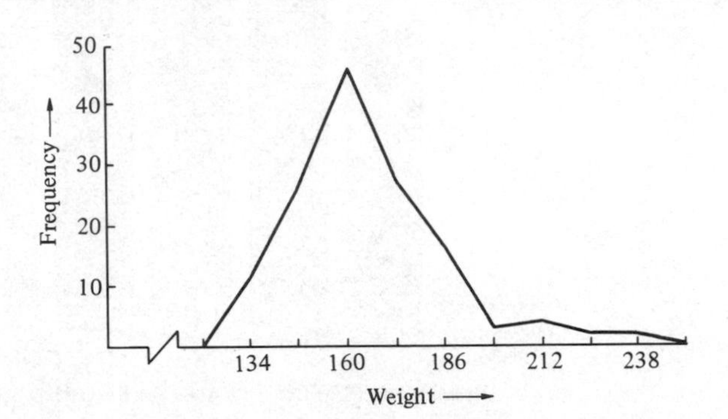

FIGURE 6.7

In addition to serving the function of providing a graphical summary of a set of data, a frequency polygon can also give us a rough idea as to the form of the distribution from which the data were obtained. For example, if you were interested in knowing whether the data might have come from a normal distribution, you could examine the frequency polygon. Is it unimodal and approximately symmetric? Does the frequency decrease as you move away from the middle? If so, the data *may* have come from a normal distribution. Is the frequency polygon flat? Is it multimodal? Is it skewed in one direction? If so, the data *may* not have come from a normal distribution. This type of examination might be called a subjective test for the form of a distribution. Some objective tests of this type will be discussed in Chapter 11.

If we wanted to compare two frequency polygons, how might we proceed? For example, if we wanted to compare the distributions of the weights of male and female statistics students by using frequency polygons, we would have a problem,

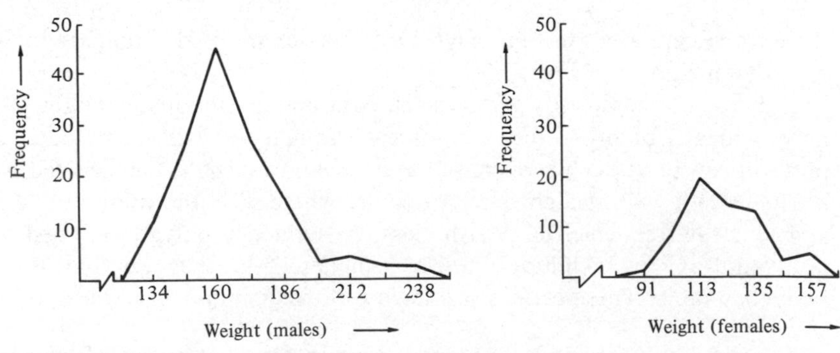

FIGURE 6.8

in that the numbers of males and females are different. If we graphed the two polygons using the same scale on the vertical (frequency) axis as in Figure 6.8, we would find them looking different primarily because of the different number of observations for each group. One way of eliminating the effect of the different numbers of observations is to use relative frequencies on the vertical axes rather than frequencies. This results in a simple change of scale. The *relative frequency* of a class is defined to be the class frequency divided by the total number of observations in the data set. For example, since the first weight class for males has a frequency of 12 and since there are 139 observations, the relative frequency for this class is 12/139, or .086. In Figure 6.9 we have graphed relative frequency polygons for the weights of males and females. As you can see, it is easier to compare the graphs in this form.

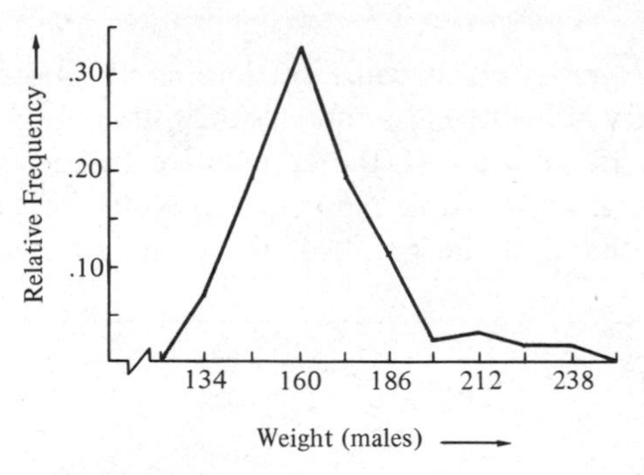

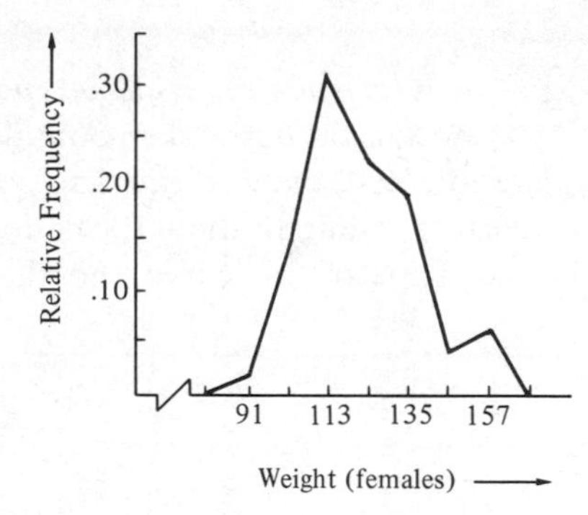

FIGURE 6.9

6.4 CUMULATIVE FREQUENCY DISTRIBUTIONS AND POLYGONS

In describing a large set of data it is sometimes useful to give the median (the 50th percentile) or some other percentiles. Finding percentiles from data that have been grouped into classes is made easier if cumulative frequency distributions are available, either in tabular or graphical form. A *cumulative frequency table* lists upper class boundaries and the cumulative frequency (i.e., the total number of observations) less than or equal to each boundary.

Example 6.4 Construct a cumulative frequency table for the weights of male statistics students given in Appendix A3.

Using the classes and frequencies given in Table 6.4, we see that there are no observations less that or equal to 127.5, 12 less than or equal to 140.5, and a *total of* 39 less than or equal to 153.5. This information gives us the first three entries in the cumulative frequency table. The complete table is given in Table 6.6.

TABLE 6.6

Upper Class Boundary	*Cumulative Frequency*
127.5	0
140.5	12
153.5	39
166.5	85
179.5	112
192.5	128
205.5	131
218.5	135
231.5	137
244.5	139

A *cumulative frequency polygon* is constructed by joining the points (x_i, y_i), where x_i is the upper class boundary of the ith class and y_i is the total number of observations less than or equal to x_i. In Figure 6.10 the cumulative frequency polygon corresponding to the information given in Table 6.6 is given. Note the similarity in the shape of this curve and the shape of the graph of the cumulative distribution

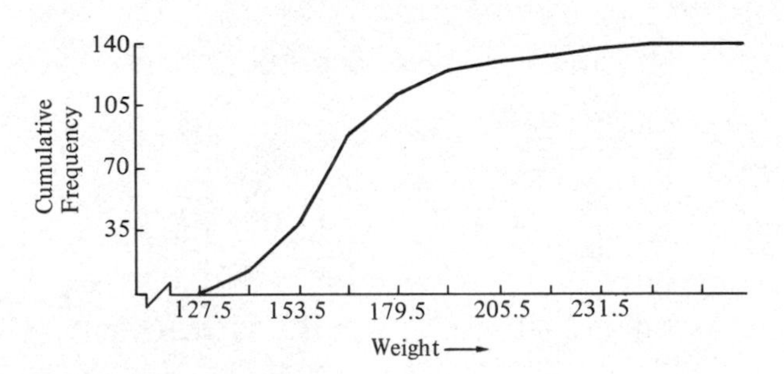

FIGURE 6.10

Cumulative Frequency Polygon for Weights of Male Students

functions for continuous random variables given in Figure 2.9 or 2.10 of Section 2.4. (In each case the curve reaches a plateau and stays there, as opposed to the behavior of the frequency polygon, which returns to the horizontal axis.)

Let us again consider the data given in Example 6.2, the weights of female students. In that example the original (i.e., ungrouped) data are given and from those data you should be able to find the sample median (see Section 5.4). How should you proceed? Since the sample median is the middle value when the observations are arranged in increasing order, you might go ahead and order the observations and then locate the $(n + 1)/2 = 31$st observation. However, with 61 values, ordering the observations is no simple task. Let us say, instead, that you use the results of the frequency

TABLE 6.7
Cumulative Frequency Distribution for Weights of Female Students

Upper Class Boundary	*Cumulative Frequency*
85.5	0
96.5	1
107.5	9
118.5	28
129.5	42
140.5	54
151.5	57
162.5	61

table given in Table 6.5 to construct the cumulative frequency table. From this table (Table 6.7) you can see that the cumulative frequency less than or equal to 118.5 is 28, while the cumulative frequency less than or equal to 129.5 is 42. What does this tell you about the location of the 31st observation? It must lie between 118.5 and 129.5 inclusive. We could now go through the original list of data and write down all 14 values in this class. In order, these would be

120	120	120	120	122	123	125
125	125	125	125	127	128.	

Since the third number in this list, corresponding to the 31st observation, is 120, this value is the sample median.

Since the purpose of grouping data is to present data in a concise fashion, the frequency or cumulative frequency tables are seldom accompanied by the actual raw data itself. Therefore, it is not possible to list the actual values contained in the median class itself. This situation is illustrated in Example 6.5.

Example 6.5 Jack Pyne, the owner of a mail-order lumber and supply business located in Seattle, was concerned about the amount of time required to fill the orders received. Jack's employees record (to the nearest minute) the times when the order was started and finished on each order. He has taken a sample of 75 filled orders from the files and found the times required to fill each one. This information is summarized in Table 6.8. Find the median for these data.

To find the median we need to determine the location of the $(n + 1)/2 = 38$th value when these times are listed in increasing order. Since the cumulative frequency up to and including 15.5 is 36 and the cumulative frequency up to and including 20.5 is 46, we know that the 38th value must lie between 15.5 and 20.5. Unfortunately, in this situation we do not know the values of the exact times, since only the summary distributions are given. It may be that all the values are bunched near 16 or they may all be bunched near 20. Since we have no way of knowing this, we might make the simplifying assumption that the observations are evenly spaced throughout the inter-

TABLE 6.8

Summary of Data for Example 6.5

Class	*Frequency*	*Upper Class Boundary*	*Cumulative Frequency*
1–5	4	5.5	4
6–10	10	10.5	14
11–15	22	15.5	36
16–20	10	20.5	46
21–25	13	25.5	59
26–30	9	30.5	68
31–35	5	35.5	73
36–40	2	40.5	75

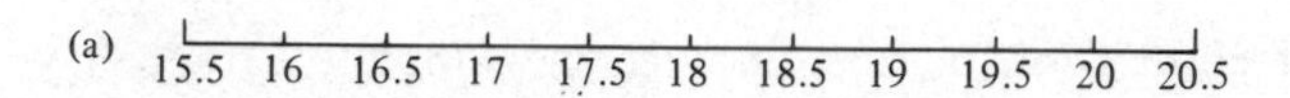

FIGURE 6.11

Location of Actual Times, Assuming Equal Spacing

val. (If the recorded times are given as integers, it would be impossible for all 10 values to be evenly spaced throughout the interval. However, since time is really a continuous quantity, the actual times could be evenly spaced.) To see how the spacing might go, consider Figure 6.11. In Figure 6.11(a) the class is divided into 10 equal parts (since there are 10 observations in the class). In Figure 6.11(b) an asterisk, representing an observation, is placed in the middle of each interval. Given that 36 observations fall below this class, the 38th one would be the second one in the class, so the median value would be 16.25. Note that had there been an even number of observations, say 76, the median would have the average of the middle two values. This would have led to finding the midpoint of the values corresponding to the 38th and 39th observations, namely 16.5.

The procedure used in finding the median in the preceding example is actually just linear interpolation. We could obtain the same results using equation (6.1). The only thing complicated in this equation is the notation. We will refer to the class containing the median as the *median class*. Let

n = total number of observations in the data
L = lower class boundary of the median class
cf = cumulative frequency up to (but not including) the median class
f = frequency of the median class
s = size of the median class
$\tilde{X}_g$ = sample median (calculated from grouped data).

Then the sample median is found from

(6.1)
$$\tilde{X}_g = L + \frac{.5n - \text{cf}}{f} \cdot s.$$

Applying this equation to Jack Pyne's data, we have

$$n = 75 \qquad f = 10$$
$$L = 15.5 \qquad s = 5,$$
$$\text{cf} = 36$$

so

$$\tilde{x}_g = 15.5 + \frac{.5(75) - 36}{10} \cdot 5$$
$$= 15.5 + \frac{1.5}{10} \cdot 5 = 16.25.$$

Had there been 76 observations instead of 75, the sample median would have been

$$\tilde{x}_g = 15.5 + \frac{.5(76) - 36}{10} \cdot 5$$
$$= 15.5 + \frac{2}{10} \cdot 5 = 16.5.$$

These two values, of course, agree with those calculated previously.

Since the median is really just the 50th percentile, you might suspect that equation (6.1) could be modified to allow you to calculate other sample percentiles. If we let x_p represent the $100 \cdot p$ percentile and in the earlier discussion replace the term "median class" with the term "percentile class," we can use equation (6.2) to find percentiles:

(6.2)
$$100 \cdot p\text{th sample percentile} = L + \frac{p \cdot n - \text{cf}}{f} \cdot s.$$

Example 6.6 Find the 80th percentile for the amount of time required for Jack's employees to fill an order.

In this case $p = .80$, so we need to find the location of the

$$p \cdot n = (.80)(75) = 60$$

observation. From Table 6.8 we see that the 60th observation will fall in the class 25.5 to 30.5, the percentile class for this problem. We identify the values in (6.2) to be

$$n = 75 \qquad \text{cf} = 59$$
$$p = .80 \qquad f = 9$$
$$L = 25.5 \qquad s = 5,$$

so

$$\text{80th sample percentile} = 25.5 + \frac{60 - 59}{9} \cdot 5 = 26.06.$$

In words, this means that 80% of all orders in the sample were filled in approximately 26 minutes or less.

In Section 2.6 we discussed finding the median and other percentiles of the distribution of a random variable by using a graph of the cumulative distribution function (see Figures 2.14 and 2.15). The same technique can be used in finding the median and percentiles from grouped sample data by using a cumulative-frequency polygon. In fact, if several percentiles are to be found, it might be advantageous to label the vertical axis in terms of cumulative relative frequency. In Figure 6.12 we

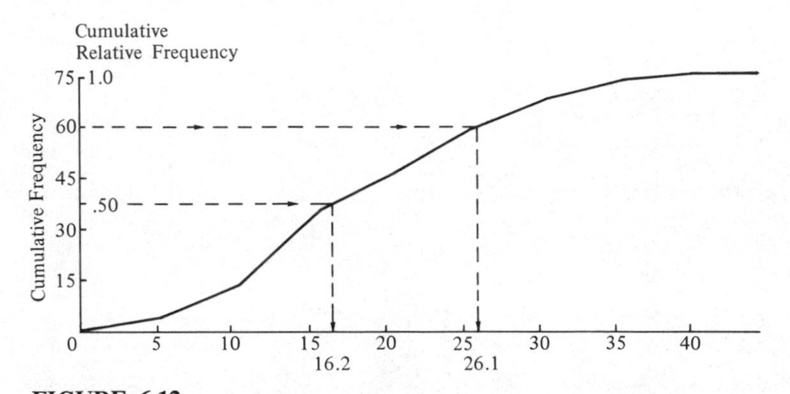

FIGURE 6.12

Cumulative Frequency Polygon for Jack Pyne's Data

give the cumulative-frequency polygon for Jack Pyne's data. The vertical axis is labeled with two scales: cumulative frequency on the left and cumulative relative frequency on the right. To locate any percentile, locate the corresponding value of p (on the right-hand scale) or equivalently the value of $n \cdot p$ (on the left-hand scale). Extend a horizontal line from that point until it intersects the polygon. Drop a vertical line from that point of intersection and read the sample percentile on the horizontal axis. If this graphical technique for finding percentiles were executed with extreme care, the results would be exactly the same as those obtained from equation (6.2). However, because of the difficulty in drawing and reading the graph precisely, this graphical technique usually gives only approximate results. In Figure 6.12 the 50th and 80th percentiles are found.

6.5 MEASURES OF CENTRAL TENDENCY FROM GROUPED DATA

In Section 2.6 we discussed three measures of central tendency for the probability distribution of a random variable, namely the mean, median, and mode of the distribution. Each of these measures has a sample counterpart. In Section 5.5 we defined the sample mean and sample median and showed how to calculate them from ungrouped data, while in Section 6.4 we showed how to calculate the sample median from grouped data. In this section we will show how to calculate the sample mean and mode from grouped data and give a brief discussion on the appropriateness of the various measures of central tendency from the viewpoint of descriptive statistics.

Definition 6.6 Given an observed set of data, the *sample mode* is the value in the data that occurs most frequently. Given a set of grouped data, the sample mode is the class midpoint of the class having the largest class frequency. This class is called the *modal class.*

Example 6.7 Noah Long, president of the Long Life Insurance Company, took a month off from his regular duties to try his hand at sales. During the first week the number of policies he sold per day were

4, 3, 4, 0, 3, 4.

The daily sales for the entire month are summarized as follows:

Number of policies sold/day:	0	1	2	3	4	5	6
Frequency:	2	4	4	6	5	2	1.

Find the mode for the first week's sales and for the entire month.

Since the number 4 occurred most often during the first week, it is the mode. For the entire month, the number 3 occurred most often; hence the mode for the month's sales is 3.

To be consistent with the convention given in Section 2.6, we will agree to say that the sample is *bimodal* if two values are tied for having the highest frequency. If three or more values are tied, we will say that no mode exists.

In this example the grouping does not combine values over a range of values into one class. The mode calculated from the grouped data is the same as would have been calculated from ungrouped data. This, of course, is not always the case. For example, in Table 6.8 (the times required to fill mail orders) we have no access to the ungrouped data and five different times are grouped into each class. The class with the highest frequency is the class 11 to 15, so the sample mode would be the corresponding class midpoint, 13. This value would most likely differ from what we would calculate from the ungrouped data had they been available.

We now turn our attention to the sample mean. Finding the sample mean for Mr. Long's first week of sales is rather easy. How would you find the sample mean for the entire month's sales? You would find the total number of policies sold and divide by the number of days devoted to selling. This latter quantity is easily found by adding all class frequencies together. To find the total number of policies sold, note, for example, that 2 policies sold per day on 4 different days gives a subtotal of 8 policies. Hence we need to take the product of the number of policies sold per day and the frequency of such an occurrence. We find

$$\text{total number of policies sold} = 0 \cdot 2 + 1 \cdot 4 + 2 \cdot 4 + \ldots + 6 \cdot 1 = 66$$

and

$$\text{total number of days} = 2 + 4 + 4 + \ldots + 1 = 24,$$

so $\bar{x} = \frac{66}{24} = 2.75$ policies per day.

How would you find the sample mean for the data on the times for Jack's employees to fill orders? (See Table 6.8.) Since the actual data are not given, you have no way of knowing where the values classified in a given class really fall in the class. As we did in Section 6.4 when we calculated the median for grouped data, we will have to make a simplifying assumption. We could assume, as we did earlier, that the values in a given class are equally spaced throughout the interval, or we could make the stronger assumption that all values are located at the class midpoint. Although both of these assumptions will lead to the same calculation of the sample mean, we will make the latter assumption for the purpose of simplifying calculations in the next section, dealing with the sample variance. If, in fact, all values in a class were located at the class midpoint, we would find the sample mean just as we did for Mr. Long's sales. Notationally, we define

$$\begin{aligned} \bar{X}_g &= \text{sample mean (calculated from grouped data)} \\ n &= \text{total number of observations in the data} \\ k &= \text{total number of classes} \\ x_i &= \text{midpoint of the } i\text{th class} \\ f_i &= \text{frequency of the } i\text{th class.} \end{aligned}$$

The sample mean for grouped data is then found from

$$\bar{X}_g = \frac{\sum_{i=1}^{k} f_i x_i}{n}, \tag{6.3}$$

where $n = \sum_{i=1}^{k} f_i$. (A common mistake made by people in finding the sample mean is to divide by k, the number of classes, rather than by n, the total number of observations! Be careful!) Note that we have written $\bar{X}_g$ using a capital letter, indicating that it is a random variable. We have not written the f_i using capital letters, although in fact they are random variables. This is due to the well-established convention of using lowercase letters. We can now find the sample mean for the data given in Table 6.8 by using equation (6.3). We have

$$\begin{aligned} \bar{X}_g &= \frac{4(3) + 10(8) + 22(13) + \ldots + 2(38)}{75} \\ &= \frac{1350}{75} = 18.0. \end{aligned}$$

In general, the value of the sample mean as calculated from grouped data will differ from the value as calculated from the original data, but the difference will usually be small. In fact, the difference will never be more than one-half the class size.

We have now discussed three measures of central tendency—the sample mean, median, and mode—which can be calculated either from ungrouped or grouped data. From the point of view of descriptive statistics, which measure should we use? Generally, in describing a set of data we want to choose a single value that is typical or representative of all the data in the set. Under various circumstances, any one of the three measures might be considered appropriate. The mode is seldom used except in a circumstance that will be discussed in Section 6.7. The choice as to whether to use the sample mean or median depends primarily on the values observed in the data. In most cases the mean is used; however, if there are a few extremely large (or small) values in the data, the mean may no longer be a "typical" value. In such a situation the median might be preferred. (One such example is that of data related to family income. The great majority of families have incomes in the range \$5000 to \$25,000, but a relatively small number of families have incomes 10 to 100 times this large. For this reason, median rather than mean incomes are often reported.) We will discuss in Section 6.7 some other elements that enter into making the choice of measure of central tendency.

Example 6.8 During the course of a televised football game between Oakland and Pittsburgh, the announcer attempted to keep the viewers informed as to significant "statistics" during the course of the game. At one point he informed them that "Oakland has averaged 8 yards on each first-down play." (The interested sports fan might like to think about his subsequent comment that "On the average Oakland has faced a second-and-two situation.") It happened, however, that early in the game a back ran for 68 yards on a first-down play. Comment on the appropriateness of his remark if, in fact, the yards gained on 12 first-down plays were

$$3, \quad 5 \quad 68, \quad 0, \quad -2, \quad 12, \quad 0, \quad 3, \quad 0, \quad -1, \quad 2, \quad 6.$$

The presence of one extremely large value indicates that the mean may not be typical. If the observations are ordered as follows:

$$-2, \quad -1, \quad 0, \quad 0, \quad 0, \quad 2, \quad 3, \quad 3, \quad 5, \quad 6, \quad 12, \quad 68,$$

we see that only 2 out of 12 times did Oakland make 8 or more yards. The value 8 is hardly typical of yards gained on first-down plays. However, the sample median of 2.5 yards is more typical.

There are many situations where people abuse statistics by doing something as simple as using an inappropriate measure of central tendency. You should keep in mind that most statistics you see and/or hear about in the media are descriptive in nature and consequently are summarizing large amounts of data. When summarizing the data, much important detail may be sacrificed. When you hear remarks such as: "Statistics can say what you want them to say. . . ." or "Here's an example of how statistics can be misleading. . . . ," you should remember that in many cases such comments result from the fact that detail is lost in summarization. An example of this happened during another football game. In this case Denver was playing Oakland and Denver led at half-time by 10–7. At that time the announcer said, "Here's an example where statistics can be misleading. Oakland had the ball for 20 minutes while Denver had it for only 10 minutes, so you'd expect Oakland to be ahead!" The time of possession may in fact be highly correlated with score, but it should not be expected to be a perfect predictor. The time of possession is actually a summary measure of all available data on each play. It could not tell you, as was the case, that Denver had a 70-yard pass play for a touchdown that consumed less than 15 seconds! A similar situation might occur for a sales representative whose manager looks at the number of calls made in an effort to measure the effectiveness of the representative. While the manager might expect a large amount of sales to follow from a large number of calls, he should keep in mind that the number of calls is merely a summary measure. There will be times when a small number of calls leads to large sales. Significant detail can be lost in summarizing data.

6.6 MEASURES OF VARIABILITY FOR GROUPED DATA

In Section 5.4 we defined the sample variance and sample standard deviation. These two quantities are the most important measures of the variability in a set of data. We would like to consider how to calculate these quantities for data that have been grouped. We will begin by considering the sales of Mr. Long during his month's work as a salesman (Example 6.7). We saw that the number of policies sold during the first week were

$$4, \quad 3, \quad 4, \quad 0, \quad 3, \quad 4.$$

To calculate s^2 for these data, we first note that $\bar{x} = 3$, so from equation (5.3), we find that

$$s^2 = \frac{(4-3)^2 + (3-3)^2 + (4-3)^2 + (0-3)^2 + (3-3)^2 + (4-3)^2}{6-1}. \tag{6.4}$$

In working out the numerical values in (6.4), you should encounter no difficulty, since the quantities to be squared are all integers. You might notice that the term "$(4-3)^2$" occurs three times and the term "$(3-3)^2$" occurs twice, so you could rearrange the right-hand side of (6.4) to read

$$\frac{3 \cdot (4-3)^2 + 2 \cdot (3-3)^2 + 1 \cdot (0-3)^2}{6-1}.$$

The idea of using multiplication for repeated terms allows us to write simple formulas for finding the variance for grouped data. Now consider the sales of Mr. Long over the entire month. Earlier we found that the sample mean for the month was 2.75 policies per day, so using the values for the month, we would find that

$$s^2 = \frac{2 \cdot (0-2.75)^2 + 4 \cdot (1-2.75)^2 + 4 \cdot (2-2.75)^2 + \ldots + 1 \cdot (6-2.75)^2}{24-1}$$

$$= \frac{58.5}{23} = 2.543.$$

The formula used to find the sample variance was

$$s^2 = \frac{\sum_{i=1}^{k} f_i(x_i - \bar{x})^2}{n-1},$$

where f_i = frequency of ith class and k = number of classes. Once again we note that the grouping in this case did not involve grouping together different values. When different values are grouped together, as in Table 6.8, for example, we generally do not know the original data values. As with the calculation of the sample mean, we will make the simplifying assumption that all values fall at the class midpoint. With this assumption we calculate the sample variance for grouped data by using

$$S_g^2 = \frac{\sum_{i=1}^{k} f_i(x_i - \bar{X}_g)^2}{n-1}. \tag{6.5}$$

The notation used here is the same as that used prior to equation (6.3). As you would expect, we define the sample standard deviation as

$$S_g = \sqrt{S_g^2}.$$

In calculating S_g^2 it is sometimes useful to put the necessary quantities in a table.

⋆***Example 6.9*** In the leftmost two columns of Table 6.9 are the classes and frequencies for the price listed in the classified ads for single-family homes under the "For Sale By Broker" heading. These homes are located in a medium-sized Midwestern city. The prices are given in thousands of dollars. Calculate the sample variance and sample standard deviation.

TABLE 6.9

Selling Price	*Frequency*	x_i	$f_i x_i$	$(x_i - \bar{X}_g)$	$f_i(x_i - \bar{X}_g)^2$
30.5–40.1	13	35.3	458.9	−10.45	1419.6325
40.2–49.8	8	45.0	360.0	−0.75	4.5000
49.9–59.5	2	54.7	109.4	8.95	160.2050
59.6–69.2	0	64.4	0	18.65	0
69.3–78.9	0	74.1	0	28.35	0
79.0–88.6	2	83.8	167.6	38.05	2895.6050
88.7–98.3	1	93.5	93.5	47.75	2280.0625
	26		1189.4		6760.0050

To calculate s_g^2 we first need to find $\bar{x}_g$, which, from (6.3), is $\sum f_i x_i / \sum f_i$. The totals of the second and fourth columns give these quantities, so we find

$$\bar{x}_g = \frac{1189.4}{26} = 45.75$$

(where we have rounded to two decimal places). Once $\bar{x}_g$ is found, we use (6.5) to find s_g^2. The fifth column gives some useful intermediate calculations, while s_g^2 is found using the totals of the second and sixth columns. We have

$$s_g^2 = \frac{6760.0050}{25} = 270.40$$

and

$$s_g = \sqrt{270.40} = 16.444.$$

In terms of dollars, the sample mean is \$45,750 and the sample standard deviation is

\$16,444. (You might give some thought as to what factors make the standard deviation so large and whether the sample mean is really the best measure of central tendency for these data.)

There are times when you might wish to calculate the sample variance by using an algebraically equivalent formula to that given in (6.5). In particular, if the sample mean is not an integer or simple decimal, it may save time to use equation (6.6).

$$S_g^2 = \frac{\left(\sum_{i=1}^{k} f_i x_i^2\right) - n(\bar{X}_g)^2}{(n-1)}. \tag{6.6}$$

Had we used this equation in Example 6.9, we would have found that

$$\sum_{i=1}^{7} f_i x_i^2 = 61{,}170.48,$$

so

$$s_g^2 = \frac{61{,}170.48 - 26(45.75)^2}{25}$$

$$= \frac{61{,}170.48 - 54{,}419.625}{25} = 270.03.$$

(The small discrepancy in the results is due to rounding error.)

6.7 MEASUREMENT SCALES

When the decennial census is taken by the federal government in the United States, a short-form questionnaire is supposed to be filled out by or for every person in the country. (Longer questionnaires are given to a sample of the population according to some systematic sampling procedure.) On the short-form questionnaire are questions relating to name, address, age, heritage, sex, number of rooms in the living unit, and so on. On this and many other questionnaires, the answers given may later be reduced to numbers for purposes of statistical summaries and for ease of storage and retrieval on computer systems. For example, a number may be assigned to sex, by

$$\text{male} \equiv 0 \qquad \text{female} \equiv 1 \tag{6.7}$$

or a number may be assigned to country in which a person is born by

$$\text{Canada} \equiv 1, \quad \text{Britain} \equiv 2, \quad \text{United States} \equiv 3, \quad \text{etc.} \tag{6.8}$$

Of course, it is easy to assign a number to age or number of rooms in a living unit. The question we would like to raise here is whether or not all numbers are the same. If they are different, how do they differ? Are some numbers more informative, in some sense, than others?

To answer these questions, we will begin by considering numbers that are assigned to categories, such as male or female, country in which born, state or province in which current residence is located, and so on. In situations such as this the number assigned to a category is simply a shorthand notation for the name of the category.

Such numbers are said to be on a *nominal* or *categorical* scale. The order of such numbers has no significance at all. The numbers assigned as in (6.7) or (6.8) could have been defined differently without destroying the purpose of the assignment. The fact that Canada is assigned a 1 while Britain is assigned a 2 does not mean that Canada is any better (or worse) a country to have been born in than Britain. Although the numbers themselves can obviously be ordered, no significance can be attached to this ordering, so calculating a sample median would be futile. It is perhaps even more obvious that a sample mean has no significance, although mathematically one could be calculated.

A second kind of measurement scale is one with which anyone who has read ratings by various agencies is familiar. AAA rates motels, Parents' Magazine rates movies, Standard and Poor rate bonds, while other agencies rate restaurants, automobiles, and so on. Consider, for example, a rating system for restaurants where an excellent restaurant is given a four-fork rating, a good one three forks, a fair one two forks, and a poor one only one fork. In this case the order is important. Assuming that the rating is done according to criteria with which you would agree, you can feel confident that a four-rated restaurant will be better than a two- or three-rated restaurant. When the order of the numbers has significance, the numbers are said to be on an *ordinal* scale. However, the differences between such numbers may not have meaning. For example, is the difference between a four and a three the same as the difference between a three and a two or a two and a one? Probably not. For another example of an ordinal scale, you might think of the teams in a football, soccer, or basketball conference or league. The teams may be ranked according to some system, but the difference in strength between the first- and second-place teams may be quite different than those between the last- and next-to-last-place teams.

Now consider two cities, Greenville and Blueville, which are to have five randomly chosen restaurants rated (one, two, three, or four forks) by a noted gourmand, Elke Sellser. After rating the restaurants in these two cities, Elke might want to summarize her findings by using a measure of central tendency. The mode might be used, but in this case the sample median would be appropriate because the ratings can be ordered. However, it would be inappropriate to use the sample mean because differences are not meaningful. A four-fork and two-fork restaurant should not be considered to "average out" to a three-fork restaurant.

A third kind of measurement scale can be illustrated by using temperatures. If you measure the temperature of three containers, each containing the same amount of water and find the temperatures to be 40, 50, and 60°C, what can you conclude about these numbers? Certainly, the order of these numbers has significance, but even more can be said. Those of you familiar with physics will remember that it takes the same amount of heat to raise a given amount of water from 40 to 50°C as from 50 to 60°C. Meaning can, in fact, be attached to differences between measurements made on such a scale. When the order of numbers and differences between numbers both have significance, the numbers are said to be on an *interval* scale. If numbers are on an interval scale, the sample mean is an appropriate measure of central tendency to use. You can see this by considering the example of temperature given earlier. If you were to combine an amount of water at 40°C with an equal amount of water at 60°C, the temperature of the resulting mixture would have a temperature of 50°C. However, there is still a deficiency in numbers on the interval scale: significance cannot be attached to ratios. For example, it is not "twice as hot" on a day when the temperature is 30°C as when it is 15°C.

The highest-order measurement scale is one in which the order of numbers, the differences between numbers, and the ratio of numbers all have significance. Such a scale is called a *ratio* scale. This is a familiar scale to all of us, since most measurements of mass, weight, length, and time are on a ratio scale. For example, say that in a census interview you determine that among the residents of a household are three people, aged 10, 30, and 60 years old. Is it true that the 60-year-old individual has lived for twice as many years as the 30-year-old? That the 30-year-old has lived three times as many years as the 10-year-old? Of course. Consequently, significance can be attached to ratios, so age (actually a time measurement) is measured on a ratio scale.

When measurements are made on a ratio scale, it is appropriate to use either the sample mean or median as a measure of central tendency. (Actually, the mode could be used, but in general the mode is used only for measurements made on a nominal scale.) Of course, if a sample mean is reasonable to calculate, it is also reasonable to calculate the sample variance (or standard deviation).

For the sake of emphasis we will summarize the discussion of this section. When measurements are made, when questionnaires are filled out, and in general whenever data are collected, numbers can be assigned to the resultant answers. However, in answer to the questions raised at the beginning of this section, the numbers may be of a different nature and not all the same because of being on different measurement scales. Whenever numbers are available, it is *mathematically possible* to calculate medians, means, or variances. However, if the measurements are not made on the appropriate scale, the values calculated may simply be nonsense. We have summarized in Table 6.10 the various types of measurement scales and the appropriate measures of central tendency which can be used with each type of measurement. (We have classified the mode as inappropriate for use with measurements on the ordinal, inter-

TABLE 6.10

Measurement Scales and Appropriate Measures of Central Tendency

Measurement Scale	*Appropriate Measure of Central Tendency*		
	Mode	*Median*	*Mean*
Nominal	Yes	No	No
Ordinal	No	Yes	No
Interval	No	Yes	Yes
Ratio	No	Yes	Yes

val, or ratio scales. Technically, it is allowable in these cases, but in practice it is seldom used.)

6.8 CODING

In the days before computers and low-cost hand calculators, the topic of coding was important, since it frequently led to a savings in time devoted to computations. We include the topic here primarily for those who may not have a hand calculator or access to a computer. The use of coding may also offer some convenience to those who do have computational aids.

We have already used a form of coding in Table 6.9 when we expressed the sale prices of homes in thousands of dollars instead of dollars (e.g., "30.5 thousand dollars" rather than "30,500 dollars"). You can see that by using coding you can have simpler figures to work with. In this case we simplified by dividing each number by 1000.

As another example, consider the bowling scores of Sherri Lane, a professional bowler. If she scores 200, 202, and 207 in three games, what is her average for these games? What would her average be if Sherri scored 240, 242, and 247? What kind of shortcuts can you use in making these calculations? (Take a moment and do the calculations.) Of course, the arithmetic here is quite easy and it is not hard to add these numbers and divide by 3. In finding the average of the first set of scores, many people would "ignore" the 200 in each score and just find the average of 0, 2, and 7. Since the average of these is 3, they would rightly conclude that the average is 203. Having calculated this average, many people would note that the second scores are all 40 higher than the first scores, so the average should be 40 higher, namely 243. This is valid reasoning and is a second example of coding.

Both of these types of coding, dividing by a constant and subtracting a constant, can be combined. If x_i represents the original data and if u_i is defined by

$$u_i = \frac{x_i - a}{b} \tag{6.9}$$

for some constants a and b, then u_i is referred to to as the coded data. If the coding is done properly, the calculations necessary to find $\bar{u}$ and s_u^2 will be easier than those necessary to find $\bar{x}$ and s_x^2. However, these quantities are related by the following equations:

$$\bar{x} = a + b\bar{u} \tag{6.10}$$

$$s_x^2 = b^2 s_u^2. \tag{6.11}$$

(Note the similarity between these relationships and those given in Theorems 2.1 and 2.2 and Theorems 2.5 and 2.6.) It is also true, of course, that

$$s_x = \sqrt{b^2 s_u^2}. \tag{6.12}$$

Example 6.10 In six games Sherri bowls 205, 240, 195, 225, 245, and 210. Use coding to find her average.

The choice of the constants a and b is not unique. We might wish to subtract 200 from each value (i.e., take $a = 200$) or we might subtract 195, 225, or some other value. Subtracting 200 from each score, for the sake of illustration, would give

$$5, \quad 40, \quad -5, \quad 25, \quad 45, \quad 10.$$

Since these numbers are all multiples of 5, we could take $b = 5$ and divide each of these new numbers by 5 to get

$$1, \quad 8, \quad -1, \quad 5, \quad 9, \quad 2.$$

We have now found the u_i values by the relationship

$$u_i = \frac{x_i - 200}{5}.$$

It is easy to see that $\bar{u} = \frac{24}{6} = 4$, so

$$\begin{aligned}\bar{x} &= a + b\bar{u} = 200 + 5\bar{u} \\ &= 220.\end{aligned}$$

Coding data can be especially helpful when calculating sample means and variances from grouped data. If the constant a is taken to be a class midpoint, either for a class of large frequency or for a class near the middle of the frequency table, and if the constant b is taken to be the class size, the u_i values (coded class midpoints) will be small integer values. This will make computations easier (even if hand calculators are used).

Example 6.11 Using the data on the cost of homes given in Table 6.9, use coding to find the sample mean and standard deviation.

Using the guidelines given earlier for choosing a and b, let us choose $a = 45.0$, the midpoint of the second class, and $b = 9.7$, the class size. In Table 6.11 we give the original class midpoints, the class frequencies, and the coded class midpoints in the first three columns. In the fourth and fifth columns we give the products $f_i u_i$ and $f_i u_i^2$.

TABLE 6.11
Coding Grouped Data

Class Midpoints, x_i	*Class Frequencies,* f_i	*Coded Class Midpoints,* u_i	$f_i u_i$	$f_i u_i^2$
35.3	13	−1	−13	13
45.0	8	0	0	0
54.7	2	1	2	2
64.4	0	2	0	0
74.1	0	3	0	0
83.8	2	4	8	32
93.5	1	5	5	25
	$\sum f_i = 26$		2	72

The sums of the second, fourth, and fifth columns can be used to find $\bar{u}_g$ and $s_{g,u}$ using equations (6.3) and (6.6). We find that

$$\bar{u}_g = \frac{\sum f_i u_i}{n} = \frac{2}{26} = .077$$

and

$$s_{g,u}^2 = \frac{(\sum f_i u_i^2) - n(\bar{u}_g)^2}{n-1}$$

$$= \frac{72 - 26(.077)^2}{25} = 2.874.$$

(Answers here were rounded to three decimal places.) Using equations (6.10) to (6.12), we get

$$\bar{x}_g = a + b\bar{u}_g = 45.0 + 9.7(.077) = 45.747$$

$$s_{g,x}^2 = b^2 s_{g,u}^2 = (9.7)^2(2.874) = 270.415$$

$$s_{g,x} = \sqrt{270.415} = 16.444.$$

These results, as they should, agree with those found in Section 6.6.

EXERCISES

★*1.* The following are measurements (the capillary basement membrane thickness) made on 42 healthy individuals. These are measurements of the average thickness of capillaries in a certain section of muscle tissue. The units of measurement are angstroms (10^{-10} meter). The measurements were taken using an electron microscope.

1270	901	1277	829	868	735	701	1006
766	942	1005	709	623	820	1005	855
794	917	895	1087	833	724	754	880
1002	872	854	652	841	860	664	1009
924	1022	735	704	947	1331	1116	859
735	836						

Using six classes, construct (a) a frequency table; (b) a frequency histogram. (*To think about:* Do these data appear to follow a normal distribution?)

★*2.* In the 1977 Sugar Bowl, Tony Dorsett of the University of Pittsburgh carried the ball 32 times and had a total of 199 rushing yards. The yards gained per carry were

4,	−2,	5,	2,	3,	−5,	4,	16,
1,	2,	1,	0,	2,	0,	22,	−2,
11,	6,	2,	10,	15,	1,	67,	−4,
4,	22,	0,	0,	4,	5,	2,	1.

If you wished to summarize this information using a frequency distribution, you would have to decide on a number of classes to use. Construct a frequency table using (a) 5 classes; (b) 15 classes. (c) Discuss the advantages and disadvantages of using 5 classes instead of 15.

★*3.* Using the random number tables, choose a random sample of 25 students from the first 100 students in the list *with* replacement. Find the corresponding ages and weights of these students. (a) Using these data, construct a frequency table and a histogram for the ages. (b) Using 5 classes construct a frequency table and a frequency polygon for the weights. (If you worked Exercise 6 of Chapter 5, you will already have the data.)

★*4.* Using the random number tables, choose a random sample of 25 students from the first 100 students in the list *without* replacement. Find the corresponding heights and grade-point averages of these students. (a) Using these data, construct a frequency table and a histogram for the heights using 5 classes. (b) Also using 5 classes, construct a frequency table and a frequency polygon for the grade-point averages. (If you worked Exercise 7 of Chapter 5, you will already have the data.)

5. Histograms and other types of graphical representation of data used in newspapers and magazines are often poorly labeled and give a distorted picture of the information they are intended to convey. In particular, the choice of scales and the failure to include a "scale break" can lead to distortion. Obtain clippings or copies of graphs from two different articles, state briefly what information these graphs are intended to convey, and give your opinion as to whether these graphs are properly or improperly used.

6. Using the frequency tables you constructed in Exercise 3, construct a cumulative frequency polygon for (a) the ages; (b) the weights. Find the median weight from (c) the ungrouped data; (d) graphically from the cumulative frequency polygon.

7. Using the frequency tables you constructed in Exercise 4, construct a cumulative relative frequency polygon for (a) heights; (b) grade-point averages. Find the median grade-point average (c) graphically from the cumulative relative frequency polygon; (d) from the frequency table using equation (6.1).

8. Using the classes and frequencies you found in Exercise 1, (a) construct a frequency polygon; (b) construct a cumulative frequency polygon. (c) Find the 25th and 75th percentiles of the CBMT graphically using the cumulative frequency polygon. (d) Find these percentiles from the frequency table using equation (6.2).

★*9.* A weekly news magazine reported the following percentiles for U.S. family incomes (in thousands of dollars) in 1976. (a) Use this information to sketch a cumulative relative frequency polygon. (b) From this graph estimate the 25th and 75th percentiles.

Family income:	4	6	8	10	12	15	20	25	50
Percentile:	7	14	22	30	38	50	69	82	98.

(*To think about:* How can you use the information given here to show that the distribution of family incomes is skewed? Could you use this information to estimate the 99th percentile?)

★*10.* The following frequency table gives asking prices for used cars as listed in the classified section of a newspaper. The prices were rounded to the nearest $10. Find the median, the 10th percentile, and the 90th percentile from this table.

Cost ($)	*Frequency*
380–1200	6
1210–2030	9
2040–2860	2
2870–3690	2
3700–4520	3

11. Find the sample mean asking price for used cars using the frequency table given in Exercise 10.

12. (a) Using the ungrouped data from Exercise 2, find the average (i.e., sample mean) number of yards per carry. Find the average number of yards per carry using the grouped data with (b) 5 classes; (c) 15 classes. (*To think about:* How can you explain any discrepancy among these answers?)

13. (a) Using the ungrouped data from Exercise 2, find the sample median number of yards per carry. Find the sample median number of yards per carry using the grouped data with (b) 5 classes; (c) 15 classes. (*To think about:* Which number would be considered more typical of the yards gained per carry, the sample mean or median? Why?)

14. Using the data as grouped in Exercise 3, find (a) the sample mean age; (b) the sample median weight.

15. Using the data as grouped in Exercise 4, find (a) the sample median height; (b) the sample mean grade-point average.

★***16.*** A portion of land in the southeastern part of Arizona was divided into 5-m^2 quadrats and the number of creosote bushes (a shrub that grows in arid regions) were counted for 20 quadrats with the following results:

No. creosote bushes/quadrat:	0	1	2	3	4	5	6
No. quadrats (frequency):	1	4	5	6	1	1	2.

Using this frequency distribution, calculate the (a) sample mean; (b) sample variance. (*To think about:* Given what you know about the mean and variance of a Poisson random variable, what does this suggest about the spatial distribution of creosote bushes?)

17. Find the sample variance and standard deviation of the asking price for used cars using the frequency table given in Exercise 10. (See also Exercise 11.)

18. Using the classes and frequencies you found in Exercise 1, calculate the sample variance and standard deviation of the CBMT.

★***19.*** The following frequency table gives the monthly rental for apartments as listed in the classified section of a newspaper. Use the information to find the sample median, mean, and standard deviation of the monthly rentals.

Rental ($)	*Frequency*
100–130	2
135–165	10
170–200	7
205–235	1
240–270	2

20. Obtain a copy of a local newspaper and make a list of the first 30 monthly rentals given in the classified section. Using the classes given in Exercise 19 (and adding classes as necessary), make a frequency distribution of these data. Using the grouped data, find the sample median, mean, and standard deviation. (*To think about:* How could you compare the results found in Exercise 19 with the results you obtained? How might differences be explained?)

21. Twenty children at a local elementary school are to be interviewed and will be asked the following: (1) What is your favorite cereal? (2) How old are you? (3) Rank the following sports 1–2–3 (with 1 = favorite): baseball, basketball, football. (4) How tall are you? (5) What color was your mother's hair this morning?

The answers to questions like these can be classified as nominal (or categorical), ordinal, interval, or ratio scale measurements. State which of these classifications pertains to questions 1 to 5, respectively. (*To think about:* For which of these questions would it be reasonable to try to compute a sample mean?)

22. Use coding to calculate the sample mean and variance of the asking price for used cars using the frequency table given in Exercise 10.

23. Use coding to calculate the sample mean and standard deviation of the monthly rental for apartments using the frequency table given in Exercise 19.

24. Use coding to calculate the sample mean and standard deviation of the following numbers:

4000 4125 4450 3900 4175 3950 4225 3975.

7

ESTIMATION

7.1 INTRODUCTION

In studying the various families of probability distributions in Chapter 4 we saw that the different members of the families were distinguished by having different parameter values. For example, different members of the family of normal distributions are distinguished by having different values of μ and/or σ^2, the parameters n and p distinguish members of the binomial family, and so on. We have seen that once these parameter values are known, it is possible to calculate probabilities associated with a particular distribution as well as to find means and variances. You may have noticed, however, that in many of the examples and exercises in Chapter 4 and in subsequent material, we have told you to *assume* that the parameter values were such and such. One reason for this is that in many cases the parameter values are unknown and by making assumptions about their values we can get numerical answers to the examples and exercises. Making such assumptions may make sense from a pedagogical viewpoint, but it certainly does not help solve practical real-world problems! If parameters are unknown, it is usually not enough to simply make assumptions about their values.

We have alluded to the problem of what to do if parameter values are unknown in earlier chapters. In Chapter 1 we discussed the relative-frequency method of assigning probabilities. We saw that to estimate the unknown probability p of an event, we could use the relative frequency of occurrence of that event in a large number of trials. In Example 4.26 we faced the problem of estimating the value of λt for a Poisson random variable, the number of raisins per cookie. In that situation we used the average number of raisins per cookie examined as an estimate of λt. In this chapter we will more formally address ourselves to the question of how to estimate unknown parameters.

7.2 POINT ESTIMATION

In Chapter 4 we discussed some special probability distributions that are observed in real-world situations. As you may have guessed, there are many other families of distributions which have proven useful in providing probabilistic models for describing behavior observed in real life in addition to those we discussed. For example, the gamma distribution, the Weibull distribution, and the beta distribution are all families of distributions indexed by two parameters. However, the methods used to estimate these parameters are more involved than we care to study in this book. In this chapter we will restrict our attention to the estimation of means (μ), variances (σ^2), and proportions (p). We will also consider estimation of differences between means or proportions and ratios of variances for different random variables ($\mu_1 - \mu_2, p_1 - p_2, \sigma_1^2/\sigma_2^2$). Since many of the statements we make will apply to all these quantities, it will be convenient to have a single symbol to represent a parameter. We will use the symbol θ as the generic representation of a parameter.

Definition 7.1 A *point estimate* of a parameter θ is a single number used as an estimate of the value of θ.

Although this definition is practically self-explanatory, you might very well ask how the number used as the point estimate is obtained. Generally, it is obtained from sample information. Let us say, for example, that a continuous random variable X has a density function that depends on the parameter θ. We might denote this density function by

$$f(x; \theta).$$

To obtain information about this distribution, and more particularly about the value of the unknown parameter θ, we need to take a sample of observations from this distribution. In this text we will generally assume that the observations obtained are independent so that we in fact will have a random sample of observations. Once we have obtained the sample of observations we will use them to obtain a point estimate of θ.

We will find that in most situations good estimators can be obtained simply by using the *sample counterpart*. For example, to estimate a population mean μ we would use a sample mean, $\bar{X}$; to estimate a proportion p, we would use a sample proportion; and so on.

Example 7.1 A random sample of five observations of students enrolled in a statistics course showed heights of 68, 73, 69, 61, and 72 inches. Use this information to give a point estimate of the mean height μ of all students enrolled in that statistics course.

Since we are interested in the mean μ, we would estimate it by using the observed value of the sample mean, $\bar{x}$. We find

$$\bar{x} = \frac{343}{5} = 68.6,$$

so the point estimate of μ is 68.6 inches.

If we were to observe a second sample of five students, we might observe heights of 62, 72, 60, 74, and 69. In this case the observed value of the sample mean $\bar{x}$, and consequently the point estimate of μ, would be $337/5 = 67.4$ inches. Does it come as a surprise to you that different samples yield different point estimates of μ? It should not, since our method of obtaining point estimates is to use the observed value of the sample mean $\bar{X}$, and of course, $\bar{X}$ is a random variable. It is generally true that a point estimate of a parameter is simply the observed value of a random variable. A random variable used to obtain point estimates of a parameter is called an *estimator* of that parameter. In Example 7.1 we would say that the random variable $\bar{X}$ is an estimator of the parameter μ. The generic representation of an estimator for a parameter θ is $\hat{\Theta}$. (Note that in keeping with our usual convention, we use a capital letter to represent a random variable. An observed value of $\hat{\Theta}$ would be denoted by $\hat{\theta}$. This observed value would be the point estimate.)

Definition 7.2 An estimator of a parameter θ is a random variable $\hat{\Theta}$ whose observed values are used as point estimates of θ.

Example 7.2 Find an estimator for the variance σ^2 of the heights of students enrolled in a statistics course.

Since we will generally use the sample counterpart of the parameter in question, we can use

$$S^2 = \frac{\sum_{i=1}^{n} (X_i - \bar{X})^2}{n - 1}$$

as an estimator for σ^2.

Example 7.3 Using the data given in Example 7.1, find a point estimate for the variance σ^2 of the heights of students enrolled in a statistics course.

The observed value of S^2 is

$$s^2 = \frac{\sum_{i=1}^{5} (x_i - \bar{x})^2}{4} = \frac{\sum_{i=1}^{5} (x_i - 68.6)^2}{4}$$

$$= \frac{89.20}{4} = 22.30,$$

which is the point estimate for σ^2.

We have said that in obtaining estimators for unknown parameters, we will generally use the sample counterpart. This approach to obtaining estimators does in fact simplify matters, but it also raises some questions: (1) This method appears to be rather intuitive; are there any objective mathematical approaches that could be used instead? (2) Are there different possible estimators for the same parameter? If so, how do we know which is better? (3) Are there any ambiguities with this approach? Detailed answers to these questions must be left to more advanced courses, but we can give some surficial answers.

In answer to question (3), there may be ambiguities with this approach. For instance, if it is known that the distribution from which you sample is symmetric, it is true that $\mu = \tilde{\mu}$; that is, the mean and median are identical. Given a sample of observations from the distribution, we might estimate the mean μ by using $\bar{x}$, but we could also estimate μ by using $\tilde{x}$, the sample median. This is reasonable since if $\mu = \tilde{\mu}$, then an estimate of $\tilde{\mu}$, namely $\tilde{x}$, is logically also an estimate of μ. This, of course, leads us to question (2), since we now have at hand two different estimators for the same parameter (and as we shall see, there are others). In general, we can say that the sample mean $\bar{X}$ is preferred over the sample median $\tilde{X}$ when in the case of symmetry both can be used as estimators of μ. The reasons for this will be discussed in Section 7.3. Finally, there are some mathematical methods that can be used as objective ways for finding estimators for parameters. Two of these methods are the *method of moments* and the *method of maximum likelihood.* In many cases, the estimators obtained by these methods turn out to be simply the sample counterparts!

• • • • •

We will now very briefly illustrate the method of obtaining maximum likelihood estimators. The method is perhaps best understood by considering a special situation. A gentleman of ill repute and feeble mind, Mr. N. Wit, has in his possession two biased silver dollars which he hopes to use to increase his fortune. One of these has a probability $p_1 = .30$ of coming up heads while the other has a probability $p_2 = .60$ of coming up heads. True to form, he has put them both in the same pocket and now cannot tell which is which. He takes one of the coins and proceeds to flip it 10 times. If he observes that it comes up heads four times, which coin has he flipped? Of course, we cannot really be sure which one it is, since with either coin it is possible to get 4 heads in 10 tries. What does your intuition tell you? You might reason that since the sample proportion of heads was .4 (which is closer to .3 than to .6), it is more apt to be coin 1. That, of course, is reasonable, but another way of arriving at the same conclusion is as follows. Define X to be the number of heads observed in 10 flips of this coin. Then X has a binomial distribution with parameters $n = 10$ and $p = p_1$ or $p = p_2$. If, in fact, p_1 is the correct value of p, what would be the probability of observing 4 heads? What would it be if p were p_2? We have from the tables of the binomial distribution

$$P[X = 4 \mid n = 10, p = p_1 = .3] = \binom{10}{4}(.3)^4(.7)^6 = .2001,$$

while

$$P[X = 4 \mid n = 10, p = p_2 = .6] = \binom{10}{4}(.6)^4(.4)^6 = .1115.$$

Since the probability is higher for $p_1 = .3$ than for $p_2 = .6$, it is not unreasonable to guess (estimate) that the correct value of p is $p_1 = .3$. That, in a nutshell, is the idea of maximum likelihood. Given observed data, estimate the parameters to be those values which make the probability the largest (i.e., a maximum). (Since for continuous random variables the probability assigned to any individual point is actually zero, this statement must be modified to say that we estimate the parameters to be those values which maximize the value of the joint probability density function, given the observed data.)

Continuing with our example, let us assume that Mr. Wit further confused things by putting an honest silver dollar in the same pocket as the two biased coins. If he chose a coin from his pocket and obtained 4 heads in 10 flips, which of the three coins would you guess that he had? (Note that .4 is halfway between .3 and .5, so the "intuitive" approach mentioned earlier will fail to be of help.) If $p = p_3 = .5$, the probability of observing 4 successes in 10 trials is

$$P[X = 4 \mid n = 10, p = p_3 = .5] = .2051,$$

so by the maximum likelihood principle we would "estimate" p to be .5.

Of course, in most estimation problems we do not have the kind of information that Mr. Wit did. In view of this, let us consider another situation where Mr. Wit is given another silver dollar by a friend who has character defects similar to his own. Assume that Mr. Wit suspects the coin is biased but has no idea whatsoever as to the value of p. If he wishes to estimate p, how should he proceed? He, of course, should obtain sample information. If he flips the coin n times and observes x heads, he might estimate the parameter p to be that value of p which makes the probability of observing x heads a maximum. In particular,

$$P[X = x] = \binom{n}{x} p^x (1 - p)^{n-x}, \tag{7.1}$$

so Mr. Wit might like to find the value of p for which (7.1) is as large as possible. For particular values of n and x, we could substitute different values of p into (7.1) until, by trial and error, we found the appropriate value of p. This is not a very practical approach and we might do better by approaching the problem mathematically. Since on the right-hand side of (7.1) we are treating n and x as fixed and are treating p as a quantity that can assume values between 0 and 1, the right-hand side is actually a function of p. We can make this functional relationship more clear by defining

$$L(p) = \binom{n}{x} p^x (1 - p)^{n-x} \qquad 0 \leq p \leq 1$$

and restating the problem to be to find the value of p that maximizes the function $L(p)$, called the likelihood function. In calculus you learn that the maximum of a function may occur where the first derivative is equal to zero. With this in mind we find $dL(p)/dp$, set this equal to zero, and solve for p.

$$\frac{dL(p)}{dp} = \binom{n}{x}[xp^{x-1}(1 - p)^{n-x} + (n - x)(1 - p)^{n-x-1}(-1)p^x] = 0.$$

Factoring out some common terms, we get

$$\binom{n}{x} p^{x-1}(1-p)^{n-x-1}[x(1-p)-(n-x)p]=0$$

$$x - xp - np + xp = 0.$$

Solving this last equation for p, we get

$$p = \frac{x}{n}.$$

[Some further work would show that this value actually maximizes the function $L(p)$.] Consequently, the maximum likelihood estimator of p is x/n. Note that this is exactly the sample proportion of successes and is the same as the estimator we would have used by following our naive "sample counterpart" approach!

Example 7.4 Let $X_1, X_2, \ldots, X_n$ be a random sample from a normal distribution having a variance known to be 1 and an unknown mean. If the sample values $x_1, x_2, \ldots, x_n$ are observed, find the maximum likelihood estimate of μ.

Following our earlier discussion, we would use as an estimate of μ that value which maximizes the joint probability density function (since these are continuous random variables), given the observed data. Since σ^2 is assumed to be 1, the joint probability density function is

$$f(x_1, x_2, \ldots, x_n) = \frac{1}{(\sqrt{2\pi})^n} \exp\left[-\frac{1}{2}\sum_{i=1}^{n}(x_i-\mu)^2\right]. \tag{7.2}$$

Since on the right-hand side of (7.2) we are treating the x_i and n as fixed and are treating μ as the variable quantity which we can adjust to maximize the right side of (7.2), we can express this relationship by defining

$$L(\mu) = \frac{1}{(\sqrt{2\pi})^n} \exp\left[-\frac{1}{2}\sum_{i=1}^{n}(x_i-\mu)^2\right].$$

We now wish to find the value of μ that maximizes the likelihood function $L(\mu)$. We do this by taking the derivative of $L(\mu)$ with respect to μ, setting this derivative equal to zero, and solving for μ as follows:

$$\frac{dL(\mu)}{d\mu} = \frac{1}{(\sqrt{2\pi})^n} \exp\left[-\frac{1}{2}\sum_{i=1}^{n}(x_i-\mu)^2\right]\left[-\frac{1}{2}\left(\sum_{i=1}^{n} 2(x_i-\mu)^1(-1)\right)\right] = 0.$$

Solving the right-hand equation gives

$$\sum_{i=1}^{n}(x_i-\mu) = 0$$

$$\sum x_i - n\mu = 0$$

$$\mu = \frac{\sum x_i}{n} = \bar{x}.$$

So the maximum likelihood estimate of μ is $\bar{x}$, the sample mean.

(Note that once again the somewhat complicated maximum likelihood method leads to the same result as the "sample counterpart" approach. Although this is not universally true, the two methods do frequently lead to the same results.)

7.3 PROPERTIES OF GOOD ESTIMATORS

In a situation where we are trying to estimate an unknown parameter that can assume values over some continuous range of values (e.g., trying to estimate p, where p can assume values between zero and 1), we must face up to the fact that we cannot hope to obtain a point estimate that is *exactly* correct. A similar situation might be found at a fair or carnival, where, for a nominal fee, Sam Barker will try to guess your weight. Of course, your true weight at any point in time is some number over a continuous range of values and Sam would face a rather hopeless task if he tried to guess your exact weight. He would obviously have a difficult time even if he tried to guess to the nearest pound! (To give himself some advantage, Sam considers a guess to be correct if it is within 5 pounds.) If we realize that we cannot estimate certain quantities exactly then, we are admitting that we will always be somewhat off in our estimates. Sometimes the estimates will be too high, sometimes they will be too low. However, we might feel that our method of estimation is a good one if the high and low estimates tend to balance out so that in the long run, on the average, our guesses are "just right." Mathematically, of course, we can find the "long-run average value" of an estimator by finding its expected value.

Definition 7.3 $\hat{\Theta}$ is said to be an *unbiased estimator* of the parameter θ if $E(\hat{\Theta}) = \theta$. An estimator that is not unbiased is said to be biased.

If an estimator is biased, the amount of bias can be measured by

$$\text{Bias } (\hat{\Theta}) = E(\hat{\Theta}) - \theta.$$

Example 7.5 If $X_1, X_2, \ldots, X_n$ is a random sample from a probability distribution having mean μ and variance σ^2, verify that $\bar{X}$, the sample mean, is an unbiased estimator of μ.

Since we have shown that the expected value of $\bar{X}$ is equal to μ in Chapter 5 [i.e., $E(\bar{X}) = \mu$], it follows from Definition 7.3 that $\bar{X}$ is an unbiased estimator for μ.

It can be shown (in fact, you were given some hints as to how to show this in Exercise 16 of Chapter 5) that the sample variance S^2 defined by

$$S^2 = \frac{\sum_{i=1}^{n} (X_i - \bar{X})^2}{n - 1}$$

is an unbiased estimator for σ^2. It is for this reason that we defined the divisor for calculating S^2 to be $n - 1$ rather than n.

Example 7.6 Let X be a random variable having a binomial distribution with parameters n and p. Show that X/n is an unbiased estimator for p.

It follows from Theorem 4.1 that $E(X) = np$, so

$$E\left(\frac{X}{n}\right) = \frac{E(X)}{n} = \frac{np}{n} = p,$$

and consequently X/n is an unbiased estimator for p.

If we consider a random sample X_1, X_2, X_3 of size 3 from a distribution having a mean μ, we know that

$$E(X_1) = E(X_2) = E(X_3) = \mu,$$

so X_1, X_2, X_3, each considered separately, are all unbiased estimators for μ. Furthermore,

$$\frac{X_1 + X_2}{2}, \quad \frac{X_1 + X_3}{2}, \quad \frac{X_2 + X_3}{2}$$

are also unbiased estimators for μ. As you can see from this, it is possible to define many different unbiased estimators for μ. How might we choose, from among several unbiased estimators, the most desirable one? We might proceed as Sam Barker did in trying to find an assistant to help him with his weight guessing. Sam had to decide between two applicants, Stan and Ollie. He had each of them guess the weights of four fairgoers, each of a different body build, but all weighing 180 pounds. Stan's guesses were 160, 190, 170, and 200 pounds, while Ollie's guesses were 165, 195, 175, and 185. Sam noted that both Stan and Ollie did about the same "on the average" (in a sense, both are unbiased estimators), but that Ollie's guesses were less variable. Since Sam was concerned about variability, Ollie got the job.

In many situations, all else being equal, an estimator having less variability would be preferred to one having more variability. Since the variance is a measure of variability, an estimator with smaller variance would be preferred to one having a larger variance.

Example 7.7 A random sample of size 3, X_1, X_2, X_3, is taken from a distribution having mean μ and variance σ^2. Find the variance of X_1, $(X_1 + X_2)/2$, and $(X_1 + X_2 + X_3)/3$.

Since X_1, X_2, and X_3 are a random sample, we know that

$$V(X_1) = V(X_2) = V(X_3) = \sigma^2$$

and that these random variables are independent. Using Corollary 3.2 and Theorem 2.6, we get

$$V(X_1) = \sigma^2$$

$$V\left[\frac{X_1 + X_2}{2}\right] = \frac{V(X_1 + X_2)}{4} = \frac{V(X_1) + V(X_2)}{4} = \frac{\sigma^2}{2}$$

$$V\left[\frac{X_1 + X_2 + X_3}{3}\right] = \frac{V(X_1 + X_2 + X_3)}{9} = \frac{V(X_1) + V(X_2) + V(X_3)}{9} = \frac{\sigma^2}{3}.$$

Each of the three quantities X_1, $(X_1 + X_2)/2$, and $(X_1 + X_2 + X_3)/3$ is an unbiased estimator for μ, but since $(X_1 + X_2 + X_3)/3$ has the smallest variance of these three, it would be the preferred estimator. We have, of course, only considered three of many possible unbiased estimators for μ based on samples of size 3. How do we find the very "best" unbiased estimator (i.e., the one with smallest variance) among all possible unbiased estimators? While the details of such an answer must be left to more advanced courses, we can say that given a sample of size n from a distribution having mean μ, the sample mean $\bar{X}$ will have the smallest variance among a large class of unbiased estimators for μ.

In trying to determine what constitutes a "good" estimator for a parameter θ, we have proposed two criteria, unbiasedness and small variance. Fortunately, in many situations the sample counterpart of a parameter will be both unbiased and have a small variance. Hence our suggestion to use the sample counterpart of a parameter as an estimator of that parameter is a reasonable one.

We might consider how to compare estimators that are biased. A quantity that combines a measure of the bias as well as the variance is the mean-squared error of an estimator.

Definition 7.4 The *mean-squared error* of an estimator $\hat{\Theta}$ is

$$\text{MSE}(\hat{\Theta}) = E(\hat{\Theta} - \theta)^2.$$

You can show (see Exercise 18) that the mean-squared error of an estimator can be expressed by

$$E(\hat{\Theta} - \theta)^2 = V(\hat{\Theta}) + [\text{Bias}\,(\hat{\Theta})]^2. \tag{7.3}$$

You can see from (7.3) that the mean-squared error combines variance and the (square of) the bias. If $\hat{\Theta}$ is an unbiased estimator of θ, the mean-squared error of $\hat{\Theta}$ will be equal to the variance of $\hat{\Theta}$.

Example 7.8 A random sample of size 10, $X_1, X_2, \ldots, X_{10}$, is taken from a distribution having mean μ and variance σ^2. Two estimators for μ are proposed: $Y_1 = (X_1 + X_{10})/2$ and $Y_2 = (X_1 + X_2 + \ldots + X_{10})/9$. Find the mean-squared error of the estimators Y_1 and Y_2.

Since $E(Y_1) = \mu$, it follows that

$$\text{MSE}(Y_1) = V(Y_1) = \frac{\sigma^2}{2}.$$

We see, however, that Y_2 is a biased estimator, since

$$E(Y_2) = \frac{E(X_1 + X_2 + \ldots + X_{10})}{9} = \frac{10\mu}{9} \neq \mu.$$

The bias is

$$\text{Bias}\,(Y_2) = E(Y_2) - \mu = \frac{10\mu}{9} - \mu = \frac{\mu}{9}$$

and the variance of Y_2 is

$$V(Y_2) = \frac{V(X_1 + X_2 + \ldots + X_{10})}{81} = \frac{10\sigma^2}{81},$$

so

$$\text{MSE}(Y_2) = \frac{10\sigma^2}{81} + \frac{\mu^2}{81} = \frac{10\sigma^2 + \mu^2}{81}.$$

There are other properties of good estimators which might be discussed (e.g., efficiency and consistency), but we will leave these discussions to other courses.

7.4 CONFIDENCE INTERVALS FOR μ WITH σ^2 KNOWN

In the previous sections we have discussed the concept of point estimation. In point estimation we give a single number as an estimate of an unknown parameter. There are many situations where giving a single number is a reasonable course of action. For example, a concessionaire who must order hot dogs to be sold at the next ballgame needs to make an estimate of the number that will be sold. The order he places must be for a single number of hot dogs. Similarly, the director of a state agency who presents a budget to the governor must present a specific single number as the final amount even though the costs of many items in the budget may be subject to chance variation. However, what about the director of revenue who must estimate the amount of money that will be available to the state government for expenditures? If the governor were to ask him for an estimate of the incoming revenue for the next fiscal year, should he give a single value for an answer? If he does give a single number, say \$2,456,572,115, what are the chances that he will have estimated correctly? The probability of his giving a precisely correct estimate is practically zero. In view of his chances of estimating correctly being so small, what might he do? He might give a range of values that he feels may ultimately be shown to contain the correct value (e.g., he might give the estimate as \$2.4 to 2.5 billion). That is, he might give an interval of values of such a size that he can feel relatively confident that the true revenue will lie within that interval. How might he increase his level of confidence? He can increase it by making the interval larger, say \$2.2 to 2.7 billion. This is the basic idea of interval estimation of a parameter. Rather than give a single number as a point estimate of a parameter, give an interval of values that will contain the parameter with high probability.

We will see that in almost all cases a confidence interval for a parameter can be found by starting with a point estimator of that parameter. The point estimator is a random variable and the probability distribution of the estimator, standardized in some appropriate way, can be used to obtain confidence intervals.

Example 7.9 A vending machine was designed to dispense 7 ounces of Fizzy Cola into 8-ounce cups. From measurements made on the amount dispensed by other machines of this type, it is felt that the actual amount dispensed, X, has a normal distribution with a standard deviation of .25 ounce. In order to estimate the mean of the amount of Fizzy Cola being dispensed, the amounts dispensed into four cups were carefully measured and found to be 6.92, 7.34, 7.26, and 6.88 ounces. Find a point estimate for μ.

To find a point estimate for μ, we simply use the sample counterpart, namely the sample mean, and observe that

$$\bar{x} = \frac{6.92 + 7.34 + 7.26 + 6.88}{4} = 7.10 \text{ ounces.}$$

Now since it was assumed that the amount dispensed into a cup is normally distributed, it follows from the discussion in Section 5.7 that the sample mean $\bar{X}$ will also have a normal distribution. We will exploit this fact to develop confidence intervals for μ. Since we know that $\bar{X}$ has a normal distribution, we can say, by standardizing $\bar{X}$, that

$$P\left[-1.96 \leq \frac{\bar{X} - \mu}{\sigma/\sqrt{n}} \leq +1.96\right] = P[z_{.025} \leq Z \leq z_{.975}] = .95, \tag{7.4}$$

where as usual the quantity z_p represents the $100 \cdot p$ percentile of the standard normal distribution. Note that the event

$$\left[-1.96 \leq \frac{\bar{X} - \mu}{\sigma/\sqrt{n}} \leq +1.96\right]$$

can be described as the simultaneous occurrence of two events,

$$\left[-1.96 \leq \frac{\bar{X} - \mu}{\sigma/\sqrt{n}}\right] \quad \text{and} \quad \left[\frac{\bar{X} - \mu}{\sigma/\sqrt{n}} \leq +1.96\right]. \tag{7.5}$$

Considering the first event defined in (7.5), we can rearrange the inequality so that μ stands alone by the following steps:

$$-1.96 \leq \frac{\bar{X} - \mu}{\sigma/\sqrt{n}}$$

$$-1.96\frac{\sigma}{\sqrt{n}} \leq \bar{X} - \mu$$

$$\mu - 1.96\frac{\sigma}{\sqrt{n}} \leq \bar{X}$$

$$\mu \leq \bar{X} + 1.96\frac{\sigma}{\sqrt{n}}. \tag{7.6}$$

In the same way the second event in (7.5) can be rearranged so that μ stands alone, the resulting inequality being

$$\bar{X} - 1.96\frac{\sigma}{\sqrt{n}} \leq \mu. \tag{7.7}$$

Since the events described by (7.6) and (7.7) are equivalent to those in (7.5), it follows that the simultaneous occurrence of these events will have the same probability as found in (7.4):

$$P\left[\bar{X} - 1.96\frac{\sigma}{\sqrt{n}} \leq \mu \leq \bar{X} + 1.96\frac{\sigma}{\sqrt{n}}\right] = .95. \tag{7.8}$$

Note that $\bar{X} - 1.96\sigma/\sqrt{n}$ and $\bar{X} + 1.96\sigma/\sqrt{n}$ are random quantities since they both depend on the random variable $\bar{X}$. Consequently, these quantities form the end points of a *random interval.* This random interval may lie entirely to the left of the constant value μ, it may lie entirely to the right of μ, or it may include the value μ. In fact, according to (7.8), there is a probability of .95 that the random interval will actually include the value μ. Just as we have distinguished throughout this book between a random variable X and the observed value of a random variable x, we must distinguish between the end points of a random interval and the observed values of the end points of a random interval. Once the data are observed and the values of the end points are calculated, they are no longer random.

Example 7.10 Twenty different random samples of size 4 were taken from a normal distribution having a known mean and variance $\mu = 5$ and $\sigma^2 = 4$. (These are actually pseudo-random samples taken by the computer.) The sample means were

3.71	5.42	5.29	4.77	5.98	5.72	4.71	2.76	5.37	5.61
4.89	6.00	4.39	5.44	7.42	4.25	4.61	5.21	4.57	3.25.

Plot the observed end points of the random intervals of the form $\bar{X} \pm 1.96(\sigma/\sqrt{n})$ on a graph.

These results are shown in Figure 7.1. The end points of the observed confidence intervals have been connected by a line and a vertical line at the value $\mu = 5$ has been drawn. In this particular example only 18 of the 20 (i.e., 90%) of the observed intervals actually contain the value μ. However, if we were to take thousands of obser-

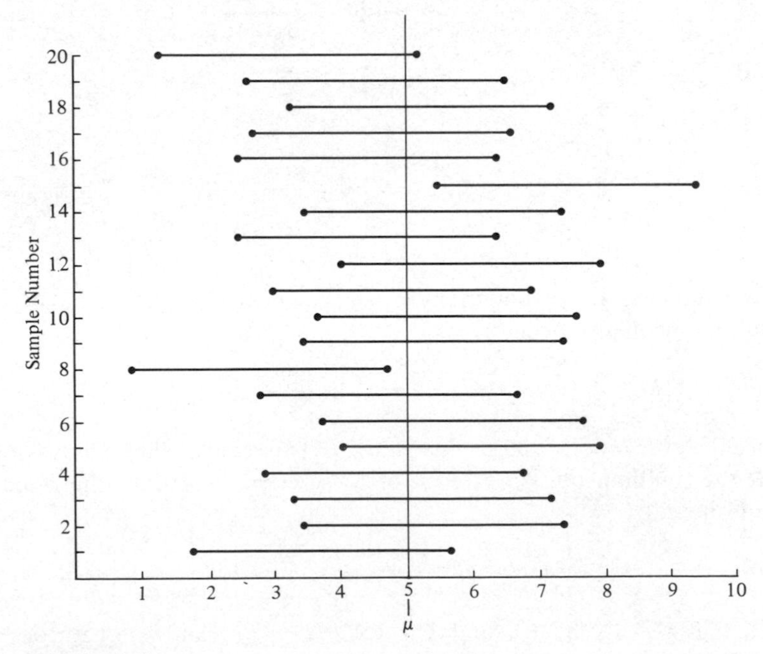

FIGURE 7.1

vations (i.e., in the long run), 95% of the observed intervals would contain the value μ. In this case the interval found using the eighth sample lies entirely to the left of μ while the interval found using the fifteenth sample lies entirely to the right of μ.

In Example 7.10 we determined some random intervals in the case where μ was actually known, and consequently we could say for certain whether or not particular observed intervals actually contained the value μ. Of course, the objective in this chapter is to obtain estimates for μ in situations where μ is *not* known. If μ is not known, then we cannot say for sure whether an observed interval really does contain the value μ. For instance, using the information in Example 7.9, we can find the end points of a random interval by

$$\bar{x} - \frac{1.96\sigma}{\sqrt{n}} = 7.10 - \frac{1.96(.25)}{\sqrt{4}} = 6.86$$

and

$$\bar{x} + \frac{1.96\sigma}{\sqrt{n}} = 7.10 + \frac{1.96(.25)}{\sqrt{4}} = 7.34$$

to get the interval (6.86, 7.34). Does this interval contain μ? We do not really know! However, what would you guess? How confident would you feel about your guess? Since 95% of all intervals found in this way will contain μ, we say that we feel 95% confident that this interval contains μ.

There is nothing very special about the value 95%. If we wanted to be more sure that a random interval would contain μ, we could use 98%, 99%, 99.9%, and so on. Of course, to find such an interval, we would have to replace the value 1.96 in (7.8) with the appropriate percentile. In general, we can write

$$(7.9) \qquad P\left[-z_{1-\alpha/2} \le \frac{\bar{X} - \mu}{\sigma/\sqrt{n}} \le z_{1-\alpha/2}\right]$$

$$= P\left[\bar{X} - z_{1-\alpha/2}\frac{\sigma}{\sqrt{n}} \le \mu \le \bar{X} + z_{1-\alpha/2}\frac{\sigma}{\sqrt{n}}\right] = 1 - \alpha,$$

so the values $\bar{X} \pm z_{1-\alpha/2}(\sigma/\sqrt{n})$ give the end points of a random interval that will contain μ with probability $1 - \alpha$.

Example 7.11 In Example 7.9 we were told that four measurements of the amount of Fizzy Cola dispensed by a vending machine had a sample mean of 7.10 ounces and that the measurements came from a normal distribution having a standard deviation of .25 ounce. (a) Find a random interval that will contain the true value of μ with probability .90. (b) What are the observed values of the end points of this interval for this sample?

Using equation (7.9), we have

$$1 - \alpha = .90,$$

so $\alpha = .10$ and $1 - \alpha/2 = .95$. Since $z_{.95} = 1.645$, it follows that

$$\left(\bar{X} - 1.645\frac{\sigma}{\sqrt{n}}, \bar{X} + 1.645\frac{\sigma}{\sqrt{n}}\right)$$

is a random interval that will contain the true value of μ with probability .90. The observed interval for this sample is

$$\left(7.10 - 1.645\left(\frac{.25}{\sqrt{4}}\right), 7.10 + 1.645\left(\frac{.25}{\sqrt{4}}\right)\right) = (6.89, 7.31).$$

Remember that the interval found in this way can be used as an estimate of the value of μ. The interval found in Example 7.11, (6.89, 7.31), is slightly smaller than the interval found earlier using the value $z_{.975} = 1.96$, namely, (6.86, 7.34). In terms of estimation, a shorter interval gives a more "precise" guess as to the value of μ than a longer interval. However, in this case the shorter interval was obtained by sacrificing the confidence we have in the estimate, a decrease from 95% to 90% in confidence. By examining equation (7.9), you will see that the only way to decrease the width of the interval while maintaining the level of confidence is to increase the sample size, n. We will pursue this thought in Section 7.12.

We can now state a formal definition of a confidence interval.

Definition 7.5 If $\hat{\Theta}_L$ and $\hat{\Theta}_U$ are the end points of a random interval satisfying

$$P[\hat{\Theta}_L \leq \theta \leq \hat{\Theta}_U] = 1 - \alpha,$$

the observed values of $\hat{\Theta}_L$ and $\hat{\Theta}_U$ for a given sample will be called a $1 - \alpha$ *confidence interval estimate* for the parameter θ.

If $X_1, X_2, \ldots, X_n$ is a random sample from a normal distribution having mean and variance μ and σ^2, it follows from our previous discussions that

$$P\left[\bar{X} - z_{1-\alpha/2}\frac{\sigma}{\sqrt{n}} \leq \mu \leq \bar{X} + z_{1-\alpha/2}\frac{\sigma}{\sqrt{n}}\right] = 1 - \alpha,$$

so the values $\bar{x} \pm z_{1-\alpha/2}(\sigma/\sqrt{n})$ give the end points of a $1 - \alpha$ confidence interval for μ.

We have mentioned in earlier chapters that not many random variables actually follow a normal distribution themselves. However, you may have noticed that our method of obtaining a confidence interval for μ was to start with the point estimator of μ, namely $\bar{X}$, and use that distribution to write a probability statement [cf. equation (7.4)]. Although the distribution of $\bar{X}$ may not be exactly normal if a sample is taken from a nonnormal distribution, the Central Limit Theorem (Section 5.8) allows us to say that if the sample size is sufficiently large, the distribution of $\bar{X}$ will be approximately normal. Consequently, for sufficiently large n, a $1 - \alpha$ confidence interval estimate for μ can be given by

$$\left(\bar{x} - z_{1-\alpha/2}\frac{\sigma}{\sqrt{n}}, \bar{x} + z_{1-\alpha/2}\frac{\sigma}{\sqrt{n}}\right). \tag{7.10}$$

Example 7.12 At the start of a new model year the employees at the Leprechaun Automobile Company receive salary and wage increases. If it is assumed that the standard deviation of the annual salary or wages of employees remains at $3200 and if a random sample of 36 employees showed a sample mean earnings of $18,540, find a 99% confidence interval estimate for μ, the mean annual earnings for all employees.

Although annual earnings would typically have a skewed distribution, the sample size is large enough to invoke the Central Limit Theorem. To use (7.10), first note that $\alpha = .01$, so

$$z_{1-\alpha/2} = z_{.995} = 2.576.$$

Using (7.10), we find that

$$\left(18{,}540 - 2.576\left(\frac{3200}{\sqrt{36}}\right), 18{,}540 + 2.576\left(\frac{3200}{\sqrt{36}}\right)\right)$$
$$= (\$17{,}166, \$19{,}914).$$

The confidence interval given in (7.10) is based on the assumption that the sample values are obtained by taking a random sample. According to the convention established in Section 5.2, we assume that sampling from a population is done with replacement, guaranteeing the independence of the observations. Since the variance of $\bar{X}$ is affected by *sampling without replacement* [cf. equation (5.13) of Section 5.8], the interval given in (7.10) would have to be appropriately modified if sampling were done without replacement.

7.5 CONFIDENCE INTERVALS FOR μ WITH σ^2 UNKNOWN

In Example 7.12 we made the assumption that the standard deviation of the annual earnings of employees at the Leprechaun Automobile Company remained at \$3200, although employees were given raises at the start of a new model year. Is that assumption reasonable? It would be reasonable if all employees were given a \$100/month raise, for example, but not if they were all given a 10% raise. There is also the possibility that raises are given on merit and not systematically. In this case it would be hard to know whether σ would increase, decrease, or remain essentially the same. We made the assumption that σ was known to be \$3200 to enable us to evaluate the end points of the interval given in (7.10). You can see that it would not be very informative to express a confidence interval estimate for μ in that situation by

$$(18{,}540 - 2.576\sigma/\sqrt{36}, 18{,}540 + 2.576\sigma/\sqrt{36}). \tag{7.11}$$

Although there are some situations where μ is unknown but σ is known, it is far more likely to be the case that σ will also be unknown. What might you try to do if you wanted a confidence interval for μ in (7.11), say, but did not know the value of σ? Given the sample values and recalling our discussions in Section 7.2 on point estimation, you might decide to replace σ by an estimate of σ, the sample standard deviation s. Good idea. However, there is one problem. You will have obtained an interval, but the confidence level would not be 99% as you originally had hoped.

Let us say, then, that we start with a random sample of size n from some normal distribution having unknown mean and unknown variance, μ and σ^2. It is true that the random variable

$$\frac{\bar{X} - \mu}{\sigma/\sqrt{n}}$$

has a standard normal distribution, but if we replace σ by the sample standard deviation S, the resulting random variable

$$\frac{\bar{X} - \mu}{S/\sqrt{n}}$$

will *not* have a standard normal distribution. This quantity is actually the quotient of two random variables. In general, it is very difficult to find the probability distribution of a quotient of random variables, but it is less of a problem when the distribution from which a sample is taken is normal.

Theorem 7.1 If $X_1, X_2, \ldots, X_n$ is a random sample from a normal distribution having mean μ and variance σ^2, the random variable

$$T = \frac{\bar{X} - \mu}{S/\sqrt{n}} \tag{7.12}$$

will have a Student's t distribution with $n - 1$ degrees of freedom.

The distribution of the random variable T defined in equation (7.12) was derived by William S. Gosset (1876–1937), a statistician for Guinness, an Irish brewery. Gosset wrote under the pen name "Student," and subsequently the distribution has been called Student's t distribution in his honor. The density function for this random variable is a complicated function given by

$$f(t) = \frac{\Gamma[(\nu + 1)/2]}{\sqrt{\nu\pi}\,\Gamma(\nu/2)}\left(1 + \frac{t^2}{\nu}\right)^{-(\nu+1)/2} \qquad -\infty < t < \infty,$$

where $\Gamma(\,\cdot\,)$ represents a gamma function. You need not be concerned with evaluating this density function, but you can see from looking at the function that the t distribution is not one, but a family of distributions indexed by a single parameter called "degrees of freedom," denoted by the symbol ν (Greek lowercase letter nu).

In Figure 7.2 we have sketched the density functions for t distributions with $\nu = 3$ and 7, and also the density function for a standard normal random variable.

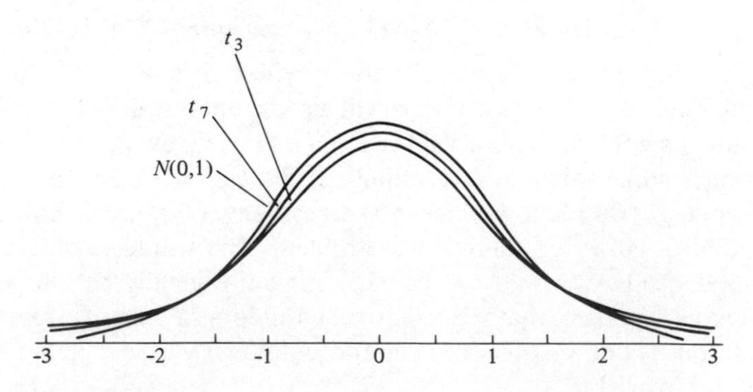

FIGURE 7.2
Density Functions for t_3, t_7, and Z

It is true, as indicated from these graphs, that as $\nu \to \infty$, the density function of the t distribution approaches the density function of a standard normal random variable. It is for this reason that the last row of the t tables, Table A6, actually gives percentiles for the standard normal distribution. The other rows of the t tables give upper percentiles for the t distribution for various degrees of freedom. Since the distributions are symmetric about zero, the lower percentiles can be easily found.

Using the fact that the random variable defined in equation (7.12) has a t distribution with $n - 1$ degrees of freedom, we can find confidence intervals for μ. More precisely, if $X_1, X_2, \ldots, X_n$ is a random sample from a normal distribution having mean μ and variance σ^2, the random variable T defined in equation (7.12) will satisfy

$$P\left[-t_{n-1,\,1-\alpha/2} \leq \frac{\bar{X} - \mu}{S/\sqrt{n}} \leq t_{n-1,\,1-\alpha/2}\right] = 1 - \alpha. \tag{7.13}$$

Just as we did in the previous section, we can solve the two inequalities given in (7.13) for μ to get

$$P\left[\bar{X} - t_{n-1,\,1-\alpha/2}\frac{S}{\sqrt{n}} \leq \mu \leq \bar{X} + t_{n-1,\,1-\alpha/2}\frac{S}{\sqrt{n}}\right] = 1 - \alpha.$$

It follows that $\bar{X} \pm t_{n-1,\,1-\alpha/2}S/\sqrt{n}$ give the end points of a random interval and, by construction, there is a probability of $1 - \alpha$ that such a random interval will contain the value μ. Consequently, an observed interval, namely,

$$\left(\bar{x} - t_{n-1,\,1-\alpha/2}\frac{s}{\sqrt{n}},\ \bar{x} + t_{n-1,\,1-\alpha/2}\frac{s}{\sqrt{n}}\right) \tag{7.14}$$

can be used as a $1 - \alpha$ confidence interval estimate for μ.

Example 7.13 The heights of first-grade children are approximately normally distributed. Given that a random sample of six first-grade children had heights (measured in inches) of

$$47.5 \quad 48 \quad 47.5 \quad 49 \quad 50 \quad 51.5,$$

find a 95% confidence interval for μ, the mean height of all first-grade children.

Since we want a 95% confidence interval, we have

$$1 - \alpha = .95,$$

so $1 - \alpha/2 = .975$. The confidence interval must be found using (7.14) since σ^2 is unknown. Since $n = 6$, we need to find

$$t_{n-1,\,1-\alpha/2} = t_{5,\,.975}.$$

From Table A6 we find $t_{5,\,.975} = 2.571$. Using $\bar{x} = 48.92$ and $s = 1.59$, we get

$$\left(48.92 - 2.571\left(\frac{1.59}{\sqrt{6}}\right),\ 48.92 + 2.571\left(\frac{1.59}{\sqrt{6}}\right)\right) = (47.25,\ 50.59)$$

You may have noticed from Table A6 that for a given percentage value, the magnitude of the percentile decreases as ν increases (e.g., from the tables you can see that $t_{5,\,.90} = 1.476 > t_{10,\,.90} = 1.372 > t_{15,\,.90} = 1.341 > t_{\infty,\,.90} = 1.282$).

As a consequence of this fact, the length of a $1 - \alpha$ confidence interval based on a t distribution when σ is unknown will generally be somewhat longer than a $1 - \alpha$ confidence interval based on a standard normal distribution when σ is known. This longer length reflects the lack of precise information about the value of σ (Exercises 32 and 46).

We would like to emphasize the fact that the confidence intervals derived in this section were derived under the assumption that the data constituted a random sample from a *normal distribution.* If the assumption of normality is violated, then the intervals obtained using (7.14) may not be $1 - \alpha$ level intervals. The associated "confidence level" may be different from $1 - \alpha$. Most authors will agree that even though the data may come from a nonnormal distribution, intervals obtained using (7.14) will be *approximate* $1 - \alpha$ confidence interval estimates for μ provided that the measurements are made on an interval or ratio scale (see Section 6.7) and the sample size is large enough. Authors do not agree, however, on what is "large enough." We would suggest as a conservative rule of thumb that if the normality assumption is violated, the sample size should be at least 30 before (7.14) can be used with confidence. For smaller sample sizes it may be possible to obtain confidence intervals using nonparametric techniques.

7.6 CONFIDENCE INTERVALS FOR $\mu_1 - \mu_2$, PAIRED OBSERVATIONS

In a large number of advertisements claims are made about the benefits of using the advertiser's product rather than a competitor's product. The benefit is sometimes expressed in terms of the difference between two values. One example of such claims can be found in advertisements for automotive products. Radial tires, for instance, are said to give you increased gas mileage as compared to ordinary tires. Certain types of motor oil are claimed to be able to give you better gas mileage than regular motor oil. Various kinds of gasoline additives are said to result in improved mileage. For the moment let us assume that we agree that any of these factors could make a difference in mileage, but we would like to estimate the amount of the difference. How could we obtain such an estimate? Say that we had four cars available for testing purposes and were interested in the effect of motor oil on gas mileage. We could proceed by putting Aristocrat oil (brand 1) into two cars and Plebeian oil (brand 2) into the other two cars, let the cars be driven for a month, and then compare the gas mileages from the different vehicles. In particular, if we defined

$$X_1 = \text{gas mileage obtained by car using Aristocrat oil}$$

$$X_2 = \text{gas mileage obtained by car using Plebeian oil,}$$

we could use the difference in the sample means, $\bar{X}_1 - \bar{X}_2$, as an estimate of $\mu_1 - \mu_2$. This potentially could be a satisfactory way of obtaining a point estimate of $\mu_1 - \mu_2$, but if conditions were not carefully controlled, the results could be of very little use. For example, if Aristocrat oil were put into compact cars driven by Ann and Heather Feathertouch while Plebeian oil was put into full-size cars driven by Harry and Larry Leadfoot, the observed differences in gas mileage could well be due to the type of car and/or driver rather than the type of oil.

As an alternative to having a separate sample of cars for each type of oil, we could consider an experiment where each car is driven with each type of oil by the same driver. For example, first drive for 1000 miles with Aristocrat oil, change oil to Plebeian, and after a suitable break-in period, drive for 1000 miles with Plebeian oil. When experimental units are subjected to two different treatments in this way, the resulting observations are said to be *paired*. In this particular example we would have four pairs of observations, each of which is a difference between the gas mileage obtained using the two oils. In general, we would consider a sample of n observations defined by

$$\begin{aligned} D_1 &= X_{1,1} - X_{2,1} = \text{difference obtained by car 1,} \\ D_2 &= X_{1,2} - X_{2,2} = \text{difference obtained by car 2,} \\ &\vdots \\ D_n &= X_{1,n} - X_{2,n} = \text{difference obtained by car } n. \end{aligned}$$

We, of course, can use the sample mean difference $\bar{D}$ as a point estimate of the population mean difference μ_D. The quantity is actually

$$\mu_D = E(D) = E(X_1 - X_2) = E(X_1) - E(X_2) = \mu_1 - \mu_2,$$

so $\bar{D}$ will, in fact, give a point estimate of $\mu_1 - \mu_2$, the difference of the two distributions.

With a point estimate in hand we can turn our attention to the problem of finding a confidence interval estimate of $\mu_1 - \mu_2$, or equivalently of μ_D. You will notice, however, that by using paired observations we have reduced a problem dealing with the means of two different distributions, the distributions of X_1 and X_2, to a problem dealing with the mean of a single distribution, the distribution of D defined by $D = X_1 - X_2$. We have already addressed ourselves to the problem of finding a confidence interval for the mean of a single distribution in Sections 7.4 and 7.5. Generally, we would not know the value of the variance of D, so we would need to use the methods of Section 7.5. If we can assume that the observations $D_1, \ldots, D_n$ comprise a random sample from a normal distribution, it follows from (7.14) that a $1 - \alpha$ confidence interval estimate for $\mu_D = \mu_1 - \mu_2$ can be given by

$$\left(\bar{d} - t_{n-1,\,1-\alpha/2}\frac{s_d}{\sqrt{n}},\ \bar{d} + t_{n-1,\,1-\alpha/2}\frac{s_d}{\sqrt{n}}\right). \tag{7.15}$$

Example 7.14 Four cars were tested for 1000 miles each using Aristocrat oil and Plebeian oil. Use the results of these tests as given in Table 7.1 to find a 95% confidence interval for $\mu_D = \mu_1 - \mu_2$, the difference in the mean mileage obtained using the two brands of oil.

Assuming the differences D are normally distributed, we use (7.15) to get the desired confidence interval. We find

$$\bar{d} = \frac{\sum d_i}{n} = \frac{0.86 + 0.69 + 1.36 - 0.33}{4} = 0.645$$

$$s_{\bar{d}}^2 = \frac{\sum (d_i - \bar{d})^2}{n - 1}$$

$$= \frac{(0.86 - 0.645)^2 + \ldots + (-0.33 - 0.645)^2}{3} = 0.503$$

$$t_{3,.975} = 3.182,$$

TABLE 7.1

	Compact 1	*Compact 2*	*Full Size 1*	*Full Size 2*
Mileage using oil 1, x_1	19.77	18.90	20.20	16.29
Mileage using oil 2, x_2	18.91	18.21	18.84	16.62
Difference $(x_1 - x_2)$	0.86	0.69	1.36	−0.33

so the end points of the confidence interval are given by $(.645 \pm 3.182\sqrt{.503}/\sqrt{4})$ or $(-.485, +1.775)$.

The use of pairing of observations minimizes the effect of extraneous factors and helps to give more precise estimates (i.e., estimators having a smaller variance or confidence intervals having a shorter length than those which could be obtained using the same sample size without pairing). Pairing can be useful in studies trying to measure the effectiveness of a treatment such as a diet. Measurements taken before a treatment is started and after the treatment has been in effect comprise a pair. Frequently, two different treatments are given to twins or siblings (in human research) or to littermates (in animal research) in an effort to control for effects due to heredity and/or environment. The measurements on these paired subjects then comprise a pair. If more than two treatments are to be studied, a generalization of the method of pairing can be used. The more general method is called the randomized complete block design and is treated in Section 10.7.

7.7 CONFIDENCE INTERVALS FOR $\mu_1 - \mu_2$, UNPAIRED OBSERVATIONS

If you knew a physiologist who was interested in learning how much children grow between first and second grades, how would you advise him to obtain sample information to estimate this growth? In view of our discussion on pairing in Section 7.6, you might advise him to go to the local elementary school, get permission from the first-grade teacher, Miss Prim, and measure the height of each of the students in first grade. You would then tell him to return to the school next year at this time, get permission from the second-grade teacher, Mr. Trim, and measure the heights of the same students. The two measurements made on each student would be paired. What objection might the physiologist raise to your suggestion? It is very likely that there would be an objection to the amount of time required for the data to be collected. This example illustrates the fact that it may not be practical from time (or perhaps cost) considerations to use pairing.

Given that it is impossible or impractical to use pairing, we would return to the idea stated at the beginning of Section 7.6, namely, to use two different samples. For the example of trying to estimate growth of children between first and second grades, we might choose a sample of children from each grade. If we let X_1 = height of a randomly chosen first-grade child and X_2 = height of a randomly chosen second-grade child, then

$$E(X_2) - E(X_1) = \mu_2 - \mu_1$$

would give the mean growth during this year. A point estimate would be given by $\bar{X}_2 - \bar{X}_1$. A subtle but important question might be raised at this point: What is the "population" in which the physiologist is interested? Is it the growth of first graders worldwide? In Europe? In North America? In the United States? In your state or province? County? City? The answer to this question is very important, since the sample chosen should be taken from the entire population of interest, not just a subset of that population. For simplicity we will assume that the interest is in students in your county and further that students in your county are relatively homogeneous with respect to socioeconomic levels, so that the members of Miss Prim's and Mr. Trim's classes constitute representative samples of children from the population of interest. If these classes have n_1 and n_2 children, respectively, the heights of these children can be written as

$$X_{1,1}, X_{1,2}, \ldots, X_{1,n_1}$$

$$X_{2,1}, X_{2,2}, \ldots, X_{2,n_2}.$$

It is not necessary, of course, for n_1 to be equal to n_2. As indicated earlier, a point estimate for $\mu_2 - \mu_1$ can be given by $\bar{X}_2 - \bar{X}_1$. To find a confidence interval for $\mu_2 - \mu_1$, we will need to know the probability distribution of $\bar{X}_2 - \bar{X}_1$. This is most easily done if we assume that the random samples come from normal distributions.

Theorem 7.2 If $X_{1,1}, X_{1,2}, \ldots, X_{1,n_1}$ and $X_{2,1}, X_{2,2}, \ldots, X_{2,n_2}$ are independent random samples from normal distributions having means μ_1 and μ_2 and variances σ_1^2 and σ_2^2, respectively, the difference in the sample means, $\bar{X}_2 - \bar{X}_1$, will be normally distributed with

$$\mu_{\bar{X}_2 - \bar{X}_1} = \mu_2 - \mu_1 \quad \text{and} \quad \sigma^2_{\bar{X}_2 - \bar{X}_1} = \frac{\sigma_2^2}{n_2} + \frac{\sigma_1^2}{n_1}. \tag{7.16}$$

Proof: The proof of this theorem follows from several theorems presented in Chapter 5. From Theorem 5.6, we know that $\bar{X}_1$ and $\bar{X}_2$ are each normally distributed with means μ_1 and μ_2 and variances σ_1^2/n_1 and σ_2^2/n_2, respectively. Since each of these is normally distributed and since $\bar{X}_2 - \bar{X}_1$ is actually a linear combination of the random variables $\bar{X}_1$ and $\bar{X}_2$, it follows from Theorem 5.5 that $\bar{X}_2 - \bar{X}_1$ has a normal distribution. Finally, the mean and variance of $\bar{X}_2 - \bar{X}_1$ can be seen to be as in (7.16) from Theorems 5.3 and 5.4.

In view of Theorem 7.2, we can say that

$$P\left[-z_{1-\alpha/2} \leq \frac{(\bar{X}_2 - \bar{X}_1) - (\mu_2 - \mu_1)}{\sqrt{\sigma_1^2/n_1 + \sigma_2^2/n_2}} \leq z_{1-\alpha/2}\right] = 1 - \alpha. \tag{7.17}$$

By rearranging the inequalities in equation (7.17) in the same way we rearranged those in (7.4), we obtain

$$P\left[(\bar{X}_2 - \bar{X}_1) - z_{1-\alpha/2}\sqrt{\frac{\sigma_1^2}{n_1} + \frac{\sigma_2^2}{n_2}} \leq \mu_2 - \mu_1 \leq (\bar{X}_2 - \bar{X}_1) + z_{1-\alpha/2}\sqrt{\frac{\sigma_1^2}{n_1} + \frac{\sigma_2^2}{n_2}}\right]$$
$$= 1 - \alpha,$$

so the quantities $(\bar{X}_2 - \bar{X}_1) \pm z_{1-\alpha/2}\sqrt{\sigma_1^2/n_1 + \sigma_2^2/n_2}$ give the end points of a random interval. These end points can be used to give confidence interval estimates for $(\mu_2 - \mu_1)$.

Example 7.15 Miss Prim's class of first graders has 22 children whose mean height is 47.75 inches, while Mr. Trim's class of second graders has 25 children whose mean height is 50.40 inches. If the standard deviations for the heights of first- and second-grade children are known to be 1.80 and 2.05 inches, respectively, find a 98% confidence interval for the mean growth, $\mu_2 - \mu_1$.

Since $z_{1-\alpha/2} = z_{.99} = 2.326$ and

$$\sqrt{\frac{\sigma_1^2}{n_1} + \frac{\sigma_2^2}{n_2}} = \sqrt{\frac{(1.80)^2}{22} + \frac{(2.05)^2}{25}} = \sqrt{.3154} = .562,$$

we get

$$(\bar{x}_2 - \bar{x}_1) \pm z_{.99}\sqrt{\frac{\sigma_1^2}{n_1} + \frac{\sigma_2^2}{n_2}} = (50.40 - 47.75) \pm 2.326(.562),$$

so a 98% confidence interval is given by (1.34, 3.96).

We should point out that the subscripts used on the variables and parameters are simply a matter of convenience. A confidence interval for $\mu_1 - \mu_2$ (rather than $\mu_2 - \mu_1$) would simply be given by

$$(\bar{x}_1 - \bar{x}_2) \pm z_{1-\alpha/2}\sqrt{\frac{\sigma_1^2}{n_1} + \frac{\sigma_2^2}{n_2}}. \tag{7.18}$$

There are some situations where the variances σ_1^2 and σ_2^2 are equal. For example, if a vending machine designed to dispense coffee is adjusted, the adjustment may affect the mean amount dispensed, but not the variance of the amount dispensed. If the variances are equal, we can write the common value as σ^2:

$$\sigma^2 = \sigma_1^2 = \sigma_2^2$$

and slightly simplify the expression of the variance of the distribution of $\bar{X}_1 - \bar{X}_2$. In particular,

$$V(\bar{X}_1 - \bar{X}_2) = \frac{\sigma_1^2}{n_1} + \frac{\sigma_2^2}{n_2} = \sigma^2\left(\frac{1}{n_1} + \frac{1}{n_2}\right),$$

so

$$P\left[-z_{1-\alpha/2} \leq \frac{(\bar{X}_1 - \bar{X}_2) - (\mu_1 - \mu_2)}{\sigma\sqrt{1/n_1 + 1/n_2}} \leq z_{1-\alpha/2}\right] = 1 - \alpha,$$

and consequently the end points of a confidence interval estimate for $\mu_1 - \mu_2$ can be given by

$$(\bar{x}_1 - \bar{x}_2) \pm z_{1-\alpha/2}\, \sigma \sqrt{\frac{1}{n_1} + \frac{1}{n_2}}. \tag{7.19}$$

Example 7.16 A vending machine designed to dispense coffee into 8-ounce cups was checked by a service technician. A sample of four cups showed an average of 7.10 ounces being dispensed. After adjustment a sample of five cups showed an average of 7.50 ounces being dispensed. Assuming that $\sigma = .20$ ounce, find a 90% confidence interval for the mean difference in the amount dispensed due to the adjustment.

Assuming that the amount dispensed follows a normal distribution, and letting $\bar{X}_1$ denote the sample mean before adjustment, we obtain, from (7.19),

$$(\bar{x}_1 - \bar{x}_2) \pm z_{.95}\, \sigma \sqrt{\frac{1}{n_1} + \frac{1}{n_2}} = (7.10 - 7.50) \pm 1.654(.20)\sqrt{\tfrac{1}{4} + \tfrac{1}{5}}$$

or $(-.66, -.18)$. This is a 90% confidence interval estimate for $\mu_1 - \mu_2$. An estimate for $(\mu_2 - \mu_1)$ would be $(.18, .66)$. Hence the effect of the adjustment was to increase the amount of coffee dispensed, the increase most likely (90% confidence) being between .18 ounce and .66 ounce.

We realize that in most situations where the means of distributions are unknown, the variances will also be unknown. In a situation where the variances are unknown but are assumed to be equal, we can replace the unknown value of σ^2 with an estimate of σ^2. We need to find a suitable estimator for σ^2, and in this instance our usual recourse to the "sample counterpart" does not quite solve our problem because in fact we have *two* sample counterparts. More explicitly, the problem we face is the following: if $X_{1,1}, X_{1,2}, \ldots, X_{1,n_1}$ is a random sample from a normal distribution having mean μ_1 and variance σ^2 and if $X_{2,1}, X_{2,2}, \ldots, X_{2,n_2}$ is an independent random sample from a second normal distribution having mean μ_2 and variance σ^2, how can we best estimate the value σ^2 common to both of these distributions? If only the first sample were available, we would, of course, estimate σ^2 by using the sample variance S_1^2. Similarly, if only the second sample were available, we would estimate σ^2 by using the sample variance S_2^2. Since both samples are available, however, it seems reasonable to combine the estimates S_1^2 and S_2^2 in some way. We might think of simply averaging the estimates by

$$\frac{S_1^2 + S_2^2}{2}.$$

If we do this, we are treating both sample variances as if they are equally good estimators of σ^2. Would you think this to be the case if S_1^2 were based on a sample size of $n_1 = 10$, say, while S_2^2 was based on a sample size of $n_2 = 100$? Certainly not! You would tend to have more confidence in an estimator based on a larger sample size. Consequently, when we combine the estimators S_1^2 and S_2^2 we give more weight to the estimator based on the larger sample size. This is done by "pooling" the information about σ^2, using "weights" equal to the degrees of freedom associated with each sample. Specifically, define the *pooled estimate* of σ^2 by

$$S_p^2 = \frac{(n_1 - 1)S_1^2 + (n_2 - 1)S_2^2}{n_1 + n_2 - 2}. \tag{7.20}$$

In order to find confidence intervals for the difference of two means, $\mu_1 - \mu_2$, when $\sigma_1^2 = \sigma_2^2 = \sigma^2$ but σ^2 is unknown, we must make use of the following theorem.

Theorem 7.3 If $X_{1,1}, X_{1,2}, \ldots, X_{1,n_1}$ and $X_{2,1}, X_{2,2}, \ldots, X_{2,n_2}$ are independent random samples from normal distributions having means μ_1 and μ_2, respectively, and a common variance σ^2, then the random variable

$$T = \frac{(\bar{X}_1 - \bar{X}_2) - (\mu_1 - \mu_2)}{S_p\sqrt{1/n_1 + 1/n_2}}$$

will have a Student's t distribution with $n_1 + n_2 - 2$ degrees of freedom.

The proof of this theorem will not be given, but we can make use of it to find confidence intervals for $\mu_1 - \mu_2$. Assuming that the hypotheses of Theorem 7.3 are satisfied, it is true that

$$P\left[-t_{n_1+n_2-2,\,1-\alpha/2} \le \frac{(\bar{X}_1 - \bar{X}_2) - (\mu_1 - \mu_2)}{S_p\sqrt{1/n_1 + 1/n_2}} \le t_{n_1+n_2-2,\,1-\alpha/2}\right] = 1 - \alpha. \tag{7.21}$$

By solving the two inequalities in (7.21) for $(\mu_1 - \mu_2)$ in much the same way as was done in solving (7.13), we obtain

$$P\left[(\bar{X}_1 - \bar{X}_2) - t_{n_1+n_2-2,\,1-\alpha/2}S_p\sqrt{\frac{1}{n_1} + \frac{1}{n_2}} \le \mu_1 - \mu_2 \right.$$
$$\left. \le (\bar{X}_1 - \bar{X}_2) + t_{n_1+n_2-2,\,1-\alpha/2}S_p\sqrt{\frac{1}{n_1} + \frac{1}{n_2}}\right] = 1 - \alpha,$$

so the quantities

$$(\bar{X}_1 - \bar{X}_2) \pm t_{n_1+n_2-2,\,1-\alpha/2}S_p\sqrt{\frac{1}{n_1} + \frac{1}{n_2}} \tag{7.22}$$

give the end points of a random interval. These end points can be used to give confidence interval estimates for $\mu_1 - \mu_2$. Note the similarity between (7.19) and (7.22). As we have observed before, if σ is unknown, confidence intervals are obtained by using the appropriate t percentiles rather than z percentiles.

Example 7.17 As in Example 7.16, assume that a vending machine designed to dispense coffee into 8-ounce cups was checked by a service technician who samples four cups before making an adjustment and five cups after making an adjustment. However, in this case assume that the value of σ^2, although not affected by adjustment, is unknown. Find a 90% confidence interval for the mean difference in the amount dispensed due to the adjustment if the samples showed the following amounts of coffee:

Before adjustment:	6.92	7.34	7.26	6.88	
After adjustment:	7.33	7.93	7.65	7.49	7.10.

Direct calculations show that $\bar{x}_1 = 7.10$, $s_1^2 = (.2338)^2$, $\bar{x}_2 = 7.50$, and $s_2^2 = (.3148)^2$. Using equation (7.20), we find that

$$s_p^2 = \frac{(4-1)(.2338)^2 + (5-1)(.3148)^2}{4+5-2}$$
$$= .0801 = (.2830)^2.$$

Being careful to use s_p and not s_p^2 in equation (7.22), we find a 90% confidence interval for $\mu_1 - \mu_2$ to be

$$(7.10 - 7.50) \pm 1.895(.2830)\sqrt{\tfrac{1}{4} + \tfrac{1}{5}}$$
$$= (-.40) \pm 1.895(.2830)(.6708).$$

Completing these calculations we find the confidence interval to be $(-.76, -.04)$.

The final case to consider in this section is the case where the values of σ_i^2 are unknown and unequal. Finding a distribution that can be used to obtain exact confidence intervals for $\mu_1 - \mu_2$ in this situation has proven to be a difficult problem. (This problem is referred to as the Behrens-Fisher problem.) While such a distribution can be found under some special conditions, a general solution has not been found. However, several approximate solutions have been proposed. One such solution is to simply use the sample variances as estimates of the unknown population variances in (7.18) and to replace the appropriate standard normal percentile with a Student's t percentile having a certain number of degrees of freedom, that is, to use

$$(7.23) \qquad (\bar{X}_1 - \bar{X}_2) \pm t_{\nu, 1-\alpha/2}\sqrt{\frac{S_1^2}{n_1} + \frac{S_2^2}{n_2}}$$

as the end points of a confidence interval for estimating $\mu_1 - \mu_2$. Different authors propose different suggestions for choosing ν. One suggestion that appears to give reasonably good results is

$$(7.24) \qquad \nu = \frac{(a_1 + a_2)^2}{a_1^2/(n_1 - 1) + a_2^2/(n_2 - 1)},$$

where $a_1 = S_1^2/n_1$ and $a_2 = S_2^2/n_2$. The value of ν so obtained will generally not be an integer, but can be rounded to an integer value. This calculation is a bit involved and another suggestion is to define

$$(7.25) \qquad \nu = \min(n_1, n_2) - 1.$$

This choice of ν seems to yield "conservative" results. That is, confidence levels designed to be equal to $1 - \alpha$ are generally larger than $1 - \alpha$. This is related to the fact that the length of the confidence interval will be somewhat longer if ν is found from (7.25) instead of (7.24). Since it is difficult to know whether or not the variances σ_1^2 and σ_2^2 are equal (or at least approximately equal), it is difficult to know whether to use (7.22) or (7.23) to obtain confidence intervals for $\mu_1 - \mu_2$. In Sections 7.9, 8.7, and 11.6 we will find objective methods for deciding whether the assumption of equal variances is reasonable. In the meantime, for simplicity in working examples or exercises, you may assume that variances are equal unless there is an explicit statement to the contrary.

Example 7.18 Find a 90% confidence interval for $\mu_1 - \mu_2$ using the information given in Example 7.17 under the assumption that the adjustment to the vending machine may have changed the variance as well as the mean of the amount of coffee dispensed.

Using the information from Example 7.17, we can find ν using equation (7.24). We have

$$a_1 = \frac{s_1^2}{n_1} = \frac{(.2338)^2}{4} = .0137,$$

$$a_2 = \frac{s_2^2}{n_2} = \frac{(.3148)^2}{5} = .0198,$$

$$\nu = \frac{(.0137 + .0198)^2}{(.0137)^2/3 + (.0198)^2/4}$$

$$= \frac{.00124}{.000161} = 7.7 \approx 8.$$

Since $t_{8,.95} = 1.860$, we find the confidence interval for $\mu_1 - \mu_2$ to be

$$(7.10 - 7.50) \pm 1.860\sqrt{\frac{(.2338)^2}{4} + \frac{(.3148)^2}{5}}$$

or $-.40 \pm .34$. If we were to take the conservative approach, we would find that

$$\nu = \min(n_1, n_2) - 1 = \min(4, 5) - 1 = 3.$$

Since $t_{3,.95} = 2.353$, we would find the confidence interval for $\mu_1 - \mu_2$ to be

$$(7.10 - 7.50) \pm 2.353\sqrt{\frac{(.2338)^2}{4} + \frac{(.3148)^2}{5}}$$

or $-.40 \pm .43$. As we indicated earlier, the conservative approach leads to a longer confidence interval.

For your convenience in problem solving we have summarized the formulas for finding confidence intervals for $\mu_1 - \mu_2$ in Table 7.2. Keep in mind that all the confidence intervals for $\mu_1 - \mu_2$ given in this section have been found under the

TABLE 7.2

Formulas for Finding Confidence Intervals for $\mu_1 - \mu_2$ (Independent Random Samples from Normal Distributions)

	$\sigma_1^2 \neq \sigma_2^2$	$\sigma_1^2 = \sigma_2^2$
σ_i^2 *Known*	$(\bar{X}_1 - \bar{X}_2) \pm z_{1-\alpha/2}\sqrt{\frac{\sigma_1^2}{n_1} + \frac{\sigma_2^2}{n_2}}$	$(\bar{X}_1 - \bar{X}_2) \pm z_{1-\alpha/2}\sigma\sqrt{\frac{1}{n_1} + \frac{1}{n_2}}$
σ_i^2 *Unknown*	$(\bar{X}_1 - \bar{X}_2) \pm t_{\nu, 1-\alpha/2}\sqrt{\frac{S_1^2}{n_1} + \frac{S_2^2}{n_2}}$ [ν given by Eq. (7.24) or (7.25)]	$(\bar{X}_1 - \bar{X}_2) \pm t_{\nu, 1-\alpha/2}S_p\sqrt{\frac{1}{n_1} + \frac{1}{n_2}}$ ($\nu = n_1 + n_2 - 2$)

assumption that independent random samples have been taken from normal distributions. Under these conditions valid confidence intervals can be found for any sample sizes. The confidence levels are exact [except in the case where equation (7.23) is used, in which case the levels are only approximate]. If the samples *do not* come from normal distributions, the procedures described in this section can be used to give *approximate* confidence intervals for $\mu_1 - \mu_2$, *provided that the sample sizes are large enough.* There has been much debate over the years as to what sample sizes might be considered to be "large enough." A conservative guideline would be to take $n_i \geq 30$. For samples of this size, the Central Limit Theorem is likely to assure that $\bar{X}_1$ and $\bar{X}_2$ are each approximately normal. If the normality assumption is violated and the sample sizes are small, the confidence level may be far from what it was intended to be. Depending on the form of the distribution from which the sample is taken, a confidence level intended to be 90%, say, could turn out to be as low as 70 or 80% or as high as 95%, for example. In summary, the assumption that the samples come from normal distributions is an important one in finding confidence intervals for $\mu_1 - \mu_2$. We will say in advance that this same assumption will be made in the next two sections, dealing with finding confidence intervals for variances. This assumption is even more crucial in these next sections.

7.8 CONFIDENCE INTERVALS FOR σ^2

We have indicated in some of the previous sections of this chapter that in many practical problems σ^2 will be unknown. In those sections we were concerned primarily with finding confidence intervals for means and we were able to simply use point estimates for σ^2. There may be situations where interest lies primarily with the variance. Consider, for example, the problem faced by Mr. Phil D. Boxx, the purchasing agent for Crunchy-O's cereal. Phil is contemplating the purchase of a machine designed to fill boxes of Crunchy-O's with 16 ounces of cereal. He knows that by making certain adjustments he can have the machine putting in 16 ounces of cereal *on the average*. However, even though the average might be correct, it is not good for the variation to be too large. (Can you think of some seasons why this might be so?) Consequently, Phil would like to get an estimate of the variance. A confidence interval estimate would probably be more informative to him than a point estimate.

In order to find a confidence interval estimate for the variance σ^2, it is necessary to find the probability distribution of the sample variance S^2 or of the sample variance when suitably normalized. In general, it is difficult to find such a distribution, but it can be found in a certain special situation.

Theorem 7.4 If $X_1, X_2, \ldots, X_n$ is a random sample from a normal distribution having mean μ and variance σ^2, the random variable

$$\frac{(n-1)S^2}{\sigma^2}$$

has a chi-square distribution with $n - 1$ degrees of freedom.

Before using the results of Theorem 7.4 to find confidence intervals for σ^2 we must give a brief discussion of the chi-square (χ^2) distribution. This is actually a family of distributions with a single parameter given the name "degrees of freedom." (Actually, the t distributions and χ^2 distributions are rather directly related, and consequently the parameter in both cases is given the same name.) A typical member of the χ^2 family is

1. Nonnegative.
2. Unimodal.
3. Skewed to the right.

Distributions corresponding to 5 and 10 degrees of freedom are shown in Figure 7.3. Since the distributions are skewed, they are asymmetric, and consequently both upper and lower percentiles must be tabled. From Table A7 you can see, for example, that

$$P[\chi_5^2 \leq \chi_{5,.05}^2 = 1.145] = .05$$

and

$$P[\chi_{10}^2 \geq \chi_{10,.95}^2 = 18.307] = 1 - P[\chi_{10}^2 \leq 18.307]$$
$$= 1 - .95 = .05.$$

The areas corresponding to these regions have been shaded in Figure 7.3.

For any ν and α, we have

$$P[\chi_{\nu,\alpha/2}^2 \leq \chi_\nu^2 \leq \chi_{\nu,1-\alpha/2}^2] = 1 - \alpha,$$

where χ_ν^2 represents a chi-square random variable having ν degrees of freedom. It follows from Theorem 7.4 that if $X_1, X_2, \ldots, X_n$ is a random sample from a normal distribution having variance σ^2, then

$$P\left[\chi_{n-1,\alpha/2}^2 \leq \frac{(n-1)S^2}{\sigma^2} \leq \chi_{n-1,1-\alpha/2}^2\right] = 1 - \alpha. \quad (7.26)$$

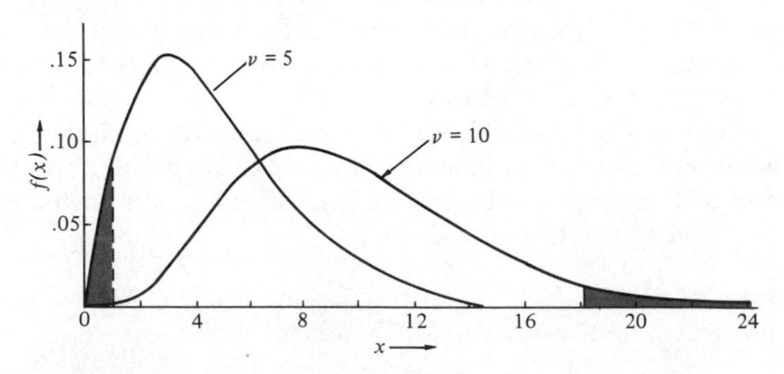

FIGURE 7.3
Chi-Square Distributions Having 5 and 10 Degrees of Freedom

If we consider the two inequalities given in brackets in (7.26), we see that each can be solved for σ^2. For example,

$$\chi^2_{n-1,\alpha/2} \leq \frac{(n-1)S^2}{\sigma^2}$$

can be solved to give

$$\sigma^2 \leq \frac{(n-1)S^2}{\chi^2_{n-1,\alpha/2}},$$

while

$$\frac{(n-1)S^2}{\sigma^2} \leq \chi^2_{n-1,1-\alpha/2}$$

can be solved to give

$$\frac{(n-1)S^2}{\chi^2_{n-1,1-\alpha/2}} \leq \sigma^2.$$

Combining these results, equation (7.26) can be rewritten as

$$P\left[\frac{(n-1)S^2}{\chi^2_{n-1,1-\alpha/2}} \leq \sigma^2 \leq \frac{(n-1)S^2}{\chi^2_{n-1,\alpha/2}}\right] = 1 - \alpha. \tag{7.27}$$

Consequently, (7.27) gives the end points of a random interval that will contain the true value of σ^2 with probability $1 - \alpha$. Therefore, we can say that

$$\left(\frac{(n-1)S^2}{\chi^2_{n-1,1-\alpha/2}}, \frac{(n-1)S^2}{\chi^2_{n-1,\alpha/2}}\right) \tag{7.28}$$

can be used to give a $1 - \alpha$ level confidence interval for σ^2.

Example 7.19 Mr. Boxx assumes that the machine designed to fill boxes of Crunchy-O's fills boxes with a random amount of cereal, that amount following a normal distribution. A machine is considered to be precise enough for production purposes if the true value of σ is less than .25 ounce. Find a 95% confidence interval for σ if six boxes were found to be filled with the following amounts:

16.52 15.53 15.74 16.37 16.21 15.63.

To find a confidence interval for σ, we first find a confidence interval for σ^2 and then take the square root of the end points of that interval. From the sample data we find $\bar{x} = 16$ and

$$\begin{aligned}(n-1)s^2 &= \sum (x_i - \bar{x})^2 = (16.52 - 16)^2 + (15.53 - 16)^2 + \ldots + (15.63 - 16)^2 \\ &= .8768.\end{aligned}$$

From Table A7 we find $\chi^2_{5,.025} = .8312$ and $\chi^2_{5,.975} = 12.8325$. Using this information in (7.28), we find a 95% confidence interval for σ^2 to be

$$\left(\frac{.8768}{12.8325}, \frac{.8768}{.8312}\right) = (.0683,\ 1.055),$$

and we find the square roots of the end points to be (.2614, 1.027). (On the basis of this information, should Mr. Boxx consider this machine to be precise enough for production purposes? No, since he would like σ to be less than .25 while the confidence interval values indicate that the actual value of σ is larger than this.)

7.9 CONFIDENCE INTERVALS FOR σ_1^2/σ_2^2

In addition to the problem of estimating the variance of one particular probability distribution, there may be situations where it is desirable to compare the variances of two different distributions. For example, Mr. Boxx, the purchasing agent mentioned in the previous section, may be faced with the problem of deciding between two different machines for filling boxes of Crunchy-O's. If, for instance, the machines are comparable in all other respects, he may decide to purchase the machine having the smaller variability in the amount actually put into the boxes. By taking a sample from each machine, he could calculate the sample variance for each and use these as point estimates of the true variances. In comparing variances for two distributions, we generally consider the ratio of the variances (for mathematical reasons that will soon become apparent). If the ratio is equal to 1, two variances are of course equal. To find confidence intervals for σ_1^2/σ_2^2, the ratio of two population variances, we will need to know the probability distribution of the ratio of sample variances suitably normalized. The following theorem in conjunction with Theorem 7.4 will allow us to find such a distribution.

Theorem 7.5 If W_1 and W_2 are independent random variables having chi-square distributions with ν_1 and ν_2 degrees of freedom, respectively, the random variable $R = (W_1/\nu_1)/(W_2/\nu_2)$ will have an F distribution with ν_1 degrees of freedom for the numerator and ν_2 degrees of freedom for the denominator.

A brief digression is perhaps in order here. The F distribution mentioned in this theorem is actually a family of distributions indexed by two parameters denoted by ν_1 and ν_2. As indicated, ν_1 is the degrees of freedom associated with the chi-square random variable in the numerator while ν_2 is the degrees of freedom associated with the chi-square random variable in the denominator. The members of the F family have characteristics similar to those of members of the chi-square family: they are (1) nonnegative, (2) unimodal, and (3) skewed to the right. In Figure 7.4 we give the density function for an F random variable with $\nu_1 = 5$, $\nu_2 = 10$. In Table A8 some upper percentiles are given. The 90, 95, and 97.5 percentiles are

$$f_{5,10,.90} = 2.52, \qquad f_{5,10,.95} = 3.33, \qquad f_{5,10,.975} = 4.24.$$

These values are shown in Figure 7.4. Although you can see from the graph that the density function is not symmetric in the usual sense, there is a kind of inverted symmetry which allows lower percentiles to be found. In Theorem 7.5 we defined a random variable R. If we now define S by

$$S = \frac{1}{R}, \tag{7.29}$$

then S will have an F distribution with ν_2 degrees of freedom for the numerator and ν_1 for the denominator. Using the relationship in (7.29), we can find lower percentiles for an F random variable. We have, by the definition of $f_{\nu_2,\nu_1,\alpha}$ and of $f_{\nu_1,\nu_2,1-\alpha}$ that

$$P[S \leq f_{\nu_2,\nu_1,\alpha}] = \alpha \quad \text{and} \quad P[R \leq f_{\nu_1,\nu_2,1-\alpha}] = 1 - \alpha, \tag{7.30}$$

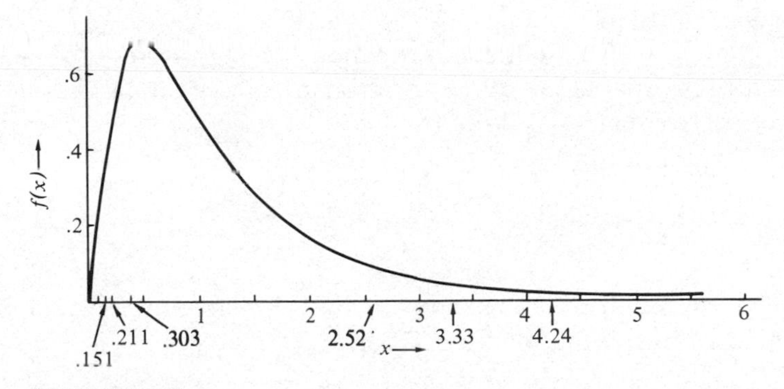

FIGURE 7.4
Density Function for an $F_{5,10}$ Random Variable

so

$$P[S \geq f_{\nu_2,\nu_1,\alpha}] = 1 - \alpha. \tag{7.31}$$

Rearranging the inequality in (7.31), we get

$$[S \geq f_{\nu_2,\nu_1,\alpha}] \Longleftrightarrow \left[\frac{1}{f_{\nu_2,\nu_1,\alpha}} \geq \frac{1}{S}\right].$$

Since $S = 1/R$, we can write

$$P\left[\frac{1}{S} \leq \frac{1}{f_{\nu_2,\nu_1,\alpha}}\right] = P\left[R \leq \frac{1}{f_{\nu_2,\nu_1,\alpha}}\right] = 1 - \alpha.$$

Comparing this last equality with the second equality in (7.30), we see that

$$P[R \leq f_{\nu_1,\nu_2,1-\alpha}] = 1 - \alpha = P\left[R \leq \frac{1}{f_{\nu_2,\nu_1,\alpha}}\right],$$

from which it follows that

$$f_{\nu_2,\nu_1,\alpha} = \frac{1}{f_{\nu_1,\nu_2,1-\alpha}}. \tag{7.32}$$

Although the details of this argument may be somewhat difficult to follow, it should not be difficult to use equation (7.32) to find lower percentiles. For example,

$$f_{5,10,.10} = \frac{1}{f_{10,5,.90}} = \frac{1}{3.30} = .303$$

$$f_{5,10,.05} = \frac{1}{f_{10,5,.95}} = \frac{1}{4.74} = .211$$

$$f_{5,10,.025} = \frac{1}{f_{10,5,.975}} = \frac{1}{6.62} = .151$$

These values are shown in Figure 7.4.

Having discussed the F distribution briefly, we now return to the problem of finding confidence intervals for σ_1^2/σ_2^2.

Theorem 7.6 If $X_{1,1}, X_{1,2}, \ldots, X_{1,n_1}$ and $X_{2,1}, X_{2,2}, \ldots, X_{2,n_2}$ are independent random samples from normal distributions having variances σ_1^2 and σ_2^2, respectively, the random variable $(S_1^2/\sigma_1^2)/(S_2^2/\sigma_2^2)$ has an F distribution with $n_1 - 1$ and $n_2 - 1$ degrees of freedom.

Proof: It follows from Theorem 7.4 that

$$\frac{(n_i - 1)S_i^2}{\sigma_i^2}$$

has a chi-square distribution with $\nu_i = n_i - 1$ degrees of freedom. Since the samples are independent, it follows from Theorem 7.5 that

$$\frac{[(n_1 - 1)S_1^2/\sigma_1^2]/(n_1 - 1)}{[(n_2 - 1)S_2^2/\sigma_2^2]/(n_2 - 1)} = \frac{S_1^2/\sigma_1^2}{S_2^2/\sigma_2^2}$$

has the stated F distribution.

To find a confidence interval for σ_1^2/σ_2^2, we note that

$$P\left[f_{n_1-1,n_2-1,\alpha/2} \le \frac{S_1^2/\sigma_1^2}{S_2^2/\sigma_2^2} \le f_{n_1-1,n_2-1,1-\alpha/2}\right] = 1 - \alpha.$$

Rearranging the inequalities in brackets to solve for σ_1^2/σ_2^2 gives

$$P\left[\frac{S_1^2/S_2^2}{f_{n_1-1,n_2-1,1-\alpha/2}} \le \frac{\sigma_1^2}{\sigma_2^2} \le \frac{S_1^2/S_2^2}{f_{n_1-1,n_2-1,\alpha/2}}\right] = 1 - \alpha,$$

so the end points of a $1 - \alpha$ confidence interval for σ_1^2/σ_2^2 can be given by

$$\left(\frac{S_1^2/S_2^2}{f_{n_1-1,n_2-1,1-\alpha/2}}, \frac{S_1^2/S_2^2}{f_{n_1-1,n_2-1,\alpha/2}}\right).$$

Using equation (7.32), the right end point can be rewritten to give

$$\left(\frac{S_1^2/S_2^2}{f_{n_1-1,n_2-1,1-\alpha/2}}, \left(\frac{S_1^2}{S_2^2}\right) f_{n_2-1,n_1-1,1-\alpha/2}\right). \tag{7.33}$$

Example 7.20 In measuring the contents of six boxes filled by one machine, Mr. Boxx determined the sample variance to be .1754 (see Example 7.19). In measuring the contents of 11 boxes filled by a second machine, he found a sample variance of .2704. Assuming that the amount dispensed follows a normal distribution for each machine, find a 95% confidence interval for σ_1^2/σ_2^2.

We will use (7.33) with

$$f_{6-1,11-1,.975} = 4.24 \quad \text{and} \quad f_{11-1,6-1,.975} = 6.62.$$

The end points of the confidence interval are

$$\left(\left(\frac{.1754}{.2704}\right)\Big/4.24, \left(\frac{.1754}{.2704}\right) \cdot 6.62\right) = (.1530, 4.294).$$

If Mr. Boxx were interested in finding a confidence interval for the difference in the mean amount put into boxes by these two machines, $\mu_1 - \mu_2$, he could use the confidence interval found here for σ_1^2/σ_2^2 to decide whether it would be appropriate to use (7.22) or (7.23) for finding a confidence interval for $\mu_1 - \mu_2$. Since the confidence interval for σ_1^2/σ_2^2 contains the value 1, it is not unreasonable to believe that $\sigma_1^2/\sigma_2^2 = 1$ (i.e., that $\sigma_1^2 = \sigma_2^2$). Consequently, (7.22) could be used to find the confidence interval for $\mu_1 - \mu_2$.

We close this section with a reminder that Theorems 7.4 and 7.6 have in their hypotheses the statement that the samples come from normal distributions. If this assumption is not satisfied, the confidence intervals obtained using (7.28) or (7.33) may not have the confidence level $1 - \alpha$.

7.10 CONFIDENCE INTERVALS FOR PROPORTIONS

In many of the opinion polls reported in the media, point estimates are given for a proportion. For example, 58% believe the president is doing a good job, 67% feel that more should be done about protecting the environment, 92% of Americans like apple pie, and so on. In many cases, however, you are not told about the total number of people interviewed or how the sample of interviewees was chosen. As we indicated in Chapter 5, if the sample was improperly chosen, the results may be meaningless. However, even if a sample is chosen randomly from a well-defined population, a point estimate may simply not be informative enough. For example, if Ms. Kay Nyne is running for the office of county animal control officer, knowing that 56% of the voters in a random sample favor her may not be as helpful to her in planning her campaign strategy as an interval estimate might be.

In most practical problems dealing with the estimation of a proportion, relatively large sample sizes are used. In view of this fact we will assume that large sample sizes are available. In particular, we will assume that the sample size n is large enough to allow the normal distribution to approximate the binomial to be used (see Section 4.10). [Small-sample techniques are available. The interested reader is referred to a book by W. J. Conover, *Practical Nonparametric Statistics* (New York: John Wiley & Sons, Inc., 1971).] If we let p denote the proportion in a large population having a certain characteristic (e.g., the characteristic of liking apple pie or of favoring a candidate for office) and if X is defined to be the number of members in a random sample of size n having that characteristic, then X will have a binomial distribution with parameters n and p if the sample is taken with replacement. (If the sample is taken without replacement but the sample size is small relative to the population size, the distribution of X will be hypergeometric, but it will be well approximated by the aforementioned binomial distribution.) It follows from the discussion of Section 4.10 that

$$(7.34) \qquad P\left[-z_{1-\alpha/2} \leq \frac{X - np}{\sqrt{npq}} \leq z_{1-\alpha/2}\right] \approx 1 - \alpha.$$

Since

$$\frac{X - np}{\sqrt{npq}} = \frac{(X/n) - p}{\sqrt{pq/n}},$$

we could rewrite (7.34) as

$$P\left[-z_{1-\alpha/2} \leq \frac{(X/n) - p}{\sqrt{pq/n}} \leq z_{1-\alpha/2}\right] \approx 1 - \alpha.$$

In either case we could attempt to solve each inequality for p. Proceeding as we have done previously, we could get

$$P\left[\frac{X}{n} - z_{1-\alpha/2}\sqrt{pq/n} \leq p \leq \frac{X}{n} + z_{1-\alpha/2}\sqrt{pq/n}\right] \approx 1 - \alpha.$$

Closer inspection, however, reveals that we have not really solved for p, since the value of p is involved in the term $\sqrt{pq/n}$. There are two choices available: (1) go back and do it right; that is, note that each inequality actually yields a quadratic in p that can be solved for p, or (2) use a point estimate for p to evaluate the offending term. We can state without going through the details that the former approach leads to a confidence interval given by

$$\text{(7.35)} \qquad \left[\left(\frac{X}{n} + \frac{z_{1-\alpha/2}^2}{2n}\right) \pm z_{1-\alpha/2}\sqrt{\frac{(X/n)(1 - X/n)}{n} + \frac{z_{1-\alpha/2}^2}{4n^2}}\right]\Big/[1 + (z_{1-\alpha/2}^2/n)].$$

A slightly less precise but simpler form can be obtained by using

$$\sqrt{\frac{X}{n}\left(1 - \frac{X}{n}\right)\Big/n}$$

to estimate $\sqrt{p(1 - p)/n}$. From

$$P\left[\frac{X}{n} - z_{1-\alpha/2}\sqrt{\frac{X}{n}\left(1 - \frac{X}{n}\right)\Big/n} \leq p \leq \frac{X}{n} + z_{1-\alpha/2}\sqrt{\frac{X}{n}\left(1 - \frac{X}{n}\right)\Big/n}\right] \approx 1 - \alpha,$$

we obtain as the end points of an approximate confidence interval for p,

$$\text{(7.36)} \qquad \frac{X}{n} \pm z_{1-\alpha/2}\sqrt{\frac{X}{n}\left(1 - \frac{X}{n}\right)\Big/n}.$$

The confidence intervals found by using (7.35) or (7.36) are only approximate confidence intervals. In fact, approximations were used in two or three different places. The binomial distribution might have been used to approximate the hypergeometric distribution, the normal to approximate the binomial, and the sample proportion to approximate the true proportion. In spite of this, experience shows these confidence intervals to be good approximations as long as the sample size n is suitably large.

Example 7.21 Find a 90% confidence interval for p, the proportion of voters favoring Ms. Nyne if 28 out of a random sample of 50 voters favored her.

Using $z_{.95} = 1.645$, $x = 28$, and $n = 50$, we obtain from (7.36),

$$\frac{28}{50} \pm 1.645\sqrt{\left(\frac{28}{50}\right)\left(\frac{22}{50}\right)\Big/50}$$

or $.56 \pm 1.645\sqrt{(.56)(.44)/50}$ [i.e., (.445, .675)].

It should be clear to you that Ms. Nyne should not feel that she has this election locked up since the confidence interval estimate for p extends well below the value .50.

Example 7.22 If, in a random sample of 1000 voters, 560 voters favor Ms. Nyne, find a 90% confidence interval for p, the proportion of all voters favoring her.

The sample proportion favoring her in this example is 56%, the same as the sample proportion in Example 7.21. However, the sample size is larger in this case, and we find the confidence interval to be

$$.56 \pm 1.645\sqrt{\frac{(.56)(.44)}{1000}}$$

or (.534, .586).

7.11 CONFIDENCE INTERVALS FOR THE DIFFERENCE BETWEEN TWO PROPORTIONS

In the previous section we considered the problem of estimating p, the proportion in a large population having a certain characteristic. It will come as no surprise to you that this proportion may be different for different subgroups of the population. For example, although 56% of the *total population* may favor Ms. Nyne for county animal control officer, it may be that 72% of all male voters might favor her while only 40% of all female voters favor her. Similarly, 68% of all voters under age 30 might favor her while only 48% of those voters 30 or older favor her. Any candidate for political office is interested in the proportion of voters in various subgroups favoring him or her, since such information can dictate campaign strategy. As another example, an advertising executive may be concerned with the proportion of consumers who buy a given product, such as Sudsy Soap, in different geographical regions. If this proportion is quite different in the South than in the Northeast, say, the executive may want to use a different advertising campaign in the two regions, say a major campaign in the region where sales are lower and a coordinated but lesser campaign in the other region. Such a strategy might be implemented if the difference in the two proportions exceeded some predetermined value.

Consider the general problem of trying to estimate the difference between two proportions. Specifically, if p_1 and p_2 are the proportions in two subpopulations having a certain characteristic and if X_1 and X_2 are the numbers of members possessing these characteristics among independent samples of sizes n_1 and n_2 from these subpopulations, respectively, the difference between p_1 and p_2, namely $p_1 - p_2$, can be estimated by using the sample counterpart. That is, a point estimate for $p_1 - p_2$ can be found from

$$\frac{X_1}{n_1} - \frac{X_2}{n_2}. \tag{7.37}$$

In order to find a confidence interval for $p_1 - p_2$, we will need to know the distribution of the difference in sample proportions (7.37). Finding the exact distribution is difficult, but finding an approximate distribution is not too hard. We pointed out in the last section that the sample proportion, X/n, has an approximately normal distribution. We also know that

$$E\left(\frac{X}{n}\right) = p \qquad \text{and} \qquad V\left(\frac{X}{n}\right) = \frac{p(1-p)}{n},$$

and, of course, these same relationships hold for each subgroup. In Section 5.7 we saw that a linear combination of independent normal random variables has a normal distribution. It follows from Theorems 5.3 to 5.5 in that section that since $(X_1/n_1) - (X_2/n_2)$ is a linear combination of independent approximately normal random variables that its distribution will be approximately normal with

$$E\left(\frac{X_1}{n_1} - \frac{X_2}{n_2}\right) = p_1 - p_2$$

and

$$V\left(\frac{X_1}{n_1} - \frac{X_2}{n_2}\right) = \frac{p_1(1-p_1)}{n_1} + \frac{p_2(1-p_2)}{n_2}.$$

By standardizing, we see that

$$P\left[-z_{1-\alpha/2} \leq \frac{\left(\frac{X_1}{n_1} - \frac{X_2}{n_2}\right) - (p_1 - p_2)}{\sqrt{\frac{p_1(1-p_1)}{n_1} + \frac{p_2(1-p_2)}{n_2}}} \leq z_{1-\alpha/2}\right] \approx 1 - \alpha,$$

so a random interval which contains $(p_1 - p_2)$ with probability $1 - \alpha$ (approximately) can be found to have end points given by

$$\left(\frac{X_1}{n_1} - \frac{X_2}{n_2}\right) \pm z_{1-\alpha/2}\sqrt{\frac{p_1(1-p_1)}{n_1} + \frac{p_2(1-p_2)}{n_2}}.$$

Of course, in attempting to evaluate the end points of this interval we face the problem of not knowing the values of p_1 and p_2. As we did in the last section, we can approximate those values by using the sample proportions. The resulting approximate confidence interval is given by

$$\left[\left(\frac{X_1}{n_1} - \frac{X_2}{n_2}\right) - z_{1-\alpha/2}\sqrt{\frac{\frac{X_1}{n_1}\left(1 - \frac{X_1}{n_1}\right)}{n_1} + \frac{\frac{X_2}{n_2}\left(1 - \frac{X_2}{n_2}\right)}{n_2}},\right.$$
$$\left.\left(\frac{X_1}{n_1} - \frac{X_2}{n_2}\right) + z_{1-\alpha/2}\sqrt{\frac{\frac{X_1}{n_1}\left(1 - \frac{X_1}{n_1}\right)}{n_1} + \frac{\frac{X_2}{n_2}\left(1 - \frac{X_2}{n_2}\right)}{n_2}}\right]. \tag{7.38}$$

Example 7.23 Observations at a store in Atlanta showed that of 325 customers buying soap, 39 bought Sudsy Soap while in a Boston store 30 of 375 bought Sudsy Soap. If p_1 and p_2 represent the proportion of consumers who buy Sudsy Soap in Atlanta and Boston, respectively, find a 90% confidence interval estimate for $p_1 - p_2$.

Using (7.38) with $z_{1-\alpha/2} = z_{.95} = 1.645$ and noting that

$$\frac{x_1}{n_1} = \frac{39}{325} = .12 \quad \text{and} \quad \frac{x_2}{n_2} = \frac{30}{375} = .08,$$

we find that

$$(.120 - .080) \pm 1.645\sqrt{\frac{(.12)(.88)}{325} + \frac{(.08)(.92)}{375}}$$

or $.04 \pm 1.645(.0228)$. This gives a confidence interval of (.002, .078).

Example 7.24 In the Medicine section of *Time* magazine it was reported that a certain drug may save or prolong the lives of hundreds of thousands of heart patients. A medical team reported that the heart death rate (on an annual basis) among 733 patients taking the drug four times a day was 4.9%. Among a comparison group of 742 patients who received a placebo, the rate was 9.5%. Treating these data as the result of independent samples, find a 99% confidence interval for the difference in the heart death rate with and without the drug.

Using (7.38) with $z_{1-\alpha/2} = z_{.995} = 2.576$, we find that

$$(.049 - .095) \pm 2.576\sqrt{\frac{(.049)(.951)}{733} + \frac{(.095)(.905)}{742}}$$

or $-.046 \pm 2.576(.0134)$. This gives a confidence interval of $(-.080, -.012)$.

Assuming that the observed differences here are due to the effect of the drug, doctors can feel rather certain (99% confident) that this drug will reduce the heart death rate by people who have suffered heart attacks by between 1.2 and 8%. Based on the lower figure and using the number of persons surviving heart attacks in a year in the United States given in the article, 400,000, the use of this drug could prolong the lives of over 4800 patients per year in the United States alone.

7.12 DETERMINATION OF SAMPLE SIZE

Given a sample from the appropriate distribution, the sample information can be used to get estimates of unknown parameters. Even the most statistically uninitiated person would agree on intuitive grounds that a larger sample would give a better (i.e., more reliable) estimate of a parameter than a smaller sample. A question often asked of a statistician is "Just how large a sample do I need to take to be able to estimate this parameter?" The statistician can give no simple answer to this question, but rather must respond with three other questions, the first of which is: How accurate do you want your estimate to be? For example, if a service technician wants a point estimate of μ, the mean amount of coffee dispensed by a vending machine, he must specify if he would like the estimate to be within .50 ounce, .10 ounce, or .01 ounce, say, of the actual mean. For the sake of example, say that the value of .10 ounce is chosen. If we give this a bit more thought, we will see that even this specification is not enough. Since $\bar{X}$, the sample mean used as a point estimate of μ, is actually a random variable, for any sample size it may or may not fall within .10 ounce of μ. There will, however, be a probability that this will happen; that is, for a given sample size, we can find

(7.39) $$P[|\bar{X} - \mu| \leq .10].$$

Because of this fact, the statistician will have to ask the service technician to specify the probability with which the estimate for μ, $\bar{X}$, will be within .10 ounce of μ. Say that the value .95 is specified. When this is done, the statistician will be able to determine the sample size necessary to attain this probability provided that the service technician can answer one more question: What is the variance of the amount of coffee dispensed into a cup? As in Example 7.16, let us assume that the amount of coffee dispensed into a cup is normally distributed with $\sigma = .20$ ounce. Summarizing, then, we are interested in the size sample necessary to have a probability of .95 that $\bar{X}$ will be within .10 ounce of μ. Since $\bar{X}$ is normally distributed, we know that

$$(7.40) \qquad P\left[-z_{.975} \leq \frac{\bar{X}-\mu}{\sigma/\sqrt{n}} \leq z_{.975}\right] = P\left[\frac{|\bar{X}-\mu|}{\sigma/\sqrt{n}} \leq z_{.975}\right] = P\left[\frac{|\bar{X}-\mu|}{\sigma/\sqrt{n}} \leq 1.96\right] = .95.$$

Dividing both sides of the inequality in (7.39) by $\sigma/\sqrt{n}$ and using (7.40), we find that

$$P\left[\frac{|\bar{X}-\mu|}{\sigma/\sqrt{n}} \leq \frac{.10}{\sigma/\sqrt{n}}\right] = P\left[\frac{|\bar{X}-\mu|}{\sigma/\sqrt{n}} \leq 1.96\right]$$

and this equality will hold if

$$\frac{.10}{\sigma/\sqrt{n}} = 1.96.$$

Solving for n, we find that

$$n = \left(\frac{1.96\sigma}{.10}\right)^2 = \left[\frac{1.96(.20)}{(.10)}\right]^2 = 15.37$$

or, using an integer value for n, 16.

More generally, we can state the problem as follows. Say that a random sample is to be chosen from a normal distribution having a known variance σ^2 and that the corresponding sample mean is to be used to estimate μ within e units with probability $1 - \alpha$. Since

$$P[|\bar{X}-\mu| \leq e] = P\left[\frac{|\bar{X}-\mu|}{\sigma/\sqrt{n}} \leq \frac{e}{\sigma/\sqrt{n}}\right],$$

the latter term will have probability $1 - \alpha$ if and only if

$$\frac{e}{\sigma/\sqrt{n}} = z_{1-\alpha/2}.$$

Consequently, the sample size necessary to meet these specifications is

$$(7.41) \qquad n = (z_{1-\alpha/2}\sigma/e)^2.$$

By examining equation (7.41), you can see why the statistician must ask three questions before the question about sample size can be answered. The quantities e, $1 - \alpha$, and σ must all be specified before n can be found. These quantities give information about the probability or degree of certainty with which $\bar{X}$ will be within e units of μ when the standard deviation of the variable of interest is σ.

Example 7.25 If the service technician wanted to estimate μ to within .025 ounce with probability .99, what sample size should be used?

We use (7.41) with $z_{1-\alpha/2} = z_{.995} = 2.576$, $e = .025$, and $\sigma = .20$ to get

$$n = \left[\frac{2.576(.20)}{.025}\right]^2 = 424.7 \approx 425.$$

The same approach can be used when the distribution is not normally distributed, provided that the resulting sample size is large enough to invoke the Central Limit Theorem. If the standard deviation is not known (as is usually the case!), it might be possible to estimate its value on the basis of a preliminary sample and then substitute the estimated value of σ in equation (7.41) to find how many additional observations are needed to meet the specifications. It may also be the case that, although the value of σ^2 is not precisely known, it can be estimated by an experienced investigator.

The discussion up to this point has been with regard to a *point estimate* of μ. One might well want an interval estimate for μ and might specify a desired length for the confidence interval. Again, assuming that a random sample is chosen from a normal distribution having a known variance, the end points of a $1 - \alpha$ confidence interval would be

$$\bar{X} - z_{1-\alpha/2}\frac{\sigma}{\sqrt{n}} \quad \text{and} \quad \bar{X} + z_{1-\alpha/2}\frac{\sigma}{\sqrt{n}}.$$

Since the length of an interval is simply the difference between the endpoints, the length is

$$l = \frac{2z_{1-\alpha/2}\sigma}{\sqrt{n}}. \tag{7.42}$$

The sample size necessary to obtain this length can be found by solving equation (7.42) for n to get

$$n = \left[\frac{z_{1-\alpha/2}\sigma}{(l/2)}\right]^2. \tag{7.43}$$

Note the similarity between equations (7.41) and (7.43).

Example 7.26 Dewey Phaster is an efficiency expert called in to study the operation in Jack Pyne's mail-order business. Dewey is willing to assume that the standard deviation in the amount of time required to fill orders is 12 minutes, but he wants to estimate the mean amount of time required, μ. If he wishes to find a 98% confidence interval estimate for μ of length 3 minutes, how many orders should he check? (Each completed order has written on it the amount of time required to fill it.)

Using equation (7.43) with $z_{1-\alpha/2} = z_{.99} = 2.326$, $l = 3$, and $\sigma = 12$,

$$n = \left[\frac{2.326(12)}{\frac{3}{2}}\right]^2 = 346.3 \approx 347.$$

The question of sample size is also asked with regard to estimation of the binomial parameter p. If we want the sample proportion, X/n, to be within e of the true value p with probability $1 - \alpha$, we can determine an expression for the necessary sample size by using the normal approximation to the binomial distribution. We have

$$P\left[\left|\frac{X}{n} - p\right| \le e\right] = P\left[\frac{|(X/n) - p|}{\sqrt{pq/n}} \le \frac{e}{\sqrt{pq/n}}\right] = 1 - \alpha$$

(approximately) if

$$\frac{e}{\sqrt{pq/n}} = z_{1-\alpha/2}.$$

Solving for n gives

$$n = \left(\frac{z_{1-\alpha/2}\sqrt{pq}}{e}\right)^2 = \left(\frac{z_{1-\alpha/2}}{e}\right)^2 pq. \tag{7.44}$$

Once again we face a familiar problem. To determine the sample size necessary to get a good estimate of p, we need to know the value of p! (And of course if we knew the value of p, we would not have to bother finding n.) In the past (Sections 7.10 and 7.11) we replaced p with a sample estimate of p, but that ploy will not work this time because we have not yet taken any sample observations. There are alternatives, however. The quantity pq *always* lies between 0 and $\frac{1}{4}$. It assumes its maximum value of $\frac{1}{4}$ when $p = \frac{1}{2}$. Consequently, the largest value of n that might be required is

$$n = \left(\frac{z_{1-\alpha/2}}{e}\right)^2 \left(\frac{1}{4}\right).$$

Hence, to be on the safe side we can use this large value. Realistically, p might well be near $\frac{1}{2}$ for many situations, in political races for one example. However, in other situations it may be possible on intuitive grounds or past experience to put an upper bound on p (if p is smaller than $\frac{1}{2}$) or a lower bound on p (if p is larger than $\frac{1}{2}$). An adequate sample size can then be determined from (7.44) using this bounding value.

Example 7.27 The campaign manager for the Democratic presidential candidate wants to estimate the proportion of voters planning to vote for the Democratic candidate in the next election. The estimate is to be within .02 with probability .95. What sample size is necessary?

Using (7.44) with $p = \frac{1}{2}$, since this is a realistic possibility, we would find

$$n = \left(\frac{1.96}{.02}\right)^2 \left(\frac{1}{2}\right)\left(\frac{1}{2}\right) = 2401.$$

Example 7.28 A medical doctor wishes to estimate the proportion of patients with a certain chronic disease who will suffer side effects from a particular drug. Similar drugs have shown side effects in 10 to 20% of the patients. How large a sample should be taken if the estimate is to be within 3% with probability .90?

A very conservative approach would be to find n using equation (7.44) with $p = \frac{1}{2}$. This would require

$$n = \left(\frac{1.645}{.03}\right)^2 \left(\frac{1}{2}\right)\left(\frac{1}{2}\right) = 751.7 \approx 752$$

observations. However, since similar drugs have shown side effects in 10 to 20% of the patients, it would not be unreasonable to use a value of, say, $p = .25$ in equation (7.44). This would require only

$$n = \left(\frac{1.645}{.03}\right)^2 (.25)(.75) = 563.7 \approx 564$$

observations, a savings of almost 200 observations. [A word of warning is in order here. If the true value of p is actually higher than the upper bound (which in this case was taken to be .25), the "savings" in sample size will be one of appearance only and not of substance, since the goal of estimating p within a certain error e with probability $1 - \alpha$ will not really have been attained.]

As a closing note we should point out that the sample sizes obtained in this section were found under the implicit assumption that samples taken from a finite population are taken *with replacement*. If they are taken without replacement, the requisite sample sizes will tend to be somewhat smaller.

7.13 ESTIMATION OF A MEAN USING STRATIFIED SAMPLING

In Figure 7.5 you can see a map of the town of Littleton. The mayor of Littleton, J. G. Gyant, would like to estimate the mean family income of the 40 families living in this town. This information is to be used for planning the city budget. The mayor is well aware that the town is divided into three neighborhoods but is not aware of the detailed information on the incomes of the families that is given in Table 7.3. Had this table been made available to Seymour Dayta, the town statistician, he would be able to find the true mean income exactly. In fact, we can see by adding the incomes together and dividing by 40 that the actual mean family income is

$$\mu = \frac{47 + 83 + \ldots + 8}{40} = \frac{980}{40} = 24.5 \text{ (thousand).}$$

Using the letter d to remind us of dollars, if d_i = the family income for the ith family,

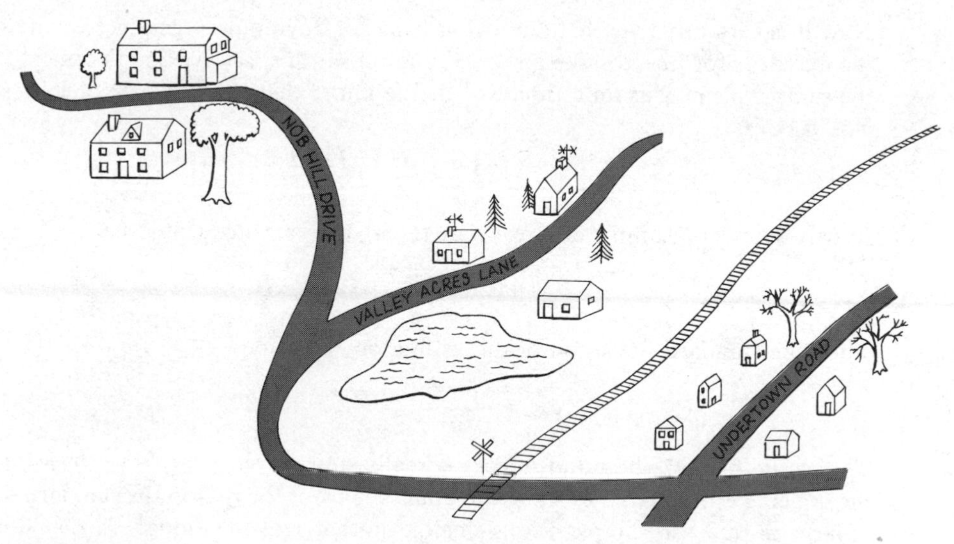

FIGURE 7.5
Map of Littleton

TABLE 7.3
Family Incomes

Incomes for:	*Income (thousands of dollars)*									
Nob Hill families	47	83	52	54	41	29	38	45	76	35
Valley Acres families	18	22	12	31	24	17	19	23	21	15
	13	26	22	27	11	16	17	23	21	22
Undertown families	8	6	10	7	8	5	9	12	7	8

the actual variance of the income could be found by

$$\sigma^2 = \frac{\sum_{i=1}^{40} (d_i - \mu)^2}{40} = 322.2.$$

[Note that had we considered the simple experiment of choosing a family at random from among the 40 families living in this town and defined the random variable X to be the income of the family chosen, X would have a probability distribution given by

$$\begin{array}{lccccccc} x: & 5 & 6 & 7 & 8 & \ldots & 76 & 83 \\ P[X = x]: & \frac{1}{40} & \frac{1}{40} & \frac{2}{40} & \frac{3}{40} & \ldots & \frac{1}{40} & \frac{1}{40}. \end{array}$$

Using this probability distribution, we would find that

$$\mu = E(X) = \sum x \cdot P[X = x] = 24.5$$

and

$$\sigma^2 = V(X) = \sum (x - \mu)^2 \cdot P[X = x] = 322.2.]$$

Now it is certainly true that an obvious way for Seymour to get an estimate of μ for the mayor is for him to take a simple random sample, say with replacement, and use the sample mean $\bar{x}$ as an estimate of μ. We know that the variance of this estimator would be

$$V(\bar{X}) = \frac{\sigma^2}{n} = \frac{322.2}{n}.$$

In particular, if a sample of size 4 were taken, the variance would be

$$V(\bar{X}) = \frac{322.2}{4} = 80.55$$

while the standard deviation of this estimator would be

$$\sigma_{\bar{X}} = \frac{\sigma}{\sqrt{n}} = \sqrt{80.55} = 8.97.$$

This value of 8.97 thousand dollars is really quite large relative to the value of the mean being estimated, 24.5 thousand dollars. Part of the reason the standard deviation is large is that the sample size is small. Another reason, though, is that due to the random nature of the sample, it is quite likely that the sample will not be representative of the city in the following sense. Intuitively, since the city is divided into three different neighborhoods in such a way that family incomes are much more similar within a given neighborhood than they are between different neighborhoods, a "representative" sample would include families from each neighborhood. Furthermore, since Valley Acres has 20 families, whereas the other neighborhoods have only 10 families, a sample of size 4 would logically appear to be most representative if it contained 1, 2, and 1 families from Nob Hill, Valley Acres, and Undertown, respectively. The probability of obtaining such a sample when a random sample of size 4 is taken with replacement can be found by using the multinomial distribution (Section 4.3):

$$P[\text{1–2–1 sample}] = \frac{4!}{1!\,2!\,1!}\left(\frac{10}{40}\right)^1\left(\frac{20}{40}\right)^2\left(\frac{10}{40}\right)^1 = .1875.$$

A sample that would seem to be most representative occurs less than 20% of the time. On the other hand, there is a rather high probability that the Nob Hill (or the Undertown) neighborhood will *not* be represented in the sample. Using the binomial distribution, we can find

$$P[\text{no Nob Hill families in sample}] = \binom{4}{0}(.25)^0(.75)^4 = .3164.$$

This is the same as the probability that no Undertown families are represented. If no Nob Hill families are in the sample, the sample mean will likely underestimate μ, while if no Undertown families are in the sample, μ will likely be overestimated. The high probability of these occurrences is one factor that leads to a large standard deviation for the estimator $\bar{X}$.

It would seem reasonable to try to get an estimator for μ which has a smaller standard deviation than that of $\bar{X}$ by choosing the sample differently. Specifically, we might choose the sample in such a way as to force it to include at least one family from each neighborhood. We could do this by randomly choosing one family from each of Nob Hill and Undertown and two families from Valley Acres. Such a sample is an example of a stratified random sample.

Definition 7.6 If a population can be subdivided into distinct groups that together comprise the entire population, these groups (subpopulations) are called *strata.*

Definition 7.7 If a population is subdivided into k strata and if independent simple random samples of size $n_i (i = 1, 2, \ldots, k)$ are chosen from each stratum, the resulting sample is called a *stratified random sample.*

We would now like to find a method of obtaining an unbiased estimator for μ based on a stratified random sample. We must first introduce some more notation. Let N_i denote the number of elements in the ith stratum, and let μ_i and σ_i^2 denote the mean and variance of a measurable quantity associated with this stratum. For example, if the neighborhoods of Littleton form the strata, we would have $N_1 = 10$, $N_2 = 20$, and $N_3 = 10$, where "Nob Hill" corresponds to stratum 1, and so on. If we denote the income of the jth family in the ith neighborhood by $d_{i,j}$, we can give expressions for μ_i and σ_i^2. For example, we could find from Table 7.3 that

$$\mu_1 = \frac{d_{1,1} + d_{1,2} + \ldots + d_{1,N_1}}{N_1} = \frac{47 + 83 + \ldots + 35}{10} = 50$$

TABLE 7.4
Information for Strata in Littleton

Stratum	*Size,* N_i	*Mean,* μ_i	*Variance,* σ_i^2	*Standard Deviation,* σ_i
Nob Hill (1)	10	50	271.0	16.46
Valley Acres (2)	20	20	25.6	5.06
Undertown (3)	10	8	3.6	1.90

and

$$\sigma_1^2 = \frac{\sum_{j=1}^{N_1} (d_{1,j} - \mu_1)^2}{N_1} = \frac{(47 - 50)^2 + (83 - 50)^2 + \ldots + (35 - 50)^2}{10} = 271.$$

For future reference we summarize this information in Table 7.4. To find an estimator for μ, we first note that μ can be expressed in terms of the quantities μ_i, N_i, and N as follows:

$$\mu = \frac{(d_{1,1} + \ldots + d_{1,N_1}) + (d_{2,1} + \ldots + d_{2,N_2}) + \ldots + (d_{k,1} + \ldots + d_{k,N_k})}{N}$$

$$= \frac{N_1\mu_1 + N_2\mu_2 + \ldots + N_k\mu_k}{N}$$

$$= \sum_{i=1}^{k} \left(\frac{N_i}{N}\right)\mu_i.$$

Since we are taking a simple random sample from each stratum, it follows that the sample mean for stratum i, namely $\bar{X}_i$, is an unbiased estimator for μ_i. Consequently, the quantity $\hat{\mu}_{\text{strat}}$ defined by

$$\hat{\mu}_{\text{strat}} = \sum_{i=1}^{k} \frac{N_i}{N}\bar{X}_i \tag{7.45}$$

will be an unbiased estimator for μ, the population mean. Once we note that this estimator is a linear combination of the independent random variables $\bar{X}_i$, we can use Theorem 5.4 to find an expression for the variance of $\hat{\mu}_{\text{strat}}$:

$$V(\hat{\mu}_{\text{strat}})_{w/} = \sum_{i=1}^{k} \left(\frac{N_i}{N}\right)^2 V(\bar{X}_i) \tag{7.46}$$

$$= \sum_{i=1}^{k} \left(\frac{N_i}{N}\right)^2 \frac{\sigma_i^2}{n_i}.$$

The expression for the variance would have to be modified somewhat if sampling were done without replacement. In view of the discussion given at the end of Section 5.6, the expression for the variance would be given by

$$V(\hat{\mu}_{\text{strat}})_{w/o} = \sum_{i=1}^{k} \left(\frac{N_i}{N}\right)^2 \frac{\sigma_i^2}{n_i}\left(\frac{N_i - n_i}{N_i - 1}\right). \tag{7.47}$$

It is obvious even to the untrained eye that calculating the variance for the estimator for μ based on a stratified random sample [using equation (7.46) or (7.47)] is more involved than finding the variance of $\bar{X}$, the estimator for μ based on a simple random sample. We shall see that although the calculations are more involved when stratification is used, better (i.e., smaller variance) estimators are obtained. For example, if four observations are taken from Littleton with $n_1 = 1, n_2 = 2$, and $n_3 = 1$, we would find, using the information in Table 7.4 and equation (7.46), that

$$V(\hat{\mu}_{\text{strat}})_{w/} = \left(\frac{10}{40}\right)^2\left(\frac{271}{1}\right) + \left(\frac{20}{40}\right)^2\left(\frac{25.6}{2}\right) + \left(\frac{10}{40}\right)^2\left(\frac{3.6}{1}\right)$$
$$= 20.36,$$

which corresponds to a standard deviation of 4.51. Comparing this value with the standard deviation of $\bar{X}$ based on a simple random sample of size 4 calculated earlier to be 8.97, we see that the standard deviation of the estimator has been cut nearly in half by the use of stratification.

Example 7.29 If Seymour decided to take a sample of size 16 from the families in Littleton with replacement, how would the standard deviations of the estimators based on a simple random sample and a stratified random sample using $n_1 = 4$, $n_2 = 8$, and $n_3 = 4$ compare?

Using simple random sampling, the variance of $\bar{X}$ would be

$$V(\bar{X}) = \frac{\sigma^2}{16} = 20.14,$$

so the standard deviation would be 4.49. For the stratified random sample, the variance would be

$$V(\hat{\mu}_{\text{strat}})_{w/} = \left(\frac{10}{40}\right)^2\left(\frac{271}{4}\right) + \left(\frac{20}{40}\right)^2\left(\frac{25.6}{8}\right) + \left(\frac{10}{40}\right)^2\left(\frac{3.6}{4}\right) \tag{7.48}$$
$$= 4.234 + 0.800 + .056 = 5.090,$$

so the standard deviation would be 2.26.

If we now look at (7.48) more closely, we can see that each stratum makes a contribution to the final variance. In fact, even a cursory examination reveals that the first stratum makes by far the largest contribution to the total variance. This is due to the fact that σ_1 is considerably larger than σ_2 or σ_3. How can we get a smaller contribution from the first stratum? That contribution is determined by

$$\left(\frac{N_1}{N}\right)^2\left(\frac{\sigma_1^2}{n_1}\right)$$

and the quantities N, N_1, and σ_1 are all properties of the population. The only quantity that can be changed is n_1. If we restrict the total sample size to 16 observations, we can increase n_1 (thus reducing the contribution to the variance of stratum 1) at the expense of decreasing n_2 or n_3 (increasing the variance contribution from those strata). Since the differences between σ_1 and σ_2 and σ_3 are so great, such a reallocation of subsample sizes will result in a net decrease in the variance. For example, if $n_1 = 6$,

$n_2 = 8$, and $n_3 = 2$, we would find

$$V(\hat{\mu}_{\text{strat}})_{w/} = \left(\frac{10}{40}\right)^2\left(\frac{271}{6}\right) + \left(\frac{20}{40}\right)^2\left(\frac{25.6}{8}\right) + \left(\frac{10}{40}\right)^2\left(\frac{3.6}{2}\right)$$

$$= 2.823 + 0.800 + .112 = 3.735.$$

This value is considerably less than the variance of 5.090 found in Example 7.29. It would be helpful if we could find a method of determining subsample sizes that would lead to the smallest possible variance of $\hat{\mu}_{\text{strat}}$ for a given total sample size. By using some methods of calculus, it can be shown that the optimal subsample sizes are found from

(7.49)
$$n_j = \frac{N_j\sigma_j}{\sum_{i=1}^{k}(N_i\sigma_i)} \cdot n,$$

where n is the total sample size.

Example 7.30 If Seymour wants to take a stratified random sample of size 16 from the families in Littleton, how should he allocate these among the three strata? What will be the variance of $\hat{\mu}_{\text{strat}}$ if sampling is done with replacement?

Using (7.49) and the information in Table 7.4, we find that

$$n_1 = \frac{N_1\sigma_1}{N_1\sigma_1 + N_2\sigma_2 + N_3\sigma_3} \cdot n = \frac{10(16.46)}{10(16.46) + 20(5.06) + 10(1.90)} \cdot 16$$

$$= \frac{164.6}{164.6 + 101.2 + 19.0}(16) = \left(\frac{164.6}{284.8}\right)(16) = 9.25.$$

Rounding to an integer we get $n_1 = 9$. Similarly, we would find that

$$n_2 = \left(\frac{101.2}{284.8}\right) \cdot 16 = 5.68 \approx 6$$

and

$$n_3 = \frac{19.0}{284.8} \cdot 16 = 1.07 \approx 1.$$

With this allocation of subsample sizes, we find the variance from (7.46) to be

$$V(\hat{\mu}_{\text{strat}})_{w/} = \left(\frac{10}{40}\right)^2\left(\frac{271}{9}\right) + \left(\frac{20}{40}\right)^2\left(\frac{26.6}{6}\right) + \left(\frac{10}{40}\right)^2\left(\frac{3.6}{1}\right) = 3.1736.$$

This variance is the smallest value that can be obtained using a stratified random sample with a total sample size of 16. Consider for a moment why Seymour would have decided on a sample size of 16. Two reasons come to mind. First, he might consider the standard deviation of $\sqrt{3.1736} = 1.78$ thousand dollars to be sufficiently small for his purposes (although he could get a smaller standard deviation with his 16 observations by simply sampling without replacement). Second, he may be constrained by a small budget. Expenditures of time and/or money must be made in order to collect data. If it costs \$10 to obtain an observation and if Seymour's budget allows only \$160 for sampling costs, it is very easy to see why he decided on a sample of size 16.

If, in fact, the cost of sampling is the same for each stratum (e.g., \$10 per observation regardless of the neighborhood), then determining the total sample size and optimal allocation is a relatively simple task. However, there may be some situations where the sampling costs may vary from stratum to stratum. For example, the time required for an interviewer to travel from one household to another could differ in a stratum consisting primarily of apartment complexes than in one consisting of single-family dwellings in a subdivision or of farmhouses in a rural area. This time differential could lead to differences in sampling costs. Returning to the problem of sampling families in Littleton, let us assume that the costs of sampling are different in each stratum. In particular, say that we have

$$c_1 = \$16, \qquad c_2 = \$9, \qquad c_3 = \$4$$

and that Seymour's budget for sampling is only \$160. How many observations should he take from each neighborhood? As one extreme, he might take 9 from Nob Hill and one each from the other neighborhoods, a total of 11 observations. The number of observations is relatively small because so much money was spent interviewing families on Nob Hill. On the other extreme, he could afford 10 observations from Undertown, 11 from Valley Acres, and 1 from Nob Hill, a total of 22 observations. Intuitively, however, neither of these extremes would be desirable since one neighborhood or another would be underrepresented. The optimal allocation of subsamples to each stratum should consider stratum size and variance as well as sampling costs. Methods of calculus can be used to show that the overall variance of $\hat{\mu}_{\text{strat}}$ is minimized if subsample sizes are found by

(7.50)
$$n_j = \frac{N_j\sigma_j/\sqrt{c_j}}{\sum_{i=1}^{k} (N_i\sigma_i/\sqrt{c_i})} \cdot n.$$

The total cost, ignoring possible fixed costs, would then be

$$C = n_1c_1 + n_2c_2 + \ldots + n_kc_k.$$

If we define h_j to be the multiplier of n in (7.50), the total cost can be expressed by

$$C = h_1nc_1 + h_2nc_2 + \ldots + h_knc_k$$
$$= \left(\sum_{i=1}^{k} h_ic_i\right)n.$$

From this we obtain the total sample size n to be

(7.51)
$$n = \frac{C}{\sum_{i=1}^{k} h_ic_i}.$$

Example 7.31 If Seymour wants to take a stratified random sample from the families in Littleton, how should he allocate observations among the three neighborhoods if his total budget allows a maximum of \$190 for sampling costs? What is the variance of the estimator $\hat{\mu}_{\text{strat}}$ if sampling is done *without* replacement?

It is perhaps easiest to summarize the necessary information in a table. In Table 7.5 the first four columns repeat information already presented earlier. The terms in column (5) and the total of that column can be used to find h_j. For example,

$$h_1 = \frac{41.15}{84.38} = .488.$$

TABLE 7.5
Information for Calculating Sample Sizes

(1) Stratum	(2) N_j	(3) σ_j	(4) c_j	(5) $N_j\sigma_j/\sqrt{c_j}$	(6) h_j	(7) h_jc_j	(8) n_j	(9) $n_j^*c_j$
1	10	16.46	16.	41.15	.488	7.808	7.808	128
2	20	5.06	9.	33.73	.400	3.600	6.400	54
3	10	1.90	4.	9.50	.112	.448	1.792	8
	40			84.38	1.000	11.856		190

The total of the terms in column (7) is used to determine the total sample size. Since $C = 190$, we find by using equation (7.51) that

$$n = \frac{190}{11.856} = 16.026,$$

which we round down to 16. Next we find the subsample sizes using equation (7.50). Expressing this using h_j, we would find that

$$n_1 = h_1 \cdot n = (.488)(16) = 7.808 \approx 8 = n_1^*$$
$$n_2 = h_2 \cdot n = (.400)(16) = 6.400 \approx 6 = n_2^*$$
$$n_3 = h_3 \cdot n = (.112)(16) = 1.792 \approx 2 = n_3^*.$$

Using n_j^* to represent the subsample sizes as rounded, the last column gives the cost of sampling in each neighborhood, the total being \$190. (Note that in general the figures will not come out quite this nicely!)

Assuming that sampling is done without replacement, we find the variance of the estimator by using equation (7.47). We have

$$V(\hat{\mu}_{\text{strat}})_{w/o} = \left(\frac{10}{40}\right)^2\left(\frac{271}{8}\right)\left(\frac{10-8}{10-1}\right) + \left(\frac{20}{40}\right)^2\left(\frac{25.6}{6}\right)\left(\frac{20-6}{20-1}\right) + \left(\frac{10}{40}\right)^2\left(\frac{3.6}{2}\right)\left(\frac{10-2}{10-1}\right)$$
$$= .470 + .786 + .100 = 1.356.$$

(Although this is *not* the smallest variance that could be obtained using 16 observations taken without replacement, it is the smallest variance that could be obtained at a cost of \$190.)

Stratified random sampling is but one method of sampling that leads to estimators having smaller variance than like estimators based on simple random sampling. If you are interested in learning more about sampling techniques, we recommend a book by W. G Cochran, *Sampling Techniques*, 3rd ed. (New York: John Wiley & Sons, Inc., 1977).

EXERCISES

1. The weights of samples of five male and five female students enrolled in a statistics course are as follows:

Males:	175	157	172	168	215
Females:	127	135	140	115	110.

(a) Find point estimates for μ_M, $\tilde{\mu}_M$, and σ_M^2 (the mean, median, and variance of the distribution of the weights of all males enrolled in the statistics course).
(b) Find point estimates for μ_F, $\tilde{\mu}_F$, and σ_F^2.

2. (a) Find a point estimate for $\mu_M - \mu_F$, the difference between the mean weights of males and females enrolled in a statistics course using the sample information given in Exercise 1.
(b) Find a point estimate for σ_M^2/σ_F^2. (*To think about:* Would you expect the weights of males or females to be more variable? What does this tell you about what you should expect the ratio σ_M^2/σ_F^2 to be?)

3. Say that you wish to estimate μ, the mean of a distribution, and are given a random sample of 25 observations, $X_1, X_2, \ldots, X_{25}$. Under what conditions might you use $\tilde{X}$, the sample median, as an estimate of μ? What advantage might there be to using $\tilde{X}$ instead of $\bar{X}$? (*Hint:* What can you say about the relative values of μ and $\tilde{\mu}$ when the distribution being sampled is symmetric? Skewed?)

4. Tony, a quality control inspector at a cereal factory, checks a sample of boxes of Frosty Wheats to see if they are adequately filled. He is interested in p, the proportion of boxes adequately filled. If he inspects 200 boxes at 9:00 A.M. and another 100 boxes at 3:00 P.M. and finds 198 and 96 boxes adequately filled, respectively, what should his estimate for p be based on the results of (a) his 9:00 A.M. inspection only?; (b) his 3:00 P.M. inspection only?; (c) both inspections combined? [*To think about:* What assumptions did you implicitly make in working part (c)?]

• **5.** Let $X_1, X_2, \ldots, X_n$ be a random sample from a Poisson distribution having a mean of λt. The joint probability distribution of these random variables is given by

$$P[X_1 = x_1, X_2 = x_2, \ldots, X_n = x_n] = \frac{e^{-n(\lambda t)}(\lambda t)^{x_1+x_2+\ldots+x_n}}{x_1!\,x_2!\ldots x_n!}.$$

Treat the product λt as a single parameter, say $\lambda t = \theta$, and find the maximum likelihood estimator for λt.

• **6.** Let $X_1, X_2, \ldots, X_n$ be a random sample from a normal distribution having a known mean of zero and an unknown standard deviation σ. The joint probability density function of these random variables is given by

$$f(x_1, x_2, \ldots, x_n) = \frac{1}{(\sqrt{2\pi})^n \sigma^n} e^{-(1/2)(\Sigma x^2/\sigma^2)}.$$

Find the maximum likelihood estimator for the parameter σ.

7. When persons having a certain kidney disease undergo dialysis, they are subject to having two major side effects (hypertension and muscle cramps). Let X represent the number of these side effects experienced by a patient undergoing dialysis. Assume that the probability distribution of X is

x (no. of side effects):	0	1	2
$P[X = x]$:	.2	.6	.2.

(a) Find $E(X)$. (b) Find $V(X)$.
Consider the next three patients arriving for treatment and let X_1, X_2, and X_3 denote the number of side effects experienced by these patients. Assume that these random variables constitute a random sample from the probability distribution of X given above. (c) Find the probability distribution of $\bar{X}$, the sample mean. (d) Find the probability distribution of $\tilde{X}$, the sample median. (*Hint:* Make a table

showing all 27 possible samples and the corresponding probability of observing each sample. For each possible sample, calculate the corresponding value of $\bar{x}$ and $\tilde{x}$. Cf. Exercise 11 of Chapter 5.)

8. Let X be as defined in Exercise 7 and let $\bar{X}$ be the sample mean based on a sample of size 3. Find $E(\bar{X})$ and $V(\bar{X})$ using $E(X)$ and $V(X)$ found in Exercise 7(a) and (b). Also find $E(\bar{X})$ and $V(\bar{X})$ using the probability distribution of $\bar{X}$ found in Exercise 7(c).

9. Using the probability distribution of $\tilde{X}$ found in Exercise 7(d), find $E(\tilde{X})$ and $V(\tilde{X})$, the mean and variance of the sample median. (To think about: Based on the results of Exercises 8 and 9, would it be better to use $\bar{X}$ or $\tilde{X}$ as an estimator of μ? Why?)

10. Repeat Exercise 7 for the case where the distribution of X is given by

x (no. of side effects):	0	1	2
$P[X = x]$:	.6	.3	.1.

(*Hint:* You can use the same table you used in Exercise 7, changing only the probability portions of that table.)

11. Let X be as defined in Exercise 10 and let $\bar{X}$ be the sample mean based on a sample of size 3. Find $E(\bar{X})$ and $V(\bar{X})$ using $E(X)$ and $V(X)$ as found in Exercises 10(a) and (b). Also find $E(\bar{X})$ and $V(\bar{X})$ using the probability distribution of $\bar{X}$ found in Exercise 10(c).

12. Using the probability distribution of $\tilde{X}$ found in Exercise 10(d), find $E(\tilde{X})$ and $V(\tilde{X})$, the mean and variance of the sample median. (*To think about:* Based on the results of Exercises 11 and 12, would it be better to use $\bar{X}$ or $\tilde{X}$ as an estimator of μ? Why?)

13. Based on what you observed in Exercises 9 and 12, under what conditions would you expect $\tilde{X}$ to be an unbiased estimator of μ?

14. Let $X_1, X_2, \ldots, X_5$ be a random sample from a distribution having mean μ and variance σ^2. Determine which of the following estimators of μ are unbiased.

(a) $(X_1 + X_3 + X_5)/3$; (b) $(X_1 + X_2 + X_3 + X_4 + X_5)/5$;
(c) $(X_1 - X_2 + X_3 - X_4 + X_5)$; (d) $(2X_2 + 4X_4)/5$.

15. Find the variance for each of the estimators defined in Exercise 14. Which of those four estimators would be considered "best"? Why? (*To think about:* Would the estimator of Exercise 14(c) or (d) be considered better? Why?)

16. (a) Show that if $X_1, X_2, \ldots, X_n$ is a random sample from a distribution having mean μ and variance σ^2, and if $\sum_{i=1}^{n} a_i = 1$, then

$$Y = a_1X_1 + a_2X_2 + \ldots + a_nX_n$$

is an unbiased estimator for μ. (b) Find an expression for the variance of Y. (*To think about:* What does this tell you about the number of possible unbiased estimators for μ?)

17. Mary, a demographer, would like to get a point estimate of the mean age μ of persons living in a small city (population 10,000). She feels that she can obtain a random sample of individuals from this city, but she is concerned that not all the responses will be accurate. She believes that a few people will give ages older than they really are but that more will give their ages as younger. To get an estimator that might compensate for these inaccuracies, she will consider using one of the two following estimators:

1. $[(X_1 + X_2 + \ldots + X_n)/n] + 2$
2. $(X_1 + X_2 + \ldots + X_n)/(n - 2)$

where X_i is the age reported by the ith individual interviewed. (a) Will these estimates be larger or smaller than the sample mean $\bar{X}$? (b) Assuming that the individuals do in fact report their ages correctly, find the expected value of each of these estimators. (c) Find the variance of each estimator. (To think about: Which estimator do you think would be better? Why? Can you think of a way of constructing an estimator better than either of these which might "adjust" for inaccurate responses?)

18. Show that $E(\hat{\Theta} - \theta)^2 = V(\hat{\Theta}) + [\text{Bias }(\hat{\Theta})]^2$. (*Hint:* Express $E(\hat{\Theta} - \theta)^2$ as $E[(\hat{\Theta} - E(\hat{\Theta})) + (E(\hat{\Theta}) - \theta)]^2$, expand the square, and then take the expected value of each term.)

19. A random sample of size n, $X_1, X_2, \ldots, X_n$ is taken from a distribution having mean μ and variance σ^2. The quantity

$$W = \frac{X_1 + X_2 + \ldots + X_n}{n - 1}$$

is taken as an estimator of μ. Find (a) the bias of W; (b) the mean squared error of W. (*To think about:* What happens to the magnitude of the bias as $n \longrightarrow \infty$?)

20. Using the standard normal tables, find

(a) $P[-.674 \leq Z \leq .674]$; (b) $P[-1.282 \leq Z \leq 1.282]$;
(c) $P[-1.645 \leq Z \leq 1.645]$; (d) $P[-1.96 \leq Z \leq 1.96]$;
(e) $P[-2.326 \leq Z \leq 2.326]$; (f) $P[-2.576 \leq Z \leq 2.576]$.

21. Using the notation $P[Z \leq z_p] = p$, find

(a) $z_{.90}$; (b) $z_{.95}$;
(c) $z_{.99}$; (d) $z_{.10}$;
(e) $z_{.05}$; (f) $z_{.01}$.

22. Using the notation $P[Z \leq z_p] = p$, assuming that $p \geq .5$ and $0 < \alpha < 1$, find

(a) $P[-z_p \leq Z \leq z_p]$; (b) $P[Z > z_p]$;
(c) $P[Z \leq z_\alpha]$; (d) $P[z_{\alpha/2} \leq Z \leq z_{1-\alpha/2}]$;
(e) $P[-z_{1-\alpha/2} \leq Z \leq z_{1-\alpha/2}]$; (f) $P[z_{\alpha/2} \leq Z \leq -z_{\alpha/2}]$.

(Your answers should be expressed in terms of p or α. Numerical answers cannot be given here.)

23. If $X_1, X_2, \ldots, X_{25}$ is a random sample from a normal distribution having unknown mean μ and σ^2 known to be 100, and if $\bar{X}$ equals 15 for this sample, find (a) a 90% confidence interval for μ; (b) a 95% confidence interval for μ.

24. A random sample of 100 accounts from persons holding Bank Americharge credit cards was taken. The (sample) mean balance was \$40.72. If the standard deviation is known to be $\sigma = \$15$, find a 95% confidence interval for the mean balance of all Bank Americharge accounts.

25. In order to estimate the mean monthly expenditures for food for a family of four living in the city of Eatintown, a random sample of 50 families was chosen. If it is known that the standard deviation of such expenditures is \$20 and if the *total* expenditures for the 50 families interviewed was \$10,600, find a 90% confidence interval for the mean monthly food expenditures for families of four. (*To think about:* Would this estimate be valid for families of four living in your hometown? Why or why not?)

26. The grade-point averages for men enrolled in a marketing course have a standard deviation of .50. A random sample of 31 women students enrolled in this course was taken and the mean grade-point average was found to be 2.98. Find a 98% confidence interval for the mean grade-point average for all women enrolled in this course. What assumptions did you make?

27. Let T_ν represent a random variable having a Student's t distribution with ν degrees of freedom. Using the notation $P[T_\nu \leq t_{\nu,p}] = p$, find from the tables

(a) $t_{8,.90}$; (b) $t_{15,.95}$;
(c) $t_{\infty,.99}$; (d) $t_{6,.10}$;
(e) $t_{20,.25}$; (f) $t_{40,.40}$.

28. Let T_ν represent a random variable having a Student's t distribution with ν degrees of freedom. Using the tables, find

(a) $P[T_{23} \leq 2.5]$; (b) $P[T_{60} > 2.0]$;
(c) $P[T_1 < -1.0]$; (d) $P[-1.86 \leq T_8 \leq 1.86]$;
(e) $P[-2.11 \leq T_{17} \leq 2.11]$; (f) $P[-2.66 \leq T_{60} \leq 2.66]$.

29. In the random sample of 31 women students referred to in Exercise 26, the sample standard deviation was found to be .56. Use this information to construct a 98% confidence interval for the mean grade-point average for all women enrolled in this course. What assumptions did you make?

★30. A fourth-grade class at the Lotta Noyes Elementary School consists of 14 boys and 6 girls. After the children went to the school nurse's office for weighing and measuring, their teacher found that the mean heights were 54.64 and 54.66 inches, with standard deviations of 2.32 and 2.64 inches for the boys and girls, respectively. Treating these data as a random sample of all fourth graders in this city, find 95% confidence intervals for the mean height of all fourth-grade (a) boys, (b) girls in the city.

★31. The mean weights of the 20 fourth-grade children described in Exercise 30 were 80.21 and 79.30 pounds, with standard deviations of 15.56 and 10.37 pounds, respectively, for boys and girls. Find approximate 98% confidence intervals for the mean weight of all fourth-grade (a) boys; (b) girls.

32. The length of any confidence interval is found by subtracting the lower end point from the upper end point. (a) Using (7.9), find an expression for the length of a $1 - \alpha$ confidence interval for the mean μ of a normal distribution when σ is known. (b) Using (7.14), find an expression for the length of a $1 - \alpha$ confidence interval for the mean μ of a normal distribution when σ is unknown. (c) Which of these lengths will tend to be larger? Why?

33. Compare the lengths of the confidence intervals found in Exercise 31(a) with those found in Exercise 31(b). What reasons can you give for the differences?

34. Good and Durdy, producers of a potting soil for house plants, want to estimate the difference in sales when the soil is packaged in boxes instead of plastic bags (each containing 3 kilograms of soil). Seven retailers in a certain area displayed the soil using boxes and plastic bags on the same shelf. Using the given results, find a 90% confidence interval for the mean difference in sales.

Retailer:	1	2	3	4	5	6	7
Boxes sold:	10	23	12	16	26	17	30
Bags sold:	28	16	18	29	23	20	31.

35. The following numbers represent the daily sales (in thousands of dollars) of stores having both a downtown and shopping center location. The stores are comparable, having approximately the same number of square feet of floor space. Find a 98% confidence interval for the mean difference between shopping center and downtown location sales.

	S&R	M.W.	JCP	K.M.	W.C.
Shopping center location:	15.2	17.0	9.4	12.8	12.6
Downtown location:	13.1	12.3	11.9	14.5	10.2.

36. After receiving a failing grade on the second hourly exam in Econ. 101, a student complained to Professor Hardknows that the exam was unfair. The student claimed that "almost everyone" did worse on the second exam than on the first exam. Professor Hardknows asked eight of the students (out of a class of 500) who were still in the classroom how they scored on the exams. The scores were as follows:

	Student							
	Jay	*Gary*	*Joe*	*Steve*	*Bob*	*Nora*	*Mark*	*Karen*
Score on exam 1	59	69	54	77	70	59	85	92
Score on exam 2	72	84	52	80	66	94	78	86

Find a 95% confidence interval for $\mu_1 - \mu_2$, where μ_i denotes the mean score on the ith exam. (*To think about:* Do the data support the student's claim that almost everyone did worse on the second exam? If it did, would it follow that the exam was unfair? Did Professor Hardknows obtain a random sample of students?)

37. Independent random samples of size $n_1 = 10$ and $n_2 = 20$ are taken from normal populations having means μ_1 and μ_2 and variances $\sigma_1^2 = 90, \sigma_2^2 = 140$. Given that the sample means are observed to be $\bar{x}_1 = 87$ and $\bar{x}_2 = 62$, find a 90% confidence interval for $\mu_1 - \mu_2$.

38. Independent random samples of size $n_1 = 9$ and $n_2 = 16$ are taken from normal populations having means μ_1 and μ_2 and a common variance, $\sigma_1^2 = \sigma_2^2 = 144$. Given that the sample means are observed to be $\bar{x}_1 = 36$ and $\bar{x}_2 = 38$, find a 95% confidence interval for $\mu_1 - \mu_2$.

39. Show that if samples are taken from two populations with common variance σ^2, the pooled sample variance S_p^2 is an unbiased estimator of σ^2.

40. The manufacturer of heart pacemakers is interested in knowing the difference in heart rates of adult males and females. It is assumed that the variances are the same for males and females. Ten males and 10 females took their (resting) heart rate with the results given in the following table. Use these results to find a 99% confidence interval for $\mu_F - \mu_M$, the mean difference in the heart rates of adult males and females. What assumptions have you made?

Female:	82	68	82	80	60	94	64	80	70	60
Male:	70	60	60	74	72	80	56	60	58	70.

★**41.** Using the information given in Exercise 30, find an 80% confidence interval for the mean difference in height between fourth-grade girls and boys. Find the interval under the assumption that (a) $\sigma_B^2 = \sigma_G^2$; (b) $\sigma_B^2 \neq \sigma_G^2$.

★**42.** Using the information given in Exercise 31, find an approximate 90% confidence interval for the mean difference in weight between fourth-grade girls and boys. Find the interval under the assumption that (a) $\sigma_B^2 = \sigma_G^2$; (b) $\sigma_B^2 \neq \sigma_G^2$.

43. Let χ_ν^2 represent a random variable having a chi-square distribution with ν degrees of freedom. Using the tables, find

(a) $P[\chi_{13}^2 \leq 9.3]$; (b) $P[\chi_{50}^2 > 28.0]$;
(c) $P[\chi_{24}^2 \leq -19.0]$; (d) $P[\chi_{11}^2 \leq 13.7]$;
(e) $P[\chi_{12}^2 \geq 28.3]$; (f) $P[9.59 \leq \chi_{20}^2 \leq 34.17]$.

44. Let χ_ν^2 represent a random variable having a chi-square distribution with ν degrees of freedom. Using the notation $P[\chi_\nu^2 \leq \chi_{\nu,p}^2] = p$, find from the tables

(a) $\chi_{5,.01}^2$; (b) $\chi_{10,.05}^2$;
(c) $\chi_{20,.25}^2$; (d) $\chi_{20,.75}^2$;
(e) $\chi_{30,.95}^2$; (f) $\chi_{60,.99}^2$.

45. Let $X_1, X_2, \ldots, X_{21}$ be a random sample from a normal distribution having a known variance, $\sigma^2 = 100$. Find the probability that the sample variance will be between (a) 77 and 120; (b) 54 and 157.

46. In comparing the lengths of confidence intervals for the mean of a normal distribution when σ is known or unknown, it can be shown that the length will be longer when σ is unknown provided that

$$S^2 > \left(\frac{z_{1-\alpha/2}}{t_{n-1,\,1-\alpha/2}}\right)^2 \sigma^2.$$

(a) Show that this statement is true (*Hint:* See Exercise 32.) (b) Show that when $n = 4$ and $1 - \alpha = .95$, the probability that the length of the confidence interval when σ is unknown will be longer than the length when σ is known is greater than .75.

★**47.** Using the information given in Exercise 30 concerning the heights of fourth-grade children, find a 95% confidence interval for the variance of heights for (a) boys; (b) girls.

★**48.** Using the information given in Exercise 31 concerning the weights of fourth-grade children, find an approximate 90% confidence interval for the variance of weights for (a) boys; (b) girls.

49. A sample of 81 cups of soda were taken from a machine designed to dispense 7.50 ounces into an 8-ounce capacity cup. The sample standard deviation was found to be .35 ounce. Find a 95% confidence interval for σ. What might you think about a manufacturer's claim that σ is equal to .20 ounce?

50. Let F_{ν_1,ν_2} represent a random variable having an F distribution with ν_1 and ν_2 degrees of freedom. Using the tables, find

(a) $P[F_{5,5} \leq 5.05]$;
(b) $P[F_{60,27} \geq 2.00]$;
(c) $P[F_{20,18} \leq 3.50]$;
(d) $P[F_{21,8} \leq .25]$;
(e) $P[F_{18,8} \leq .33]$;
(f) $P[F_{20,7} \leq .40]$.

51. Let F_{ν_1,ν_2} represent a random variable having an F distribution with ν_1 and ν_2 degrees of freedom. Using the notation $P[F_{\nu_1,\nu_2} \leq f_{\nu_1,\nu_2,p}] = p$, find from the tables

(a) $f_{4,6,.95}$;
(b) $f_{4,6,.05}$;
(c) $f_{8,10,.975}$;
(d) $f_{10,8,.025}$;
(e) $f_{12,12,.995}$;
(f) $f_{12,12,.005}$.

52. A random sample $X_1, \ldots, X_5$ is taken from a normal distribution having $\sigma_X^2 = 100$ and an independent random sample $Y_1, \ldots, Y_{10}$ is taken from a normal distribution having $\sigma_Y^2 = 100$. Find the probability that the ratio of sample variances, S_X^2/S_Y^2, will lie between .166 and 3.63.

53. Consider the same situation as in Exercise 52, but assume that $\sigma_Y^2 = 400$. Find the probability that the ratio of sample variances, S_X^2/S_Y^2, will be (a) less than or equal to 1.18; (b) less than or equal to .064.

★***54.*** Using the information given in Exercise 30 concerning the heights of fourth-grade children, find a 90 percent confidence interval for the ratio of variances of the heights of boys and girls, σ_B^2/σ_G^2.

★***55.*** Using the information given in Exercise 31 concerning the weights of fourth-grade children, find an approximate 95% confidence interval for the ratio of variances of the weights of boys and girls, σ_B^2/σ_G^2.

56. Find a 90% confidence interval for p, the probability of a "success" if 64 successes were observed in 100 trials.

★***57.*** A certain type of thumbtack was dropped 100 times and landed "point up" 55 times. Find a 95% confidence interval for the probability that this type of thumbtack will land point up when dropped.

58. The week before an election a candidate for the legislature, Denny Krat, takes a poll to estimate the number of voters in his district who plan to vote for him. Mr. Krat finds that 212 out of 400 randomly chosen voters plan to vote for him. On the basis of this poll, should Mr. Krat feel that he has the election well in hand? Explain your reasoning.

59. The personnel manager at a large manufacturing plant, Mr. Hiram N. Firem, is concerned about absenteeism. In examining records he finds that 20 of a sample of 100 workers were absent 1 or more days last month. Find a 98% confidence interval for the proportion of workers who were absent 1 or more days last month. (*To think about:* Would this be a good indication of the rate of absenteeism that might be expected 6 months from now? Why?)

60. A bank official wishes to estimate the proportion of checking accounts that have an error on their monthly statement. A random sample of 100 accounts is examined and 4 are found to contain errors. (a) Find a 99% confidence interval for this proportion using the usual normal approximation. (b) Are the results obtained reasonable? Why or why not?

• *61.* Using methods of calculus show that the quantity $p(1 - p)$ achieves a maximum at the value $p = \frac{1}{2}$.

62. In an effort to determine the difference in the proportion of men and women $(p_M - p_W)$ favoring a proposed constitutional amendment, 100 men and 100 women are interviewed. If 60 men and 55 women favor the amendment, find a 90% confidence interval for the difference $p_M - p_W$. (*To think about:* Would this result be valid if 100 married couples were interviewed? Why?)

63. Mr. Firem (Exercise 59) wishes to compare the rate of absenteeism for "new workers" and "old timers." A worker is considered to be "new" if he has worked for the company 1 year or less. Six of 36 new workers and 16 of 64 old workers were absent 1 or more days last month. Find a 95% confidence interval for the difference $p_N - p_O$.

64. The Watzit Manufacturing Company has the option of buying one of two machines for producing a new line of watzits. The first machine costs $10,000 more than the second machine. All else being equal, the second machine will be purchased. One hundred watzits are produced by each machine with 5 from the first and 8 from the second machine being defective. (a) Find a 98% confidence interval for the difference in proportions, $p_1 - p_2$. (b) What machine do you think should be purchased? Why?

65. The marketing director of the Water's Salt Company wishes to learn about the packaging preference of purchasers of Water's Salt. In a supermarket the salt is sold in rectangular boxes and in the same supermarket a month later the salt is sold in cylindrical containers. The price remained the same as that of other brands. Of 280 customers buying salt the first month, 28 bought Water's Salt, while 77 of 350 customers bought Water's Salt the second month. What conclusions might the marketing director draw?

66. To get a point estimate of a mean μ of a normal distribution having a variance $\sigma^2 = 50$, how large a sample should be taken if the point estimate is to be within 4 units of μ with probability .90?

67. Referring to Exercise 25, how many families should be interviewed if it is desired to have a 90% confidence interval for the mean monthly expenditure of food for a family of four which has a length of $5?

68. How large of a sample should be taken in order to be 98% confident that the sample proportion will be within .03 of a true proportion p?

69. A pollster, George Giddyup, wants to estimate the proportion of people in favor of the resignation of a controversial cabinet member. How many people should be interviewed to be 90% confident that the sample proportion will be (a) within .02 of the true population proportion?; (b) within .01?

★*70.* A news article gave the results of a poll on presidential performance. At the end of the article it stated "As with any sample survey, the results of the AP-NBC News polls could differ from the results of interviews with all Americans because of chance variations in the sample. For polls with 1600 interviews, the results should vary no more than three percentage points either way because of sample errors. That is, if 20 polls were conducted with the same questionnaire, the results would vary from these results by no more than three percentage points at least 19 times." Comment on this statement.

71. In Exercise 38 you found a 95% confidence interval for $\mu_1 - \mu_2$. (a) What is the length of the confidence interval? (b) What would the length be if n_1 were 10 and n_2 were 15? (c) What would the length be if n_1 were 12 and n_2 were 13?

72. Say you are interested in finding a 90% confidence interval for the difference $\mu_1 - \mu_2$ and that you know the values of the variances (cf. Exercise 37). In particular, assume that $\sigma_1^2 = (10)^2$ and $\sigma_2^2 = (40)^2$. Assume that you have money enough to take a total of 50 observations which must be divided between the two groups. What will the length of the confidence interval be if you divide the 50 observations so that (a) $n_1 = 10$, $n_2 = 40$; (b) $n_1 = 20$, $n_2 = 30$; (c) $n_1 = 25$, $n_2 = 25$? If, instead, you know that $\sigma_1^2 = (20)^2$ and $\sigma_2^2 = (30)^2$, what will the length of the confidence interval be if (d) $n_1 = 10$, $n_2 = 40$; (e) $n_1 = 20$, $n_2 = 30$; (f) $n_1 = 25$, $n_2 = 25$? (*To think about:* Can you see a relationship between the sample size and standard deviation in finding confidence intervals having shortest length?)

73. In each of the following situations, give one or more methods for subdividing the population into strata.

(a) An agricultural economist wishes to sample farms in a given state to determine the average yield/acre of corn.

(b) A counselor wants to find the average number of hours worked per week on outside jobs by students at a large university.

(c) The chairman of a large chain of stores wants to estimate the average hourly wage of employees.

74. Ann Landers frequently asks readers to send in their opinions on a given topic (e.g., "If you had it to do over again, would you marry the same person?"). The results of these opinions are published in a future column. Are these opinions based on a random sample? A stratified random sample? A representative sample? What conclusions could be drawn from such results?

Use the following information in Exercises 75 to 80. The town of Largerton has 1000 families living in one of three subdivisions. Five hundred families live in Sunrise Estates (located near the town's elementary school), 300 families live in Rock Terrace (located near the high school), and 200 families live in Sunset Park. Assume that you are interested in estimating the mean annual family income. The overall standard deviation of the annual family income (in thousands of dollars) is 12. Assume further that you plan to sample 100 families. Let Sunrise Estates, Rock Terrace, and Sunset Park be strata 1, 2, and 3, respectively.

75. If a single simple random sample is to be taken, $\bar{X}$ can be used as an estimator of μ. What is the standard deviation of this estimator if the sample is taken (a) with replacement?; (b) without replacement?

76. If the cost of sampling is the same within each subdivision, (a) how should the subsamples be allocated to the three subdivisions if $\sigma_1 = 3$, $\sigma_2 = 8$, and $\sigma_3 = 6$? (b) Find the standard deviation of the stratified estimator of μ using the allocation found in (a) if sampling is done with replacement. (c) What would it be if sampling were done without replacement?

77. If the cost of sampling is the same within each subdivision, (a) how should the subsamples be allocated to the three subdivisions if $\sigma_1 = \sigma_2 = \sigma_3 = 6$? (b) Find the standard deviation of the stratified estimator of μ using the allocation found in (a) if sampling is done with replacement. (c) What would it be if sampling were done without replacement? [*To think about:* Would your answer to (a) change if $\sigma_1 = \sigma_2 = \sigma_3 = 8$? 10?]

78. If a simple random sample of size n_i is taken from stratum i and if the strata sample mean incomes (in thousands of dollars) are found to be $\bar{X}_1 = 15$, $\bar{X}_2 = 19.2$, and $\bar{X}_3 = 21.3$, find $\hat{\mu}_{\text{strat}}$.

79. If the costs of sampling within the strata are $c_1 = \$1.00$/unit, $c_2 = \$1.69$/unit, and $c_3 = \$1.21$/unit, (a) how should the subsamples be allocated to the three subdivisions if $\sigma_1 = 3$, $\sigma_2 = 8$, $\sigma_3 = 6$? (b) Find the standard deviation of the stratified estimator of μ using the allocation found in (a) if sampling is done with replacement. (c) What would it be if sampling were done without replacement? (d) What will be the total cost of sampling the 100 units? (e) How many units could be sampled if the total budget for sampling is \$100?

80. If the costs of sampling within the strata are $c_1 = \$1.00$/unit, $c_2 = \$1.69$/unit, and $c_3 = \$1.21$/unit, (a) how should the subsamples be allocated to the three subdivisions if $\sigma_1 = \sigma_2 = \sigma_3 = 6$? (b) Find the standard deviation of the stratified estimator of μ using the allocation found in (a) if sampling is done with replacement. (c) What would it be if sampling were done without replacement? (d) What will be the total cost of sampling the 100 units? (e) How many units could be sampled if the total budget for sampling is \$100?

8

PARAMETRIC TESTS OF HYPOTHESES

8.1 INTRODUCTION

Almost every package you see in a grocery store (and in many other types of stores as well) has a statement on it concerning quantitative aspects of the contents. A box of Cheerios may be labeled "15 ounces," a box of Kleenex may say "200 tissues," a bottle of Coke may be labeled "1 liter," and so on. You could no doubt add many other products to this list with very little effort. Now ask yourself whether these statements are all *precisely* true. If you remember some of our earlier discussions concerning continuous quantities such as weight or volume, you will answer this question in the negative. You tend to forget this for products that are canned, but it is easier to see this for products that are bottled. If you look at a carton of bottles of soda pop, you will see slightly different levels in the bottles. Of course, they do not all contain exactly the advertised amount of 16 ounces, say. If not, then what does the statement on the carton mean? One reasonable interpretation would be that the *average contents* is equal to the advertised amount. This kind of reasoning also holds for packages that have contents measured by count rather than weight or volume, such as tissues, paper towels, vitamin tablets, and so on. If the actual contents of these packages are random variables, the statement of their contents may be interpreted as a statement about the mean of the distribution of these random variables, or perhaps about some other aspect of their distribution. Such statements may be called statistical hypotheses.

Definition 8.1 A *statistical hypothesis* is a statement about the probability distribution of a random variable.

In many cases a statistical hypothesis is a statement about a parameter of the probability distribution of a random variable. If we define the random variable X to be the weight of a randomly chosen box of Crunchy-Munchy cereal, each of the following statements is a statistical hypothesis:

X has a normal distribution.
X has a normal distribution with mean $\mu = 16$ ounces.
X has a mean $\mu = 16$ ounces.
X has a normal distribution with standard deviation $\sigma = .25$ ounce.
X has a standard deviation smaller than .25 ounce.
X has a normal distribution with $\mu = 16$ ounces and $\sigma = .25$ ounce.

You can see that some of these hypotheses are rather general while others are more specific.

We all know, of course, that the mere fact that a statement is put into print does not guarantee that the statement is true. Consequently, just because a box of Crunchy-Munchy cereal is labeled "16 ounces," it does not necessarily follow that the box will actually contain 16 ounces. Most of us have had the experience of buying a package that we have found to be half-empty, so our experience tells us that we may not always get the advertised amount. If we think of the advertised amount as a statistical hypothesis (e.g., "the mean is 16 ounces"), we can see that this hypothesis need not be true. As a matter of fact, virtually any statistical hypothesis can be false. Since this is the case, we might like to have a statistical hypothesis that is true to replace the false one. One way to do this is to make an alternative statement which is the logical complement to the original hypothesis. For example, the logical alternative to "X has a normal distribution" is "X does not have a normal distribution." The logical alternative to "X has a mean $\mu = 16$ ounces" is "X has a mean $\mu \neq 16$ ounces." We can denote the original hypothesis by H_0 (read "*h* naught" or "*h* zero" and called the null hypothesis) and the alternative hypothesis by H_a (some authors use H_1 instead of H_a). There are ways of defining an alternative hypothesis other than the logical alternative, but in most cases we consider the logical alternative to be adequate. Now, given a null and an alternative hypothesis, both will not be true simultaneously. It would be desirable to have an objective method to allow us to decide whether to believe that the null hypothesis or the alternative hypothesis is true. The development of such methods is the purpose of this chapter.

There are some packages which are labeled with statements that are, by our definition, statistical hypotheses. A package of standard 60-watt light bulbs that we bought was labeled "avg. life 1000 hours/avg. light output 855 lumens" (these values are about the same for most major brands). A package of "long-life" 60-watt light bulbs we bought was labeled "1500 avg. hours" and "608 avg. lumens." Now consider a situation where we have a bulb taken from one of these two packages but because of the printing on the top of the bulb being smeared, say, we cannot tell by looking at it whether it is a standard or a long-life bulb. What simple experiment could we perform to try to make an educated guess as to its type? We could either measure its lifetime or its light output. Since we are more apt to have instruments for measuring time (e.g., clock and/or calendar) than for measuring light output, we could measure the value of X, the lifetime of this bulb. Basically, we would like to distinguish between the following hypotheses:

$$H_0 : \mu = 1000 \qquad H_a : \mu = 1500$$

where μ is the mean of the random variable X. How would you use the observed value of X (i.e., the observed lifetime) to decide whether the null or alternative hypothesis is more reasonable? Which would you guess to be true if the lifetime were observed to be 900 hours? 1600 hours? 1125 hours? 1350 hours? Most people in this situation would guess H_0 to be true if X were observed to be less than 1250 hours and guess H_a to be true if X were greater than or equal to 1250 hours. In the absence of any other information, this would seem to be a reasonable procedure for deciding between these two hypotheses. This illustrates the general idea of testing hypotheses which we will develop as we go through this chapter: State the null and alternative hypotheses; choose a random variable whose observed value will be used to decide whether to believe the null or alternative hypothesis; choose a region such that the null hypothesis will be rejected if and only if the observed value of the random variable falls in that region; actually observe the random variable and make the decision called for by the agreed-upon rule.

8.2 TYPE I AND TYPE II ERRORS

As you are reading through this book, you should be getting an understanding of the nature of random variables. Implicit in the very name "random variable" is the idea that their values cannot be perfectly predicted. Although you may be able to make a statement such as "90% of the time the observed value of a random variable X will lie between a and b," you also know that the other times the values will be somewhat smaller or larger than these values. In view of this unpredictability, when we depend on the value of a random variable for deciding between a null and an alternative hypothesis, we must be prepared to be led astray by an extreme value. To see how we might be led astray, let us make some further assumptions about the distribution of X, the lifetime of a 60-watt light bulb mentioned in the previous section. In particular, let us assume that the lifetime is normally distributed with a standard deviation of 200 hours. (We must emphasize that this assumption of normality is being made because of the ease of finding probabilities from a normal distribution. For pedagogical reasons, therefore, this is a convenient assumption. The actual distribution *may not be* normal. In Chapter 11 we will discuss methods whereby hypotheses about the distribution of a random variable can be tested.) Using the hypotheses

$$H_0 : \mu = 1000 \qquad H_a : \mu = 1500$$

stated in the previous section, we can see that if the null hypothesis is true, the mean will be 1000 hours, whereas it will be 1500 hours if the alternative hypothesis is true.

In Figure 8.1(a) we have shown the distribution of X when H_0 is true, and in Figure 8.1(b) we have the distribution when H_a is true. As you have learned from your study of the normal distribution, most of the time the values of X will fall within 2 standard deviations of the mean. Consequently, most values will fall between 600 and 1400 when the value of the mean is actually 1000. This does not bode well, however, when we think of the procedure we were going to use to make an educated guess as to whether H_0 or H_a is true. We said that

$$\text{if } X \text{ is observed} < 1250 \text{ hours, believe } H_0\ (\mu = 1000) \text{ true, but}$$

$$\text{if } X \text{ is observed} \geq 1250 \text{ hours, believe } H_a\ (\mu = 1500) \text{ true.}$$

You can see from Figure 8.1(a) that it is quite possible for μ to actually be equal to 1000 but for X to be greater than or equal to 1250. In such a case we would make an error by wrongly believing H_a to be true. We could be led astray by the random variable X taking on a value somewhat larger than usual. You can also see from Figure 8.1(b) that it is possible for μ to actually be equal to 1500 (i.e., for H_a to be true) but for X to be less than 1250. In this case an error would be made by wrongly believing H_0 to be true. Whenever we depend on the value of a random variable to help us decide which of two statistical hypotheses is apt to be true, we must be prepared for the possibility of making a wrong judgment.

In the example we have been discussing, the random variable X, the lifetime of the light bulb, has been used to help decide between two statistical hypotheses. We will refer to any procedure used to decide which of two statistical hypotheses is true as a *statistical test of hypotheses*. A random variable whose observed value is used to decide which of two statistical hypotheses should be accepted as true will be called a *test statistic*. The random variable X, the lifetime of the light bulb, is an example of a test statistic.

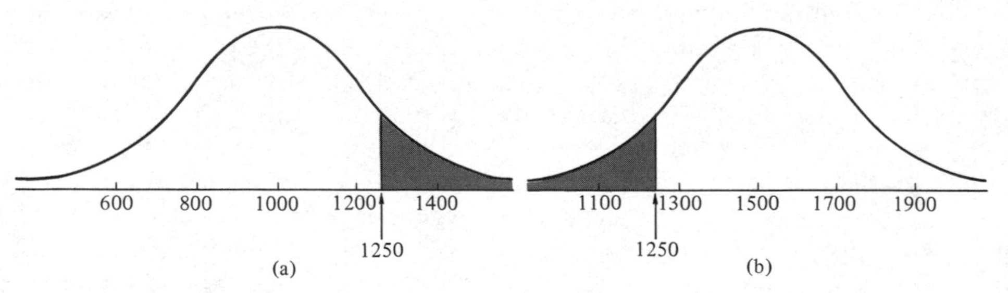

FIGURE 8.1

Definition 8.2 The *critical region* of a statistical test of hypotheses is the set of all values of the test statistic which calls for the rejection of the null hypothesis.

In the light-bulb example, the critical region is defined by the set

$$C = [1250, \infty).$$

In almost all the tests we consider the critical region will be an interval of values or a union of two intervals. We frequently refer to the point(s) that separate the critical region and noncritical region as the *critical value*(*s*). Here the value 1250 would be the critical value.

We saw earlier that errors can be made in decisions made by following a statistical test procedure. Formally, we distinguish between two types of errors.

Definition 8.3 An error made by rejecting a null hypothesis when the null hypothesis is in fact true is called a *type I error*.

Definition 8.4 An error made by accepting a null hypothesis when the alternative hypothesis is in fact true is called a *type II error*.

In a statistical test of hypotheses there are two ways a correct decision can be made as well as two ways that an error can be made. (Of course, if a statistical test of hypotheses is worth its salt, there will be more times when a correct decision is made than a wrong decision!) These four possibilities are shown in Figure 8.2.

Now ask yourself when a type I error will be made in testing the hypotheses about the mean lifetime of the light bulbs. A type I error will be made if X is greater than or equal to 1250 hours when μ is actually 1000 hours. In our prior discussion we pointed out that it is *possible* to make a type I error, but we can be even more precise and find the *probability* of making a type I error. The probability will be equal to the shaded area in Figure 8.1(a). Similarly, the shaded area in Figure 8.1(b) corresponds to the probability of making a type II error. The probability of making a type I error will be denoted by α, while the probability of making a type II error will be denoted by β.

		True State of Affairs	
		H_0 true	*H_a true*
Decision	*Accept H_0 as true*	OK	Type II error
	Reject H_0	Type I error	OK

FIGURE 8.2

Example 8.1 Find the values of α and β if the hypotheses $H_0 : \mu = 1000$ and $H_a : \mu = 1500$ are tested by the criteria stated earlier in this section.

We can get a rough idea of the value of α (or β) by studying the graphs in Figure 8.1, but to get the exact answer we must proceed as follows:

$$\begin{aligned}\alpha &= P[\text{type I error}] = P[\text{reject } H_0 \,|\, H_0 \text{ true}]\\ &= P[\text{test statistic is in the critical region} \,|\, H_0 \text{ true}]\\ &= P[X \geq 1250 \,|\, H_0 \text{ true}]\\ &= P[X \geq 1250 \,|\, \mu = 1000].\end{aligned}$$

Since we have assumed X is normally distributed with $\sigma = 200$, we find the value of α by standardizing X and then using the standard normal tables:

$$P[X \geq 1250 \mid \mu = 1000] = P\left[\frac{X - \mu}{\sigma} \geq \frac{1250 - 1000}{200}\right] = P\left[Z \geq \frac{250}{200} = 1.25\right]$$
$$= 1 - .8944 = .1056$$

(i.e., $\alpha = .1056$). Similarly, we find β by

$$\begin{aligned}\beta &= P[\text{type II error}] = P[\text{accept } H_0 \mid H_a \text{ true}] \\ &= P[\text{test statistic is not in the critical region} \mid H_a \text{ true}] \\ &= P[X < 1250 \mid \mu = 1500] \\ &= P\left[\frac{X - \mu}{\sigma} < \frac{1250 - 1500}{200}\right] = P[Z < -1.25] = .1056.\end{aligned}$$

You should be able to see that the values of α and β are determined by the choice of the critical region. In Figure 8.3 are graphs of the distribution of X when (a) $\mu = 1000$ and (b) $\mu = 1500$. The critical region has been changed to values of X which are 1300 or more. The shaded regions correspond to α and β. What happens to the relative values of α and β when the critical value is increased? What would happen to these values if the critical value were decreased to 1150 or 1200, say? You can see that as one goes up, the other goes down.

Since it is impossible to make both α and β smaller simultaneously by simply changing the critical value, we might want to consider which of the two should be made smaller or if they should be made equal. The answer to such a question would depend on the consequences of each type of error. For example, if the consequences of committing a type I error were more severe than those of committing a type II error, we would want to make α smaller than β. In the light-bulb example the situation is somewhat artificial and there are no obvious consequences of making a wrong decision. The next examples illustrate situations where the consequences of the two types of errors are quite different.

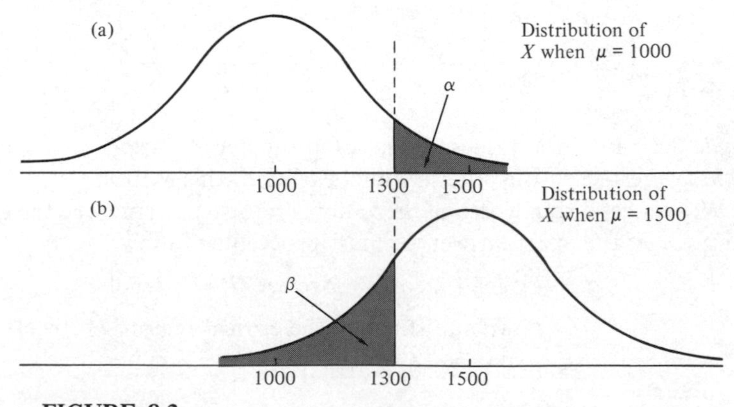

FIGURE 8.3

Example 8.2 (*Criminal Law*) If a person is accused of a crime and brought to trial, the jury must decide between two hypotheses,

$$H_0 : \text{Defendant is innocent} \qquad \text{and} \qquad H_a : \text{Defendant is guilty.}$$

What are the consequences of a type I error? of a type II error?

We can write the consequences of each type of decision in a table such as the one in Figure 8.2. This table is given in Figure 8.4. You can see that the consequence of a type I error is the conviction of an innocent person, and the consequence of a type II error is the acquittal of a guilty person.

		True State of Affairs	
		H_0 true (defendant innocent)	*H_a true (defendant guilty)*
Decision Made by the Jury	*Accept H_0 (find defendant innocent)*	Defendant acquitted (correct decision)	Defendant acquitted (type II error)
	Reject H_0 (find defendant guilty)	Defendant convicted (type I error)	Defendant convicted (correct decision)

FIGURE 8.4

In the example of criminal law, a type I error, the conviction of an innocent person, is considered by our society to be more severe than a type II error. Although we would ideally like for both of these errors to have zero probability, we know this is impossible. To make the probability of committing a type I error small, the trial is to be conducted under the assumption that the person is innocent (i.e., H_0 is true) unless there is *convincing* evidence to the contrary. This same convention has been adopted for the testing of statistical hypotheses. The null and alternative hypotheses are generally set up so that (1) the consequences of a type I error are the more severe, and (2) the null hypothesis is assumed to be true unless there is convincing evidence to the contrary. [Having said that, we must now qualify remark (1). For one thing, this assumes that the consequences of each type of error are known. For another, the severity of the consequences of an error may be viewed differently by different parties. We will discuss the idea of "point of view" further as we proceed.]

Example 8.3 (*The Picture-Happy Tourist*) While vacationing in Yellowstone National Park, Westman Kodiak sees a very young bear cub. He would like to get a memorable picture for the folks back home and decides that the picture will be more memorable if it has his daughter Polly feeding the cub. Before sending Polly over to the cub, he remembers that mother bears sometimes look askance at human contact with their cubs. If Westman sets up

$$H_0 : \text{mother bear is nearby} \qquad \text{and} \qquad H_a : \text{mother bear is not nearby}$$

(see Figure 8.5), what are the consequences of a type I and type II error?

FIGURE 8.5

In this case we can also make a table similar to the one given in Figure 8.2. As you look at this table (Figure 8.6), try to understand the logic behind each entry in the table. In this situation a type I error could have serious consequences for Polly while a type II error has relatively mild consequences. (Do you see the reason for the question mark after the "correct decision" of rejecting H_0 when H_a is actually true? While this decision is correct from the point of view of the hypotheses set up, it does not take into account the park rules about not bothering the animals!)

		True State of Affairs	
		H_0 true (Mother is nearby)	*H_a true (Mother is not nearby)*
Decision Made by Westman	*Accept H_0 (believe Mother is nearby)*	Take picture of cub alone (correct decision)	Take picture of cub alone—miss a memorable picture (type II error)
	Reject H_0 (believe Mother is not nearby)	Send Polly by cub—anger Mother (type I error)	Send Polly by cub—get picture [correct decision (?)]

FIGURE 8.6

You will notice that in this example the null and alternative hypotheses were set up so that the consequences of a type I error are more severe than those of a type II error. If Westman is rational, he will act as if H_0 is true unless there is convincing evidence to the contrary.

In considering Example 8.1 and in the ensuing discussion, we saw that α and β will both change as the critical value changes. As α decreases, β increases, and vice versa. It is possible, then, by choosing the critical region appropriately, to make the value of α as small as you like. Unfortunately, this is done at the expense of making β larger, perhaps larger than is tolerable. [In the example of criminal law (Example 8.2) it is possible to make $\alpha = 0$ by finding all defendants innocent. If this were done, what would happen to the value of β? Is this tolerable?]

There is a way of making both α and β smaller simultaneously in tests of statistical hypotheses. The way that this can be done is to gather more information, that is, to take a larger sample size. In Example 8.1 we assumed that there was only one light bulb available for testing. We might assume that we have a large shipment of light bulbs, say 100 gross, which are unmarked and are either standard bulbs or extra life bulbs having a mean lifetime of 1200 hours (not to be confused with the "long-life" bulbs described earlier). In a situation like this, we certainly have the option of testing a larger number of bulbs. If we were to test n light bulbs and measure the lifetimes $X_1, X_2, \ldots, X_n$ of each, what would we use as a test statistic? Since we are interested in the mean μ, we would use the sample mean $\bar{X}$ as the test statistic. This illustrates a rule of thumb that is generally safe to follow:

> In testing hypotheses about a parameter θ, the sample counterpart of θ will generally be a good test statistic.

Another way of expressing this rule is to say that a test statistic for a parameter θ will generally be the same statistic used as a point estimate of θ. For example, $\bar{X}$ would be used for tests about μ, S^2 for tests about σ^2, X/n for tests about p, and so on.

We would now like to investigate the interrelationships among α, β, the critical region, and the sample size n. Let us add some more assumptions to the situation just described, where we have a large shipment, 100 gross, of unmarked light bulbs which are either standard bulbs ($\mu = 1000$) or "extra life" bulbs ($\mu = 1200$). Assume that standard bulbs can be sold for 50 cents each while extra life bulbs can be sold for 60 cents. If we believe μ to be 1000 hours when the actual value is 1200 hours, we lose a potential profit of

$$(10 \text{ cents/bulb}) \times 100 \times 144 \text{ bulbs} = \$1440.$$

On the other hand, if we believe μ to be 1200 hours and sell the bulbs as "extra life" bulbs when the actual value is 1000 hours (i.e., the bulbs are in fact standard bulbs), we risk a lawsuit and the loss of goodwill of unhappy customers. The consequences of these decision errors are quite different. Furthermore, it is difficult to assign a dollar value to "loss of goodwill." Nevertheless, let us assume that the management decides that the latter error is more serious. In this case we would set up

$$H_0 : \mu = 1000 \text{ hours} \quad \text{and} \quad H_a : \mu = 1200 \text{ hours}. \tag{8.1}$$

Since the cost of sampling light bulbs is not as large as that of sampling very expensive items, for example space rockets, there is no problem in taking several observations. We might want to find the sample size n and critical value c to obtain specified values of α and β.

Example 8.4 Find the sample size n and critical value c necessary to make $\alpha = .001$ and $\beta = .005$ in testing the hypotheses (8.1). Assume that the lifetimes of light bulbs are normally distributed with $\sigma = 200$ hours.

As indicated earlier, we will use $\bar{X}$, the sample mean, as a test statistic. Since the individual lifetimes are normally distributed, it follows that $\bar{X}$ will be normally distributed. Intuitively, we would reject H_0 only if $\bar{X}$ is sufficiently large, say $\bar{X} \geq c$. Using the definitions of α and β, we have

$$\alpha = .001 = P[\text{reject } H_0 \mid H_0 \text{ true}] = P[\bar{X} \geq c \mid \mu = 1000] \tag{8.2}$$

$$\beta = .005 = P[\text{accept } H_0 \mid H_a \text{ true}] = P[\bar{X} < c \mid \mu = 1200]. \tag{8.3}$$

By standardizing $\bar{X}$ we get, in place of (8.2) and (8.3),

$$\alpha = .001 = P\left[\frac{\bar{X} - \mu}{\sigma/\sqrt{n}} \geq \frac{c - \mu}{\sigma/\sqrt{n}} \middle| \mu = 1000\right] = P\left[Z \geq \frac{c - \mu}{\sigma/\sqrt{n}} = \frac{c - 1000}{200/\sqrt{n}}\right] \tag{8.4}$$

$$\beta = .005 = P\left[\frac{\bar{X} - \mu}{\sigma/\sqrt{n}} < \frac{c - \mu}{\sigma/\sqrt{n}} \middle| \mu = 1200\right] = P\left[Z < \frac{c - \mu}{\sigma/\sqrt{n}} = \frac{c - 1200}{200/\sqrt{n}}\right], \tag{8.5}$$

where Z represents a standard normal random variable. Now we know that if

$$P[Z \geq a] = .001,$$

then $a = z_{.999}$ (i.e., a is the 99.9 percentile of the standard normal distribution). Using this fact along with (8.4), we deduce that

$$\frac{c - 1000}{200/\sqrt{n}} = z_{.999} = 3.090. \tag{8.6}$$

Similar reasoning applied to (8.5) yields

$$\frac{c - 1200}{200/\sqrt{n}} = z_{.005} = -2.576. \tag{8.7}$$

Considering (8.6) and (8.7) together, we see that we have two equations in the two unknowns c and n. Solving each equation for c and equating the results allows us to solve for n as follows:

$$c = 1200 - 2.576\left(\frac{200}{\sqrt{n}}\right) = 1000 + 3.090\left(\frac{200}{\sqrt{n}}\right).$$

Therefore,

$$200 = (3.090 + 2.576)\left(\frac{200}{\sqrt{n}}\right)$$

$$\sqrt{n} = \frac{(5.666)(200)}{200}$$

$$n = (5.666)^2 = 32.10.$$

So we could take a sample size of 32 light bulbs and find $\bar{X}$, the mean lifetime of these bulbs. We can then find the critical value c by substituting $n = 32$ into (8.6) to get

$$c = 1000 + 3.090\left(\frac{200}{\sqrt{32}}\right) = 1109.25.$$

Of course, it would be possible to make α and β both smaller still by taking an even larger value of n. However, one must consider a law of diminishing returns here since the cost and time involved in the sampling process will soon outweigh the benefits of correctly classifying the bulbs. (To consider an extreme situation, if you tested all 14,400 bulbs, you would be virtually assured of knowing the correct type of light bulb. Unfortunately, you would not have any bulbs left to sell!)

In situations where testing is expensive in terms of time and/or money, a limit may be placed on the sample size. In such situations the sample size n and the probability of a type I error α are specified in advance, and from these values the critical value c and the probability of a type II error β can be found.

Example 8.5 Find the critical value c and the value of β if the hypotheses (8.1) are tested with a sample size $n = 25$ and $\alpha = .001$. Assume that the lifetimes of light bulbs are normally distributed with $\sigma = 200$ hours.

As in Example 8.4, we use $\bar{X}$ as a test statistic and will reject H_0 for values of $\bar{X}$ that are large, say $\bar{X} \geq c$. Using the definition of α and standardizing $\bar{X}$, we find that

$$\begin{aligned}\alpha = .001 &= P[\text{reject } H_0 \mid H_0 \text{ true}] = P[\bar{X} \geq c \mid \mu = 1000] \\ &= P\left[\frac{\bar{X} - \mu}{\sigma/\sqrt{n}} \geq \frac{c - \mu}{\sigma/\sqrt{n}} \middle| \mu = 1000\right] = P\left[Z \geq \frac{c - 1000}{200/\sqrt{25}}\right],\end{aligned}$$

where Z represents a standard normal random variable. As in equation (8.6), we have

$$\frac{c - 1000}{200/\sqrt{25}} = 3.090$$

$$c = 1000 + 3.090\left(\frac{200}{\sqrt{25}}\right) = 1123.6.$$

Given this critical value, we can find β by

$$\begin{aligned}\beta &= P[\text{accept } H_0 \mid H_a \text{ true}] = P[\bar{X} < c \mid \mu = 1200] \\ &= P\left[\frac{\bar{X} - \mu}{\sigma/\sqrt{n}} < \frac{c - \mu}{\sigma/\sqrt{n}} \middle| \mu = 1200\right] = P\left[Z < \frac{1123.6 - 1200}{200/\sqrt{25}}\right] \\ &= P[Z < -1.91] = .0281.\end{aligned}$$

These last two examples illustrate the relationships among n, c, α, and β. In fact, if any two of these four quantities are specified, the other two quantities will be determined. In practice, however, the most common situations are to have α and n specified or to have α and β specified. [In the beginning of this section, we made the assumption that the distribution of the lifetime of these light bulbs was normal. This assumption was made to allow us to calculate α and β when testing two hypotheses using a single observed lifetime (a sample of size 1). The assumption of normality must be valid in that case to ensure the accuracy of the values of α and β. However, because of the Central Limit Theorem, the assumption of normality is not so important to the calculations of α and β when the sample sizes are larger, as in Example 8.4.]

8.3 ONE- AND TWO-SIDED ALTERNATIVE HYPOTHESES

The examples of hypothesis-testing problems used in the previous sections were simple in the sense that we assumed that the value of the mean μ was exactly one of two values. Those hypotheses were also "simple" in the statistical sense. A *simple hypothesis* is one that completely specifies the distribution of the test statistic. The null hypothesis "$H_0 : \mu = 1000$" given in (8.1) is a simple hypothesis, since if it is true (and if the other assumptions of normality and known variance are true), the test statistic $\bar{X}$ will have a normal distribution, specifically

$$\bar{X} \sim N\left(\mu = 1000, \frac{\sigma^2}{n}\right).$$

Such a complete specification is not always the case, as you can see by the following example.

Example 8.6 Summer Daze works for a consumer protection group and is concerned with protecting the interests of consumers. She is currently interested in seeing whether or not the makers of Bubbly Cola are properly filling their 1-liter bottles. She wishes to test the null hypothesis

$$H_0 : \mu \leq 1000 \text{ ml}$$

against the alternative hypothesis

$$H_a : \mu > 1000 \text{ ml}.$$

Past experience shows that the contents of a randomly chosen bottle of Bubbly Cola is a random variable having an approximately normal distribution with a standard deviation of 4 ml. Should she reject H_0 (i.e., conclude that customers are in fact getting their money's worth) if a sample of 25 bottles had a mean content of 1001.6 ml?

Before solving this problem we would first like to point out how these hypotheses differ from the simple hypotheses given in (8.1). Since we are again concerned with tests of hypotheses about a population mean μ, we will use the sample mean $\bar{X}$ as the test statistic. If H_0 is true, we know that $\bar{X}$ will have a normal distribution with variance σ^2/n [i.e., $(4)^2/25$ for this example], but we *do not* know precisely the value of μ. Knowing that H_0 is true does not give us specific information about the exact value of μ; it only tells us that μ is less than or equal to 1000. In this case the null hypothesis does not completely specify the distribution of $\bar{X}$. Any hypothesis that does not completely specify the distribution of a test statistic is called a *composite hypothesis*. In Example 8.6 both H_0 and H_a are composite hypotheses.

When considering a composite null hypothesis, we face a small problem in finding the probability of committing a type I error. You should be able to see that the critical region for this test (i.e., values of $\bar{X}$ that call for rejection of H_0) will be all values greater than or equal to some constant c. Furthermore, since H_0 will be rejected only if there is strong evidence that it is not true, we know that c must be a value larger than 1000. (For reasons to be explained later, we will see that $c = 1001.03$ is a good choice.) To find the probability of a type I error, we would have to find

$$\begin{aligned} \text{(8.8)} \qquad P[\text{type I error}] &= P[\text{reject } H_0 \mid H_0 \text{ true}] \\ &= P[\bar{X} \geq c \mid \mu \leq 1000] \\ &= P\left[\frac{\bar{X} - \mu}{\sigma/\sqrt{n}} \geq \frac{c - \mu}{\sigma/\sqrt{n}} \,\middle|\, \mu \leq 1000\right] \\ &= P\left[Z \geq \frac{1001.03 - \mu}{4/\sqrt{25}} \,\middle|\, \mu \leq 1000\right]. \end{aligned}$$

You can see that since the exact value of μ is not specified, we cannot give a unique answer for the probability in (8.8). The answer will depend on what the value of μ really is. You can see from the graphs in Figure 8.7 that the probability in (8.8) will depend on the value of μ, and also that this probability is the largest when μ is actually equal to 1000. In view of this fact, when testing a composite null hypothesis, we will define

$$\alpha = \text{maximum } P[\text{type I error}].$$

Note that in the hypothesis stated in Example 8.6 the null hypothesis

$$H_0 : \mu \leq 1000$$

was given to include the value "1000." It really would not have changed the problem significantly had we written

$$H_0 : \mu < 1000, \qquad H_a : \mu \geq 1000,$$

but we will follow the convention of always including the equal sign with the null hypothesis. With this convention we will find that for problems in this text the value of α (i.e., the maximum probability of a type I error) will occur when equality actually holds for the parameter specified by the null hypothesis. Using the critical value $c = 1001.03$, we find that

$$\begin{aligned}\alpha &= \text{maximum } P[\text{type I error}] \\ &= \text{maximum } P[\bar{X} \geq c \mid \mu \leq 1000] \\ &= P[\bar{X} \geq c \mid \mu = 1000] \\ &= P\left[\frac{\bar{X} - \mu}{\sigma/\sqrt{n}} \geq \frac{1001.03 - 1000}{4/\sqrt{25}}\right] = P[Z \geq 1.2875] \approx .10.\end{aligned}$$

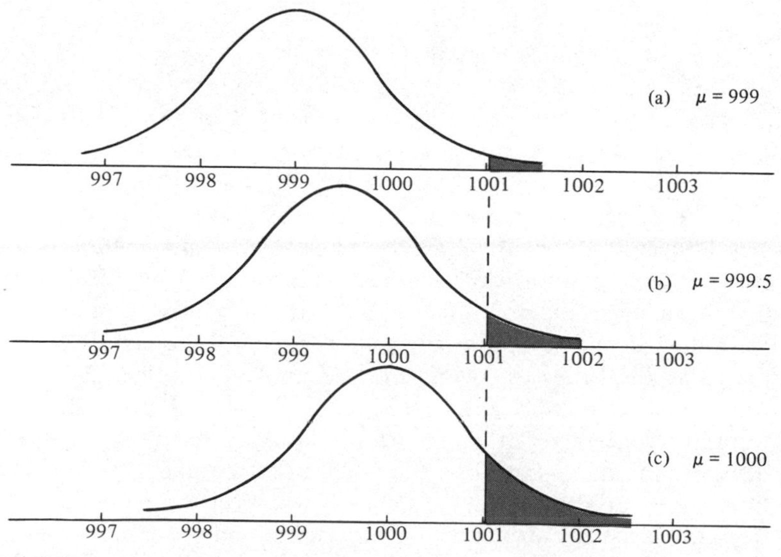

FIGURE 8.7

Reversing the problem, if it is desired to have the value of α equal to .10, the critical value should be 1001.03.

Example 8.6 (*continued*) Returning now to the solution of the problem, if α is taken to be .10, Summer should reject H_0, since the observed value of the test statistic, $\bar{x} = 1001.6$, is in the critical region for this test. This means that she can conclude that the buyers of Bubbly Cola 1-liter bottles are getting their money's worth.

Example 8.7 Summer wishes to perform another test for "500-milliliter" bottles of Bubbly Cola. Assuming that the standard deviation of the amount of cola in a bottle is 4 milliliters and that 25 bottles will be sampled, what critical value should she use if α is to be equal to .05?

The hypotheses in this case would be

$$H_0 : \mu \leq 500 \qquad \text{and} \qquad H_a : \mu > 500.$$

Since this is a composite null hypothesis, we find that

$$\begin{aligned}\alpha &= \text{maximum } P[\text{reject } H_0 \,|\, \mu \leq 500] \\ &= P[\bar{X} \geq c \,|\, \mu = 500] \\ &= P\left[\frac{\bar{X} - \mu}{\sigma/\sqrt{n}} \geq \frac{c - 500}{4/\sqrt{25}}\right] \\ &= P\left[Z \geq \frac{c - 500}{.8}\right].\end{aligned}$$

To make this last probability equal to .05, we must solve

$$\frac{c - 500}{.8} = z_{.95} = 1.645$$

for c. We find that

$$c = 500 + 1.645(.8) = 501.316.$$

As a note on terminology, we should point out that the value of α chosen for a test is often referred to as the *significance level* of the test. Generally, α is chosen to be small. For no reason other than custom, the values of α (significance levels) most often used in tests of hypotheses are .10, .05, and .01.

We mentioned in Section 8.2 that the "severity" of different types of errors may be viewed differently by different people. A similar question of point of view arises in stating statistical hypotheses. As an employee for a consumer protection group, Summer (Examples 8.6 and 8.7) is concerned primarily with the consumer. She will complain to the Bubbly Cola company if the bottles are underfilled but will not complain if they are overfilled. Her point of view is certainly different than that of her uncle Olaf, who is the quality control engineer at the Bubbly Cola plant. Olaf would be concerned if the amount put into the bottles was either too much or too little. If too much is put into the bottles, the company will lose profits, while if too little is put into the bottles, the company will receive complaints and adverse publicity from the consumer protection group. In checking to see if a bottling machine is in proper adjustment, he may choose a sample of "1-liter" bottles, measure their contents, and use the results to test

(8.9) $$H_0 : \mu = 1000 \text{ ml} \quad \text{vs.} \quad H_a : \mu \neq 1000 \text{ ml}.$$

You can see from this illustration that if a person is concerned only with values of a parameter that are extreme in one direction, hypotheses will be set up differently than if he is concerned with values that are extreme in both directions (i.e., with values that are either larger or smaller than some standard). If you are trying to decide whether an alternative hypothesis should be "one-sided" or "two-sided," you must ask yourself whether you are concerned with deviations in one direction from a standard or with deviations in either direction from a standard.

Now let us consider how a test of the hypotheses in (8.9) might be performed. As in previous examples, we are interested in testing hypotheses about a mean μ. The test statistic we should use then would be the sample mean $\bar{X}$, in this case the sample mean contents of "1-liter" bottles of Bubbly Cola. Intuitively, if $\bar{X}$ is "near" the value 1000, the sample data support the null hypothesis that $\mu = 1000$, but if $\bar{X}$ is too much smaller than 1000 (say $\bar{X} \leq c_1$) or too much larger than 1000 (say $\bar{X} \geq c_2$), this would be evidence against H_0. Consequently, a reasonable test would be

$$\text{reject } H_0 \text{ if } \bar{X} \leq c_1 \text{ or if } \bar{X} \geq c_2$$

$$\text{do not reject } H_0 \text{ if } c_1 < \bar{X} < c_2.$$

To find specific values of c_1 and c_2 for this problem, we would have to make some further assumptions about the significance level desired (α), the standard deviation of the amount of soda in a bottle, and the sample size.

Example 8.8 Olaf has determined from past experience that the amount of soda dispensed into a "1-liter" bottle is approximately normally distributed with a standard deviation of 4 milliliters. To test

$$H_0 : \mu = 1000 \text{ ml} \quad \text{vs.} \quad H_a : \mu \neq 1000 \text{ ml}$$

he plans to measure the contents of 16 bottles. What critical values c_1 and c_2 should he use if α is to be .05?

If a sample of 16 is chosen, $\bar{X}$ will have a normal distribution with a standard deviation $\sigma/\sqrt{n} = 4/\sqrt{16} = 1$. Furthermore, since this is a simple null hypothesis, we know that if H_0 is true, then

$$\bar{X} \sim N(\mu_{\bar{X}} = 1000, \sigma_{\bar{X}}^2 = 1).$$

A sketch of the distribution of $\bar{X}$ is given in Figure 8.8. We wish to find c_1 and c_2 so that

(8.10)
$$\begin{aligned} \alpha = .05 &= P[\text{reject } H_0 \mid H_0 \text{ true}] \\ &= P[\bar{X} \leq c_1 \text{ or } \bar{X} \geq c_2 \mid \mu = 1000] \\ &= P[\bar{X} \leq c_1 \mid \mu = 1000] + P[\bar{X} \geq c_2 \mid \mu = 1000]. \end{aligned}$$

[Note that since the events $(\bar{X} \leq c_1)$ and $(\bar{X} \geq c_2)$ are mutually exclusive events, we can say that the probability of the union of these two events will be equal to the sum of the probabilities given in (8.10).] The simplest way to make the sum in (8.10) equal

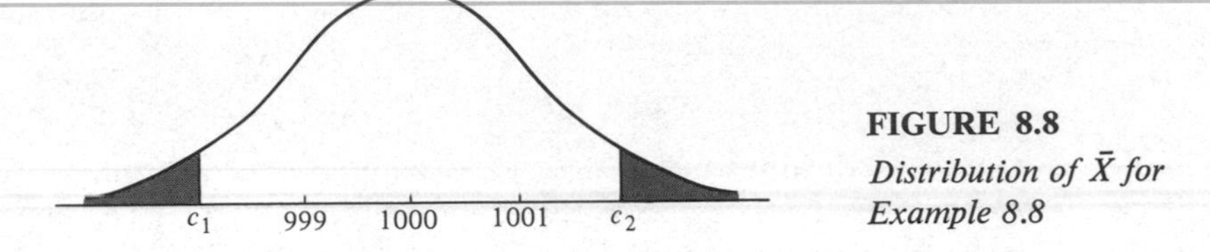

FIGURE 8.8
Distribution of $\bar{X}$ for Example 8.8

to .05 is to make each term equal to .025. (In general, to make the sum equal to α, make each term equal to $\alpha/2$.) Using this approach gives

$$P[\bar{X} \leq c_1 \mid \mu = 1000] = P\left[\frac{\bar{X} - \mu}{\sigma/\sqrt{n}} \leq \frac{c_1 - 1000}{4/\sqrt{16}}\right] = P\left[Z \leq \frac{c_1 - 1000}{4/\sqrt{16}}\right] = .025,$$

so

$$\frac{c_1 - 1000}{4/\sqrt{16}} = z_{.025} = -1.96 \tag{8.11}$$

$$c_1 = 1000 - 1.96(1) = 998.04.$$

Similarly,

$$P[\bar{X} \geq c_2 \mid \mu = 1000] = P\left[\frac{\bar{X} - \mu}{\sigma/\sqrt{n}} \geq \frac{c_2 - 1000}{4/\sqrt{16}}\right] = P\left[Z \geq \frac{c_2 - 1000}{4/\sqrt{16}}\right] = .025,$$

so

$$\frac{c_2 - 1000}{4/\sqrt{16}} = z_{.975} = +1.96, \tag{8.12}$$

from which

$$c_2 = 1000 + 1.96(1) = 1001.96.$$

Closer examination of Figure 8.8 will show you that c_1 and c_2 are simply the 2.5 and 97.5 percentiles of the distribution of $\bar{X}$. The values of c_1 and c_2 were found by standardizing $\bar{X}$ to a standard normal random variable and setting the standardized constants equal to the 2.5 and 97.5 percentiles of the standard normal distributions [equations (8.11) and (8.12)]. From this you can see that in general there are two equivalent ways of describing the critical values for a test about μ with a two-sided alternative.

1. Reject H_0 if $\bar{X} \leq c_1$ or $\bar{X} \geq c_2$, where c_1 and c_2 are the $100(\alpha/2)$ and $100(1 - \alpha/2)$ percentiles of the distribution of $\bar{X}$.
2. Reject H_0 if $Z = (\bar{X} - \mu)/(\sigma/\sqrt{n}) \leq z_{\alpha/2}$ or if $Z \geq z_{1-\alpha/2}$.

The first method uses the unstandardized $\bar{X}$ as test statistic, while the second uses the standardized $\bar{X}$, namely Z, as test statistic. The first method is perhaps a bit easier if several tests of the same hypotheses might be conducted (e.g., Olaf may test to see if the bottling machine is working properly every hour), but the second method is easier if a particular test is to be conducted only once.

Example 8.9 The Grow Quik Plant company has used a certain mix of soil and fertilizers to start tomato plants from seeds. After a fixed number of days, say d days, the average height of the plants has been 7.6 centimeters (about 3 inches), with a standard deviation of 1.2 cm. Daisy La Fleur, a company agronomist, wishes to try a new soil and fertilizer mix to see if there is any change in the growth of the plants. Assuming that the heights after d days are approximately normally distributed and that $\sigma = 1.2$ cm for the new mixture, what conclusions should be drawn if a box of 25 tomato plants showed a mean height of 7.9 cm using $\alpha = .10$?

A key word in the statement of this problem is the word "change." Of course, a change would occur if the new mix led to mean heights either smaller than the "standard" 7.6 cm or larger than that value. Consequently, it is appropriate to use a two-sided alternative hypothesis. In particular, set up

$$H_0 : \mu = 7.6 \text{ cm} \qquad \text{vs.} \qquad H_a : \mu \neq 7.6 \text{ cm}.$$

We know that the sample size is $n = 25$ and that the significance level is $\alpha = .10$. The unstandardized test statistic would be $\bar{X}$, the sample mean, while the standardized statistic is

$$Z = \frac{\bar{X} - \mu}{\sigma/\sqrt{n}}.$$

The critical values will be either the 5 and 95 percentiles of the distribution of $\bar{X}$ or the 5 and 95 percentiles of the standard normal distribution. Using the latter we would reject H_0 if the observed standardized value is less than or equal to $z_{.05}$ or greater than or equal to $z_{.95}$ that is, reject H_0 if

$$z \leq z_{.05} = -1.645 \qquad \text{or} \qquad z \geq z_{.95} = +1.645.$$

The observed value of z is

$$z = \frac{\bar{x} - \mu}{\sigma/\sqrt{n}} = \frac{7.9 - 7.6}{1.2/\sqrt{25}} = 1.25. \tag{8.13}$$

Since this value is not in the critical region, we do not reject H_0. [Note that in (8.13) we substituted for μ the value 7.6, the value of μ when H_0 is in fact true.]

In most of the examples we do from this point on we will find it more convenient to use standardized random variables for test statistics and percentiles of the standardized distributions for critical values. This procedure will work for one-sided or two-sided alternatives.

Example 8.10 A researcher from the agronomy department has suggested that Daisy switch to a particular soil and fertilizer mix for starting tomato plants, claiming that the slight differential in costs would be offset by hardier plants. (Assume that the hardiness of a plant is reflected by its height.) Should Daisy be convinced of his claim (and switch to this mix) if a flat of 36 plants had a mean height of 7.95 cm after d days? As in the previous example, assume that the heights are approximately normally distributed with $\sigma = 1.2$ cm. Use $\alpha = .01$.

The first thing we must do in this problem is to decide whether to use a one-sided or two-sided alternative. The key word here is "taller." Since the researcher's claim is true only if the mean is larger than the standard value of 7.6 cm, Daisy will want to use a one-sided alternative. This brings her to two choices. Should she use

$$H_0 : \mu \leq 7.6 \qquad \text{vs.} \qquad H_a : \mu > 7.6$$

or

$$H_0 : \mu \geq 7.6 \qquad \text{vs.} \qquad H_a : \mu < 7.6?$$

Recall that by convention the null hypothesis is not rejected unless there is *convincing* evidence that it is not true. Note that using the first set of hypotheses, Daisy will believe the researcher's claim *only if* there is convincing evidence that his claim *is true.* Using the second set of hypotheses, she will believe the researcher's claim *unless* there is convincing evidence that his claim *is not true.* Daisy is aware of the fact that many innovators tend to make exaggerated claims about their innovations. Consequently, she decides not to act in a naive manner but to require the researcher to convince her that the claim is true. She will set up

$$H_0 : \mu \leq 7.6 \qquad \text{vs.} \qquad H_a : \mu > 7.6. \tag{8.14}$$

Continuing with the question of how to choose the null and alternative hypotheses, note that if H_0 is "$\mu \leq 7.6$," a type I error would mean that an inferior and more expensive mix would be used by the company. A type II error would mean that a standard mix would be retained instead of using a mix that would lead to somewhat hardier plants. Although both errors may be undesirable, most companies would feel that the expenses involved in unnecessary change are worse than missing a slight improvement over the current method. (We will see in a later section that greater improvements are likely to be detected than slight improvements.) Consequently, by setting up the hypotheses in this way, the type I error is the more severe.

Intuitively, we see from (8.14) that H_0 should be rejected only if the sample mean $\bar{X}$ is "large." To get $\alpha = .01$ we would find c so that

$$\begin{aligned}\alpha = .01 &= \max P[\text{reject } H_0 \,|\, H_0 \text{ true}] \\ &= P[\bar{X} \geq c \,|\, \mu = 7.6].\end{aligned}$$

Hence c is the 99 percentile of the distribution of $\bar{X}$ when $\mu = 7.6$. Equivalently, H_0 should be rejected if the standardized sample mean exceeds the 99 percentile of the standard normal distribution, namely

$$z_{.99} = 2.326.$$

The observed value is

$$z = \frac{\bar{x} - \mu}{\sigma/\sqrt{n}} = \frac{7.95 - 7.6}{1.2/\sqrt{36}} = 1.75.$$

Since this is not in the critical region, Daisy should not accept the researcher's claim.

Note that the researcher did, in fact, present evidence to support his claim. (The sample mean height of 7.95 cm is certainly larger than the current standard of 7.6 cm.) However, the evidence was not convincing evidence. This example illustrates a helpful rule of thumb for deciding upon null and alternative hypotheses. If an innovator makes a claim to you, adopt a "show-me" attitude and believe that claim only if there is convincing evidence that it is true.

8.4 PARAMETRIC VERSUS NONPARAMETRIC TESTS

In the examples that we've considered up to this point in this chapter we have assumed that the measurement being made (e.g., amount of soda in a bottle, or height of a plant) was a random variable that had a normal distribution. With this assumption we were able to conclude that the test statistic $\bar{X}$, used for tests of hypotheses about a mean μ, also had a normal distribution. This fact allowed us to find critical regions for the test that would lead to specified values of α. Any test procedure like this which is dependent upon specific assumptions about the family of distributions from which the random variables come is called a *parametric test*. Since the test procedures used in the previous sections were based on the assumption that the measurements of interest came from a normal distribution, these would be called parametric tests. Most of the other test procedures that will be presented in this chapter will depend on this same assumption—that the measurements of interest come from a normal distribution. Consequently, these other test procedures will also be parametric tests.

In our initial discussions about the normal distribution (Section 4.8) we warned that not all "measurements" follow a normal distribution. The measurements referred to are measurements of time, length, height, weight, and so on, and are measurements of continuous quantities. Of course, if measurements are count measurements (e.g., number of days an employee is absent, number of defective items produced by a machine in 1 hour, etc.), they cannot follow a normal distribution exactly, since these measurements are discrete. An important question to ask is the following: "What will happen if I use a parametric test procedure based on the assumption that the data given come from a normal distribution, say, when in fact that data do *not* come from a normal distribution?" Unfortunately, no simple answer can be given.

What can go wrong if a parametric test is used when the underlying assumptions are not satisfied? One important thing is that the significance level may be affected. For example, you may follow a procedure where α, the probability of a type I error, will be .05, say, if the underlying assumptions are satisfied. However, if these assumptions are not satisfied, following this procedure may lead to a significance level of .10 or .15, say. You might be a lot more likely to make a type I error than you think you are!

Some test procedures are not very sensitive to mild deviations from the underlying assumptions. Although the significance level may change slightly from the nominal value, the changes will not be drastic. For example, a test might have a nominal significance level of .05 when the assumptions are satisfied and a level between .04 and .06 when there are moderate deviations from the assumptions. Such a test procedure is said to be *robust*. In the next section we will discuss a "*t* test" used for testing hypotheses about a mean μ. This test assumes that the data are measurements from a normal distribution. However, the *t* test is a robust procedure in that if the measurements come from a continuous distribution which is nonnormal, say from a distribution that is slightly skewed, the significance level will not be seriously affected. A word of warning is in order here. A healthy or "robust" individual can ignore some of the requisites for good health (adequate sleep, diet, exercise) some of the time, but if too many requisites are ignored too often, even the healthiest person will suffer.

In the same way even a robust test procedure will be adversely affected by gross departures from the assumptions

It is unfortunately true that not all test procedures are robust. For example, the F test (Section 8.7) for comparing the variance of two independent normal distributions is very sensitive to departures from the assumption of normality. What alternatives, then, do we have for situations where it is inappropriate to use parametric test procedures because underlying assumptions are not satisfied? We can use "nonparametric" test procedures. *Nonparametric test procedures* are test procedures whose validity does not depend on specific assumptions about the family of probability distributions from which the measurements are taken. Nonparametric tests of hypotheses will be the subject of Chapter 11. The purpose of this present section is to make you aware of the fact that there is a choice to be made between parametric and nonparametric procedures. The choice is not an arbitrary one, but one to be based on whether certain underlying assumptions are satisfied.

In the remainder of this chapter we will remind you of the assumptions necessary to use various parametric procedures We can say in advance that the primary assumption is that the data come from a normal distribution. There are some relatively simple questions you can ask to see if the assumption of normality is reasonable:

1. Are the measurements discrete or continuous? If discrete, they cannot *exactly* follow a normal distribution.
2. Are the measurements made on a nominal or ordinal scale? If so, they cannot be normal.
3. Does a histogram of the data follow approximately a bell shape? If so, the measurements *may* be normal. (An affirmative answer here might be sufficient for justifying the use of robust procedures.)

A fourth but less simple question is:

4. Do the data "pass" a test for normality? If so, parametric test procedures requiring the assumption of normality may be used. If not, nonparametric procedures are more appropriate.

Some tests for normality will be discussed in Chapter 11.

8.5 TESTS FOR A MEAN

At the end of Section 8.1 we gave a general outline of steps to be followed in a hypothesis-testing problem. We are now in a position to give a more specific outline of steps. These steps are given in Table 8.1. Following the steps outlined here will help you in setting up and testing appropriate hypotheses. Be careful, however, to check to see if underlying assumptions for a given test procedure are satisfied (or "close" to being satisfied).

A word of warning is in order here. In testing a null hypothesis against an

TABLE 8.1
Steps to Follow in Hypothesis-Testing Problems

Step	*Comments*
1. State null and alternative hypotheses.	First decide whether to use a one- or two-sided alternative. If one-sided, carefully decide on the direction of the inequality.
2. Choose α, the significance level.	The most common values are .05 or .01, depending on the seriousness of committing a type I error. Other values can be used.
3. Determine the sample size, n.	This may be determined by time and/or cost considerations. The size may be chosen to achieve certain error probabilities.
4. Choose a suitable test statistic.	For parametric tests this will generally be the sample counterpart of the parameter being tested. Generally, the test statistic will be standardized.
5. Find the critical region.	This will usually be determined by percentiles of the standardized test statistic.
6. Find the observed value of the test statistic and determine whether or not to reject H_0.	If the (standardized) test statistic is in the critical region, reject H_0. If not, do not reject H_0.

alternative hypothesis we are using sample information as "evidence." If the evidence is heavily weighted against H_0, we say "reject H_0" or "accept H_a," while if it is not heavily weighted against H_0, we say "do not reject H_0" or perhaps "accept H_0." Generally speaking, we cannot be sure that a right decision has been made. For example, failure to reject H_0 does not mean that H_0 is true. It may be that a type II error has been made. Furthermore, although we know the probability of committing a type I error, we generally do not know the probability of committing a type II error.

In the previous sections we gave some examples of tests of hypotheses about the mean of a normal distribution having known variance. How would we proceed if the assumption of normality did not hold? Why was this assumption necessary? We made the assumption that the observations came from a normal distribution so that we could be sure that the test statistic $\bar{X}$ would have a normal distribution. Then standardizing $\bar{X}$ allowed us to find critical values by using percentiles of the standard normal distribution. Fortunately, the assumption that the individual observations come from a normal distribution is not necessary in order for $\bar{X}$ to have an approximately normal distribution. You will recall from Section 5.9 that according to the Central Limit Theorem the sample mean $\bar{X}$ will be approximately normally distributed *if the sample size is large enough* for practically any underlying distribution. We said at that time that if the underlying distribution is continuous, unimodal, and not too skewed, $\bar{X}$ will be approximately normal for n as small as 5 or 10. In most other situations a sample size of $n \geq 30$ is large enough to invoke the Central Limit Theorem. We can summarize this discussion as follows:

In order to test hypotheses about the mean μ of a distribution, if σ^2 is known, then $\bar{X}$ will be a suitable test statistic. Furthermore, for n sufficiently large, $\bar{X}$ will have an approximately normal distribution.

Example 8.11 The weights of college-age males are nonnormal, being skewed to the right. The standard deviation is known to be about 25 pounds. A dictician, Ella Slender, has conjectured that the mean weight of males at a certain small college is different from the national mean of 165 pounds. Is her conjecture justified if a random sample of 16 males from this college had a sample mean of 156.3 pounds? Take α to be .05.

Even though the distribution of weights is not normal, the distribution is continuous, unimodal, and only moderately skewed. Consequently, a sample of size 16 should be large enough to invoke the Central Limit Theorem. Since Ella is interested in a "difference," a two-sided alternative is appropriate. Following the steps of the outline, we get

1. $H_0 : \mu = 165$ vs. $H_a : \mu \neq 165$.
2. $\alpha = .05$.
3. $n = 16$.
4. The standardized test statistic will be $(\bar{X} - \mu)/(\sigma/\sqrt{n})$, where μ is taken to be 165, the value when H_0 is true.
5. Since the sample size is large enough to invoke the Central Limit Theorem, the test statistic will be approximately distributed as a standard normal random variable. Since it is a two-sided alternative, the critical values will be $z_{\alpha/2} = z_{.025} = -1.96$ and $z_{1-\alpha/2} = z_{.975} = +1.96$.
6. The observed value of the test statistic is

$$z = \frac{\bar{x} - \mu}{\sigma/\sqrt{n}} = \frac{156.3 - 165}{25/\sqrt{16}} = -1.392.$$

Since this value is not in the critical region (see Figure 8.9), do not reject H_0. While there is some evidence that Ella's conjecture is correct, it is not convincing evidence.

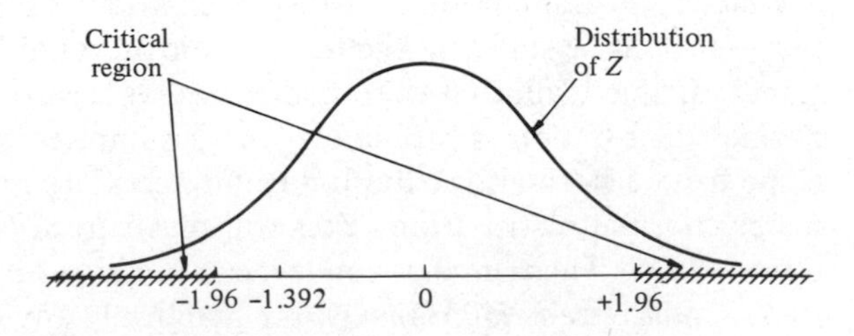

FIGURE 8.9

The procedure used for tests about the mean of a distribution when σ^2 is known is, in a sense, nonparametric. The procedure will work for *any* underlying distribution, provided that the sample size n is large enough to invoke the Central Limit Theorem.

If the underlying distribution is normal, the test procedure described here works for any sample size.

As you can see, the use of this test procedure does depend on σ being known. If σ is unknown, as is the case in many real problems, then we must use a different test statistic. If you review Section 7.5, paying particular attention to Theorem 7.1, you will see how we handled a similar problem in trying to estimate a mean μ when σ^2 is unknown. Recall that if $X_1, X_2, \ldots, X_n$ is a random sample from a normal distribution having mean μ, then

$$T = \frac{\bar{X} - \mu}{S/\sqrt{n}} \tag{8.15}$$

will have a Student's t distribution with $n - 1$ degrees of freedom. Consequently, if measurements come from a normal distribution, a suitable test statistic for tests about a mean is T as defined in (8.15). The critical values for such a test will be the appropriate percentiles of a t random variable with $n - 1$ degrees of freedom. The observed value of the test statistic is found by substitution into (8.15) with the value μ replaced by the numerical value specified by the null hypothesis.

Example 8.12 Daisy La Fleur, agronomist for the Grow Quik Plant Company (Examples 8.9 and 8.10), wants to see if a newly developed variety of tomato plant grows more quickly than the variety that has been used. Using the same soil and fertilizer mix that led to an average height of 7.6 cm after d days, she planted 10 seeds. Later, the heights of these plants were found to be

7.9	9.1	6.6	8.1	7.9
7.5	8.5	7.8	8.1	8.5.

Assuming that the heights of these plants follow approximately a normal distribution with an unknown variance, set up and test appropriate hypotheses about the mean height of the new variety of plants. Use a .05 significance level.

Since Daisy is interested in knowing whether the new variety grows "more quickly" than the present variety, a one-sided alternative hypothesis is appropriate. The burden of proof is put on the new variety if we set up

1. $H_0 : \mu \leq 7.6$ cm vs. $H_a : \mu > 7.6$ cm.

Following the remaining steps of the outline, we get

2. $\alpha = .05$.
3. $n = 10$.
4. Since we have assumed the heights $X_1, X_2, \ldots, X_{10}$ to have come from a normal distribution with unknown variance, the appropriate test statistic will be $T = (\bar{X} - \mu)/(S/\sqrt{n})$. Here the value of μ will be taken to be 7.6, the value of μ when equality holds in H_0.
5. The test statistic has a t distribution with $n - 1 = 9$ degrees of freedom. Intuitively, we would reject H_0 only if $\bar{X}$ is large, so we reject H_0 only if the

observed value of t is large, in particular greater than or equal to the 95 percentile (i.e., reject if $t \geq t_{9,.95} = 1.833$).

6. From the given data, we find

$$\bar{x} = \frac{7.9 + 9.1 + \ldots + 8.5}{10} = 8.0$$

$$s^2 = \frac{\sum_{i=1}^{n} (x_i - \bar{x})^2}{n - 1} = \frac{\sum_{i=1}^{10} (x_i - 8.0)^2}{9} = .444,$$

so

$$t = \frac{8.0 - 7.6}{.667/\sqrt{10}} = 1.90.$$

Since this value is in the critical region (see Figure 8.10), Daisy can conclude that the new variety of tomato plants are taller after d days than the variety currently being used.

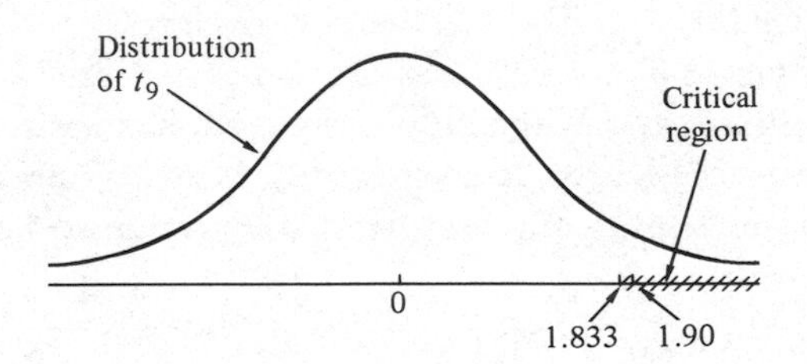

FIGURE 8.10

In order for the random variable T defined by (8.15) to follow a Student's t distribution exactly, it is necessary that the sample data come from a normal distribution. Studies have shown, however, that the distribution of T will approximately follow a Student's t distribution even if the sample data come from a nonnormal distribution. The approximation will be best, of course, when the departure from normality is not too great. It is very difficult to give precise guidelines here as to what constitutes "too great" a departure from normality. A crude rule of thumb would be to use the following guidelines: If the sample data come from a normal distribution, T has exactly a Student's t distribution for any sample size; if they come from a continuous distribution that is unimodal and symmetric, T will have approximately a Student's t distribution for n greater than 5 to 10, say; if the sample data come from a distribution that is unimodal and only moderately skewed, T will have approximately a Student's t distribution for n greater than 15 to 20; for most other distributions from which the sample data might have come, T will have approximately a Student's t distribution for n greater than 30 to 50. In the terminology of Section 8.4, a test for a parameter μ using T as the test statistic would be robust in terms of the assumption of normality. As we warned you there, however, do not put too much trust in robustness! If you wish to test for a mean μ and you have a small set of data from a decidedly nonnormal distribution, perform the test by using nonparametric procedures rather than a t test. The Wilcoxon signed rank test described in Chapter 11 is one possible choice.

Example 8.13 All airlines are concerned about the number of vacant seats on regularly scheduled flights. The All-American Airlines is no exception to this concern. They found that on their Dallas to Los Angeles run, the mean number of vacant seats was 30.2. Two months after instituting a special reduced fare, they wanted to see if the mean number of vacant seats had been reduced to 20 or less (a reduction of 10 vacant seats being necessary to offset the fare reduction). Treating the next 41 flights as a random sample of flights to be made over the next several months, what conclusion should they draw if the mean and variance of the number of vacant seats in these next 41 flights are 17.3 and 30.25? Use $\alpha = .05$.

The number of vacant seats/flight is a discrete, not continuous, random variable and hence cannot be normally distributed However, for a sample size this large we can rely on the robustness of the t-test procedure to give us good approximate results. A one-sided alternative is appropriate here and following the steps of the outline we get:

1. $H_0 : \mu \geq 20$ vs. $H_a : \mu < 20$.
2. $\alpha = .05$.
3. $n = 41$.
4. Since the variance is unknown and the sample size is relatively large, we will use $T = (\bar{X} - \mu)/(S/\sqrt{n})$ as the test statistic. The value of μ will be taken to be 20.
5. The test statistic has approximately a t distribution with $n - 1 = 40$ degrees of freedom. Intuitively, we would reject H_0 only if $\bar{X}$ is small, so we reject H_0 only if the observed value of t is small. Since $\alpha = .05$, reject H_0 if $t \leq t_{40,.05} = -1.684$.
6. From the given information, we find that

$$t = \frac{17.3 - 20}{\sqrt{30.25}/\sqrt{41}} = -3.15.$$

Since this value is in the critical region (Figure 8.11), the null hypothesis should be rejected. Assuming that the fare reduction was the only factor affecting the change in number of vacant seats, they can reasonably conclude that the increased revenue from additional passengers has more than offset the loss of revenue from fare reductions. (Can you think of factors other than fare reduction that might affect the number of vacant seats?)

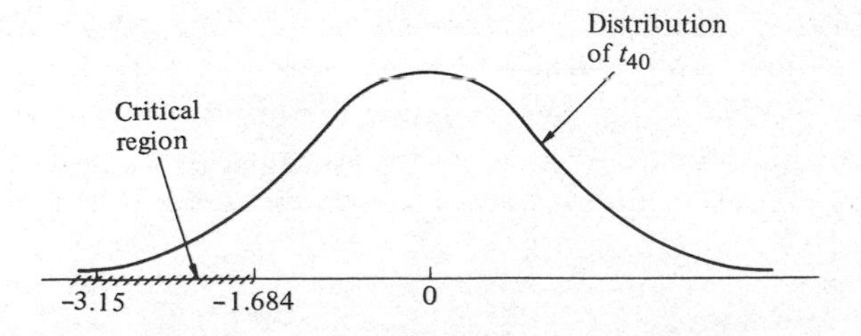

FIGURE 8.11

8.6 TESTS FOR DIFFERENCES BETWEEN MEANS

In Sections 7.6 and 7.7 we discussed the problem of obtaining interval estimates of the difference between two means, $\mu_1 - \mu_2$. You saw at that time that there are several different methods of finding the confidence intervals for $\mu_1 - \mu_2$, the appropriate method depending on the experimental design and/or what was known about the variances. In a similar way there are different test statistics that can be used to test hypotheses about a difference between means. The appropriate choice of test statistic parallels the appropriate choice of method of finding confidence intervals. A basic underlying assumption for all these tests is that the sample data come from normal distributions or that the sample sizes are large enough to invoke the Central Limit Theorem or to rely on the robustness of the t distribution. If this basic assumption is not satisfied, nonparametric procedures should be used. If it is satisfied, the appropriate test statistic can be found by asking the questions in Table 8.2. The

TABLE 8.2

Questions for Finding Appropriate Test Statistic

Question	*Answer*	*Action*
1. Are the data paired?	No	Ask question 2.
	Yes	Use $T = \frac{\bar{D} - \mu_D}{S_D/\sqrt{n}}$ as the test statistic ($n - 1$ degrees of freedom).
2. Are σ_1^2 and σ_2^2 known?	No	Ask question 3.
	Yes	Use $Z = \frac{(\bar{X}_1 - \bar{X}_2) - (\mu_1 - \mu_2)}{\sqrt{\sigma_1^2/n_1 + \sigma_2^2/n_2}}$ as the test statistic.
3. Are σ_1^2 and σ_2^2 equal?	No	Use $T = \frac{(\bar{X}_1 - \bar{X}_2) - (\mu_1 - \mu_2)}{\sqrt{S_1^2/n_1 + S_2^2/n_2}}$ [degrees of freedom given by (7.24) or (7.25)].
	Yes	Use $T = \frac{(\bar{X}_1 - \bar{X}_2) - (\mu_1 - \mu_2)}{S_p\sqrt{1/n_1 + 1/n_2}}$ as the test statistic ($n_1 + n_2 - 2$ degrees of freedom).

notation used in this table is the same as that used in Chapter 7, namely $\bar{X}_1$, S_1^2, and n_1 all refer to the first sample, which is taken from a distribution having mean and variance μ_1 and σ_1^2, and so on. In calculating the observed value of the appropriate test statistic, substitute for $\mu_1 - \mu_2$ or μ_D the constant specified by the null hypothesis. The flowchart given in Flowchart 8.1 shows this same information in a slightly different form.

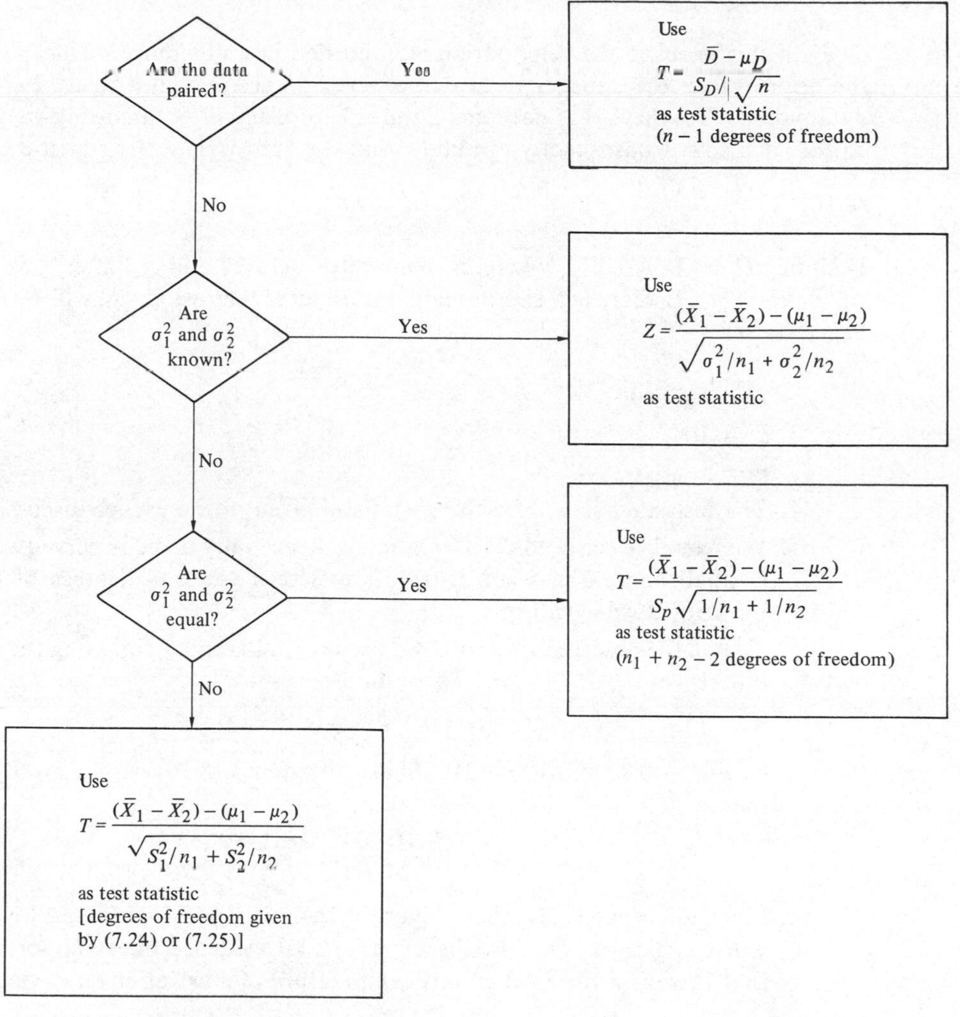

FLOWCHART 8.1

⋆***Example 8.14*** A veterinarian was interested in knowing if there is a rise in the body temperature of sows after farrowing. Sensors were attached to five sows so that body temperature could be monitored both before and after farrowing. Assume that these temperatures are approximately normally distributed. Set up and test appropriate hypotheses using a. 01 level of significance if the observed temperatures were:

	Sow				
Temperature (°C)	*1*	*2*	*3*	*4*	*5*
Before farrowing	38.3	38.4	38.0	38.7	38.2
After farrowing	39.4	39.6	38.6	39.5	39.0

In this problem the veterinarian is interested in a difference of means. To find the appropriate test statistic we can consider the questions in Table 8.2. Are these data paired? Yes they are, since temperatures before and after farrowing were obtained for each sow. Consequently, we know that the appropriate test statistic will be

$$T = \frac{\bar{D} - \mu_D}{S_D/\sqrt{n}}. \tag{8.16}$$

Defining $D_i = X_{B,i} - X_{A,i}$, where B represents "before" and A "after," we see that $\mu_D = \mu_B - \mu_A$. If there is a rise in temperature after farrowing, μ_D will be negative; hence we set up

1. $H_0 : \mu_D \geq 0$ vs. $H_a : \mu_D < 0$.
2. $\alpha = .01$.
3. $n = 5$.
4. The test statistic will be the T statistic given in the earlier discussion.
5. We would reject H_0 only if $\bar{D}$ is small, hence only if the observed value of T is small. Since T has a t distribution with $n - 1 = 4$ degrees of freedom, the critical value will be $t_{4,.01} = -3.747$.
6. To find the observed value of T we need to find $\bar{D}$ and S_D by using the observed differences $d_1, d_2, \ldots, d_5$. These differences are

   $$-1.1 \qquad -1.2 \qquad -0.6 \qquad -0.8 \qquad -0.8,$$

   so $\bar{d} = -0.90$ and $s_D = .245$. Substituting into (8.16) with $\mu_D = 0$, we find that

   $$t = \frac{-0.90 - 0}{.245/\sqrt{5}} = -8.2.$$

   Since this value is less than the critical value of -3.747, the null hypothesis will be rejected. That is, the experimental evidence does support the idea that there is a rise in the body temperature of sows after farrowing.

Example 8.15 Mary Land, a realtor in a large Eastern city, believes that the cost of housing in a certain subdivision has gone up more than \$15,000 over the last 10 years. She believes that the standard deviation of the cost of single-family dwellings 10 years ago was \$4000 but that it is currently \$8000. In checking a random sample of old records she has found that 30 houses in this subdivision sold for an average of \$22,875 10 years ago. Furthermore, a random sample of 25 houses in this subdivision sold last year for an average of \$40,345. Is her belief supported by the data?

If μ_1 and μ_2 represent the mean selling prices of homes in this subdivision last year and 10 years ago, respectively, Mary is interested in the quantity $\mu_1 - \mu_2$. In answering the questions in Table 8.2, we find that the data are not paired, but the variances are known. While the selling price of homes is most likely skewed to the right, and hence not normal, we would not expect the skewness to be too marked within a given subdivision. Since the sample sizes are moderately large, we can invoke the Central Limit Theorem to say that

$$Z = \frac{(\bar{X}_1 - \bar{X}_2) - (\mu_1 - \mu_2)}{\sqrt{\sigma_1^2/n_1 + \sigma_2^2/n_2}} \tag{8.17}$$

will have a standard normal distribution (approximately). If we adopt a "show-me" attitude toward her claim, we will set up

1. $H_0 : \mu_1 - \mu_2 \leq 15{,}000$ vs. $H_a : \mu_1 - \mu_2 > 15{,}000$.
2. In this problem the significance level was not specified. In the absence of information about the consequences of a type I error, we can take $\alpha = .05$.
3. There are two samples here, $n_1 = 25$ and $n_2 = 30$.
4. The test statistic should be Z, as defined in (8.17).
5. Since we would reject H_0 only if $\bar{X}_1 - \bar{X}_2$ is large, we would reject H_0 if Z is large. In particular, reject H_0 if the observed value of Z is at least $z_{.95} = 1.645$.
6. To find the observed value of Z, we substitute into (8.17) using 15,000 for $\mu_1 - \mu_2$. We find that

$$z = \frac{(40{,}345 - 22{,}875) - (15{,}000)}{\sqrt{(8000)^2/25 + (4000)^2/30}} = \frac{17{,}470 - 15{,}000}{\sqrt{3{,}093{,}333}}$$

$$= \frac{2470}{1759} = 1.404.$$

Since this does not exceed the critical value 1.645 (see Figure 8.12), do not reject H_0. There is certainly some evidence that Mary's belief is correct, but the evidence is not quite strong enough to convince a skeptic using $\alpha = .05$. (Note, however, that someone less skeptical using a significance level of .10 would have been convinced!)

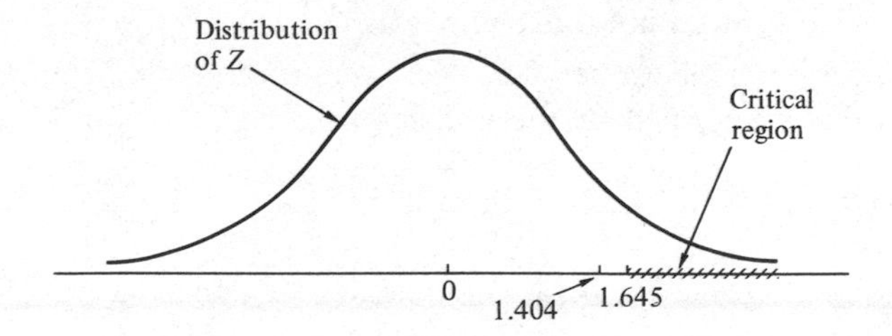

FIGURE 8.12

Example 8.16 Henry J. Ford owns a mail-order auto parts store. He has two workers, Edsel and Lincoln, who are given orders to fill. The orders are assigned to the workers randomly. Henry is going to give both men a raise and will give equal raises if it appears that both men work at about the same speed; otherwise, the faster worker will be given a larger raise. To judge how fast the men work, Henry has looked at the times required for each man to fill 30 orders. These orders were randomly selected from among the orders filled during the last month. What should Henry do if the sample mean and standard deviation for Edsel were 19.2 minutes and 4.3 minutes while the figures for Lincoln were 22.3 minutes and 5.5 minutes? Use $\alpha = .10$.

If we let μ_1 and μ_2 denote the mean times for Edsel and Lincoln, respectively, Henry would be interested in knowing if $\mu_1 = \mu_2$ or, equivalently, if $\mu_1 - \mu_2 = 0$. A two-sided alternative would be in order here. In answering the questions in Table 8.2, we find it easy to answer the first two questions in the negative. However, the answer to the third question is not so clear (i.e., it is not clear from the information given in the problem whether or not $\sigma_1^2 = \sigma_2^2$). However, if the workers really are of about the same ability, the variances for the amount of time it takes each to fill an order should be about the same. (Of course, the statement that "$\sigma_1^2 = \sigma_2^2$" is really another statistical hypothesis which you will learn how to test in the next section.) On the assumption that the variances are equal, we would use

$$T = \frac{(\bar{X}_1 - \bar{X}_2) - (\mu_1 - \mu_2)}{S_p\sqrt{1/n_1 + 1/n_2}} \tag{8.18}$$

as the test statistic. Note that although the times for filling orders may not be normal, the sample sizes are large enough to be able to assume that T will have approximately a Student's t distribution with $n_1 + n_2 - 2$ degrees of freedom. Following the outline for hypothesis testing, we find:

1. $H_0 : \mu_1 - \mu_2 = 0$ vs. $H_a : \mu_1 - \mu_2 \neq 0$.
2. $\alpha = .10$.
3. The two sample sizes are $n_1 = 30$, $n_2 = 30$.
4. The test statistic should be T, as defined in (8.18).
5. Since this is a two-sided alternative, we would reject H_0 if $\bar{X}_1 - \bar{X}_2$ is too much smaller or larger than zero, or equivalently if the observed absolute value of T is at least $t_{58,.95} = 1.672$. (This value was found by interpolation, but it obviously does not differ much from $t_{60,.95} = 1.671$.)
6. To find the observed value of T, substitute into (8.18), replacing $(\mu_1 - \mu_2)$ by the value 0 specified by the null hypothesis. First, however, find S_p by

$$S_p = \sqrt{\frac{(n_1 - 1)S_1^2 + (n_2 - 1)S_2^2}{n_1 + n_2 - 2}} = \sqrt{\frac{29(4.3)^2 + 29(5.5)^2}{58}}$$
$$= \sqrt{24.37} = 4.94.$$

Using this value, we observe that

$$t = \frac{(19.2 - 22.3) - 0}{4.94\sqrt{\frac{1}{30} + \frac{1}{30}}} = \frac{-3.1}{4.94\sqrt{.0666}} = -2.43.$$

Since this value is in the critical region (see Figure 8.13), the null hypothesis should be rejected. Henry should conclude that there is a difference in the speed at which these men work. Furthermore, since Edsel's sample average time was smaller, the evidence is that he works faster and so Edsel should get a larger raise.

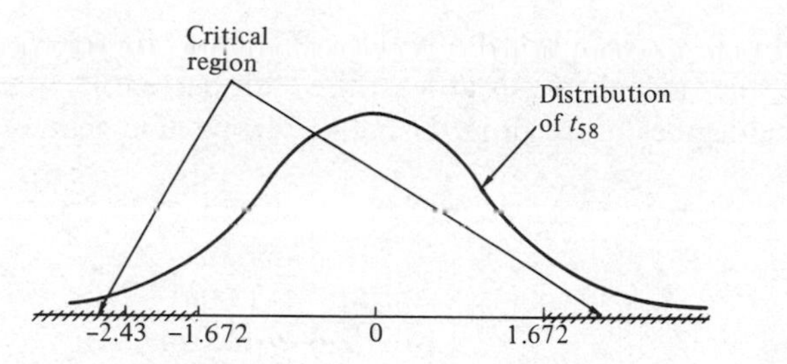

FIGURE 8.13

Example 8.17 Henry's auto parts store is currently organized so that parts are stored in bins arranged alphabetically (accelerator cables, brake shoes, carburetors, etc.). Henry's sister Henrietta also owns a parts store, but in her store, parts are arranged according to a different scheme, by part number. To see if his workers would be more efficient with the new scheme, he has asked Edsel to work in Henrietta's store. After a week to learn the scheme, Edsel filled 34 orders chosen randomly from Henry's orders. The sample mean and standard deviation of the time required to fill the orders were 17.6 minutes and 5.9 minutes. Based on this information and the information in Example 8.16, does it appear to be worthwhile for Henry to reorganize his store?

If we let μ_1 and μ_2 denote the mean time required for Edsel to fill an order under Henry's scheme and Henrietta's scheme, respectively, the difference $\mu_1 - \mu_2$ would give the time differential for Edsel under the two schemes. In this problem a one-sided alternative would be appropriate, since Henry would want to reorganize only if times for filling orders were shorter. In finding the appropriate test statistic using the questions in Table 8.2, we see that the data are not paired and the variances are not known. It is not clear whether the variances are equal. For the moment let us assume that they are not equal. Then the appropriate test statistic would be

$$T = \frac{(\bar{X}_1 - \bar{X}_2) - (\mu_1 - \mu_2)}{\sqrt{S_1^2/n_1 + S_2^2/n_2}}. \tag{8.19}$$

This random variable should have approximately a Student's t distribution with degrees of freedom ν found from (7.24) or (7.25). Recall that in (7.24) we had

$$\nu = \frac{(a_1 + a_2)^2}{(a_1^2/(n_1 - 1)) + (a_2^2/(n_2 - 1))}$$

with $a_i = S_i^2/n_i$ and in (7.25) we had $\nu = \min(n_1, n_2) - 1$. For the sake of illustration we use (7.24), although the effect of the difference in the degrees of freedom for these sample sizes is not great. Following the outline for hypothesis testing, we find:

1. $H_0 : \mu_1 - \mu_2 \leq 0$ vs. $H_a : \mu_1 - \mu_2 > 0$.
2. The significance level was not specified in the statement of the problem. Since reorganization would be a major undertaking, Henry would want a very small probability of a type I error, so let us take $\alpha = .01$.
3. The sample sizes are $n_1 = 30$ (Henry's store) and $n_2 = 34$ (Henrietta's store).
4. The test statistic should be T, as defined in (8.19).

5. Since this is a one-sided alternative, H_0 would be rejected only if $\bar{X}_1 - \bar{X}_2$ is large, that is, only if T is at least the 99th percentile of a t random variable having ν degrees of freedom. Using (7.24), we find that

$$a_1 = \frac{S_1^2}{n_1} = \frac{(4.3)^2}{30} = .616, \qquad a_2 = \frac{S_2^2}{n_2} = \frac{(5.9)^2}{34} = 1.024,$$

so

$$\nu = \frac{(.616 + 1.024)^2}{(.616^2/29) + (1.024^2/33)}$$

$$= \frac{2.690}{.0449} = 59.91 \approx 60.$$

Consequently, the critical value is $t_{60,.99} = 2.390$.

6. To find the observed value of T, substitute into (8.19), replacing $\mu_1 - \mu_2$ by the constant specified in H_0, namely 0. We find that

$$t = \frac{(19.2 - 17.6) - 0}{\sqrt{(4.3)^2/30 + (5.9)^2/34}} = \frac{1.6}{\sqrt{1.64}} = 1.25.$$

Since this value is not in the critical region, do not reject H_0. Although there is some evidence that Edsel fills orders faster under this new parts organization scheme, the evidence is not convincing enough for Henry to want to reorganize.

8.7 TESTS ABOUT VARIANCES

In the earlier sections of this chapter we have been primarily concerned with the means of random variables. There are situations where the amount of variability is of as much or more concern. For example, in any production process where containers are being filled, the mean amount of a product put into the container is of interest. However, the variability of the amount might also be of interest. This may be due to aesthetic considerations or for more practical reasons. Consider a machine that is supposed to put glue into "250-milliliter" containers. If the amount dispensed into the bottles is quite variable, a large proportion of containers will tend to overflow, creating cleaning and possibly safety problems.

Example 8.18 Elmer is a purchasing agent for a glue manufacturing company. The machines currently used to fill 250-ml bottles actually dispense a random amount of glue into the bottles. The amount is normally distributed, with a standard deviation of 4 ml. Elmer is considering purchase of a new machine which fills bottles at a slightly faster rate than the machine currently in use. However, he will only make the purchase if the new machine is less variable in filling bottles than is the current machine. What should he do if a sample of 81 bottles filled by the new machine had a standard deviation of 3.15 ml? Use $\alpha = .01$.

This example illustrates an example where a person is concerned with variability. (Elmer is less concerned with the mean amount being dispensed, since the mean can easily be controlled by relatively simple adjustments.) Before we can solve the problem, however, we will need to find a suitable test statistic. Since this problem calls for a test of hypotheses about a variance σ^2, the logical candidate for the test statistic is the sample variance S^2. To find appropriate critical regions, we will have to know something about the probability distribution of S^2. Recall from Section 7.8 that if a random sample $X_1, X_2, \ldots, X_n$ is taken from a normal distribution having variance σ^2, then

$$\frac{(n-1)S^2}{\sigma^2} \tag{8.20}$$

will have a χ^2 distribution with $n - 1$ degrees of freedom. Consequently, if we standardize S^2 as in (8.20), it will be possible to use percentiles of the chi-square distribution as critical values. To see how this is done, let us proceed through the outline of the test procedure.

Example 8.18 *(continued)*

1. $H_0 : \sigma \geq 4$ vs. $H_a : \sigma < 4$ (or, equivalently, $H_0 : \sigma^2 \geq 16$ vs. $H_a : \sigma^2 < 16$). Note that by setting up the hypotheses this way, Elmer puts the burden of proof on the new machine. He will reject H_0 only if there is convincing proof that σ is less than 4.
2. $\alpha = .01$.
3. $n = 81$.
4. The test statistic could be either S^2 or the standardized version of S^2 as defined by (8.20).
5. The null hypothesis will be rejected only if S^2 or, equivalently, $(n-1)S^2/\sigma^2$ is sufficiently small. To get the desired significance level, we will want

$$\begin{aligned} \alpha = .01 &= \max P[\text{reject } H_0 \mid H_0 \text{ true}] \\ &= \max P[S^2 \leq c \mid \sigma^2 \geq 16] \\ &= P[S^2 \leq c \mid \sigma^2 = 16] \\ &= P[(n-1)S^2/\sigma^2 \leq (n-1)c/\sigma^2 \mid \sigma^2 = 16]. \end{aligned}$$

   Since, under the assumption of normality, $(n-1)S^2/\sigma^2$ will have a χ^2_{n-1} distribution, it follows that the critical value will be the .01 percentile for a χ^2_{80} random variable. In particular,

$$P\left[\frac{(n-1)S^2}{\sigma^2} \leq \chi^2_{80,.01}\right] = P[\chi^2_{80} \leq \chi^2_{80,.01} = 53.54] = .01,$$

   the critical value is 53.54. The null hypothesis will be rejected if $(n-1)S^2/\sigma^2 \leq 53.54$.
6. For this problem the observed value of the standardized test statistic is

$$\frac{(n-1)s^2}{\sigma^2} = \frac{(80)(3.15)^2}{(4)^2} = 49.61.$$

   Since this value is in the critical region (Figure 8.14), it follows that H_0 should be rejected. Consequently, Elmer would feel that there is sufficient evidence to reject the null hypothesis. According to the criterion he has established, he should purchase the new machine. (You may have noticed that the graph in Figure 8.14 is rather symmetric. This is due to the fact that for a large

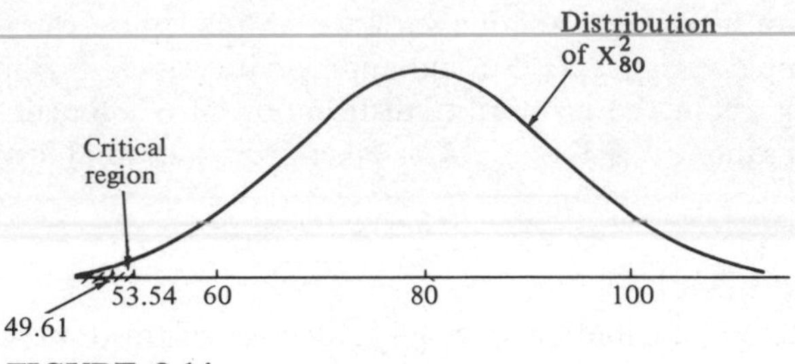

FIGURE 8.14

number of degrees of freedom, the chi-square distribution is only slightly skewed.)

In general, a test of

$$H_0 : \sigma^2 \geq \sigma_0^2 \quad \text{vs.} \quad H_a : \sigma^2 < \sigma_0^2,$$

where σ_0^2 is some specified constant value, will reject H_0 at a significance level α if

$$\frac{(n-1)S^2}{\sigma^2} \tag{8.21}$$

is less than or equal to the critical value $\chi^2_{n-1,\alpha}$. If the hypotheses are

$$H_0 : \sigma^2 \leq \sigma_0^2 \quad \text{vs.} \quad H_a : \sigma^2 > \sigma_0^2,$$

the null hypothesis will be rejected if the test statistic (8.21) is greater than or equal to $\chi^2_{n-1,1-\alpha}$. If the alternative hypothesis is two-sided, that is, in testing

$$H_0 : \sigma^2 = \sigma_0^2 \quad \text{vs.} \quad H_a : \sigma^2 \neq \sigma_0^2,$$

the null hypothesis will be rejected if the test statistic (8.21) falls at or below $\chi^2_{n-1,\alpha/2}$ or at or above $\chi^2_{n-1,1-\alpha/2}$. Keep in mind, however, that these tests are based on the assumption that the observations come from a normal distribution.

In Elmer's situation he wanted to compare an unknown variance (that of a new machine) with a known variance (that of the machine currently in use). In other situations it may be necessary to compare two variances, both of which are unknown.

Example 8.19 Ben West is the owner of the West Wrist Watch Works and needs to decide which of two movements to buy for use in a new model watch. One movement is made by the Good Time Company and the other by the Time Lee Company. The movements cost about the same, so Ben will buy the movements that keep the best time. This will be determined by the variability in the amount of time gained in 1 week after a preliminary adjustment. Ben bought 10 of each type of movement and found that after a week the watches using the movements from the Good Time Company (company 1) had gained

.42 −1.87 .23 −.30 −.10 −1.27 .03 −1.48 −2.68 .79

minutes each. (Here a negative value indicates time lost.) The watches using movements from the Time Lee Company (company 2) gained

2.42 −1.73 2.17 .45 1.01 −.91 .62 −.53 1.45 −2.16

minutes. If he finds that the watches keep time equally well, he will compare them on a second criterion—how loud they tick. What decision should Ben make?

To solve this problem, we need to find a suitable test statistic. This problem calls for comparison of two variances and we found in Section 7.9 that such a comparison is best made by considering the ratio of the variances, say σ_1^2/σ_2^2. It follows that a suitable candidate for the test statistic is the ratio of sample variances, S_1^2/S_2^2. To find critical regions for test procedures using this ratio as a test statistic, it will be necessary to know the distribution of S_1^2/S_2^2. We saw in Section 7.9 that if $X_{1,1}, X_{1,2}, \ldots, X_{1,n_1}$ and $X_{2,1}, X_{2,2}, \ldots, X_{2,n_2}$ are independent samples taken from normal distributions having variances σ_1^2 and σ_2^2, respectively, then

$$F = \frac{S_1^2/\sigma_1^2}{S_2^2/\sigma_2^2} \tag{8.22}$$

will have an F distribution with $n_1 - 1$ and $n_2 - 1$ degrees of freedom. Consequently, if we standardize the ratio S_1^2/S_2^2 as in equation (8.22), we can use percentiles of the F distribution as critical values. Let us now proceed through the outline of a test procedure.

Example 8.19 (*continued*)

1. $H_0 : \sigma_1^2/\sigma_2^2 = 1$ vs. $H_a : \sigma_1^2/\sigma_2^2 \neq 1$. Note that if the null hypothesis is true, the two movements will have the same variability.
2. $\alpha = .05$. This value was not specified by the problem. In the absence of further information about the consequences of each type of error, this is a reasonable figure to use.
3. $n_1 = n_2 = 10$.
4. The test statistic will be S_1^2/S_2^2, standardized as in (8.22).
5. The null hypothesis will be rejected if the test statistic is either too large or too small, since the alternative is two-sided. The critical values will be percentiles of the F distribution chosen in such a way as to make α equal to .05. That is,

$$\begin{aligned}\alpha = .05 &= P[\text{reject } H_0 \mid H_0 \text{ true}]\\ &= P[(S_1^2/S_2^2 \leq c_1) \text{ or } (S_1^2/S_2^2 \geq c_2) \mid H_0 \text{ true}]\\ &= P[S_1^2/S_2^2 \leq c_1 \mid H_0 \text{ true}] + P[S_1^2/S_2^2 \geq c_2 \mid H_0 \text{ true}].\end{aligned}$$

The desired significance level can be obtained by setting each term equal to $\alpha/2 = .025$:

$$P[S_1^2/S_2^2 \leq c_1 \mid H_0 \text{ true}] = \frac{\alpha}{2} = .025 \tag{8.23}$$

and

$$P[S_1^2/S_2^2 \geq c_2 \mid H_0 \text{ true}] = \frac{\alpha}{2} = .025. \tag{8.24}$$

Since

$$\frac{S_1^2/\sigma_1^2}{S_2^2/\sigma_2^2} = \frac{S_1^2}{S_2^2} \cdot \frac{\sigma_2^2}{\sigma_1^2},$$

it follows that the ratio S_1^2/S_2^2 can be "standardized" by multiplying by σ_2^2/σ_1^2. This quantity will have an F distribution. Consequently, (8.23) can be rewritten as

$$P\left[\frac{S_1^2}{S_2^2}\frac{\sigma_2^2}{\sigma_1^2} \leq c_1\frac{\sigma_2^2}{\sigma_1^2}\middle| H_0 \text{ true}\right] = P\left[F_{n_1-1,n_2-1} \leq c_1\frac{\sigma_2^2}{\sigma_1^2}\middle| H_0 \text{ true}\right] = \frac{\alpha}{2}.$$

This last equality will be true only if

$$c_1\left(\frac{\sigma_2^2}{\sigma_1^2}\right) = f_{n_1-1,n_2-1,\alpha/2}.$$

Hence we see that the critical values when using the standardized test statistic defined in (8.22) will simply be percentiles of the F distribution; that is, H_0 will be rejected if (8.22) is greater than or equal to

$$f_{n_1-1,n_2-1,1-\alpha/2} = f_{9,9,.975} = 4.03$$

or is less than or equal to

$$f_{n_1-1,n_2-1,\alpha/2} = f_{9,9,.025} = \frac{1}{f_{9,9,.975}} = \frac{1}{4.03} = .248.$$

6. For this problem we calculate $s_1^2 = 1.284$ and $s_2^2 = 2.474$. In standardizing the ratio S_1^2/S_2^2 we must use the ratio σ_2^2/σ_1^2. When H_0 is true, however, this ratio will be equal to 1. Consequently, the observed value of the standardized test statistic is

$$\left(\frac{s_1^2}{s_2^2}\right)\left(\frac{\sigma_2^2}{\sigma_1^2}\right) = \left(\frac{1.284}{2.474}\right)(1) = .519.$$

Since this ratio is not in the critical region (see Figure 8.15), H_0 should not be rejected. In view of the fact that H_0 was not rejected, what should Ben do? Since the evidence is not strong enough to reject the assumption that the variability for the two movements is the same, he should compare them on a second criterion. (What action would he have taken in this case had the critical value been .520 instead of .248?)

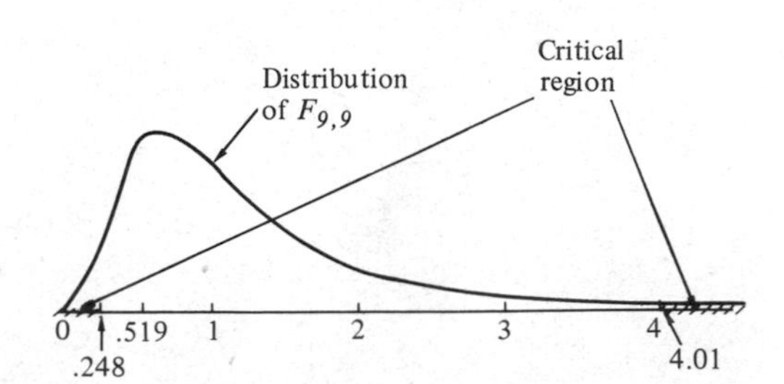

FIGURE 8.15

In this example it was desired to see if the variances σ_1^2 and σ_2^2 were equal (i.e., if the ratio of variances was equal to 1). In other situations it might be desirable to see if the ratio of variances is equal to some constant other than 1, say a constant k. By going through arguments similar to those in Example 8.19, you can see that a test of hypotheses

$$H_0 : \frac{\sigma_1^2}{\sigma_2^2} = k \qquad \text{vs.} \qquad H_a : \frac{\sigma_1^2}{\sigma_2^2} \neq k,$$

where k is some specified constant, will use as a test statistic

$$F = \frac{S_1^2}{S_2^2} \cdot \frac{1}{k} \tag{8.25}$$

and the null hypothesis will be rejected at a significance level α if the observed value of F falls at or below $f_{n_1-1,n_2-1,\alpha/2}$ or is at or above $f_{n_1-1,n_2-1,1-\alpha/2}$. The same test statistic would be used for one-sided alternatives with critical values chosen accordingly. Specifically, if

$$H_0 : \frac{\sigma_1^2}{\sigma_2^2} \leq k \qquad \text{vs.} \qquad H_a : \frac{\sigma_1^2}{\sigma_2^2} > k \tag{8.26}$$

are being tested, H_0 will be rejected if F [defined by (8.25)] exceeds $f_{n_1-1,n_2-1,1-\alpha}$, while if

$$H_0 : \frac{\sigma_1^2}{\sigma_2^2} \geq k \qquad \text{vs.} \qquad H_a : \frac{\sigma_1^2}{\sigma_2^2} < k, \tag{8.27}$$

then this null hypothesis will be rejected if F is less than $f_{n_1-1,n_2-1,\alpha}$. [Note that one-sided alternatives can always be put in the form of that of (8.26) by appropriately labeling the populations as "1" or "2." However, it is not necessary to do this.]

Example 8.20 Happy Joe, sales representative for the Good Time Company, claims that their electronic movement is less than one-third as variable as their standard spring-driven movement. Ben has requested some electronic movements for testing, believing that if Joe's claim is correct, the extra expense of electronic movements will be well spent. However, to discourage Joe from making exaggerated claims, Ben will not use these electronic movements if the testing does not convincingly support Joe's claim. What should Ben do if the 12 movements that Joe left for testing gained

−.04	−.61	.14	−.19	−.18	.30
−.55	−.16	−.20	−.32	.60	.49

minutes after 1 week? Use a significance level of .05. Also use the data on the standard movements that Ben found in Example 8.19.

Care must be taken here in setting up the null and alternative hypotheses. If we let σ_E^2 represent the variance of electronic movements and σ_S^2 that of spring movements, then Joe's claim is that $\sigma_E^2 < \frac{1}{3}\sigma_S^2$ or equivalently $\sigma_E^2/\sigma_S^2 < \frac{1}{3}$. Since Ben will believe Joe's claim only if the test results convincingly support it, Ben would set up

1. $H_0 : \sigma_E^2/\sigma_S^2 \geq \frac{1}{3}$ vs. $H_a : \sigma_E^2/\sigma_S^2 < \frac{1}{3}$. Continuing with the hypothesis testing outline, we find:
2. $\alpha = .05$.
3. $n_E = 12, n_S = 10$.
4. Assuming that the amount of time gained by each of these movements

follows a normal distribution, the appropriate test statistic is $F = (S_E^2/S_S^2)(1/k)$ [see equation (8.25)].

5. Since these hypotheses are of the form (8.27), the null hypothesis will be rejected if F is less than or equal to

$$f_{n_E-1,n_S-1,\alpha} = f_{11,9,.05} = \frac{1}{f_{9,11,.95}} = \frac{1}{2.90} = .3448.$$

6. Using the data given in the problem, we find that $S_E^2 = .1436$. We see that the constant k is equal to $\frac{1}{3}$, so $1/k = 3$. In Example 8.19 we found $S_S^2 = 1.284$, so the observed value of the test statistic is

$$f = \left(\frac{s_E^2}{s_S^2}\right)\left(\frac{1}{k}\right) = \left(\frac{.1436}{1.284}\right)(3) = .3354.$$

Since this is less than the critical value, H_0 will be rejected. Ben is satisfied that Joe's claim is correct and so he will purchase these electronic movements.

In Sections 7.7 and 8.6 we discussed problems of estimation and tests of hypotheses for the difference of two means, $\mu_1 - \mu_2$, for normal distributions. The appropriate choice of interval estimate or of test statistic in the situation where the population variances σ_1^2 and σ_2^2 are unknown depends on whether these unknown variances are equal or unequal. For example, if the variances are equal, the appropriate test statistic would be T defined by (8.18) using the pooled estimate S_p^2 of the common variance. On the other hand, if the variances are unequal, the appropriate test statistic would be given by (8.19). One way to decide on which approach might be appropriate is to follow the suggestion made in Section 7.9. An equivalent way is the following:

1. Test $H_0 : \sigma_1^2/\sigma_2^2 = 1$ vs. $H_a : \sigma_1^2/\sigma_2^2 \neq 1$ using a level of significance of .10 (or perhaps .20).
2. If H_0 is not rejected, assume that the variances are equal and proceed accordingly [e.g., use T as defined by (8.18) as test statistic], but if
3. H_0 is rejected, assume that the variances are unequal. [This would imply that T as defined by (8.19) should be used as the test statistic.]

Example 8.21 In Example 8.16 Henry J. Ford was trying to decide whether to give equal raises to his two employees, Edsel and Lincoln. It was assumed that the variances of the amount of time required for each man to fill an order were the same. (This allowed us to use the pooled estimate of the variance, S_p^2, in a test statistic.) Is the assumption of equal variances reasonable? Use the information that the sample variance based on 30 orders was (4.3 minutes)^2 for Edsel and was (5.5 minutes)^2 for Lincoln.

To see if this assumption is reasonable, we perform the following test of hypotheses.

1. $H_0 : \sigma_1^2/\sigma_2^2 = 1$ vs. $H_a : \sigma_1^2/\sigma_2^2 \neq 1$.
2. $\alpha = .10$.

3. $n_1 = n_2 = 30$.
4. From (8.24) we see that since $k = 1$, the appropriate test statistic is $F = S_1^2/S_2^2$.
5. Since the test statistic will have an F distribution when H_0 is true, it follows that H_0 will be rejected if the observed value of the test statistic falls at or below $f_{29,29,.05} = 1/f_{29,29,.95}$ or at or above $f_{29,29,.95}$. From Table A8 we see that $f_{30,29,.95} = 1.85$ and $f_{24,29,.95} = 1.90$, so we can use $f_{29,29,.95} \approx 1.86$ and $1/f_{29,29,.95} \approx .538$.
6. As in Example 8.16, we let Edsel's data be denoted by subscript 1. The observed value of the test statistic is

$$f = \frac{s_1^2}{s_2^2} = \frac{(4.3)^2}{(5.5)^2} = .611.$$

Since this is not in the critical region, H_0 is not rejected. According to the criteria we set up, the assumption of equal variances is reasonable and the appropriate test statistic to be used for tests about the difference of the two means would be T, as defined by (8.18). This was the test statistic used in Example 8.16.

8.8 TESTS FOR PROPORTIONS

In any manufacturing process there is always a possibility of an imperfect product due to equipment or human error. For example, packages may not be filled to within specified tolerances, screws may be made without a slot, battery cases may have cracks, light bulbs may be improperly sealed, and so on. A certain small proportion of defective products may be tolerable, but if the proportion exceeds a specified amount, perhaps 1 or 5 or 10%, the process may have to be stopped and some adjustments made. (The "adjustment" may be as simple as turning a dial or waking up the operator, or it may require a complex overhaul or the hiring of new personnel.) A quality control engineer might be interested in testing a hypothesis about the proportion of defective products being produced.

Individuals involved in politics are also frequently concerned with proportions. Prior to election a candidate for office will be vitally concerned with the proportion of the electorate in his or her favor. After election the concern may be with the proportion of constituents for or against a particular issue. Hence a politician may be interested in testing hypotheses about proportions.

Many other examples could be given of individuals interested in tests about proportions. The key to performing such tests is finding a suitable test statistic and using the probability distribution of that statistic to find critical regions for the test. As in other testing situations, it is possible to use the sample counterpart as the test statistic. To test a hypothesis about a proportion p, where p can represent the probability of a "success" for some dichotomous trial, n independent observations can be taken. The number of successes in these n trials, call it X, will have a binomial distribution. Either X or X/n can be used as a test statistic. If n is less than or equal to 20, the binomial tables could be used to find critical values. If n is greater than 20, an appropriate approximate distribution, usually the normal distribution, can be used to find critical values. We will illustrate the former case by an example to point out one difficulty encountered in performing tests of hypotheses using a discrete random variable (in this case a binomial random variable) as a test statistic. Our main emphasis, however, will be on using the normal approximation for test purposes.

Example 8.22 The Cherry Berry Cookie Company packages cookies 20 to a box. Gerry, the quality control supervisor, would like the proportion of broken cookies to be 20% or less, but she is not willing to stop the packaging process unless there is convincing evidence that this proportion is being exceeded. If Gerry opens a single box at random from those being produced and counts the number of broken cookies, what critical value should she use if she wants α, the probability of a type I error, to be equal to .01? (Assume that cookies in a box are broken or not broken independently of one another.)

To work this problem we can go through the first five steps of the hypothesis testing outline, since our primary interest is in determining a critical value.

1. Since Gerry wishes to believe that p, the proportion of broken cookies, is less than or equal to .20 unless there is evidence to the contrary, the appropriate null and alternative hypotheses are
$$H_0: p \leq .20 \qquad \text{vs.} \qquad H_a: p > .20.$$
2. $\alpha = .01$.
3. Since one box of 20 cookies is to be inspected, $n = 20$.
4. The appropriate test statistic is either X/n, the sample proportion of broken cookies, or X, the total number of broken cookies among the 20.
5. Intuitively, the null hypotheses will be rejected only if X/n is greater than or equal to some critical value c, that is, if
$$\frac{X}{n} \geq c,$$
or, equivalently, if X is greater than or equal to some constant c^*. The two criteria are equivalent, since
$$\frac{X}{n} \geq c \Longleftrightarrow X \geq nc \equiv c^*.$$
Since the null hypothesis is composite, we find c^* from
$$\text{(8.27)} \qquad \alpha = \max P[\text{reject } H_0 \mid H_0 \text{ true}] = \max P[X \geq c^* \mid p \leq .20] = P[X \geq c^* \mid p = .20].$$
Since we want α to be equal to .01, we must find a constant c^* to satisfy
$$P[X \geq c^* \mid p = .20] = .01.$$

Examining the cumulative binomial tables with $n = 20$ and $p = .20$, we find that

$$
\begin{aligned}
& P[X \geq 10 \mid n = 20, p = .20] = .0026 \\
& P[X \geq \ 9 \mid n = 20, p = .20] = .0100 \\
(8.28) \qquad & P[X \geq \ 8 \mid n = 20, p = .20] = .0322 \\
& P[X \geq \ 7 \mid n = 20, p = .20] = .0867.
\end{aligned}
$$

It follows that the appropriate choice of c^* is 9, so the null hypothesis would be rejected if

$$X \geq c^* = 9 \qquad \text{or} \qquad \frac{X}{n} \geq \frac{c^*}{n} = c = \frac{9}{20} = .45.$$

In the context of this problem, then, Gerry should stop the packaging process only if the number of broken cookies in the box is nine or more.

One difficulty encountered with tests using a discrete random variable is the inability to easily obtain a particular prespecified significance level. For example, in the previous problem, what critical value would Gerry use if she wanted $\alpha = .05$? From (8.28) we see that a critical value of 8 will make the significance level equal to .0322 (which is too small), while a value of 7 will make it .0867 (which is too large). Furthermore, since X is a discrete random variable, there is no middle ground between these two values! There are some methods of circumventing this problem, but they are beyond the scope of this book.

What is your feeling about the inspection procedure for controlling the proportion of broken cookies employed by the quality control supervisor? Do you think there is a good chance that a high true proportion of broken cookies, say $p = .25$ or .30, might go undetected by this procedure? (See Exercise 38.) How could the procedure be improved? Do you think it could be improved by examining more cookies? It seems logical that obtaining more information by examining more cookies will result in an improved procedure and this, in fact, is generally the case. When examining boxes of cookies, the costs of sampling are relatively small, so taking a larger sample should present no difficulty. (Of course, there are situations such as firing off rockets, doing large animal research, and so on, where costs are high and large samples cannot easily be obtained.)

Example 8.23 Gerry, the quality control supervisor at the Cherry Berry Cookie Company, will choose 20 boxes of 20 cookies each for examination to see if the proportion of broken cookies is at a satisfactory level. If she counts the total number of broken cookies, what critical value should she choose to have a significance level of .01?

The first four steps of the hypothesis-testing outline will be the same as in Example 8.22. Following the same reasoning that led to (8.27), we wish to find a constant c^* to satisfy

$$P[X \geq c^* \mid p = .20] = .01.$$

Since X will have a binomial distribution with $n = 400$ and $p = .20$, we can use the normal approximation to the binomial distribution to get the critical value (approximately). Using subscripts to distinguish between the binomial random variable and its normal approximation, we find that

$$P[X_B \geq c^* \,|\, n = 400, p = .20] \approx P[X_N \geq c^* \,|\, \mu = np = 80, \sigma^2 = npq = 64]$$

$$= P\left[\frac{X_N - np}{\sqrt{npq}} \geq \frac{c^* - 80}{8}\right] = .01.$$

Since $P[Z \geq z_{.99} = 2.326] = .01$, it follows that c^* can be found by solving

$$\frac{c^* - 80}{8} = 2.326$$

(i.e., $c^* = 98.6$). This is not an integer, but using a value of 99 will give a critical value for which the significance level is approximately .01. According to the criterion set up, Gerry should stop the packaging process only if 99 or more of 400 cookies are broken. (By working Exercise 42 and comparing the results with Exercise 38, you will see that this inspection procedure is better than the one described in Example 8.22, where only 20 cookies were inspected.)

Whenever the sample size is large enough to justify using the normal approximation to the binomial distribution (see Section 4.10), the critical values for tests about a proportion p can be found by using percentiles of the standard normal distribution. In testing

$$H_0 : p \leq p_0 \qquad \text{vs.} \qquad H_a : p > p_0$$

with a significance level α, reject H_0 if X satisfies

$$\frac{X - np_0}{\sqrt{np_0 q_0}} \geq z_{1-\alpha}. \tag{8.29}$$

[An equivalent expression for (8.29) can be obtained by dividing the numerator and denominator of the left-hand side of (8.29) by n to get

$$\frac{(X/n) - p_0}{\sqrt{p_0 q_0 / n}} \geq z_{1-\alpha}.$$

Similar expressions could be given in the next two cases to be discussed as well.] In testing $H_0 : p \geq p_0$ against $H_a : p < p_0$, reject H_0 if X satisfies

$$\frac{X - np_0}{\sqrt{np_0 q_0}} \leq z_\alpha.$$

Finally, a test with a two-sided alternative, $H_0 : p = p_0$ vs. $H_a : p \neq p_0$, would reject H_0 if X satisfies

$$\frac{X - np_0}{\sqrt{np_0 q_0}} \geq z_{1-\alpha/2} \qquad \text{or} \qquad \frac{X - np_0}{\sqrt{np_0 q_0}} \leq z_{\alpha/2}.$$

Example 8.24 Congressman Jay Porkbarol favors construction of a dam on the Plugadup river at a site in his district. He will vote in favor of funding the Plugadup dam project unless he can be convinced that over 60% of his constituents oppose construction of the dam. Using a .05 level of significance, how should he vote if a random sample of 600 voters in his district showed that 375 opposed construction of the dam?

In view of Congressman Porkbarol's predisposition in favor of funding this project, the hypotheses that he would set up would be

1. $H_0 : p \leq .60$ vs. $H_a : p > .60$, where p is the proportion of voters opposing the funding of the project.
2. $\alpha = .05$.
3. $n = 600$
4. The number of voters in the sample of 600 opposing the funding of the project would be a hypergeometric random variable if sampling was done without replacement. However, since the sample size of 600 is small relative to the total number of voters in a congressional district, the binomial approximation to the hypergeometric distribution should be good. Since the sample size is large enough to justify using the normal approximation to the binomial distribution, the appropriate test statistic is $(X - np_0)/\sqrt{np_0q_0}$, where $p_0 = .60$.
5. From (8.29) we see that the null hypothesis should be rejected if the observed value of the test statistic is greater than or equal to the critical value of $z_{1-\alpha} = z_{.95} = 1.645$.

   The observed value of the test statistic is

$$\frac{x - np_0}{\sqrt{np_0q_0}} = \frac{375 - 600(.60)}{\sqrt{600(.60)(.40)}} = \frac{375 - 360}{\sqrt{144}} = 1.25.$$

   Since this value does not exceed the critical value of 1.645, the null hypothesis should not be rejected. In view of the criteria he set up, Congressman Porkbarol should vote in favor of funding the project. (You can see, of course, that the criteria set forth by the congressman reflect his predisposition. The same sample results would be interpreted quite differently by a group opposed to this project!)

In an effort to determine whether constituents in all parts of his district felt the same about the Plugadup dam project, the congressman might take two samples of voters. One sample might come from those counties in and around the proposed dam site, while the second might come from counties farther away. (He might also compare urban areas with rural areas or use some other comparison.) How could sample information so obtained be used in this case? In particular, let p_1 and p_2 represent the proportion of voters in the first and second regions, with X_1 and X_2 being the number in the samples who oppose construction. If there is no difference in these proportions, then p_1 and p_2 will be equal. Consequently, it would be desirable to test

$$H_0 : p_1 - p_2 = 0 \quad \text{vs.} \quad H_a : p_1 - p_2 \neq 0. \tag{8.30}$$

To test these hypotheses it will be necessary to find a suitable test statistic. Using the sample counterpart we see that

$$\frac{X_1}{n_1} - \frac{X_2}{n_2},$$

the difference in sample proportions, is a logical candidate. Of course, to be able to find critical values, it is necessary to find the probability distribution of this difference or an approximation to that distribution. If both sample sizes are large enough to justify using the normal approximation to the binomial, we can say that

$$\left(\frac{X_1}{n_1}\right) \sim N\left(p_1, \frac{p_1 q_1}{n_1}\right) \quad \text{and} \quad \left(\frac{X_2}{n_2}\right) \sim N\left(p_2, \frac{p_2 q_2}{n_2}\right).$$

Since a linear combination of normal random variables will have a normal distribution (Section 5.7), it follows that *if the samples are chosen independently*, then

(8.31)
$$\left[\left(\frac{X_1}{n_1} - \frac{X_2}{n_2}\right) - (p_1 - p_2) \middle/ \sqrt{\frac{p_1 q_1}{n_1} + \frac{p_2 q_2}{n_2}}\right]$$

will have a standard normal distribution (approximately). The standardized test statistic then is given by (8.31) and the null hypothesis would be rejected if the observed value of this test statistic were to be at least $z_{1-\alpha/2}$ or at most $z_{\alpha/2}$ [for a two-sided alternative as in (8.30)].

Now say that samples of size $n_1 = 400$ and $n_2 = 200$ were observed with $x_1 = 240$ and $x_2 = 135$. What would the observed value of the test statistic (8.31) be if $H_0 : p_1 - p_2 = 0$ is in fact true? The numerator would be

$$\left(\frac{x_1}{n_1} - \frac{x_2}{n_2}\right) - (p_1 - p_2) = \left(\frac{240}{400} - \frac{135}{200}\right) - (0)$$
$$= -.075.$$

Notice that we substituted zero for $p_1 - p_2$, since if H_0 is in fact true, the value of $p_1 - p_2$ will be zero. Now how do we evaluate the denominator of (8.31)? If the null hypothesis is actually true, p_1 and p_2 will be equal, but their common value is not specified! It could be that $p_1 = p_2 = .10$ or .37 or .62 or any other value between zero and 1. Since we do not know what the value of this parameter is, what can we do? We can replace it by a point estimate of the common value. This will make the approximation of the test statistic still more approximate, but the approximation is still a good one.

Let us turn our attention for the moment to the problem of finding a point estimate of the common value p. (Say that $p_1 = p_2 = p$.) If p_1 and p_2 are equal, X_1/n_1 and X_2/n_2 are both point estimates of the same parameter value. We could combine these estimates by looking at the *total number* of successes divided by the *total number* of trials, namely, $(X_1 + X_2)/(n_1 + n_2)$. We can denote this estimate of p by $\hat{p}$:

(8.32)
$$\hat{p} = \frac{X_1 + X_2}{n_1 + n_2}.$$

If $p_1 = p_2 = p$, the denominator of (8.31) would be

$$\sqrt{\frac{p_1 q_1}{n_1} + \frac{p_2 q_2}{n_2}} = \sqrt{\frac{pq}{n_1} + \frac{pq}{n_2}} = \sqrt{pq\left(\frac{1}{n_1} + \frac{1}{n_2}\right)}.$$

An estimate of this quantity would be given by

$$\sqrt{\hat{p}\hat{q}\left(\frac{1}{n_1}+\frac{1}{n_2}\right)}.$$

Using the observed values given earlier, we would find

$$\hat{p}=\frac{x_1+x_2}{n_1+n_2}=\frac{240+135}{400+200}=.625,$$

so

$$\sqrt{\hat{p}\hat{q}\left(\frac{1}{n_1}+\frac{1}{n_2}\right)}=\sqrt{(.625)(.375)(\tfrac{1}{400}+\tfrac{1}{200})}=.0436.$$

Summarizing this discussion, we can say that the appropriate test statistic for testing hypotheses $H_0: p_1-p_2=0$ vs. $H_a: p_1-p_2\neq 0$ would be

$$\left(\frac{X_1}{n_1}-\frac{X_2}{n_2}\right)\bigg/\sqrt{\hat{p}\hat{q}\left(\frac{1}{n_1}+\frac{1}{n_2}\right)}, \tag{8.33}$$

where $\hat{p}$ is defined by (8.32). This same test statistic would be used for one-sided alternatives as well, namely,

$$H_0: p_1-p_2\leq 0 \quad \text{vs.} \quad H_a: p_1-p_2>0$$

or

$$H_0: p_1-p_2\geq 0 \quad \text{vs.} \quad H_a: p_1-p_2<0.$$

The appropriate critical values would be percentiles of the standard normal distribution.

Example 8.25 Does Congressman Porkbarol have reason to believe that constituents living near the proposed dam site feel the same about the Plugadup dam project as those living farther away if 240 of 400 voters living near the proposed site are opposed to the dam, while 135 of 200 of those away from the site are opposed? Use a .05 level of significance.

1. In this case it is appropriate to use a two-sided alternative,
$$H_0: p_1-p_2=0 \quad \text{vs.} \quad H_a: p_1-p_2\neq 0.$$
2. $\alpha=.05$.
3. $n_1=400, n_2=200$.
4. Since the sample sizes are relatively large, the normal approximation will be appropriate, so the test statistic given in (8.33) is appropriate.
5. The null hypothesis will be rejected if the observed value of the test statistic is $z_{\alpha/2}=z_{.025}=-1.96$ or less or $z_{1-\alpha/2}=z_{.975}=+1.96$ or more.
6. The observed value of the test statistic is
$$\left(\frac{x_1}{n_1}-\frac{x_2}{n_2}\right)\bigg/\sqrt{\hat{p}\hat{q}\left(\frac{1}{n_1}+\frac{1}{n_2}\right)}=\frac{-.075}{.0436}=-1.72.$$
Since this value is not in the critical region, H_0 should not be rejected. The congressman can believe that constituents near and away from the proposed dam site feel about the same about the project.

Example 8.26 Antony, the owner of Antony's Pizzeria, is interested in knowing the effect on sales of a coupon good for 50 cents toward the purchase of any size pizza. Antony sponsors a "Birthday Club," where children can get a free small pizza during the week of their birthday. He randomly selects 300 families from the Birthday Club list and sends a coupon to half these families. The coupon expires in 2 weeks and is good only for the family to which it was sent. After 2 weeks have passed he found that 47 coupons had been returned. Calls to the 150 families who did not receive coupons revealed that 31 of them had bought at least one pizza from Antony during the past 2 weeks. Can Antony conclude that the coupon helped increase sales among Birthday Club families? Take $\alpha = .05$.

1. In this case a one-sided alternative is appropriate. Since Antony would not want to send out coupons unless there was evidence that they really helped sales, the hypothesis should be

$$H_0 : p_1 - p_2 \leq 0 \qquad \text{vs.} \qquad H_a : p_1 - p_2 > 0,$$

where p_1 represents the proportion of families receiving coupons who bought at least one pizza from Antony and p_2 corresponds to the proportion of "noncoupon" families buying at least one pizza.

2. $\alpha = .05$.
3. $n_1 = n_2 = 150$.
4. The test statistic would be given by (8.33).
5. In view of the alternative hypothesis being that $p_1 - p_2$ is positive, H_0 would be rejected only if the test statistic is $z_{1-\alpha} = z_{.95} = 1.645$ or more.
6. The value of $\hat{p}$ [equation (8.32)] is

$$\hat{p} = \frac{47 + 31}{150 + 150} = .26,$$

so the observed value of the test statistic would be

$$\left(\frac{x_1}{n_1} - \frac{x_2}{n_2}\right)\Big/\sqrt{\hat{p}\hat{q}\left(\frac{1}{n_1} + \frac{1}{n_2}\right)} = (.3133 - .2067)/\sqrt{(.26)(.74)(\tfrac{1}{150} + \tfrac{1}{150})}$$

$$= \frac{.1066}{.0506} = 2.11.$$

Since this value is in the critical region, H_0 should be rejected. Antony can conclude that the coupon offer had a positive effect on sales for Birthday Club families. (Could he conclude from this that sending coupons to the public at large would have a positive effect on sales? Why?)

8.9 *p* VALUES

The test procedures that we have discussed in this chapter have all been concerned with deciding between a null hypothesis and an alternative hypothesis. The decision rule (i.e., rule used to decide between these) is based on whether the observed value of a test statistic does or does not fall into a critical region. The approach we have used has been to treat the problem simply in terms of whether or not the test statistic is in the critical region, with no regard for *how far into* or *how far away from* the critical region the observed value actually is. Such information would, in many situations, be more informative than simply a statement that "H_0 was rejected when α was .05," for example.

Consider the following illustration. Douglas Furr is a worker in a tree nursery where white pine seedlings are being grown. Under normal conditions these seedlings will grow 25 centimeters on the average over a certain time period, the actual growth approximately following a normal distribution with a standard deviation of 3.2 cm. Doug is experimenting with different nutrients to add to the soil to increase growth. If a particular nutrient combination appears to be promising, he will present the results to his supervisor, who will then oversee more extensive tests. A group of 16 seedlings given a particular nutrient has been growing for the required time. What report should Doug give his supervisor if he observes a sample mean growth of 26.30 if he tests

$$H_0 : \mu \leq 25 \qquad \text{vs.} \qquad H_a : \mu > 25$$

using $\alpha = .05$? What if the sample mean were 25.8? 26.33? 27.2?

The appropriate test statistic would be either $\bar{X}$, the sample mean, or the standardized sample mean. In either case the null hypothesis would be rejected for values that are "too large." In terms of the standardized test statistic, the critical value would be $z_{1-\alpha} = z_{.95} = 1.645$, while in terms of $\bar{X}$ the critical value would be found from

$$\frac{c - \mu}{\sigma/\sqrt{n}} = \frac{c - 25}{3.2/\sqrt{16}} = 1.645,$$

so $c = 25 + 1.645(.8) = 26.316$. (See Example 8.6 in Section 8.3 for a similar problem.) It is an easy matter now to compare each of the sample means to the critical value of 26.316. These values are all shown in Figure 8.16. You can see that the value 26.30 does not fall into the critical region, so, according to the decision rule set up, Doug would report to his supervisor that this particular nutrient combination is "not promising." Similarly, a sample mean of 25.8 would be considered "not promising," while the means of 26.33 and 27.2 would be reported as "promising."

If you were Doug's supervisor, would you be at all troubled by his method of

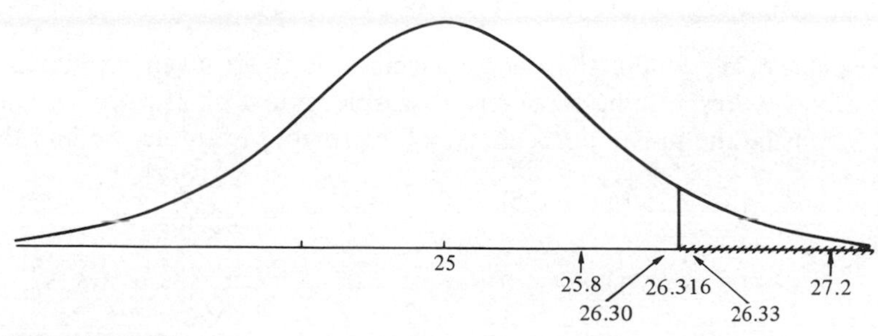

FIGURE 8.16

reporting? Consider the fact that sample means as close as 26.30 and 26.33 were classified differently (one being labeled as "not promising" and the other as "promising"), while sample means as far apart as 25.8 and 26.3 or as 26.33 and 27.2 were classified the same! Also, wouldn't you rather give priority in further tests to the nutrient combination that led to a sample mean of 27.2 than the one that had a mean of 26.33?

You should be able to see from this illustration that there are situations where it would be helpful to know how far into a critical region an observed test statistic falls rather than just that it falls into a critical region. Similarly, if it fails to fall in the critical region, it may be helpful to have an idea of how close it actually came to the critical region. One way of getting a numerical measure of this is to find the p value corresponding to the observed value of a test statistic.

Definition 8.5 The *p value* corresponding to the observed value x of a test statistic X is the maximum probability that the test statistic X will assume a value as or more extreme than x when H_0 is true.

The notion of "extreme" is determined by the alternative hypothesis, as we shall see in the following examples.

Example 8.27 Find the p value corresponding to an observed sample mean of 27.2 when testing the hypotheses

$$H_0 : \mu \leq 25 \qquad \text{vs.} \qquad H_a : \mu > 25,$$

where μ represents the mean of a normal distribution having a standard deviation of 3.2. Assume that the sample mean is based on 16 observations.

This is exactly the situation given for the study of the growth of white pine seedlings. Since in an ordinary test of these hypotheses H_0 would be rejected for values of $\bar{X}$ that are "too large," the statement in Definition 8.5 of "a value as or more extreme than x" would be interpreted here as "a value as large or larger than 27.2." That is,

$$\begin{aligned} p &= \max P[\bar{X} \geq 27.2 \mid H_0 \text{ true}] \\ &= P[\bar{X} \geq 27.2 \mid \mu = 25] \\ &= P\left[\frac{\bar{X} - \mu}{\sigma/\sqrt{n}} \geq \frac{27.2 - 25}{3.2/\sqrt{16}} = 2.75\right] = P[Z \geq 2.75] = .0030. \end{aligned}$$

Example 8.28 Under the same conditions as those given in Example 8.27, find the p values corresponding to observed sample means of 25.8, 26.30, and 26.33.

Using the same approach as in the previous example, we find that

$$p = P[\bar{X} \geq 25.8 \mid \mu = 25] = P\left[\frac{\bar{X} - \mu}{\sigma/\sqrt{n}} \geq \frac{25.8 - 25}{3.2/\sqrt{16}} = 1\right] = .1587$$

$$p = P[\bar{X} \geq 26.3 \mid \mu = 25] = P\left[\frac{\bar{X} - \mu}{\sigma/\sqrt{n}} \geq \frac{26.3 - 25}{3.2/\sqrt{16}} = 1.625\right] = .0521$$

$$p = P[\bar{X} \geq 26.33 \mid \mu = 25] = P\left[\frac{\bar{X} - \mu}{\sigma/\sqrt{n}} \geq \frac{26.33 - 25}{3.2/\sqrt{16}} = 1.6625\right] = .0482.$$

Each of these p values calculated in the last two examples is shown as the shaded regions of Figure 8.17. Also indicated on these graphs is the critical value 26.316 corresponding to a significance level of $\alpha = .05$. You should be able to see from these graphs that

> H_0 will be rejected at a significance level α if and only if the p value corresponding to the observed value of the test statistic is less than or equal to α.

Since the p value corresponding to $\bar{x} = 27.2$ is .0030, we can see immediately that H_0 would be rejected if α were equal to .05. In fact, H_0 would be rejected if α were .01 or even .005, but not if it were .001.

In performing tests of hypotheses using p values, the six-step outline used throughout this chapter would be modified somewhat. Steps 1 to 4, the statement of the null and alternative hypotheses, choice of significance level, sample size(s), and test statistic all remain the same. Step 5 would be modified to read: "Find the p value," and step 6 would say "reject H_0 if the p value is less than or equal to α."

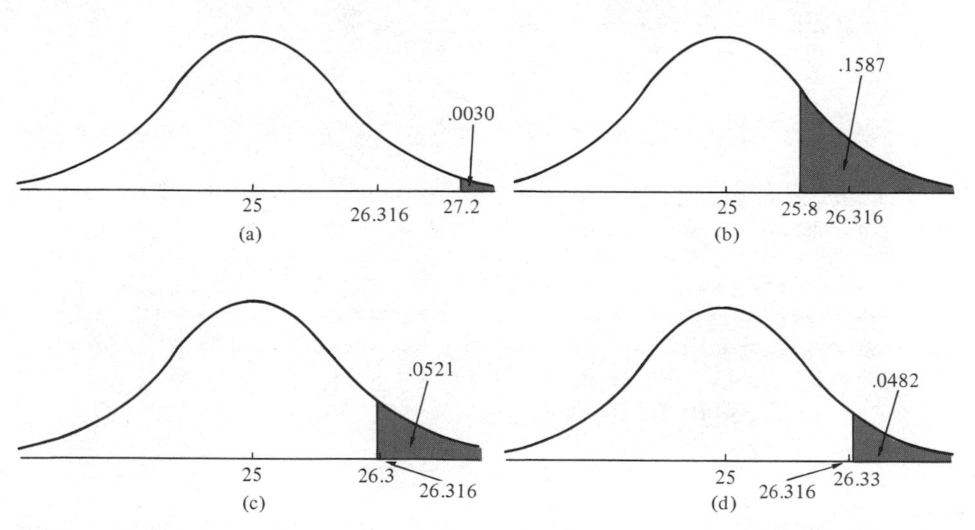

FIGURE 8.17

Example 8.29 To test

$$H_0 : \mu \geq 40 \qquad \text{vs.} \qquad H_a : \mu < 40,$$

where μ represents the mean of a normal distribution having a known standard deviation of 6.5, a random sample of 25 observations was chosen. What would be the p value corresponding to a sample mean of 41? Of 37.4? In each case would H_0 be rejected if α were .05? .01?

In this case the null hypothesis would be rejected for values of $\bar{X}$ that are "too small," so the statement in Definition 8.5 of "a value as or more extreme than x" would be interpreted here as "a value as small or smaller than x." In the first case we would find that

$$p = P[\bar{X} \leq 41 \mid \mu = 40] = P\left[\frac{\bar{X} - \mu}{\sigma/\sqrt{n}} \leq \frac{41 - 40}{6.5/\sqrt{25}} = 0.77\right] = .7793,$$

while in the second case

$$p = P[\bar{X} \leq 37.4 \mid \mu = 40] = P\left[\frac{\bar{X} - \mu}{\sigma/\sqrt{n}} \leq \frac{37.4 - 40}{6.5/\sqrt{25}} = -2\right] = .0228.$$

These p values are shown as the shaded regions in Figure 8.18. It should not be hard for you to see that H_0 would not be rejected for $\alpha = .05$ or $.01$ when $\bar{x} = 41$ is observed. However, since for $\bar{x} = 37.4$, $p = .0228$ is less than $\alpha = .05$, H_0 would be rejected in this case, but not when $\alpha = .01$.

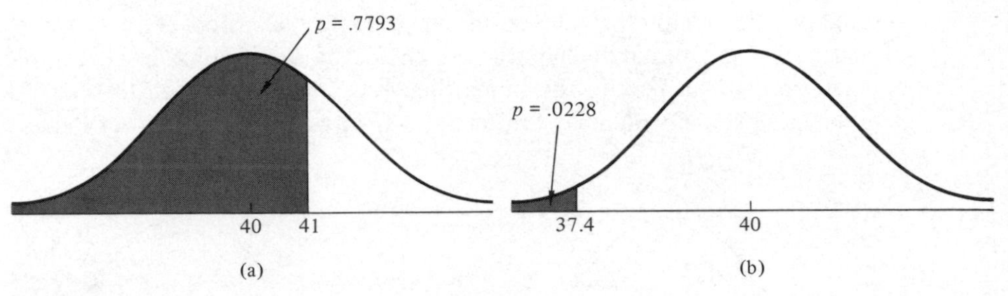

FIGURE 8.18

Example 8.30 Consider the same situation as in Example 8.29, but assume this time that σ is unknown. Specifically, to test

$$H_0 : \mu \geq 40 \qquad \text{vs.} \qquad H_a : \mu < 40,$$

a random sample of 25 observations was chosen from a normal distribution. The sample mean and sample standard deviation were found to be $\bar{x} = 37.4$ and $s = 6.35$. Find the p value. Would H_0 be rejected if α were .05? .01?

In this situation the appropriate test statistic is

$$T_{n-1} = \frac{\bar{X} - \mu}{S/\sqrt{n}},$$

which has a t distribution with 24 degrees of freedom. The observed value of the test statistic is

$$t = \frac{37.4 - 40}{6.35/\sqrt{25}} = -2.047.$$

The probability that T_{24} will be less than or equal to this value is

$$p = P(T_{24} \leq -2.047).$$

Because of the limited number of percentiles given in the t tables, we cannot find this probability exactly. We can say, however, that p will lie between .025 and .050, since

$$t_{24,.025} = -2.064 < -2.047 < -1.711 = t_{24,.050}.$$

In fact, p will be close to .025. Consequently, H_0 would be rejected if $\alpha = .05$ but not if $\alpha = .01$.

Example 8.31 In an effort to see if "1-liter" bottles of Froot Jooce are being properly filled, Joy chose 25 bottles and carefully measured the contents. From past experience she knows that the standard deviation of the amount in the bottles is about 8 ml. If she were testing

$$H_0 : \mu = 1000 \text{ ml} \quad \text{vs.} \quad H_a : \mu \neq 1000 \text{ ml},$$

what would the p value be if she determined the sample mean to be 998 ml?

For any two-sided alternative, the null hypothesis will be rejected if the test statistic is either too "large" or too "small." Consequently, we must consider extremes to be values on *either side* of the value specified by the null hypothesis. In this example the sample mean is 2 units below the hypothesized mean ($998 - 1000 = -2$). The p value would be the probability that the test statistic will *differ from* the hypothesized value by 2 or more units. That is,

$$p = P[\bar{X} - \mu \leq -2 \text{ or } \bar{X} - \mu \geq +2].$$

Although it has not been assumed that the amount of juice being put into the bottles follows a normal distribution, since the sample size is 25, we can invoke the Central Limit Theorem and find

$$\begin{aligned} p &= P[(\bar{X} - \mu \leq -2) \text{ or } (\bar{X} - \mu \geq +2)] = 2 \cdot P[\bar{X} - \mu \leq -2] \\ &= 2 \cdot P\left[\frac{\bar{X} - \mu}{\sigma/\sqrt{n}} \leq \frac{-2}{8/\sqrt{25}}\right] = 2 \cdot P[Z \leq -1.25] \\ &= 2(.1056) = .2112. \end{aligned}$$

The shaded regions in Figure 8.19 correspond to the p value for this test.

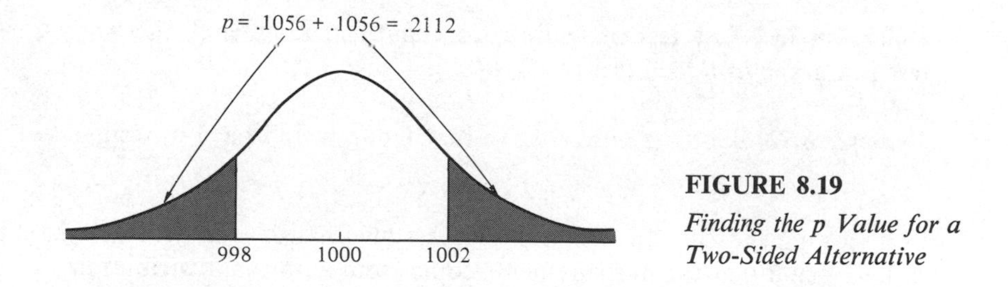

FIGURE 8.19
Finding the p Value for a Two-Sided Alternative

(Note that the p value would have been the same had the sample mean been observed to be 1002 ml.)

In finding the p value for two-sided alternatives when the distribution of the test statistic is not symmetric, take twice the probability of observing a value of the test statistic as or more extreme than the one observed given that H_0 is true.

8.10 POWER OF TESTS OF HYPOTHESES

In Section 7.12 we discussed the problem of how to determine a sample size to meet certain criteria in estimation of a parameter. For example, how large must the sample size n be in order to have a .90 probability, say, of estimating μ to within 2 units?

Similar questions can be asked in hypothesis-testing situations. We gave some attention to this matter in Section 8.2 when we discussed the interrelationships among α, β, n, and c for testing one simple hypothesis against another. Relationships such as these exist for tests involving composite hypotheses, as we shall see.

Let us consider once again the situation of the nursery worker, Douglas Furr, who is experimenting with white pine seedlings. You will recall that under normal conditions these seedlings would grow, on the average, 25 cm in a particular time interval, the actual growth following a normal distribution with a standard deviation of 3.2 cm. Doug was experimenting with different nutrient combinations in an effort to find a combination that might lead to increased growth. We assumed that he would be interested in testing the hypotheses

$$H_0 : \mu \leq 25 \qquad \text{vs.} \qquad H_a : \mu > 25$$

using $\alpha = .05$. In the last section it was assumed that a sample size of $n = 16$ was used. Of course, this particular sample size need not be used; rather, a larger or smaller one might be used. The question that might be asked here is: What is the best sample size to use? As in Section 7.12, the question of what sample size should be used can be answered only after answers to other questions are given.

Keep in mind that *for any sample size n* it is possible to construct a test of hypotheses for which α will be equal to a predetermined value. That is, it is always possible to control the probability of making a type I error. However, as the sample size n increases, the probability of committing a type II error in some sense decreases. One way of seeing the relationship between sample size and the probability of committing errors is to plot power curves.

Definition 8.6 In a test of hypotheses about a parameter θ, the *power curve* of the test is a graph of θ vs. $P[\text{reject } H_0 \mid \theta]$.

Example 8.32 Plot the power curve that Doug would have in testing

$$H_0 : \mu \leq 25 \qquad \text{vs.} \qquad H_a : \mu > 25$$

using a sample of size 16 with $\alpha = .05$. Assume that the standard deviation is known to be 3.2 and that the measurements come from a normal distribution.

In this case the parameter of interest is μ, so the power curve will be a graph of μ vs. $P[\text{reject } H_0 \mid \mu]$. Since in this case there is a one-sided alternative with the alternative hypothesis of "large" values μ, H_0 will be rejected when $\bar{X}$, the test statistic, is greater than or equal to some critical value c. The value of c is determined by using the desired level of significance and the appropriate percentile of the standardized test statistic $Z = (\bar{X} - \mu)/(\sigma/\sqrt{n})$. We find

$$\begin{aligned} \alpha = .05 &= P[\bar{X} \geq c \mid \mu = 25] \\ &= P\left[\frac{\bar{X} - \mu}{\sigma/\sqrt{n}} \geq \frac{c - \mu}{\sigma/\sqrt{n}} \middle| \mu = 25\right] \\ &= P\left[Z \geq \frac{c - 25}{3.2/\sqrt{16}}\right]. \end{aligned}$$

This last equality is true only if

$$\frac{c - 25}{3.2/\sqrt{16}} = z_{.95} = 1.645,$$

so $c = 25 + 1.645(3.2/\sqrt{16}) = 26.316$ (cf. Figure 8.16).

We can now easily find one point on the power curve. Since

$$P[\text{reject } H_0 \mid \mu = 25] = \alpha = .05,$$

we know that the power curve passes through the point (25, .05). Furthermore, because we have found the critical value, we can find other points on the power curve. For instance, when $\mu = 26$,

$$\begin{aligned} P[\text{reject } H_0 \mid \mu = 26] &= P[\bar{X} \geq c \mid \mu = 26] \\ &= P[\bar{X} \geq 26.316 \mid \mu = 26] = P\left[\frac{\bar{X} - \mu}{\sigma/\sqrt{n}} \geq \frac{26.316 - 26}{3.2/\sqrt{16}}\right] \\ &= P[Z \geq .395] = .3464, \end{aligned}$$

so the curve passes through (26, .3464). In fact, for this example we have

$$\begin{aligned} P[\text{reject } H_0 \mid \mu] &= P[\bar{X} \geq 26.316 \mid \mu] \\ &= P\left[\frac{\bar{X} - \mu}{\sigma/\sqrt{n}} \geq \frac{26.316 - \mu}{3.2/\sqrt{16}}\right] = P\left[Z \geq \frac{26.316 - \mu}{3.2/\sqrt{16}}\right], \end{aligned}$$

so the power curve goes through the points $(\mu, P[Z \geq (26.316 - \mu)/.8])$. Some of these values are given here and are shown in Figure 8.20.

μ:	25	26	26.316	27	28
$P[Z \geq (26.316 - \mu)/.8]$:	.0500	.3464	.5000	.8037	.9823.

You can see that the largest probability of rejecting H_0 when H_0 is true, in this case when $\mu \leq 25$, occurs at the value $\mu = 25$. This maximum value is, of course, equal to α (cf. the discussion in Section 8.3 concerning composite alternatives). This graph can also be used to find the probability of a type II error for any *specific value*

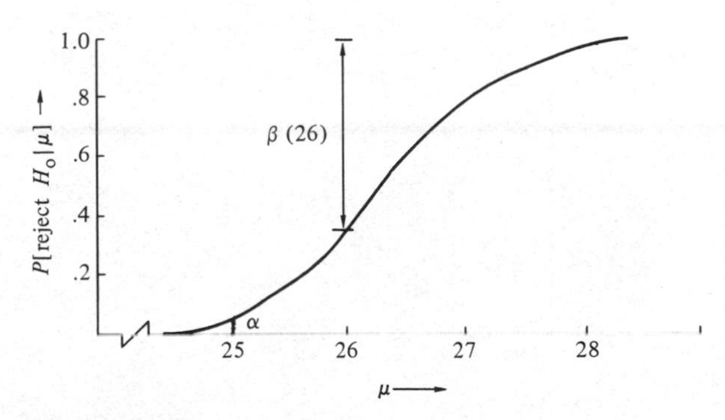

FIGURE 8.20
Power Curve for n = 16

of μ in H_a. For example,

$$P[\text{accept } H_0 \mid \mu = 26] = 1 - P[\text{reject } H_0 \mid \mu = 26]$$
$$= 1 - .3464 = .6536.$$

This value is denoted on the graph by $\beta(26)$.

Example 8.33 Plot the power curve for testing the hypotheses of the previous example when a sample of size 36 is used.

Since the level of significance is to remain at .05, it follows that the power curve will pass through (25, .05). To find other points on the power curve, it is necessary to find the critical value. This is found from

$$\alpha = .05 = P[\bar{X} \geq c \mid \mu = 25]$$
$$= P\left[\frac{\bar{X} - \mu}{\sigma/\sqrt{n}} \geq \frac{c - 25}{3.2/\sqrt{36}}\right] = P\left[Z \geq \frac{c - 25}{3.2/\sqrt{36}}\right],$$

so

$$\frac{c - 25}{3.2/\sqrt{36}} = z_{.95} = 1.645$$

or $c = 25.877$. For a given value of μ, then, the power curve can be found by using

$$P[\text{reject } H_0 \mid \mu] = P[\bar{X} \geq 25.877 \mid \mu]$$
$$= P\left[Z \geq \frac{25.877 - \mu}{3.2/\sqrt{36}}\right].$$

For example, for $\mu = 26$ we have

$$P[\text{reject } H_0 \mid 26] = P\left[Z \geq \frac{25.877 - 26}{.533} = -.231\right] = .5925.$$

For the values

μ:	25	25.877	26	27	28
$P[Z \geq (25.877 - \mu)/.533]$:	.0500	.5000	.5925	.9778	.99997

a graph of the power curve is shown in Figure 8.21. Note that while α remains the

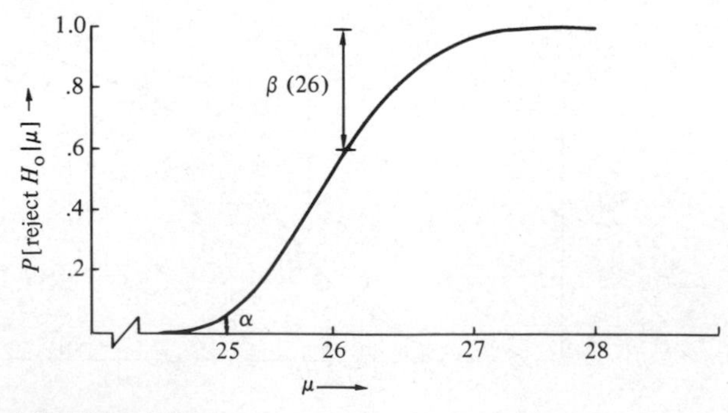

FIGURE 8.21
Power Curve for n = 36

same as when the sample size was 16, the probability of committing a type II error when $\mu = 26$ has been reduced to

$$P[\text{accept } H_0 \mid \mu = 26] = 1 - P[\text{reject } H_0 \mid \mu = 26]$$
$$= 1 - .5925 = .4075.$$

[This is shown on the graph in Figure 8.21 as $\beta(26)$.]

As you might expect, for a fixed level of significance, the probability of committing a type II error for a specific alternative (such as $\mu = 26$) decreases as the sample size increases. In a sense, the test with a larger sample size is better able to discriminate between a true mean of 25 and a true mean of 26 than a test based on a smaller sample size. This ability to discriminate is referred to as "power." A test whose power curve lies above the power curve of another for all values of the parameter corresponding to the alternative hypothesis is said to be more powerful. In Figure 8.22 are both of the power curves given in Figures 8.20 and 8.21. As you would

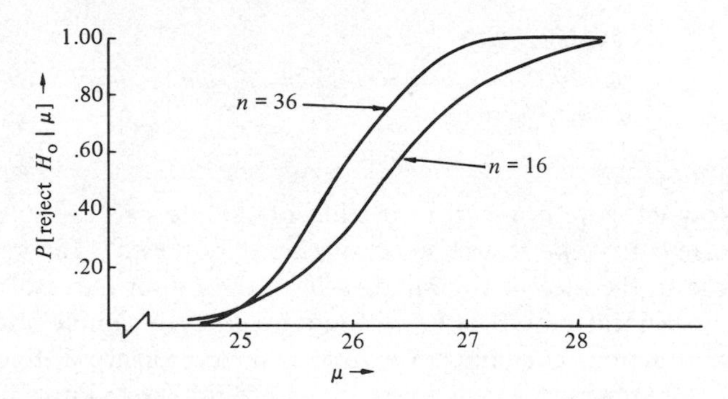

FIGURE 8.22
Power Curves for $n = 16$ and $n = 36$

expect, for all values of μ greater than 25 (values corresponding to the alternative hypothesis), the power curve for the sample size of 36 lies above the power curve for the sample size of 16. What happens, however, for values of μ less than 25? You can see that in this case the power curve corresponding to $n = 36$ lies *below* the one corresponding to $n = 16$. Is this good or bad? It is good since it means that for the larger sample size you are less likely to reject H_0 for values of μ less than 25 (i.e., when H_0 is true).

Now consider the question of what the power curve would look like for a "perfect" test. If a test were perfect, H_0 would *always* be rejected when H_a is actually true but would *never* be rejected when H_0 is actually true. For the hypotheses we have been considering in this section a "perfect" or "ideal" test would satisfy

$$P[\text{reject } H_0 \mid \mu] = 1 \qquad \text{if } \mu > 25$$
$$P[\text{reject } H_0 \mid \mu] = 0 \qquad \text{if } \mu \leq 25.$$

The corresponding power curve is called the ideal power curve and is shown in Figure 8.23 along with the power curves for $n = 16$ and $n = 36$. What do you think the power curve would look like for a sample size of 50? 100? If you guessed that it would get closer to the ideal as the sample size increased, you would be correct.

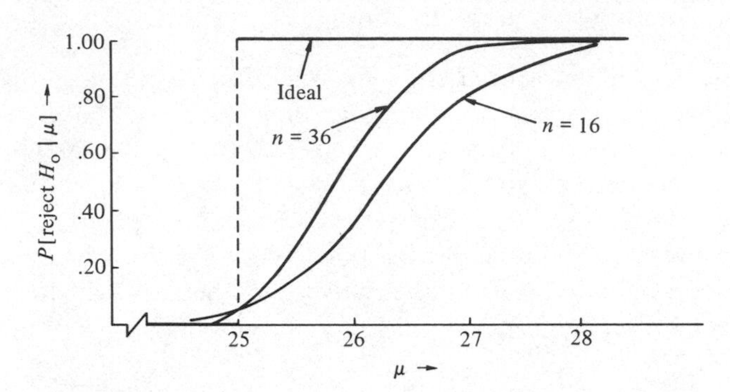

FIGURE 8.23

Power Curves for n = 16, n = 36, and the Ideal Power Curve

Now let us return to the question of sample size. How large a sample should an experimenter take to test some specific hypotheses? The answer will depend on how close to the ideal it is desired to have the power curve of the test. To get very close to the ideal power curve will require a large sample size. However, in most practical situations one does not wish to make very sharp distinctions between values that are close together. Doug, for example, would like to know if a particular nutrient combination increases growth over the standard value of 25 cm. However, he would not care to further investigate a combination that had a mean $\mu = 25.01$, since such an increase is of no practical importance. A test that was "too powerful" would be rather sure to detect such differences. Statistically, the value of 25.01 is greatet than 25, but such a difference is of no practical significance. In determining sample size, one must ask the questions "What kind of differences do you wish to detect?" and "How sure do you want to be that you will detect such differences?" For example, Doug may decide that he wants to detect a difference of 2 cm or more (i.e., a growth of 27 cm or more) with a probability of at least .90. In other words, if μ is actually 27, he would like to reject H_0 with probability .90. Assuming that he still wants α to be .05, he would need to find n by solving the following two equations for n and c:

$$P[\text{reject } H_0 \mid \mu = 25] = P[\bar{X} \geq c \mid \mu = 25] = P\left[\frac{\bar{X} - \mu}{\sigma/\sqrt{n}} \geq \frac{c - 25}{3.2/\sqrt{n}}\right] = .05$$

and

$$P[\text{reject } H_0 \mid \mu = 27] = P[\bar{X} \geq c \mid \mu = 27] = P\left[\frac{\bar{X} - \mu}{\sigma/\sqrt{n}} \geq \frac{c - 27}{3.2/\sqrt{n}}\right] = .90.$$

It follows that

$$\frac{c - 25}{3.2/\sqrt{n}} = z_{.95} = 1.645$$

and

$$\frac{c - 27}{3.2/\sqrt{n}} = z_{.10} = -1.282.$$

These equations can be solved for n and c just as was done in Example 8.4 of Section 8.2. Solving each equation for c gives

$$c = 25 + 1.645\left(\frac{3.2}{\sqrt{n}}\right) \tag{8.34}$$

$$c = 27 - 1.282\left(\frac{3.2}{\sqrt{n}}\right).$$

Equating these two, we can find n from

$$27 - 1.282\left(\frac{3.2}{\sqrt{n}}\right) = 25 + 1.645\left(\frac{3.2}{\sqrt{n}}\right)$$

$$2\sqrt{n} = (1.645 + 1.282)(3.2)$$

$$n = 21.93 \approx 22.$$

Substituting into (8.34) we find that the critical value should be

$$c = 25 + 1.645\left(\frac{3.2}{\sqrt{22}}\right) = 26.122.$$

If you were to graph the power curve for this test, you would find that for values of μ equal to 27 *or more*, the probability of rejecting H_0 is *at least* .90, thus satisfying Doug's requirements (cf. Exercise 51).

Power curves can be found for tests of hypotheses with two-sided alternatives as well as for those with one-sided alternatives. Since the critical region for two-sided alternatives consists of two distinct regions, finding the probability of rejecting H_0 is a bit more involved than for one-sided alternatives.

Example 8.34 Olaf, the quality control engineer at the Bubbly Cola plant, is concerned about the amount of cola being put into "1-liter" bottles (cf. Example 8.8). To test

$$H_0 : \mu = 1000 \qquad \text{vs.} \qquad H_a : \mu \neq 1000,$$

Olaf will measure the contents of 16 bottles. He assumes that the amount put into a bottle approximately follows a normal distribution with a standard deviation of 4 ml and wishes to have $\alpha = .05$. Plot the power curve for this test and also plot the ideal power curve for this test.

To get the power curve, we will need to find $P[\text{reject } H_0 \mid \mu]$ for various values of μ. Since this is a test with a two-sided alternative, H_0 will be rejected if $\bar{X}$ is either too "small" or too "large." In Example 8.8 it was found that the critical values should be 998.04 and 1001.96. Using these values, we find that

$$\begin{aligned}
P[\text{reject } H_0 \mid \mu] &= P[\bar{X} \leq 998.04 \text{ or } \bar{X} \geq 1001.96 \mid \mu] \\
&= P\left[\frac{\bar{X}-\mu}{\sigma/\sqrt{n}} \leq \frac{998.04-\mu}{4/\sqrt{16}}\right] + P\left[\frac{\bar{X}-\mu}{\sigma/\sqrt{n}} \geq \frac{1001.96-\mu}{4/\sqrt{16}}\right] \\
&= P[Z \leq 998.04 - \mu] + P[Z \geq 1001.96 - \mu].
\end{aligned}$$

Of course, when $\mu = 1000$, $P[\text{reject } H_0 \mid \mu]$ will be .05. The values of this probability for other values of μ are given in Table 8.3. By plotting the values of μ against the probabilities in the last column of Table 8.3, we get the power curve shown in Figure 8.24. By looking at the entries in the second and third columns, you can see that the power curve will be symmetric about the value $\mu = 1000$.

TABLE 8.3

μ	$P[Z \leq 998.04 - \mu]$	$P[Z \geq 1001.96 - \mu]$	$P[\text{reject } H_0 \mid \mu]$
996	.9793	.0000	.9793
997	.8508	.0000	.8508
998	.5160	.0000	.5160
999	.1685	.0014	.1699
1000	.0250	.0250	.0500
1001	.0014	.1685	.1699
1002	.0000	.5160	.5160
1003	.0000	.8508	.8508
1004	.0000	.9793	.9793

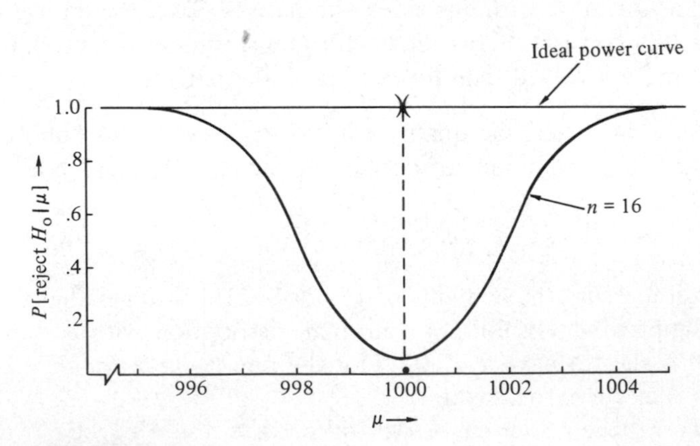

FIGURE 8.24
Power Curves for a Two-Sided Alternative

An ideal power curve would correspond to a test that will always reject for values of μ different from 1000 and never reject if μ equals 1000. That is,

$$P[\text{reject } H_0 \mid \mu] = \begin{cases} 1 & \text{if } \mu \neq 1000 \\ 0 & \text{if } \mu = 1000. \end{cases}$$

This ideal curve is plotted in Figure 8.24.

In this situation as well as in the one-sided alternative case, tests based on larger sample sizes will have power curves closer to the ideal than tests based on smaller sample sizes. Would you recommend to Olaf that he take a large sample to be sure that his test will have a power curve close to the ideal? If this were to be the case, there would be a very high probability that $H_0(\mu = 1000)$ will be rejected even for values of μ quite close to 1000, say 999.5 or 1000.5. If Olaf's plan for adjusting the filling machine is to stop for adjustment when H_0 is rejected, do you think he would want to stop for adjustment if the true mean is really quite near μ? Of course not. He knows that the machinery is not perfect and is willing to allow some tolerance on either side of the target value of 1000 ml. Consequently, he will not want to take too large a sample. The size of his sample will depend on the size of the differences he wants to detect and how certain he will be to detect those differences.

Example 8.35 Olaf wants to devise a procedure to decide whether or not a machine is properly filling 1-liter (1000-ml) bottles. What sample size should he use if he wishes to have a probability of .05 of needlessly adjusting the machine when it is, in fact, working properly but to have a probability of at least .90 of detecting it when the amount of filling is off by 2.5 ml or more?

The conditions set forth in this example actually require three equations to be satisfied:

$$\begin{aligned} P[\text{reject } H_0 \mid \mu &= 1000] = .05 \\ P[\text{reject } H_0 \mid \mu &= 997.5] = .90 \\ P[\text{reject } H_0 \mid \mu &= 1002.5] = .90. \end{aligned}$$

By using these three equations, we can solve for the three unknowns n, c_1, and c_2, where c_1 and c_2 are the critical values. We know that H_0 will be rejected if $\bar{X} \leq c_1$ or $\bar{X} \geq c_2$. By standardizing, we find that

$$P[\bar{X} \leq c_1 \text{ or } \bar{X} \geq c_2 \mid \mu = 1000]$$

$$= P\left[\frac{\bar{X} - \mu}{\sigma/\sqrt{n}} \leq \frac{c_1 - 1000}{4/\sqrt{n}}\right] + P\left[\frac{\bar{X} - \mu}{\sigma/\sqrt{n}} \geq \frac{c_2 - 1000}{4/\sqrt{n}}\right] = .05.$$

Since it is desired to detect a difference of 2.5 ml in either direction, we can assume that the values of c_1 and c_2 will be symmetrically located relative to 1000. Hence we can take

$$\frac{c_1 - 1000}{4/\sqrt{n}} = z_{.025} = -1.96 \tag{8.35}$$

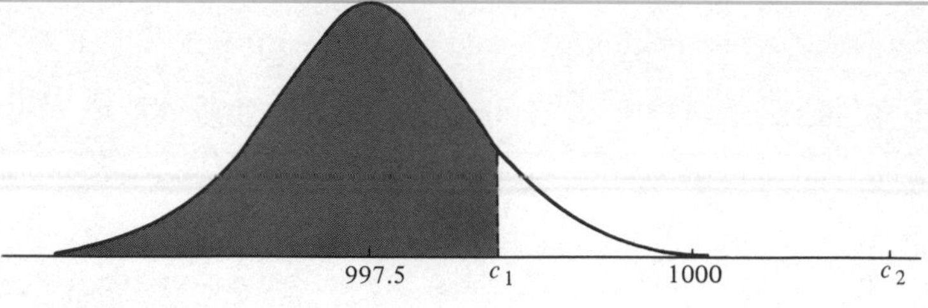

FIGURE 8.25

and

$$\frac{c_2 - 1000}{4/\sqrt{n}} = z_{.975} = +1.96.$$

A second equation can be found from

$P[\bar{X} \leq c_1 \text{ or } \bar{X} \geq c_2 \mid \mu = 997.5]$

$$= P\left[\frac{\bar{X} - \mu}{\sigma/\sqrt{n}} \leq \frac{c_1 - 997.5}{4/\sqrt{n}}\right] + P\left[\frac{\bar{X} - \mu}{\sigma/\sqrt{n}} \geq \frac{c_2 - 997.5}{4/\sqrt{n}}\right] = .90.$$

A bit of reflection will convince you that when the mean μ is actually equal to 997.5, the distribution of $\bar{X}$ will look roughly like that given in Figure 8.25. There is a small probability that $\bar{X}$ will exceed c_2, but the actual value is negligible. Consequently, virtually all of the probability will come from $\bar{X}$ being less than c_1 (the shaded region on the graph). It follows that

$$\frac{c_1 - 997.5}{4/\sqrt{n}} = 1.282.$$

Solving this equation and the first equation in (8.35) simultaneously, we find that

$$c_1 = 997.5 + 1.282\left(\frac{4}{\sqrt{n}}\right)$$

$$c_1 = 1000 - 1.96\left(\frac{4}{\sqrt{n}}\right),$$

so

$$n = \left[\frac{(1.96 + 1.282)(4)}{2.5}\right]^2 = 26.9 \approx 27.$$

Consequently, Olaf should take a sample of 27 bottles. If the sample mean is less than

$$c_1 = 1000 - 1.96\left(\frac{4}{\sqrt{27}}\right) = 998.5 \text{ ml}$$

or greater than

$$c_2 = 1000 + 1.96\left(\frac{4}{\sqrt{27}}\right) = 1001.5 \text{ ml},$$

the machine should be adjusted. Otherwise, no adjustment should be made.

EXERCISES

1. Statistical hypotheses are generally stated in pairs, a "null" hypothesis and an "alternative" hypothesis. Frequently, the alternative hypothesis is the logical complement of the null hypothesis. Give the logical alternative to each of the following null hypotheses:

   (a) X has a mean $\mu = 1000$ ml.
   (b) X has a mean $\mu \geq 32$ oz.
   (c) The standard deviation of X is no more than 1 mm.
   (d) X follows a Poisson distribution.

2. Mr. Denny Krat has 1 week left before the election and still has \$2000 left in his campaign fund. He knows that either $p \leq .50$ or $p > .50$, where p is the proportion of voters in his district who will vote for him. He plans to interview a sample of voters. If the number in the sample favoring him is "large," he will assume that $p > .50$ and will not spend the \$2000; if the number is "small," he will assume that $p \leq .50$ and will spend the money on his campaign. Discuss the consequences of a wrong decision if (a) p is really greater than .50; (b) p is really less than or equal to .50. (c) In view of the answers to parts (a) and (b), which hypotheses should he call the null hypothesis?

3. In most communities, if a person is bitten by an animal and if the animal cannot be found, it is assumed that the animal is rabid. The bitten person is usually given a series of rabies shots. It is clear that either the animal is rabid or the animal is not rabid. A doctor's treatment would be different for these two cases. If the animal were known to be rabid, shots would be given, while if the animal were known to be not rabid, only the wound would be treated. Discuss the consequences of a wrong decision if (a) the animal is really rabid; (b) the animal is really not rabid. (c) In view of the answers to parts (a) and (b), which hypothesis should be called the null hypothesis?

4. Vicki has recently purchased a new dress costing \$125. She wants to wear the dress to dinner, but she does not want to get it wet if it rains, since the rain will ruin the dress. She can carry her \$2 plastic raincoat with her to keep the dress dry in case of rain, but would rather not take it unnecessarily since it does not match the dress. Assume that she considers the two hypotheses H_0: it will rain tonight and H_a: it will not rain tonight. Make a table similar to the one in Figure 8.4 showing the consequences of the possible errors.

5. Vicki will decide whether to accept H_0 or H_a of the previous exercise by using a weather-predicting device in her home. If the device predicts rain, she will accept H_0, but not otherwise. This device correctly predicts rain 80% of the time and correctly predicts no rain 70% of the time. For example, the probability that it will predict rain given that it actually does rain is .80. Determine the values of α and β for this situation.

6. Assume that in Exercise 2 Mr. Krat decides to interview 100 voters and that he will assume $p > .50$ if and only if X, the number in the sample favoring him, is 55 or greater. Find the probability that he will assume $p > .50$ when the true value of p is actually (a) .40; (b) .50. Find the probability that he will assume $p \leq .50$ when the true value of p is actually (c) .50; (d) .60.

The following situation will be considered in Exercises 7 to 11: A random sample of n observations is to be taken from a normal distribution having a known variance $\sigma^2 = 400$. To test H_0: $\mu = 20$ against the alternative H_a: $\mu = 30$, $\bar{X}$ will be calculated.

7. If $n = 16$, find α and β if the critical region is chosen to be values of $\bar{X}$ greater than or equal to (a) 25; (b) 27. For a fixed sample size, what is the effect on α and β when the critical value is increased?

8. Repeat Exercise 7 when the sample size is $n = 25$. Comparing these results with those of Exercise 7, for a fixed critical region, what is the effect on α and β when the sample size is increased?

9. If the sample size is $n = 16$, what should the critical region be if it is desired to have (a) $\alpha = .10$?; (b) $\alpha = .05$? In each of these cases, find the corresponding value of β.

10. If the critical region is $C = \{\bar{X} \geq 24\}$, what sample size should be chosen to make (a) $\alpha = .05$?; (b) $\alpha = .01$? In each of these cases find the corresponding value of β.

11. Find the critical region and the sample size necessary to make (a) $\alpha = .05$, $\beta = .05$; (b) $\alpha = .05$, $\beta = .10$.

12. Assume that you are a worker with a consumer advocate group. *From your point of view*, state appropriate null and alternative hypotheses for the following situations:

(a) A box of Crunchie-Munchies is labeled "10 oz." You wish to test hypotheses about the mean weight μ.
(b) A package of Wheeze Weeds Cigarettes states "10.2 mg of tar, average per cigarette." You wish to test hypotheses about the mean tar contents μ.
(c) A package of Sure-Gro petunia seeds claims that the germination rate is 80%. You wish to test hypotheses about the proportion p of seeds that germinate.
(d) The Kurl-E-Q Shampoo company makes shampoo for dry, regular, and oily hair. The shampoos differ in their acidity, the "regular" shampoo claiming an acidity rating of 5. (Higher or lower ratings correspond to shampoo for "dry" or "oily" hair.) You wish to test hypotheses about the mean acidity rating μ for the regular shampoo.

★13. In tests of hypotheses as well as in determining confidence intervals, it is frequently assumed that observed data come from a normal distribution. Construct a histogram for each of the following sets of data and tell whether the assumption of normality appears to be reasonable to you.

(a) The following are yards gained by Tony Dorsett on running plays in the 1977 Sugar Bowl (Pitt. vs. Georgia) (see Exercise 2 of Chapter 6): 4, −2, 5, 2, 3, −5, 4, 16, 1, 2, 1, 0, 2, 0, 22, −2, 11, 6, 2, 10, 15, 1, 67, −4, 4, 22, 0, 0, 4, 5, 2, 1.
(b) The following are lifetimes (in days) of male mice following a radiation dose of 300 rads at age 5 to 6 weeks: 655, 624, 662, 193, 430, 734, 202, 638, 757, 229, 691, 800, 240, 747, 855, 259, 778, 868, 321, 136, 873, 434, 376, 910, 496, 616, 1015.

In all hypothesis-testing problems, follow the general outline given in Table 8.1 of Section 8.5 and *explicitly* state all relevant information.

14. In order to test $H_0: \mu \geq 20$ vs. $H_a: \mu < 20$, a random sample of 36 observations was taken from a normal distribution having a standard deviation of 60. Determine whether H_0 should be rejected with $\alpha = .05$ if $\bar{x}$ is observed to be (a) 21; (b) 5. (*To think about:* Would this test procedure be appropriate if the sample came from a nonnormal distribution?)

15. Some manufacturers attempt to meet their packaging claims by erring in favor of the buyer. For example, an internationally known cereal manufacturer claims that their goal is to have at least 95% of all of their packages have contents that exceed the advertized amount. Assuming that the amount of cereal in a "16-oz" box follows a normal distribution with $\sigma = 0.5$ ounce, (a) show that their goal will be met if the true mean μ is 16.83 ounces or more. (b) The quality control supervisor has devised the following scheme to decide whether or not to adjust the filling machine: Sample 25 boxes and adjust the machine if the sample mean is 16.65 ounces or less. What null and alternative hypotheses have been implicitly set up? (c) What is the value of α?

16. Assume that the quality control supervisor of Exercise 15 uses the following scheme to decide whether or not to adjust the filling machine: Sample 25 boxes and adjust the machine if the sample mean is either 16.47 ounces or less or 17.19 ounces or more. (a) What null and alternative hypotheses have been implicitly set up? (*Hint:* Is the alternative one sided or two-sided?) (b) What is the value of α? (*To think about:* Do you think it would be better to follow the scheme described in this exercise or the previous exercise? Why?)

17. In order to test $H_0: \mu = 50$ vs. $H_a: \mu \neq 50$, a random sample of four observations was taken from a normal distribution. Determine whether H_0 should be rejected with $\alpha = .05$ if the sample mean and standard deviation are found to be (a) 54.1 and 2.4; (b) 58.5 and 5.6. (*To think about:* Would this procedure be appropriate if the sample came from a nonnormal distribution?)

18. A restaurateur, Dolesome Mel, owns two restaurants. In one coffee is sold for 40 cents with unlimited refills available. Mel knows that patrons of this restaurant who order coffee drink, on the average, 14 ounces of coffee and that the standard deviation is 6 ounces. In the second restaurant, coffee is sold for 4 cents with unlimited refills. Mel would like to know if the reduced price of coffee has a significant effect on the amount drunk per patron ordering coffee. Set up and test appropriate hypotheses using $\alpha = .01$ if a sample of 36 coffee drinkers in the second restaurant drank 16.4 ounces on the average. Assume that $\sigma = 6$ ounces. (*To think about:* In this context, is it reasonable to assume that the amount of coffee consumed per coffee drinker follows a normal distribution?)

19. McRonald's Hamburger Shoppe advertises "QUARTER-POUND HAMBURGERS" (precooked weight). Assume you will patronize this establishment only if you can be *convinced* that you are getting *at least* a quarter-pound. Set up and test appropriate hypotheses using $\alpha = .05$ if a sample of six hamburgers showed precooked weights of 3.96, 4.04, 4.08, 3.92, 4.16, and 4.14 oz. On the basis of the criterion set up in this problem, would you patronize this establishment? (*To think about:* Do you think that this criterion is too stringent? How might you modify the hypotheses to give McRonald's some benefit of the doubt?)

20. The number of people entering a Thrifty-Nifty discount store on a Monday is approximately normally distributed with a mean of 600 and a standard deviation of 150. The store manager Sandy Sures is interested in knowing whether an ad in the Sunday Shopper will increase the number of people entering the store. What conclusion should the manager draw if on four Mondays (following the appearance of ads) the mean number of people entering the store was (a) 592?; (b) 810? What assumptions did you make?

21. In the preceding exercise, Sandy knows that an increase of 100 or more customers must take place to offset the cost of the ad. With this in mind, rework Exercise 20.

22. Using the data from Exercise 34 of Chapter 7, test the null hypothesis that there is no difference in the mean sales when soil is packaged in boxes or plastic bags. Use $\alpha = .05$.

23. The manager of a store having a shopping center location claims that the average daily sales are greater than the average daily sales of a store of comparable size at a downtown location. Do the data given in Exercise 35 of Chapter 7 support this claim? Use $\alpha = .05$ and test appropriate hypotheses.

24. In order to test H_0: $\mu_1 - \mu_2 \leq 10$ vs. H_a: $\mu_1 - \mu_2 > 10$, where μ_1 and μ_2 represent the means of two normal distributions having known variances of $\sigma_1^2 = 625$ and $\sigma_2^2 = 121$, independent random samples were taken from the two distributions. The first sample consisted of 100 observations with $\bar{x}_1 = 127.3$ and the second sample consisted of 44 observations with $\bar{x}_2 = 105.1$. What conclusion should be drawn if $\alpha = .01$? (*To think about:* How would the analysis change if the distributions were known to be nonnormal?)

★**25.** Some of the mice irradiated at age 5 to 6 weeks (see Exercise 13) were conventional mice exposed to a normal environment, others were germ-free mice. Let μ_1 and μ_2 denote the mean lifetime of the conventional and germ-free mice, respectively. Use the following data to test H_0: $\mu_1 \geq \mu_2$ vs. H_a: $\mu_1 < \mu_2$ at the $\alpha = .01$ level, assuming that the lifetimes are normally distributed.

Lifetimes (conventional mice):	200	245	318	495	628
Lifetimes (germ-free mice):	430	638	691	747	778.

(*To think about:* In view of the results of Exercise 13, is the assumption of normality reasonable? If not, what alternatives are there for testing these hypotheses?)

26. Mr. Rick Shaw owns a fleet of 12 taxicabs. He plans to buy 6 Goodstone steel-belted radials and 6 Firerich steel-belted radials to put on the rear wheels of his cabs. The tires will be checked at 500-mile intervals and will be considered to be worn out when the tread wear bars show. He can either (A) put one new tire on each of the 12 cabs or (B) put one of each brand on the rear tires of 6 cabs. From a statistical viewpoint, which would be the preferred procedure? Why?

27. If Mr. Shaw (Exercise 26) followed plan (A) and observed the following data, can he conclude that one brand is better than the other? Explain. (Use $\alpha = .05$.)

Mileage on Goodstone tires:	32,000	31,500	38,500	37,000	40,000	37,000
Mileage on Firerich tires:	34,500	31,000	39,000	38,500	41,000	38,000.

28. If Mr. Shaw (Exercise 26) followed plan (B) and if he observed the following mileage data, can he conclude that one brand is better than the other? Explain. (Use $\alpha = .05$.)

	Cab					
	1	*2*	*3*	*4*	*5*	*6*
Goodstone tires	32,000	31,500	38,500	37,000	40,000	37,000
Firerich tires	34,500	31,000	39,000	38,500	41,000	38,000

(*To think about:* Would it be reasonable to expect the mileages to be exactly normally distributed? Approximately?)

29. To test $H_0: \sigma^2 \geq 100$ vs. $H_a: \sigma^2 < 100$, where σ^2 is the variance of a normal distribution, a random sample of 21 observations was taken from this distribution. Using $\alpha = .05$, what should be concluded if the sample variance was found to be (a) 50?; (b) 160?

30. A soda machine dispenses Fizzy Cola into 8-ounce cups, the amount dispensed being approximately normal with a variance of (.25 ounce)^2. The distributors of Fizzy Cola are considering the possibility of installing a new machine if it can be shown to be less variable in dispensing than the current one. Using $\alpha = .01$, set up and test appropriate hypotheses if a sample of 30 cups had a variance of .0382 oz^2.

31. Mr. Shaw (Exercise 26) has kept records on gas mileage achieved by his cabs in city driving and has found the mean to be 14.3 mpg with a standard deviation of 2.1 mpg. He would like to expand his business to include intercity rental service and is interested in knowing whether there is a difference in the variability in mileage in highway driving compared with city driving. His sister, Ava, has allowed him to examine some records from her company, Ava's Car Rental. Most of the miles on Ava's cars are highway miles. Based on 61 tankfuls of gas, he calculated a sample standard deviation of 2.6 mpg. What conclusion should he draw? (Use $\alpha = .05$.)

32. To test $H_0: \sigma_1^2/\sigma_2^2 = 1$ vs. $H_a: \sigma_1^2/\sigma_2^2 \neq 1$, where σ_1^2 and σ_2^2 are variances of two normal distributions, independent random samples of size 10 were taken from these distributions. Using $\alpha = .05$, what should be concluded if the sample variances were found to be (a) $S_1^2 = 16.2$, $S_2^2 = 5.1$; (b) $S_1^2 = 16.2$, $S_2^2 = 86.7$?

33. To test $H_0: \sigma_1^2 \geq 2\sigma_2^2$ vs. $H_a: \sigma_1^2 < 2\sigma_2^2$, where σ_1^2 and σ_2^2 are variances of two normal distributions, independent random samples of sizes $n_1 = 16$ and $n_2 = 25$ were taken from these distributions. Using $\alpha = .01$, what should be concluded if the sample variances were found to be (a) $S_1^2 = 86.7$, $S_2^2 = 43.1$?; (b) $S_1^2 = 6.5$, $S_2^2 = 43.1$?

★34. The data gathered on students in a statistics course having a large enrollment can be used to test hypotheses concerning the variability of measurements for males and females. Measurements were given by 139 males and 61 females. In each of the following situations, test $H_0: \sigma_M^2 = \sigma_F^2$ vs. $H_a: \sigma_M^2 \neq \sigma_F^2$ at the .05 level of significance.

(a) For heights of students, $S_M = 2.7$ inches and $S_F = 2.6$ inches.
(b) For weights of students, $S_M = 20.6$ pounds and $S_F = 15.5$ pounds.
(c) For heart rates of students, $S_M = 10.9$ beats/minute and $S_F = 11.1$ beats/minute.

(*To think about:* If H_0 were *known* to be true in every case and if you performed 20 tests like those above, how many times would you expect to wrongly reject H_0? Why?)

★35. Using the data given in Exercise 25, assuming normality, test the hypothesis H_0: $\sigma_1 \leq \sigma_2$ against the alternative H_a: $\sigma_1 > \sigma_2$ using a .05 level of significance.

36. Using the data given in Exercise 27, assuming normality, test the hypothesis H_0: $\sigma_G^2 = \sigma_F^2$ against the alternative H_a: $\sigma_G^2 \neq \sigma_F^2$ at the .01 level of significance.

37. Using the data in Exercise 30 of Chapter 7 concerning the heights of boys and girls in a fourth-grade class at the Lotta Noyes Elementary School, check the reasonableness of the assumption that the variability in heights is the same for fourth-grade boys and girls in general. Do the same for the weights of fourth-grade boys and girls using the data given in Exercise 31 of Chapter 7. (*To think about:* What do these results tell you about the appropriateness of the method you used to find confidence intervals in those two problems?)

38. If the quality control supervisor (Example 8.22 of Section 8.8) rejects $H_0: p \leq .20$ in favor of $H_a: p > .20$ if and only if 9 or more cookies in a sample of 20 are broken, what is the probability that H_0 will *not* be rejected when in fact the true value of p is .25?; .30?

39. Test the hypothesis $H_0: p \leq .20$ against the alternative $H_a: p > .20$ using $\alpha = .01$ if p represents the proportion of successes and if in 100 trials (a) 18, (b) 30 successes were observed.

40. The manufacturer of an insecticide, Fly-Die, claims that a single spray will kill at least 90% of the flies sprayed. As purchasing agent for the Noc M. Dead Slaughterhouse, you wish to test this claim. Set up and test appropriate hypotheses using a .05 level of significance. What conclusion should you draw if 822 of 900 flies sprayed were killed?

41. Congressman Porkbarol favors construction of the Plugadup dam and states that a majority of his constituents also favor the construction. If you are skeptical of claims made by politicians and take a sample of his constituents to test his claim, what would you conclude if 215 out of 400 persons surveyed favored construction of the dam?

42. If the quality control supervisor (Example 8.23 of Section 8.8) rejects $H_0: p \leq .20$ in favor of $H_a: p > .20$ if and only if 99 or more cookies in a sample of 400 are broken, what is the probability that H_0 will *not* be rejected when in fact the true value of p is .25?; .30? (*To think about:* Compare these results with those of Exercise 38. Which inspection procedure appears to be better? Why?)

43. To test the hypothesis H_0: $p_1 - p_2 = 0$ against the alternative H_a: $p_1 - p_2 \neq 0$ using $\alpha = .01$, where p_1 and p_2 represent the probability of a success for binomial random variables X_1 and X_2, respectively, you observe $n_1 = 100$ and $n_2 = 200$ trials. What conclusions should you draw if you observe (a) $x_1 = 80$, $x_2 = 170$ successes; (b) $x_1 = 25$, $x_2 = 70$ successes?

★**44.** Let p_1 represent the proportion of voters living in the city (urban) and p_2 the proportion of voters living outside the city (rural) who voted for the democratic candidate in a past presidential election. Test $H_0: p_1 - p_2 = 0$ vs. $H_a: p_1 - p_2 \neq 0$ at the .01 level of significance by treating the following data as random samples of urban and rural voters: 10,187 out of 20,690 voters in Columbia, Missouri, and 6686 out of 12,070 "out county" voted for the democratic candidate. (*To think about:* Is it reasonable to consider these data representative of the entire United States? Of Missouri?)

45. Burt Howdy, a sportscaster, implied during a telecast that a baseball player was a better batter in his home park than on the road, citing his batting average of .292 at home and .272 on the road this season. Treating the data from this season as a random sample of batting appearances during his lifetime, test $H_0: p_H - p_R \leq 0$ against $H_a: p_H - p_R > 0$. For simplicity, assume that he batted 250 times at home and 250 times on the road. Use $\alpha = .05$. (*To think about:* Would X = number of hits in n at bats follow a binomial distribution? What about the assumption of independence? What about streaks and slumps?)

46. In a study of brand recognition, interest centered on whether or not the proportion of men who correctly identified a product manufactured under a given name exceeded the proportion of women correctly identifying the product. Assume that in a random sample of 50 men, 40 correctly identified a product manufactured under the brand name "Homelite" while 28 of a random sample of 50 women made a correct identification. Set up and test appropriate hypotheses using a .05 level of significance.

47. In testing $H_0: \mu \leq 50$ vs. $H_a: \mu > 50$, a random sample of 36 observations was taken from a normal distribution having a standard deviation of 90. (a) If $\mu = 50$, what is the probability of observing a sample mean $\bar{X}$ larger than 70? (b) If in testing H_0, $\bar{X}$ is observed to be 70, what is the p value? (c) If $\alpha = .05$, should H_0 be rejected? (d) If $\alpha = .10$, should H_0 be rejected?

48. Referring to Exercise 14, find the p value if $\bar{x}$ is observed to be (a) 21; (b) 5.

49. Referring to Exercise 20, find the p value if the average number of people entering the store was (a) 592; (b) 810.

50. Mr. Krat (Exercises 2 and 6) sets up $H_0: p \leq .50$ with the alternative $H_a: p > .50$ and decides to reject H_0 if $X \geq 55$, where X is the number in a sample of 100 voters favoring him. (a) Plot the power function for this test. (b) What is the significance level for this test?

51. Plot the power curve for a test of $H_0: \mu \leq 25$ vs. $H_a: \mu > 25$, where μ is the mean of a normal distribution having a known standard deviation of 3.2. Assume that a random sample of 22 observations is used and that α is equal to .05 (cf. the discussion in Section 8.10).

52. In order to test $H_0: \mu \geq 20$ vs. $H_a: \mu < 20$, a random sample of 36 observations was taken from a normal distribution having a standard deviation of 60 (cf. Exercise 14). If $\alpha = .05$ for this test, (a) determine the value c such that H_0 will be rejected if $\bar{X} \leq c$. (b) Plot the power function for this test. (c) Repeat part (a) if a random sample of 100 observations is taken. (d) Plot the power function for this test on the same axes used in part (b).

53. In order to test $H_0: \mu = 20$ vs. $H_a: \mu \neq 20$, a random sample of 36 observations was taken from a normal distribution having a standard deviation of 60. If $\alpha = .05$ for this test, (a) determine the values c_1 and c_2 so that H_0 will be rejected if either $\bar{X} \leq c_1$ or $\bar{X} \geq c_2$. (b) Plot the power function for this test. (c) Repeat part (a) if a random sample of 100 observations is taken. (d) Plot the power function for this test on the same axes used in part (b).

9

REGRESSION AND CORRELATION

9.1 INTRODUCTION

In Chapter 3 we discussed joint probability distributions for two (or more) random variables. We saw that in some cases two random variables X and Y were independent while in other cases they were not independent. If the joint probability distribution for X and Y is known, it is possible to determine whether or not X and Y are independent by using methods described in Chapter 3. If the joint distribution is not known (as is usually the case), it is possible to devise statistical methods for testing the null hypothesis of independence. Some of these methods will be discussed in Section 11.13.

In many situations random variables X and Y are dependent and it is often possible to use information about one of these variables, say X, to make predictions about the other variable. It may also be possible to measure the degree of association between random variables. The former case (prediction) is covered under the general heading of *regression analysis* while the latter (association) comes under the general heading of *correlation analysis*.

You are no doubt familiar with examples where two variables are related. For example, a person's height and weight, the amount of rainfall and the yield of corn in a particular state, and the volume of sales and the amount of money spent on advertising are all pairs of variables which are related. As another example, in the study of the laws of motion in physics you learn that the distance traveled by a free-falling object can be found from the formula

$$d = \tfrac{1}{2}gt^2. \tag{9.1}$$

The variables d, the distance traveled, and t, the time of fall, are related through this formula (g is simply a constant, equal to 9.8 m/sec^2 for objects freely falling to the earth). The relationship between the variables d and t is *deterministic* in the sense that if wind, air resistance, and other disturbing factors can be ignored (e.g., on the moon), the value of d completely determines the value of t, and vice versa. If the experiment were to be repeated a number of times, the results would always be the same. However, we will not concern ourselves with deterministic relationships, but with *stochastic* (statistical) relationships. These are relationships where the value of one variable *does not* completely determine the value of another.

Example 9.1 Billy Joe is interested in finding the height of a bridge over the Tallahatchee River. He decides to drop an object from the bridge, time its fall, and determine the height by using equation (9.1). He is going to perform this experiment on a windy day. Is the relationship between the height of the bridge and the time of the fall deterministic or stochastic?

Since the wind can have an effect on the time of fall, the relationship will be stochastic. If the actual height of the bridge was 80 meters, say, then under ideal conditions the time of the fall would be

$$t = \sqrt{\frac{2d}{g}} = 4.04 \text{ seconds.}$$

However, on a windy day the time required would not necessarily be this value, and in fact would not always be the same value for different repetitions of this experiment.

We have seen that a distinction can be made between deterministic and stochastic relationships between random variables. Another distinction that should be made is between association and causality. Frequently, people will learn of an association between variables and assume that a "cause–effect" relationship must exist between them. This may be the case, but it is not necessarily so. For example, if a student is chosen at random from a large class and if X represents that student's grade on the first hourly exam of the semester and Y represents the grade on the second hourly exam, you might expect that a relationship exists between X and Y. These no doubt have some degree of association (e.g., low values of X tend to go with low values of Y and high values with high values), but there is not a cause–effect relationship. You would no doubt agree that a D on the first exam does not "cause" a D on the second, nor does an A on the first "cause" an A on the second. "That's obvious," you say? Well in this example it may be, but in others it may not be. For instance, in a medical study doctors examined a number of adult women and found that women with a large number of children tended to have higher blood pressure than women with a small number (or no) children. Here it would be easy to jump to the conclusion that raising children causes higher blood pressure. Now while many mothers might tend to agree with that conclusion, a more careful analysis of this medical data showed that a primary factor in higher blood pressure is age. Even among women having the same number of children, the older women tended to have higher blood pressure than the younger women. This relationship could be hidden in a less carefully performed analysis. Women having had few children (0, 1, or 2, say) will tend to be younger ones who may still bear other children while women having had more children (3, 4, or more, say) will tend to be older ones. Generally speaking, statistical analysis of data can be helpful in determining whether or not a strong relationship exists between variables. If a strong relationship does exist, it is not for the statistician to judge whether or not a cause–effect relationship exists. Such judgment should be left to experts in the particular substantive field. (At this point the statistician can also be of help to the expert by assisting in the design of the experiment used to determine whether or not a cause–effect relationship exists. It is extremely important to keep all other variables that might affect response fixed. Then the suspected causative variable is varied and the response is measured.)

In some cases a very clear relationship between variables exists. Consider, for example, a greenhouse experiment performed by Peter. Peter is raising peppers and wants to determine the amount of water to put on the pepper plants to obtain maximum yield. It is obvious that the amount of water put on the plants will be directly related to the yield. In fact, if Peter never waters the plants (i.e., if the amount of water is zero), there will be no yield. Let us assume that Peter decides to experiment with pepper plants by putting different amounts of water on them. To be specific, assume that he takes eight groups of five plants each. For the first group, he will put on no water; for the second, he will put on 50 ml of water daily; for the third, 100 ml of water daily; and for the eighth, 350 ml of water daily (see Figure 9.1).

Notationally, we can let x_i denote the amount of water put on the ith plant. (Note that the value of x_i is controlled by Peter and hence is *not* a random variable.) We can also let Y_i denote the yield of the ith plant, measured by the weight of ripe peppers picked from the plant. If we know that $x_i = 0$, we can be rather certain that Y_i will also be zero. What do you think will happen when $x_i = 50$? Since this is not a large amount of water, you might expect some yield, but you might think it will be small. This will likely be the case. However, remember that there are five different plants that will receive 50 ml of water daily. Will all these plants have *exactly* the same yield? Of course not! For this reason you can see that Y_i is actually a random variable. After the experiment is completed, an observed value y_i will have

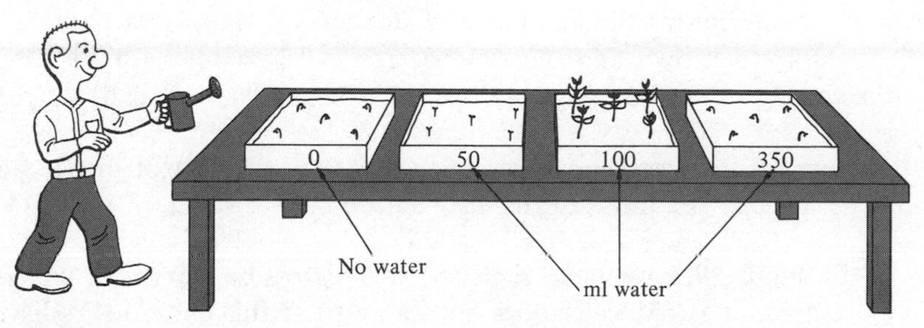

FIGURE 9.1
Peter's Experiment

been found, but this value cannot be completely predicted using only the knowledge of x_i, the amount of water put on the plant. The random variable Y corresponding to $x = 50$ will have a probability distribution, a mean, and a variance. Of course, each of these quantities may be different when the random variable Y corresponds to $x = 100$. In particular, when $x = 100$, we know that more water is being given, so it way well be that the mean yield will be higher in this case than when $x = 50$. To indicate the dependence of the mean (or variance) of Y on the value of x, we will use the notation

$$\mu_{Y|x}$$

to indicate the mean of the random variable Y for a given value of x. Similarly, $\sigma^2_{Y|x}$ will denote the variance of the random variable Y for a given value of x. When convenient, we will also write these as $\mu_{Y_i|x_i}$ and $\sigma^2_{Y_i|x_i}$.

When Peter finishes his experiment he will observe values y_i. While we would expect the observed values to be near the means $\mu_{Y_i|x_i}$, we would not expect them to be exactly equal to these values. In Figure 9.2 we have shown what the results of

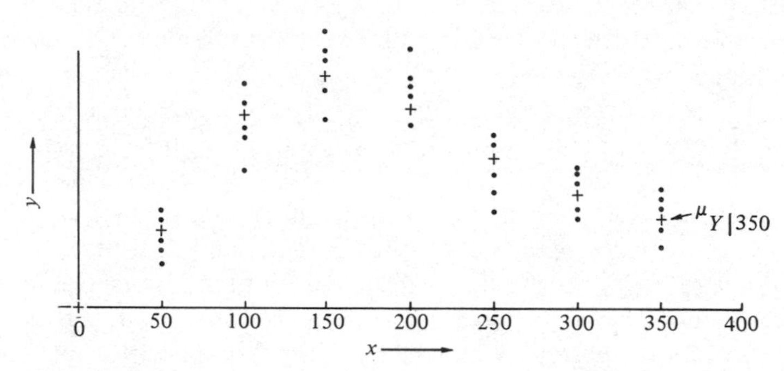

FIGURE 9.2
Possible Results of Peter's Experiment

Peter's experiments may have been like. The "+" signs indicate $\mu_{Y|x}$, the actual means of the distribution of Y for each different x, and the dots indicate the observed values. (Note that as the amount of water put on the plants becomes too large, the plants start to drown and the yield decreases.) It would not be difficult to draw a curve through the means $\mu_{Y|x}$ for each x. Such a curve is called a regression curve.

Definition 9.1 A *regression curve* is a curve passing through the points $(x, \mu_{Y|x})$, that is, through the mean of the distribution of Y given x.

In Figure 9.3 we have sketched the regression curve for Peter's experiment. We are assuming that Peter does not know what this curve looks like, for if he did, he would easily be able to answer his question about the amount of water leading to maximum yield. The optimal amount would be about 150 ml per day.

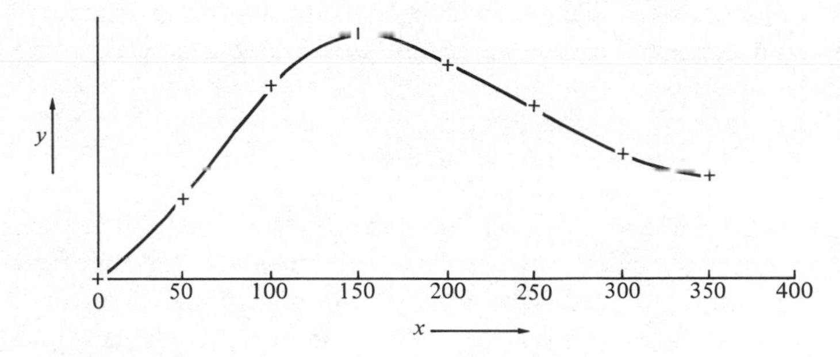

FIGURE 9.3
The Regression Curve for Peter's Experiment

This example illustrates one important use of a regression curve, namely, for purposes of optimization. A regression curve can also be used for purposes of prediction. Of course, to really make use of the regression curve, we need to be able to describe it mathematically. A regression curve is simply the graph of a special function of the variable x. It could be of any mathematical form involving trignometric and exponential functions, for example. Simpler mathematical forms would be polynomials. For instance, the regression curve in some situation might be described by the function

$$\mu_{Y|x} = \beta_0 + \beta_1 x + \beta_2 x^2 + \ldots + \beta_k x^k,$$

a polynomial of degree k. If k equals 1, the regression curve would be linear, say

$$\mu_{Y|x} = \beta_0 + \beta_1 x. \tag{9.2}$$

Although this last form is quite simple mathematically, it is seldom true in practice that a regression curve is *exactly* linear. However, if we restrict our attention to a particular range of x values, the regression curve may be approximately linear. For instance, in Figure 9.3 you can see that for x between 50 and 100, the regression curve is almost a straight line. The same is true (for a different straight line) for values of x between 200 and 300. In this chapter we will concentrate on problems where the regression curve is linear [i.e., of the form (9.2)]. In Section 9.7 we will briefly discuss the situation where the regression curve is a polynomial.

Now that we know how a regression curve is defined and some uses of a regression curve, the next question for us to ask is how we find it mathematically. If we knew the probability distribution of Y for each x, then by using methods from Section 2.5 we could find the mean of Y for each x. A graph of these values would be the regression curve. If X happened to be a random variable (as opposed to the situation in Peter's experiment where the values of x could be controlled) and if the joint probability distribution of X and Y were known, then by using methods described in Chapter 3, the conditional expectation of Y given x, $E(Y|x)$ or $\mu_{Y|x}$, could be found. A graph of x versus $\mu_{Y|x}$ would be the regression curve. Unfortunately, in most real-world situations these distributions are unknown, so the regression curve cannot be found exactly. In such a case the regression curve must be estimated. The next sections of this chapter will be devoted to the problem of estimation of the regression curve and certain associated tests of hypotheses.

9.2 ESTIMATING A SIMPLE LINEAR REGRESSION LINE

We saw in the previous section that by restricting attention to a portion of the range of x values, the regression curve may be approximately linear over that range. A particular situation where such a linear relationship may exist is related to the growth of children. If we were to consider age (x) and height (y), we would find that for boys (or girls) of ages between 6 and 12 years, the graph of x versus $\mu_{Y|x}$ (age versus mean height for that age) would be approximately linear. A similar relationship exists for height (x) and weight (y) if we restrict attention to heights between 36 and 48 inches, say. In Table 9.1 we have given heights and weights of boys having heights within

TABLE 9.1
Heights and Weights for Seven Boys

Height, x:	36	38	40	42	44	46	48
Weight, y:	30	35	35	42	51	48	53

this range. We have said that the regression curve is approximately linear for heights within this range. Does this statement agree with your intuition and experience? You certainly would agree that as children grow in stature they also grow in weight, but it may not be totally clear that the relationship is linear, even over this restricted range. To get an idea as to whether or not this linearity assumption is reasonable, it is helpful to make a scatter diagram. A *scatter diagram* is simply a two-dimensional plot of the data points (x_i, y_i). A scatter diagram for the data given in Table 9.1 is shown in Figure 9.4. Although this is only a small amount of data, you can see from the scatter diagram that the linearity assumption is reasonable.

In Figure 9.5 are some examples of what scatter diagrams might look like in other situations. In Figure 9.5(a) and (b) you see that the assumption of linearity may be reasonable over a portion of the range of x, but certainly not over the entire

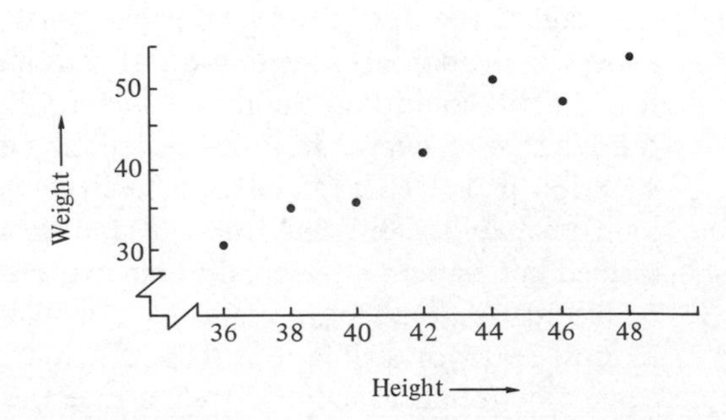

FIGURE 9.4
Scatter Diagram for Heights and Weights of Boys

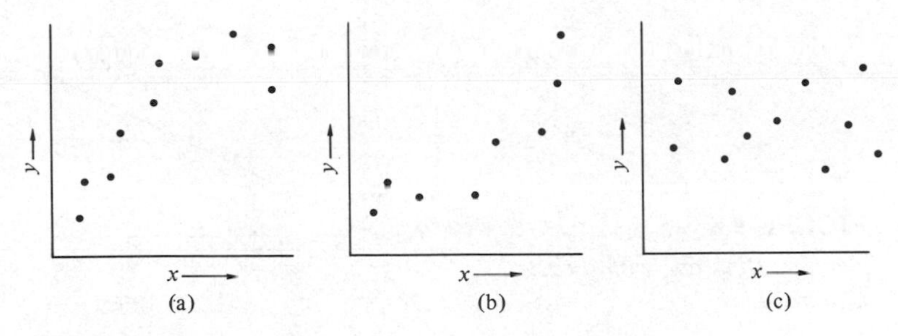

FIGURE 9.5
Other Scatter Diagrams

range. In Figure 9.5(c) it would appear that no relationship between x and y values exists. If a scatter diagram does not indicate at least an approximately linear relationship, it would be inappropriate to use the methods to be discussed in this section for estimating the regression curve.

Once it has been determined that the assumption of linearity is reasonable, the next step is to find the regression line. The functional form of the regression line is

$$\mu_{Y|x} = \beta_0 + \beta_1 x,$$

where β_0 and β_1 are parameters. The "y intercept" is given by β_0 and the slope is given by β_1. Now remember that for a given value of x, the quantity $\mu_{Y|x}$ is the mean value of Y. The actual observed value of Y, namely y, will generally vary on one side or the other of the mean. If this variation were not present, it would be a simple matter to determine the regression line and hence the values of the parameters β_0 and β_1. However, since this variation is present, we cannot hope to find the true regression line exactly. Instead, we find a "sample" regression line, that is, a line based on the sample values (x_i, y_i), having functional form

$$b_0 + b_1 x,$$

and then we use b_0 as a point estimate of β_0 and b_1 as a point estimate of β_1.

Now look at the scatter diagram in Figure 9.4. Can you visualize a straight line that passes through or near the plotted points? Such a line is referred to as an "eyeball" fitted line. This line is based on the sample observations, but the problem with using an eyeball line is that it is too subjective. Other persons, using the same technique, would tend to get slightly different lines than the one you got. We would rather have an objective method for determining a line that "best fits" a particular set of data. We will define a line to be of best fit if it is the line of "least squares." In Figure 9.6 we have repeated the scatter diagram of heights and weights of boys. The least-squares line has been drawn in. You can see that the line misses the points by some amount. On the graph we have indicated the "error" or the amount of miss for the fifth point by e_5. Mathematically, this error is equal to

$$e_5 = y_5 - (b_0 + b_1 x_5).$$

This "error" can be measured for each point, (x_i, y_i), $i = 1, 2, \ldots, n$. The *method*

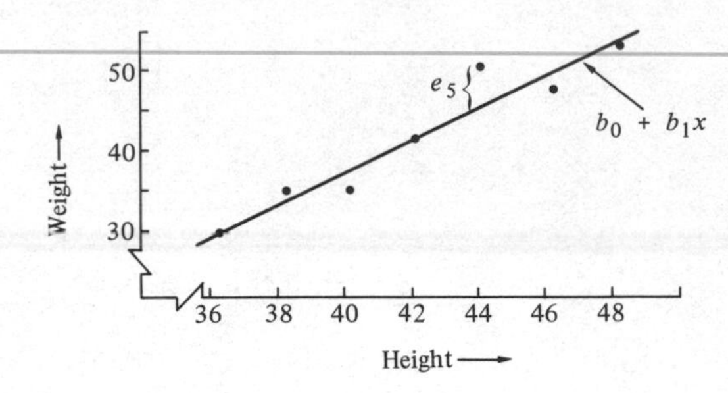

FIGURE 9.6
Scatter Diagram with Least-Squares Line

of least squares is a mathematical method for choosing the quantities b_0 and b_1 in such a way that the sum of the *squared errors* is least (i.e., is a minimum). In other words, b_0 and b_1 are chosen so that the sum

$$\sum_{i=1}^{n} e_i^2 = \sum_{i=1}^{n} [y_i - (b_0 + b_1 x_i)]^2 \tag{9.3}$$

is as small as possible. This sum will be as small as possible if we take b_0 and b_1 to be

$$b_1 = \frac{\sum_{i=1}^{n} (x_i - \bar{x})(y_i - \bar{y})}{\sum_{i=1}^{n} (x_i - \bar{x})^2} \tag{9.4}$$

$$b_0 = \bar{y} - b_1 \bar{x}. \tag{9.5}$$

Example 9.2 Plot a scatter diagram and find the least-squares line for the following data:

x_i:	1	2	3	4	5
y_i:	3	6	7	8	11.

The scatter diagram for these data is shown in Figure 9.7. To find b_0 and b_1, it is helpful to arrange the numbers in a table such as we have done in Table 9.2. From this table it is easy to find the sums necessary to calculate b_0 and b_1. From (9.4) we find that

$$b_1 = \frac{\sum (x_i - \bar{x})(y_i - \bar{y})}{\sum (x_i - \bar{x})^2} = \frac{18}{10} = 1.8,$$

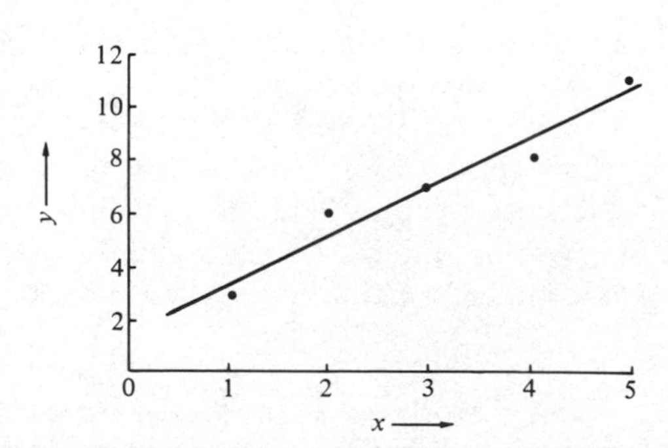

FIGURE 9.7
Scatter Diagram and Least-Squares Line for Example 9.2

TABLE 9.2

	x_i	y_i	$(x_i - \bar{x})$	$(y_i - \bar{y})$	$(x_i - \bar{x}) \cdot (y_i - \bar{y})$	$(x_i - \bar{x})^2$
	1	3	−2	−4	8	4
	2	6	1	−1	1	1
	3	7	0	0	0	0
	4	8	1	1	1	1
	5	11	2	4	8	4
Σ	15	35	0	0	18	10

and from (9.5) we find that

$$b_0 = \bar{y} - b_1\bar{x} = 7 - (1.8)(3) = 1.6.$$

The least-squares line is

$$b_0 + b_1 x = 1.6 + 1.8x.$$

In this example $b_0 = 1.6$ is a point estimate of β_0 and $b_1 = 1.8$ is a point estimate of β_1. The line $1.6 + 1.8x$ is an estimate of the true regression line $\beta_0 + \beta_1 x$. This estimated regression line is shown in Figure 9.7.

Using equations (9.4) and (9.5) and the format in Table 9.2 for calculating b_0 and b_1 is most convenient when $\bar{x}$ and $\bar{y}$ come out to be integer-valued or simple fractions. If this is not the case, the calculations of the deviations from the mean, $x_i - \bar{x}$ and $y_i - \bar{y}$, become more cumbersome. Furthermore, the possibility of significant roundoff error increases. In such situations there are alternative procedures that you can use. One possibility is to use a proven computer program, punch the data on cards using the appropriate format, and let the computer find the values of b_0 and b_1. This is the approach that would be used for most "real" problems with a large amount of data. A second alternative is to use a superduper hand calculator with a regression routine built in. Both of these alternatives, however, presuppose access to the needed hardware and software. A third alternative is to perform calculations using a simple hand calculator and a formula that is algebraically equivalent to (9.4). One equivalent formula is

$$b_1 = \frac{n \sum (x_i y_i) - \sum (x_i) \sum (y_i)}{n \sum (x_i^2) - (\sum x_i)^2}. \tag{9.6}$$

In the next example we will find b_1 using both equations (9.4) and (9.6).

Example 9.3 Find the estimated regression line for the height–weight data for boys given in Table 9.1.

The estimated regression line is simply the least-squares line. To find this line, we construct Table 9.3 to more easily find the necessary sums. This table is useful for finding b_1 by using equation (9.4). Using the sums given in this table in equations (9.4) and (9.5), we find that

$$b_1 = \frac{\sum (x_i - \bar{x})(y_i - \bar{y})}{\sum (x_i - \bar{x})^2} = \frac{222}{112} = 1.982$$

$$b_0 = \bar{y} - b_1\bar{x} = 42 - (1.982)(42) = -41.244.$$

TABLE 9.3

	x_i	y_i	$(x_i - \bar{x})$	$(y_i - \bar{y})$	$(x_i - \bar{x}) \cdot (y_i - \bar{y})$	$(x_i - \bar{x})^2$
	36	30	−6	−12	72	36
	38	35	−4	−7	28	16
	40	35	−2	−7	14	4
	42	42	0	0	0	0
	44	51	2	9	18	4
	46	48	4	6	24	16
	48	53	6	11	66	36
Σ	294	294	0	0	222	112

Consequently, the estimated regression line is

$$b_0 + b_1 x = -41.244 + 1.982x.$$

(Note that b_1 has been given to three decimal places. Such rounding, although necessary, may cause some minor errors in subsequent calculations.)

To find b_1 by using (9.6), we can construct a table similar to Table 9.3. In Table 9.4 we give the quantities necessary to find b_1.

TABLE 9.4

	x_i	y_i	$x_i y_i$	x_i^2
	36	30	1,080	1,296
	38	35	1,330	1,444
	40	35	1,400	1,600
	42	42	1,764	1,764
	44	51	2,244	1,936
	46	48	2,208	2,116
	48	53	2,544	2,304
Σ	294	294	12,570	12,460

Substituting into (9.6), we find that

$$b_1 = \frac{(7)(12{,}570) - (294)(294)}{(7)(12{,}460) - (294)^2}$$

$$= \frac{1554}{784} = 1.982.$$

This estimated regression line was plotted along with the corresponding scatter diagram in Figure 9.6.

We can now formally state the important results of this section as a theorem.

Theorem 9.1 Given n pairs of points, $(x_1, y_1), (x_2, y_2), \ldots, (x_n, y_n)$, the straight line $y = b_0 + b_1 x$ for which the sum

$$\sum e_i^2 = \sum [y_i - (b_0 + b_1 x_i)]^2$$

is a minimum is found by defining

$$b_1 = \frac{\sum (x_i - \bar{x})(y_i - \bar{y})}{\sum (x_i - \bar{x})^2} \qquad \text{and} \qquad b_0 = \bar{y} - b_1 \bar{x}.$$

[*Note:* It is common for authors to define

$$\hat{y}_i = b_0 + b_1 x_i$$

to be the "fitted value" of y for a given x_i. Although this notation is somewhat inconsistent with that introduced in Section 7.2, the quantity $\hat{y}_i$ is really an estimate of $\mu_{Y|x_i}$ or, equivalently, $\beta_0 + \beta_1 x_i$. Using this notation, an alternative expression for $\sum e_i^2$ is

$$\sum (y_i - \hat{y}_i)^2.]$$

• • • • •

Proof: Recalling that a maximum or minimum for a function may occur at a point where the first derivative equals zero, we begin by defining a function of two variables, $f(b_0, b_1)$, by

$$f(b_0, b_1) = \sum_{i=1}^{n} [y_i - (b_0 + b_1 x_i)]^2.$$

Note that in this setting we treat b_0 and b_1 as variables and the x_i and y_i as constants. We would like to find values of b_0 and b_1 for which the first derivative of f is equal to zero. However, since f is a function of two variables, we must consider partial derivatives, obtain two equations, and solve the two equations for zero simultaneously. We then have

$$\frac{\partial f(b_0, b_1)}{\partial b_0} = \sum 2[y_i - (b_0 + b_1 x_i)]^1(-1) \tag{9.7}$$

$$\frac{\partial f(b_0, b_1)}{\partial b_1} = \sum 2[y_i - (b_0 + b_1 x_i)]^1(-x_i). \tag{9.8}$$

Setting the partial derivatives (9.7) and (9.8) both equal to zero and simplifying somewhat, we obtain

$$\sum [y_i - (b_0 + b_1 x_i)] = 0 \tag{9.9}$$

$$\sum [x_i y_i - (b_0 x_i + b_1 x_i^2)] = 0. \tag{9.10}$$

We can easily solve (9.9) for b_0 by

$$\sum y_i - \sum b_0 - \sum b_1 x_i = 0$$

$$\sum y_i - n b_0 - b_1 \sum x_i = 0$$

$$b_0 = \frac{\sum y_i}{n} - b_1 \frac{\sum x_i}{n}$$

$$b_0 = \bar{y} - b_1 \bar{x}.$$

Substituting this value of b_0 into (9.10) gives

$$\sum [x_i y_i - (\bar{y} - b_1\bar{x})x_i - b_1 x_i^2] = 0$$
$$\sum [x_i y_i - x_i\bar{y} + b_1 x_i\bar{x} - b_1 x_i^2] = 0$$
$$\sum [x_i(y_i - \bar{y}) - b_1 x_i(x_i - \bar{x})] = 0$$
$$\sum x_i(y_i - \bar{y}) - b_1 \sum x_i(x_i - \bar{x}) = 0.$$

If we now solve this last equation for b_1, we will get

(9.11) $$b_1 = \frac{\sum x_i(y_i - \bar{y})}{\sum x_i(x_i - \bar{x})}.$$

This expression for b_1 is not quite in the form as given in Theorem 9.1. However, there are several different algebraically equivalent forms for b_1. These different forms arise because of the fact that

$$\sum (x_i - \bar{x}) = 0 \quad \text{and} \quad \sum (y_i - \bar{y}) = 0.$$

(See Exercise 14 and also Exercise 8 of Chapter 5.) Using this, we see that

(9.12) $$\begin{aligned}\sum (x_i - \bar{x})(y_i - \bar{y}) &= \sum [x_i(y_i - \bar{y}) - \bar{x}(y_i - \bar{y})] \\ &= \sum x_i(y_i - \bar{y}) - \sum \bar{x}(y_i - \bar{y}) \\ &= \sum x_i(y_i - \bar{y}) - \bar{x}\sum (y_i - \bar{y}) \\ &= \sum x_i(y_i - \bar{y}) - \bar{x}(0) \\ &= \sum x_i(y_i - \bar{y}).\end{aligned}$$

In the same way, it is true that

(9.13) $$\sum (x_i - \bar{x})(y_i - \bar{y}) = \sum y_i(x_i - \bar{x})$$

and

(9.14) $$\sum (x_i - \bar{x})^2 = \sum (x_i - \bar{x})(x_i - \bar{x}) = \sum x_i(x_i - \bar{x}).$$

Using (9.12) and (9.14) in equation (9.11), we get

$$b_1 = \frac{\sum (x_i - \bar{x})(y_i - \bar{y})}{\sum (x_i - \bar{x})^2}.$$

Although we have obtained formulas for b_0 and b_1 as stated in the theorem, the proof of the theorem is not quite complete. We have simultaneously solved two equations based on setting the first (partial) derivative equal to zero. It is possible (mathematically) that such a procedure could lead to maximizing, rather than minimizing, the sum of squared deviations $\sum e_i^2$. Although we will not go through the details here, we can say that further analysis would show that these solutions do, in fact, minimize the sum of squared deviations.

9.3 PROPERTIES OF THE LEAST-SQUARES ESTIMATORS

In the development up to this point, we have considered pairs of points (x_i, y_i) where we have considered y_i to be the observed value of a random variable Y_i. Since the values of b_0 and b_1 found by (9.4) and (9.5) depend on these y_i values, they, too, are the observed values of random variables. To make this fact more clear, we can use our usual convention of denoting random variables by uppercase letters and observed values of random variables by lowercase letters. Rewriting (9.4) and (9.5) using this notation, we get

$$B_1 = \frac{\sum (x_i - \bar{x})(Y_i - \bar{Y})}{\sum (x_i - \bar{x})^2} \tag{9.15}$$

$$B_0 = \bar{Y} - B_1\bar{x}. \tag{9.16}$$

From these equations it should be clear to you that B_0 and B_1 are random variables. Since the observed values of these random variables, b_0 and b_1, are used to estimate the parameters β_0 and β_1, it follows that B_0 and B_1 are estimators of β_0 and β_1 (see Definition 7.2). In this section we would like to consider properties of these estimators. In order to endow these estimators with desirable properties, it will be necessary to make certain assumptions about the random variables Y_i. We will introduce these assumptions as necessary.

We will first show that

B_0 and B_1 are linear combinations of the random variables Y_i.

By using the algebraic equivalence given in (9.13), it is easy to see that

$$B_1 = \frac{\sum (x_i - \bar{x})Y_i}{\sum (x_i - \bar{x})^2}.$$

Defining

$$c_i = \frac{x_i - \bar{x}}{\sum (x_i - \bar{x})^2},$$

we have

$$B_1 = \sum \left[\frac{x_i - \bar{x}}{\sum (x_i - \bar{x})^2}\right] Y_i = \sum c_i Y_i, \tag{9.17}$$

so B_1 can be written as a linear combination of the random variables Y_i (see Definition 5.7). Substituting this expression for B_1 into (9.16), we get

$$\begin{aligned} B_0 &= \bar{Y} - (\textstyle\sum c_i Y_i)\bar{x} \\ &= \sum \left(\frac{1}{n}\right) Y_i - \sum c_i \bar{x} Y_i \\ &= \sum \left(\frac{1}{n} - c_i\bar{x}\right) Y_i \\ &= \sum a_i Y_i, \end{aligned}$$

where

$$a_i = \left(\frac{1}{n} - c_i\bar{x}\right) = \left(\frac{1}{n} - \frac{(x_i - \bar{x})\bar{x}}{\sum (x_i - \bar{x})^2}\right).$$

Consequently, B_0 can also be written as a linear combination of the random variables Y_i.

If it is true that the regression curve is really linear, that is, if

Assumption 1 $E(Y_i | x_i) = \mu_{Y_i | x_i} = \beta_0 + \beta_1 x_i$,

it will be true that

B_0 and B_1 are unbiased estimators of β_0 and β_1.

The concept of unbiased estimators was discussed in Section 7.3. You will recall that unbiasedness is a desirable property for estimators to have. The proof of the fact that these estimators are unbiased follows directly from properties of expected values and, in particular, from Theorem 5.3. We need to show that

$$E(B_0) = \beta_0 \qquad \text{and} \qquad E(B_1) = \beta_1.$$

We will use the fact that $\sum (x_i - \bar{x}) = 0$. Using this equality, it follows that

$$\sum c_i = \sum \left[\frac{(x_i - \bar{x})}{\sum (x_i - \bar{x})^2} \right] = \frac{\sum (x_i - \bar{x})}{\sum (x_i - \bar{x})^2} = 0. \tag{9.18}$$

It also follows from (9.14) that

$$\sum c_i x_i = \sum \left[\frac{x_i (x_i - \bar{x})}{\sum (x_i - \bar{x})^2} \right] = \frac{\sum x_i (x_i - \bar{x})}{\sum (x_i - \bar{x})^2} = 1. \tag{9.19}$$

Using (9.17), (9.18), (9.19), and Assumption 1, it is not hard to see that

$$\begin{aligned} E(B_1) &= E(\sum c_i Y_i) = \sum c_i E(Y_i) \\ &= \sum c_i(\beta_0 + \beta_1 x_i) \\ &= \beta_0 \sum c_i + \beta_1 \sum c_i x_i \\ &= \beta_0(0) + \beta_1(1) = \beta_1. \end{aligned}$$

This shows that B_1 is an unbiased estimator of β_1. The proof that B_0 is an unbiased estimator of β_0 is left to Exercise 15.

If we make the assumption that the Y_i are independent, that is, if

Assumption 2 the random variables $Y_1, Y_2, \ldots, Y_n$ are independent,

it is not hard to get expressions for the variances of the estimators B_0 and B_1. Since each of these estimators is a linear combination of the Y_i, it follows from Theorem 5.4 that

$$V(B_0) = V(\sum a_i Y_i) = \sum a_i^2 \, V(Y_i) \tag{9.20}$$

and

$$V(B_1) = V(\sum c_i Y_i) = \sum c_i^2 \, V(Y_i). \tag{9.21}$$

Now it certainly may be the case that the variances

$$V(Y_i) = V(Y | x_i) = \sigma^2_{Y | x_i}$$

are different for each i. If you consider Peter's pepper-growing experiment (Figure 9.2), you will see that for different values of x_i (i.e., for different amounts of water), the variability in y, the yield, will be different. However, in some situations these variances may be the same. If they are the same, the common value of the variance can be denoted by σ^2. If we assume that this is the case, that is, if we make

Assumption 3 $V(Y_i) = \sigma^2$ for all i,

we can get simpler expressions for $V(B_0)$ and $V(B_1)$. In particular, we can show that

$$V(B_0) = \sigma^2\left[\frac{1}{n} + \frac{\bar{x}^2}{\sum (x_i - \bar{x})^2}\right] \tag{9.22}$$

and

$$V(B_1) = \frac{\sigma^2}{\sum (x_i - \bar{x})^2}. \tag{9.23}$$

That equation (9.23) is true follows from (9.21), Assumption 3, and the fact that

$$\sum c_i^2 = \sum\left[\frac{x_i - \bar{x}}{\sum (x_i - \bar{x})^2}\right]^2 = \sum \frac{(x_i - \bar{x})^2}{[\sum (x_i - \bar{x})^2]^2} = \frac{1}{\sum (x_i - \bar{x})^2}. \tag{9.24}$$

The proof that (9.22) is true is left to Exercise 16.

We have now found expressions for the mean and variance of each of the estimators B_0 and B_1. Even more can be said, however. Not only are B_0 and B_1 point estimates of β_0 and β_1, they are the *best* point estimators among a large class of estimators. They are best in the sense of having the smallest variance. This result is stated in the following theorem, which we state without proof.

Theorem 9.2 (*Gauss–Markov Theorem*) If Assumptions 1, 2, and 3 hold, then B_0 and B_1 as defined by (9.4) and (9.5) are unbiased estimators of β_0 and β_1. Furthermore, these estimators have the smallest variance among the class of all unbiased estimators of β_0 and β_1 which are linear combinations of the random variables Y_i.

(Note that B_0 and B_1 are often called BLUE. This does not refer to their emotional state, but is an acronym for Best Linear Unbiased Estimate.)

In order to be able to find confidence intervals for β_0 and β_1 and/or to test hypotheses about these parameters, it is helpful to make an assumption about the probability distribution of the random variables Y_i. If we make

Assumption 4 Y_i has a normal distribution for each i,

it will be possible to make such inference. In particular, it follows from Theorem 5.5 that if Assumption 4 holds, then

B_0 and B_1 will be normally distributed random variables.

That this statement is true follows from the fact shown earlier in this section that B_0 and B_1 are linear combinations of the independent random variables Y_i.

An interesting mathematical result is that if Assumptions 1 to 4 all hold, the parameters β_0, β_1, and σ^2 can be estimated by the method of maximum likelihood (Section 7.2). The maximum likelihood estimates of β_0 and β_1 turn out to be exactly the same as the least-squares estimates given in (9.4) and (9.5). The maximum likelihood estimate of σ^2 is given by

$$\frac{\sum (Y_i - \hat{Y}_i)^2}{n}.$$

Unfortunately, this estimate is biased, so it is generally not used as an estimate of σ^2. Instead, the quantity

$$S^2_{Y|x} = \frac{\sum (Y_i - \hat{Y}_i)^2}{n-2} \tag{9.25}$$

is used because it is an unbiased estimator for σ^2.

9.4 INFERENCE IN SIMPLE LINEAR REGRESSION

In the last section we gave assumptions about the random variables Y_i which allowed us to deduce several properties of the least-squares estimators β_0 and β_1. For easy reference and summary we will again state those assumptions here.

Assumption 1 $E(Y_i \mid x_i) = \mu_{Y_i|x_i} = \beta_0 + \beta_1 x_i$.

Assumption 2 $Y_1, Y_2, \ldots, Y_n$ are independent random variables.

Assumption 3 $V(Y_i) = \sigma^2$ for $i = 1, 2, \ldots, n$.

Assumption 4 Y_i has a normal distribution for $i = 1, 2, \ldots, n$.

Unless explicitly stated otherwise, we will assume that each of these assumptions holds in the problems through the remainder of this chapter.

Example 9.4 Mr. Havitall, owner of the Havitall Hardware store, is interested in studying the relationship of size of an advertisement and profits. Over a period of several weeks he has placed ads of various sizes in the local newspaper. The costs of the ads (x) and the amount of profit for the week following the appearance of the ad (Y) are given in Table 9.5. Based on this information, find point estimates of β_0

TABLE 9.5
Havitall Sales Data

Cost of ads, x:	0	75	150	225	300
Profit, y:	325	362	351	382	460

and β_1. Also, *in the context of this problem*, interpret the meaning of β_0 and β_1. (Of course, a cost of zero corresponds to no ad being placed that week.)

Direct calculations show that

$$\sum x_i = 750, \quad \sum x_i^2 = 168{,}750, \quad \sum x_i y_i = 303{,}750, \quad \sum y_i = 1880.$$

Using equations (9.6) and (9.5), we find that

$$b_1 = \frac{(5)(303{,}750) - (750)(1880)}{(5)(168{,}750) - (750)^2}$$

$$= \frac{108{,}750}{281{,}250} = .387$$

$$b_0 = 376 - (.387)(150) = 317.95.$$

These values are the point estimates of β_1 and β_0, respectively. In the context of this problem, β_0, the Y intercept, is the amount of the profit when $x = 0$ (i.e., when no money is spent on advertising). The slope, β_1, would correspond to the change in profit per amount spent on advertising. If the slope is positive, the change in profit would be an increase.

In this problem the true values of β_0 and β_1 were unknown and consequently had to be estimated. We know, of course, that a point estimate of any parameter is likely to be in error. For this reason, it is generally more informative to give an interval estimate of parameters since, in some sense, this type of estimate takes error into account. Consider for a moment why Mr. Havitall might be interested in knowing the value of β_1. What would be the ramifications if β_1 were known to be less than 0? Greater than 0? Some reflection on the matter will allow you to see that if β_1 is less than zero, the increased sales will not offset the cost of advertising, leading to a net loss in profit. If β_1 exceeds the value zero, increased sales will more than offset the cost of advertising and lead to increased profit.

This illustration points out the fact that it may be desirable to be able to find confidence intervals and/or test hypotheses for parameters in regression problems. To do this, it is necessary to know the probability distribution of the random variables B_0 and B_1.

Theorem 9.3 If Assumptions 1 to 4 hold, the random variables B_0 and B_1 are each normally distributed with means

$$\mu_{B_0} = E(B_0) = \beta_0, \qquad \mu_{B_1} = E(B_1) = \beta_1$$

and variances

$$\sigma^2_{B_0} = V(B_0) = \sigma^2\left(\frac{1}{n} + \frac{\bar{x}^2}{\sum (x_i - \bar{x})^2}\right)$$

and

$$\sigma^2_{B_1} = V(B_1) = \frac{\sigma^2}{\sum (x_i - \bar{x})^2},$$

respectively.

The proof of this theorem follows directly from the properties of B_0 and B_1 given in the previous section.

Since B_0 and B_1 are each normally distributed, it follows that each can be standardized. That is,

$$\frac{B_0 - \beta_0}{\sigma_{B_0}} \quad \text{and} \quad \frac{B_1 - \beta_1}{\sigma_{B_1}} \tag{9.26}$$

each have a standard normal distribution. Knowing these distributions allows us to find confidence intervals for β_0 or β_1 using the techniques developed in Chapter 7. For example,

$$B_0 \pm z_{1-\alpha/2}\sigma_{B_0} \quad \text{and} \quad B_1 \pm z_{1-\alpha/2}\sigma_{B_1} \tag{9.27}$$

would give the end points of $1 - \alpha$ confidence intervals for β_0 and β_1, respectively. Similarly, the statistics in (9.26) could be used as test statistics for testing hypotheses about β_0 or β_1 following the techniques developed in Chapter 8.

Example 9.5 If Mr. Havitall knows that σ^2, the variance of the weekly profit for each value of x, is equal to $(\$25)^2$, find a 95% confidence interval for β_1 using the data given in Example 9.4.

To solve this problem, we need to use (9.27). We also need to find σ_{B_1}. Using (9.23), we find that

$$\sigma_{B_1} = \sqrt{\frac{\sigma^2}{\sum (x_i - \bar{x})^2}} = \frac{\sigma}{\sqrt{\sum (x_i^2) - (\sum x_i)^2/n}} = \frac{25}{\sqrt{(168{,}750) - [(750)^2/5]}}$$

$$= \frac{25}{\sqrt{56{,}250}} = .105.$$

Since $z_{1-\alpha/2} = z_{.975} = 1.96$, we have

$$.387 \pm 1.96(.105)$$

or $(.181, .593)$ as a 95% confidence interval for β_1.

Unfortunately, in most problems of this type, the true value of σ^2 is really unknown. In these situations it becomes necessary to estimate this value in order to make inference. We were confronted with a similar situation in Section 7.5 when we tried to obtain confidence intervals for a mean μ when σ^2 was unknown. We saw there that when sampling from a normal distribution, if σ is replaced by S, the resulting statistic of interest, namely, $(\bar{X} - \mu)/(S/\sqrt{n})$, has a t distribution with $n - 1$ degrees of freedom. You will also remember that S^2 is defined by

$$\frac{\sum (X_i - \bar{X})^2}{n - 1},$$

which we can think of as follows:

$$\frac{\sum [(i\text{th observed value}) - (\text{estimate of mean of } i\text{th value})]^2}{\text{number necessary to get an unbiased estimate of } \sigma^2}.$$

An analogous formula can be used in a regression problem leading to the estimator for σ^2 given in (9.25), namely,

$$S_{Y|x}^2 = \frac{\sum (Y_i - \hat{Y}_i)^2}{n - 2}.$$

If this estimator is used in place of σ^2 in equations (9.22) and (9.23), we obtain estimators of $\sigma_{B_0}^2$ and $\sigma_{B_1}^2$ as follows:

$$S_{B_0}^2 = S_{Y|x}^2\left[\frac{1}{n} + \frac{\bar{x}^2}{\sum (x_i - \bar{x})^2}\right] \tag{9.28}$$

$$S_{B_1}^2 = \frac{S_{Y|x}^2}{\sum (x_i - \bar{x})^2}. \tag{9.29}$$

Since it can be shown that

$$\frac{B_0 - \beta_0}{S_{B_0}} \quad \text{and} \quad \frac{B_1 - \beta_1}{S_{B_1}} \tag{9.30}$$

each has a t distribution with $n - 2$ degrees of freedom, it follows that confidence intervals for β_0 or β_1 can be found from

$$B_0 \pm t_{n-2, 1-\alpha/2} S_{B_0} \quad \text{and} \quad B_1 \pm t_{n-2, 1-\alpha/2} S_{B_1}. \tag{9.31}$$

Likewise, the statistics in (9.30) could be used as test statistics for testing hypotheses about β_0 or β_1.

A direct calculation of $\sum (Y_i - \hat{Y}_i)^2$ for use in finding $S^2_{Y|x}$ would be rather tedious, since $\hat{Y}_i$ would have to be found for each i. By using alternative formulas which are algebraically equivalent, some of this tedium can be avoided. The following are some equivalent forms that can be used.

$$\sum (Y_i - \hat{Y}_i)^2 = \sum Y_i^2 - B_0 \sum Y_i - B_1 \sum x_i Y_i \tag{9.32}$$

$$= \sum (Y_i - \bar{Y})^2 - \frac{[\sum (x_i - \bar{x})(Y_i - \bar{Y})]^2}{\sum (x_i - \bar{x})^2} \tag{9.33}$$

$$= \sum (Y_i - \bar{Y})^2 - B_1^2 \sum (x_i - \bar{x})^2. \tag{9.34}$$

Some further computational simplifications can be obtained by using

$$\sum (x_i - \bar{x})^2 = \sum x_i^2 - n\bar{x}^2 = \sum x_i^2 - \frac{(\sum x_i)^2}{n}.$$

Example 9.6 Estimate the value of σ^2 for the variance of weekly profit using the data of Example 9.4. Use this value to find a 95% confidence interval for β_1.

Using the data of Example 9.4, we see that

$$\sum y_i^2 = 717{,}394.$$

Using this value in conjunction with (9.32), we find that

$$s^2_{Y|x} = \frac{\sum y_i^2 - b_0 \sum y_i - b_1 \sum x_i y_i}{n-2}$$

$$= \frac{717{,}394 - (317.950)(1880) - (.387)(303{,}750)}{5-2}$$

$$= 698.917.$$

This is an estimate of σ^2. Using (9.29), we have

$$s_{B_1} = \sqrt{\frac{s^2_{Y|x}}{\sum (x_i - \bar{x})^2}} = \frac{26.437}{\sqrt{56250}} = .111,$$

so the end points of a 95% confidence interval for β_1 are found from (9.31) to be

$$b_1 \pm t_{3,.975} s_{B_1} \quad \text{or} \quad .387 \pm (3.182)(.111)$$

or .034 and .740.

In this example we found a confidence interval for β_1. Of course, hypotheses about β_1 could also be tested. A commonly tested null hypothesis is the hypothesis that $\beta_1 = 0$. Why is this hypothesis tested? Consider the implications of this fact in terms of the linear regression curve. If $\beta_1 = 0$, then

$$\mu_{Y|x} = \beta_0 + \beta_1 x$$

simplifies to

$$\mu_{Y|x} = \beta_0.$$

This would mean that the mean of the distribution of Y for each x is the same value, namely, β_0. Consequently, the quantity x has no effect on the mean of Y. You can see that knowledge of this fact would be helpful in practical problems.

Example 9.7 Using the Havitall data, test the hypotheses

$$H_0 : \beta_0 \leq 300 \qquad \text{vs.} \qquad H_1 : \beta_0 > 300.$$

Use a .05 level of significance.

You might first note that in the context of this problem, if the null hypothesis is true, then with no advertising the weekly profits accruing to Mr. Havitall are \$300 or less. The appropriate test statistic for this problem is $(B_0 - \beta_0)/S_{B_0}$ [see (9.30)], which has a t distribution with $(n - 2)$ degrees of freedom. The null hypothesis would be rejected if the test statistic is greater than or equal to the critical value of

$$t_{n-2,1-\alpha} = t_{3,.95} = 2.353.$$

Since the value of the test statistic is

$$t = \frac{b_0 - \beta_0}{s_{B_0}} = \frac{317.950 - 300}{26.437\sqrt{\frac{1}{5} + (150)^2/56{,}250}}$$

$$= \frac{17.950}{20.478} = .877,$$

which does not exceed the critical value, H_0 cannot be rejected.

When considering the parameters β_0 and β_1, there may be times when it is desired to test hypotheses about these parameters and there may be other times when estimates are desired. In the next two cases we consider, the primary interest is in the area of estimation. We alluded to these cases earlier in Peter's pepper-growing problem. There are two aspects in which Peter might be interested. First, for a fixed value of x, what is an estimate of the *mean response* for all plants that might be grown using a value of x (amount of water); second, what is an estimate of the *response of an individual* y value? The answers to these questions are very much related. In fact, if we were to consider only point estimates, the answers would be identical! To see why this is so, consider a more familiar example. Imagine yourself sitting in a large auditorium with several hundred people. If one person is going to be randomly chosen from this group and you are asked in advance of the choosing to guess that person's height, what would you guess? If the group consisted of ordinary men and women, you would guess (estimate) a single number, say 68 inches, which might appear to be an "average" number for the group you observe. Now consider what you would do if a number of persons, say 10, were to be chosen and you were asked to guess the average (mean) height for the group of 10. What would your guess be in this case? It would again be the same number; that is, your point estimate would be the same in each case! On the other hand, ask yourself the following question: In which of these two situations are you more likely to be close to correct? From the work that you have done earlier in this book, it should be clear to you that you would likely be closer in guessing the mean, since there is less variability in estimating a mean than in estimating a single value. A similar thing is true in a regression problem and the difference in variability is reflected in the length of an interval estimate of a mean response or of an individual response.

Specifically, if we wish to estimate the mean response for a given x value (i.e., to estimate $\mu_{Y|x} = \beta_0 + \beta_1 x$), we would use as a point estimator the quantity

$$\hat{Y} = B_0 + B_1 x.$$

We have seen that both B_0 and B_1 are linear combinations of the random variables $Y_1, Y_2, \ldots, Y_n$ [see equation (9.17) and following], so it follows that $\hat{Y}$ is also a linear combination of these random variables. In fact, using c_i and a_i as defined in Section 9.3, we can write

$$\hat{Y} = \sum (a_i + c_i x) Y_i.$$

It is important to note that the coefficient $(a_i + c_i x)$ depends on x, so it should not be surprising that the value of x enters into the calculation of the variance of $\hat{Y}$. Further, since $\hat{Y}$ is a linear combination of independent normal random variables, $\hat{Y}$ will be normally distributed. Some algebraic computations would show that

$$\hat{Y} \sim N\left[\mu_{\hat{Y}} = \beta_0 + \beta_1 x, \sigma^2_{\hat{Y}} = \sigma^2 \left(\frac{1}{n} + \frac{(x - \bar{x})^2}{\sum (x_i - \bar{x})^2}\right)\right].$$

Following the usual procedure for obtaining confidence intervals, a $1 - \alpha$ confidence interval for the mean response, $\beta_0 + \beta_1 x$, when σ^2 is known would be given by

$$\hat{Y} \pm z_{1-\alpha/2}\sigma\sqrt{\frac{1}{n} + \frac{(x - \bar{x})^2}{\sum (x_i - \bar{x})^2}}.$$

If σ^2 is unknown, it must be estimated by using $S^2_{Y|x}$, leading to a confidence interval

$$\hat{Y} \pm t_{n-2,1-\alpha/2} S_{Y|x} \sqrt{\frac{1}{n} + \frac{(x - \bar{x})^2}{\sum (x_i - \bar{x})^2}}. \tag{9.35}$$

We again would like to emphasize how the value of x enters into the picture. As x deviates from $\bar{x}$, the width of the confidence interval increases. The width is smallest when $x = \bar{x}$.

Example 9.8 Find a 90% confidence interval for the mean profit that Mr. Havitall can expect if he buys advertising space costing \$225. What would the interval be if he bought space costing \$150?

From the results of previous examples, we find

$$b_0 = 317.95, \qquad b_1 = .387, \qquad s_{y|x} = \sqrt{698.917}$$

$$n = 5, \qquad \bar{x} = 150, \qquad \sum (x_i - \bar{x})^2 = 56{,}250.$$

We have

$$\hat{y}_{(225)} = b_0 + b_1(225) = 405.025.$$

Since $t_{3,.95} = 2.353$, substituting into (9.35) gives

$$405.025 \pm 2.353\sqrt{698.917}\sqrt{\frac{1}{5} + \frac{(225 - 150)^2}{56{,}250}}$$

or $405.025 \pm 2.353(26.437)\sqrt{.3}$. Consequently, the end points are 370.95 and 439.10.

In the second part of this problem, $x = \bar{x} = 150$, so

$$\hat{y}_{(150)} = b_0 + b_1(150) = 376,$$

and the end points of the interval are given by

$$376 \pm 2.353(26.437)\sqrt{\tfrac{1}{5} + 0}$$

or 348.18 and 403.82.

We could actually go further than we have in this example and construct a graph that shows the least-squares line and the end points of a 90% confidence interval for the mean response for each x. This is done in Figure 9.8. Note that on the horizontal axis we only graphed values up to \$300, the largest amount Mr. Havitall actually spent on advertising. Could we not find estimates for values of x above 300?

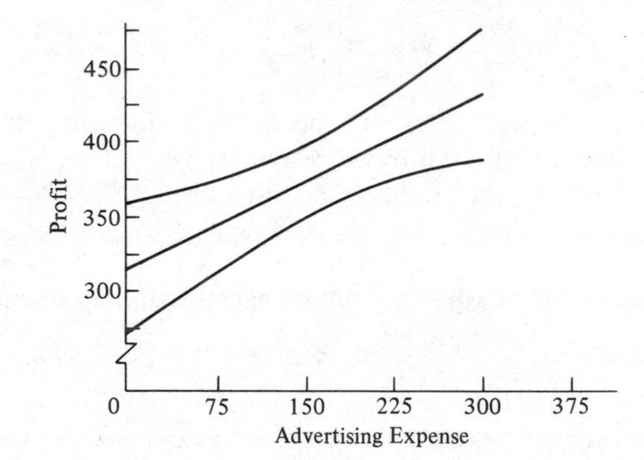

FIGURE 9.8
90% Confidence Interval Curves for Mean Response

Mathematically, we could, of course, merely by substituting into the appropriate formulas. However, such extrapolation beyond the range of observed values is a risky business which we strongly advise against. Why? Because initially we assumed that the cost of ads and the amount of profit would be approximately linearly related over the given range of x values. The linear relationship will very likely not continue to hold over a larger range of values. In fact, owing to the limitations of the size of his store and staff, Mr. Havitall knows that after some point increased expenditures on advertising will lead to decreased rather than increased profit.

Rather than consider the mean response, a person may be interested in predicting the value of the next individual response for a given level of x. As we indicated earlier, making such a prediction accurately is more difficult than predicting or estimating a mean response. In this case there are two sources of variability, variability in estimating the true mean and variability of the individual response about the mean value. A point estimate of the value of the response for a given value of x is given by

$$\hat{Y} = B_0 + B_1 x$$

while a $1 - \alpha$ "prediction interval" is given by

$$\hat{Y} \pm z_{1-\alpha/2}\sigma\sqrt{1 + \frac{1}{n} + \frac{(x - \bar{x})^2}{\sum (x_i - \bar{x})^2}}$$

if σ is known or by

$$\hat{Y} \pm t_{n-2,1-\alpha/2}S_{Y|x}\sqrt{1 + \frac{1}{n} + \frac{(x - \bar{x})^2}{\sum (x_i - \bar{x})^2}} \tag{9.36}$$

if σ is unknown. Note that, as we indicated, the width of these confidence intervals will be greater than those intervals for the mean response.

Example 9.9 Find a 95% prediction interval for the amount of profit Mr. Havitall will get when he places an ad of a size costing \$75.

Using (9.36) and previously obtained information, we find that

$$\hat{Y}_{(75)} = b_0 + b_1(75) = 346.98$$

and

$$346.98 \pm 3.182(26.437)\sqrt{1 + \frac{1}{5} + \frac{5625}{56250}}$$

or 346.98 ± 95.91. That is, the end points of a 95% prediction interval are 251.07 and 442.89.

We will close this section on inference in simple linear regression by noting that inference can be made about the value of σ^2 as well as the other parameters we have considered. Interval estimates can be found by using $S^2_{Y|x}$ and the techniques given in Section 7.8, and hypotheses can be tested using the techniques of Section 8.7. In each case the appropriate degrees of freedom to use with the chi-square random variable is $n - 2$.

9.5 USE OF RESIDUALS IN EXAMINING MODEL ASSUMPTIONS

In our present state of technology, we have many powerful tools to help us accomplish difficult tasks. Most homes have many of these tools: power drills, food grinders, vacuum cleaners, and many others are very common. Although most of these tools are simple to use and may be very beneficial, their very simplicity can make them that much more dangerous. A child may try to "fix" a toy with the drill or "clean" the fish tank with the vacuum cleaner, for example, because it looks so easy when the parents use the tools. Analogously, regression analysis, especially when implemented through use of a computer program, is a powerful tool for data analysis and statistical inference. With prepackaged statistical programs, it is a tool that is easily used. However, it is dangerous in the sense that if improperly used, misleading and erroneous conclusions can be drawn! When is it improper to use the techniques that we have presented in this chapter? When the basic underlying assumptions 1 to 4 summarized at the beginning of Section 9.4 fail to hold. In this section we will show you how the "residuals" can be used in examining the reasonableness of model assumptions.

Definition 9.2 If $y_1, y_2, \ldots, y_n$ are the observed values of a response variable and $\hat{y}_1, \hat{y}_2, \ldots, \hat{y}_n$ are the fitted values corresponding to the values $x_1, x_2, \ldots, x_n$, respectively, the differences

$$(y_i - \hat{y}_i) \qquad i = 1, 2, \ldots, n$$

are said to be the *residuals*.

Example 9.10 In Table 9.6, x represents the initial speed of an automobile (in km/hr) and y represents the distance (in meters) required to bring the automobile to a full stop. Find the residuals if a linear model is used.

The least-squares line that best fits these data is given by

$$\hat{y} = b_0 + b_1 x = -20 + 1.20x.$$

Using this equation to find the fitted values, the residuals are easily found. For example, when $x = 50$, we have

$$y - \hat{y} = 31.8 - 40.0 = -8.2.$$

The complete set of fitted values and residuals are given in the third and fourth rows of Table 9.6, respectively.

TABLE 9.6

Stopping Distance Data for Example 9.10

x:	10	20	30	40	50	60	70	80	90	100
y:	3.4	7.3	15.3	21.0	31.8	46.5	59.2	72.4	90.5	112.6
$\hat{y}$:	−8.0	4.0	16.0	28.0	40.0	52.0	64.0	76.0	88.0	100.0
$y - \hat{y}$:	+11.4	+3.3	−.7	−7.0	−8.2	−5.5	−4.8	−3.6	2.5	12.6

Mathematically, it is always possible to find the best-fitting least-squares line for any set of data, as long as there are two or more distinct x values. However, any statistical inference done using the resulting estimates b_0 and b_1 may be invalid if the underlying assumptions 1 to 4 fail to hold. In Example 9.10, for instance, we were able mathematically to find the least-squares line that best fits the data. Before attempting to do statistical inference, however, it is necessary to check the assumptions. The first assumption is that of linearity. Does it appear as if the true regression curve is linear? That is, does it appear as if

$$\mu_{Y|x} = \beta_0 + \beta_1 x$$

for some parameters β_0 and β_1? To get an idea about this, we can plot a scatter diagram with the least-squares line plotted on it, or we can plot the residuals (on the vertical axis) against the x values (on the horizontal axis). The latter graph is referred to as a *residual plot*. Both of these graphs are shown in Figure 9.9. When you examine the graphs, does it appear to you as if the true regression curve is linear? While a linear fit is not too bad, the observed y values do not appear to be randomly scattered about the least-squares line. There is a distinct pattern to the points. The first two

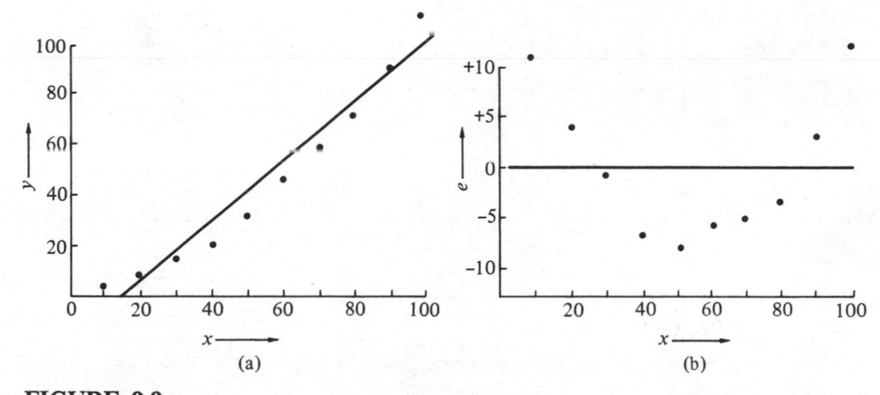

FIGURE 9.9

Scatter Diagram (a) and Residual Plot (b)

and last two points lie above the least-squares line while the remaining points lie below it. A similar and more obvious pattern can be seen in the residual plot. These patterns indicate that perhaps a parabola (i.e., a second-degree polynomial) might describe the true regression line better than a straight line. You might have guessed this if you remembered from studying physics that stopping distance and initial speed are related through a second-degree equation.

A visual examination of the scatter diagram or the residual plot can give you a subjective method for checking the assumption of linearity. If there are repeated observations at one or more of the x values, a statistical test of the assumption of linearity can be performed. This test is referred to as a "lack-of-fit" test and is based on the F distribution. We will not present the details of the test here, but would refer you to a textbook on regression techniques. [For example, see J. Neter and W. Wasserman, *Applied Linear Statistical Models* (Homewood, Ill.: Richard D. Irwin, Inc., 1974).]

We have seen that a residual plot can be used to check on the assumption of linearity. These plots can be used to check on the other assumptions, as well. For example, if the Y_i values are independent and if the assumption of linearity is appropriate, a residual plot should show "random scatter" around the line corresponding to 0 residual. This scatter is illustrated in Figure 9.10(a). If there appears to be an apparent lack of randomness as in Figure 9.10(b), this may indicate dependence among the Y_i variables. As in the check for linearity, this graphical check is only subjective. An objective statistical test of this assumption can be performed by using a "runs test." This test will be discussed in Chapter 11. The assumption of independence is generally regarded as the most important one in making inferences about the regression relationships.

In earlier discussions of Peter's pepper-growing experiment (Sections 9.1 and 9.3), we pointed out that when x, the amount of water put on the plants, is small, the variability in the yield will be smaller than with moderate amounts of water. It is certainly true that in many situations the variability in the response variable Y

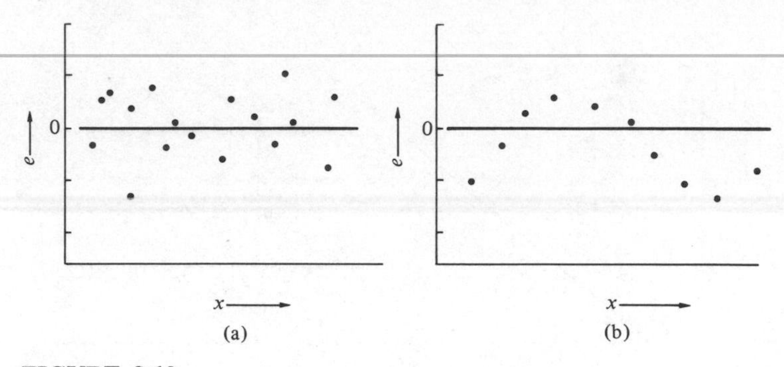

FIGURE 9.10
Residual Plots for the Assumption of Independence

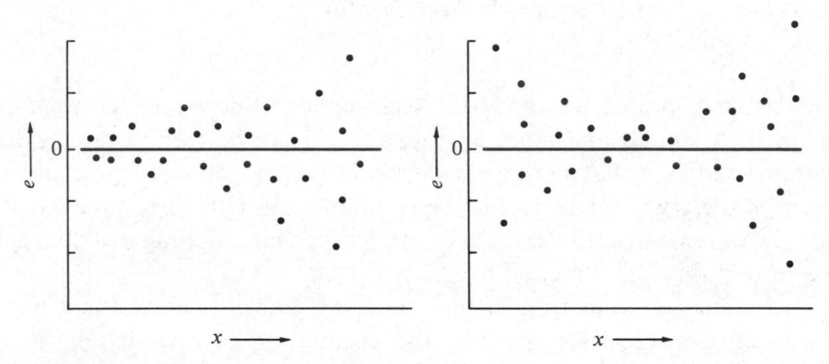

FIGURE 9.11
Residual Plots with Variance Showing a Dependence on x

will vary with x. This dependency on x may be difficult to detect with a small number of observations, but with a larger number of observations the dependency may be more obvious. In Figure 9.11 are residual plots showing cases where the variance would appear to depend on x. One way of obtaining a statistical test for the assumption of a common variance for each x value would be to divide the data into two sets and to test the null hypothesis that the variances for each set are equal. If the assumption of normality was reasonable, an appropriate test for equality of variances would be based on the F distribution, as described in Section 8.7.

If the first three assumptions are satisfied, the fourth assumption of normality for each of the Y_i can be checked by using the residuals. In this case a histogram of the residuals could be plotted and a visual check of the histogram would give some indication of the reasonableness of the normality assumption. A histogram such as the one in Figure 9.12(a) would indicate that the assumption is reasonable, whereas one such as Figure 9.12(b) would not. A statistical test for this assumption could be performed using one of the "goodness-of-fit" techniques discussed in Chapter 11.

A final use of residuals and residual plots is for the purpose of checking for "outliers." An outlier is an observation that lies extremely far from the other observa-

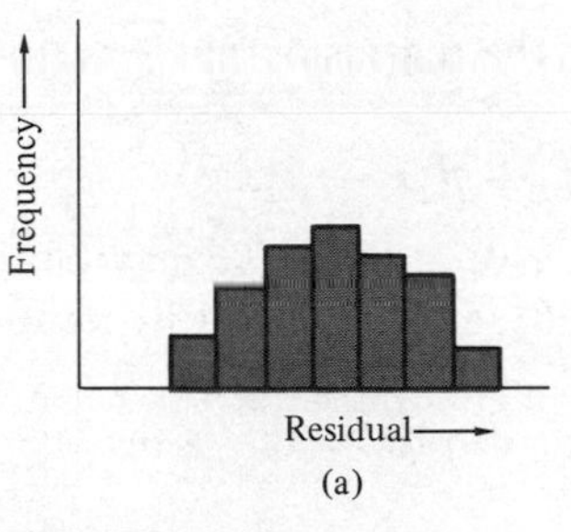

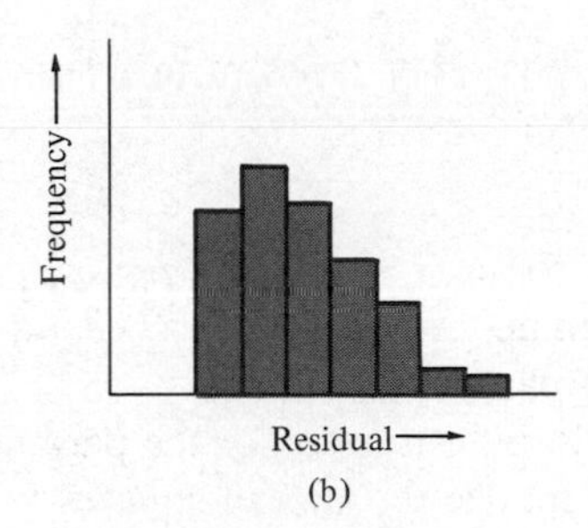

FIGURE 9.12
Histograms of Residuals

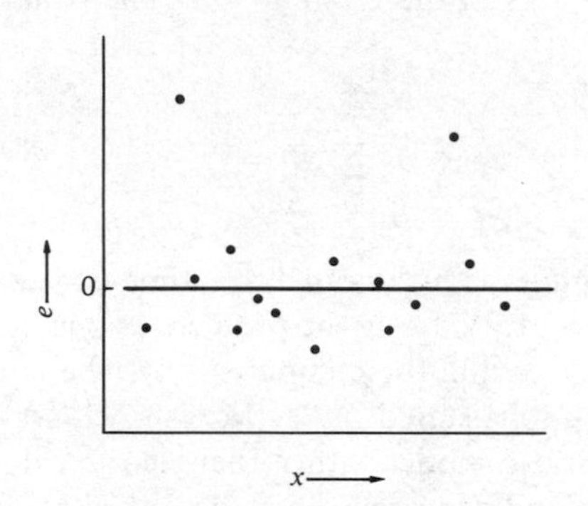

FIGURE 9.13
Residual Plot Showing Two Outliers

tions. In Figure 9.13 you can see two outliers. When an outlier is found, it is a good idea to look at the original data and see if you have a valid data point. It may be that a nonsampling error (see Section 5.3) has been made or that some error was made in the performance of the experiment. (For example, Peter's assistant may have mistakenly watered a particular plant with 500 ml of water instead of 50.0 ml.) In such an instance the value should be corrected, if possible, or deleted from the data. However, if no explanation can be found, it would be wrong to change or delete the value! Such changes would invalidate any statistical inference performed.

In closing we should point out that the set of residuals $e_1, e_2, \ldots, e_n$ are not independent. This can be seen from the fact that the sum of the residuals must equal zero, that is,

$$\sum e_i = \sum (y_i - \hat{y}_i) = 0.$$

However, if the Y_i themselves are independent and if the sample size is moderately large, the amount of dependence is small.

9.6 POLYNOMIAL REGRESSION

We saw in the last section that there are situations where the regression curve may not be linear but may be quadratic in x. This may be the case in problems dealing with acceleration (such as free-falling objects) or deceleration (such as stopping distances) as well as other types of problems. If the regression curve is quadratic, it will be of the form

$$\mu_{Y|x} = \beta_0 + \beta_1 x + \beta_2 x^2.$$

As in the case where the regression curve is linear, it may be desirable to estimate the unknown parameters β_0, β_1, and β_2, and/or to test hypotheses about the values of the parameters.

Point estimates for the parameters can be found by the method of least squares. To minimize the sum of squares

$$\sum_i [y_i - (b_0 + b_1 x_i + b_2 x_i^2)]^2,$$

it is necessary to solve three equations in three unknowns. The equations are as follows:

$$\begin{aligned} nb_0 + b_1 \sum x_i + b_2 \sum x_i^2 &= \sum y_i \\ b_0 \sum x_i + b_1 \sum x_i^2 + b_2 \sum x_i^3 &= \sum x_i y_i \\ b_0 \sum x_i^2 + b_1 \sum x_i^3 + b_2 \sum x_i^4 &= \sum x_i^2 y_i. \end{aligned}$$

Solving these three equations simultaneously can be a time-consuming task. The process can be made somewhat easier by viewing the problem as one in matrix algebra, but can be made still more easy by having the computer solve the equations.

If we again make the assumptions stated in Section 9.3, with a modification of Assumption 1 to reflect the quadratic model rather than the linear, the estimators B_0, B_1, and B_2 will have the same kind of properties as described in that section. The details may change somewhat, but the logic remains the same.

Example 9.11 Find the least-squares estimators b_0, b_1, and b_2 for the stopping distance data given in Table 9.6. Using these estimates, find the fitted values $\hat{y}$ and plot the residuals.

We found these estimates by using a computer program. A portion of the computer output for this problem is shown in Figure 9.14. In the regression equation "X1" and "X2" are interpreted as x and x^2, respectively. The observed values, fitted

```
THE REGRESSION EQUATION IS
Y = .6883 + .1703 X1 + .00934 X2

                                           ST. DEV.        T-RATIO =
           COLUMN        COEFFICIENT       OF COEF.        COEF./S.D.
             --              .69             1.44              .48
X1           C2            .1703            .0599             2.84
X2           C3          .009337           .000531           17.58
```

y_i:	3.0	7.8	15.3	21.4	32.0	46.3	58.8	72.4	90.9	112.1
$\hat{y}_i$:	3.3	7.8	14.2	22.4	32.5	44.5	58.4	74.1	91.6	111.1
$y_i - \hat{y}_i$:	-.3	.0	1.1	-1.0	-.5	1.8	.4	-1.7	-.7	1.0

FIGURE 9.14

Sample Computer Output Giving Least-Squares Estimates

values, and residuals are given after the computer output. This information is used for the residual plot given in Figure 9.15. The scatter of the residuals about the line zero appears (subjectively) to be random. The comparison between this plot and the one given in Figure 9.9(b) is indeed striking. A second-degree polynomial appears to describe the regression curve better than a straight line. That it does fit better is consistent with the physics of the situation.

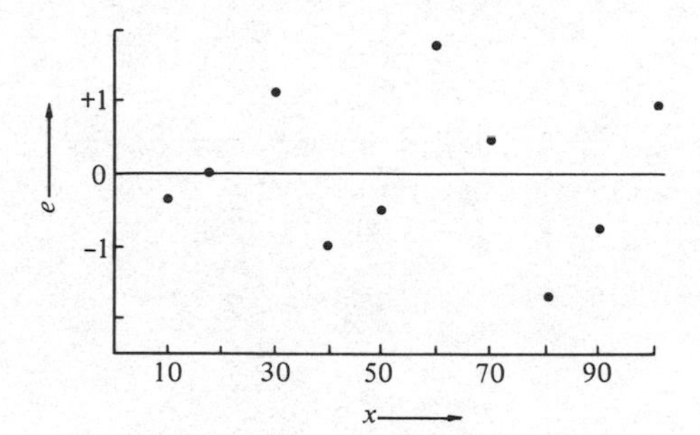

FIGURE 9.15
Residual Plot for Example 9.11

★ ***Example 9.12*** Researchers in the area of fisheries and wildlife are interested in studying growth rates of fish. In one study the amount of growth (Y) was examined as a function of total length at the beginning of the study period (x). Both measurements are given in millimeters. The species of fish investigated was bluegill (*Lepomis macrochirus*). Using the data given in Table 9.7, plot a scatter diagram. Determine whether the regression curve would be better described by a linear or a quadratic function and find the least-squares estimates of the parameters. Also plot the estimated regression curve.

TABLE 9.7
Annual Growth Increment (y) *for Initial Body Length* (x) *for Bluegill*

x (*Initial Length*)	y (*Growth*)	x (*Initial Length*)	y (*Growth*)
48	21	138	22
52	19	138	19
51	18	130	26
53	22	140	21
69	32	160	13
71	36	157	11
69	31	156	16
75	29	161	17
101	37	173	3
107	31	168	6
100	30	172	1
104	36	178	0

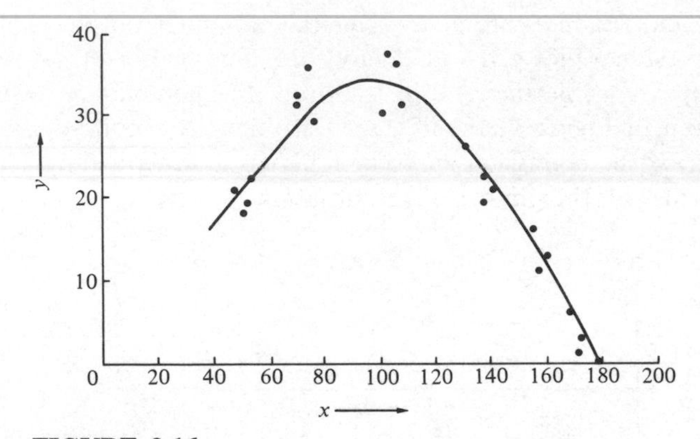

FIGURE 9.16
Scatter Diagram and Estimated Quadratic Regression Curve for Bluegill Growth Data

The scatter diagram for these data is given in Figure 9.16. It is not hard to see from this diagram that a regression curve that is quadratic would be a better choice than one that is linear. A computer analysis of the data yields the following least-squares estimates:

$$b_0 = -16.8793, \qquad b_1 = 1.0309, \qquad b_2 = -0.0053.$$

The fitted values then would be obtained from

$$\hat{y} = -16.8793 + 1.0309x - 0.0053x^2.$$

This curve is also plotted on the scatter diagram. (Note, by the way, that while a quadratic function may provide an adequate model for the regression curve over the range of x values used, we cannot use this model outside this range. If we do, then for an initial length of, say, 200 mm ($x = 200$), the predicted "growth" would be -22.7 mm (i.e., the fish is predicted to shrink in length!). (Talk about fish stories!) This is another illustration of the dangers of extrapolation beyond the range of the x values for which observed y's are obtained.)

The two examples given in this section illustrate the need for fitting regression curves which are not simply linear in x. In these examples second-degree polynomials (i.e., quadratic curves) are adequate, but in other situations a higher-order degree may be necessary. If the true regression curve is of the form

$$\mu_{Y|x} = \beta_0 + \beta_1 x + \beta_2 x^2 + \ldots + \beta_k x^k, \tag{9.37}$$

then, given adequate data, least-squares estimates of the parameters $\beta_0, \beta_1, \ldots, \beta_k$ can be found. If Assumptions 2 to 4 of Section 9.4 also hold, then tests of hypotheses and interval estimates relating to these parameters can be performed and found.

The estimates for the parameters would rarely be found by hand these days for real-world problems. The work most likely would be performed by a computer. As we indicated at the start of Section 9.5, the computer is a useful tool that can be "dangerous" if misused. Remember that the computer output is only valid if the underlying assumptions stated earlier are (at least approximately) satisfied. These assumptions should be checked when polynomial regression or the yet-to-be-discussed multiple regression is used as well as when simple linear regression is used.

9.7 MULTIPLE REGRESSION

In our introductory remarks in Section 9.1, we indicated that a purpose of regression analysis was to be able to predict the value of a response variable Y for given levels of an "independent" variable x. For example, Peter might want to predict the yield of peppers for a given value of amount of water, while Mr. Havitall might want to predict profit for a given value of the amount of money spent on an ad. In both of these cases, the investigator is interested in the effect of one specific "independent" variable on the response. Can you think of other quantities, in addition to water, which might affect the yield of peppers? What about the amount of sunlight, nitrogen, potassium, or phosphates? Clearly, there are many other factors that could affect the yield! If all other factors except the amount of water are kept constant, it is appropriate to express the mean yield, $\mu_{Y|x}$, as a function of x only. However, in actual field experiments rather than greenhouse experiments, it may be desirable or necessary to include other "independent" variables. If these variables are denoted by $x_1, x_2, \ldots, x_k$, it may be possible to express the mean response of Y by

$$\mu_{Y|x_1,x_2,\ldots,x_k} = \beta_0 + \beta_1 x_1 + \beta_2 x_2 + \ldots + \beta_k x_k. \tag{9.38}$$

[Notice the similarity between (9.38) and (9.37), the regression curve for the case of polynomial regression.]

In order to get point estimates for the parameters $\beta_0, \beta_1, \ldots, \beta_k$, the methods of least squares can be used. In order to get the estimates, it is necessary to simultaneously solve the following system of equations:

$$\begin{aligned} nb_0 + b_1 \sum x_{1i} + b_2 \sum x_{2i} + \ldots + b_k \sum x_{ki} &= \sum y_i \\ b_0 \sum x_{1i} + b_1 \sum x_{1i}^2 + b_2 \sum x_{1i}x_{2i} + \ldots + b_k \sum x_{1i}x_{ki} &= \sum x_{1i}y_i \\ &\vdots \\ b_0 \sum x_{ki} + b_1 \sum x_{1i}x_{ki} + b_2 \sum x_{2i}x_{ki} + \ldots + b_k \sum x_{ki}^2 &= \sum x_{ki}y_i. \end{aligned}$$

As we indicated in the case of polynomial regression, this system is generally not solved by hand but rather by using a computer. However, if the values of the x's can be chosen properly, then many of the terms of the form $\sum x_{ji}$ or $\sum x_{ji}x_{hi}$ will equal zero, giving an easier system of equations to solve.

★***Example 9.13*** In a study to investigate the effect of frequency of training (x_1) and intensity of training (x_2) on physical work capacity (y), sedentary males from the faculty and staff of a large university were randomly assigned to one of six training groups. There were three different levels of "frequency of training," namely, 1, 3, or 5 days/week and two levels of "intensity of training," namely, maintaining the

TABLE 9.8
Data for Example 9.13

Subject	x_1	x_2	y	*Subject*	x_1	x_2	y
1	5	150	134	19	5	120	182
2	5	150	312	20	5	120	235
3	5	150	196	21	5	120	135
4	5	150	249	22	5	120	171
5	5	150	222	23	5	120	193
6	5	150	170	24	5	120	203
7	3	150	93	25	3	120	278
8	3	150	306	26	3	120	174
9	3	150	160	27	3	120	111
10	3	150	183	28	3	120	110
11	3	150	296	29	3	120	119
12	3	150	209	30	3	120	121
13	1	150	253	31	1	120	55
14	1	150	165	32	1	120	98
15	1	150	93	33	1	120	71
16	1	150	127	34	1	120	54
17	1	150	135	35	1	120	84
18	1	150	146	36	1	120	186

heart rate, while exercising, at 120 or 150 beats/minute. The *total* amount of time of exercising for all groups was 50 minutes/week. Using the data given in Table 9.8, find the least-squares estimates for β_0, β_1, and β_2, assuming that the mean response (gain in physical work capacity) is given by

$$\mu_{Y|x_1,x_2} = \beta_0 + \beta_1 x_1 + \beta_2 x_2.$$

A computer analysis of these data gave the values

$$b_0 = -108.2, \qquad b_1 = 19.48, \qquad b_2 = 1.609,$$

so predicted values of increase in work capacity will be given by

$$\hat{y} = -108.2 + 19.48x_1 + 1.609x_2. \tag{9.39}$$

(For example, a participant exercising 5 days/week at a 150-beats/minute level of intensity would be predicted to have a gain in physical work capacity of

$$\hat{y} = -108.2 + 19.48(5) + 1.609(150) = 230.55).$$

In this example the investigator was interested in studying the effect on physical work capacity of two different quantities, frequency and intensity of training. It seems intuitively plausible that each of these quantities could have some effect on physical work capacity. You might wonder, however, what would have happened if the investigator had considered only one of these quantities at a time, ignoring the other one. Using only frequency of training, a simple linear regression analysis would give

$$\hat{y} = 109.0 + 19.48x_1 \tag{9.40}$$

as the least-squares line. In this case the sum of squares of the residuals about the regression line or sum of squares of error is

$$\text{SSE}(X_1) = \sum (y - \hat{y})^2 = 133{,}303.$$

(The X_1 in parentheses indicates which independent variable is being included in the model.) We will see later that this quantity represents "unexplained" variation (i.e., variation in the y's not explained by the fact that the means of the y distributions lie on a particular regression line). If the investigator had considered only the intensity of training, a simple linear regression analysis would give

$$\hat{y} = -49.78 + 1.609x_2 \tag{9.41}$$

as the least-squares line. In this case the sum of squares of the residuals about the regression line is

$$\text{SSE}(X_2) = \sum (y - \hat{y})^2 = 148{,}752.$$

As you can see, it is possible to consider these two "independent" variables singly. However, when both variables are treated together, more of the variability is explained. In particular, when multiple regression is used, the sum of squares of residuals is

$$\text{SSE}(X_1, X_2) = \sum (y - \hat{y})^2 = 112{,}326,$$

which is lower than for either of the sums based on simple linear regression. This example illustrates the fact that when more "independent" variables are used in a model, more variability can be explained. This is also reflected in the fact that better predictions can be made when more variables are included. A word of warning is in order here: Virtually any additional "independent" variable can be included in a model and will result in a decrease in the residual sum of squares. Such variables should only be retained in a model if subsequent tests of hypotheses about the value of the coefficient β_i show it to be significantly different from zero. For instance, in the study on physical work capacity the investigator might have used a third variable x_3, the number of years of education of a participant. Assuming that the means of the y responses are given by

$$\mu_{Y|x_1,x_2,x_3} = \beta_0 + \beta_1 x_1 + \beta_2 x_2 + \beta_3 x_3,$$

the investigator would find that $\text{SSE}(X_1, X_2, X_3)$ is smaller than $\text{SSE}(X_1, X_2)$. However, in this problem it would not be likely that a subsequent test of $H_0 : \beta_3 = 0$ would call for rejection of this null hypothesis. (Do you see why?) In such a case, the third term should not be included in the model.

In the example on physical work capacity, you may have noticed that the least-squares estimates b_1 and b_2 of the parameters β_1 and β_2 were the same whether both variables were included in a multiple regression situation (9.39), or whether one variable at a time was considered in a simple linear regression situation, (9.40) or (9.41). We must emphasize that *this will not always happen*. It did happen here because of the special way the x_i values were chosen. Since this was a carefully designed study, the investigator could easily control the values of the x_i's. In many studies such control is not possible, and the least-squares estimates of the parameters β_i will depend very much on the model used. In addition, extra care must be taken in interpreting the results of regression analysis if the x_i are not controlled by the experimenter.

An experiment closely related to Peter's pepper-growing experiment would be one where Peter had plots of peppers in various locations around a county or state and where he measured the yield (y) along with the rainfall (x) for each plot. Unless Peter has more influence with Mother Nature than you or we do, it should be obvious that he does not have control over the values of x used. In such a situation x is actually the observed value of a random variable X and the joint probability distribution of X and Y must be considered. Fortunately, with only minor modifications in the assumptions stated in Section 9.3, the subsequent statements concerning estimation, hypothesis testing, and prediction still hold true when the x values are observed values of a random variable X. One additional condition that must hold is that the distribution of the random variable X not involve the parameters β_0, β_1, or σ^2. In the next example, both X_1 and X_2 are random variables.

★***Example 9.14*** In a study of selling price of calves at a feeder calf sale, the selling price (dollars per hundred weight) (y), the weight of the animal (x_1), and the grade of

TABLE 9.9

Data for Example 9.14

Price/100 lb, y	*Weight,* x_1 *(lb)*	*Grade,* x_2	*Price/100 lb,* y	*Weight,* x_1 *(lb)*	*Grade,* x_2
36.75	736	0	33.50	339	1
35.50	506	0	36.00	517	1
28.00	394	0	33.00	396	1
31.00	419	0	30.50	382	1
32.75	504	0	36.00	479	1
35.25	599	0	35.25	515	1
31.50	442	0	37.50	606	1
33.50	505	0	30.50	336	1
32.50	512	0	33.00	425	1
31.25	400	0	34.75	498	1
35.00	558	0	36.25	594	1
34.25	350	0	35.25	375	1
33.25	365	0	35.50	366	1
24.25	421	0	36.25	345	1
20.75	393	0	37.25	422	1
21.00	343	0	39.75	664	1
23.00	449	0	36.75	450	1
27.00	536	0	37.75	703	1
23.00	453	0	33.25	552	1
26.75	480	0	36.50	471	1
25.00	413	0	34.25	542	1
26.00	463	0	32.00	354	1
20.00	357	0	39.00	664	1
26.25	429	0	36.75	440	1
25.25	443	0	39.50	450	1

the animal (x_2) were found for 50 calves. The grade was determined to be either "good" ($x_2 = 0$) or "choice" ($x_2 = 1$) as judged by a specialist. Using the data given in Table 9.9, find the least-squares line (or plane) and the sum of squares of error when (a) only x_1 is used, (b) only x_2 is used, and (c) both x_1 and x_2 are used in the model.

These data were analyzed using a computer program. The results are given in Table 9.10. You should be able to see that in this problem the values of x_1, the weights, cannot easily be controlled by an experimenter and are, in fact, observed values of

TABLE 9.10
Results for Example 9.14

Variables Included	*Least-Squares Estimates*	*Sum of Squares of Error*
x_1	$20.46 + .0248x_1$	$\text{SSE}(X_1) = 1063$
x_2	$28.75 + 6.69x_2$	$\text{SSE}(X_2) = 798$
x_1, x_2	$18.91 + .0214x_1 + 6.25x_2$	$\text{SSE}(X_1, X_2) = 581$

a random variable. The values of x_2 were also random, but for the purposes of this study, 25 calves of each grade were selected. As is typical in such problems, the coefficients of x_1 and x_2 are different when only one variable is included in the model than when both are. You can also see that there is a substantial reduction in the error sum of squares when both variables are included in the model.

In closing this section we would like to point out that care must be exercised in choosing "independent" variables to be included in the multiple regression model (9.38). The variables should be related to the real-world situation being analyzed. For example, in studying crop yield per acre for different farms, one might consider variables such as rainfall, farm size, and nitrogen in the soil, but variables such as the height of the farmer or the number of children in his family may not be relevant. Another factor to be considered in choosing variables is the interrelationship among the "independent" variables. If you are studying gas mileage of automobiles, for example, you may wish to consider engine size and total weight as factors. However, these two variables are rather highly interrelated, since heavy cars generally have bigger engines and bigger engines contribute to the weight of a car! When variables are closely related or highly correlated, the error sum of squares when both are included won't be much smaller than when only one variable is included. It is better to include variables that are measuring quite different things. For further details on this subject, we again refer you to a text on regression analysis.

9.8 THE CORRELATION COEFFICIENT

In Sections 9.1 and 9.7 we indicated that it is possible for both X and Y to be random variables. If both are random variables, it would be of interest to know of their joint probability distribution. If this joint distribution itself is unknown, it might be possible to estimate parameters associated with this distribution. One such parameter is the correlation coefficient ρ, defined in Section 3.5 by

$$\rho = \frac{\text{Cov}(X, Y)}{\sigma_X \sigma_Y}.$$

Using properties of the covariance of X and Y, equivalent mathematical forms for ρ are the following:

$$\rho = \frac{E[(X - \mu_X)(Y - \mu_Y)]}{\sigma_X \sigma_Y} = \frac{E(XY) - E(X)E(Y)}{\sigma_X \sigma_Y}. \tag{9.42}$$

The correlation coefficient can be shown to satisfy the following properties:

1. $-1 \leq \rho \leq 1$.
2. $|\rho| = 1$ if and only if X and Y are linearly related [i.e., if for some constants β_0 and β_1 ($\beta_1 \neq 0$), $Y = \beta_0 + \beta_1 X$]. If $\beta_1 > 0$, then ρ will equal $+1$, while if $\beta_1 < 0$, ρ will equal -1.

The next example illustrates the second property.

Example 9.15 If X is a discrete random variable with probability distribution

x:	-1	0	1
$P[X = x]$:	.3	.4	.3,

and if the random variable Y is defined by $Y = 1 + 2X$, find the joint probability distribution of X and Y and the correlation coefficient ρ.

TABLE 9.11

Joint Probability Distribution for Example 9.15

		y		
		−1	*1*	*3*
x	*−1*	.3	0	0
	0	0	.4	0
	1	0	0	.3

Remembering that the joint probability distribution can be given in a table (as in Section 3.1) it should not be hard to see that the joint distribution is as given in Table 9.11. From this table it is easy to find the marginal distributions and to obtain

$$E(X) = 0, \quad E(Y) = 1, \quad E(X^2) = .6, \quad E(Y^2) = 3.4, \quad E(XY) = 1.2.$$

It follows that $\sigma_X^2 = .6$ and $\sigma_Y^2 = 2.4$, so

$$\rho = \frac{E(XY) - E(X)E(Y)}{\sigma_X \sigma_Y} = \frac{1.2 - (0)(1)}{\sqrt{(.6)(2.4)}} = \frac{1.2}{\sqrt{1.44}} = 1.$$

It is important to emphasize that ρ *is a measure of the strength of the linear relationship* between two random variables. It is possible for X and Y to be related by a nonlinear relationship and to have ρ equal to zero, for example (see Exercise 33).

If ρ is unknown, it is possible to get a point estimate for ρ by using the sample counterpart to (9.42). The sample correlation coefficient is defined by

$$(9.43) \qquad r = \frac{\sum (x_i - \bar{x})(y_i - \bar{y})/(n-1)}{s_X s_Y} = \frac{[\sum (x_i y_i) - n\bar{x}\bar{y}]/(n-1)}{s_X s_Y}$$

$$= \frac{[(\sum x_i y_i) - n\bar{x}\bar{y}]}{[\sqrt{(\sum x_i^2 - n\bar{x}^2)(\sum y_i^2 - n\bar{y}^2)}]}.$$

(There are some mathematical reasons for dividing the numerator by $n - 1$. We will not go into those reasons here.) The sample correlation coefficient r satisfies properties analogous to those stated earlier for ρ, namely:

1. $-1 \leq r \leq 1$.
2. $|r| = 1$ if and only if the sample pairs are linearly related, that is, if for some constants b_0 and b_1 ($b_1 \neq 0$), $y_i = b_0 + b_1 x_i$ for all i. If $b_1 > 0$, then r will equal $+1$, while if $b_1 < 0$, r will equal -1.

It is also true that r is a measure of the strength of the linear relationship between the observed random variables, (x_i, y_i), $i = 1, 2, \ldots, n$.

Example 9.16 The first two columns of Table 9.12 give engine size (in cubic inches) and EPA estimates of gas mileage for 18 cars selected from among a larger set of cars all of the same model year. Find the sample correlation coefficient.

By using equation (9.43) and the sums in Table 9.12, we find the sample correlation coefficient to be

$$r = \frac{\sum (x_i y_i) - n\bar{x}\bar{y}}{\sqrt{[\sum x_i^2 - n\bar{x}^2][\sum y_i^2 - n\bar{y}^2]}}$$

$$= \frac{99{,}323.7 - 18(4206/18)(433.0/18)}{\sqrt{[1{,}060{,}534 - 18(4206/18)^2][10{,}535.14 - 18(433.0/18)^2]}}$$

$$= \frac{-1853.967}{\sqrt{(77{,}732)(119.084)}} = -.609.$$

We have said that r is a measure of the strength of the linear relationship between the observed values in a sample. In what sense does it measure this relationship? How does it fit in with the linear regression problem discussed earlier in this chapter? We will attempt to answer these questions in the remainder of this section.

Consider once again the gas mileage data given in Example 9.16. In particular, look at the y values plotted on the vertical line in Figure 9.17(a). On this line we have also shown the sample mean $\bar{y} = 24.06$ by a short horizontal line. As you look at this line you can see that the y_i values are scattered about the sample mean $\bar{y}$. This should come as no surprise, since the values of a random variable are typically

TABLE 9.12
Data for Example 9.16

Engine Size, x_i	Gas Mileage (EPA est.), y_i	$x_i y_i$	x_i^2	y_i^2
200	29.9	5,980.0	40,000	894.01
171	27.3	4,668.3	29,241	745.29
121	25.8	3,121.8	14,641	665.64
182	23.4	4,258.8	33,124	547.56
258	21.7	5,598.6	66,564	470.89
351	20.4	7,160.4	123,201	416.16
225	26.7	6,007.5	50,625	712.89
198	26.0	5,148.0	39,204	676.00
250	25.6	6,400.0	62,500	655.36
250	25.2	6,300.0	62,500	635.04
232	24.7	5,730.4	53,824	610.09
258	24.3	6,269.4	66,564	590.49
167	24.1	4,024.7	27,889	580.81
171	23.0	3,933.0	29,241	529.00
304	22.1	6,718.4	92,416	488.41
182	22.1	4,022.2	33,124	488.41
360	21.0	7,560.0	129,600	441.00
326	19.7	6,422.2	106,276	388.09
Σ 4206	433.0	99,323.7	1,060,534	10,535.14

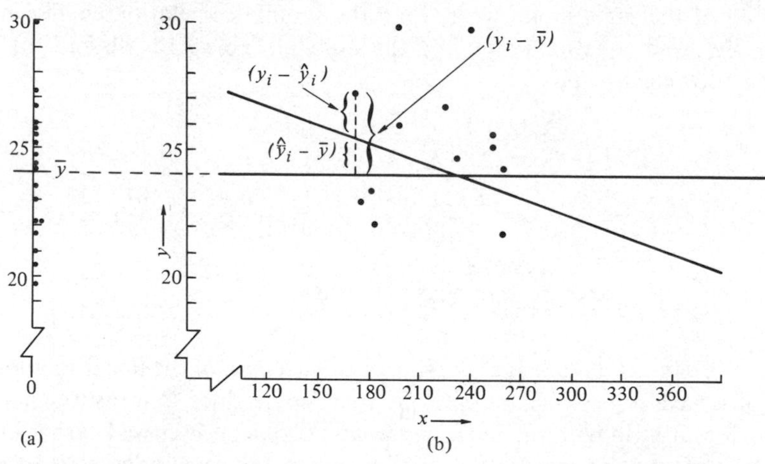

FIGURE 9.17
Graphs of Data from Example 9.16

scattered about the sample mean. However, it still might be of interest to try to explain *why* there is as much scatter as there is. A way of measuring the total scatter is by using the sum of squares of the observed values about their mean. In this example the sum is

$$\sum (y_i - \bar{y})^2 = \sum (y_i^2) - n\bar{y}^2 = 119.084.$$

This sum is frequently denoted by "SST." Why is this sum so large? Surely you can think of several factors that can cause variability in gas mileage: the type of oil used, the type of tires used, weather conditions, whether or not the driver is a teenager and/or whether the driver's parents are in the car, and so on. Of course, another obvious factor is the engine size. How much of the variability can be accounted for by the engine size factor? To get an idea about this, look at the scatter diagram in Figure 9.17(b). On this graph we have plotted the least-squares line

$$\hat{y} = 29.629 - .0239x.$$

A horizontal line corresponding to $\bar{y} = 24.06$ is also plotted on this graph. How can we explain the variability of the y_i values about the mean line $\bar{y}$? We can account for some of the variability by noting that the points y_i are scattered about the least-squares line. For example, when $x_i = 351$, the estimated value for y_i is $\hat{y}_i = 21.24$. Intuitively, we expect the value of y_i to be relatively small because this car has a relatively large engine! This explains to some degree why the y_i value of 20.4 is so far from the mean. Mathematically, we can write

$$y_i - \bar{y} = (y_i - \hat{y}_i + \hat{y}_i - \bar{y}) = (y_i - \hat{y}_i) + (\hat{y}_i - \bar{y}). \tag{9.44}$$

This equation shows that for each i, the deviation from the mean $\bar{y}$ can be written as the sum of the deviation of y_i from its predicted value $\hat{y}_i$ and the deviation of the predicted value $\hat{y}_i$ from the mean $\bar{y}$. This partitioning can be seen in Figure 9.17(b). Furthermore, if we square the terms on the left- and right-hand sides of (9.44) and sum these terms, an interesting algebraic relationship holds:

$$\begin{aligned} \text{SST} &= \sum (y_i - \bar{y})^2 = \sum [(y_i - \hat{y}_i) + (\hat{y}_i - \bar{y})]^2 \\ &= \sum [(y_i - \hat{y}_i)^2 + 2(y_i - \hat{y}_i)(\hat{y}_i - \bar{y}) + (\hat{y}_i - \bar{y})^2] \\ &= \sum (y_i - \hat{y}_i)^2 + 2\sum (y_i - \hat{y}_i)(\hat{y}_i - \bar{y}) + \sum (\hat{y}_i - \bar{y})^2. \end{aligned}$$

However, you can show (Exercise 37) that $\sum (y_i - \hat{y}_i)(\hat{y}_i - \bar{y}) = 0$, so

$$\begin{aligned} \text{SST} &= \sum (y_i - \hat{y}_i)^2 + \sum (\hat{y}_i - \bar{y})^2 \\ &= \text{SSE} + \text{SSR}, \end{aligned} \tag{9.45}$$

where SSE is as previously defined and $\text{SSR} = \sum (\hat{y}_i - \bar{y})^2$. SSR is called the sum of squares due to the linear regression of Y on X. In other words, the total sum of squares, SST, can be attributed to two factors, SSE (which is the sum of squares of residuals about the estimated regression line) and SSR. This partitioning then offers an explanation to our original question about why SST is as large as it is. Quantitatively, the *proportion* of the variability in y explained by the regression on x is given by

$$\text{SSR/SST}.$$

It should be obvious that since SSR is a sum of squares, it must always be greater than or equal to zero. Furthermore, because of equation (9.45), the ratio must be less than or equal to 1. The closer the proportion is to 1, the more variability is explained by the regression. If the proportion is 0, none of the variability of the y's is explained by the regression on x, while if the proportion is 1, all of the variability of the y's is explained by the regression on x. Algebraically, we can see that

$$\begin{aligned} \text{SSR} &= \sum (\hat{y}_i - \bar{y})^2 = \sum [(b_0 + b_1 x_i) - \bar{y}]^2 \\ &= \sum \{[(\bar{y} - b_1\bar{x}) + b_1 x_i] - \bar{y}\}^2 \\ &= \sum [b_1(x_i - \bar{x})]^2 = b_1^2 \sum (x_i - \bar{x})^2. \end{aligned}$$

Using this relationship and using equation (9.4) for b_1, we see that

$$\begin{aligned} \frac{\text{SSR}}{\text{SST}} &= \frac{b_1^2 \sum (x_i - \bar{x})^2}{\sum (y_i - \bar{y})^2} \\ &= \left[\frac{\sum (x_i - \bar{x})(y_i - \bar{y})}{\sum (x_i - \bar{x})^2}\right]^2 \frac{\sum (x_i - \bar{x})^2}{\sum (y_i - \bar{y})^2} \\ &= \left[\frac{\sum (x_i - \bar{x})(y_i - \bar{y})/(n-1)}{\sqrt{\dfrac{\sum (x_i - \bar{x})^2 \sum (y_i - \bar{y})^2}{(n-1)(n-1)}}}\right]^2 = \left[\frac{\sum (x_i - \bar{x})(y_i - \bar{y})/(n-1)}{s_x s_y}\right]^2 \\ &= r^2. \end{aligned}$$

In other words, the square of the sample correlation coefficient, r^2, gives us the proportion of the variability in y explained by the *linear* regression relationship. It is in this sense that r is a measure of the strength of the linear relationship between x and y. These statements answer the question raised after Example 9.16.

The scatter diagrams in Figure 9.18 and the corresponding least-squares lines along with the values of r and r^2 illustrate how r (and, more important, r^2) measure

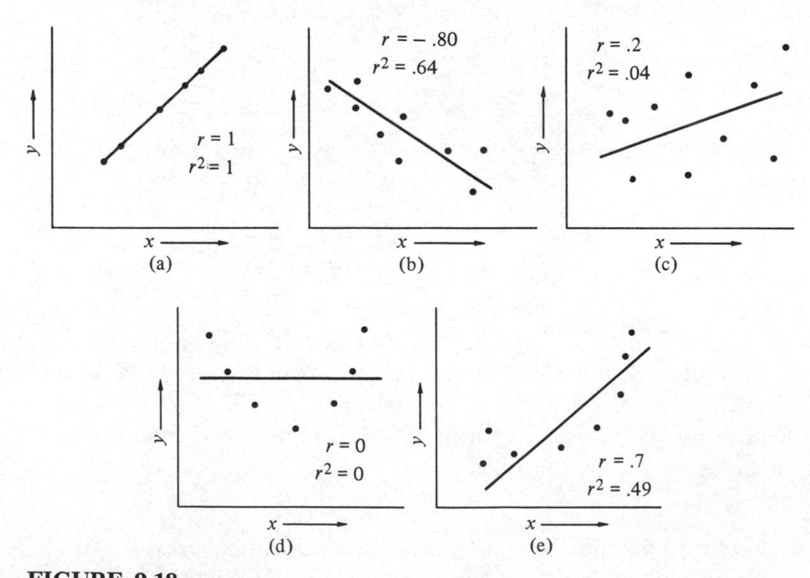

FIGURE 9.18

Scatter Diagrams with Associated r Values

the strength of the linear relationship between x and y in a sample. These diagrams also illustrate some other properties of r: first, the sign of r will be positive if and only if the slope of the least-squares line, b_1, is positive. Second, even if r^2 has some moderately large value, as in Figure 9.18(e), the linear model may not be appropriate. Finally, a value of r at or near zero (c or d) may not mean that no relationship exists. Remember, r measures only the strength of the *linear relationship.*

9.9 THE BIVARIATE NORMAL DISTRIBUTION

We have seen that the normal distribution plays an important role in the study of the probability distributions of single random variables. In the study of the joint probability distribution of two or more random variables, we find that a "joint normal" distribution plays an equally important role. This distribution is called a bivariate normal distribution if two random variables are considered.

How can we know if two random variables will have a joint probability distribution that is bivariate normal? If enough data are available, we might perform a statistical goodness-of-fit test (similar to methods which will be discussed in Chapter 11). We can also make use of the fact that if X and Y together follow a bivariate normal distribution, each of X and Y singly must follow an ordinary univariate normal distribution. Given this criterion, you should be able to see, for example, that if X represents the height of a husband and Y the height of a wife in a randomly selected married couple, the assumption of bivariate normality may be reasonable. This is so since each of X and Y taken alone should follow a normal distribution. Similarly, the heights of adult brothers and sisters may follow a bivariate normal distribution. However, if adult brother–sister pairs are considered, and if X represents the number of children in the brother's family and Y the number in the sister's family, the assumption of bivariate normality would be unreasonable. Since X and Y are both discrete random variables assuming only small integer values, neither, when taken alone, would be even approximately normal.

Another method that can be used to get an idea about the reasonableness of the bivariate normal assumption is to make a scatter diagram. If the sample observations do come from a bivariate normal distribution, the points will roughly lie within an elliptical (or circular) region with a heavier concentration near the middle, and with a gradually decreasing concentration away from the middle. In Figure 9.19 some scatter diagrams are given. Can you tell for which of these the assumption of bivariate normality is reasonable? It is reasonable in Figure 9.19(a), (b), and (c), but not in (d).

If X and Y do, in fact, have a bivariate normal distribution, the joint probability density function will have the following form:

$$(9.46)\quad f(x, y) = \frac{1}{2\pi\sigma_X\sigma_Y\sqrt{1-\rho^2}} \exp\left\{-\frac{1}{2(1-\rho^2)}\left[\left(\frac{x-\mu_X}{\sigma_X}\right)^2 - 2\rho\left(\frac{x-\mu_X}{\sigma_X}\right)\left(\frac{y-\mu_Y}{\sigma_Y}\right) + \left(\frac{y-\mu_Y}{\sigma_Y}\right)^2\right]\right\}.$$

Although this function looks rather complicated, it is worth studying. First you

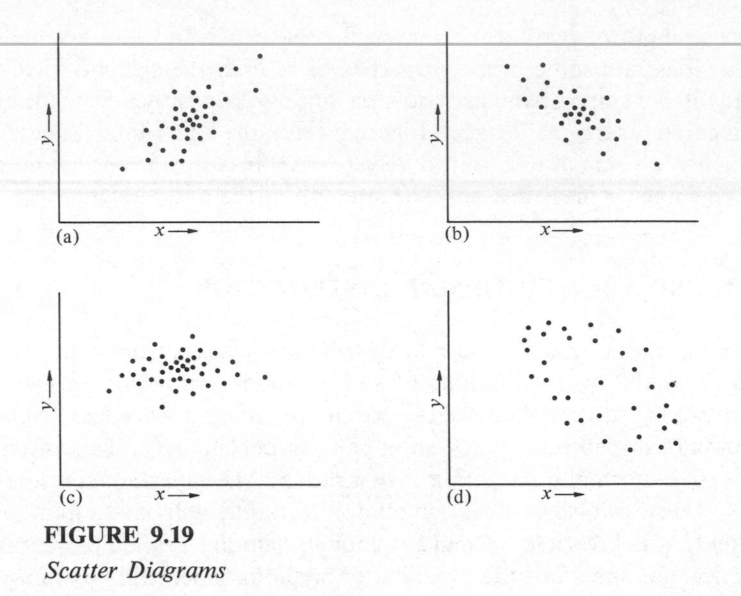

FIGURE 9.19
Scatter Diagrams

will notice that there are five parameters, denoted by $\mu_X, \mu_Y, \sigma_X, \sigma_Y$, and ρ. These parameters represent exactly what you would guess (assuming that you would be willing to guess!). Before guessing, however, it might be helpful to turn back to Chapter 3 and review Sections 3.2, 3.3, and 3.5. Remember that marginal and conditional distributions can be found from the joint distribution. If the joint distribution is bivariate normal, then the marginal distribution of X is normal with mean μ_X and variance σ_X^2, and similarly $Y \sim N(\mu_Y, \sigma_Y^2)$. Also, the correlation coefficient for X and Y will be given by ρ. Hence the five parameters of the bivariate normal distribution represent the means and standard deviations of the marginal distributions of X and Y and also the correlation coefficient between X and Y.

A graph of the bivariate normal density (9.46) is also informative. In Figure 9.20(a) such a graph is given. As you might expect, the peak of this probability mound occurs when $x = \mu_X$ and $y = \mu_Y$. To get an idea as to the form of the conditional distribution of Y for a given x value, consider a slice through this mound parallel to the y axis at the given value of x. A cutaway view of this is given in Figure 9.20(b). A similar slice made at a given y value and parallel to the x axis is shown in Figure 9.20(c). These views indicate, as is the case, that the conditional distributions of Y given x (and of X given y) also follow a normal distribution. (It seems that no matter how you slice it, it comes up normal!) Of course, to completely specify the distribution, we must give its mean and variance. It can be shown that

$$Y \mid x \sim N\left(\mu_Y + \rho\frac{\sigma_Y}{\sigma_X}(x - \mu_X), \sigma_Y^2(1 - \rho^2)\right) \tag{9.47}$$

and

$$X \mid y \sim N\left(\mu_X + \rho\frac{\sigma_X}{\sigma_Y}(y - \mu_Y), \sigma_X^2(1 - \rho^2)\right).$$

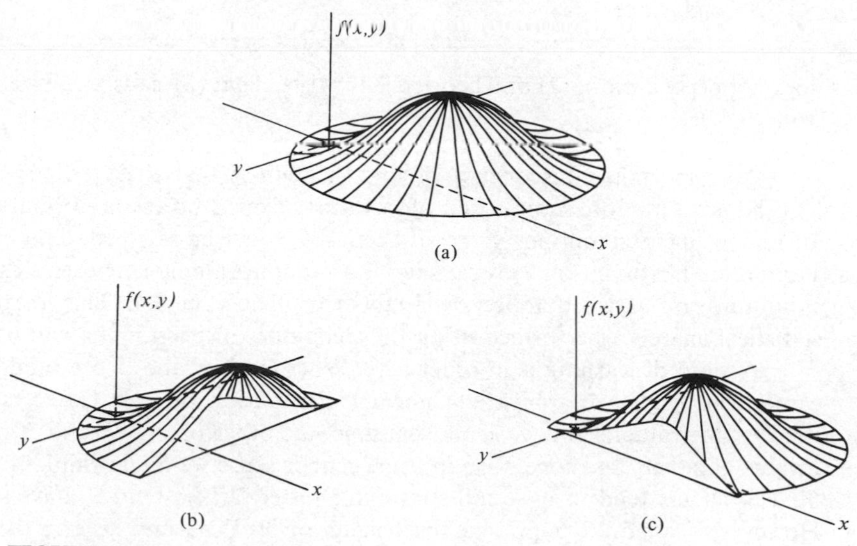

FIGURE 9.20

The Bivariate Normal Distribution

There are some very important results which can be deduced from the fact that the conditional distribution of Y given x is of the form given in (9.47). We will state these results in a theorem.

Theorem 9.4 If X and Y have a bivariate normal distribution with parameters $\mu_X, \mu_Y, \sigma_X, \sigma_Y$, and ρ, then:

1. The regression function $E(Y|x)$ is a linear function of x.
2. For each x, $V(Y|x)$ does not depend on x.
3. For each x, the conditional distribution of $Y|x$ is normal.

Analogous statements can be made for the conditional distribution of X given y.

Proof: The proof of this theorem follows directly from (9.47). The conditional mean is

$$E(Y|x) = \mu_Y + \rho\frac{\sigma_Y}{\sigma_X}(x - \mu_X).$$

By rearranging terms, this can be written as

$$E(Y|x) = \left(\mu_Y - \rho\frac{\sigma_Y}{\sigma_X}\mu_X\right) + \rho\frac{\sigma_Y}{\sigma_X}x.$$

If we define β_0 to be $(\mu_Y - \rho(\sigma_Y/\sigma_X)\mu_X)$ and β_1 to be $\rho(\sigma_Y/\sigma_X)$, we see that

$$E(Y|x) = \beta_0 + \beta_1 x,$$

so that it is a linear function of x.

Second, since

$$V(Y|x) = \sigma_Y^2(1 - \rho^2)$$

does not depend on x, (2) of Theorem 9.4 is true. That (3) is also true follows directly from (9.47).

An important consequence of this theorem is that if $(X_1, Y_1), (X_2, Y_2), \ldots, (X_n, Y_n)$ are a random sample of n observations from a bivariate normal distribution, all four of the assumptions given in Section 9.3 will be satisfied. The independence assumption (Assumption 2) is satisfied if a random sample is chosen. That the other conditions are satisfied follows directly from Theorem 9.4. This means that any statistical inference performed using the techniques of Section 9.4 will be valid.

A word of warning is in order here. When both X and Y are random, as is the case when they have a bivariate normal distribution, it is not always possible to think of the value of one variable "causing" the other one. If X and Y represent the heights of fathers and sons, then in some genetic sense we might think of X "causing" Y. Tall fathers tend to have tall sons and shorter fathers tend to have shorter sons. However, if X and Y represent the heights of brothers and sisters, then neither X nor Y can be thought of as "causing" the other. In this case you can see that the variables are related, will be positively correlated in fact, but do not cause each other. Both are related, in a very real sense, to another factor, namely, having common parents.

9.10 INFERENCE FOR ρ

If it is known that X and Y come from a bivariate normal distribution, it follows from Theorem 9.4 of the previous section that there is no question about the regression line being linear. Although in some situations it may be of interest to estimate the parameters β_0 and β_1 of the regression line, it may be the case that interest lies in estimating ρ. This is especially true in situations where a cause–effect relationship does not exist. In such a case we can think of ρ as a measure of the degree of association between X and Y.

You will recall from our discussions of ρ in Sections 3.5 and 9.8 that, if X and Y are independent, ρ will be zero. However, in general, it need not be the case that ρ being zero means that X and Y are independent (see Exercise 33). However, *if X and Y have a bivariate normal distribution*, then when $\rho = 0$, X and Y will be independent. [You can see that this is true by considering the joint density given in (9.46). When $\rho = 0$, the joint density can be written as the product of the marginals of X and Y.] This fact allows us to perform a statistical test for independence of X and Y by testing the null hypothesis that $\rho = 0$. In the remainder of this section we will present methods for obtaining confidence intervals for ρ and performing tests of hypotheses about ρ. Based on the following theorem, we can find a test statistic for testing $H_0: \rho = 0$.

Theorem 9.5 If $(X_1, Y_1), (X_2, Y_2), \ldots, (X_n, Y_n)$ are a random sample from a bivariate normal distribution and if $\rho = 0$, then

$$r\sqrt{\frac{n-2}{1-r^2}} \tag{9.48}$$

has a t distribution with $n-2$ degrees of freedom.

The proof of this theorem is beyond the scope of this book. We can see, however, that a suitable test statistic for testing hypotheses about $\rho = 0$ is given by (9.48).

Example 9.17 In Example 9.16 we found the sample correlation coefficient between engine size (x) and estimated gas mileage (y) for 18 cars to be $r = -.609$. Assuming that these values come from an approximately bivariate normal distribution, test $H_0: \rho = 0$ vs. $H_a: \rho \neq 0$ using $\alpha = .01$.

Following the steps for hypothesis testing given in Chapter 8, we have:

1. $H_0: \rho = 0, H_a: \rho \neq 0$.
2. $\alpha = .01$.
3. $n = 18$.
4. Since the conditions of Theorem 9.5 are (approximately) satisfied, the appropriate test statistic will be $r[(n-2)/(1-r^2)]^{1/2}$.
5. This test statistic has a t distribution with $n - 2 = 16$ degrees of freedom. We would reject H_0 if the absolute value of the test statistic is greater than or equal to $t_{16,.995} = 2.921$.
6. The observed value of the test statistic is

$$r\sqrt{\frac{n-2}{1-r^2}} = -.609\sqrt{\frac{16}{1-(-.609)^2}} = -3.071.$$

Since this value lies in the critical region H_0 would be rejected. Equivalently, the null hypothesis that X and Y are independent would be rejected.

When ρ is not equal to zero, the exact probability distribution of r is very complicated, so it is not easy to find simple probability statements that can be used to find confidence intervals for ρ. By using a mathematical transformation, it is possible to find approximate confidence intervals for ρ. This transformation was developed by R. A. Fisher and is generally referred to as Fisher's Z transformation. To distinguish this from the standard normal, we will use the notation Z'. Defining

$$Z' = \frac{1}{2}\ln\left(\frac{1+r}{1-r}\right), \tag{9.49}$$

it can be shown that for n sufficiently large, Z' will have an approximately normal distribution with

$$\mu_{Z'} = \frac{1}{2}\ln\left(\frac{1+\rho}{1-\rho}\right)$$

and variance

$$\sigma_{Z'}^2 = \frac{1}{n-3}.$$

(Many authors say that this approximation is good for $n \geq 25$; others say that n

should be greater than 50. We, too, will recommend $n \geq 25$ as a reasonable rule of thumb.) Since

$$\frac{Z' - \mu_{Z'}}{\sqrt{1/(n-3)}} \tag{9.50}$$

is approximately standard normal, we can use the techniques of Chapter 7 to obtain a $1 - \alpha$ confidence interval for $\mu_{Z'}$. The end points of this interval would be given by

$$Z' \pm z_{1-\alpha/2}\sqrt{\frac{1}{n-3}}. \tag{9.51}$$

The inverse of the transformation (9.49) applied to these end points would give end points for a confidence interval for ρ. You may be relieved to find that you do not have to carry out the transformation or inverse transformation by hand, since transformed values of r are given in Table A9. The procedure used for finding confidence intervals is perhaps best illustrated by an example.

Example 9.18 The engine sizes (x) and estimated gas mileages (y) for 12 additional cars were determined, bringing the total sample size to 30. The observed value of the sample correlation coefficient was $r = -.642$. Find an approximate 95% confidence interval for ρ.

To find this interval, we follow these steps:

1. Find the transformed value of r. From Table A9 we find the value corresponding to $r = +.642$ to be $z' = .7616$. Because of the form of the transformation, the value $-z'$ corresponds to the value $-r$, so we get $z' = -.7616$.
2. Find the end points of a confidence interval for $\mu_{Z'}$ using (9.51). Since $\alpha = .05$ and $n = 30$, we find that

$$-.7616 \pm 1.96\sqrt{1/27}$$

or $(-1.139, -.3844)$.
3. Find the inverse transformed values for these end points. We look for the values .3844 and 1.139 in the body of the table and find the corresponding value of r in the margins. The value .3844 is just halfway between .3838 and .3850, which correspond to r values of .366 and .367. Here we can take the inverse value to be .3665. The value 1.139 appears in the table and corresponds to an r value of .814. Considering the sign of the end points, we find a 95% confidence interval for ρ to be $(-.814, -.3665)$.

Fisher's transformation can also be used for testing hypotheses about the value of ρ if ρ is different from zero. The statistic given in (9.50) can be used as the test statistic.

A graphical method can also be used to find confidence intervals for ρ when samples are taken from a bivariate normal distribution. The advantages of using a graph are that numerical calculations as in (9.51) need not be done and that graphs can be used for small sample sizes. The disadvantages are that a different graph is needed for each α, not all sample sizes can be included on the graph, and they can be read to at most two decimal places of accuracy.

Example 9.19 Using the chart in Table A10, find a 95% confidence interval for ρ if a sample of size 10 taken from a bivariate normal distribution had a sample correlation of $r = +.40$.

Using the chart corresponding to 95%, lay a straightedge vertically on the chart at the value $r = +.40$. Note the places where the straightedge intersects the curves labeled $n = 10$. Follow these points horizontally until they intersect the vertical (ρ) scale. These points of intersection give the end points of the interval. Doing this, we read the values $-.30$ and $+.78$. (There is some doubt as to the accuracy of the second digit.)

In closing this section we would like to emphasize these points: If a sample is taken from a bivariate normal distribution, a test of $\rho = 0$ is equivalent to a test for independence of the variables. Any value of ρ, no matter how close to ± 1, only indicates the degree of association between the variables. No cause–effect relationship is implied. Finally, the techniques used for testing hypotheses and finding confidence intervals are valid only if the assumption of bivariate normality is approximately satisfied. If this assumption is not satisfied, then nonparametric techniques as discussed in Chapter 11 should be used.

EXERCISES

1. A car stops at a gas station after driving M miles since the last fill-up. (a) Is the relationship between the number of gallons G required to fill the tank and M stochastic or deterministic? (b) Assume that gasoline cost c cents/gallon. Is the relationship between the number of gallons G and the total cost of filling up, C, stochastic or deterministic?

2. The distribution of a discrete random variable Y for different values of x is given below. Find $E(Y|x)$ for $x = 1, 2, 3$, and 4 and graph x vs. $E(Y|x)$. Is the regression curve linear?

	y			
	0	*1*	*2*	*3*
$P[Y = y \| x = 1]$:	.1	.2	.3	.4
$P[Y = y \| x = 2]$:	.1	.2	.4	.3
$P[Y = y \| x = 3]$:	.1	.4	.3	.2
$P[Y = y \| x = 4]$:	.4	.3	.2	.1

[*Hint:* Do the points $(x, E(Y|x))$ fall *exactly* on a straight line?]

• *3.* The probability density function for a continuous random variable Y for different values of x is given below. Find $E(Y|x)$ for $x = 1, 2, 3, 4$ and graph x vs. $E(Y|x)$. Is the regression curve linear?

$$f(y|x) = \frac{x + 2y}{x + 1} \qquad 0 \leq y \leq 1, \quad 0 \leq x \leq 4.$$

[*Hint:* Do the points $(x, E(Y|x))$ fall *exactly* on a straight line?]

•4. Repeat Exercise 3 for the case where

$$f(y|x) = \frac{2(x + xy)}{3x} \qquad 0 \leq y \leq 1, \quad 0 < x \leq 4.$$

5. The following data give heights (x) measured in inches and weights (y) measured in pounds of children at Lotta Noyes Elementary School. The first set of data is for girls, the second set is for boys. Plot scatter diagrams for each of these sets of data. Does it appear that the regression curve would be linear?

Girls											
	x:	44	46	48	50	52	54	56	58	60	62
	y:	46	59	53	61	64	67	80	95	102	125

Boys											
	x:	44	46	48	50	52	54	56	58	60	62
	y:	51	50	60	54	60	80	120	97	95	107.

★6. The following data give heights (x) measured in inches and weights (y) measured in pounds of students enrolled in a statistics course. For each height x there are two weight observations. The first set of data is for girls, the second set is for boys. Plot scatter diagrams for each of these sets of data. Does it appear that the regression curve would be linear?

Girls											
	x:	60	61	62	63	64	65	66	67	68	69
	y:	100	115	140	125	110	160	115	127	135	140
		100	115	107	112	130	125	145	120	135	160

Boys											
	x:	65	66	67	68	69	70	71	72	73	74
	y:	130	157	130	160	165	155	145	172	175	177
		155	135	160	165	165	170	160	148	170	179.

★7. In a study of the characteristics of Rawlings and Spalding major league baseballs, the weight (x_1), circumference (x_2), and coefficient of restitution (y) were measured for samples of both kinds of baseballs. The following tables give this information for 20 of each brand.

Rawlings baseballs:

x_1:	5.07	5.08	5.09	5.08	5.20	5.17	5.07	5.24	5.12	5.16
x_2:	9.04	9.03	9.03	9.03	9.04	9.05	9.00	9.07	9.01	9.00
y:	52.6	52.8	52.7	54.7	53.7	53.7	48.1	50.9	50.9	54.5

x_1:	5.18	5.19	5.24	5.27	5.26	5.10	5.28	5.29	5.27	5.09
x_2:	9.02	9.02	9.01	9.00	9.00	9.01	9.00	9.02	9.04	9.01
y:	53.7	53.8	53.8	51.8	52.7	53.6	51.0	51.0	51.8	52.1.

Spalding baseballs:

x_1:	4.88	5.02	5.03	5.04	4.84	5.06	5.17	4.98	5.13	5.08
x_2:	9.04	9.02	9.13	9.03	9.04	9.13	9.13	9.03	9.13	9.07
y:	51.8	52.8	50.0	48.3	48.3	50.8	52.5	51.7	51.7	48.3

x_1:	5.09	5.08	5.04	4.89	4.97	4.92	5.19	5.20	5.08	5.10
x_2:	9.05	9.15	9.05	9.04	9.05	9.08	9.08	9.13	9.06	9.09
y:	50.0	50.0	52.5	49.2	50.9	51.7	51.7	50.0	51.8	51.8.

(a) For the Rawlings baseballs, make a scatter diagram of weight (x_1) vs. coefficient of restitution (y).

(b) For the Spalding baseballs, make a scatter diagram of circumference (x_2) vs. coefficient of restitution (y).

★*8.* In a report entitled "A Population Analysis of the Yellowstone Grizzly Bears" by J. Craighead, J. Varney, and F. Craighead, data are given on the age structure of the grizzly bear population. In particular, one can find the number of grizzly bears in one year (x) and the number of cubs born the next year (y). Make a scatter diagram using the information in the table below. What kind of relationship, if any, appears to exist?

x:	154	169	166	155	177	185	187	202	175	181	195
y:	35	30	39	40	24	40	32	30	32	28	21

9. To see that the sum of squares really is a minimum when b_0 and b_1 are found using equations (9.4), calculate $\sum e_i^2$ as defined in equation (9.3) using the data from Example 9.2 with (a) $b_0 = 1.6$ and $b_1 = 1.8$; (b) $b_0 = 1.7$ and $b_1 = 1.8$; (c) $b_0 = 1.6$ and $b_1 = 1.9$; (d) any value of b_0 and b_1 of your choice. Compare the sums.

10. Given the following set of data, (a) make a scatter diagram; (b) find the least-squares line; (c) plot this line on the graph with the scatter diagram.

x:	1	2	3	4	5
y:	17	19	24	27	33.

★*11.* Using the data given in Exercise 8, find the least-squares line that best fits these data. Using this line, how many cubs would you expect to be born next year if the grizzly bear population this year is found to be (a) 160; (b) 190? [*To think about:* What would be the predicted number of cubs born next year using this equation if the grizzly bear population is 0?; 325? What does this tell you about extrapolating beyond the range of observed x values?]

12. There are various noise levels at the Watzit plant and the operations supervisor, A. Kumagain, wishes to check the time required for workers to assemble a watzit under the varying conditions. The noise levels X, measured in decibels (dB), and the time required for assembly Y, measured in minutes, for eight workers is given in the following table.

x:	10	10	30	30	50	50	70	70
y:	11	14	14	18	18	19	24	22.

(a) Plot a scatter diagram. (b) Find the least-squares line and graph it. (c) What time might you expect it to take to assemble a watzit if the noise level is 40 dB?

★*13.* In an experiment performed at the agricultural experiment station of a large university, a particular variety of corn was planted in 15 different small plots (10 by 10 inches). The 15 plots were randomly divided into three sets of five plots each, receiving the equivalent of 0, 50, and 100 pounds/acre of nitrogen. Insofar as possible, all other factors were kept constant. After a short period the plants were harvested and dry matter (measured in grams/plot) was determined. The data for these 15 plots, which are a subset of a larger set of data, are as follows:

x (lb/acre)	*y (grams/plot)*	*x (lb/acre)*	*y (grams/plot)*
0	60	50	78
0	90	50	98
0	81	100	114
0	82	100	107
0	78	100	103
50	104	100	94
50	95	100	122
50	92		

(a) Plot a scatter diagram for the data. Does it appear as if the regression curve may be linear?
(b) Find the least-squares line and graph it on the scatter diagram.
(c) *In the context of this problem*, what meaning can be attached to the values of β_0 and β_1?

14. Show that for any set of numbers $x_1, x_2, \ldots, x_n$, $\sum (x_i - \bar{x}) = 0$. (*Hint:* Use properties of summations. Also cf. Example 2.20 and Exercise 8 of Chapter 5.)

15. Show that if the regression curve is linear [i.e., if $E(Y_i) = \beta_0 + \beta_1 x_i$], then $B_0 = \bar{Y} - B_1\bar{x}$ is an unbiased estimator of β_0. [*Hint:* Use the fact that $E(B_1) = \beta_1$. Also find $E(\bar{Y})$.]

16. Show that (9.22) of Section 9.3 is true. That is, show that if the Y_i are independent and if $V(Y_i|x_i) = \sigma^2$ for all x_i, then

$$V(B_0) = V(\sum a_i Y_i) = \sigma^2\left[\frac{1}{n} + \frac{\bar{x}^2}{\sum (x_i - \bar{x})^2}\right].$$

[*Hint:* Expand $\sum a_i^2 = \sum [(1/n) - c_i\bar{x}]^2$ and use equations (9.18) and (9.24).]

17. In a problem where the regression curve is assumed to be linear, the following quantities were calculated: $n = 7$, $\sum x_i = 21$, $\sum y_i = 35$, $\sum (x_i - \bar{x})(y_i - \bar{y}) = -63$, $\sum (x_i - \bar{x})^2 = 28$, and $\sum (y_i - \bar{y})^2 = 162$. Find (a) the equation for the regression line; (b) an estimate of $\sigma^2_{Y|x}$; (c) an estimate of $\sigma^2_{B_0}$; (d) an estimate of $\sigma^2_{B_1}$.

18. Using the data given in Exercise 10, find estimates of $\sigma^2_{Y|x}$, $\sigma^2_{B_0}$, and $\sigma^2_{B_1}$.

★19. Using the grizzly bear data (Exercises 8 and 11), find estimates of $\sigma^2_{Y|x}$, $\sigma^2_{B_0}$, and $\sigma^2_{B_1}$. [*To think about:* What assumptions have you made about $V(Y|x)$ for various values of x? If this assumption is reasonable for x values between 150 and 200, will it necessarily be reasonable for values outside this range?]

20. Using the watzit data in Exercise 12, assuming that the variance in Y is the same for each sound level x, find estimates of $\sigma^2_{Y|x}$, $\sigma^2_{B_0}$, and $\sigma^2_{B_1}$.

21. Assuming that the Y values in Exercise 17 are normally distributed, (a) find a 95% confidence interval for β_0. (b) Test H_0: $\beta_1 \geq -2.0$ against H_a: $\beta_1 < -2.0$ using a .05 level of significance.

22. (a) Using the data of Exercise 10, test $H_0: \beta_0 = 15$ against $H_a: \beta_0 \neq 15$ using a .01 level of significance. (b) Find a 90% confidence interval for β_1. (You may wish to use the results of Exercise 18.)

★**23.** Using the grizzly bear data of Exercise 8, test the null hypothesis that $\beta_1 \geq 0$ using a .05 level of significance. (*To think about:* What might a negative slope imply about the self-regulating nature of the grizzly bear population in a protected environment?) You may wish to use the results of Exercises 11 and 19.

24. (a) Using the watzit data (Exercise 12), find a 99% confidence interval for β_0. (b) Test $H_0: \beta_1 \leq 0$ against $H_a: \beta_1 > 0$ at the .01 level of significance. (*To think about:* What might a positive slope imply about the effect of noise on the time required for assembly of a watzit?) You may wish to use the results of Exercise 20.

25. Show that if the fitted value of Y for a given value x, $\hat{Y}$, is expressed as a linear combination of the random variables Y_i by

$$\hat{Y} = \sum (a_i + c_i x) Y_i,$$

where a_i and c_i are defined in (9.17) and following in Section 9.3, then

$$\sigma_{\hat{Y}}^2 = \sigma^2 \left[\frac{1}{n} + \frac{(x - \bar{x})^2}{\sum (x_i - \bar{x})^2}\right]$$

(cf. Section 9.4).

★**26.** Using the corn data given in Exercise 13, (a) find a 90% confidence interval for the mean amount of dry matter that would be expected if the equivalent of 25 pounds of nitrogen/acre were used per plot. (b) Find a general expression for this interval if the equivalent of x pounds of nitrogen/acre are used. (c) Find a 95% prediction interval for the amount of dry matter in an individual plot if that plot is to receive the equivalent of 75 pounds of nitrogen/acre.

★**27.** Using the data from Exercise 8 and the least-squares line from Exercise 11, find the residuals, $y_i - \hat{y}_i$. Using these values, make a residual plot. Examine the residual plot. Does it appear that the assumption of a common variance (Assumption 3) is reasonable?

28. In the following data, x represents the number of homework assignments handed in by a statistics student and Y indicates whether the student passed an hourly examination. (Here $y = 1$ indicates passing while $y = 0$ indicates failing.)

x:	0	1	2	3	4	5	6	7
y:	0	0	1	0	1	1	1	1.

The least-squares line fitting these points is given by

$$\hat{y} = .083 + .155x.$$

(a) Find the residuals, $y_i - \hat{y}_i$, and make a residual plot. (b) Examine the residual plot. Is the assumption of linearity (Assumption 1) reasonable? (*To think about:* Can the mean of Y for a given x be greater than 1? What is $\hat{Y}$ when $x = 6$?) (c) Is the assumption that the conditional distribution of Y given x is normally distributed reasonable? Why or why not?

★**29.** Using the corn data and the least-squares line found in Exercise 13, (a) find the residuals, $y_i - \hat{y}_i$, and make a residual plot. (b) Examine the residual plot. Is the assumption of a common variance for the distribution of Y for each x reasonable? Explain.

★30. In an experiment similar to the one described in Exercise 13, it was desired to study the effect of a growth-regulating compound (Am Chem 66-329) on the corn growth. Here, too, the response was measured in grams/plot. Growth regulator equivalent to the amount of 0, .5, and 2 pounds/acre was applied to 15 randomly selected plots. The resulting data are shown in the table.

x (lb/acre)	*y (grams/plot)*	*x (lb/acre)*	*y (grams/plot)*
0	80	.5	98
0	87	.5	112
0	83	2	78
0	80	2	98
0	85	2	74
.5	121	2	81
.5	110	2	82
.5	98		

(a) Make a scatter diagram for these data. (b) Find and plot the least-squares regression line $\hat{y} = b_0 + b_1x$. (c) Find SSE $= \sum (y_i - \hat{y}_i)^2$. (d) Make a residual plot. (e) Examine the residual plot. Does the assumption of linearity appear to be reasonable? Does the assumption of a common variance appear to be reasonable? Explain. (f) Would it be proper to perform statistical inference of β_0 and/or β_1? Why or why not?

★31. (a) Verify that the least-squares quadratic regression curve for the data in Exercise 30 is given by

$$\hat{y} = 83.0 + 66.2x - 33.2x^2.$$

(b) Plot this curve on the scatter diagram of Exercise 30(a). (c) Find the residuals $(y_i - \hat{y}_i)$ and calculate SSE $= \sum (y_i - \hat{y}_i)^2$. Compare this value with the answer found in Exercise 30(c). (d) Make a residual plot. (e) Examine the residual plot. Does the assumption of a common variance appear to be reasonable? Explain. (f) Would it be proper to perform statistical inference on β_0 and/or β_1? Why or why not?

★32. The data in Table 9.13 were obtained from a newspaper article relating to "fast-food" restaurants. (a) Plot a scatter diagram of price (y) and weight of meat (x). (b) Find and plot the least-squares line $\hat{y} = b_0 + b_1x_1$. (c) Make a residual plot. (d) Plot a scatter diagram of price (y) and weight of bun and condiments (x_2). (e) Find and plot the least-squares line $\hat{y} = b_0 + b_2x_2$. (f) Make a residual plot. (g) Verify that the least-squares plane for the multiple regression model is

$$\hat{y} = .084 + .244x_1 + .0402x_2.$$

33. If X has probability distribution

x:	-1	0	1
$P[X = x]$:	.3	.4	.3

and if $Y = X^2$, find the joint probability distribution of X and Y. Find the correlation coefficient ρ (cf. Example 9.15 and the following discussion).

34. Using the data and the results of Exercise 10, find the values of r and r^2.

35. Using the information given in Exercise 17, find the values of r and r^2.

TABLE 9.13
Data for Exercise 32

Name	*Price*	*Cooked Weight of Meat (oz)*	*Weight of Bun and Condiments (oz)*
Burger King Hamburger	$ .39	1.38	2.12
Dairy Queen Junior Burger	.40	.84	2.07
Dairy Queen Big Brazier	.80	2.21	2.97
Hardee's Hamburger	.35	1.26	2.45
Jack-in-the-Box Hamburger	.35	1.09	3.18
McDonald's Hamburger	.35	1.05	2.07
Sonic Mini Kiddy Burger	.55	.81	2.41
Sonic Burger	.85	2.38	4.97
Wendy's Single	.85	2.31	4.76
Burger King Whopper	.99	2.59	6.58
Burger King Double Whopper	1.59	4.76	6.44
Dairy Queen Super Brazier	1.49	4.48	3.36
Hardee's Big Deluxe	1.09	2.73	5.04
Hardee's Big Twin	.85	2.63	3.60
Jack-in-the-Box Jumbo Jack	.99	2.56	6.47
Jack-in-the-Box Double Jumbo Jack	1.49	5.01	7.31
McDonald's Quarter Pounder	.80	2.52	2.77
McDonald's Big Mac	.85	2.14	4.48
Sonic Super Burger	1.45	5.04	6.30
Wendy's Double	1.45	4.62	4.73
Wendy's Triple	1.95	7.14	4.66

★**36.** Find the sample correlation coefficient between (a) the weight (x_1) and the coefficient of restitution (y) for the Rawlings baseballs and (b) the circumference (x_2) and the coefficient of restitution (y) for the Spalding baseballs using the data given in Exercise 7.

37. Show that $\sum (Y_i - \hat{Y}_i)(\hat{Y}_i - \bar{Y}) = 0$ [see (9.45) of Section 9.8]. [*Hint:* Use the fact that $\hat{y}_i = b_0 + b_1 x_i = \bar{y} + b_1(x_i - \bar{x})$.]

★**38.** (a) Using the data from Exercise 32, find the proportion of variation in the price of hamburgers (Y) explained by the linear regression on x_1, the weight of the meat. (b) What proportion is accounted for by x_2, the weight of the bun and condiments? (*To think about:* Which of these two factors is the most important in determining the price of a hamburger?)

39. Find the proportion of variation in Y, the amount of time required for workers to assemble a watzit, explained by the regression on x, the noise level at the Watzit plant (see Exercise 12).

40. Let X and Y represent the adult heights of a brother and a sister in a family. Assume that X and Y have a bivariate normal distribution with parameters $\mu_X = 69.5$ inches, $\mu_Y = 64.5$ inches, $\sigma_X = 2.5$ inches, $\sigma_Y = 2.4$ inches, and $\rho = .6$.

(a) Plot the regression line x vs. $E(Y|x)$.
(b) Plot the regression line $E(X|y)$ vs. y.
(c) Find the conditional variances $V(Y|x)$ and $V(X|y)$.

(d) What is the conditional distribution of the heights of sisters having a brother 68 inches tall? 72 inches tall?

41. In the following situations discuss whether or not it might be reasonable to assume that X and Y follow a bivariate normal distribution. Give reasons for your answers.

(a) A car is chosen at random from the registration lists in your state. $X =$ age of the car chosen, $Y =$ mileage on the odometer of the car chosen.

(b) A person is chosen at random from the membership of Eager-Eaters Anonymous. $X =$ weight of person chosen when he joined EEA, $Y =$ current weight of person chosen.

(c) A sociologist chooses a family at random from a low-rent housing project. $X =$ amount of food stamps bought by the family in the last year, $Y =$ amount of federal income tax paid by the family in the last year.

(*To think about:* If a sample of 20 observations were chosen in each of the situations, and if r were calculated, would you expect r to be positive or negative? In which, if any, of these situations would it be appropriate to test $H_0 : \rho = 0$ using the techniques of this chapter? In the other situations nonparametric techniques would be more appropriate.)

★***42.*** A representative sample of (systolic) blood pressure and weight readings were taken on 15 individuals with the following results:

Blood pressure:	138	160	166	153	190	132	154	160
Weight:	122	178	142	150	185	111	153	180
Blood pressure:	153	185	120	166	156	136	152	
Weight:	120	144	123	157	147	132	179.	

(a) Plot a scatter diagram of weight and blood pressure. (b) Find the sample correlation coefficient r. (c) Test the null hypothesis that $\rho = 0$ against a two-sided alternative. Use $\alpha = .01$. (d) Find an approximate 90% confidence interval for ρ using Fisher's transformation. (e) Find an approximate 95% confidence interval for ρ using the graphical method.

★***43.*** In the days before stretch socks and tube socks, grandmother would try to determine the proper sock size for her grandchildren by wrapping the sock around their fist. If the toe of the sock touched the heel, she declared the sock to be the right size. (For obvious reasons, grandmother considered it improper to try on socks.) What assumption was grandmother making about the relationship between fist "circumference" and foot length? Using the data in the table, find the sample correlation coefficient r. Find a 95% confidence interval for ρ. (*To think about:* What do these results tell you about grandmother's technique?)

	Grandchild							
	Shelly	*Mike*	*Rick*	*Suzy*	*Pete*	*Jody*	*Derek*	*Scott*
X (fist circumference)	8.25	9.25	9.00	7.50	7.75	8.00	7.00	6.50
Y (foot length)	9.25	9.25	8.25	8.00	8.00	7.00	7.00	6.25

44. In an effort to predict a person's success in sales, a personnel manager has administered two tests, one measuring persistence and the other measuring personality.

The manager studied a sample of records of sales personnel after 6 months' working and found the following results.

Persistence score:	50	61	54	73	70	77	61	42	47	45
Personality score:	34	78	87	90	46	69	77	70	66	83
Amount of sales (thousands):	12	36	42	40	32	35	37	35	29	42.

(a) Find the sample correlation between persistence and sales. (b) Test $\rho = 0$ at the .05 level of significance. (c) Find a 95% confidence interval for ρ using the graphical method. (d) Find the sample correlation between personality and sales. (e) Test $\rho = 0$ at the .05 level of significance. (f) Find a 99% confidence interval for ρ using the graphical method.

★**45.** (a) The sample correlation coefficient between the circumference (x_2) and the coefficient of restitution (y) for Rawlings baseballs (Exercise 7) was found to be $r = -.0247$. Find a 95% confidence interval for ρ using Fisher's transformation. (b) The sample correlation coefficient between the weight (x_1) and the coefficient of restitution (y) for Spalding baseballs was found to be $r = +.2203$. Find a 99% confidence interval for ρ using Fisher's transformation.

10

ANALYSIS OF VARIANCE

10.1 INTRODUCTION

In Chapters 7 and 8 we discussed the problem of statistical inference for the means of two populations. A commonly asked question is whether or not the means are equal. In one exercise we saw that the owner of a small taxi company, Rick Shaw, was interested in knowing whether the mean mileage on two different brands of tires was the same. In another a store manager, Sandy Sures, was interested in knowing whether the mean number of customers entering the store was affected by the appearance of an ad in a paper (Exercises 20 and 26 of Chapter 8). In similar situations it is possible for an individual to be interested in comparing more than two means. For example, Rick might want to compare the mean mileage on five different brands of tires. If there is no significant difference among the means, he might buy the least expensive brand. Likewise, Sandy might want to compare the mean number of customers entering the store under four different conditions: no ads, a newspaper ad, a radio ad, or a television ad. Sandy might choose the advertising method leading to the highest mean. As another example, an efficiency expert, Dewey Phaster, might want to compare the mean productivity of workers under three systems of coffee breaks: one 30-minute break, two 15-minute breaks, or three 10-minute breaks per shift. The system leading to the highest productivity might be used.

You can see from these examples that there are many situations where an investigator is interested in comparing means from different populations. In its simplest form, the statistical method used to compare the means is an extension of the method used in Chapter 8 for comparing two means when the data are unpaired. In that method two independent samples are chosen from the two populations of interest, while in the extension we choose an independent sample from each population of interest. We also saw in Chapter 8 that another method can be used when the data are paired. We will find that this method can be generalized to a comparison of several means. The statistical methods used are based on *analyzing the variation* in the observed values through partitioning schemes similar to the scheme used in Section 9.8. Consequently, the general method of statistical analysis used is referred to as *analysis of variance*. The acronym ANOVA is also used to refer to this method.

10.2 ONE-WAY CLASSIFICATION

So that you may better see the need for the procedures about to be introduced, we will describe a particular problem. We will refer to this problem or an extension of it throughout this chapter.

Carl Brundum is the owner of the Grindstone Works. This factory produces grinding wheels for grinding and sharpening purposes. Throughout the day the employees who make the wheels are allowed to take coffee breaks. Carl, of course, hopes that since the breaks relieve boredom, they will ultimately lead to higher productivity. Dewey Phaster, his efficiency expert, proposes a study of four different methods of giving coffee breaks. The methods are as follows:

1. Two 15-minute breaks at times chosen by the employer.
2. Three 10-minute breaks at times chosen by the employer.
3. Two 15-minute breaks at times chosen by the individual workers.
4. Three 10-minute breaks at times chosen by the individual workers.

Dewey and Carl would like to find which (if any) of these methods of giving coffee breaks leads to the highest mean productivity as measured by the number of wheels produced per shift by individual employees (see Figure 10.1). Dewey proposes to randomly select 20 employees and have 5 each work under the four different systems. For each employee the number of wheels produced in a shift will be found.

FIGURE 10.1
Coffee Breaks Lead to Higher Productivity

If we define

$$\mu_1 = \text{expected number of wheels produced/shift by employees given breaks under method 1,}$$

and define μ_2, μ_3, and μ_4 similarly, then Dewey would like to know if all these means are the same, that is, if

$$\mu_1 = \mu_2 = \mu_3 = \mu_4$$

or if one (or more) of these means differ from the others. The data for the 20 workers are given in Table 10.1. Also given in the table are the sums (i.e., total production for

TABLE 10.1

Production Figures from the Grindstone Works

	Method			
	1	*2*	*3*	*4*
Production	80	81	78	93
	86	82	82	90
	83	88	84	91
	85	83	82	93
	81	86	84	88
Sample mean	83	84	82	91
Sum	415	420	410	455
Sum of squares	34,471	35,314	33,644	41,423

five workers using a particular type of break), the sample means, and the sum of squares (which could be used to get an estimate of the variance of the production under each of the methods of giving coffee breaks). Given these data, how might Dewey proceed to test to see if all four means are equal? In Chapter 8 we found several methods for testing for the equality of means from *two* populations. Dewey might consider testing the means in pairs (i.e., test to see if $\mu_1 = \mu_2$, test to see if $\mu_2 = \mu_3$, if $\mu_3 = \mu_4$, etc.). There are two problems with this approach, however. First, it would require several tests rather than just one single test. Second, if each individual test were conducted using a level of significance of .05, say, then the overall level of significance would be much higher. For instance, if $H_0 : \mu_1 = \mu_2 = \mu_3$ is actually true, and if the three null hypotheses $\mu_1 = \mu_2$, $\mu_2 = \mu_3$, and $\mu_1 = \mu_3$ are each tested using a .05 level of significance, then it can be shown that the probability of *accepting all three* hypotheses when H_0 is true is at least .85. [If the tests were independent, this probability would be $(.95)^3 = .857$.] Hence, the probability of rejecting H_0 when it is true using this approach could be as large as .15, not .05!

While the technique of using several pairwise tests of means does not appear to be the best way to test the null hypothesis of the equality of several means, a brief review of the pairwise test might be in order as we seek a better way to test this hypothesis. First remember that we assumed that we had independent samples from each population. We then said that if it is reasonable to assume that the variances are equal and the distributions are normal, a suitable test statistic for tests about $\mu_1 - \mu_2$ would be given by (8.18). If under the null hypothesis $\mu_1 - \mu_2 = 0$, this test statistic becomes

$$T = \frac{\bar{X}_1 - \bar{X}_2}{S_p\sqrt{(1/n_1) + (1/n_2)}}.$$

Remember that S_p^2 is the pooled estimate of the common variance σ^2. This pooled estimate will play a major role in the sequel.

At this point it is appropriate to give some definitions and to introduce some notation. Since, as we will see later, the techniques used in this chapter are similar to techniques used in regression analysis, the notation will be similar.

Definition 10.1 A *factor* is an independent variable that may affect the distribution of the response random variable.

A factor may be either quantitative or qualitative in nature. The independent variables considered in our study of regression were mainly quantitative (e.g., amount of water put on pepper plants). In the study of analysis of variance they may be quantitative but they may also be qualitative. For instance, the difference between the time of a coffee break being chosen by the employer or employee is qualitative. In a study of sales of Water's Salt, the factor "shape of box" would be a qualitative factor (cf. Exercise 65 of Chapter 7).

Definition 10.2 A *factor level* is a particular form that a factor assumes. If a single factor is being considered, the factor levels are called *treatments*.

In the study being conducted in the Grindstone Works, there are four factor levels or treatments being considered. We will see later that multifactor studies can also be considered. In such cases a combination of different factor levels is *also called a treatment*.

Now say that we are studying a problem with one factor and t factor levels or treatments. We can define μ_i to be the mean response of the variable of interest under treatment i and μ (with no subscript) to be the overall mean or the average of the treatment means:

$$\mu = \sum_{i=1}^{t} \frac{\mu_i}{t}. \tag{10.1}$$

Using μ, we can express the means μ_i as

$$\mu_i = \mu + \tau_i, \tag{10.2}$$

where τ_i is a parameter measuring the treatment effect. Note that since $\tau_i = \mu_i - \mu$, it must be true that

$$\sum_{i=1}^{t} \tau_i = 0 \tag{10.3}$$

(see Exercise 2).

In view of (10.2) and (10.3), it should not be hard to see that the hypotheses

(10.4) $$H_0 : \mu_1 = \mu_2 = \ldots = \mu_t$$

(10.5) $$H_0 : \tau_1 = \tau_2 = \ldots = \tau_t = 0$$

are equivalent. Therefore, a test of the hypothesis that the treatment means μ_i are all equal is equivalent to a test of the hypothesis that the treatment effects are all equal to zero.

In order to test the hypotheses (10.4) or (10.5), we need to obtain sample data and find a suitable test statistic, a statistic whose distribution is known when the null hypothesis is true. Notationally, we will denote the sample observations by

$$Y_{i,j} \qquad i = 1, 2, \ldots, t \text{ and } j = 1, 2, \ldots, n.$$

This assumes that for each treatment, a random sample of n observations is taken. (If the sample sizes are different for each treatment, the analysis becomes a bit more complicated. We will not consider the case of unequal sample sizes in detail.) For the data in Table 10.1, for example, we have $y_{1,1} = 80$ [i.e., the first observation ($j = 1$) under the first treatment ($i = 1$) was 80 wheels produced]. Similarly, $y_{1,2} = 86$, $y_{2,1} = 81$, and so on. In terms of the parameters introduced earlier, we can write

$$E(Y_{i,j}) = \mu_i = \mu + \tau_i.$$

This information, although important, is not enough to allow us to find a parametric test for (10.4). We must make some further assumptions. In particular, we make:

Assumption 1 The observations $Y_{i,j}$ are independent.

Assumption 2 For each i, the variance of $Y_{i,j}$ is σ^2.

Assumption 3 For each i, $Y_{i,j}$ has a normal distribution.

(Note the similarity between these assumptions and those made in Section 9.3.) If these assumptions all hold, the distributions of the random variables $Y_{i,j}$ may be as given in Figure 10.2. Note that the means of these distributions may differ, but the variances are all the same.

If this common variance σ^2 is unknown, how can it be estimated? We faced a similar situation in Section 7.7, where we tried to estimate the common variance of two populations. We saw there that we could estimate σ^2 by the sample variance of the first population, S_1^2, or by the sample variance of the second population, S_2^2. However, further reflection showed that it would be better to combine or "pool" these estimates and use

$$S_p^2 = \frac{(n_1 - 1)S_1^2 + (n_2 - 1)S_2^2}{n_1 + n_2 - 2}$$

as an estimate of σ^2. We do a similar thing here in estimating the common variance of t populations. In this case the pooled estimate would be

(10.6) $$S_p^2 = \frac{(n_1 - 1)S_1^2 + (n_2 - 1)S_2^2 + \ldots + (n_t - 1)S_t^2}{n_1 + n_2 + \ldots + n_t - t}$$

where

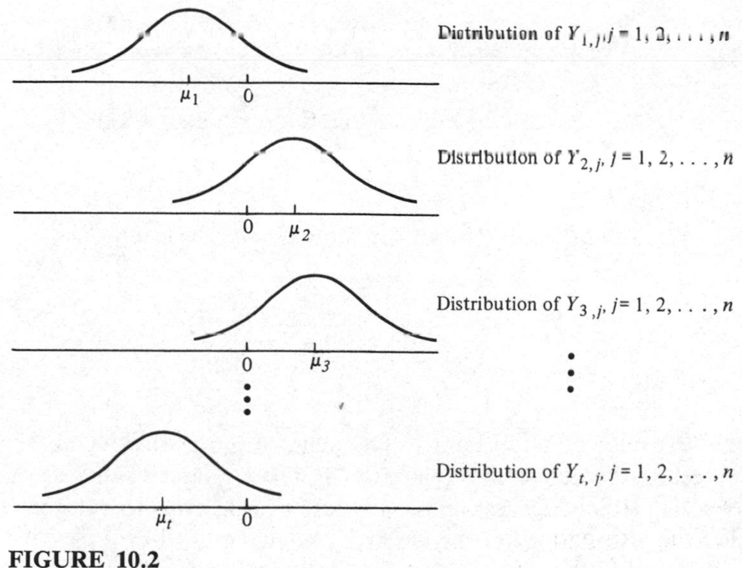

FIGURE 10.2
Distributions of $Y_{i,j}$

(10.7) $$S_i^2 = \frac{\sum_{j=1}^{n_i} (Y_{i,j} - \bar{Y}_{i\cdot})^2}{(n_i - 1)}.$$

(Here we have used the shorthand notation $\bar{Y}_{i\cdot}$ to represent the mean of the ith sample. We let $Y_{i\cdot}$ represent the sum $\sum_j Y_{i,j}$ and $\bar{Y}_{i\cdot} = Y_{i\cdot}/n_{i\cdot}$.) Since we have made the simplifying assumption that all sample sizes are equal, (10.6) and (10.7) can be simplified somewhat. Furthermore, for computational purposes we can express the sum of squares in (10.7) in an equivalent form [cf. (5.3) and (5.4) of Section 5.4]:

$$(n - 1)S_i^2 = \sum_{j=1}^{n} (Y_{i,j} - \bar{Y}_{i\cdot})^2 = \sum_{j=1}^{n} Y_{i,j}^2 - n\bar{Y}_{i\cdot}^2.$$

Using this form, we can write

(10.8) $$S_p^2 = \frac{\sum_{i=1}^{t}\left[\sum_{j=1}^{n} Y_{i,j}^2 - n\bar{Y}_{i\cdot}^2\right]}{t(n - 1)}.$$

Example 10.1 Using the production figures from the Grindstone Works given in Table 10.1, find the pooled estimate of the common variance σ^2.

If we find each sample variance separately, we get

$$s_1^2 = \frac{\sum_{j=1}^{n} (y_{1,j} - \bar{y}_{1\cdot})^2}{n - 1} = \frac{\sum_{j=1}^{n} (y_{1,j} - 83)^2}{5 - 1} = 6.5$$

$$s_2^2 = \frac{\sum_{j=1}^{n} (y_{2,j} - \bar{y}_{2\cdot})^2}{n - 1} = \frac{\sum_{j=1}^{n} (y_{2,j} - 84)^2}{5 - 1} = 8.5$$

$$s_3^2 = \frac{\sum_{j=1}^{n} (y_{3,j} - \bar{y}_{3\cdot})^2}{n-1} = \frac{\sum_{j=1}^{n} (y_{3,j} - 82)^2}{5-1} = 6.0$$

$$s_4^2 = \frac{\sum_{j=1}^{n} (y_{4,j} - \bar{y}_{4\cdot})^2}{n-1} = \frac{\sum_{j=1}^{n} (y_{4,j} - 91)^2}{5-1} = 4.5,$$

so $s_p^2 = 6.375$.

We would, of course, get the same answer by using (10.8),

$$s_p^2 = \frac{\sum_{i=1}^{4} \left[\sum_{j=1}^{5} Y_{i,j}^2 - 5(\bar{Y}_{i\cdot})^2 \right]}{4(5-1)} = \frac{144{,}852 - 144{,}750}{16}$$

$$= 6.375.$$

We must be careful not to lose sight of our goal, which is to find a test statistic for testing the null hypothesis given in (10.4), namely, that all treatment means μ_i are equal. (If it is any reassurance, we are over halfway to our goal!) We now consider what the distributions of the variables $Y_{i,j}$ would be like *if in fact H_0 were to be true.* In this case all the treatment means will be equal, so the distributions would be as shown in Figure 10.3. With this in mind, let us consider the distribution of the sample means $\bar{Y}_{1\cdot}$, $\bar{Y}_{2\cdot}$, and so on. We know from Theorem 5.6 of Section 5.6 that if a random sample is taken from a normal distribution, the sample mean will have a normal distribution. Consequently, we know that

$$\bar{Y}_{i\cdot} \sim N\left(\mu_i, \frac{\sigma^2}{n}\right),$$

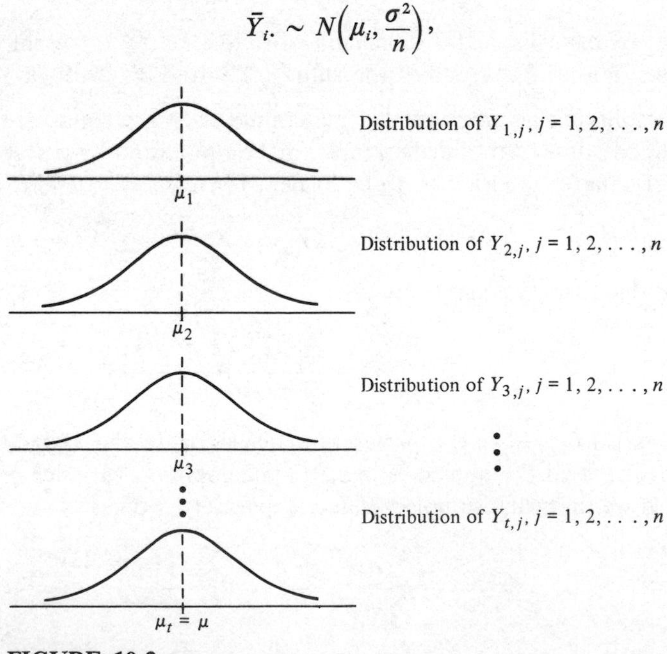

FIGURE 10.3

Distributions of $Y_{i,j}$ When $\mu_1 = \mu_2 = \ldots = \mu_t = \mu$

for each $i = 1, 2, \ldots, t$. However, when all the means μ_i are equal, the random variables $\bar{Y}_{i\cdot}$ all have the very same probability distribution. Since these random variables are independent, it follows that $\bar{Y}_{1\cdot}, \bar{Y}_{2\cdot}, \ldots, \bar{Y}_{t\cdot}$ are a random sample from a $N(\mu, \sigma^2/n)$ distribution. (Here we let μ denote the common mean.) With this fact established, we now can say that the sample variance of the $\bar{Y}_{i\cdot}$'s can be used as an estimate of the population variance, σ^2/n. Consequently, n times the sample variance will give an estimate of σ^2; that is,

$$nS_{\bar{Y}}^2 = \frac{n\sum_{i=1}^{t}[\bar{Y}_{i\cdot} - \bar{Y}_{\cdot\cdot}]^2}{t-1} = \frac{n\sum_{i=1}^{t}\left[\bar{Y}_{i\cdot} - \left(\sum_{i=1}^{t}\bar{Y}_{i\cdot}\right)\Big/t\right]^2}{t-1}$$

is an estimate of σ^2 based on t observations. In the case of an equal number of observations per treatment, we may view $\bar{Y}_{\cdot\cdot}$ (the sample mean of all observations) as the sample mean of the $\bar{Y}_{i\cdot}$'s (i.e., $\sum_i \bar{Y}_{i\cdot}/t$).

Now notice that we have found two different estimators for the parameter σ^2. One estimator, S_p^2, is an unbiased estimator for σ^2 *no matter what the values of the* μ_i *happen to be*. The second estimator, $nS_{\bar{Y}}^2$, is an unbiased estimator for σ^2 *provided that the means* μ_i *are all equal* [*i.e., if the null hypothesis* (10.4) *is true*]. Since we have assumed that the random variables $Y_{i,j}$ are normally distributed, it can be shown (by a theorem similar to Theorem 7.6 of Section 7.9) that the quantity

$$\frac{nS_{\bar{Y}}^2}{S_p^2} \tag{10.9}$$

will have an F distribution with $(t - 1)$ and $t(n - 1)$ degrees of freedom, when the means μ_i are all equal.

The next important question to ask is what happens if the null hypothesis in (10.4) is *not true* (i.e., when the means μ_i are not equal)? It can be shown that in this case the random variable $nS_{\bar{Y}}^2$ is not an unbiased estimator for σ^2. In fact,

$$E(nS_{\bar{Y}}^2) = \sigma^2 + \frac{n\sum_{i=1}^{t}(\mu_i - \mu)^2}{t-1} = \sigma^2 + \frac{n\sum_{i=1}^{t}\tau_i^2}{t-1}.$$

where μ and τ_i are as defined in (10.1) and (10.2). Hence, if H_0 is not true, then $nS_{\bar{Y}}^2$ will be a biased estimator for σ^2 with the bias being positive. Now (at long last) we see a candidate for a test statistic for testing the null hypothesis in (10.4), namely, $nS_{\bar{Y}}^2/S_p^2$. We can summarize our discussion as follows:

> If Assumptions 1, 2, and 3 given earlier in this section hold, the common variance σ^2 can always be estimated unbiasedly by S_p^2. If the null hypothesis given in (10.4) is true, then $nS_{\bar{Y}}^2$ will also be unbiased for σ^2 and the ratio
>
> $$\frac{nS_{\bar{Y}}^2}{S_p^2} \sim F_{t-1,\,t(n-1)}.$$
>
> If the null hypothesis is not true, the numerator will be biased, so the ratio will tend to be larger than otherwise expected. Consequently, we can use the ratio (10.9) as a test statistic. If the ratio exceeds $f_{t-1,\,t(n-1),\,1-\alpha}$, then reject H_0. Otherwise, do not reject H_0.

Before illustrating this test procedure by an example, we will give an alternative computational formula for $nS_{\bar{Y}}^2$.

$$(10.10) \qquad nS_{\bar{Y}}^2 = \frac{1}{t-1}\left[\sum_{i=1}^{t} \frac{\left(\sum_{j=1}^{n} y_{i,j}\right)^2}{n} - \frac{(\sum_i \sum_j y_{i,j})^2}{nt}\right].$$

Example 10.2 Using the production figures from the Grindstone Works given in Table 10.1, test

$$H_0: \mu_1 = \mu_2 = \ldots = \mu_4 \qquad \text{vs.} \qquad H_a: \text{not all } \mu_i \text{ are equal}$$

using a .05 level of significance.

The appropriate test statistic here is given in (10.9). We found s_p^2 in Example 10.1 to be 6.375, so we need here to find $ns_{\bar{Y}}^2$. Using (10.10), we get

$$ns_{\bar{Y}}^2 = \frac{1}{4-1}\left[\frac{723{,}750}{5} - \frac{(1700)^2}{5(4)}\right] = 83.333.$$

Consequently, the ratio is

$$\frac{83.333}{6.375} = 13.07.$$

We must compare this observed value of the test statistic with

$$f_{4-1,\,4(5-1),\,.95} = 3.24.$$

Since the test statistic exceeds this critical value, H_0 is rejected and we conclude that not all the means are equal.

In the context of this problem, Dewey the efficiency expert, can claim that there is, in fact, a significant difference between mean production based on different methods of giving coffee breaks. A further analysis is required for him to ascertain which methods have mean productions different from the others. We will address this question in Section 10.4.

We will close this section with two final comments. First, if only two treatments are being considered (i.e., if $t = 2$), the test based on the test statistic (10.9) is equivalent to the t test given by (8.18). This follows since algebraically the test statistic (10.9) is the square of the test statistic given by (8.18) (when $\mu_1 = \mu_2$) and because the square of a T random variable having ν degrees of freedom is the same as an F random variable having 1 and ν degrees of freedom. Second, if the sample sizes $n_1, n_2, \ldots, n_t$ are unequal, it is still possible to obtain a suitable test statistic for testing the null hypothesis of equal means μ_i. Specifically, use

$$\frac{\sum_{i=1}^{t} n_i(\bar{Y}_{i\cdot} - \bar{Y}_{\cdot\cdot})^2/(t-1)}{\sum_{i=1}^{t}\sum_{j=1}^{n_i}(Y_{i,j} - \bar{Y}_{i\cdot})^2 \Big/ \left[\left(\sum_{i=1}^{t} n_i\right) - t\right]},$$

where

$$\bar{Y}_{\cdot\cdot} = \sum_{i=1}^{t}\sum_{j=1}^{n_i} Y_{i,j} \Big/ \left(\sum_{i=1}^{t} n_i\right),$$

as a test statistic and reject H_0 only if the observed value exceeds $f_{\nu_1,\nu_2,1-\alpha}$, where $\nu_1 = t - 1$ and $\nu_2 = \left(\sum_{i=1}^{t} n_i\right) - t$.

10.3 ESTIMATION OF PARAMETERS AND PARTITIONING THE SUM OF SQUARES

In the last section we assumed that the response random variable $Y_{i,j}$, corresponding to the jth observation of the ith treatment, would have its mean expressed by

$$E(Y_{i,j}) = \mu_i.$$

If there are t treatments under consideration, there are t parameters, $\mu_1, \mu_2, \ldots, \mu_t$. which we may wish to estimate. If μ_i is expressed by

$$\mu_i = \mu + \tau_i$$

as in (10.2), we might want to estimate the parameters $\mu, \tau_1, \tau_2, \ldots, \tau_t$. At first this may appear to be a bit different than estimating the μ_i, since now we are considering $t + 1$ parameters. However, in light of (10.3), we know that $\sum \tau_i = 0$, so once τ_1, $\tau_2, \ldots$, and τ_{t-1} are specified, τ_t is uniquely determined (cf. Example 4.8 of Section 4.3). Consequently, in either case there are t parameters to estimate. By using the method of least squares (as discussed in Section 9.2), we can show that the point estimates are given by their sample counterparts. In particular, we would have

$$\hat{\mu}_i = \bar{Y}_{i\cdot}$$
$$\hat{\mu} = \bar{Y}_{\cdot\cdot}$$
$$\hat{\tau}_i = \bar{Y}_{i\cdot} - \bar{Y}_{\cdot\cdot}\,.$$

Furthermore, if we make Assumptions 1 and 2 as stated in the last section ($Y_{i,j}$ independent and having common variance), these estimators are BLUE (Best Linear Unbiased Estimates). In order to find confidence intervals or to test hypotheses about these parameters, we would make the additional assumption of normality of the $Y_{i,j}$. We will not pursue these ideas further here, however.

In our study of regression and correlation analysis we raised the question of why there appeared to be so much variability in the Y values. A similar question can be raised here. For example, if Carl, the owner of the Grindstone Works, were to see the production data given in Table 10.1 and if he were unaware of the fact that the employees worked under different coffee break methods, he might wonder why there was so much variability in the amounts produced. If he would graph these production values on a scale as in Figure 10.4(a), he could see the variability in these values. One way of measuring this variability is to calculate the sum of squared deviations around the overall mean. Denoting this mean by $\bar{Y}_{\cdot\cdot}$, the total sum of squares is

$$\text{SST} = \sum_{i=1}^{t}\sum_{j=1}^{n}(Y_{i,j} - \bar{Y}_{\cdot\cdot})^2 = \sum_{i=1}^{t}\sum_{j=1}^{n} Y_{i,j}^2 - nt(\bar{Y}_{\cdot\cdot})^2 = 352.$$

In an effort to explain why this sum of squares is as large as it is, Dewey might suggest that the production figures for each method of giving coffee breaks be graphed sepa-

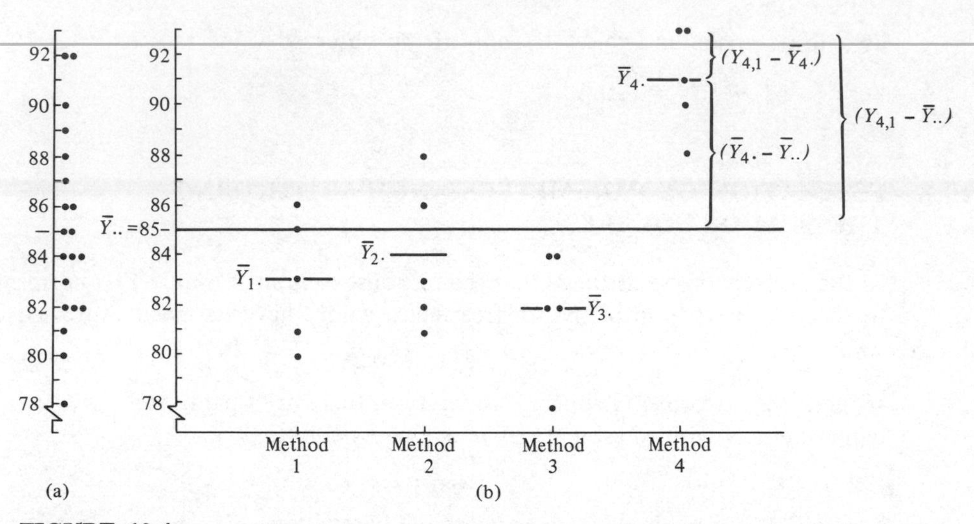

FIGURE 10.4
Graphs of Production Data from the Grindstone Works

rately. These graphs are given in Figure 10.4(b). (Note that there is no significance to the positions to which the methods are assigned on the horizontal axis, since the method names are simply labels. These values could be permuted with no essential change in the discussion to follow. This was not the case, however, in regression analysis.) A horizontal line is shown corresponding to the overall mean $\bar{Y}_{..}$ and short horizontal lines corresponding to the treatment means $\bar{Y}_{i\cdot}$ are also shown. You can see that algebraically

$$(Y_{i,j}-\bar{Y}_{..})=(Y_{i,j}-\bar{Y}_{i\cdot})+(\bar{Y}_{i\cdot}-\bar{Y}_{..}) \tag{10.11}$$

for all i and j. This relationship is shown in Figure 10.4(b) for the particular values $i=4$ and $j=1$. If we now consider the sum of squares SST using the relationship in (10.11), we find that

$$\begin{aligned}\text{SST} &= \sum_i\sum_j(Y_{i,j}-\bar{Y}_{..})^2=\sum_i\sum_j[(Y_{i,j}-\bar{Y}_{i\cdot})+(\bar{Y}_{i\cdot}-\bar{Y}_{..})]^2\\ &= \sum_i\sum_j(Y_{i,j}-\bar{Y}_{i\cdot})^2+2\sum_i\sum_j(Y_{i,j}-\bar{Y}_{i\cdot})(\bar{Y}_{i\cdot}-\bar{Y}_{..})+\sum_i\sum_j(\bar{Y}_{i\cdot}-\bar{Y}_{..})^2.\end{aligned}$$

It is not too hard to see that the middle term sums to zero (see Exercise 11), so the total sums of squares, SST, can be written as

$$\text{SST}=\sum_i\sum_j(Y_{i,j}-\bar{Y}_{i\cdot})^2+\sum_i\sum_j(\bar{Y}_{i\cdot}-\bar{Y}_{..})^2. \tag{10.12}$$

We define the first term on the right-hand side of (10.12) to be the "sum of squares due to error," SSE (sometimes referred to as sum of squares within treatments, since it measures the random variation of observations treated similarly). The second term we name to be the "sum of squares due to treatments," SSTR (since it measures the variability of the treatment means from the overall mean). That is,

$$\text{SST}=\text{SSE}+\text{SSTR},$$

where

$$\text{SSE} = \sum_i \sum_j (Y_{i,j} - \bar{Y}_{i\cdot})^2 \quad \text{and} \quad \text{SSTR} = \sum_i \sum_j (\bar{Y}_{i\cdot} - \bar{Y}_{\cdot\cdot})^2.$$

Some algebraically equivalent forms that can be used for computational purposes are

$$\text{(10.13)} \qquad \text{SSE} = \sum_i \Big[\sum_j Y_{i,j}^2 - n\bar{Y}_{i\cdot}^2\Big] = \sum_i \sum_j Y_{i,j}^2 - \Big(\sum_i (Y_{i\cdot})^2/n\Big)$$

and

$$\text{(10.14)} \qquad \text{SSTR} = n\sum_i (\bar{Y}_{i\cdot} - \bar{Y}_{\cdot\cdot})^2 = \frac{\sum_i (Y_{i\cdot})^2}{n} - \frac{(Y_{\cdot\cdot})^2}{nt}.$$

For the Grindstone data, we have

$$\text{SSE} = \sum_i \sum_j Y_{i,j}^2 - \Big(\sum_i (Y_{i\cdot})^2/n\Big) = 144{,}852 - (723{,}750/5) = 102,$$

and

$$\text{SSTR} = \frac{\sum_i (Y_{i\cdot})^2}{n} - \frac{(Y_{\cdot\cdot})^2}{nt} = \frac{723{,}750}{5} - \frac{2{,}890{,}000}{20} = 250,$$

so SST $= 102 + 250 = 352$.

We see two important facts here. First, the total variability, as measured by the total sum of squares, SST, can be expressed as the sum of two terms. One term, SSTR, gives the variability of the treatment means, $\bar{Y}_{i\cdot}$, around the overall mean $\bar{Y}_{\cdot\cdot}$ [see Figure 10.5(a)]. The other term, SSE, gives the variability of the individual values, $Y_{i,j}$, around the treatment means $\bar{Y}_{i\cdot}$. A visual representation of this fact for the Grindstone data is given in Figure 10.5(b). (To avoid the graph being too cluttered, we show only the data for method 2.) The second important fact is that the terms SSTR and SSE are essential ingredients in the test statistic (10.9) developed in the last section for testing the null hypothesis (10.4), namely, that all treatment means are equal. More precisely, the test statistic (10.9) could be expressed as

$$\text{(10.15)} \qquad \frac{\text{SSTR}/(t-1)}{\text{SSE}/t(n-1)}.$$

It is convenient to display the total sums of squares with the component parts in a a table referred to as an analysis of variance table or an ANOVA table. In such a table it is also conventional to show the degrees of freedom (i.e., the divisor), associated with each sum of squares. It is interesting to note that the "degrees of freedom" can be partitioned just as the total sum of squares can be partitioned. The ratio (10.15), the value of the test statistic for testing the null hypothesis (10.4), is also generally displayed in an ANOVA table. In Table 10.2 is shown the general form for an ANOVA table when one factor is being analyzed (a "one-way" ANOVA) and when the sample sizes for each treatment are equal. (A similar table could be given for the case where sample sizes are unequal.) Note that the expected mean square for treatments is found under the assumption that the treatment effects are not all zero. If they are all zero, then, as indicated in the last section, the expected value will simply be σ^2. In Table 10.3 we give the ANOVA table for the Grindstone data, summarizing the analysis given earlier.

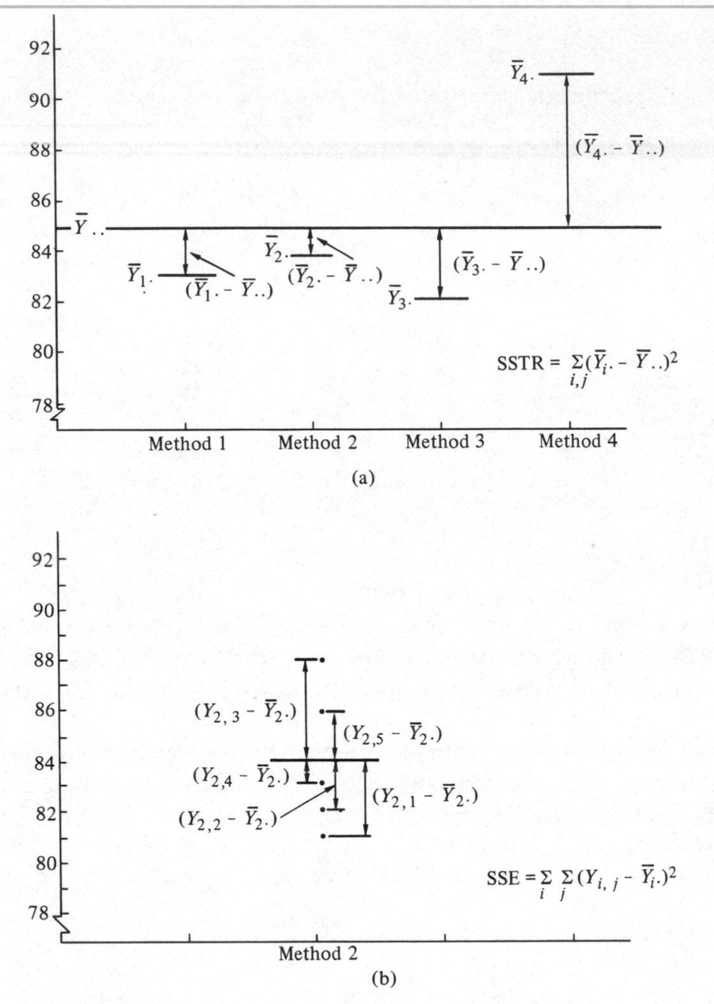

FIGURE 10.5

TABLE 10.2

General Form of a One-Way ANOVA Table

Source of Variation	*Degrees of Freedom (df)*	*Sums of Squares (SS)*	*Mean Square (MS)*	*Expected Mean Square [E(MS)]*	*F Ratio*
Among treatments	$t - 1$	SSTR	MSTR = SSTR/$(t - 1)$	$\sigma^2 + \frac{n \sum_i \tau_i^2}{t - 1}$	$\frac{\text{MSTR}}{\text{MSE}}$
Within treatments	$t(n - 1)$	SSE	MSE = SSE/$(t(n - 1))$	σ^2	
Total	$nt - 1$	SST			

TABLE 10.3

ANOVA Table for Grindstone Data

Source of Variation	*df*	*SS*	*MS*	*E(MS)*	*F Ratio*
Among treatments	3	250	83.333	$\sigma^2 + \frac{5}{3}\sum \tau_i^2$	13.07
Within treatments	16	102	6.375	σ^2	
Total	19	352			

Example 10.3 A class of 20 third-grade children was randomly divided into five groups of four students each for the purpose of studying the effectiveness of five different methods of teaching mathematics. The methods were (1) individualized instruction, (2) TV instruction, (3) traditional lecture, (4) team teaching, and (5) tape-recorded instruction. After 6 weeks, each child was given a standardized examination. The scores attained and some relevant sums are given in Table 10.4. The

TABLE 10.4

Data for Methods of Math Teaching

	Method 1	*Method 2*	*Method 3*	*Method 4*	*Method 5*
Test scores	93	73	75	89	59
	97	77	84	81	64
	92	67	80	76	55
	85	76	70	75	67
Mean	91.75	73.25	77.25	80.25	61.25
Sum	367	293	309	321	245
Sum of squares	33,747	21,523	23,981	25,883	15,091

math supervisor for the school, Adam Quick, would like to know if there is a difference in the mean scores attained by students receiving different methods of instruction. Test the null hypothesis of no difference in the mean scores attained under each method. Use $\alpha = .10$.

The quantities necessary to perform this test are summarized in the ANOVA table given in Table 10.5. The requisite sums of squares are found using equations (10.13) and (10.14), namely,

$$\text{SSE} = 120{,}225 - \frac{479{,}085}{4} = 453.75$$

$$\text{SSTR} = \frac{479{,}085}{4} - \frac{2{,}356{,}225}{20} = 1960.$$

TABLE 10.5

ANOVA Table for Methods of Math Teaching

Source of Variation	df	SS	MS	E(MS)	F Ratio
Among treatments	4	1960	490	$\sigma^2 + \frac{4}{4}\sum \tau_i^2$	16.20
Within treatments	15	453.75	30.25	σ^2	
Total	19	2413.75			

The F ratio of 16.20 must be compared to the appropriate F percentile. Since $\alpha = .10$, we find the $1 - \alpha = .90$ percentage point of the $F_{4,15}$ distribution. This value is found from Table A8 to be 2.36. The observed F ratio exceeds this value, so the null hypothesis of equal treatment means is rejected.

10.4 MULTIPLE COMPARISONS

We have seen how a test of the hypotheses

$$H_0 : \mu_1 = \mu_2 = \ldots = \mu_t \quad \text{vs.} \quad H_a : \text{not all } \mu_i \text{ are the same}$$

can be performed by using the ratio (10.15) as a test statistic. If these hypotheses are tested and if H_0 cannot be rejected at a particular level of significance, we say that based on the observed data there does not appear to be a significant difference between the treatment means. In most instances no further analysis is called for. On the other hand, if H_0 is rejected, there is evidence that the treatment means are not all the same. Generally, an investigator is not satisfied in knowing that differences exist, but wishes to know specifically *which* means differ from others. In the Grindstone Works example, for instance, Carl would not be satisfied in knowing only that different methods of assigning coffee breaks lead to different mean production. He would like to know which method (or methods) lead to the highest production. Consequently, once there is evidence that means may in fact be different, it may be desirable to simultaneously test

$$\begin{array}{lll} H_0 : \mu_1 - \mu_2 = 0 & \text{vs.} & H_a : \mu_1 - \mu_2 \neq 0 \\ H_0 : \mu_1 - \mu_3 = 0 & \text{vs.} & H_a : \mu_1 - \mu_3 \neq 0 \\ \vdots & & \vdots \\ H_0 : \mu_3 - \mu_4 = 0 & \text{vs.} & H_a : \mu_3 - \mu_4 \neq 0, \end{array} \tag{10.16}$$

that is, to compare all treatment means pairwise. There are other comparisons that might also be made. For example, Carl might want to compare the average production under coffee break methods 1 and 2 (times chosen by the employer) with the average production under coffee break methods 3 and 4 (times chosen by the individual worker). Here he could consider

$$H_0 : \frac{\mu_1 + \mu_2}{2} = \frac{\mu_3 + \mu_4}{2} \quad \text{vs.} \quad H_a : \frac{\mu_1 + \mu_2}{2} \neq \frac{\mu_3 + \mu_4}{2}.$$

An equivalent way of stating H_0 is

$$H_0 : \tfrac{1}{2}\mu_1 + \tfrac{1}{2}\mu_2 - \tfrac{1}{2}\mu_3 - \tfrac{1}{2}\mu_4 = 0.$$

Referring to the example on methods of math teaching, Adam might be interested in comparing all pairs of treatment means as in (10.16), but he also might like to compare average scores under methods involving personal contact (methods 1, 3, and 4) with the impersonal methods using TV and tape-recorded instruction (methods 2 and 5). Consequently, he might test

$$H_0 : \tfrac{1}{3}\mu_1 + \tfrac{1}{3}\mu_3 + \tfrac{1}{3}\mu_4 - \tfrac{1}{2}\mu_2 - \tfrac{1}{2}\mu_5 = 0$$

vs.

$$H_a : \tfrac{1}{3}\mu_1 + \tfrac{1}{3}\mu_3 + \tfrac{1}{3}\mu_4 - \tfrac{1}{2}\mu_2 - \tfrac{1}{2}\mu_5 \neq 0.$$

Whenever several (i.e., multiple) tests or comparisons of treatment means are made simultaneously, care must be exercised in finding a significance level, choosing the hypotheses to be tested, and interpreting the results. In choosing the hypotheses to be tested, for example, it is generally better to choose them before the data are analyzed rather than to have them be suggested by the data. Our intention in this section is to introduce you to some ideas about multiple comparisons but not to go into too much detail. For more detail we again refer you to a text on analysis of variance.

There are several methods that can be used for multiple comparisons. Some of these are Bonferroni, Duncan's multiple range test, Tukey's method for multiple comparisons, and Scheffe's method for multiple comparisons. Each of these has advantages and disadvantages under different circumstances. Since the Bonferroni method is perhaps the easiest method to apply and since it uses techniques previously developed, we will briefly discuss the method as it would be applied to comparing several pairs of means. The results that the Bonferroni method gives are conservative in the sense that if c comparisons are made, if an "overall" α level is chosen, and if all null hypotheses are actually true, the probability is *at least* $1 - \alpha$ that none of the null hypotheses will be wrongly rejected.

The Bonferroni method of multiple comparisons is basically repeated applications of the test for the difference of two means given in (8.18). In that case we assumed that the two population variances were unknown but equal and used S_p^2 [defined in (7.20)] as an estimate of the variance. In the Bonferroni method of multiple comparisons, we proceed in a similar way with two critical differences. First, the common variance for all treatment means is estimated by the mean-squared error defined earlier by

$$\text{MSE} = \frac{\text{SSE}}{t(n-1)}.$$

[This is equivalent to S_p^2 as defined in (10.8), the pooled variance estimate based on t samples rather than two samples.] Second, the critical value will be given by

$$t_{t(n-1),\,1-\alpha^*/2}.$$

Here α^* is defined to be α, the overall level of significance, divided by c, the number of comparisons to be made. For example, in testing

$$H_0 : \mu_i - \mu_j = 0 \qquad \text{vs.} \qquad H_a : \mu_i - \mu_j \neq 0$$

for c different pairs of means, H_0 will be rejected if

$$\frac{|\bar{Y}_{i\cdot} - \bar{Y}_{j\cdot}|}{\sqrt{(\text{MSE})(1/n + 1/n)}} \geq t_{t(n-1),1-\alpha/2c},$$

or, equivalently, if

$$|\bar{Y}_{i\cdot} - \bar{Y}_{j\cdot}| \geq t_{t(n-1),1-\alpha/2c}\sqrt{(\text{MSE})\left(\frac{1}{n} + \frac{1}{n}\right)}.$$

Consequently, to use the Bonferroni method of multiple comparisons, we proceed as follows. First, decide on an overall α level. Frequently, this α value will be the same as was used in the original F test of the null hypothesis that all treatment means are equal. Second, decide on the number of comparisons to be made. If there are t treatments and all distinct pairs of treatment means are to be compared, the number of comparisons will be $\binom{t}{2}$. Third, find the critical t value $t_{t(n-1),1-\alpha/2c}$ from the t tables. If the $1 - \alpha/2c$ percentage point is not given in the tables, it might be necessary to use interpolation. Fourth, use this t value to calculate

$$t_{t(n-1),1-\alpha/2c}\sqrt{\text{MSE}\left(\frac{1}{n} + \frac{1}{n}\right)}. \tag{10.17}$$

Fifth, for each pair of means being considered, reject $H_0 : \mu_i - \mu_j = 0$ whenever $|\bar{Y}_{i\cdot} - \bar{Y}_{j\cdot}|$ is at least as great as the value given in (10.17).

Example 10.4 In Example 10.3, the null hypothesis of no difference in the mean scores attained under different methods of math teaching was rejected. Use Bonferroni's method to find which treatment means may differ from the others.

Following the steps just outlined, we first choose α. We will use $\alpha = .10$, since this was the value used in the F test in Example 10.3. Second, since there are five treatments under consideration, we will make

$$c = \binom{5}{2} = 10$$

comparisons. The critical t value will be

$$t_{t(n-1),1-\alpha/2c} = t_{5(4-1),1-.10/2(10)} = t_{15,.995} = 2.947.$$

Next we calculate, from (10.17),

$$t_{t(n-1),1-\alpha/2c}\sqrt{\text{MSE}\left(\frac{1}{n} + \frac{1}{n}\right)} = 2.947\sqrt{30.25\left(\frac{1}{4} + \frac{1}{4}\right)} = 11.46.$$

Finally we compare the absolute difference $|\bar{Y}_{i\cdot} - \bar{Y}_{j\cdot}|$ with the constant 11.46 for each i, j. The easiest way to do this is to write the sample means $\bar{Y}_{i\cdot}$ in increasing order and compare differences with 11.46. In this case we have

$$\bar{Y}_{5\cdot} = 61.25, \qquad \bar{Y}_{2\cdot} = 73.25, \qquad \bar{Y}_{3\cdot} = 77.25,$$
$$\bar{Y}_{4\cdot} = 80.25, \qquad \bar{Y}_{1\cdot} = 91.75.$$

Since the difference $|\bar{Y}_{5\cdot} - \bar{Y}_{2\cdot}| = 12.0 > 11.46$, we would say that μ_5 differs significantly from μ_2. Since the sample means are in increasing order, it should be immediately obvious that μ_5 is significantly different from μ_3, μ_4, and μ_1. Comparing $\bar{Y}_{2\cdot}$ with $\bar{Y}_{3\cdot}$, we see that the absolute difference does not exceed 11.46, so μ_2 and μ_3 are not considered significantly different. Likewise, μ_2 and μ_4 are not significantly different, but μ_2 and μ_1 are significantly different, since $|\bar{Y}_{2\cdot} - \bar{Y}_{1\cdot}| = 18.50 > 11.46$. Continuing with the other comparisons, we ultimately conclude that μ_2, μ_3, and μ_4 are not significantly different, while μ_5 is significantly smaller than the rest, while μ_1 is significantly larger than the rest. (Note that in the context of this problem, Adam, the school math supervisor, might conclude that for third-grade students in his school district the student test scores will be highest when individualized instruction is given and lowest when tape-recorded instruction is given. The other three methods of instruction lead to test scores that are the same, for all practical purposes.) It is customary to indicate sets of means that are not significantly different by drawing a line under the corresponding sample means when these sample means are ordered. Here we would write

$$61.25 \qquad \underline{73.25 \qquad 77.25 \qquad 80.25} \qquad 91.75.$$

Example 10.5 Using the data from the Grindstone Works (Tables 10.1 and 10.3), determine which, if any, of the methods of assigning coffee breaks leads to different mean values of productivity.

Based on the results of Example 10.2 (and the table given in Table 10.3), the null hypothesis that all treatment means are equal was rejected with $\alpha = .05$. Following the outline of steps for multiple comparisons, we will take the overall α level to be .05. Since there are four treatments, we can compare $\binom{4}{2} = 6$ different pairs of means; hence we take $c = 6$. The critical t value will be

$$t_{t(n-1),1-\alpha/2c} = t_{4(5-1),1-.05/2(6)} = t_{16,.9958} \approx 3.027.$$

(This value was found by interpolating between $t_{16,.9950} = 2.921$ and $t_{16,.9975} = 3.252$.) Next, we calculate from (10.17)

$$t_{t(n-1),1-\alpha/2c}\sqrt{\text{MSE}\left(\frac{1}{n}+\frac{1}{n}\right)} = 3.027\sqrt{(6.375)(\tfrac{1}{5}+\tfrac{1}{5})} = 4.83.$$

Now we compare the absolute differences $|\bar{Y}_{i\cdot} - \bar{Y}_{j\cdot}|$ with the constant 4.83. Arranging the sample means $\bar{Y}_{i\cdot}$ in increasing order, we have

$$\bar{Y}_{3\cdot} = 82, \qquad \bar{Y}_{1\cdot} = 83, \qquad \bar{Y}_{2\cdot} = 84, \qquad \bar{Y}_{4\cdot} = 91.$$

It is not hard to see that that $\bar{Y}_{4\cdot}$ differs from the other sample means by more than 4.83. The other differences do not exceed 4.83, so we conclude that μ_4 is significantly greater than the other three treatment means, while μ_1, μ_2, and μ_3 do not significantly differ from each other. We could indicate this by

$$\underline{82 \qquad 83 \qquad 84} \qquad 91.$$

Based on this analysis, Carl could be told that method 4, three 10-minute coffee breaks at times chosen by the individual workers, is the method of assigning coffee breaks that leads to highest productivity.

10.5 TWO-WAY CLASSIFICATION WITHOUT INTERACTION

In Definition 10.1 of Section 10.2 we defined a *factor* as an independent variable that may affect the distribution of a response variable. In many studies it is desirable to consider two or more different factors at one time. Such considerations can be more efficient than conducting separate studies for each different factor. Some examples of two-factor studies are the breaking strength of concrete as a function of the proportion of cement (factor A) and the curing time (factor B); the mileage obtained on tires as a function of tire type (factor A) and climate (factor B); the amount of impurity in a particular chemical as a function of filtration method (factor A) and type of technician (factor B); blood pressure as a function of age category (factor A) and smoking habits (factor B); and so on.

If there are a levels of factor A and b levels of factor B, and if a complete study is done where observations are made for each level of factor A with each level of factor B, a total of $a \cdot b$ different factor combinations will be considered. It is possible to think of each combination of factor levels as a "treatment" and to analyze the problem as a one-way analysis of variance problem. However, treating the factors separately is a refinement of the one-way ANOVA procedure that often allows us to identify which individual factors have a greater influence on the response. If you look back at the Grindstone production data and the methods of coffee breaks discussed at the beginning of Section 10.2, you will see that the four "treatments" can be considered as the possible combinations of two levels of "factor A," choice of time determined by employer or employee, and two levels of "factor B," frequency of breaks (two or three per shift). Earlier we analyzed the data as a one-way ANOVA problem. In this section we will analyze it as a two-way problem.

Since there are two factors to consider now, we will have to use an additional subscript in our notation. We will let $Y_{i,j,k}$ represent the kth sample observation corresponding to the ith level of factor A and the jth level of factor B. The expected value of $Y_{i,j,k}$ could be denoted by

$$E(Y_{i,j,k}) = \mu_{i,j} \qquad i = 1, 2, \ldots, a, \quad j = 1, 2, \ldots, b,$$

or we could specifically identify factor effects for factors A and B by writing

$$\mu_{i,j} = \mu + \alpha_i + \beta_j. \tag{10.18}$$

This expression is analogous to (10.2) for the one-way model. By analogy with (10.3) we also impose the restrictions that

$$\sum_{i=1}^{a} \alpha_i = 0 \qquad \text{and} \qquad \sum_{j=1}^{b} \beta_j = 0. \tag{10.19}$$

(We apologize to you for reusing the symbols "α" and "β" with meanings different from those used earlier. We trust that you will be able to distinguish α_i, denoting the effect of the ith level of factor A, from α, denoting the level of significance of statistical tests, by the context of the problem. This type of notation is rather firmly entrenched in the statistical literature and we feel that it is better in the long run to use established notation.)

If the true expected value of the random variables $Y_{i,j,k}$ is actually of the form (10.18), it is possible to obtain point estimates of the unknown parameters μ, α_i ($i = 1, 2, \ldots, a$), and β_j ($j = 1, 2, \ldots, b$) by the method of least squares. Using the "dot notation" introduced earlier, we find the estimates to be

$$\hat{\mu} = \bar{Y}_{...}$$
$$\hat{\alpha}_i = \bar{Y}_{i..} - \bar{Y}_{...}$$
$$\hat{\beta}_j = \bar{Y}_{.j.} - \bar{Y}_{....}$$

Generally, in problems of this type, interest is not in point estimation but in testing the null hypothesis of equal treatment means. That is, we would generally be interested in testing

$$H_0 : \mu_{1,1} = \mu_{1,2} = \ldots = \mu_{1,b} = \ldots = \mu_{a,b}$$

vs.

$$H_a : \text{not all treatment means are equal.}$$

In view of the restriction given in (10.19), the null hypothesis is equivalent to

$$(10.20) \qquad H_{0(1)} : \quad \alpha_1 = \alpha_2 = \ldots = \alpha_a = 0$$

and

$$H_{0(2)} : \quad \beta_1 = \beta_2 = \ldots = \beta_b = 0.$$

To find appropriate test statistics for testing the null hypotheses in (10.20), we could proceed as we did in the one-way ANOVA problem. We assume, as we did earlier, that for each i and j the random variables $Y_{i,j,k}$, $k = 1, 2, \ldots, n$, form a random sample of size n from a normal distribution with mean $\mu_{i,j}$ and a *common variance* σ^2. We then obtain a pooled estimate of this common variance which is unbiased whether or not H_0 is true. This estimate is

$$(10.21) \quad \text{MSE} = \frac{\text{SSE}}{abn - a - b + 1}$$
$$= \sum_{i,j,k} [Y_{i,j,k} - (\bar{Y}_{i..} + \bar{Y}_{.j.} - \bar{Y}_{...})]^2/(abn - a - b + 1)$$
$$= \frac{\{\sum_{i,j,k} Y_{i,j,k}^2 - [(\sum_i Y_{i..}^2)/bn] - [(\sum_j Y_{.j.}^2)/an] + [(Y_{...})^2/abn]\}}{abn - a - b + 1}.$$

We also obtain independent estimates of σ^2 which are unbiased only if H_0 is true, and which have a positive bias (and hence will tend to be larger than otherwise expected) if H_0 is not true. In particular, one estimate of σ^2 is related to factor A and the other to factor B. The sums of squares necessary for obtaining these estimates can be obtained by partitioning the total sum of squares as follows:

$$\begin{aligned}
\text{SST} &= \sum_{i,j,k} (Y_{i,j,k} - \bar{Y}_{...})^2 \\
&= \sum_{i,j,k} [(Y_{i,j,k} - \bar{Y}_{i..} - \bar{Y}_{.j.} + \bar{Y}_{...}) + (\bar{Y}_{i..} - \bar{Y}_{...}) + (\bar{Y}_{.j.} - \bar{Y}_{...})]^2 \\
&= \sum_{i,j,k} [Y_{i,j,k} - (\bar{Y}_{i..} + \bar{Y}_{.j.} - \bar{Y}_{...})]^2 + \sum_{i,j,k} (\bar{Y}_{i..} - \bar{Y}_{...})^2 + \sum_{i,j,k} (\bar{Y}_{.j.} - \bar{Y}_{...})^2 \\
&= \text{SSE} + \text{SSA} + \text{SSB}.
\end{aligned}$$

An easier computational formula for SSE is given in (10.21), while the following computational formulas for SST, SSA, and SSB can be used:

$$\text{SST} = \sum_{i,j,k} Y_{i,j,k}^2 - \frac{(Y_{...})^2}{abn}$$

$$\text{SSA} = \frac{\sum_i Y_{i..}^2}{bn} - \frac{(Y_{...})^2}{abn}$$

$$\text{SSB} = \frac{\sum_j Y_{.j.}^2}{an} - \frac{(Y_{...})^2}{abn}.$$

This information, along with the test statistics for testing the null hypotheses (10.20), can be displayed in an ANOVA table as in Table 10.6.

TABLE 10.6

General Form of ANOVA Table for Two-Way ANOVA Without Interaction

Source of Variation	*Degrees of Freedom*	*Sum of Squares*	*Mean Square*	*Expected Mean Squares*	*F Ratio*
Factor A	$a - 1$	SSA	MSA = SSA/$(a - 1)$	$\sigma^2 + \frac{nb \sum \alpha_i^2}{a - 1}$	MSA/MSE
Factor B	$b - 1$	SSB	MSB = SSB/$(b - 1)$	$\sigma^2 + \frac{na \sum \beta_j^2}{b - 1}$	MSB/MSE
Error	$abn - a - b + 1$	SSE	MSE = SSE/$(abn - a - b + 1)$	σ^2	
Total	$abn - 1$	SST			

Example 10.6 In the manufacture of microelectronic components and semiconductor devices, it is necessary to perform quality control inspection using a microscope. Mae Fynedit is the quality control supervisor for the Micro-Corp, a company that specializes in the manufacture of these products. She has set up an experiment to study the effect of illumination and of magnification on the time (in minutes) required for an inspector to find the flaws in a sample component. In this experiment, two levels of illumination (low and high) and four levels of magnification (5X, 8X, 14X, and 26X) were used. Using the data given in Table 10.7, find the two-way ANOVA

TABLE 10.7
Data for Example 10.6

Factor A (Illumination)	*Factor B (Magnification)* 5×	8×	14×	26×
Low	.49	.29	.39	.51
	.45	.31	.33	.45
	.50	.36	.30	.45
Mean	.48	.32	.34	.47
Sum	1.44	.96	1.02	1.41
Sum of squares	.6926	.3098	.3510	.6651
High	.32	.25	.24	.32
	.39	.22	.32	.39
	.44	.28	.28	.31
Mean	.35	.25	.28	.34
Sum	1.05	.75	.84	1.02
Sum of squares	.3701	.1893	.2384	.3506

table assuming no interaction, and test the null hypotheses (10.20) that all factor effects are zero, using a .025 level of significance.

Before attempting to construct the ANOVA table, it is necessary to be sure to understand the notation being used for the two-way analysis. In this particular problem, we have

$$a = \text{number of levels of factor A} = 2$$

$$b = \text{number of levels of factor B} = 4$$

$$n = \text{number of observations for each factor combination} = 3.$$

Also to be sure that you understand the subscripting method and the dot notation (not to be confused with the dot–dash notation of Morse Code), you should verify from Table 10.7 that

$$Y_{1,1,1} = .49, \quad Y_{1,2,2} = .31, \quad Y_{2,2,1} = .25, \quad Y_{2,4,2} = .39,$$

$$Y_{1,1\cdot} = 1.44, \quad Y_{1,4\cdot} = 1.41, \quad Y_{2,1\cdot} = 1.05, \quad Y_{2,3\cdot} = .84,$$

$$Y_{1\cdot\cdot} = 4.83, \quad Y_{2\cdot\cdot} = 3.66, \quad Y_{\cdot 1\cdot} = 2.49, \quad Y_{\cdot 3\cdot} = 1.86,$$

$$Y_{\cdots} = 8.49, \quad \sum_{i,j,k} Y_{i,j,k}^2 = .6926 + .3098 + \ldots + .3506 = 3.1669.$$

With this notation understood, you can use the computational formulas to verify that

$$\text{SST} = 3.1669 - \frac{(8.49)^2}{24} = .1636$$

$$\text{SSA} = \frac{36.7245}{12} - \frac{(8.49)^2}{24} = 0.570$$

$$\text{SSB} = \frac{18.4887}{6} - \frac{(8.49)^2}{24} = .0781.$$

The final sum of squares needed, SSE, can be found by

$$\text{SSE} = \text{SST} - \text{SSA} - \text{SSB} = .0285.$$

All of this information is summarized in Table 10.8. Using this information we can test the null hypothesis of no factor effects by comparing the F ratios with the appropriate F percentiles using $\alpha = .025$. For factors A and B, respectively, we would use

$$f_{1,19,.975} = 5.92 \quad \text{and} \quad f_{3,19,.975} = 3.90.$$

Since the F ratios both exceed the respective critical values, both factors, illumination and magnification, can be said to have a significant effect on the time required for an inspector to find flaws in a sample component.

TABLE 10.8

ANOVA Table for Example 10.6

Source of Variation	*df*	*SS*	*MS*	*E(MS)*	*F Ratio*
A	1	.0570	.0570	$\sigma^2 + 12\sum \alpha_i^2$	38.0
B	3	.0781	.0261	$\sigma^2 + 2\sum \beta_j^2$	17.4
Error	19	.0285	.0015	σ^2	
Total	23	.1636			

Two remarks are in order at this point. After Mae, the quality control supervisor, determines that each factor does have a significant effect on inspection time, she would no doubt want to follow the multiple comparison procedures discussed in Section 10.4 to determine which factor B levels differ from each other. Second, since two F tests were done here, the overall level of significance is larger than the level of significance for each test separately. Consequently, if two tests are performed using a .025 level of significance, for example, the overall level of significance will be greater than .025, but no more than .05 (cf. the discussion about the Bonferroni method for multiple comparisons in Section 10.4).

Example 10.7 Analyze the data from the Grindstone Works by using a two-way analysis of variance procedure with factor A representing the choice of breaks (by employer or worker) and factor B representing the frequency of breaks (either two or three times per shift). Test the null hypotheses given in (10.20) to see if any factor effects differ significantly from zero. Use a level of significance of .025 for each test.

In this problem there are two levels of factor A and two levels of factor B with five observations for each factor combination, so

$$a = 2, \quad b = 2, \quad \text{and} \quad n = 5.$$

The data in Table 10.1 can be used to find the two-way ANOVA table, but it might be helpful for you to see these same data presented in a two-way format. This is done in Table 10.9. Using these data, we can find the appropriate ANOVA table (Table 10.10). In order to test the null hypotheses of zero-factor effects, it is necessary to find the appropriate F percentiles. Since there are two tests to be performed, each

TABLE 10.9
Data From the Grindstone Works Presented in a Two-Way Format

Factor B (Frequency of Breaks)	*Factor A (Choice of Breaks)*			
	Employer		*Worker*	
	Two	*Three*	*Two*	*Three*
	80	81	78	93
	86	82	82	90
	83	88	84	91
	85	83	82	93
	81	86	84	88
Mean	83	84	82	91
Sum	415	420	410	455
Sum of squares	34,471	35,314	33,644	41,423

TABLE 10.10
ANOVA Table for Example 10.7

Source of Variation	*df*	*SS*	*MS*	*E(MS)*	*F Ratio*
A	1	45	45	$\sigma^2 + 10 \sum \alpha_i^2$	4.20
B	1	125	125	$\sigma^2 + 10 \sum \beta_j^2$	11.67
Error	17	182	10.71	σ^2	
Total	19	352			

with significance level of .025, and since the degrees of freedom are the same, the critical value in each case will be $f_{1,17,.975} = 6.04$. Since the F ratio for factor A is less than this value, we cannot reject the null hypothesis that $\alpha_1 = \alpha_2 = 0$ [i.e., that the method of choosing breaks (employer or worker choice) has no effect]. On the other hand, the F ratio for factor B does exceed this critical value, so it appears as if the frequency of breaks does have an effect on productivity.

We will close this section with a word of warning. The analysis described in this section is appropriate *only when the mean for a given treatment combination can be described by* (10.18), that is, when

$$\mu_{i,j} = \mu + \alpha_i + \beta_j.$$

In such a case, the effect due to each factor is said to be "additive." If there is doubt as to whether the mean can be described in this relatively simple way, it would be better to consider the methods described in the next section, where an "interaction" term is included. We will see that the Grindstone data, analyzed in this section for illustrative purposes, are actually better analyzed by using a two-way analysis of variance with interaction.

10.6 TWO-WAY CLASSIFICATION WITH INTERACTION

In the last section we assumed that when two factors A and B are considered, the mean response for an experimental unit receiving a treatment combination of the *i*th level of factor A and the *j*th level of factor B would be

$$E(Y_{i,j,k}) = \mu + \alpha_i + \beta_j.$$

When we use this model, we implicitly assume that each factor contributes to the overall mean in the same way, no matter what level the other factor is at. That is, we assume that the contribution to the overall mean by a given level of factor A is in no way influenced by the given level of factor B. If this is true, we say there is no "interaction" and the effects are said to be additive. However, in many cases factors do interact. You may be familiar with this effect from your study of chemistry, for example. The reaction when chemical A is added to a mixture may be determined by or affected by the amount of chemical B added. As another example, we might consider the effect on the harmony of a community picnic of the level of factor A (absence or presence of the Hatfield family) and the level of factor B (absence or presence of the McCoy family). If we assumed that the traditional feud was in effect, you could believe that there could be an interaction effect on the harmony at the picnic! If the feud were not in effect, it could be possible for each family's presence to increase the level of harmony with no interaction. These two cases are illustrated graphically in Figure 10.6(a) and (b). In these graphs, level 1 for each factor indicates the family's absence while factor 2 indicates their presence. The possibility also exists that if the two families have become good friends, the overall level of harmony when both families are present together would be higher than expected if there were no interaction. This case is illustrated in Figure 10.6(c). If there is an interaction effect, it can be either positive or negative. These graphs illustrate a point that holds in general. If there is no interaction, the lines connecting the mean response for levels of factor B, say, for each level of factor A will be parallel. If there is interaction, the lines will not be parallel [see Figure 10.6(a) and (c)].

If an interaction effect is present, the mean response for an experimental unit receiving a treatment combination of the *i*th level of factor A and the *j*th level of factor B would be

$$E(Y_{i,j,k}) = \mu + \alpha_i + \beta_j + (\alpha\beta)_{i,j}, \tag{10.22}$$

where $(\alpha\beta)_{i,j}$ is a term that is the effect of the interaction between the *i*th level of factor A and the *j*th level of factor B. The parameters μ, α_i, β_j, and $(\alpha\beta)_{i,j}$ can be estimated by the method of least squares. The point estimates are given by

$$\hat{\mu} = \bar{Y}_{...}$$

$$\hat{\alpha}_i = \bar{Y}_{i..} - \bar{Y}_{...}$$

$$\hat{\beta}_j = \bar{Y}_{.j.} - \bar{Y}_{...}$$

$$\widehat{(\alpha\beta)}_{i,j} = \bar{Y}_{i,j.} - \bar{Y}_{i..} - \bar{Y}_{.j.} + \bar{Y}_{...}.$$

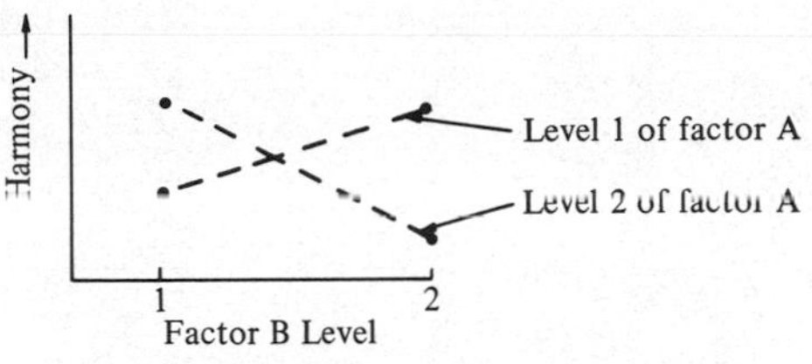

(a) Feud in effect (negative interaction)

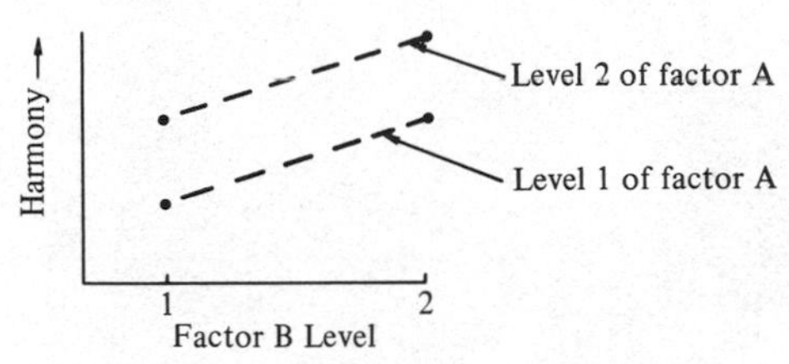

(b) Feud not in effect (no interaction)

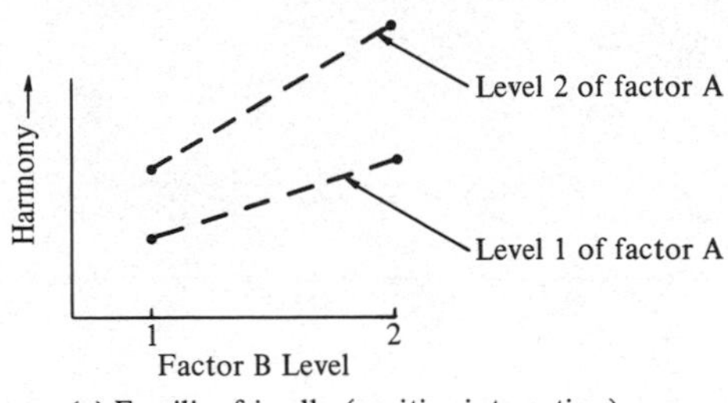

(c) Families friendly (positive interaction)

FIGURE 10.6
Illustration of the Effect of Interaction

The hypotheses of interest in this situation would be

(10.23)
$$H_0: \text{ all treatment means are equal}$$
vs.
$$H_a: \text{ not all treatment means are equal.}$$

If we impose restrictions analogous to (10.3) and (10.19), namely

$$\sum_{i=1}^{a} \alpha_i = 0, \quad \sum_{j=1}^{b} \beta_j = 0, \quad \sum_{i=1}^{a} (\alpha\beta)_{i,j} = 0, \quad \text{and} \quad \sum_{j=1}^{b} (\alpha\beta)_{i,j} = 0,$$

then the null hypothesis (10.23) will be true if and only if each of the following hypotheses is true:

(10.24)
$$H_{0(1)}: \quad \alpha_1 = \alpha_2 = \ldots = \alpha_a = 0$$
$$H_{0(2)}: \quad \beta_1 = \beta_2 = \ldots = \beta_b = 0$$
$$H_{0(3)}: \quad (\alpha\beta)_{1,1} = (\alpha\beta)_{1,2} = \ldots = (\alpha\beta)_{a,b} = 0.$$

As we did in Section 10.5, we can test these hypotheses by using appropriate F ratios, provided that Assumptions 1 to 3 of Section 10.2 are satisfied. In this case the sums of squares necessary for finding these ratios can be found by using the following formulas:

$$\text{SST} = \sum_{i,j,k} (Y_{i,j,k} - \bar{Y}_{...})^2$$
$$= \sum_{i,j,k} (Y_{i,j,k})^2 - \frac{(Y_{...})^2}{nab}$$
$$\text{SSE} = \sum_{i,j,k} (Y_{i,j,k} - \bar{Y}_{i,j\cdot})^2$$
$$= \sum_{i,j,k} (Y_{i,j,k})^2 - \frac{\sum_{i,j} (Y_{i,j\cdot})^2}{n}$$
$$\text{SSA} = nb \sum_i (\bar{Y}_{i..} - \bar{Y}_{...})^2 = \frac{\sum_i (Y_{i..})^2}{nb} - \frac{(Y_{...})^2}{nab}$$
$$\text{SSB} = na \sum_j (\bar{Y}_{\cdot j\cdot} - \bar{Y}_{...})^2 = \frac{\sum_j (Y_{\cdot j\cdot})^2}{na} - \frac{(Y_{...})^2}{nab}$$
$$\text{SSAB} = \text{SST} - \text{SSE} - \text{SSA} - \text{SSB}.$$

This information, along with the test statistics for testing the null hypotheses (10.24), can be displayed in an ANOVA table as in Table 10.11.

TABLE 10.11

General Form of ANOVA Table for Two-Way ANOVA with Interaction

Source of Variation	*Degrees of Freedom*	*Sum of Squares*	*Mean Square*	*Expected Mean Square*	*F Ratio*
Factor A	$a - 1$	SSA	$\text{MSA} = \frac{\text{SSA}}{a-1}$	$\sigma^2 + \frac{nb \sum \alpha_i^2}{a-1}$	$\frac{\text{MSA}}{\text{MSE}}$
Factor B	$b - 1$	SSB	$\text{MSB} = \frac{\text{SSB}}{b-1}$	$\sigma^2 + \frac{na \sum \beta_j^2}{b-1}$	$\frac{\text{MSB}}{\text{MSE}}$
Interaction	$(a - 1)(b - 1)$	SSAB	$\text{MSAB} = \frac{\text{SSAB}}{(a-1)(b-1)}$	$\sigma^2 + \frac{n \sum (\alpha\beta)_{i,j}^2}{(a-1)(b-1)}$	$\frac{\text{MSAB}}{\text{MSE}}$
Error	$ab(n - 1)$	SSE	$\text{MSE} = \frac{\text{SSE}}{ab(n-1)}$	σ^2	
Total	$nab - 1$	SST			

Example 10.8 Analyze the data from the Grindstone Works by using a two-way analysis of variance procedure. Assume that the treatment means are given by (10.22) (i.e., assume that there may be an interaction effect). Test the null hypotheses given in (10.24) to see if any factor effects or interaction effects differ significantly from zero. Use a level of significance of .025 for each test.

We first note, as we did in Example 10.7, that

$$a = 2, \qquad b = 2, \qquad \text{and} \qquad n = 5.$$

Using the data as displayed in Table 10.9, you can verify that

$$\text{SST} = 144{,}852 - \frac{(1700)^2}{20} = 352$$

$$\text{SSE} = 144{,}852 - \frac{723{,}750}{5} = 102$$

$$\text{SSA} = \frac{1{,}445{,}450}{10} - \frac{2{,}890{,}000}{20} = 45$$

$$\text{SSB} = \frac{1{,}446{,}250}{10} - \frac{2{,}890{,}000}{20} = 125$$

$$\text{SSAB} = 352 - 102 - 45 - 125 = 80.$$

Using these numbers we can complete the ANOVA table and use the resulting figures to test the hypotheses. From Table 10.12 we see that the F ratios must each be com-

TABLE 10.12

Two-Way ANOVA Table Including Interaction for the Grindstone Data

Source of Variation	*df*	*SS*	*MS*	*E(MS)*	*F Ratio*
A	1	45	45	$\sigma^2 + 10 \sum \alpha_i^2$	7.06
B	1	125	125	$\sigma^2 + 10 \sum \beta_j^2$	19.61
Interaction	1	80	80	$\sigma^2 + 5 \sum (\alpha\beta)_{i,j}^2$	12.55
Error	16	102	6.375	σ^2	
Total	19	352			

pared with the 97.5 percentile of an $F_{1,16}$ distribution, namely, $f_{1,16,.975} = 6.12$. Since each of the ratios exceeds this critical value, each of the null hypotheses in (10.24) would be rejected and we would conclude that each factor and the interaction effects are significantly different from zero.

Note that the decision to reject the null hypothesis of no interaction effect tells us that we probably made the right choice in analyzing these data using the interaction model (10.22) rather than the "no interaction" model (10.18). On the other hand, this would have been the right choice even if we could not reject the null hypothesis of no interaction! The reason for this is that to be conservative, we should use the interaction model unless there is rather strong evidence (based on past research, for example) that no interaction exists.

Often it is desirable to use graphical means to get an idea (albeit somewhat subjective) as to whether interaction may exist, and if it does exist, what form it

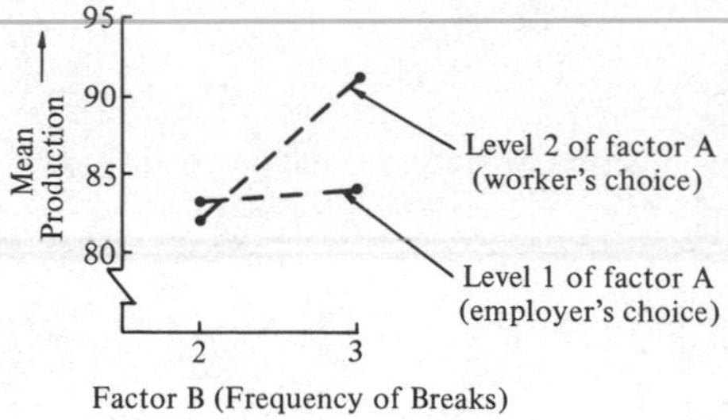

FIGURE 10.7

Observed Mean Production for Factor Level Combinations in Grindstone Data Example

might take (positive or negative). This can be done by drawing graphs similar to the ones in Figure 10.6. In Figure 10.7 we have such a graph. The points connected by line segments correspond to the observed mean production for different levels of factor A, while the levels of factor B are shown on the horizontal axis (although A and B could be interchanged with no essential change in the meaning). The fact that these lines are not close to parallel is not surprising in view of the fact that the F test showed the interaction effect to be significantly different from zero. In Figure 10.8

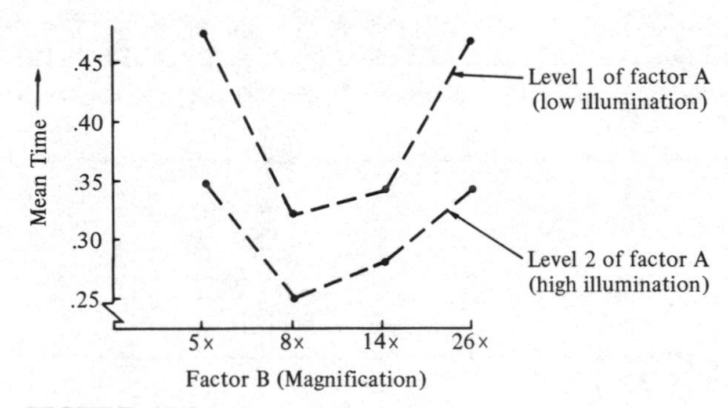

FIGURE 10.8

Observed Mean Time for Correct Identification for Factor Level Combinations in Micro-Corp Data Example

we have drawn a graph for the Micro-Corp data of Example 10.6. In this case the lines are not too far from being parallel, so we would not be too surprised if a test did not show the interaction effect to be significantly different from zero.

Example 10.9 Analyze the data from the Micro-Corp (Table 10.7) using a two-way model with interaction. Test the null hypotheses given in (10.24) to see if any factor effects or interaction effects differ significantly from zero. Use a .025 level of significance for each test.

By using the data from Table 10.7, we obtain the ANOVA table shown in Table 10.13. To test for the effect of factor A (level of illumination), we need to compare

TABLE 10.13
ANOVA Table for Example 10.9

Source of Variation	*df*	*SS*	*MS*	*E(MS)*	*F Ratio*
A	1	.057037	.057037	$\sigma^2 + 12\sum \alpha_i^2$	41.481
B	3	.078112	.026037	$\sigma^2 + 2\sum \beta_j^2$	18.936
Interaction	3	.006413	.002138	$\sigma^2 + \sum (\alpha\beta)_{i,j}^2$	1.555
Error	16	.022000	.001375	σ^2	
Total	23	.163562			

the F ratio with $f_{1,16,.975} = 6.12$. Since 41.481 exceeds this value, we conclude that the effect of the level of illumination is significantly different from zero. Similarly, comparing with $f_{3,16,.975} = 4.08$, we conclude that the effect of level of magnification is significant, but that there is not a significant interaction effect. (This last conclusion is consistent with the subjective analysis about interaction that we made based on Figure 10.8.)

10.7 RANDOMIZED COMPLETE BLOCK DESIGN

In Sections 7.6, 7.7, and 8.6 we discussed the problems of estimation and hypothesis testing for the difference of two means. In Section 7.6 we discussed a problem concerning the effect of two different oils, Aristocrat and Plebeian, on gas mileage. We would now like to consider a more general version of the same problem. Specifically, let us assume that we have four different types of oil (Aristocrat, Plebeian, Royal, and Super Slick) and we wish to determine which, if any, of these oils will lead to better gas mileage. Assume further that we have 20 cars from a company car pool in which to test these oils. How should we proceed?

If you review the discussion in Section 7.6, you will see that one procedure would be to randomly choose five cars to be tested with Aristocrat Oil, five with Plebeian Oil, and so on. When the different treatments are assigned to the experimental units (i.e., the units receiving the treatments) in a completely random way, the method of assignment is referred to as a *completely randomized design.* Such a design is illustrated in Figure 10.9. Can you think of any shortcomings with such a design? If the fleet of company cars consisted of cars of the same make, model, year, engine size, and so on, this design is satisfactory. However, as you can see from Figure 10.9, this need not be the case. In fact, the company car pool has four each of five different types of cars. If the treatments are assigned at random, it could happen, for example, that mostly compact cars receive Royal Oil while mostly station wagons receive Super Slick Oil. Since this situation would not be desirable we would like to have a method of assigning treatments that would avoid this type of problem. You can see that this could easily be done by assigning at random a treatment (a different

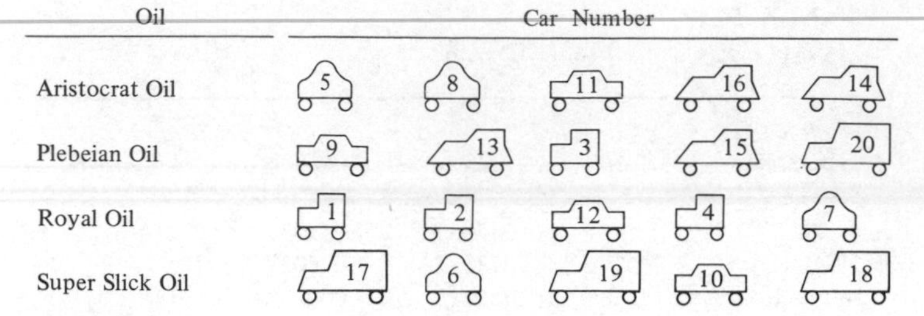

FIGURE 10.9

Treatments Assigned in a Completely Random Way

oil) to each of the compacts, each of the station wagons, and so on. Such an assignment is referred to as a *randomized complete block design* and is illustrated in Figure 10.10. The objective of this design is to group together experimental units which are as alike as possible into a group called a block. Then the treatments are assigned randomly to elements within the block. In this way the observed differences in responses (e.g., gas mileage) of units within a block will be due to differences in treatment effects (plus some random variation) rather than to inherent differences in the experimental units.

Oil	Car Number				
Aristocrat Oil	1	5	10	13	20
Plebeian Oil	3	7	11	16	17
Royal Oil	2	6	9	14	19
Super Slick Oil	4	8	12	15	18

FIGURE 10.10

Treatments Assigned by Blocks

Whenever the experimental units scheduled to receive treatments are heterogeneous, it is advantageous to group together those which are most similar into blocks. Such grouping can be done in medical experiments where patients having similar diseases, age, sex, and so on, are grouped together in a block to receive one of several treatments being studied. In other situations one might group siblings together, while in animal experiments littermates may be grouped together.

If treatments are assigned according to a randomized complete block design, the analysis of variance procedure used to test to see if there is a significant difference between treatments is the same as is used in a two-way ANOVA without interaction. (This analysis is appropriate under the assumption that there really is no interaction between blocks and treatments in such an analysis.) The expected value of the response of the experimental unit receiving the ith treatment in the jth block would be given by

$$E(Y_{i,j}) = \mu + \tau_i + \beta_j$$

where τ_i represents the effect of the ith treatment and β_j the effect of the jth block. Typically, we would like to test the hypotheses

$$(10.25) \quad H_0: \ \tau_1 = \tau_2 = \ldots = \tau_t = 0 \quad \text{vs.} \quad H_a: \ \text{not all } \tau_i \text{ are equal}$$

by using an F test. [We might also wish to see if any of the block effects differ significantly from zero, that is, to test

$$(10.26) \quad H_0: \ \beta_1 = \beta_2 = \ldots = \beta_b = 0 \quad \text{vs.} \quad H_a: \ \text{not all } \beta_j \text{ are equal.}$$

We would perform a test of (10.26) to see if the method of blocking actually was effective.] The general form for the ANOVA table used here, except for notational differences, is the same as that given in Table 10.6. For your convenience we present that table again with appropriate notational changes in Table 10.14. The table shown

TABLE 10.14

General Form of ANOVA Table for Randomized Complete Block Design

Source of Variation	*Degrees of Freedom*	*Sum of Squares*	*Mean Square*	*Expected Mean Square*	*F Ratio*
Treatments	$t-1$	SSTR	MSTR $= \dfrac{\text{SSTR}}{t-1}$	$\sigma^2 + \dfrac{b\sum\tau_i^2}{t-1}$	$\dfrac{\text{MSTR}}{\text{MSE}}$
Blocks	$b-1$	SSB	MSB $= \dfrac{\text{SSB}}{b-1}$	$\sigma^2 + \dfrac{t\sum\beta_j^2}{b-1}$	$\dfrac{\text{MSB}}{\text{MSE}}$
Error	$tb-t-b+1$	SSE	MSE $= \dfrac{\text{SSE}}{tb-t-b+1}$	σ^2	
Total	$tb-1$	SST			

is for the situation where each treatment is assigned once to experimental units within a block. The necessary sums of squares can be calculated using the following computational formulas:

$$\text{SSTR} = \frac{\sum_i (Y_{i\cdot})^2}{b} - \frac{(Y_{\cdot\cdot})^2}{tb}$$

$$\text{SSB} = \frac{\sum_j (Y_{\cdot j})^2}{t} - \frac{(Y_{\cdot\cdot})^2}{tb}$$

$$\text{SST} = \sum_{i,j} (Y_{i,j})^2 - \frac{(Y_{\cdot\cdot})^2}{tb}$$

$$\text{SSE} = \text{SST} - \text{SSTR} - \text{SSB}.$$

If each treatment is assigned to n experimental units within a block rather than just one, some minor modifications must be made in these formulas and in the ANOVA table (cf. Table 10.6).

Example 10.10 The data in Table 10.15 represent the gas mileages obtained by company cars using four different types of oil. Construct an ANOVA table and perform the appropriate F test to see if there is a significant difference in gas mileage obtained by cars using the different oils. Use $\alpha = .05$.

Using the given data and the computational formulas, we find that

$$\text{SSTR} = 17{,}802 - 17{,}760.8 = 41.2$$
$$\text{SSB} = 19{,}788 - 17.760.8 = 2027.2$$
$$\text{SST} = 19{,}852 - 17{,}760.8 = 2091.2$$
$$\text{SSE} = 2091.2 - 41.2 - 2027.2 = 22.8.$$

TABLE 10.15
Data for Example 10.10

		Block (Type of Car)					
		Sub-compact	*Compact*	*Inter-mediate*	*Full Size*	*Station Wagon*	*Sum $Y_{i\cdot}$*
Type of Oil	*Aristocrat*	43	37	24	19	17	140
	Plebeian	44	39	26	19	21	149
	Royal	46	43	30	22	19	160
	Super-Slick	43	37	28	20	19	147
	Sum $Y_{\cdot j}$	176	156	108	80	76	
		$Y_{\cdot\cdot} = 596$	$\sum Y_{i,j}^2 = 19{,}852$				

Using these sums of squares, we can construct the ANOVA table given in Table 10.16. We test the null hypothesis of no treatment effect (10.25) by comparing the F

TABLE 10.16
ANOVA Table for Example 10.10

Source	*df*	*SS*	*MS*	*E(MS)*	*F Ratio*
Treatments	3	41.2	13.7333	$\sigma^2 + \frac{5\sum \tau_i^2}{3}$	7.2281
Blocks	4	2027.2	506.8000	$\sigma^2 + \sum \beta_j^2$	266.7368
Error	12	22.8	1.9	σ^2	
Total	19	2091.2			

ratio with $f_{3,12,.95} = 3.49$. Since 7.2281 exceeds this critical value, H_0 would be rejected. We would conclude that there is a significant difference in the gas mileages obtained under the different treatments (i.e., using these different types of oil).

We will end this section with a final example. This example illustrates the usefulness of the ANOVA procedure in research and development.

⋆***Example 10.11*** The data in Table 10.17 were obtained from an experiment carried out by a large animal feed manufacturing company. The purpose of the experiment was to evaluate three experimental pig starter rations fed to early weaned pigs. In order to control for any possible genetic differences among the pigs, three littermates were randomly chosen from among the pigs in a litter for each of nine litters. In this case the sets of three littermates form blocks and the treatments, different experimental starter rations, were randomly assigned to these three little pigs. (Note that this example describes a real experiment and is really not a fairy tale!) The measured

TABLE 10.17
Data for Example 10.11

		Block (litter)									Sum
		1	2	3	4	5	6	7	8	9	$Y_{i.}$
Treatment	Mix A	33.5	35.0	44.0	34.0	40.5	40.0	36.0	44.0	35.5	342.5
	Mix B	26.5	32.0	32.0	29.0	34.0	37.0	40.0	38.0	30.0	298.5
	Mix C	31.0	28.0	38.5	31.5	37.0	35.0	38.5	46.5	33.0	319.0
	Sum $Y_{.j}$	91.0	95.0	114.5	94.5	111.5	112.0	114.5	128.5	98.5	

$Y_{..} = 960 \qquad \sum Y_{i,j}^2 = 34{,}779$

response was weight gained after a certain period of time. Test to see if there is a significant difference among the different starter rations. Use a .10 level of significance.

The ANOVA table for this problem is given in Table 10.18. To test the null

TABLE 10.18
ANOVA Table for Example 10.11

Source	df	SS	MS	E(MS)	F Ratio
Treatments	2	107.7222	53.8611	$\sigma^2 + \frac{9\sum \tau_i^2}{2}$	7.0574
Blocks	8	415.8333	51.9792	$\sigma^2 + \frac{3\sum \beta_j^2}{8}$	6.8108
Error	16	122.1111	7.6319	σ^2	
Total	26	645.6666			

hypothesis of no treatment effect (10.25), we need to compare the F ratio with $f_{2,16,.90} = 2.67$. Since the ratio exceeds this value, H_0 is rejected and it is concluded that there is evidence that the treatment means are indeed different. (At this point we generally would perform multiple comparisons to see which mixes led to means that differed from the others. We might also test (10.26) to see if the blocking were effective. You can see that the block effect is significantly different from zero at a .005 level since the F ratio exceeds $f_{8,16,.995} = 4.52$.)

EXERCISES

1. In Section 10.2 it was stated that a factor may be either quantitative or qualitative in nature. State whether the factors listed here are quantitative or qualitative. The response random variable of interest is the amount of sales of Munchi-Crunchies cereal in a week. Some factors that might affect the sales are:
(a) The color of the boxes (red, blue, or yellow).
(b) The size of the boxes (8, 12, 16, or 20 ounces).
(c) The inclusion or noninclusion of a prize.
(d) The height of the boxes (20, 25, or 30 centimeters).
(e) The location of the display [first (top), second, or third shelf].

2. Using equations (10.1) and (10.2) of Section 10.2, show that

$$\sum \tau_i = 0.$$

3. Independent random samples were taken from three normal distributions having different means but the same variance σ^2. The sample observations were:

Sample 1:	2.46	.23	2.14		
Sample 2:	4.56	2.84	5.08	4.52	
Sample 3:	3.59	1.41	3.85	2.26	1.54.

(a) Find the sample variances S_1^2, S_2^2, and S_3^2. (b) Find S_p^2, the pooled estimate of σ^2.

4. Independent random samples were taken from three normal distributions having a common variance σ^2. The sample observations were:

Sample 1:	3.36	1.08	2.36	0.24
Sample 2:	1.68	1.52	1.28	0.72
Sample 3:	0.04	3.40	1.44	1.56.

(a) Find the sample variances S_1^2, S_2^2, and S_3^2. (b) Find S_p^2, the pooled estimate of σ^2, by using (10.6) of Section 10.2. (c) Find S_p^2 by using (10.8).

5. Assuming that the normal distributions in Exercise 4 all have the same mean, find an estimate of σ^2 using (10.10).

★*6.* The price/earnings ratio (PE ratio) is found by dividing the price of one share of common stock in a specific company by that company's earnings (or profits) per one share of common stock. Thus the PE ratio reflects how much investors are willing to pay for $1's worth of current earnings. It is an indication of investors' expectations of the future and of the relative expensiveness of a share of common stock. A sample of eight oil companies, utilities, and banks, respectively, were selected and their PE ratings were determined. The data follow.

Oil companies:	10.50	4.29	2.80	6.38	2.94	4.34	3.92	3.63
Utilities:	8.10	6.82	5.35	5.18	4.18	5.36	5.50	5.43
Banks:	9.21	5.54	11.63	8.08	11.77	4.03	7.87	5.71.

If μ_1, μ_2, and μ_3 denote the mean values of the PE ratings for all oil companies, utilities, and banks, test

$$H_0: \mu_1 = \mu_2 = \mu_3 \quad \text{vs.} \quad H_a: \text{not all means are equal}$$

by using (10.9) as the test statistic. Use a .025 level of significance. Explicitly state your conclusion in the context of this problem.

★**7.** An economics instructor was interested in investigating differences that might exist between different sections of the same class taught by him at different times of the day (7:30 A.M., 11:30 A.M., and 2:30 P.M.). The measured response is the final grade in the class. A random sample of 9 was taken from each class.

7:30 A.M.:	80	78	63	65	65	63	60	70	73
11:30 A.M.:	80	70	78	68	72	55	72	72	72
2:30 P.M.:	58	63	58	72	60	65	50	68	63.

If μ_1, μ_2, and μ_3 denote the mean final grade in the 7:30, 11:30, and 2:30 classes, respectively, test the null hypothesis that all means are equal against the alternative hypothesis that they are not all equal by using (10.9) as the test statistic. Use a .05 level of significance. Explicitly state your conclusion in the context of this problem. (*To think about:* Ideally, for the purposes of this study, how should students be assigned to the different class times? Is such an assignment possible? What affect might this have on the interpretation of the results of this study?)

8. At the end of Section 10.2 we stated that if only two treatments are being considered, then the test of $H_0: \mu_1 = \mu_2$ vs. $H_a: \mu_1 \neq \mu_2$ based on the test statistic (10.9) is equivalent to the t test based on (8.18). With this in mind, verify that

$$(t_{2(n-1),\,1-\alpha/2})^2 = f_{1,\,2(n-1),\,1-\alpha}$$

by finding the appropriate percentiles in Tables A6 and A8 when (a) $n = 3$, $\alpha = .05$; (b) $n = 4$, $\alpha = .10$. (c) Using the notation of this chapter, show that if $\mu_1 = \mu_2$ and $n_1 = n_2 = n$, the test statistic in (8.18) can be written as

$$T = \frac{\bar{Y}_{1\cdot} - \bar{Y}_{2\cdot}}{\sqrt{\dfrac{\sum_i \sum_j (Y_{i,j} - \bar{Y}_{i\cdot})^2}{2(n-1)}\left(\dfrac{1}{n} + \dfrac{1}{n}\right)}}.$$

Note that this test statistic will have a Student's t distribution with $2(n-1)$ degrees of freedom. (d) Show that the square of the test statistic, T^2, is algebraically equivalent to the test statistic in (10.9), which can be written as

$$\frac{n[(\bar{Y}_{1\cdot} - \bar{Y}_{\cdot\cdot})^2 + (\bar{Y}_{2\cdot} - \bar{Y}_{\cdot\cdot})^2]}{\sum_i \sum_j (Y_{i,j} - \bar{Y}_{i\cdot})^2/2(n-1)}.$$

Note that this test statistic will have an F distribution with 1 and $2(n-1)$ degrees of freedom.

9. (a) Using the data from Exercise 7, find least-squares estimates of μ, μ_i, and τ_i, $i = 1, 2, 3$. (b) Under what conditions will the least-squares estimators be BLUE?

(c) Find the residuals $(y_{i,j} - \bar{y}_{i\cdot})$ and make a residual plot. (d) Based on a visual examination of the residual plot, does the assumption of a common variance for each distribution appear reasonable?

★*10.* (a) Using the data from Exercise 6, find least-squares estimates of μ, μ_i, and τ_i, $i = 1, 2, 3$. (b) Under what conditions will the least-squares estimators be BLUE? (c) Find the residuals $(y_{i,j} - \bar{y}_{i\cdot})$ and make a residual plot. (d) Based on a visual examination of the residual plot, does the assumption of a common variance for each distribution appear reasonable?

11. Show that $\sum_i \sum_j (Y_{i,j} - \bar{Y}_{i\cdot})(\bar{Y}_{i\cdot} - \bar{Y}_{\cdot\cdot}) = 0$. [This fact allows the total sum of squares, SST, to be expressed as in (10.12).]

12. Construct a one-way analysis of variance table using the data from Exercise 6.

★*13.* Construct a one-way ANOVA table using the data given in Exercise 7 (cf. Table 10.2 of Section 10.3).

★*14.* A question of concern in recent years is how noise affects human beings in their working environments. A study was designed to compare the effect of three conditions—no noise; soft, "relaxing" music; and TV playing—on a person's ability to perform a mental task. The task consisted of rote memory of seven to eight nonsense syllables, all of the same length (i.e., 45 characters) per trial. Nonsense syllables were used to minimize any prior learning of phrases that would bias performance times, which were measured in minutes. Thirty subjects were randomly divided among the three groups. The performance times were:

No noise:	1.96	.94	1.62	1.09	2.26	1.82	1.61	1.71	1.61	1.46
Soft "relaxing" music:	1.98	1.15	1.08	1.05	1.97	1.32	2.14	3.17	1.63	1.73
TV:	2.06	1.61	1.46	1.62	2.86	2.51	2.66	2.60	2.23	1.98.

(a) Construct a one-way ANOVA table using these data. (b) State what assumptions must be satisfied in order for the test statistic to have an F distribution. (c) Explicitly state what conclusions can be drawn in testing for equality of treatment means using a .05 level of significance. (d) Find the residuals $(y_{i,j} - \bar{y}_{i\cdot})$ and make a residual plot. (e) Based on a visual examination of the residual plot, does the assumption of a common variance for each distribution appear reasonable?

★*15.* (a) Using a one-way analysis of variance procedure, analyze the data of Exercise 13 of Chapter 9 by considering the different levels of nitrogen as the different "treatments." (b) Analyze the data of Exercise 30 of Chapter 9 by considering the different levels of growth regulating compound as the different treatments.

★*16.* Based on the results of the one-way analysis of variance performed in Exercise 6 (PE ratios for different types of companies), is a multiple comparison follow-up appropriate? If so, compare all treatments pairwise using an overall level of significance of .06.

★*17.* Based on the results of the one-way analysis of variance performed in Exercise 7 (class performances at different times of the day), is a multiple comparison follow-up appropriate? If so, compare all treatments pairwise using an overall level of significance of .06.

★*18.* Based on the results of the one-way analysis of variance performed in Exercise 14 (the effect of noise on performance), is a multiple comparison follow-up appropriate? If so, compare all treatments pairwise using an overall level of significance of .06.

★*19.* In Table 9.8 of Section 9.7, data are given on the gain in physical work capacity for individuals exercising under different regimens. These data can be analyzed by using a two-way analysis of variance procedure assuming no interaction. Factor A can correspond to frequency of training (three levels) and factor B to intensity of training (two levels). (a) Construct a two-way ANOVA table for this problem (cf. Table 10.6 of Section 10.5.) (b) Test the null hypothesis (10.20); that is, test to see if the factor effects differ significantly from zero, using a .025 level of significance.

★*20.* During the past several years, farmers have invested record sums in new field machinery. One major question confronting agricultural economists is: How does the current book value of field machinery on a per acre basis vary with the number of tillable acres (factor A) and type of farm (factor B)? The farms were classified by number of tillable acres (of 100-acre size groups) and type of farm (mainly livestock, roughly evenly divided between livestock and crop, and mainly crop). The data in the table represent a portion of an actual experiment carried out in a Midwestern state. The entries in the 2 × 2 table are current book value (in dollars) of

	Type of Farm		
Number of Tillable Acres	*Livestock*	*Livestock–Crop*	*Crop*
200–299	63.56	79.06	80.90
	39.87	71.58	58.03
	93.98	61.28	80.27
300–399	55.92	57.69	40.44
	81.48	43.44	81.03
	64.47	41.32	83.27
400–499	66.05	45.06	20.13
	42.84	22.18	55.19
	25.86	38.99	48.28
500–599	28.60	15.83	37.69
	39.32	36.81	23.32
	7.36	35.68	30.51

field machinery on a per acre basis. (These data were gathered by county extension agents from farmers who were willing to cooperate.) Assume that responses have means given by

$$E(Y_{i,j,k}) = \mu + \alpha_i + \beta_j + (\alpha\beta)_{i,j}$$

so that a two-way analysis of variance with interaction is appropriate. (a) Construct an ANOVA table for this problem (cf. Table 10.11 of Section 10.6). (b) To get a visual picture of a possible interaction effect, construct graphs similar to those in Figures 10.7 and 10.8 of Section 10.6. Plot the mean response on the vertical axis and the levels of factor B on the horizontal axis. Connect the points correponding to each level of factor A. (c) Using a .025 level of significance, perform tests to see whether either of the factor effects or the interaction effect differs significantly from zero. [*To think about:* Was your visual impression obtained from part (b) consistent with the test result in part (c)? Is it likely that the data from this problem are a true random sample of farmers within the state? Why or why not?]

21. In the United States, hospitals are usually under one of three types of ownership and control—private nonprofit hospital, government hospital, and private for-profit hospital. Health economists are interested in determining whether type of ownership (factor A) is important in determining a hospital's occupancy rate (the average percentage of beds occupied over the course of a year). Another factor (factor B) which has been felt to possibly affect occupancy rates is the complexity of the facilities available. Hospitals were classified into three levels (low, medium, high) with respect to this factor. The data in the table represent a hypothetical sample of occupancy rates from 54 hospitals (6 of each factor-level combination). (a) Assuming a two-way model with interaction, construct an ANOVA table for this problem. (b) To get a visual picture of a possible interaction effect, construct graphs similar to those in Figures 10.7 and 10.8 of Section 10.6. Plot the mean

Complexity Level	*Nonprofit Private*		*Government*		*Profit Private*	
Low	77.4	83.6	65.2	60.3	63.7	71.3
	79.3	80.5	74.3	71.2	55.8	74.6
	75.4	72.4	61.4	64.8	69.7	54.3
Medium	73.6	71.0	65.3	81.0	66.2	78.6
	79.5	81.3	69.2	84.3	70.8	81.2
	83.4	81.8	74.8	68.1	69.3	72.1
High	79.3	77.6	82.3	76.3	71.1	73.0
	69.4	84.5	79.8	84.6	58.3	75.4
	74.6	83.1	69.7	67.5	61.4	64.0

response on the vertical axis and the levels of factor B on the horizontal axis. Connect the points corresponding to each level of factor A. (c) Using a .025 level of significance, perform tests to see whether either of the factor effects or the interaction effects differ significantly from zero.

★*22.* It is desired to study the yield of rubber (in grams) from three varieties of guayule bush (a northern Mexico rubber-producing bush). Since these bushes were planted in four different locations, it is reasonable to treat the locations as blocks. By doing this it is possible to keep soil type, amount of rainfall, and other environmental

	Location			
Variety	*1*	*2*	*3*	*4*
1	4.06	6.42	4.43	6.64
	3.75	4.72	7.31	5.92
2	6.85	3.88	6.71	5.82
	4.94	6.22	6.67	5.08
3	2.96	2.03	5.41	0.48
	2.71	5.08	0.87	1.97

factors as constant as possible within a block. The data obtained are shown in the table. (a) Since in this problem we have randomized complete blocks, construct an ANOVA table using these data (cf. Table 10.14 of Section 10.7). (b) Test the null hypothesis of no treatment effect using a .05 level of significance. (c) To see if blocking has been effective, test the null hypothesis of no block effect using a .05 level of significance.

★23. The economics instructor of Exercise 7 might be able to improve the study on the effect of the different times of the day on student performance. It might be possible to group the students according to past achievement as measured by grade point average (low, medium, or high). These groups would be the blocks for the study and three students, say, from each group would be assigned to each of the class times. Assume that the resulting responses (scores on final exams) for such a study were as given in the table. (a) Construct an ANOVA table using these data. (b) Test the null hypothesis of no treatment effect using a .05 level of significance. (c) To see if blocking has been effective, test the null hypothesis of no block effect using a .05 level of significance. (*To think about:* Can you see any problem for an instructor attempting to implement such a study? Would you yourself be willing to be randomly assigned to one of the three time periods?)

	Past Achievement		
Class Time	*Low*	*Medium*	*High*
7:30 A.M.	63	63	80
	60	65	78
	65	73	70
11:30 A.M.	68	72	72
	55	70	80
	72	72	78
2:30 P.M.	58	58	72
	50	63	68
	60	63	65

★24. A serious concern of the agricultural industry is with insecticide research and the evaluation of field damage from insects. Corn was planted in two different fields.

	Field	
Insecticide	*1*	*2*
1	1	3
	3	3
	2	2
2	3	4
	3	5
	4	5
3	5	5
	4	5
	5	6

Each field was divided into thirds and one of three different types of insecticide was randomly assigned to each third. At an appropriate point in time, the roots of randomly selected plants were examined for rootworm damage. The amount of damage (the response) was assessed by a rating system that assigned a value of 1 through 6 to quantify the degree of damage (1 corresponding to no damage and 6 to severe damage). A researcher used an analysis of variance procedure to evaluate the performance of the insecticides. (a) Using the data given in the table, duplicate that analysis. That is, construct an analysis of variance table and perform appropriate tests. Use a .05 level of significance. (b) Explicitly state which assumptions must be satisfied in order for the test statistic to follow an *F* distribution. (c) One of these assumptions is clearly violated. Which one is violated? Explain.

11

NONPARAMETRIC TESTS OF HYPOTHESES

11.1 INTRODUCTION

In Section 8.4 we discussed some differences between parametric and nonparametric tests of hypotheses. You should go back and reread that section before continuing. We will summarize that discussion and add some to it here.

In finding the critical regions for the tests of hypotheses presented in Chapter 8, we used our knowledge about the probability distribution of the test statistic and the desired level of significance. In almost every case an underlying assumption about the distribution from which our sample observations came was made, the assumption that the distribution was normal. This allowed us to say that the test statistic had a certain probability distribution (e.g., a t distribution, χ^2 distribution, or F distribution). We know, however, that in many cases such an assumption cannot be made. For example, if the measurement scale for the observations is nominal or ordinal (see Section 6.7), the distribution cannot be normal. If the distribution is skewed or multimodal, it also cannot be normal.

What are the possible consequences of using a parametric test based on the underlying assumption of normality when that assumption is false? The main consequence is that the level of significance may be different than what you think it is. If the procedure is "robust," as may be the case when the sample size is large and the Central Limit Theorem can be invoked, for example, the difference in the true level of significance and the hoped for level may be small. However, in other cases the difference may be large. For example, the true level of significance may be .10 or .15 instead of a desired level of .05.

What can we do if we know or suspect that the underlying assumption of normality cannot be made? We can use test procedures that use less restrictive assumptions, which are more likely to be satisfied. These test procedures generally do not depend on specific distributional assumptions about the probability distribution from which the measurements are taken and are called "nonparametric" or "distribution-free" tests. (A distinction is sometimes made between these two terms, but we will not concern ourselves with that distinction.)

We already mentioned the possible consequences of using a parametric test when it is inappropriate to do so, namely having a test whose true level of significance is different than the desired level. You might well ask about the possible consequences of using a nonparametric procedure when a parametric procedure would be justified. In such a case the test is less "powerful" (see Section 8.10); that is, the probability of rejecting H_0 for a given value of a parameter θ corresponding to H_a being true is less with the nonparametric test than with a parametric test. However, the level of significance will be the same in both cases. Some of these different consequences are depicted in Figure 11.1. In light of our convention of setting up the null hypothe-

		True State of Affairs	
		Nonparametric procedure Appropriate	*Parametric procedure Appropriate*
Decision of Investigator	*Use nonparametric procedure*	Right choice	Correct level of significance but a slight loss of power
	Use parametric procedure	Unknown level of significance, possibly too large	Right choice

FIGURE 11.1
Consequences of Choices of Test Procedure

sis so that a type I error is more serious, it would be better, if in doubt, to maintain the desired level of significance at the risk of some loss of power. Consequently, a conservative approach to the choice of a parametric or a nonparametric test procedure would be to choose a nonparametric procedure if any doubt exists about the appropriateness of a parametric procedure. This would be especially true when the sample size is small, since it is more difficult to know from the data whether an assumption such as normality is appropriate.

11.2 TESTS FOR THE MEDIAN

You will recall from the discussion in Section 2.6 that the mean, μ, and the median, $\tilde{\mu}$, as well as the mode can be used as measures of central tendency for the probability distribution of a random variable. Remember that if a distribution is symmetric, the mean and median will be equal, while if it is skewed, they will be unequal. If the distribution is skewed to the right, for example, the mean will be larger than the median. If the distribution is highly skewed, the mean may be considerably larger than the median and consequently not as "typical" or "representative" of the value that the random variable may assume. For this and other reasons, the median, $\tilde{\mu}$, is often used as a measure of central tendency and nonparametric tests about "location" are frequently related to the median rather than to the mean. In this section we will present some nonparametric test procedures for testing

$$H_0: \tilde{\mu} = \tilde{\mu}_0 \quad \text{vs.} \quad H_a: \tilde{\mu} \neq \tilde{\mu}_0 \tag{11.1}$$

or analogous one-sided hypotheses.

Will Nockon is the owner of the Nockon Wood Company, a small company that turns out dowel rods. The procedure used to cut dowels to certain nominal lengths is somewhat crude, so the actual lengths of the rods vary quite a bit. When the procedure is working properly, the distribution of the length of pine dowel rods, which are nominally 100 centimeters in length, is as shown in Figure 11.2(b). As you can see, the distribution is skewed to the right, with the median at 100 cm. The process can go out of adjustment by having the lengths of the rods longer or shorter than usual. If this happens, the distribution of the lengths of the rods could look like one of those given in Figure 11.2(a) or (c). To check on the process, Will has a simple testing procedure where he compares the lengths of a sample of rods with a standard meter-long dowel (see Figure 11.3). He simply notes whether a sample rod is longer (+) or shorter (−) than the standard. How can he use this information to test the null hypothesis that the process is in adjustment; that is, how can he test

$$H_0: \tilde{\mu} = 100 \text{ cm} \quad \text{vs.} \quad H_a: \tilde{\mu} \neq 100 \text{ cm?} \tag{11.2}$$

If you look at Figure 11.2, you should be able to see that when the dowels are too long, he will find a large number of + signs, while if the dowels are too short, he will find a small number of + signs. This suggests that

$$X = \text{number of + signs from a sample of size } n \tag{11.3}$$

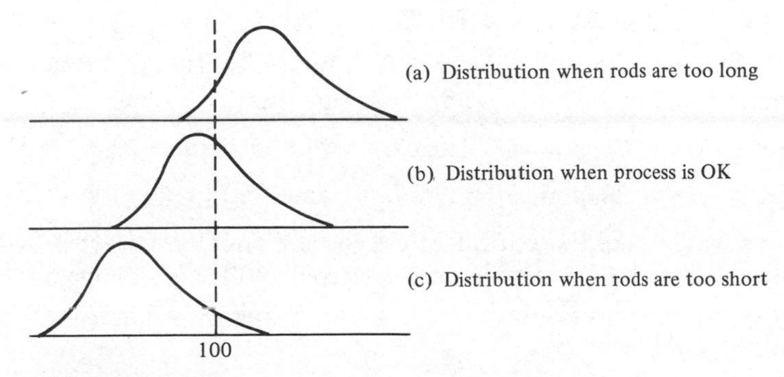

FIGURE 11.2

Possible Distributions of Lengths of "100-cm" Dowel Rods

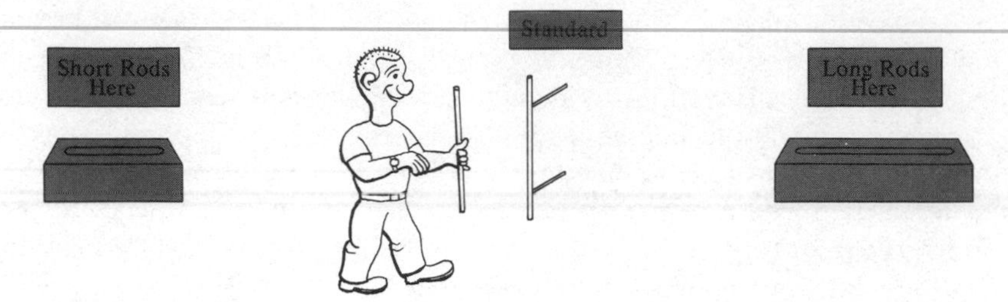

FIGURE 11.3
Will's Inspection Procedure

can be used as a test statistic. H_0 would be rejected if X were either too large or too small, the exact critical values being determined by the desired level of significance. To find these values is not difficult once we know the probability distribution of the test statistic X. Assuming that the probability of getting a dowel rod of *exactly the same length* as the standard is zero, each dowel will be classified as + (long) or − (short). This suggests that X will have a binomial distribution. It remains to determine the value of p, the probability of a "success" (getting a +), when H_0 is true. Since when H_0 is true, $\tilde{\mu} = 100$ (i.e., the median is 100), by the very definition of median (Definition 2.12 of Section 2.6), p must equal .5. Consequently, the test statistic will have a binomial distribution with $p = .5$ when H_0 is true. Actually, the hypotheses (11.1) could be expressed as

$$H_0: \ p = .5 \qquad \text{vs.} \qquad H_a: \ p \neq .5$$

and could be tested by simply using the methods of Section 8.8.

Example 11.1 Given that Will's inspection procedure showed 17 +'s out of 20 rods inspected, test the hypotheses in (11.2) using a .05 level of significance.

Remember that when $p = .5$, the binomial distribution is symmetric. Since we are considering a two-sided alternative, we would find critical values c_1 and c_2 such that for X as defined in (11.3),

$$P[X \leq c_1 \,|\, n = 20, p = .5] = .025 \qquad \text{and} \qquad P[X \geq c_2 \,|\, n = 20, p = .5] = .025.$$

Unfortunately, since the binomial distribution is discrete, we cannot get probabilities exactly equal to .025, but we do find from Table A2 that

$$P[X \leq 5 \,|\, n = 20, p = .5] = .0207, \qquad P[X \geq 15 \,|\, n = 20, p = .5] = .0207,$$
$$P[X \leq 6 \,|\, n = 20, p = .5] = .0577, \qquad \text{and} \qquad P[X \geq 14 \,|\, n = 20, p = .5] = .0577,$$

so we would take the critical values to be 5 and 15. The observed value of 17 is in the critical region, so H_0 would be rejected. Will should conclude that the process is out of adjustment. [Note that if we were to calculate a p-value (Section 8.9) for this test, it would be

$$2(P[X \geq 17 \,|\, n = 20, p = .5]) = 2(.0013) = .0026.$$

Remember that when test statistics have discrete distributions, it is often not possible to find a critical region leading to a predetermined level of significance. For this reason it is customary to express test results in terms of p values for many nonparametric tests.]

Note that had n been greater than 20, we would use the normal approximation to the binomial to determine critical values.

The procedure used for testing the hypotheses in this last example is called the *sign test*. In general, to use the sign test, we need a random sample from the population of interest. The variable of interest (e.g., the length of dowel rods) must be measured on at least an ordinal scale and should be continuous in the neighborhood of the median. (This assures that the variable of interest will be either greater than or less than the hypothesized median, so that outcomes can be classified in exactly one of two ways. If the measurements are such that some values may be equal to the hypothesized median, we say that a "zero" has occurred. One way of treating zeros is to ignore them, thus reducing the sample size.) If the sample observations are $X_1, X_2, \ldots, X_n$, a sign test can be used to test

$$H_0: \tilde{\mu} = \tilde{\mu}_0 \quad \text{vs.} \quad H_a: \tilde{\mu} \neq \tilde{\mu}_0$$

or analogous one-sided hypotheses. The test statistic will be

$$X = \text{number of sample values exceeding } \tilde{\mu}_0$$

or, equivalently,

$$X = \text{number of positive signs among the set}$$
$$\{X_1 - \tilde{\mu}_0, X_2 - \tilde{\mu}_0, \ldots, X_n - \tilde{\mu}_0\}.$$

Assuming that there are no zeros, when H_0 is true, X will have a binomial distribution with parameters n and $p = .5$.

Example 11.2 The data in Table 11.1 represent average rates of productivity increase for a sample of 25 industries in a recent year. Test the hypotheses

$$H_0: \tilde{\mu} \leq 2.0\% \quad \text{vs.} \quad H_a: \tilde{\mu} > 2.0\%$$

(i.e., test to see if there is convincing evidence that the median of the average rate of productivity increase was more than 2.0%). Use a .05 level of significance.

The test statistic in this problem will be X, the number of positive signs among the set $\{X_1 - 2.0, X_2 - 2.0, \ldots, X_n - 2.0\}$. In this case there is one "zero," since the last value is equal to 2.0. This value will be ignored and we will count the number of + signs among the remaining 24 values. Since n is greater than 20 and since the

TABLE 11.1

Data for Example 11.2

2.1	2.6	2.1	2.1	.5	2.4	2.3	2.1	7.1	5.3	2.4	2.6	2.7
2.1	2.8	1.5	2.8	1.8	2.5	2.8	2.6	.2	.4	6.5	2.0	

hypotheses are equivalent to

$$H_0: \; p \leq .5 \qquad \text{vs.} \qquad H_a: \; p > .5,$$

we perform the test as we did in Section 8.8. In particular, we use

$$Z = \frac{X - np_0}{\sqrt{np_0 q_0}}$$

as the standardized test statistic and, in view of the one-sided alternative, reject H_0 if the observed z value is $z_{.95} = 1.645$ or more. Here we have

$$z = \frac{19 - 24(.5)}{\sqrt{24(.5)(.5)}} = 2.858,$$

so we reject H_0 and conclude that the median of the average gain in productivity was more than 2.0%.

The sign test certainly has some very desirable features. For example, the assumptions that must be satisfied in order to use it are not very restrictive, the calculations required to find the number of + signs are quite simple, and the tables needed to find critical values are ones with which you are already familiar. As is often the case, however, these desirable features are obtained at a cost. The sign test is not as "powerful" as other tests, because some information is being ignored. In particular, by considering only the sign of the difference $X_i - \tilde{\mu}_0$, we are ignoring the magnitude of the difference. In the previous example it is easy to see that there is quite a difference between

$$2.1 - 2.0 = +0.1 \qquad \text{and} \qquad 7.1 - 2.0 = +5.1,$$

but the sign test only makes use of the sign and not the magnitude. Of course, in some situations such as Will's inspection procedure of Example 11.1, only the signs are available. However, if the magnitudes are available, we can get a more powerful test procedure by using the Wilcoxon signed rank test. The assumptions that must be satisfied to use the sign test, the Wilcoxon signed rank test, and, for purposes of comparison, a t test are given in Table 11.2. In studying the conditions you can see

TABLE 11.2

Assumptions That Must Be Satisfied to Use Various Tests

Sign Test	*Signed Rank Test*	*t Test*
1. Independent observations	1. Independent observations	1. Independent observations
2. Distributions continuous around the median	2. Distributions continuous everywhere	2. Distributions continuous everywhere
3. Distributions need not be symmetric	3. Distributions must be symmetric	3. Distributions must be normal
4. Distributions may be different but have the same median	4. Distributions must be the same, hence have the same median	4. Distributions must be the same, hence have the same mean and variance

that they become more restrictive as you read from left to right. While there are some similarities, some differences are that the sign test can be used on data that are strictly ordinal, while the Wilcoxon signed rank test and the t test can be used on interval data. Another major difference is that for the sign test the distributions need not be symmetric, for the signed rank test they must be symmetric but not normal, while for the t test they must not only be symmetric but normal. However, the signed rank test is robust relative to mild departures from symmetry. In the following we will motivate and then outline the test procedure to be used for a signed rank test. Details as to the derivation of the distribution of the test statistic will be left to the next section.

Consider a random sample of observations taken from one of the three distributions shown in Figure 11.4. You might think of the distributions as corresponding to the lengths of birch dowel rods of nominal length 50 cm. Note that in this case we are assuming the distribution to be symmetric. If [as in Figure 11.4(b)] the null hypotheses that $\tilde{\mu} = 50$, say, is true, then we would expect the observed values (indicated here by dots) to be symmetrically scattered about the value 50. Values a large distance from 50 are as likely to be above 50 as below 50. However, if the true median is larger than 50 [as in Figure 11.4(a)], values that are a large distance from the hypothesized value of 50 are more apt to be above than below 50. On the other hand, if the true median is smaller than 50 [as in Figure 11.4(c)], values that are a large distance from the hypothesized value of 50 are more apt to be below 50 than above. With this in mind, say that we consider the distances from the hypothesized median, say $\tilde{\mu}_0$, of each element of a random sample coming from one of the distributions (a), (b), or (c). These values would be

$$|X_1 - \tilde{\mu}_0|, |X_2 - \tilde{\mu}_0|, \ldots, |X_n - \tilde{\mu}_0|.$$

If these values are ranked from smallest to largest (the smallest receiving a rank of 1), and if we define

$T^+ =$ sum of ranks corresponding to observations for which differences are positive,

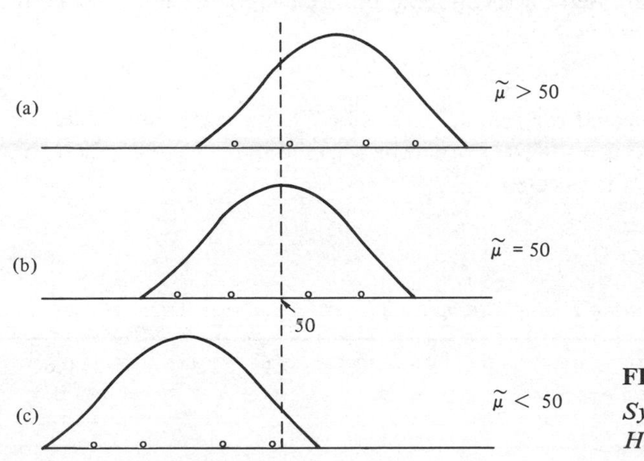

FIGURE 11.4
Symmetric Distributions Having Different Medians

T^+ could be used as a test statistic. Considering Figure 11.4(a), you can see that larger differences (higher ranks) will tend to be positive, so T^+ will tend to be large if $\tilde{\mu}$ is greater than the hypothesized value. Similarly, from Figure 11.4(c) you can see that T^+ will tend to be small if $\tilde{\mu}$ is less than the hypothesized value. Consequently, in testing

(11.4) $H_0: \tilde{\mu} \leq \tilde{\mu}_0$ vs. $H_a: \tilde{\mu} > \tilde{\mu}_0$, reject H_0 if T^+ is "large"

(11.5) $H_0: \tilde{\mu} \geq \tilde{\mu}_0$ vs. $H_a: \tilde{\mu} < \tilde{\mu}_0$, reject H_0 if T^+ is "small"

(11.6) $H_0: \tilde{\mu} = \tilde{\mu}_0$ vs. $H_a: \tilde{\mu} \neq \tilde{\mu}_0$, reject H_0 if T^+ is "large" or "small."

Just how "large" or "small" T^+ must be is determined by the level of significance and the sample size. Critical values when $\tilde{\mu} = \tilde{\mu}_0$ are given in Table A11.

If the distribution of the observed values $X_1, \ldots, X_n$ is truly continuous, these values would all be distinct. However, because of round off in measurement it is possible for "ties" to occur in the data as well as "zeros." As with the sign test, we will ignore zeros. When there are ties, we average the ranks that would have been assigned to the values if they had not been tied. With these conventions in mind, we would test any of (11.4) to (11.6) as follows.

1. Find the absolute differences $|X_1 - \tilde{\mu}_0|, \ldots, |X_n - \tilde{\mu}_0|$.
2. Rank these absolute differences.
3. Calculate T^+.
4. Compare T^+ with the critical value determined from Table A11.

Example 11.3 Will has measured a sample of 10 birch dowel rods and has found their lengths to be

49.5 49.0 48.7 49.5 50.4 48.6 49.3 50.6 48.0 50.0

Test: $H_0: \tilde{\mu} \geq 50$ vs. $H_a: \tilde{\mu} < 50$. Use $\alpha = .05$.

We first find the absolute differences, noting the sign of the difference (see Table 11.3). Ignoring the zero, we rank the remaining values. Note that since the two differences "0.5" were tied for second and third place, they each received the "average rank" of 2.5. We now find that

$$T^+ = 1 + 4 = 5.$$

Since H_0 will be rejected for "small" values of T^+, we see from Table A11 that for a sample of size 9 and $\alpha = .05$, H_0 would be rejected if T^+ were 8 or less. Since for this example $T^+ = 5$, H_0 is rejected.

TABLE 11.3

Values for Calculating T^+ for Example 11.3

Absolute difference:	0.5	1.0	1.3	0.5	0.4	1.4	0.7	0.6	2.0	0
Sign of difference:	−	−	−	−	+	−	−	+	−	
Rank:	2.5	6	7	2.5	1	8	5	4	9	

Note that when n ranks are assigned, the total sum of all ranks will be equal to $n(n + 1)/2$. If we define

T^- = sum of ranks of observations for which differences are negative, then

$$T^+ + T^- = \frac{n(n+1)}{2}. \tag{11.7}$$

There are times when it may be easier to find T^- and then use (11.7) to obtain T^+ by subtraction.

Since the assumptions necessary to perform a signed rank test are satisfied whenever the normality assumption necessary to perform a t test is satisfied, we could use a signed rank test instead of a t test. If the normality assumption really does hold, the t test will be slightly more powerful. However, in a large number of problems, both test procedures will lead to the same decision about rejecting H_0. This is illustrated by the next example.

Example 11.4 In Example 8.12 of Section 8.5, Daisy La Fleur wanted to see if a newly developed variety of tomato plant would grow more quickly than the variety that had been used. Using the data given in the first row of Table 11.4, she tested

$$H_0: \mu \leq 7.6 \quad \text{vs.} \quad H_a: \mu > 7.6$$

TABLE 11.4

Values for Calculating T^+ for Example 11.4

Height of plant, X_i:	7.9	9.1	6.6	8.1	7.9	7.5	8.5	7.8	8.1	8.5
$\lvert X_i - 7.6 \rvert$:	0.3	1.5	1.0	0.5	0.3	0.1	0.9	0.2	0.5	0.9
Sign of difference:	+	+	−	+	+	−	+	+	+	+
Rank:	3.5	10	9	5.5	3.5	1	7.5	2	5.5	7.5

using $\alpha = .05$. The null hypothesis was rejected because the test statistic was in the critical region (although just barely). Test the same hypotheses using the signed rank test.

Because the distribution is assumed to be symmetric, the mean and median are equal, hence the hypotheses could be expressed as

$$H_0: \tilde{\mu} \leq 7.6 \quad \text{vs.} \quad H_a: \tilde{\mu} > 7.6.$$

Comparing these with (11.4), we see that H_0 will be rejected only if T^+ is too large. Consequently, we would find the critical value by using Table A11. For a sample size of 10, we find that the critical value is 45. The observed value of T^+ is most easily found from (11.7) to be

$$T^+ = \frac{10(11)}{2} - T^- = 55 - 10 = 45.$$

Since this value is in the critical region, H_0 would be rejected. (The same decision was reached in this case as when the t test was used.)

If the sample size exceeds 15, then Table A11 cannot be used to find critical values. However, a normal approximation to the distribution of T^+ can be used in such a case. In particular, for large n,

$$Z = \frac{T^+ - (n(n+1))/4}{\sqrt{n(n+1)(2n+1)/24}} \sim N(0, 1) \tag{11.8}$$

approximately. H_0 would be rejected if the observed value of Z is in the critical region determined by the appropriate percentile z_p.

Example 11.5 Julie Rhee is the personnel manager for the Gold Ring Company, a large company specializing in a line of jewelry products. She is interested in the median income of workers of this company and in particular wants to test

$$H_0 : \tilde{\mu} = \$17{,}000 \quad \text{vs.} \quad H_a : \tilde{\mu} \neq \$17{,}000$$

at a .05 level of significance. She randomly selected the files of 20 workers and found the salaries (in thousands of dollars) to be as given in Table 11.5. What conclusions can she draw?

Although the distribution of salaries is typically skewed, we rely on the robustness of the signed rank test and use T^+ as a test statistic. Since the alternative is two-sided and since the normal approximation to the distribution of T^+ is required, we use $\pm z_{.975} = \pm 1.96$ as the critical values. We have $T^+ = 131.5$, so from (11.8) we

TABLE 11.5
Values for Calculating T^+ for Example 11.5

Salary, X_i	$\lvert X_i - 17 \rvert$	*Sign of Difference*	*Rank*
18.2	1.2	+	8
17.3	.3	+	3
16.9	.1	−	1
17.5	.5	+	4.5
15.6	1.4	−	10
25.8	8.8	+	16
35.2	18.2	+	17
16.8	.2	−	2
53.6	36.6	+	20
44.8	27.8	+	19
15.8	1.2	−	8
16.5	.5	−	4.5
16.3	.7	−	6
38.6	21.6	+	18
21.4	4.4	+	14
14.3	2.7	−	13
12.1	4.9	−	15
19.6	2.6	+	12
14.9	2.1	−	11
15.8	1.2	−	8

find the observed value of Z to be

$$z = \frac{131.5 - (20)(21)/4}{\sqrt{20(21)(41)/24}} = .989.$$

Since this value is not in the critical region, H_0 cannot be rejected at the .05 level of significance. It is reasonable to say that the median salary is $17,000.

11.3 PROBABILITY DISTRIBUTION OF THE WILCOXON SIGNED RANK STATISTIC

In this section we will show how the probability distribution for the signed rank statistic T^+ can be derived. If you are not concerned with the "whys" of test procedures but only in the "hows," you may wish to go directly to the next section.

If the conditions necessary to perform a signed rank test are satisfied (Table 11.2), we know that the distribution of the observed variable X is symmetric about its median. Because of this symmetry, an observed value is just as likely to be d or more units above the median as d or more units below the median. Consequently, when the differences $|(X_i - \tilde{\mu})|$ are considered, the largest of these differences is as likely to be a positive difference as a negative difference. In terms of the signed rank test, we can say that the sign associated with the largest absolute difference (rank n) has a probability of $\frac{1}{2}$ of being either positive or negative. The very same argument holds true for each of the other ranks: they are equally likely to be assigned to positive or negative differences.

Once we know that when differences about the true median are considered the ranked values are equally likely to be positive as negative, we can find the exact probability distribution of T^+ by constructing a table similar to Table 5.1 of Section 5.5. In that case we found the exact probability distribution of $\bar{X}$. In particular, consider a random sample of size 3. The ranks assigned to the absolute differences $D_i = |X_i - \tilde{\mu}|$ must be 1, 2, and 3. You can see that the largest value that T^+ can assume is 6, which will occur when all three differences are positive, the probability that all three are positive is $(\frac{1}{2})^3$, so

$$P[T^+ = 6] = \tfrac{1}{8}.$$

In fact, each of the assignments of $+$ or $-$ signs to the ranks 1, 2, and 3 is equally likely, having probability $\frac{1}{8}$. We make use of this fact in constructing the first two columns of Table 11.6. In the last column we give the value of T^+ for each configuration of signs. You can see from the table that there are two ways that T^+ can equal 3, so $P[T^+ = 3] = \frac{2}{8}$. The entire probability distribution for T^+ when $n = 3$ is

t^+:	0	1	2	3	4	5	6
$P[T^+ = t^+]$:	$\frac{1}{8}$	$\frac{1}{8}$	$\frac{1}{8}$	$\frac{2}{8}$	$\frac{1}{8}$	$\frac{1}{8}$	$\frac{1}{8}$.

Using this distribution, it is not hard to see that the cumulative probabilities are given by

t^+:	0	1	2	3	4	5	6
$P[T^+ \leq t^+]$:	.125	.250	.375	.625	.750	.875	1.00.

TABLE 11.6
Possible Configuration of Signs with Ranks, n = 3

Rank 1	Rank 2	Rank 3	Probability of Signs	T^+
+	+	+	$\frac{1}{8}$	6
+	+	−	$\frac{1}{8}$	3
+	−	+	$\frac{1}{8}$	4
+	−	−	$\frac{1}{8}$	1
−	+	+	$\frac{1}{8}$	5
−	+	−	$\frac{1}{8}$	2
−	−	+	$\frac{1}{8}$	3
−	−	−	$\frac{1}{8}$	0

The critical values in Table A11 are obtained from cumulative probabilities such as these.

You will be asked to show in Exercise 6 that when $n = 4$, the probability distribution of T^+ is

t^+:	0	1	2	3	4	5	6	7	8	9	10
$P[T^+ = t^+]$:	$\frac{1}{16}$	$\frac{1}{16}$	$\frac{1}{16}$	$\frac{2}{16}$	$\frac{2}{16}$	$\frac{2}{16}$	$\frac{2}{16}$	$\frac{2}{16}$	$\frac{1}{16}$	$\frac{1}{16}$	$\frac{1}{16}$.

From this distribution we can see how critical values can be found. For example, when testing the null hypothesis (11.5), if $\alpha = .10$,

$$P[T^+ \leq 0] = \tfrac{1}{16} < .10$$

while

$$P[T^+ \leq 1] = \tfrac{2}{16} > .10,$$

so $t^+ = 0$ is the critical value for the lower tail. You may be able to see by considering the distributions of T^+ when $n = 3$ and $n = 4$ that the distribution is symmetric.

11.4 TESTS FOR DIFFERENCES IN LOCATION (PAIRED OBSERVATIONS)

In Sections 7.6 and 8.6 we discussed estimation and test procedures for the difference of two means, $\mu_1 - \mu_2$, when the data are paired. We saw that inference could be performed using the Student's t distribution if we assumed that the differences in the sample pairs

$$D_i = X_{1,i} - X_{2,i}$$

were normally distributed. If the assumption of normality is unreasonable, but it can be assumed that the differences are a random sample from a continuous symmetric distribution, the Wilcoxon signed rank test can be used to test hypotheses about the median difference $\tilde{\mu}_D$. (Since symmetry implies $\tilde{\mu}_D = \mu_D$, this would be equivalent to tests about $\mu_D = \mu_1 - \mu_2$.) The test procedure is as described in Section 11.2, using D_i in place of "X_i."

TABLE 11.7
Values for Calculating T^+ for Example 11.6

Estimated price, $X_{1,i}$:	\$20.53	20.62	19.96	20.70	20.88	21.09	20.69	21.02	20.54	20.85
Actual price, $X_{2,i}$:	20.35	20.31	20.19	20.46	20.61	20.59	20.68	20.63	20.57	20.31
Difference, D_i:	.18	.31	−.23	.24	.27	.50	.01	.39	−.03	.54
$\lvert D_i - 0 \rvert$:	.18	.31	.23	.24	.27	.50	.01	.39	.03	.54
Sign of $(D_i - 0)$:	+	+	−	+	+	+	+	+	−	+
Rank:	3	7	4	5	6	9	1	8	2	10

Example 11.6 The data in the first two rows of Table 11.7 represent estimated price ($X_{1,i}$) per hundred weight for hogs (estimated by some complicated scheme before the actual sale) and the actual price ($X_{2,i}$) as determined after the actual sale. To determine whether the estimates are satisfactory, it is desired to test to see if $\mu_1 - \mu_2$ equals zero or not, or equivalently to test

$$H_0 : \tilde{\mu}_D = 0 \qquad \text{vs.} \qquad H_a : \tilde{\mu}_D \neq 0.$$

Use the Wilcoxon signed rank test with $\alpha = .05$ to test these hypotheses.

Since the alternative hypothesis is two-sided, we would reject H_0 if T^+ is either too small or too large. With $n = 10$ and $\alpha = .05$, we determine the critical values from Table A11 to be 8 and 47. That is, reject H_0 if $T^+ \leq 8$ or $T^+ \geq 47$. The observed value of T^+ is most easily found by using (11.7):

$$T^+ = \frac{n(n+1)}{2} - T^- = \frac{10(11)}{2} - 6 = 49.$$

Since this value is in the critical region, H_0 is rejected. We would conclude that there is a difference in the means of the distributions of estimated and actual price (at a .05 level of significance). The estimates may not be considered satisfactory, since they tend to overestimate the actual price.

If it is not reasonable to make the assumption that the differences D_i have a symmetric distribution and/or if only the signs of the relevant differences are available, the sign test may be the appropriate test to use. If the hypothesized median difference for a problem is $\tilde{\mu}_0$, the appropriate test statistic to use would be

$$X = \text{number of } + \text{ signs among the set } \{D_1 - \tilde{\mu}_0, D_2 - \tilde{\mu}_0, \ldots, D_n - \tilde{\mu}_0\}.$$

Example 11.7 Chubby McSwine is the president of a large local chapter of Eager Eaters Anonymous. He has checked the weights of all members before Thanksgiving and again at the first meeting in January. Based on the information about 12 members given in Table 11.8, he would like to test the null hypothesis that the median weight

TABLE 11.8
Data for Example 11.7

Weight before:	156	129	178	147	225	186	230	139	210	167	208	192
Weight after:	152	125	161	147	214	185	250	136	223	165	210	196
Difference, D_i:	4	4	17	0	11	1	−20	3	−13	2	−2	−4
$(D_i - 5)$:	−1	−1	12	−5	6	−4	−25	−2	−18	−3	−7	−9
Sign of difference:	−	−	+	−	+	−	−	−	−	−	−	−

loss over the holiday period was 5 or more pounds:

$$H_0: \tilde{\mu}_D \geq 5 \qquad \text{vs.} \qquad H_a: \tilde{\mu}_D < 5.$$

Test these hypotheses using $\alpha = .05$.

Past experience has indicated that the distribution of weight losses may be skewed, so the sign test would be appropriate.

The test statistic, the number of + signs, is equal to 2. We would reject H_0 only if X is less than or equal to 2, since

$$P[X \leq 2 \mid n = 12, p = .5] = .0193 \qquad \text{and} \qquad P[X \leq 3 \mid n = 12, p = .5] = .0730.$$

The observed value of the test statistic is in the critical region, so H_0 is rejected. Chubby must conclude that the median weight loss was less than 5 pounds.

11.5 TEST FOR DIFFERENCES IN LOCATION (UNPAIRED OBSERVATIONS)

We have discussed in earlier sections the fact that pairing (or blocking in the case of more than two "treatments") can be helpful in eliminating the effect of extraneous factors when checking for a difference in the means of distributions. We know, however, that it is not always possible to use pairing, but that in many cases we must take independent observations from the populations of interest. In Section 7.7, for example, we saw that to estimate how much children grow between first and second grades (if no past records were available) would require a year to obtain the data (if pairing were used) or would require two independent samples. To do inference, we assumed that the samples were independent random samples from normal distributions. To pool variances, we had to make the assumption that the variances were equal. You will remember that if two normal distributions have the same variance but different means, the graphs of the pdf's would be identical in shape but shifted from each other. In this section we will make a less stringent assumption. In particular, we will assume that the data consist of independent random samples whose probability density functions are identical in shape, but (perhaps) shifted from each other. This shift will have the effect of making the distributions have different medians. These

TABLE 11.9
Comparison of Assumptions for Two-Sample Tests

Rank Sum Test	*Two-Sample t Test*
1. Independent random samples	1. Independent random samples
2. Both samples come from continuous distributions having the same general form	2. Both sample come from normal distributions with the same variance
3. Medians may be different	3. Means may be different

assumptions are shown in Table 11.9. The test that will be discussed in this section is called the Wilcoxon rank sum test.

The two sample sizes need not, of course, be equal. We will use the convention of labeling the smaller of the two samples with a "1" subscript. Hence the sample sizes will be denoted by n_1 and n_2 ($n_1 \leq n_2$) and the samples themselves by

$$X_{1,1}, X_{1,2}, \ldots, X_{1,n_1}$$
$$X_{2,1}, X_{2,2}, \ldots, X_{2,n_2}.$$

With this notation in mind, scan the graphs in Figure 11.5. In Figure 11.5(a) you see the situation where the first distribution is to the left of the second, so that $\tilde{\mu}_1 < \tilde{\mu}_2$. In Figure 11.5(b) and (c) we show the distributions when $\tilde{\mu}_1 = \tilde{\mu}_2$ and $\tilde{\mu}_1 > \tilde{\mu}_2$. The circles along the axis of the first distribution represent actual sample observations for four observed values. The stars along the axis of the second distribution represent five observed values from this distribution. In each case we have superimposed both sets of observed values on a single axis to the right of the distributions. As you look at these configurations you should be able to see that when the medians are the same [as in Figure 11.5(b)], the circles and stars (representing the observations from the two distributions) are fairly well intermixed. When the medians are separated, the circles and stars become more separated. This suggests the following scheme for testing

$$H_0: \tilde{\mu}_1 = \tilde{\mu}_2 \quad \text{vs.} \quad H_a: \tilde{\mu}_1 \neq \tilde{\mu}_2 \tag{11.9}$$

or analogous one-sided hypotheses. Combine the two sets of sample data and rank the combined set (assigning a rank of 1 to the smallest value). Find the sum of the ranks corresponding to the observations in the smaller sample, call this sum T_1; that is, define

$$T_1 = \text{sum of ranks corresponding to values in the smaller sample.}$$

If this statistic is used as a test statistic for the hypotheses (11.9), you can see from Figure 11.5 that

if $\tilde{\mu}_1 < \tilde{\mu}_2$, T_1 will tend to be "small" and
if $\tilde{\mu}_1 > \tilde{\mu}_2$, T_1 will tend to be "large."

How "small" or "large" T_1 will have to be before rejecting the null hypothesis of equal medians will depend on the level of significance and the probability distribution

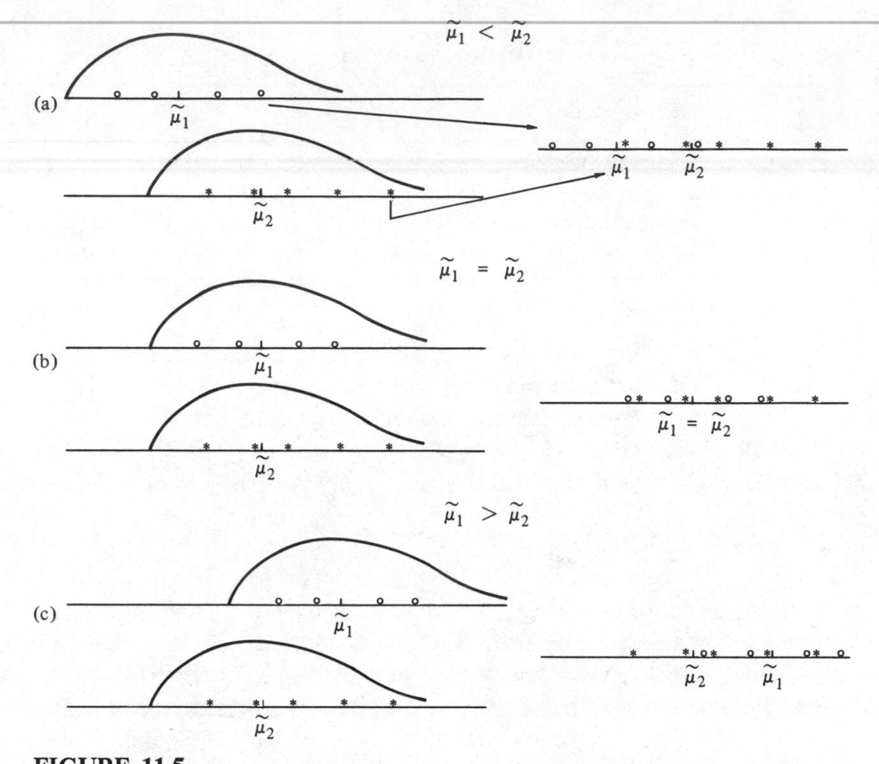

FIGURE 11.5
Possible Locations for Two Distributions

of T_1. In Table A12 you will find lower and upper critical values for various values of n_1 and n_2.

We will not go into detail as to how the probability distribution of T_1 is found when H_0 is true, but we will make a few comments. First, the smallest value that T_1 can assume occurs when all the values from the first sample lie to the left of those of the second sample [an extreme case of Figure 11.5(a)]. The ranks then will be 1, 2, . . . , n_1, so

$$T_1 = 1 + 2 + \ldots + n_1 = \frac{n_1(n_1 + 1)}{2}.$$

Likewise, the largest value T_1 can assume is when all the values of the first sample are the largest, yielding

$$T_1 = (n_2 + 1) + (n_2 + 2) + \ldots + (n_2 + n_1) = n_1 n_2 + \frac{n_1(n_1 + 1)}{2}. \quad (11.10)$$

Also, there are a total of $n_1 + n_2$ different ranks that can be assigned, and of these n_1 are to be assigned to the first sample. To find the number of different sets of ranks that could be assigned to the first sample, we use combinations. There are

$$\binom{n_1 + n_2}{n_1} = \frac{(n_1 + n_2)!}{n_1!\, n_2!}$$

different ways of assigning ranks. Using the fact that when H_0 is true, each of these combinations is equally likely, the probability distribution of T_1 can be found (see Exercise 14).

If the assumption that the distributions are continuous is satisfied, all the observed values will be different and no ties will occur. However, because of measurement limitations and rounding, ties may occur. In such instances the average of the ranks that would have been assigned to the values if no ties were present is assigned to each tied value. If there are many ties, a correction term is recommended. For details we refer you to a text on nonparametric statistics. If the sample sizes are large, a normal approximation to the distribution of T_1 can be used. In particular,

$$\text{(11.11)} \qquad \frac{T_1 - n_1(n_1 + n_2 + 1)/2}{\sqrt{n_1 n_2(n_1 + n_2 + 1)/12}} \sim N(0, 1) \quad \text{(approximately)}$$

if $n_2 > 10$ and $n_1 \geq 5$. (For values of n_1 and n_2 not satisfying these conditions and which are not given in Table A12, more detailed tables must be consulted.) A continuity correction factor (cf. Section 4.10) can be used here to improve the approximation.

Example 11.8 Using the data as shown in Figure 11.5, find the value of T_1 for each of the situations shown.

Since T_1 is the sum of the ranks of the observations in the smaller sample, we find the ranks corresponding to the "circles." We find that

$$\text{in (a), } T_1 = 1 + 2 + 4 + 6 = 13,$$
$$\text{in (b), } T_1 = 1 + 3 + 6 + 7 = 17,$$
$$\text{and in (c), } T_1 = 3 + 5 + 7 + 9 = 24.$$

Example 11.9 In Example 7.17 we assumed that a vending machine designed to dispense coffee into 8-ounce cups was adjusted by a service technician. To see if the adjustment made a significant difference in the amount dispensed, test

$$H_0: \tilde{\mu}_1 = \tilde{\mu}_2 \qquad \text{vs.} \qquad H_a: \tilde{\mu}_1 \neq \tilde{\mu}_2,$$

where the subscript 1 indicates the median before adjustment and corresponds to the smaller sample. Use the data from Example 7.17 and $\alpha = .10$.

The data (measured in ounces) are given in Table 11.10. The ranks for the

TABLE 11.10
Data for Example 11.9

(Rank):	2	6	4	1	
Before adjustment:	6.92	7.34	7.26	6.88	
(Rank):	5	9	8	7	3
After adjustment:	7.33	7.93	7.65	7.49	7.10

combined sample are shown above the observed value. From this you can see that

$$T_1 = 2 + 6 + 4 + 1 = 13.$$

To find the critical values, note that $\alpha = .10$ and use Table A12. For samples of size $n_1 = 4$ and $n_2 = 5$, we find lower and upper critical values to be 12 and 28. Since the value 13 is not in the critical region, H_0 is not rejected. Although there is some indication that the adjustment made a difference in the amount dispensed, the evidence (based on these sample sizes) is not convincing at a .10 level of significance. (The results were "close" to significant, however, since a more detailed analysis shows that the p value is .112.)

★***Example 11.10*** In Table 11.11 data are presented on the lifetimes (in days) of some male mice who were exposed to a radiation dose of 300 rads at age 5 to 6 weeks. Those in the first group were conventional mice exposed to a normal environment and those in the second group were germ-free mice. Use the data to test $H_0: \tilde{\mu}_1 \geq \tilde{\mu}_2$ vs. $H_a: \tilde{\mu}_1 < \tilde{\mu}_2$ at a .01 level of significance.

TABLE 11.11

Data for Example 11.10

Rank:	6	11	4	8	1	2	7	12				
Lifetimes: (conventional mice)	414	594	206	428	51	163	420	621				
Rank:	3	5	10	9	17	13	14	16	15	18	20	19
Lifetimes: (germ-free mice)	195	301	434	430	778	655	658	757	737	807	1015	868

In view of the form of the histograms you constructed in Exercise 13 of Chapter 8 based on such lifetimes, it may well be that the lifetimes are not normally distributed. In view of this fact, it would be better to use a rank sum test rather than a two-sample t test. The appropriate test statistic here would be T_1, the sum of the ranks corresponding to the smaller sample. Since the alternative hypothesis is $\tilde{\mu}_1 < \tilde{\mu}_2$, we would reject H_0 only if T_1 is "small." In particular, we would reject H_0 if the observed value of T_1 were found to be no larger than the first percentile. Since $n_1 = 8$ and $n_2 = 12$, we will use the normal approximation and reject if the standardized value of T_1 is no larger than $z_{.01} = -2.326$. The observed value of T_1 can be found from Table 11.11 to be 51, so from (11.11) we get

$$z = \frac{51 - 8(21)/2}{\sqrt{8(12)(21)/12}} = -2.55.$$

Since this value is in the critical region, H_0 would be rejected. We would conclude that the median lifetime is less for the conventional mice than for the germ-free mice.

11.6 TEST FOR DIFFERENCES IN VARIABILITY

We have discussed in earlier chapters the importance of being able to test for variability. We mentioned, for example, that variability is an important aspect for quantities such as the amount of a certain quantity (e.g., cereal, soda, glue, etc.) being put into a container or in the amount of time gained (or lost) by different watch movements. In comparing variances of two different distributions as in Section 8.7, we made the assumption that the samples come from normal distributions. If this assumption is satisfied, the appropriate test statistic has an F distribution. Unfortunately, this test procedure is not very robust with respect to departures from normality. For this reason, if the sample sizes are small and/or there is doubt as to the validity of the normality assumption, it is advisable to use nonparametric procedures.

The appropriate choice of a nonparametric procedure for testing for differences in variability depends on the medians of the distributions of interest. Different tests are applicable depending on whether the medians are the same or different, known or unknown. The first test that we will describe is appropriate if the medians are the same. (It can be adapted to the case where the medians are different, but known.) More specifically, if we assume that the null hypothesis is

$$H_0 : \text{distributions are the same}$$

and if we wish to test this against the alternative

$$H_a : \text{distributions differ in variability}$$

(but have the same median), and if the distributions are continuous (so that theoretically we can ignore the possibility of ties), a Siegel–Tukey test can be used. To see how an appropriate test statistic can be found, consider the distributions shown in Figure 11.6. As we did in the last section, we let circles denote observations from the first distribution and stars denote observations from the second distribution. In Figure 11.6(a) the distributions are identical and the observations from the two distributions, shown together on the axis to the right, are interspersed. In Figure 11.6(b), however, the first distribution has a greater amount of variability than the second. In this situation the first observations (circles) are at the extremes of the combined samples. You can see that if we assigned ranks in the usual way (1 to the smallest, etc.), the first distribution would have both the smallest and largest ranks, which would tend to balance each other out. If we assign "scores" to the ordered combined samples according to the scheme shown in Figure 11.7, a small total "score" will correspond to the distribution having the greater amount of variability, and vice versa. (Other authors may suggest slightly different scoring schemes, but the basic idea is the same.)

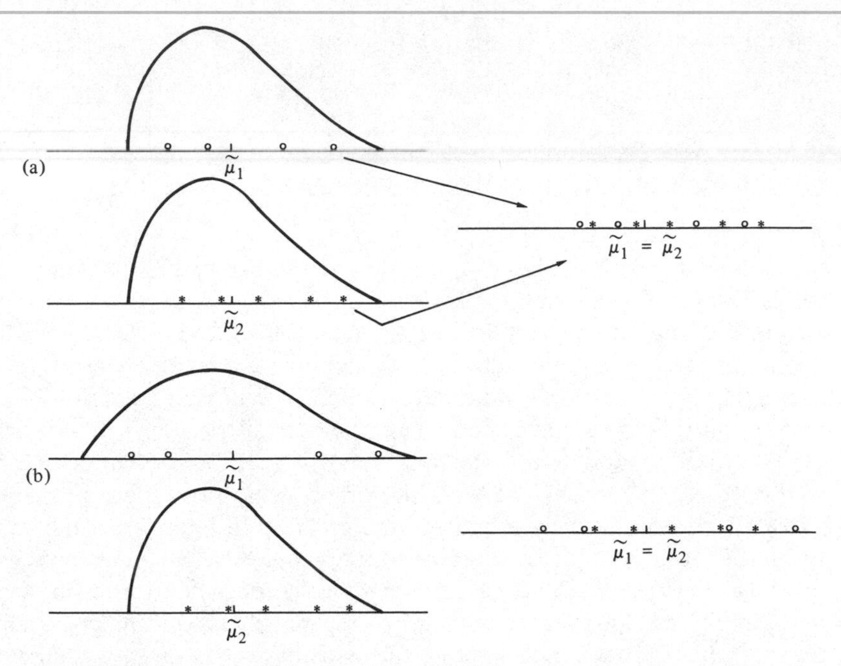

FIGURE 11.6

Distributions with Same Median and Same (a) or Different (b) Variability

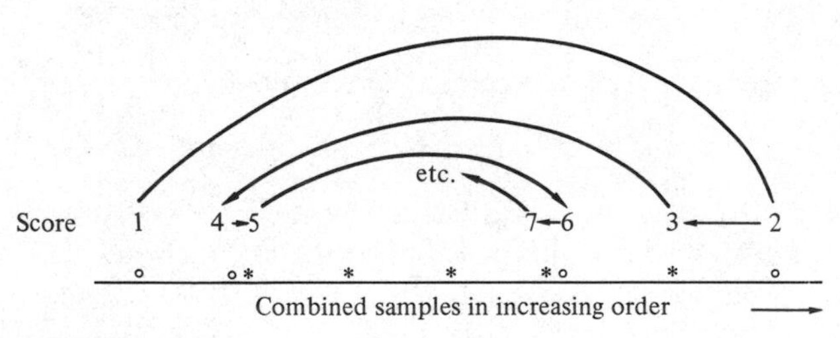

FIGURE 11.7

Assigning Scores to Combined Samples

Example 11.11 Find the total score assigned to sample 1 when the data are as in Figure 11.6.

Using the alternating scoring scheme as shown in Figure 11.7, you can see that the total score for the first sample in Figure 11.6(a), call it S_1, is

$$S_1 = 1 + 5 + 7 + 3 = 16.$$

In Figure 11.6(b), the total score for the first sample is

$$S_1 = 1 + 4 + 6 + 2 = 13.$$

If S_1, the total score of the smaller sample, is used as a test statistic and if the null hypothesis is actually true, the distribution of S_1 is exactly the same as the distribution of T_1, the rank sum statistic introduced in the last section. Consequently, Table A12 can be used to determine the critical values for certain values of n_1 and n_2. If $n_1 \geq 10$ and $n_2 > 10$, the normal approximation analogous to the one given in (11.11) can be used.

Example 11.12 Ben West, owner of the West Wrist Watch Works, wanted to know whether the variability in time gained by movements made by the Good Time Company differed from the variability in time gained by movements made by the Time Lee Company (see Example 8.19 of Section 8.7). What conclusion can be drawn if after 1 week watches using the movements from the Good Time Company (company 1) had gained

$$.42 \quad -1.87 \quad .23 \quad -.30 \quad -.10 \quad -1.27 \quad .03 \quad -1.48 \quad -2.68 \quad .79$$

minutes each, while watches using the movements from the Time Lee Company (company 2) had gained

$$2.42 \quad -1.73 \quad 2.17 \quad .45 \quad 1.01 \quad -.91 \quad .62 \quad -.53 \quad 1.45 \quad -2.16$$

minutes. Use $\alpha = .05$.

In this case it is not unreasonable to assume that the median is the same, namely zero, for both movements (i.e., for each company about half the movements run fast and the rest run slow). The other assumptions necessary to use the Siegel–Tukey test seem reasonable, so we can test

H_0: distributions of time gained is the same for both movements

vs.

H_a: distributions differ in variability

by using the total score for the first sample as the test statistic. Since the alternative is two-sided, H_0 will be rejected if S_1 is either too small or too large. From Table A12 we see that for $n_1 = 10$, $n_2 = 10$, $p = .025$, and $1 - p = .975$, the lower and upper critical values are 78 and 132. To find the observed value of the test statistic, the combined samples need to be ordered and then scores assigned. This has been done in Figure 11.8. We have placed a circle around the values corresponding to the first sample. We find the sum to be 126. Since this value is not in the critical region, H_0

Score	1	4	5	8	9	12	13	16	17	20
Time	(-2.68)	-2.16	(-1.87)	-1.73	(-1.48)	(-1.27)	-.91	-.53	(-.30)	(-.10)
Score	19	18	15	14	11	10	7	6	3	2
Time	(.03)	(.23)	(.42)	.45	.62	(.79)	1.01	1.45	2.17	2.42

FIGURE 11.8

Assignment of Scores for Example 11.12

is not rejected. There is not sufficient evidence to say that these distributions differ in variability. (Note that the same conclusion was reached in Example 8.19.)

The Siegel–Tukey test is easy to use and has the advantage of not requiring extra tables. Unfortunately, the test procedure is not a very good one if the medians of the two distributions are unknown and unequal. In such a situation it would be better to use the Moses test. This test requires that each sample be randomly subdivided into equal-size groups. In making the groups be of equal size, there may be some "leftover" observations that must be ignored. Because of this random subdivision procedure, it is possible to get different answers (either reject H_0 or do not reject H_0) with the same data. In spite of disadvantages like these, it is still better to use the Moses test rather than the Siegel–Tukey test if the medians are unknown and unequal. For more detail on the Moses test, we refer you to a text on nonparametric statistics.

11.7 TEST FOR RANDOMNESS

Since we introduced the concept of a random sample in Chapter 5, we have repeatedly made statements to the effect "let $X_1, X_2, \ldots, X_n$ be a random sample from a distribution . . . ." In the study of regression and analysis of variance, we have assumed that the response variables "$Y_1, Y_2, \ldots$" or "$Y_{i,j}$" are independent. In each case we assume that the observed values in the sample data are independent (i.e., the value for one variable does not affect the value of another). The assumption of independence is necessary to simplify probability calculations and without this assumption, most of the procedures for statistical inference would not be valid. When the variables are independent, the observed values are influenced only by "random factors" and not by values assumed by other random variables. Consequently, a test for this characteristic is called a test for randomness.

Consider for a moment how a lack of randomness could come into being. One factor that could affect randomness is time. When data are collected, it is frequently collected over some period of time rather than all at once. It may be that changes take place over time which affect the values. To see this, consider the following illustration. Mindy, a saleswoman for Cosmic Cosmetics, started her job last November. She wanted to compare two different methods of selling cosmetics: method 1 consisted simply of going door to door, while method 2 consisted of first leaving a Cosmic Cosmetic Catalog at a home and then returning the next day to try to make a sale. Mindy spent her first 4 weeks of selling using method 1 and the next 4 weeks using method 2. The resulting weekly sales (in dollars) were

420 436 448 503 471 525 639 652.

Would you think that these values could be considered random samples of four values for each of the methods? There are several ways that time could affect the "randomness" of these observations. First, there is a general training effect. As Mindy gets more experience, her sales expertise can be expected to increase. Second, there is a specific effect of adapting to a particular method. (Another factor is that the approaching Christmas season would tend to increase sales.) If you look more carefully at the data, you will see that there is an increasing trend within each set of 4 weeks as well as for the combined set. It would be inappropriate to treat either group of four as a random sample from a particular distribution.

The previous illustration may give you some ideas as to how you might test for randomness. For instance, if the data are ordered, not by magnitude but according to time of collection for example, and if a pattern is apparent, a lack of randomness is indicated. Of course, for this idea to be useful, we must be able to quantify what we mean by a "pattern." To this end, consider a waiting line where 6 boys and 6 girls are waiting to go into the Bijou Theater. If you observed the line and were curious as to whether boys and girls were in a random order, you might note the order. Three (of 2156) possibilities are

B B B B B B G G G G G G

B B G B G G G B B B G G

B G B G B G B G B G B G.

(The first ordering might be for a group of elementary school children, the second for junior high, and the third for high school or college students.) Which of these seem to be random and which seem to be in a pattern? The first ordering indicates a pattern of grouping or clustering while the third is an alternating pattern (possibly couples waiting together) which is too thorough a mixing to indicate randomness. The middle pattern would appear to be random. A quantitative method for distinguishing among these patterns is to count the number of "runs" in each sequence. A run is defined in a way similar to a streak in competitive events. Specifically, a *run* is a number (one or more) of consecutive events of the same kind. A run is terminated by an event of a second kind. In the first movie-line sequence, there is one run of six B's followed by a second run of six G's. In the second case we have six different runs, which we could indicate by underlining as follows:

$\underline{B\ \ B}\quad \underline{G}\quad \underline{B}\quad \underline{G\ \ G\ \ G}\quad \underline{B\ \ B\ \ B}\quad \underline{G\ \ G}.$

In the third case there are 12 runs each of length 1. You can see that either too few or too many runs may indicate a lack of randomness.

If we have a sequence of observations, n_1 of one type and n_2 of a second type ($n_1 \leq n_2$), and we wish to see if these different types are in a random order, we can use the total number of runs in a sequence as a test statistic. Specifically, if R is the total number of runs, then in testing

$$H_0: \text{sequence is random} \quad \text{vs.} \quad H_a: \text{sequence is not random},$$

reject H_0 if R is either too large or too small, since this is essentially a two-sided alternative. For testing

$$H_0: \text{sequence is random} \quad \text{vs.} \quad H_a: \text{sequence shows clustering}$$

(a one-sided alternative), reject H_0 if R is too small, while for testing

$$H_0: \text{sequence is random} \quad \text{vs.} \quad H_a: \text{sequence is too mixed},$$

reject H_0 if R is too large.

The critical values, of course, depend on the level of significance and the probability distribution of R. It is somewhat tedious, although not difficult, to find the probability distribution of R. In particular, if the sequence consisting of n_1 observations of one type (say A's) and n_2 of another type (say B's) is truly random, then each of the

$$\binom{n_1 + n_2}{n_1} = \frac{(n_1 + n_2)!}{n_1!\,n_2!}$$

different orderings are equally likely. If each of these orderings is written out and if the number of runs for each ordering is found, the probability distribution for R is found by

$$P[R = r] = (\text{no. of orderings having } r \text{ runs}) \Big/ \binom{n_1 + n_2}{n_1}.$$

The distribution of R will be symmetric only if n_1 and n_2 are equal. The lower and upper critical values for selected values of n_1 and n_2, $2 \leq n_1 \leq n_2 \leq 20$, are given in Table A-13. For sample sizes sufficiently large, a normal approximation to the distribution of R can be used. In particular, if we let $N = n_1 + n_2$, then

$$\frac{R - (1 + 2n_1n_2/N)}{\sqrt{\dfrac{2n_1n_2(2n_1n_2 - N)}{N^2(N+1)}}} \sim N(0, 1) \quad \text{(approximately)}.$$

This approximation can be improved somewhat by using a continuity correction term.

Example 11.13 In the production of items that need to be made with a high degree of precision, there is a high probability of a defective item being produced. If among 20 such items made by the same individual in a week's time, the items were found to be defective (D) or not defective (N) as follows,

D D N D N N D D D N

N N N D D N N N N N,

does it appear as if the defectives and nondefectives occur in random order? Use a .05 level of significance.

In this case we would test H_0: "defectives and nondefectives occur randomly" against a two-sided alternative. We see that there are 8 D's and 12 N's in the sequence, so we have $n_1 = 8$ and $n_2 = 12$. The test statistic here will be R, the total number of runs. Since the alternative is two-sided, H_0 would be rejected if R is either too large or too small. From Table A13 we find the critical values to be 6 and 16. The observed value of R is 8 runs, so H_0 is not rejected. There is not sufficient evidence to reject the null hypothesis of randomness.

An important application of the test for randomness is in the examination of residuals as discussed in regression (Chapter 9) and analysis of variance (Chapter 10). If all the underlying assumptions are satisfied, a residual plot should show essentially random scatter around the line corresponding to zero residual. (Here you might wish to review Sections 9.4 and 9.5.) If the pattern is not random, this indicates that one or more of the assumptions is not satisfied and that the analysis performed is not valid.

In the examination of residuals for randomness, we would look at the pattern of positive residuals (+) and negative residuals (−).

Example 11.14 Use the residual plot in Figure 9.9(b) of Section 9.5 and test to see if the residuals are random by using a runs test. Use $\alpha = .10$.

In examining this plot we see two positive residuals, followed by six negative residuals, followed by two more positive residuals. We see that there are a total of four positive residuals and six negative residuals, so we have $n_1 = 4$ and $n_2 = 6$. The test statistic will be R, the number of runs. From Table A13 we find that when $n_1 = 4$ and $n_2 = 6$, the critical values are 3 and 9. The observed value of R is 3, so H_0 is rejected. That is, we reject the null hypothesis of randomness of the residuals and conclude that a pattern may exist.

This test objectively reinforces the subjective judgment made in Chapter 9 that this residual plot indicates a lack of randomness. Of course, this test does not make use of the rather clear parabolic pattern that we see visually. (Another nonparametric test, the "runs up and down" test would make better use of this pattern.) Consequently, the test does have its shortcomings. Another factor that we should mention relative to application of the runs test to the residuals is that the set of residuals is not independent (see comments at the end of Section 9.5). Since the dependence is relatively weak, however, the effect on the runs test will be minimal if the sample size is moderately large.

11.8 GOODNESS-OF-FIT TESTS

In previous chapters we have often made assumptions about the distributions from which random samples have been taken. The most commonly assumed distribution has been the normal distribution, since with this assumption we were able to say that a certain test statistic would have a t distribution, a chi-square distribution, or an F distribution, for example. We have assumed in other situations that samples have come from multinomial, Poisson, uniform, or other distributions. In this section we will discuss two ways in which the sample observations can be used to test to see if they may have actually come from the assumed distribution.

Consider for a moment how you might attempt to solve the following problem. Say that you are playing a simple game with a younger sister (or brother) which involves the throwing of a single die. After a period of time she becomes frustrated and complains that the die is unfair. It is up to you to convince her that the die really is fair and that you are winning solely because of superior ability. How could you proceed? You know that rolling the die only a few times would not give much information, so you might roll it a large number of times, say 60 times. If the die is really fair, you'd "expect" (in both the everyday sense and the statistical sense) to get each face 10 times. However, you would not be surprised if the "observed" number of times you got each face differed slightly from 10. Large discrepancies between the "observed" and "expected" might convince you that sis was right after all! To get a quantitative measure of the differences between observed and expected values, we could consider the sum of these differences squared. A better way is to divide these square differences by the expected values. The resulting quantity

$$X^2 = \sum_{i=1}^{6} \frac{(o_i - e_i)^2}{e_i} \tag{11.12}$$

could be used as a test statistic. Here e_i is the expected number of times that an i should appear and o_i is the observed number of times it actually did appear ($i = 1, 2, \ldots,$ or 6).

The number coming up on a die follows a multinomial distribution and a test statistic similar to the one given in (11.12) can be used to test the null hypothesis that a set of data is actually a sample from a particular multinomial distribution. More explicitly, say that $X_1, X_2, \ldots, X_n$ is a random sample from some multinomial distribution corresponding to an experiment having k distinct outcomes (see Section 4.3). To test

$$H_0: p_1 = p_{1o}, p_2 = p_{2o}, \ldots, p_k = p_{ko} \tag{11.13}$$

against the alternative that two or more of these equalities fail to hold, we define

$$o_i = \text{number of times the } i\text{th outcome occurs in } n \text{ trials}$$

and

$$e_i = np_{io} = \text{expected number of times the } i\text{th outcome would occur if } H_0 \text{ is actually true,}$$

and we use as a test statistic

$$X^2 = \sum_{i=1}^{k} \frac{(o_i - e_i)^2}{e_i}.$$

The null hypothesis would be rejected only if the test statistic is too large, since small values of X^2 correspond to good agreement with the hypothesized probabilities. It is difficult to find the exact probability distribution of the test statistic X^2, but for large n the distribution is well approximated by a chi-square distribution with $k - 1$ degrees of freedom. (A conservative rule of thumb is to have n large enough so that e_i is approximately 5 or more for all i.) Consequently, H_0 would be rejected at significance level α whenever X^2 exceeds the $\chi^2_{k-1, 1-\alpha}$ percentage point. Since this test is used to see how well the data fit or agree with the hypothesized distribution, it is called a "goodness-of-fit" test.

Example 11.15 John Luvs is in charge of the accounts receivable department of the Misery-Luvs Company. An accountant wishes to check on the department and asks John his ideas about outstanding accounts. John replies that he believes that accounts should be about as follows

Less than 3 months delinquent	50%
3–6 months delinquent	25%
6–9 months delinquent	15%
More than 9 months delinquent	10%.

TABLE 11.12
Calculations for Example 11.15

	o_i	$e_i = np_{io}$	$o_i - e_i$	$(o_i - e_i)^2/e_i$
	27	30	-3	.3000
	19	15	4	1.0667
	11	9	2	.4444
	3	6	-3	1.5000
Total	60	60	0	3.3111

The accountant, knowing that it is impractical to examine all accounts, chooses a random sample of 60 accounts and finds 27, 19, 11, and 3 accounts in these categories. Can the accountant conclude that John's ideas are accurate?

In this problem we need to test

$$H_0: \ p_1 = .50, \quad p_2 = .25, \quad p_3 = .15, \quad p_4 = .10$$

against the alternative that two or more of these values are incorrect. If we use a .05 level of significance, we will reject H_0 if the observed value of the test statistic X^2 exceeds $\chi^2_{4-1,\,1-.05} = \chi^2_{3,\,.95} = 7.81473$. An easy way to calculate the test statistic is to make a table as we have done in Table 11.12. (Note that the "$o_i - e_i$" column must sum to zero, thus providing a check on your arithmetic.) Since the observed value of the test statistic is 3.3111, which is not in the critical region, H_0 is not rejected. Based on these data, John's ideas would appear to be accurate.

One characteristic of the multinomial distribution is that exactly one of k distinct outcomes must be observed on each trial. This is not true in the case of the Poisson distribution (no theoretical upper limit to the number of possible outcomes) or the normal distribution (a continuum of possible outcomes), for example. In each of these cases, however, it is possible to group possible outcomes so that the outcomes must fall into one of k distinct groups. Grouping together different outcomes may also be necessary so that the expected number in each group is large enough for the chi-square approximation to be good. Consequently, grouping outcomes together allows us to use a goodness-of-fit test procedure for distributions other than just the multinomial.

The null hypothesis (11.13) is "simple" (cf. Section 8.3) in the sense that the values of the proportions p_{io} are completely specified by the null hypothesis. If the hypothesis is not simple, it may be necessary to estimate certain parameters from the data in order to determine the values p_{io}. The effect of this estimation is to decrease the number of degrees of freedom for the approximate chi-square distribution of the test statistic. Specifically, if there are k distinct outcomes possible and if d parameters are estimated, the number of degrees of freedom for the chi-square distribution is $k - d - 1$.

★***Example 11.16*** A portion of land in southeastern Arizona was divided into 5-m² quadrats and the number of creosote bushes were counted for 20 quadrats with the following results:

No. creosote bushes/quadrat:	0	1	2	3	4	5	6
Observed frequency:	1	4	5	6	1	1	2.

Test the null hypothesis that the spatial distribution of these plants follows a Poisson distribution. Take $\alpha = .05$.

Since the parameters of the Poisson have not been specified, it will be necessary to estimate them from the data. Specifically, we need to estimate λt, the mean number of occurrences for an area of 5 m^2. This can be estimated by using the sample mean. We find that

$$\widehat{\lambda t} = \bar{x} = \frac{(1)(0) + (4)(1) + (5)(2) + \ldots + (2)(6)}{1 + 4 + 5 + \ldots + 2} = \frac{53}{20} = 2.65.$$

Using this estimate and interpolating in Table A3, we find the probabilities and expected values to be as given in Table 11.13. Here we see an example of two reasons

TABLE 11.13

Probabilities and Expected Values for Example 11.16

No. bushes:	0	1	2	3	4	5	6 or more
$p_{io}(P[X = i \mid \lambda t = 2.65])$:	.0707	.1873	.2480	.2190	.1451	.0770	.0529
$e_i = np_{io}$:	1.414	3.746	4.960	4.380	2.902	1.540	1.058

that grouping is necessary. Since theoretically it is possible for a Poisson random variable to assume any integer value, to have only k distinct possible outcomes it is necessary to combine values. For this reason we have made the last class to be "6 or more." However, even so, not all outcomes have expected values of about 5. If we group together 0 and 1 and also group 4, 5, and 6 or more, this condition will be approximately satisfied. We can now perform a goodness-of-fit test by using $k = 4$ possible outcomes. The observed and expected values are combined as necessary. You can see from Table 11.14 that the observed value of the test statistic is 1.0136. This value must be compared with the 95th percentile of a chi-square distribution

TABLE 11.14

Calculations for Example 11.16

Group	o_i	e_i	$o_i - e_i$	$(o_i - e_i)^2/e_i$
0–1	5	5.160	−.16	.0050
2	5	4.960	.04	.0030
3	6	4.380	1.62	.5992
4 or more	4	5.500	−1.50	.4091
Total	20	20	0	1.0136

having

$$k - d - 1 = 4 - 1 - 1 = 2$$

degrees of freedom. From Table A7 we find $\chi^2_{2,.95} = 5.99146$ and see that the observed value of the test statistic is not in the critical region. We cannot reject the null hypothesis that the spatial distribution of the plants follows a Poisson distribution.

If a set of data is thought to have come from a normal distribution with unspecified mean and standard deviation, the unknown mean and standard deviation can be calculated from the data (at a "cost" of 2 degrees of freedom) and a chi-square goodness-of-fit test can be performed (although the significance level may only be approximately equal to the nominal value). In order to obtain k distinct categories into which the data can be placed, it is necessary to artificially define classes. One way of doing this is to choose classes having equal probability associated with them and such that each class has an expected number of about 5 associated with it. (If the data are already grouped and if the original data are not available, the mean and standard deviation must be estimated from the grouped data and a procedure similar to that of the previous example must be followed.)

Example 11.17 Perform a chi-square goodness-of-fit test using a .05 level of significance to test the null hypothesis that the following set of data comes from a normal distribution. The data have been put in increasing order for convenience.

220	223	230	237	247	249	251	252	256
274	277	280	303	324	337	339	346	366
376	385	401	407	410	425	485.		

The first step here is to find the sample mean and standard deviation. These values are $\bar{x} = 316$ and $s = 75.00$. In Figure 11.9 we show the density function for a normal distribution having $\mu = 316$ and $\sigma = 75$. Since there are 25 observations, we would like to divide the possible values for the normal distribution into five classes, each having a probability of .2 associated with them, and consequently expected values of

$$e_i = 25(.2) = 5.0.$$

The distribution can be properly divided by finding the 20th, 40th, 60th, and 80th percentiles of this distribution. You can verify that these values (shown on the graph)

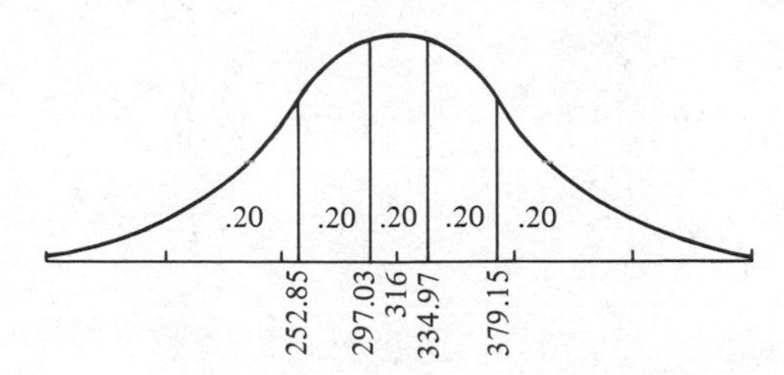

FIGURE 11.9
Normal Curve for Example 11.17

TABLE 11.15
Calculations for Example 11.17

Group	o_i	e_i	$o_i - e_i$	$(o_i - e_i)^2/e_i$
Less than 252.85	8	5	3	1.80
252.85–297.03	4	5	−1	.20
297.03–334.97	3	5	−2	.80
334.97–379.15	4	5	−1	.20
Greater than 379.15	6	5	1	.20
Total	25	25	0	3.20

are

$$x_{.20} = 316 - .842(75) = 252.85, \qquad x_{.40} = 316 - .253(75) = 297.03,$$
$$x_{.80} = 316 + .842(75) = 379.15, \qquad x_{.60} = 316 + .253(75) = 334.97.$$

Once these classes and expected values are established, it is a simple matter to compute the value of the test statistic. The necessary calculations for this are shown in Table 11.15. The value of the test statistic, 3.20, must be compared with the 95th percentile of a chi-square distribution having

$$k - d - 1 = 5 - 2 - 1 = 2$$

degrees of freedom. This value is 5.99146, so H_0 is not rejected.

The procedure just discussed for testing for normality is relatively easy to carry out, but unfortunately it is not a very powerful procedure (see Section 8.10). By this we mean that even if the sample comes from a nonnormal distribution, the probability of rejecting the null hypothesis of normality is relatively small unless the sample size is rather large. A somewhat more powerful procedure that can be used for tests about the distributional form for continuous random variables (not just normal random variables) is the Kolmogorov–Smirnov test. The Kolmogorov–Smirnov test is based on comparing the cumulative distribution function of the hypothesized distribution with the "sample cumulative distribution function." (This contrasts with the chi-square goodness-of-fit procedure, which in some sense compares sample data with the probability density function.) Specifically, the test statistic is denoted by D, the largest absolute difference between the hypothesized and sample cumulative distribution functions. The critical values for various sample sizes and significance levels are given in Table A14.

If $X_1, X_2, \ldots, X_n$ is a random sample from some distribution, the *sample cumulative distribution function* is defined by

$$F_n(x) = \frac{\text{no. of sample observations} \leq x}{n}.$$

A graph of this function will be a step function similar to those for the CDF of discrete random variables (Section 2.4). In order to test

$$H_0: \text{sample comes from a distribution having CDF } F_0(x)$$

against the alternative that it comes from some other distribution, we use as a test statistic

$$D = \text{largest absolute difference between } F_n(x) \text{ and } F_0(x).$$

The null hypothesis will be rejected at a specified level of significance if D is too large.

Example 11.18 Use a Kolmogorov–Smirnov test with a .05 level of significance to test the null hypothesis that the following data are a random sample from a uniform distribution over the interval from 0 to 1. (The data have been put in increasing order for convenience.)

.100 .135 .253 .354 .376 .467 .520 .863 .876 .973

Since the distribution is specified to be uniform (0, 1), we can see (by arguments similar to those given in Example 2.14 of Section 2.4) that

$$F_0(x) = x \qquad 0 \leq x \leq 1.$$

This function is shown in Figure 11.10 along with the sample distribution function.

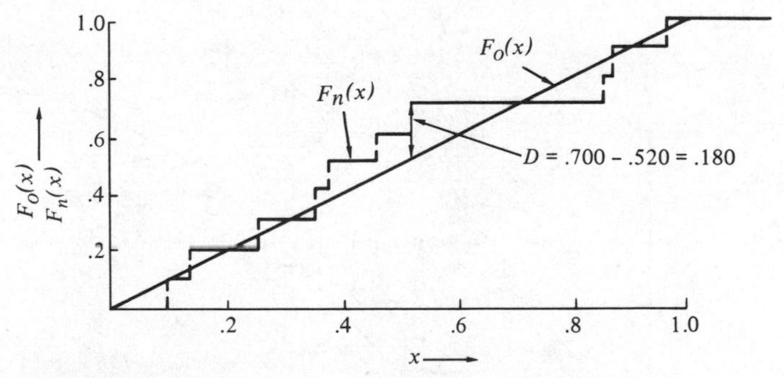

FIGURE 11.10
Distribution Functions for Example 11.18

You can also see from the graphs that the largest absolute difference, the value of the test statistic, is .180. From Table A14 we see that for $n = 10$ and $\alpha = .05$, the critical value is .409. Since D is less than this value, H_0 is not rejected (i.e., we do not reject the null hypothesis that the sample came from a uniform distribution).

You can see from the graphs in Figure 11.10 that since $F_0(x)$ is a continuous nondecreasing function and since $F_n(x)$ is a step function, the largest differences must occur right where the jumps or steps occur. The absolute differences may be largest just before the jump occurs or right after it occurs. Consequently, instead of making a graph, we could make a table showing the absolute differences $|F_n(x) - F_0(x)|$ just before and after each step. The steps, of course, occur at each observed data value.

The process (when done by hand instead of a computer subroutine) is tedious whether done graphically or in a table.

Example 11.19 Use a Kolmogorov–Smirnov test with a .05 level of significance to test the null hypothesis that the data given in Example 11.17 come from a normal distribution.

Since the particular normal distribution is not specified, it is necessary to estimate the mean and standard deviation. These values were found earlier to be $\bar{x} = 316$ and $s = 75$. (Note that estimating these parameters from the data will make the test conservative. That is the actual level of significance will be smaller than the nominal value.) Graphing the sample distribution function is not difficult, but graphing the normal distribution function is more challenging. We need to standardize and use the standard normal tables. For example,

$$F_0(220) = P[X \leq 220 \mid \mu = 316, \sigma = 75]$$
$$= P\left[Z \leq \frac{220 - 316}{75}\right]$$
$$= F_Z(-1.28) = .1003.$$

A graph of the distribution functions is shown in Figure 11.11. Calculations show

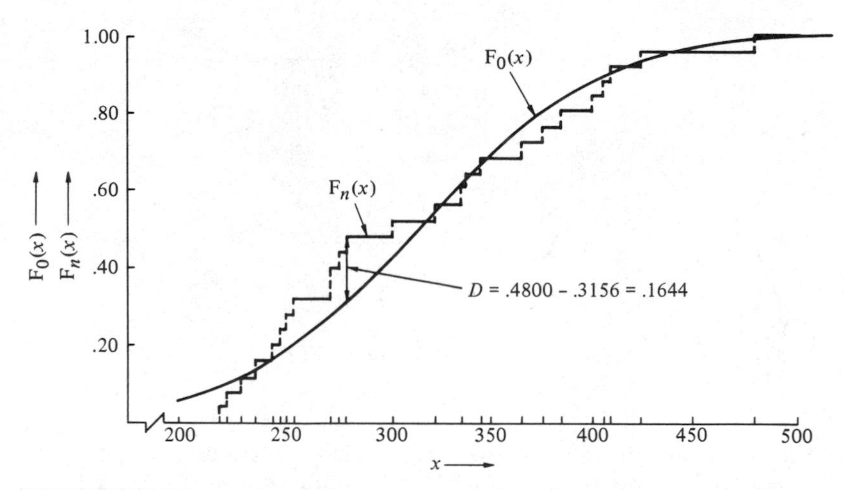

FIGURE 11.11

that the maximum absolute difference is $D = .1644$. The critical value for $n = 25$ and $\alpha = .05$ is found from Table A14 to be .264. Consequently, the null hypothesis is not rejected; that is, there is not sufficient evidence to reject the hypothesis that the sample observations come from a normal distribution.

We will close the section with two comments. First in deciding between the chi-square or Kolmogorov–Smirnov test for a goodness-of-fit test of data to some hypothesized distribution, you must consider the type of data at hand. If the data are categorical (nominal), a chi-square procedure must be used. If it is ordinal, interval, or ratio, the Kolmogorov–Smirnov test is more appropriate. Second, if primary interest lies in testing data for normality, then rather than use a "general-purpose" test such as the Kolmogorov–Smirnov, it is better to use a test designed specifically for this purpose. If you are interested in such tests we would refer you to W. J. Conover, *Practical Nonparametric Statistics* (New York: John Wiley & Sons, Inc., 1971), for references.

11.9 ONE-WAY ANALYSIS OF VARIANCE

In Section 11.5 we presented the rank sum test as a nonparametric test procedure for tests about the difference in location of two populations. The samples were assumed to be independent samples from the two populations. In this section we will present a method that can be used to test whether k different populations have the same location or whether one or more populations differ in location from the others. The test procedure that we will use is the Kruskal–Wallis test, a nonparametric analog of the one-way ANOVA test procedure discussed in Section 10.2. In that situation we assumed that the responses followed a normal distribution. If it is not certain that the normality assumption holds, the nonparametric Kruskal–Wallis procedure may be more appropriate.

We will assume that we have available k independent samples (of possibly different sizes) from k populations. Under the null hypothesis we assume that all populations are identical (and consequently all have the same median). We will assume that if the populations differ, they differ in location, and consequently the alternative hypothesis will be that the medians differ (cf. assumptions in Table 11.9 and the graphs in Figure 11.5 in Section 11.5). Specifically, say that we have k independent random samples

$$\begin{array}{c} x_{1,1}, x_{1,2}, \ldots, x_{1,n_1} \\ x_{2,1}, x_{2,2}, \ldots, x_{2,n_2} \\ \vdots \\ x_{k,1}, x_{k,2}, \ldots, x_{k,n_k} \end{array}$$

(with $n_1 \leq n_2 \leq \ldots \leq n_k$) and we wish to test

$$H_0: \tilde{\mu}_1 = \tilde{\mu}_2 = \ldots = \tilde{\mu}_k \qquad \text{vs.} \qquad H_a: \text{not all medians are equal.}$$

To find a suitable test statistic, we will consider the ranks assigned to the observations in each sample. If the null hypothesis is true, the distributions are all the same and the various possible assignments of ranks to the sample observations should all be equally likely. Since the possible ranks that could be assigned to $x_{1,1}$, for example, are

$$1, 2, \ldots, N \qquad (N = n_1 + n_2 + \ldots + n_k),$$

and since these are all equally likely, the *expected value of the rank* assigned to $x_{1,1}$ would be $(N + 1)/2$. Since this is true for each sample value, the expected mean rank for all n_1 values in the first sample would also be $(N + 1)/2$. Consequently, if we define

$$\bar{T}_1 = \text{mean rank of sample observations from first sample,}$$

we would expect that $\bar{T}_1$ would not deviate too much from its expected value (assuming that H_0 is true). Likewise, we would expect $\bar{T}_2, \bar{T}_3, \ldots \bar{T}_k$, to also be close to $(N+1)/2$. One way of measuring the overall deviation from the expected mean rank is to use

$$\sum_{i=1}^{k} n_i\left(\bar{T}_i - \frac{N+1}{2}\right)^2. \tag{11.14}$$

Note, however, that if H_0 is not true so that the population medians are different, the quantity (11.14) will tend to be larger than otherwise expected. Consequently, we could use (11.14) as a test statistic. As you might imagine, however, there are many different sample sizes (n_i) and populations (k) which would have to be considered to make complete tables for the probability distribution of this statistic. Therefore, it is highly desirable to find a good approximation to the exact distribution. Such an approximation can be found if (11.14) is multiplied by an appropriate constant, namely, $12/N(N+1)$. In particular,

$$H = \frac{12}{N(N+1)}\sum_{i=1}^{k} n_i\left(\bar{T}_i - \frac{N+1}{2}\right)^2 \tag{11.15}$$

will have an approximately chi-square distribution with $k - 1$ degrees of freedom. The approximation is relatively good if each sample size n_i is 5 or more. For $k = 3$ and $n_1 \leq n_2 \leq n_3 \leq 5$, critical values for the test statistic H are given in Table A15.

Example 11.20 Samples of size 5 were taken from each of three populations, with results as shown in Table 11.16. Using these data, test the null hypothesis that the population medians are all equal against the alternative that at least one median differs from the other. Use $\alpha = .05$.

The appropriate test statistic is the random variable H as defined in (11.15). In Table 11.16 we show the ranks associated with each value as well as the sum and

TABLE 11.16
Data for Example 11.20

Sample 1		*Sample 2*		*Sample 3*	
Value	*Rank*	*Value*	*Rank*	*Value*	*Rank*
.61	5	.59	4	1.08	12
.19	2	.78	7	1.14	14
.79	8.5	1.07	11	.77	6
.04	1	1.10	13	1.28	15
.46	3	.84	10	.79	8.5
	$T_1 = 19.5$		$T_2 = 45$		$T_3 = 55.5$
	$\bar{T}_1 = 3.9$		$\bar{T}_2 = 9$		$\bar{T}_3 = 11.1$

the average of the ranks for each sample. Note that we treat ties by assigning the average of the ranks that would have been assigned had the values not been tied. We find

$$H = \frac{12}{15(15+1)}\{[5(3.9-8)^2] + [5(9-8)^2] + [5(11.1-8)^2]\}$$
$$= 6.855.$$

This value must be compared with the critical value. Since $n_1 = n_2 = n_3 = 5$ and $\alpha = .05$, we find the critical value from Table A15 to be 5.78. Since the observed value is in the critical region, H_0 is rejected and we would conclude that the medians are not the same.

There are two algebraically equivalent formulas that can be used to calculate the test statistic H. We have

$$H = \frac{12}{N(N+1)} \sum_{i=1}^{k} \frac{[T_i - n_i(N+1)/2]^2}{n_i} \tag{11.16}$$
$$= \left[\frac{12}{N(N+1)} \sum_{i=1}^{k} \frac{T_i^2}{n_i}\right] - 3(N+1).$$

Example 11.21 The kangaroo rat inhabits desolate arid regions of the southwestern United States, surviving on a diet of dry seeds, an occasional succulent leaf, and no drinking water. Environmental physiologists have studied this rodent under laboratory conditions. One such study focused on the animal's response to heat and cold and dietary restrictions. The measured response was percent of hepatic fatty acid compared to the total fatty acid in the liver. To obtain this measurement it was necessary to kill the animals, so different animals had to be used for each set of conditions. Assume that 17 animals were available for the study and that 6 were randomly assigned to treatment 1 (warm conditions, abundant food), 6 to treatment 2 (cold conditions, abundant food), and 5 to treatment 3 (cold conditions, limited food). If the results are as shown in Table 11.17, could a researcher conclude that the median response differs for these three treatments? Use $\alpha = .05$.

TABLE 11.17
Data for Example 11.21

Treatment 1 (W-A)		*Treatment 2 (C-A)*		*Treatment 3 (C-L)*	
%	*Rank*	%	*Rank*	%	*Rank*
.80	12	1.39	17	.52	5
.57	6	.66	9	.58	7.5
.71	11	1.25	16	.58	7.5
.36	3	.88	13	.28	1
.44	4	.93	14	.33	2
.68	10	1.07	15		
	$T_1 = 46$		$T_2 = 84$		$T_3 = 23$

Using the Kruskal–Wallis test to test the null hypothesis of no differences among the medians, we find the value of the test statistic H by using equation (11.16). We have

$$H = \left[\frac{12}{17(18)}\left(\frac{46^2}{6} + \frac{84^2}{6} + \frac{23^2}{5}\right)\right] - 3(18) = 10.097.$$

Since the sample sizes are larger than the tabled values, we use the chi-square approximation to the distribution of H and reject the null hypothesis if H is at least $\chi^2_{3-1,.95} = 5.99$. In this case the null hypothesis is rejected.

In most situations, if the null hypothesis of equal medians is not rejected, there is not convincing evidence that the medians are different and the matter is not pursued further. However, if there is evidence that the medians are in fact different, an investigator is generally interested in knowing specifically *which* medians differ from the others. In order to determine this, multiple comparisons (Section 10.4) are generally performed. For instance, in the previous example an investigator might be interested in knowing whether one median differed from the other two or whether all three were different. As in the parametric analysis, there are many ways to perform multiple comparisons. As we did earlier, we could use Bonferroni's method here. In this case we would perform a rank sum test for each pair of medians we wish to compare. If we make c comparisons, we would use a level of significance of $\alpha/2c$ for each pair, thus guaranteeing an overall level of significance of no more than α. You can verify that with an overall level of significance of .05, the median of treatment 2 differs from that of both treatments 1 and 3, but the medians of treatments 1 and 3 are not significantly different (see Exercise 32).

As with other nonparametric test procedures based on ranks, the underlying distribution of the response is assumed to be continuous so that theoretically ties cannot occur. If, because of roundoff, ties do occur and are numerous, a correction for ties should be used. See S. Siegel, *Nonparametric Statistics* (New York: McGraw-Hill Book Company, 1956), for more detail on this.

11.10 TEST FOR EQUALITY OF POPULATION PROPORTIONS

In the last section we considered the problem of comparing k different populations, with interest focused primarily on the medians of those populations. To perform a Kruskal–Wallis test of the null hypothesis that the medians are all equal, it is necessary to be able to rank the observations from each population; consequently, the measurement scale for the observations must be at least ordinal. In this section we will discuss a method for comparing k populations whose observed values can be measured only on a categorical (nominal) scale. The method used will be based on comparing the proportion of observed values in each of the categories.

As an example, say that Kay Marrt is the director of marketing for a small chain of variety stores. The items sold in these stores can be classified into one of five categories: clothing, housewares, hardware, sporting goods, or miscellaneous. Kay wishes to know if the proportion of items sold in each of these categories is the same in each of the four stores in the chain. To gather information on these proportions she chose one day at random and found the number of items sold in each

TABLE 11.18
Sales Data for Four Variety Stores

		Category					
		Clothing	*House wares*	*Hard ware*	*Sporting Goods*	*Miscel-laneous*	*Store Total*
Store	*1*	510	363	410	242	895	2,420
	2	388	265	305	245	837	2,040
	3	860	620	689	488	1,703	4,360
	4	1,362	936	1,092	741	2,649	6,780
Category Total		3,120	2,184	2,496	1,716	6,084	15,600 *Grand Total*

category for each store. The values that she obtained are given in Table 11.18. How can these data be used to answer her question about the proportions? Let us first put her question in the form of a statistical hypothesis. Let

$p_{i,j}$ = proportion of items sold in the ith store which fall into the jth category.

[For example, $p_{1,2}$ is proportion of items sold in the first store which fall into the second (housewares) category.] The null hypothesis of interest is

$$H_0: p_{1,j} = p_{2,j} = p_{3,j} = p_{4,j} \qquad \text{for } j = 1, 2, \ldots, 5, \tag{11.17}$$

while the alternative would be that one or more of these equalities fail to hold. There is a strong similarity between the null hypotheses (11.17) and (11.13) of Section 11.9. In that section we considered a goodness-of-fit test of observed proportions to specified (hypothesized) proportions, while in the present case we wish to compare four different sets of proportions whose values are unspecified. In view of the similarity in hypotheses, it should not be surprising that the test statistic is similar. In fact, we use essentially the same test statistic, namely,

$$X^2 = \frac{\sum_{i,j} (o_{i,j} - e_{i,j})^2}{e_{i,j}}. \tag{11.18}$$

(Here $o_{i,j}$ and $e_{i,j}$ represent the observed and expected numbers in the ith row and jth column.) In this case, while the proportions in a given category are assumed to be the same for each store, their common values are unknown and hence must be estimated from the data. For example, we may wish to estimate the proportion of sales that are categorized as clothing. Since we assume that the proportion is the same for each store, that is, $p_{1,1} = p_{2,1} = p_{3,1} = p_{4,1}$ [see (11.17)], we can denote this common value by p_1 and get an overall estimate for this value. We see that the total number of clothing sales for all stores was 3120, so the overall estimate of p_1 is

$$\hat{p}_1 = \frac{3120}{15{,}600} = .20.$$

Consequently, the estimated expected number of clothing items for the first store is

$$e_{1,1} = n_{1.}\hat{p}_1 = 2420(.20) = 484,$$

where $n_{1}.$ represents the total sales for the first store. (Note that we have a notational dilemma here. We would like to use $o_{i,j}$ for the mnemonic of associating o with "observed." However, in calculating expectations for multinomial random variables, we like to use np. We trust you will not be confused by our trying to get the best of both worlds. If we define $n_{i,j} = o_{i,j}$, then

$$n_{i\cdot} = \sum_j n_{i,j} = \sum_j o_{i,j} = \text{total number of observations for } i\text{th population},$$

$$n_{\cdot j} = \sum_i n_{i,j} = \sum_i o_{i,j} = \text{total number of observations for } j\text{th category},$$

and

$$n_{\cdot\cdot} = \sum_{i,j} n_{i,j} = \sum_{i,j} o_{i,j} = \text{grand total number of observations.)}$$

In general, then,

$$e_{i,j} = n_{i\cdot}\hat{p}_j = n_{i\cdot}\left(\frac{n_{\cdot j}}{n_{\cdot\cdot}}\right). \tag{11.19}$$

The test statistic X^2 given in (11.18) has an approximately chi-square distribution (if the expected values are approximately 5 or more) with the number of degrees of freedom equal to

$$(k - 1)(c - 1) = (4 - 1)(5 - 1) = 12$$

(where k = number of populations and c = number of categories). The null hypothesis (11.17) will be rejected if the test statistic is too large. Of course, X^2 is the appropriate test statistic for testing the more general null hypothesis

$$H_0: p_{1,j} = p_{2,j} = \ldots = p_{k,j} \qquad j = 1, 2, \ldots, c$$

against the alternative that one or more of these inequalities fail to hold.

Example 11.22 Using the data Kay gathered (Table 11.18), test the null hypothesis that the proportion of items sold in each category is the same for all four stores against the alternative that they are not all the same. Use a .05 level of significance.

An easy way to display the quantities necessary to calculate the value of the test statistic X^2 is to rewrite a table containing the observed values and placing the expected values parenthetically next to these. This has been done in Table 11.19. The expected values were calculated using equation (11.19). We find X^2 to be

$$X^2 = \frac{(510 - 484)^2}{484} + \frac{(363 - 338.8)^2}{338.8} + \ldots + \frac{(2649 - 2644.2)^2}{2644.2} = 17.9766.$$

[Note that computationally it is often easier to find X^2 by using the form

$$X^2 = \left[\sum_{i,j} \frac{(o_{i,j})^2}{e_{i,j}}\right] - n_{\cdot\cdot},$$

which is algebraically equivalent to equation (11.18).] We need to compare the observed value of the test statistic with the 95th percentile of a chi-square distribution having $(k - 1)(c - 1) = 12$ degrees of freedom. This value is 21.0261. Since the observed value of the test statistic does not exceed this critical value, H_0 is not rejected. The proportion of items sold in a given category can be considered to be the same for all four stores.

TABLE 11.19

Data for Example 11.22 with Expected Numbers Shown

		Category					
		Clothing	House-wares	Hard-ware	Sporting Goods	Miscel-laneous	Store Total
Store	1	510(484)	363(338.8)	410(387.2)	242(266.2)	895(943.8)	2,420
	2	388(408)	265(285.6)	305(326.4)	245(224.4)	837(795.6)	2,040
	3	860(872)	620(610.4)	689(697.6)	488(479.6)	1,703(1700.4)	4,360
	4	1,362(1356)	936(949.2)	1,092(1084.8)	741(745.8)	2,649(2644.2)	6,780
Category Total		3,120	2,184	2,496	1,716	6,084	15,600 Grand Total

Example 11.23 An economist studied workers in two geographical regions. The economist interviewed 500 workers from region 1 and 300 workers from region 2. Those workers were classified as unemployed, blue collar, white collar, or professionals. Use the data given in Table 11.20 to test the null hypothesis that in the two regions the proportion of workers in each classification is the same against the alternative that some proportions differ. Use a .01 level of significance.

The expected values for each category are shown parenthetically in Table 11.20. In this case $k = 2$ and $c = 4$, so the test statistic, X^2, will have an approximate chi-square distribution with $(k - 1)(c - 1) = 3$ degrees of freedom. The critical value is $\chi^2_{3,.99} = 11.3449$, while the observed value of the test statistic is

$$X^2 = \left(\sum \frac{o_{i,j}^2}{e_{i,j}}\right) - n.. = 815.3193 - 800 = 15.3193.$$

TABLE 11.20

Data for Example 11.23

		Employment Category				
		Unemployed	Blue Collar	White Collar	Professional	Total
Region	1	86(100)	131(125)	179(187.5)	104(87.5)	500
	2	74(60)	69(75)	121(112.5)	36(52.5)	300
Total		160	200	300	140	800

Consequently H_0 will be rejected. The economist can conclude that the proportions of workers in each category differs between these regions.

11.11 ANALYSIS OF RANDOMIZED COMPLETE BLOCK DESIGN

In Section 10.7 we discussed the advantages of using a randomized complete block design instead of the completely randomized design. In particular, we saw that by grouping experimental units that are alike into a group called a block, the observed differences in responses within each block would more likely be due to differences in the treatments and not to differences in the experimental units themselves. For this reason we saw that to test for the effect of different brands of oil on gas mileage, it was better to construct "blocks" consisting of similar cars than to simply randomly assign oils to various cars. You may recall that the analysis carried out in that problem depended on the assumption of normality. In this section we will present a method of testing to see if the responses to the different treatments are the same, which does not depend on the normality assumption. This nonparametric test procedure is called Friedman's test.

The basic underlying assumptions for Friedman's test are that the responses between blocks are independent, the data within blocks are ordinal, and within each block the treatment effects are all the same, so that the distribution of the responses is the same. The alternative of interest is that the location of one (or more) of the response distributions differs from the others. In terms of the oil example, if for the compacts, say, there is no difference among the treatment effects, the distribution of the mileages will be the same for all four brands of oil. If we let Y_A, Y_P, Y_R, and Y_S represent the mileages when Aristocrat, Plebeian, Royal, and Super Slick oils (respectively) are used in compact cars and if the distributions of these mileages really are the same, each possible assignment of ranks to these four responses should be equally likely. In such a case, what is the *expected rank* that would be assigned to Y_A? If each rank is equally likely, we would find

$$E(\text{rank for } Y_A) = 1(\tfrac{1}{4}) + 2(\tfrac{1}{4}) + 3(\tfrac{1}{4}) + 4(\tfrac{1}{4}) = \tfrac{5}{2}.$$

More generally, if there were k different oils being compared, we would find

$$E(\text{rank for } Y_A) = 1\left(\frac{1}{k}\right) + 2\left(\frac{1}{k}\right) + \dots + k\left(\frac{1}{k}\right) = \frac{k+1}{2}.$$

Likewise, the expected rank for Y_P, Y_R, and Y_S, for example, would each be $(k + 1)/2$. The same expected rank would be found for every block considered, so if b blocks were used, the expected sum of the ranks for Y_A, say, over all b blocks would be

$$E(\text{sum of ranks for } Y_A \text{ taken over all } b \text{ blocks}) = \frac{b(k+1)}{2},$$

with a similar result holding for each other oil being considered. This idea is summarized in Table 11.21. Knowing what sum of ranks to expect when H_0 is true gives us a means for finding a test statistic. In particular, we compare the sum of ranks for each treatment with the expected sum of ranks when H_0 is true. Large deviations from the expected sum would indicate that H_0 is not true. In particular, we can define

$$S = \sum_{i=1}^{k} \left[R_i - \frac{b(k+1)}{2}\right]^2, \tag{11.20}$$

where R_i is the sum of the ranks assigned to the ith treatment. If we used S as our test statistic, we would reject H_0 if S is too large, since this indicates a large amount

TABLE 11.21
Expected Ranks When H_0 Is True

		Block Number				
		1	2	...	b	Expected sum of ranks over all b blocks
Treatment	1 (*Aristocrat Oil*)	$\left(\frac{k+1}{2}\right)$	$\left(\frac{k+1}{2}\right)$	...	$\left(\frac{k+1}{2}\right)$	$b(k+1)/2$
	2 (*Bard Oil*)	$\left(\frac{k+1}{2}\right)$	$\left(\frac{k+1}{2}\right)$	...	$\left(\frac{k+1}{2}\right)$	$b(k+1)/2$
	⋮	⋮				⋮
	k (*Kens Oil*)	$\left(\frac{k+1}{2}\right)$	$\left(\frac{k+1}{2}\right)$	...	$\left(\frac{k+1}{2}\right)$	$b(k+1)/2$
		↑ Expected ranks within block 1, etc.				

of disagreement with what we would expect if H_0 were true. Because of the form of this statistic used for large-sample approximations, it is common to use a function of S as a test statistic. In particular, we define

$$Q = \frac{12S}{bk(k+1)} = 12\left[\sum \left(R_i - \frac{b(k+1)}{2}\right)^2\right] \Big/ bk(k+1) \tag{11.21}$$

and reject H_0 if Q is too large.

We can summarize the preceding discussion as follows. If we have k different treatments each of which is assigned to one of k experimental units within a block, and if we have a total of b blocks, then to test

(11.22) H_0: the distributions of the responses to the k treatments within a block are the same

against the alternative that the distributions differ in location, we first rank the observed responses within each block. Next, find the sum of the ranks associated with each treatment and call this sum R_i, $i = 1, 2, \ldots, k$. Then calculate the value of Q defined by (11.21). Find the critical value from Table A16 corresponding to the appropriate values of α, b, and k. For large b, we can use the fact that Q has (approximately) a chi-square distribution with $(k-1)$ degrees of freedom and reject H_0 if Q exceeds $\chi^2_{k-1,\,1-\alpha}$. For computational convenience, we can use the fact that

$$Q = 12\left[\left(\sum_{i=1}^{k} R_i^2\right) - b^2(k+1)^2k/4\right] \Big/ bk(k+1).$$

Example 11.24 Use the data on car mileages for different brands of oil given in Table 10.15 of Section 10.7 to test the null hypothesis (11.22). Use a .01 level of significance.

The data are repeated in Table 11.22 with the modification that the mileages (responses) within each type of car (block) are ranked. The ranks are shown parenthe-

TABLE 11.22
Data for Example 11.24

		Type of Car (Block)					
		Sub-compact	Compact	Inter-mediate	Full Size	Station Wagon	R_i
Type of Oil	Aristocrat	43(1.5)	37(1.5)	24(1)	19(1.5)	17(1)	6.5
	Plebeian	44(3)	39(3)	26(2)	19(1.5)	21(4)	13.5
	Royal	46(4)	43(4)	30(4)	22(4)	19(2.5)	18.5
	Super-Slick	43(1.5)	37(1.5)	28(3)	20(3)	19(2.5)	11.5

tically and the sum of ranks is shown in the last column. Since $k = 4$ and $b = 5$, we find

$$S = \sum (R_i - 12.5)^2 = 74 \qquad \text{and} \qquad Q = 8.88.$$

Since $b = 5$, we cannot use Table A16 to find the critical value, but rather we must use the chi-square approximation. The critical value is

$$\chi^2_{4-1,.99} = 11.3449,$$

but Q does not exceed this value, so H_0 is not rejected. (Note, however, that it would have been rejected had α been .05.)

Example 11.25 Ellen Swee is the development manager for the Swee-Tooth Candy Company. Ellen wishes to test market a new candy bar to be chosen from three candidates: Chocolate Goo, Slickery Licorice, and Caramel Sticks. Since these candies are designed to appeal to children, she wishes to have some children rank the three

TABLE 11.23
Data for Example 11.25

		Child (Block)									
		Amy	Bob	Carl	Dawn	Ed	Fay	Gene	Hans	Iris	R_i
Type of Candy	Chocolate Goo	1	1	1	2	1	2	1	1	1	11
	Slickery Licorice	2	3	3	1	3	2	3	2	3	22
	Caramel Sticks	3	2	2	3	2	2	2	3	2	21

candies. If one candy receives better ratings than the others, that candy will be test-marketed; otherwise, the decision about which to market will be made on other grounds. Nine children were randomly selected, and each was asked to rank the three candy bars. (Note that the response measured here is actually a rank.) Given the data in Table 11.23, test the null hypothesis that the distribution of the responses to each candy bar is the same (i.e., that all are equally rated) against the alternative that at least one is rated differently than the others. Use a .05 level of significance.

Each child ranked the three candy bars (with a rank of 1 indicating the favorite). Since Fay considered all three bars equally good, she assigned a rank of 2 to each. Since the data given here are already in terms of ranks, we proceed directly to the calculation of the test statistic Q. Since $k = 3$ and $b = 9$, we find

$$Q = \frac{12\left\{\sum_{i=1}^{3}\left[R_i - \left(\frac{9(4)}{2}\right)\right]^2\right\}}{9(3)(4)} = 8.222.$$

From Table A16 we find that when $\alpha = .05$, the critical value is 6.222, so H_0 would be rejected. Further examination of the data shows that Chocolate Goo received the best ranks and hence is the logical candidate for test marketing.

11.12 RANK CORRELATION AND INDEPENDENCE

In Chapters 3 and 9 we discussed joint probability distributions for two random variables X and Y. We have seen that in some situations there is a cause–effect relationship between two variables (as when $X =$ amount of annual rainfall and $Y =$ crop yield), while in other situations there may only be an association but not a cause–effect relationship between them (as when $X =$ height of brother and $Y =$ height of sister within a family). It is not the statistician's place to decide whether or not a cause–effect relationship exists between variables, but the statistician can provide measures of association.

A commonly used measure of association is ρ, the correlation coefficient, defined in Section 3.5. If the joint distribution for two random variables X and Y is known, ρ can be calculated. Some of the important properties of the correlation coefficient were given in Section 9.8. Briefly these properties are:

1. $-1 \leq \rho \leq 1$.
2. $|\rho| = 1$ if and only if X and Y are linearly related.

The correlation ρ is a *measure of the strength of the linear relationship* between X and Y. It is possible for ρ to be zero and for a strong nonlinear relationship to exist. However, it is always true that if X and Y are independent, ρ will equal zero [equation (3.33) of Section 3.5].

The sample correlation coefficient r, defined by

$$r = \frac{\sum (x_i - \bar{x})(y_i - \bar{y})}{\sqrt{\sum (x_i - \bar{x})^2 \sum (y_i - \bar{y})^2}} \tag{11.23}$$

[cf. equation (9.43) of Section 9.8], can be used as a point estimate of ρ. In order to find confidence intervals for ρ, we had to make the assumption that the data consisted of independent observations from a bivariate normal distribution. This distribution has the special property that $\rho = 0$ if and only if X and Y are independent, so a test of the hypothesis that $\rho = 0$ is equivalent to a test for independence. Unfortunately, the assumption of bivariate normality cannot be made for many sets of data. For example, if the data for the X variable and/or Y variable are only ordinal, the assumption of bivariate normality cannot hold. In this section we will present two ways of measuring and testing for association between random variables when the assumption of bivariate normality cannot be made.

★*Example 11.26* A researcher was interested in the relationship between the number of rabbits in a litter (X) and the mean weight of the baby rabbits in the litter (Y). The study was done using rabbits from a particular strain. The data from the experiment are given in Table 11.24. Calculate the sample correlation coefficient r. Can the assumption of bivariate normality be made for these data?

Direct calculations show that $r = -.750$. Since the litter sizes are small integer values, it follows that these values cannot follow a normal distribution; hence X and Y together cannot be bivariate normal.

TABLE 11.24

Data for Example 11.26

Doe	*Litter Size, X*	*Mean Weight, Y (grams)*	*Doe*	*Litter Size, X*	*Mean Weight, Y (grams)*
A	2	47.0	K	4	47.7
B	3	42.3	L	3	47.7
C	6	39.7	M	6	37.1
D	5	40.3	N	1	52.7
E	4	41.4	O	2	42.3
F	7	39.4	P	1	59.2
G	8	40.5	Q	6	48.8
H	1	54.2	R	3	51.2
I	2	49.4	S	2	51.0
J	5	43.1	T	3	49.0

Since the assumption of bivariate normality is not reasonable for the data in the previous example, it would be inappropriate to test the null hypothesis that X and Y are independent using the methods of Chapter 9. A measure of association that can be used to test this hypothesis is Spearman's rank correlation coefficient r_S. This quantity is obtained by finding the sample correlation coefficient using the ranks of the observed values rather than the values themselves. Specifically, if we define

$$R(X_i) = \text{rank of } X_i \text{ among the set } X_1, X_2, \ldots, X_n$$

and

$$R(Y_i) = \text{rank of } Y_i \text{ among the set } Y_1, Y_2, \ldots, Y_n,$$

then Spearman's rank correlation coefficient is defined by

$$r_S = \frac{\sum [R(X_i) - \overline{R(X)}][R(Y_i) - \overline{R(Y)}]}{\sqrt{\{\sum [R(X_i) - \overline{R(X)}]^2\}\{\sum [R(Y_i) - \overline{R(Y)}]^2\}}} \tag{11.24}$$

[cf. equation (11.23)]. Since the ranks must be all values from 1 through n, it follows that

$$\overline{R(X)} = \overline{R(Y)} = \frac{1 + 2 + \ldots + n}{n} = \frac{n + 1}{2},$$

so (11.24) can be simplified and can be expressed by any of the following algebraically equivalent formulas:

$$(11.25) \qquad r_S = \frac{\sum [(R(X_i) - (n + 1)/2)(R(Y_i) - (n + 1)/2)]}{\sqrt{\sum (R(X_i) - (n + 1)/2)^2][\sum (R(Y_i) - (n + 1)/2)^2]}}$$

$$= \frac{\sum R(X_i)R(Y_i) - n(n + 1)^2/4}{\sqrt{[(\sum R(X_i)^2) - n(n + 1)^2/4][\sum R(Y_i)^2 - n(n + 1)^2/4]}}.$$

If there are no tied observations, either of the two following forms can be used:

$$r_S = \frac{\sum [R(X_i) - (n + 1)/2)(R(Y_i) - (n + 1)/2)]}{n(n^2 - 1)/12}$$

$$= 1 - \left\{\frac{6 \sum [R(X_i) - R(Y_i)]^2}{n(n^2 - 1)}\right\}.$$

⋆***Example 11.27*** Using the data on litter size and mean weight of baby rabbits given in Table 11.24, find the value of Spearman's rank correlation coefficient.

To find r_S we must first find the ranks of the X's (litter size) and Y's (mean weights). In Table 11.25 we show the original data with the ranks for each X and each Y shown parenthetically. For tied observations we have again used the convention of assigning the average of the ranks that would have been assigned to the tied observations. Using these ranks, we find

$$\sum R(X_i)R(Y_i) = (5.5)(11) + (9.5)(7.5) + \ldots + (9.5)(14) = 1681.5,$$

$$\sum R(X_i)^2 = 2855, \qquad \sum R(Y_i)^2 = 2869.5.$$

Substituting into (11.25), we find that

TABLE 11.25

Rabbit Data Showing Ranks

Doe	*Litter Size, X*	*Mean Weight, Y*	*Doe*	*Litter Size, X*	*Mean Weight, Y*
A	2(5.5)	47.0(11)	K	4(12.5)	46.7(10)
B	3(9.5)	42.3(7.5)	L	3(9.5)	47.7(12)
C	6(17)	39.7(3)	M	6(17)	37.1(1)
D	5(14.5)	40.3(4)	N	1(2)	52.7(18)
E	4(12.5)	41.4(6)	O	2(5.5)	42.3(7.5)
F	7(19)	39.4(2)	P	1(2)	59.2(20)
G	8(20)	40.5(5)	Q	6(17)	48.8(13)
H	1(2)	54.2(19)	R	3(9.5)	51.2(17)
I	2(5.5)	49.4(15)	S	2(5.5)	51.0(16)
J	5(14.5)	43.1(9)	T	3(9.5)	49.0(14)

$$r_S = \frac{1681.5 - 20(21)^2/4}{\sqrt{[2855 - 20(21)^2/4][2869.5 - 20(21)^2/4]}} = -.797.$$

Spearman's rank correlation coefficient and the ordinary sample correlation have several properties in common (as you might expect in view of the way r_S was defined). For example, r_S will always be between -1 and $+1$. If for a set of sample observations $|r| = 1$, then $|r_S|$ will equal 1. However, it is possible for the ranks of the X's and the Y's to be in perfect agreement so that $r_S = 1$, while the original sample pairs (x_i, y_i) are not linearly related so that $r \neq 1$ (see Exercise 40). Nevertheless, the ordinary sample correlation and r_S both measure association in the sense that if large values of X tend to go with large values of Y, then r will tend to be positive and large X ranks will tend to go with large Y ranks, so r_S will also tend to be positive. Similarly, if large X values tend to go with small Y values, both r and r_S will tend to be negative. Furthermore, because of the way that r_S is defined, it is a *measure of the strength of the linear relationship between the ranks of the* X's *and* Y's.

As we have said before, if the random variables X and Y are actually independent, ρ, the correlation coefficient, will be zero, and consequently both r and r_S will tend to be near zero. If we wanted to test

$$H_0: X \text{ and } Y \text{ are independent} \tag{11.26}$$

against the alternative that they are not independent, it would be reasonable to use r as a test statistic, rejecting H_0 if r is either too "large" or too "small." Unfortunately, even when H_0 is true, it is not easy to find the probability distribution of r (which is necessary to find critical values) unless we make the assumption of bivariate normality. Of course, our purpose here is to avoid that assumption, so we must look for a different candidate for a test statistic. The candidate we will choose is r_S, Spearman's rank correlation coefficient.

If $(X_1, Y_1), (X_2, Y_2), \ldots, (X_n, Y_n)$ is a bivariate random sample from a continuous joint (bivariate) distribution such that the data are at least ordinal, then when the null hypothesis of independence (11.26) is true, the distribution of r_S can easily be found. Consequently, r_S can be used as a test statistic for testing (11.26) with the critical values found from Table A17. The assumption of continuity is made to (theoretically) avoid the problem of tied observations. If ties are extensive, a correction for ties can be made (see Siegel, *Nonparametric Statistics*). The probability distribution of r_S is relatively easy to find when H_0 is actually true since for any given ranking of the X values, say, all $n!$ possible rankings of the Y values are equally likely (see Exercise 45). In testing

$$H_0: X \text{ and } Y \text{ are independent} \quad \text{vs.} \quad H_a: X \text{ and } Y \text{ are not independent}$$

at a significance level α, reject H_0 if the absolute observed value of r_S is greater than or equal to the critical value corresponding to $p = \alpha/2$. One-sided alternatives can also be tested. For example, in testing

$$H_0: X \text{ and } Y \text{ are independent} \quad \text{vs.} \quad H_a: \text{a positive relationship exists},$$

at a significance level α, reject H_0 if the observed value of r_S is greater than or equal to the critical value corresponding to $p = \alpha$. Since the distribution of r_S is symmetric about zero, the alternative hypothesis that a negative relationship exists between X and Y would be accepted if r_S is less than or equal to the negative of the critical value corresponding to $p = \alpha$.

★***Example 11.28*** Using the data on litter size (X) and mean weight of baby rabbits (Y) given in Table 11.24, test the null hypothesis that X and Y are independent against the alternative that a negative relationship exists. Use a .01 level of significance.

Note that intuitively we would expect that if any relationship exists between X and Y, it would be negative, so we have chosen this as the alternative hypothesis. In Example 11.26 we found the observed value of r_S to be $-.797$. From Table A17 we find that for $n = 20$ and $\alpha = .01$, the critical value is $-.522$, so H_0 is rejected and we would conclude that a negative relationship probably exists.

We know, of course, that r_S measures the strength of the linear relationship between the ranks of the X and Y observations. It may be that a nonlinear relationship exists for which r_S is typically close to zero. In such a case there may be a high probability of failing to reject the null hypothesis of independence when that hypothesis is false (i.e., of committing a type II error).

If the sample size n is greater than 30, the quantity

$$r_S\sqrt{n-1}$$

is approximately standard normal. Consequently, the appropriate critical values can be found by using $(z_{1-\alpha}/\sqrt{n-1})$. For example, if $n = 37$ and $\alpha = .05$, the critical value for r_S would be

$$\frac{z_{.95}}{\sqrt{37-1}} = \frac{1.645}{6} = .2742.$$

In many investigations the bivariate data obtained are nominal (categorical) data and are not ordinal. If the data are not ordinal, ranks cannot be assigned to the responses, and consequently Spearman's rank correlation coefficient cannot be calculated. For example, a personnel manager for a large company may randomly select 200 workers, say, and classify each worker in two ways: by the department in which the worker is employed and by job satisfaction (satisfied, neutral, or dissatisfied). In this situation Spearman's rank correlation coefficient cannot be calculated. However, the personnel manager may wish to know if the two responses are independent. If they are independent, efforts at improving employees' feelings of job satisfaction could be the same for all departments. If they are dependent, different efforts might be made for different departments. We will present here a method by which the null hypothesis of independence between the two classifications can be tested.

Example 11.29 The personnel manager of the Tooze Company, Teresa Crowd, randomly selected 200 workers and classified them according to department (management, production, or shipping) and job satisfaction. The results are shown in Table 11.26. Test the null hypothesis

$$H_0: \text{classifications are independent}$$

against the alternative that they are not independent. Use a .05 level of significance.

TABLE 11.26

Classification of Responses of Workers at the Tooze Company

		Job Satisfaction			
		Satisfied	Neutral	Dissatisfied	Total
Department	Management	20	11	5	36
	Production	69	27	12	108
	Shipping	43	10	3	56
	Total	132	48	20	200

Before we can test this hypothesis we must, of course, find a suitable test statistic. You may notice a similarity between this table and the tables in Section 11.11, where we considered a test for equality of proportions for different populations. It will not be surprising, then, that the test statistic is the same for the two situations. Here, too, we consider the agreement between the observed and expected number of observations in each cell in the table. The expected values are found by the following logic. Say that a worker is chosen at random and that H_0 is true (i.e., the classifications are independent). If we let $p_{i,j}$ represent the probability that a randomly chosen worker will be classified in the ith row and jth column, then, for example,

$$\begin{aligned} p_{1,1} &= P[(\text{worker is in management}) \cap (\text{worker is satisfied})] \\ &= P[\text{worker is in management}] \cdot P[\text{worker is satisfied}]. \end{aligned}$$

Consequently, the expected number would be

$$n_{..}p_{1,1} = n_{..}\{P[\text{worker is in management}] \cdot P[\text{worker is satisfied}]\}.$$

Of course, these last probabilities are really unknown, so they must be estimated from the data. Using the notation of Section 11.11, we have

$$\hat{P}[\text{worker is in management}] = \frac{n_{1.}}{n_{..}} = \frac{36}{200} = .18$$

and

$$\hat{P}[\text{worker is satisfied}] = \frac{n_{.1}}{n_{..}} = \frac{132}{200} = .66.$$

The estimated expected number is then

$$e_{1,1} = n_{..}\left(\frac{n_{1.}}{n_{..}}\right)\left(\frac{n_{.1}}{n_{..}}\right) = 200(.18)(.66) = 23.76$$

and in general

$$e_{i,j} = n_{..}\left(\frac{n_{i.}}{n_{..}}\right)\left(\frac{n_{.j}}{n_{..}}\right) = \frac{n_{i.}n_{.j}}{n_{..}}$$

[cf. equation (11.19)]. We use as a test statistic

$$\begin{aligned} X^2 &= \sum \frac{(o_{i,j} - e_{i,j})^2}{e_{i,j}} \qquad (11.27) \\ &= \left(\sum \frac{o_{i,j}^2}{e_{i,j}}\right) - n_{..}, \end{aligned}$$

which has (approximately) a chi-square distribution (if the expected values are approximately five or more) with the number of degrees of freedom equal to

$$(r - 1)(c - 1),$$

where $r =$ number of rows in the table and $c =$ number of columns in the table. (This number of degrees of freedom is determined by the number of cells in the table and the number of parameters that must be estimated.) The null hypothesis of independence between classifications will be rejected at significance level α if the test statistic X^2 is greater than or equal to $\chi^2_{(r-1)(c-1),1-\alpha}$.

Example 11.29 *(continued)* We can now calculate the test statistic X^2 and compare it with the critical value

$$\chi^2_{(3-1)(3-1),1-.05} = \chi^2_{4,.95} = 9.48773.$$

The expected values for each cell are shown parenthetically in Table 11.27. Using these values, we find the test statistic to be

$$X^2 = \frac{(20 - 23.76)^2}{23.76} + \frac{(11 - 8.64)^2}{8.64} + \ldots + \frac{(3 - 5.60)^2}{5.60} = 5.11.$$

Since this value is not in the critical region, the null hypothesis of independence is not rejected.

TABLE 11.27

Data for Example 11.28 with Expected Numbers

		Job Satisfaction			
		Satisfied	*Neutral*	*Dissatisfied*	*Total*
Department	*Management*	20(23.76)	11(8.64)	5(3.60)	36
	Production	69(71.28)	27(25.92)	12(10.80)	108
	Shipping	43(36.96)	10(13.44)	3(5.60)	56
	Total	132	48	20	200

This type of analysis for independence of different classifications is called *contingency table* analysis. It is possible to generalize this procedure to three or more types of classifications rather than just two.

EXERCISES

1. Bill Nockon, a district sales manager for the Nockon Wood Co., believes that the median number of calls made by salespersons in his district is 6 per day. A sample of 20 daily reports showed the following number of calls:

10 9 7 3 9 5 8 3 7 6

5 2 13 8 4 7 9 11 10 14.

Test $H_0 : \tilde{\mu} = 6$ vs. $H_a : \tilde{\mu} \neq 6$ using a sign test with a .05 level of significance.

2. In an effort to investigate the readiness of boys compared to the readiness of girls for school, a psychologist interviewed 11 sets of brother–sister twins. Each child was classified as "very immature" (V), "immature" (I), "ready" (R), or "advanced" (A). Assigning a + sign if the brother of the pair of twins is more ready than the sister and a—sign if the sister is more ready, use a sign test to test the null hypothesis that boys and girls are equally ready for school using the given data. Use $\alpha = .10$.

Twin set:	1	2	3	4	5	6	7	8	9	10	11
Boy:	V	I	R	R	V	A	R	I	I	I	I
Girl:	R	R	R	I	I	R	A	I	A	R	R.

3. Fourteen people were weighed before and after a special carbohydrate diet. It is only of interest to know if the diet was effective in reducing weight. Set up and test appropriate hypotheses using a sign test and a .10 level of significance using the given data.

Weight Before	*Weight After*	*Weight Before*	*Weight After*
189	180	235	213
238	224	230	220
224	212	178	191
231	251	210	224
223	200	249	230
202	186	195	180
256	240	281	260

4. According to a book about birds, the median number of eggs laid by the tiny golden-crowned kinglet is 8. An ornithologist would like to see if this median number is the same in a region destroyed by fire a few years ago. The number of eggs found in 10 nests in this region were 9, 10, 7, 6, 5, 4, 5, 7, 6, and 9. Test H_0: $\tilde{\mu} = 8$ against a two-sided alternative using a Wilcoxon signed rank test and a .05 level of significance.

5. Test the hypotheses in Exercise 1 using a Wilcoxon signed rank test.

6. Find the probability distribution of T^+, the sum of the positive ranks in a sample, when $n = 4$ and when the assignment of + or − signs to each rank is equally likely. [*Hint:* Each possible assignment of signs to the four ranks has probability $(1/2)^4$. Follow the technique used in Table 11.6 of Section 11.3.]

7. Test the hypotheses in Exercise 3 using a Wilcoxon signed rank test.

8. A group of army recruits was tested before and after a long march. Captain

Recruit Number	*Before Score*	*After Score*	*Recruit Number*	*Before Score*	*After Score*
1	13.7	12.7	9	15.7	14.1
2	15.5	16.0	10	10.0	10.3
3	14.4	13.2	11	11.4	9.9
4	11.3	10.5	12	9.0	9.7
5	11.1	11.5	13	9.5	8.8
6	10.5	9.2	14	12.2	11.3
7	8.0	11.9	15	11.4	9.3
8	13.7	11.0			

Marvel believes that due to fatigue the "after" scores will tend to be lower than the "before" scores. Based on the given data, is there convincing evidence to support the Captain's belief using a .05 level of significance? Set up and test appropriate hypotheses using a Wilcoxon signed rank test.

9. To compare the academic effectiveness of East Junior High and West Junior High, an experiment was designed using 10 sets of identical twins. Each twin had completed the sixth grade and each set had obtained their schooling in the same classroom at each grade level. One twin from each set was randomly assigned to each junior high. At the end of ninth grade each child was given an achievement test. Using the given data test the null hypothesis that the two schools are the same in academic effectiveness, as measured by this achievement test, against the alternative that they are not equally effective. Use a Wilcoxon signed rank test with $\alpha = .05$.

Twin pair:	1	2	3	4	5	6	7	8	9	10
East score:	67	80	65	70	86	50	63	81	86	60
West score:	39	75	69	55	74	52	56	72	89	47.

10. An educational researcher is interested in testing whether third graders' attitudes toward reading are changed by having stories read to them daily as opposed to just giving them free time to read their own stories. Twenty students were randomly divided between the two programs. After a period of time each child's attitude toward reading was rated on a scale ranging from 1 (very poor) to 20 (excellent). Using the given data, perform a Wilcoxon rank sum test to see if there is any significant difference between the groups at the $\alpha = .10$ level.

Stories read to them:	1	4	7	8	8	10	12	14	16	18
Read their own stories:	2	3	3	4	6	8	9	13	17	19.

11. Mike Pike is a regional distributor of fishing equipment. He is interested in knowing whether sales personnel with college degrees have higher sales than those without college degrees. A sample of 8 persons with college degrees and 12 without was chosen and their sales (in thousands of dollars) over a certain time period were found. Using the given data, set up and test $H_0 : \tilde{\mu}_1 \leq \tilde{\mu}_2$ vs. $H_a : \tilde{\mu}_1 > \tilde{\mu}_2$, where the subscript 1 corresponds to personnel having college degrees. Use a rank sum test with $\alpha = .05$.

Sales (with degree):	67	103	96	97	76	84	76	102				
Sales (without degree):	74	80	76	85	83	78	99	86	83	90	78	72.

12. A study reported on in a psychiatry journal compared people having above-average and below-average IQ in terms of the percent of abstract responses on a particular test. Using the given data, which are similar to that reported, test the null hypothesis that there is no difference in the percent of abstract responses for the two groups. Use a rank sum test with $\alpha = .01$. (*To think about:* Might a one-sided alternative be appropriate here? If so, what might the alternative be?)

Above-average IQ:	69	54	38	40	27	54	20	22	19
Below-average IQ:	8	19	23	11	4	8	19	20	4.

13. Let n_1 represent the smaller and n_2 the larger of two sample sizes. If T_1 is the sum of the ranks assigned to the smaller sample when the combined samples are ranked, show that the largest value that T_1 can assume is

$$n_1 n_2 + \frac{n_1(n_1 + 1)}{2}.$$

14. Let T_1 be defined as in Exercise 13. If the populations from which the samples are chosen are the same, so that each different assignment of ranks to the n_1 observations in the smaller sample is equally likely, find the probability distribution of T_1 when $n_1 = 2$ and $n_2 = 3$. [*Hint:* List all $\binom{5}{2}$ possible assignment of ranks to the two values. Each of these assignments is equally likely.]

15. Both ecologists and people in marketing are concerned with ways of encouraging consumers to buy detergents having a low phosphate content. In an experiment, two comparable grocery stores were used. One store (test store) made cards stating the percentage of phosphates for each detergent and taped these cards to the shelves. The second store (control store) displayed detergents in the usual way without any extra information. Thirty purchasers of detergents were observed in each store and the detergent bought was classified as to phosphate content. Let $\tilde{\mu}_1$ denote the median phosphate level for detergents bought in the first store. Using the given data, test $H_0: \tilde{\mu}_1 \geq \tilde{\mu}_2$ vs. $H_a: \tilde{\mu}_1 < \tilde{\mu}_2$ using a rank sum test with a .05

	Level of Phosphate Content				
	Very Low	*Low*	*Medium*	*High*	*Very High*
Test store (1)	8	4	13	4	1
Control store (2)	4	2	10	6	8

level of significance. [*Hint:* Since the 60 observations must be classified into only five ordered classes, there are a large number of ties. For example, 12 values are tied for the smallest ranks, so each tied observation will receive the average rank of 6.5. Six values are tied for the next ranks, ranks 13 through 18 inclusive, so each of these will receive the average rank of 15.5, etc. The sum of ranks T_1 will be found by taking $8(6.5) + 4(15.5) + \cdots$.]

16. Jerry Nutsan, purchasing agent for the Nutsan Bolts factory, is considering purchase of a new machine for making bolts having "2.5 cm" diameter. The new machine, as well as the old, makes bolts having the right diameter "on the average," but Jerry is concerned about the variability. If the variability of the new machine can be shown to be significantly less than that of the old, he will buy the new one. Careful measurements have given the following data. What should Jerry do? Base your answer on a Siegel–Tukey test with $\alpha = .05$.

New machine:	2.52	2.48	2.47	2.51	2.55	2.49	2.51
	2.48	2.49	2.51	2.51	2.52	2.50	2.48
	2.49						
Old machine:	2.54	2.47	2.53	2.53	2.54	2.48	2.46
	2.47	2.46	2.53	2.47	2.57	2.53	2.45
	2.46.						

17. In Exercise 11, the median sales for persons with and without a college degree were compared. Using these same data, use a Siegel–Tukey test with $\alpha = .05$ to test the null hypothesis that there is no difference in variability between these groups.

18. An analysis of variance problem was performed to compare two means. The residuals determined were the following:

Sample 1:	−11	−10	−6.5	−5	−1.5	4	7	7	7	9
Sample 2:	−25.5	−24.5	−19	−9	8	10	12	14	15.5	18.5.

(The residuals are listed in increasing order for convenience.) (a) Make a residual plot. (b) Use a Siegel–Tukey test with $\alpha = .01$ to test the null hypothesis of no difference in variability.

19. Many elm trees have been planted along the north side of Elm Boulevard. Mr. Palm, who specializes in tree diseases, has gone down the street and classified each tree as healthy (H) or diseased (D). The results were:

H H H H H D D D H H H H H H

D H D D D D H H D D D D D D.

Use a runs test to test the null hypothesis that the diseased trees occur randomly against the alternative that diseased trees tend to cluster together. Use a .01 level of significance.

20. A communication line between a central computer and a data transmission unit may undergo "line drops." A line drop occurs when a "bit," which can be either a 1 or 0, is reversed in value. In a test run, a sequence of 31 bits, all 1's, was transmitted to the computer, but the sequence actually received by the computer was

0 0 1 1 1 1 1 1 1 1 1 1 0 0 0 0

1 1 1 1 1 0 0 0 0 0 1 1 1 1 1.

Test the null hypothesis that line drops occur at random. Use a runs test with $\alpha = .05$.

21. A lab technician performed a series of analyses using the same specimens. The analyses were done at half-hour intervals throughout the day (starting at 8 A.M.). A least-squares line was fitted to the data and the following residuals were found:

.4 1.0 .8 .6 −1.3 −.7 −.8 −.2

−.3 −.8 −.5 −1.8 1.2 1.4 .8 .2.

(a) Make a residual plot. (Plot time on the x-axis.) (b) Use a runs test to see if the residuals occur in a random order relative to the value zero. Use $\alpha = .05$.

22. Consider the possible orderings of two A's and three B's. List all 10 of these orderings and find the number of runs in each ordering. Assuming that each ordering is equally likely, find the probability distribution of R, the number of runs.

★***23.*** A die was tossed 60 times with the following results:

Face up:	1	2	3	4	5	6
Frequency:	6	12	15	9	10	8.

Use a chi-square goodness-of-fit test to test the null hypothesis that the die is fair. Use a .01 level of significance.

★*24.* According to Mendel's theory of inheritance, cross breeding of two plants, one having round yellow seeds and the other wrinkled green seeds, will give second-generation plants having seeds distributed in the following four categories in the ratio of 9: 3: 3: 1. C_1 = round and yellow, C_2 = wrinkled and yellow, C_3 = round and green, and C_4 = wrinkled and green. An experimenter found that of 4530 seeds,

2504 853 881 292

were classified into categories 1 to 4, respectively. Use a chi-square goodness-of-fit test to see if these data support Mendel's theory. Take $\alpha = .10$.

25. The manager of the Wood Pile, a local lumberyard, claims that 10% of his customers will buy cherry lumber, 25% will buy oak, and 65% will buy pine. A series of purchases was observed and of 200 customers, 16, 44, and 140 bought cherry, oak, and pine, respectively. In light of these results, what do you think of the manager's claim? Set up and test appropriate hypotheses.

★*26.* The following data give the number of deaths from the kick of a horse per army corps per year for 10 Prussian Army Corps for 20 years (a total of 200 observations).

No. deaths/corps/year:	0	1	2	3	4
Observed frequency:	109	65	22	3	1.

(a) Estimate the mean number of deaths/corps/year from the data. (b) Perform a chi-square goodness-of-fit test of the null hypothesis that the data follow a Poisson distribution (cf. Exercise 34 of Chapter 4).

27. In Exercise 18 two sets of residuals are given. Combine these residuals into a single sample of size 20 and perform a chi-square goodness-of-fit test to test the null hypothesis that the data come from a normal distribution. (*Hint:* First find the sample mean and variance and use these as estimates of μ and σ^2. Divide the normal distribution with this mean and variance into four regions each of probability .25. Find the observed and expected numbers in each region.) Use $\alpha = .10$.

28. Repeat Exercise 27 using a Kolmogorov–Smirnov test instead of a chi-square goodness-of-fit test.

29. Cables consisting of several strands of wire were manufactured under three different processes. To compare strength, samples of wires from each process were subjected to testing. The following data give the number of kilograms the wires could support before breaking.

Sample Number	*Process 1*	*Process 2*	*Process 3*
1	348	338	342
2	345	350	354
3	350	335	339
4	344	333	342
5	342	345	340
6	346	331	330
7	349	354	343
8	345	339	341
9	341	339	351
10	347	343	349

Perform a Kruskal–Wallis test to see if the different processes lead to cables having the same strength. Use $\alpha = .01$.

30. In a drug-screening experiment 13 mice were used to determine if certain drugs are able to control tumor growth. Cancerous cells were implanted in each mouse so that a tumor would grow. Eight mice (four in each treatment group) were randomly selected to receive treatment with drug 1 and drug 2. The five remaining mice constituted a control group. After a fixed period of time the tumors were removed and weighed, the weight of the tumor being considered indicative of the effectiveness of the drugs in arresting or reducing the cancer. Given the following data on actual tumor weights (in milligrams), use a Kruskal–Wallis test to compare the median tumor weights for the three groups. Use $\alpha = .05$.

Drug 1:	4	8	6	7	
Drug 2:	3	5	4	9	
Control:	8	12	17	16	14.

★31. The data on statistics students in Appendix A3 give the year in school for each student as well as their grade-point averages (GPA). Only three students from this set are freshman (year 1). Treating this data set as a sample from a larger population, take as a sample all freshman students, the first 7 sophomore, junior, and senior students (total sample size of 24). Find the corresponding grade-point averages and use a Kruskal–Wallis test to compare the median grade-point averages for the four different years in school. Use $\alpha = .10$.

32. Referring to Example 11.21 of Section 11.10, for purposes of performing multiple comparisons, perform a rank sum test for each pair of medians. Make use of the fact that for an overall α of .05 when c, the number of comparisons, is equal to three, and when $n_1 = n_2 = 6$, the lower and upper critical values for the rank sum test statistic corresponding to $\alpha/2c$ are 24 and 54, respectively. Which treatment medians differ from the others?

33. In a marketing study, single coupons (good on purchase of a cereal product) and joint-offer coupons (good on purchase of a cereal product and fresh fruit) were each sent to 180 randomly selected households. The redemption rate was found

	Redeemed	*Unredeemed*
Single coupon	61	119
Joint coupon	83	97

for each type of coupon. Perform a chi-square test for equality of proportions to determine whether there is a difference in the redemption rate for the two types of coupons. Use a .05 level of significance.

34. Three machines produce plastic fishing bobbers of different sizes. To compare the rate of defectives for each machine, 100 bobbers from each machine were sampled. Each bobber was tested and classified as good, slightly defective, or bad. The findings were as shown on the following page. Use a chi-square test to see if the proportion in each category is the same for each size of bobber. Use $\alpha = .10$.

	Good	Slight Defect	Bad
Large bobbers	82	14	4
Medium bobbers	84	10	6
Small bobbers	89	6	5

35. A researcher has suspected that the occurrence of cerebral vascular accidents is different for people with different diastolic blood pressure levels. Three groups of 50 people, having low, normal, or high blood pressure levels, were studied until death. It was noted whether or not they had a cerebral vascular accident. Using the given data, test appropriate hypotheses to see if the researcher's suspicions can be substantiated.

		Cerebral Vascular Accident	
		Yes	*No*
Diastolic	*Low*	1	49
Blood	*Normal*	4	46
Pressure	*High*	15	35

36. Show that the following algebraic equality holds:

$$\sum \frac{(o_{i,j} - e_{i,j})^2}{e_{i,j}} = \left(\sum \frac{o_{i,j}^2}{e_{i,j}}\right) - n_{\cdot\cdot\cdot}$$

(*Hint:* Expand the quantity on the left-hand side. Use the fact that $\sum o_{i,j} = \sum e_{i,j} = n_{\cdot\cdot\cdot}$)

★37. The data given here was collected by the U.S. Department of Agriculture. The values given are retail food prices in capitals of various countries. Use a Friedman's test to see if retail food prices tend to differ in different countries. Use a .01 level of significance.

	Food						
Capital	*Sirloin (lb)*	*Broilers (lb)*	*Eggs (doz)*	*Cheese (lb)*	*Milk (qt)*	*Potatoes (lb)*	*Oranges (doz)*
Bonn	\$ 6.13	\$1.10	\$1.28	\$1.55	\$ .56	\$.12	\$2.89
Mexico City	1.45	.97	.60	3.27	.27	.16	.27
Stockholm	6.23	1.88	1.92	2.69	.43	.24	2.95
Tokyo	15.87	1.88	1.04	2.68	1.03	.51	2.85

38. In a marketing study it was desired to know how various factors influenced consumer choices of large appliances. Eight people who recently purchased large appliances were asked to rank the following four factors (1 = most influential, . . . , 4 = least influential): endorsement by celebrities, brand loyalty, results of government testing, recommendation of a friend. Using the given results, use a Friedman's test to see if these factors are rated differently. Use $\alpha = .05$.

Factor	Purchaser							
	Joe	*Kate*	*Laura*	*Mal*	*Nell*	*Otto*	*Peg*	*Quin*
Endorsement	4	3	2	4	4	3.5	3	4
Brand loyalty	1	2	1	1.5	2	1	2	2
Testing results	3	4	4	3	2	3.5	4	3
Friend	2	1	3	1.5	2	2	1	1

39. Suppose that an experimenter is interested in evaluating the relative effectiveness of four drugs in reducing pain in patients experiencing migraine headaches. Seven patients were selected and were given each drug for a month at a time. At the end of each month, the patients were asked what proportion of headaches (in the course of the month) never reached the "severe pain" stage. Using the given data, compare the effectiveness of the drugs using Friedman's test ($\alpha = .05$).

Drug	*Patient*						
	Rod	*Sue*	*Tim*	*Una*	*Vic*	*Wanda*	*Yolanda*
A	.3	.6	.3	.1	.2	.2	.2
B	.3	.4	.5	.2	.3	.5	.3
C	.7	.8	.7	.5	.6	.4	.7
D	.6	.7	.8	.9	1.0	.9	1.0

40. Given the following (x, y) pairs, calculate both r, the ordinary correlation coefficient, and r_S, Spearman's rank correlation coefficient.

x:	0	1	2	3	4
y:	0	2	3	8	17.

★***41.*** Using the first 10 males in the data set in Appendix A3, list their weights (x) and pulse rates (y). Find the value of Spearman's rank correlation coefficient r_S.

★***42.*** Using the first 15 females in the data set in Appendix A3, list their weights (x) and pulse rates (y). Find the value of r_S, Spearman's rank correlation coefficient. Using this value, test the null hypothesis that weight and pulse rate are independent for females. Use $\alpha = .01$.

★***43.*** The following data give the number of pounds of meat sold in a week by a grocery store (x) and the number of items advertised that week (y). Calculate Spearman's rank correlation coefficient and test the null hypothesis that X and Y are independent against the alternative that they are positively related. Use $\alpha = .10$.

Week:	1	2	3	4	5	6
Pounds sold, x:	7730	11,028	8152	8208	8751	8524
No. items, y:	5	8	9	8	7	10.

★***44.*** The following data were obtained from a sample of mutual fund companies. The variable X represents total net assets (in millions) and Y represents the offering price on a particular date. Test the null hypothesis that X and Y are independent using a .05 level of significance.

Mutual fund:	B	Ch	Co	D	F	I	L	P	U
Net assets, x:	51.1	314.0	30.0	26.7	2.2	41.9	14.0	42.4	14.7
Offering price, y:	9.44	7.06	11.74	8.57	12.50	15.63	7.62	11.19	12.01.

45. Consider a sample of $n = 4$ (x, y) pairs. Assuming that for any ranking of the x's, all $n!$ possible rankings of the y's are equally likely, find the probability distribution of r_S.

★46. One year 340 Boy Scout troops were sampled and information as to their sponsoring organization was obtained. Seven years later in a follow-up study it was determined whether the sponsors had changed or not in the intervening time. Using these data, test to see if the type of sponsoring organization and change in sponsor are independent. Use $\alpha = .01$.

	Change in Sponsor	
Type of Sponsoring Organization	*No*	*Yes*
Youth-oriented	156	70
Partially youth-oriented	50	31
Non-youth-oriented	12	21

47. A sample of 300 students was chosen for purposes of comparing major area of studies and interest in politics. Do the following data support the null hypothesis of independence? Use $\alpha = .05$.

	Major Area		
Interest in Politics	*Natural Sciences*	*Journalism*	*Social Sciences*
Low	56	12	66
Medium	32	32	34
High	22	16	30

48. People who have a fear of heights are classified as being acrophobic. This fear affects both males and females. A sample of 200 persons showed the tabled results. Do these data support the null hypothesis that acrophobia is independent of an individual's sex? Use $\alpha = .10$.

	Acrophobic	*Not Acrophobic*
Female	18	85
Male	14	83

12

BAYESIAN INFERENCE

12.1 INTRODUCTION

In the material presented up to this point, virtually all inference has been based solely on sample information. Estimation of parameters, for example, has been done by using "sample counterparts." There are many statisticians who feel that other information is also available and that this information should be combined with sample information in order to make inferences. This information is generally based on past experience, but it is usually somewhat subjective in nature. The mathematical device used for combining past or "prior" information with sample information is Bayes' theorem, and consequently the resulting inference is referred to as Bayesian inference. Before reading on in this section, you should review Sections 1.7 and 1.11 on subjective probability and Bayes' theorem.

Have you ever flipped a coin to see if it would come up heads or tails? Most people have done this many times in their lives, so it would be rather surprising if you had not done this experiment a number of times. You probably have not kept careful records on the outcomes of these experiments, but you probably have the feeling that virtually all coins you have flipped have been "fair" (i.e., the probability of the coin coming up heads is approximately .5). Now take a coin from your pocket and examine it. If it is an "ordinary" coin, what do you think the probability will be that it will come up heads if you flip it? Although you have no sample information *for this particular coin*, you do have some prior information about the behavior of similar coins, so you would probably conjecture that the probability will be .5. (You may argue that you've based your answer on the "equally likely" principle and not on past experience. However, the judgment that heads and tails are, in fact, "equally likely" is also a subjective judgment based to some degree on past experience.) We might think of this subjective conjecture as reflecting prior information. Now say that you decide to gather some objective sample information about p, the probability that this particular coin will come up heads when flipped. More precisely, assume you flip the coin 5 times and observe 3 heads. What would your estimate of p be? If you were to use the ordinary estimation procedure discussed in Chapter 7, the estimate would be the sample proportion

$$\frac{x}{n} = \frac{3}{5} = .6.$$

Do you really believe that this estimate is very close to the correct value? Of course not! At this point you have a choice. You can

1. ignore the sample information and believe that $p = .5$,
2. ignore the prior information and believe the sample estimate of .6, or
3. combine both the prior and sample information.

The third option would seem to be a reasonable one, and this option will be investigated in this chapter.

12.2 PRIOR, CONDITIONAL, AND POSTERIOR DISTRIBUTIONS

In Example 1.40 we discussed a situation where an inspector for the Watzit Manufacturing Company was interested in knowing the proportion of defectives being produced by a watzit maker. We will repeat that example here with a slight modification.

Example 12.1 The inspector at the Watzit Manufacturing Company, Dee Syder, inspects items coming out of a watzit maker. From past experience Dee knows that on some days 1% of the items produced are defective, while on other days 5 or 10% are defective. More explicitly, 60% of the time 1% of the items produced are defective, while 5 and 10% are bad 30 and 10% of the time, respectively. If he inspects two items today and finds both to be defective, what is the probability that the watzit maker is currently producing 1% defectives? 5%? 10%?

As in Chapter 1, it is helpful to visualize a watzit maker (Figure 12.1) which has an internal dial that is set to determine the rate of defectives. (While such a dial does not really exist, thinking of such a dial can be helpful in working through the problem.) Based on his prior knowledge only, Dee would suspect that the dial is set a 1%, since this has been the setting about 60% of the time. On the other hand, having inspected two items and finding both defective, the sample information indicates that the current rate of defectives may be high (i.e., the dial may be set at 5 or 10%). By using Bayes' Theorem, both the prior information and sample information can be combined. Using the same notation as in Example 1.40, we can define events

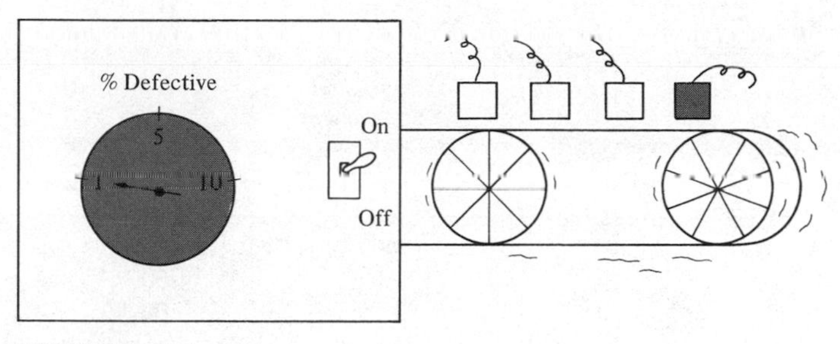

FIGURE 12.1
Watzit-Maker Set at 1% Defective Rate

A_1, A_2, A_3, and B by

$$A_1 = \text{dial is set at } 1\%$$
$$A_2 = \text{dial is set at } 5\%$$
$$A_3 = \text{dial is set at } 10\%$$
$$B = \text{two inspected items are both befective.}$$

From the information given in the problem, we have

$$P(A_1) = .60, \qquad P(A_2) = .30, \qquad P(A_3) = .10.$$

Assuming independence among items produced by this machine, we find that

$$\begin{aligned} P(B \mid A_1) &= P[\text{both items are defective} \mid \text{dial is set at } 1\%] \\ &= P[\text{first item is defective} \mid \text{dial is set at } 1\%] \cdot \\ &\quad\; P[\text{second item is defective} \mid \text{dial is set at } 1\%] \\ &= (.01)(.01) = .0001. \end{aligned}$$

Similarly, $P(B \mid A_2) = .0025$ and $P(B \mid A_3) = .0100$. Putting this information into Bayes' formula, we get

$$\begin{aligned} P(A_1 \mid B) &= \frac{P(A_1 \cap B)}{P(B)} = \frac{P(A_1) \cdot P(B \mid A_1)}{P(A_1) \cdot P(B \mid A_1) + P(A_2) \cdot P(B \mid A_2) + P(A_3) \cdot P(B \mid A_3)} \\ &= \frac{(.60)(.0001)}{(.60)(.0001) + (.30)(.0025) + (.10)(.0100)} = \frac{.00006}{.00181} \\ &= .0331. \end{aligned}$$

Likewise, we find that $P(A_2 \mid B) = .4144$ and $P(A_3 \mid B) = .5525$.

The information needed to work this example can be most easily summarized in a table similar to Table 1.1. This summary is given in Table 12.1. One difference between these tables is that here the first column is headed "proportion" rather than "setting." Note that the first two columns actually give a discrete probability distribution over the possible values of p. Another difference in the tables is that the event A_i has been replaced with $p = p_i$. Finally, note that the first and last columns together

TABLE 12.1

Proportion, p_i	Prior Probability, $P(p = p_i)$	Conditional Probability of Observed Event, $P(B \mid p_i)$	Product, $P(p = p_i) \cdot P(B \mid p_i)$	Posterior Probability, $P(p = p_i) \cdot P(B \mid p_i)/\sum$ $= P(p = p_i \mid B)$
.01	.60	.0001	.00006	.0331
.05	.30	.0025	.00075	.4144
.10	.10	.0100	.00100	.5525
	1.00		$\sum = .00181$	1.0000

give a revised probability distribution over the possible values of p. The initial distribution is referred to as the *prior distribution*, since it gives the probability distribution prior to obtaining sample information, while the revised probability distribution is referred to as the *posterior distribution*, since it gives the probability distribution after (post) the sample information is obtained. This terminology is used in the table headings. Note that here too the sum of the fourth column, denoted by $\sum$, is equal to $P(B)$.

The posterior probability distribution found by Bayes' theorem is an updated version of the prior distribution. It reflects the current state of belief of the inspector (in this case), combining both prior and sample information currently available. If more sample information became available, the "old" posterior distribution would be used as a "new" prior, and this distribution would be revised to get the latest version of the posterior distribution.

Example 12.2 In Example 1.40 Dee inspected a *single item* and found it to be defective, thus revising his prior distribution to

$$\begin{array}{llll} p_i: & .01 & .05 & .10 \\ P[p = p_i]: & .1935 & .4839 & .3226. \end{array}$$

If he now inspects a second item and finds that it, too, is defective, what should his new posterior distribution be?

To solve this problem, we can construct a table like Table 12.1. The observed event in this case, which we will denote by B, is "an inspected item is found to be defective." The conditional probabilities are

$$P[B \mid p = .01] = .01, \qquad P[B \mid p = .05] = .05, \qquad P[B \mid p = .10] = .10.$$

The last column of Table 12.2 gives the new posterior probabilities. If you examine the last column of Tables 12.1 and 12.2, you will see that (except for some roundoff error) they are the same. Even though the posterior distributions were obtained in slightly different ways (in one case the prior was revised once, after two items were inspected, while in the second case the prior was revised twice, after the inspection of each of two items) the same amount of information ultimately was available to the inspector, so it is reasonable to have the final posterior probability distributions equal.

TABLE 12.2

Proportion, p_i	Prior Probability, $P[p = p_i]$	Conditional Probability of Observed Event, $P[B \mid p_i]$	Product, $P[p = p_i] \cdot P[B \mid p_i]$	Posterior Probability, $P[p = p_i \mid B]$
.01	.1935	.01	.00194	.0332
.05	.4839	.05	.02420	.4144
.10	.3226	.10	.03226	.5524
	1.000		$\sum = .05840$	1.0000

You will notice that in these problems we have treated the quantity p as a random variable, whereas in previous chapters we treated p as a parameter, and consequently as a constant. This is one major difference between Bayesian inference and "classical" inference. In Bayesian inference, parameters are treated like random variables instead of like constants. Notationally, as we did before, we could use the symbol theta as the generic representation of a parameter, but we use an uppercase theta Θ to denote that it is a random variable. In our brief introduction to Bayesian inference we will only be considering two parameters: p, the parameter of the binomial distribution, and μ, the mean of a normal distribution. If the parameter Θ can assume only a discrete set of values, say $\theta_1, \theta_2, \ldots \theta_k$, and if a certain event B is observed, the posterior probability distribution for Θ is most easily found by constructing a table that is analogous to Table 12.1 or 12.2. The table headings would be as in Table 12.3.

In order to find conditional probabilities, it is usually necessary to use one of the special probability distributions described in Chapter 4. In later examples and exercises we will use the binomial, negative binomial, and hypergeometric distributions.

TABLE 12.3

Parameter Value θ_i	Prior Probability, $P[\Theta = \theta_i]$	Conditional Probability, $P[B \mid \theta_i]$	Product, $P[\Theta = \theta_i] \cdot P[B \mid \theta_i]$	Posterior Probability, $P[\Theta = \theta_i \mid B]$

Example 12.3 Assume that the prior distribution over the values of p, the proportion of defectives produced by a watzit maker, is as given initially in Example 12.1, namely,

$$\begin{array}{rccc} p_i\colon & .01 & .05 & .10 \\ P[p = p_i]\colon & .60 & .30 & .10. \end{array}$$

During the first shift the inspector checked 20 items and found only 1 defective. What should the posterior probability distribution over the values of p be?

In this case the sample information is that 1 item in a sample of 20 was found to be defective. If we denote by the random variable X the number of defective items in a sample of 20, X will have a binomial distribution. Consequently,

$$P[B \mid p_i] = P[X = 1 \mid n = 20, p = p_i]$$
$$= \binom{20}{1}(p_i)^1(1 - p_i)^{19}.$$

When $p_i = .05$ or $.10$, this conditional probability can be found from the binomial tables. When $p_i = .01$, this probability can be found by direct calculations to be .1652. The posterior probability distribution for p can be found by constructing Table 12.4. Notice that since the sample information ($x/n = 1/20 = .05$) most strongly supports $p = .05$, the probability that $p = .05$ was increased from .30 in the prior distribution to .4730 in the posterior distribution.

TABLE 12.4

Parameter Value, p_i	*Prior Probability,* $P[p = p_i]$	*Conditional Probability,* $P[B \mid p_i]$	*Product,* $P[p = p_i] \cdot P[B \mid p_i]$	*Posterior Probability,* $P[p = p_i \mid B]$
.01	.60	.1652	.09912	.4141
.05	.30	.3774	.11322	.4730
.10	.10	.2702	.02702	.1129
	1.00		$\sum = .23936$	1.0000

12.3 PRIOR DISTRIBUTION FOR THE BINOMIAL PARAMETER p

As you think back to the examples of the first two sections, what aspect about them struck you as most unrealistic? The name of the item being produced, a "watzit," is of course a bit whimsical. However, if you think of this as simply representing some manufactured item, you would probably agree that it is realistic to say that some proportion of the items will be defective while the others would be acceptable. What is unrealistic is to assume that the proportion of defectives must be constrained to one of three values! Actually, the proportion could take on any value between zero and 1 (see Figure 12.2). This being the case, we must treat p as a continuous random variable rather than a discrete random variable. Consequently, to describe the prior distribution of p we will need to use a continuous probability density function.

In Chapter 2 when we first discussed continuous probability density functions for a random variable X, we said that they would allow us to assign probabilities to events like $a \leq X \leq b$. It is also true that the value of the density function will be large in regions where X is likely to assume values and small in regions where X is less likely to assume values. Consider for example the graph of a pdf for a random

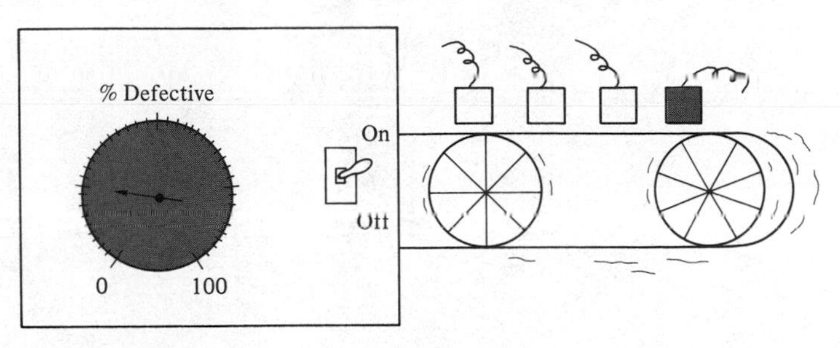

FIGURE 12.2

Watzit Maker with a Continuous Defective Rate Dial

variable X shown in Figure 12.3. Even though no vertical scale is shown, so that exact probabilities cannot be calculated, you should be able to see that X is more likely to fall between 2 and 4 than between 0 and 2, for example, and that this latter event is more likely than for X to fall between 8 and 10. Keeping this notion in mind, we can see in general terms how we might like to choose a prior density function over the possible values of p: the density function should be large in regions where we expect p to be and smaller where we do not expect p to be.

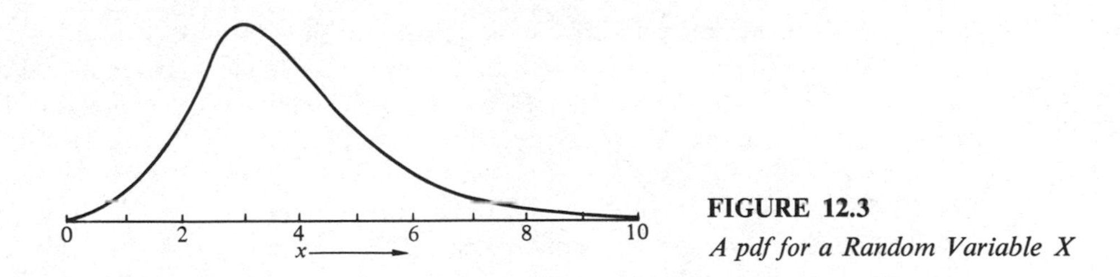

FIGURE 12.3

A pdf for a Random Variable X

Example 12.4 Consider three situations where we are interested in a probability p:

1. An amateur silversmith makes a medallion with an eagle on one side. Let p represent the probability that the medallion will land eagle side up when flipped.
2. The USA and USSR are planning a series of joint scientific satellite launchings. Let p represent the probability of a successful launch.
3. A machine produces No. 10 1-inch bolts. Let p represent the probability that a bolt produced by this machine will be defective.

Which of the density functions in Figure 12.4 best describes a reasonable prior density for p for each of these situations?

Although we do not really know how good this amateur silversmith is at making medallions, our intuition tells us that p might be near .5, although we would not be too surprised if p were as low as .3 or as high as .7. In the absence of other information,

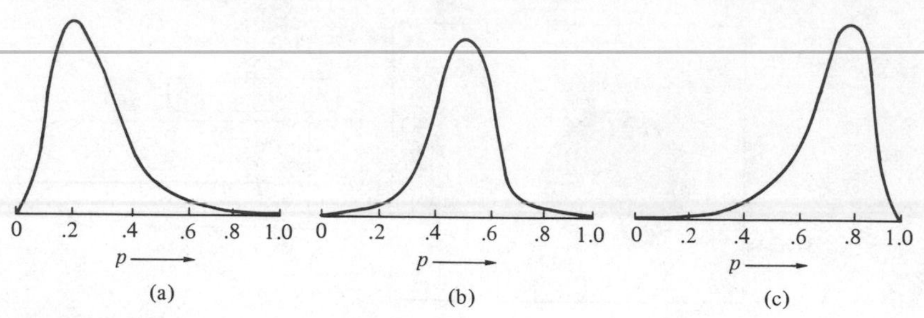

FIGURE 12.4

it would seem reasonable to choose the prior given in Figure 12.4(b) over (a) or (c).

The space programs of the USA and USSR have been relatively successful, although both have had some dramatic failures. We would expect the probability of a successful launch to be greater than .5, so (c) would be the prior among the three which is the most reasonable.

If a machine producing bolts is operating in a reasonably satisfactory condition, the probability that it will produce a defective bolt should be quite small, certainly less than 20%, say. Consequently, the prior density function in (a) would be the most reasonable choice of prior densities among these three.

At this point you may have some feeling for how to make a rough sketch of a prior distribution for a parameter p in certain circumstances. It would be desirable to be able to quantify these feelings by giving a mathematical expression for these prior density functions. A family of density functions that can be used for this purpose is the beta family. This is a two-parameter family of distributions and by appropriate choice of these parameter values we can obtain density functions that look like those given in Figure 12.4, as well as many others. For reasons that will become apparent later, we will denote the parameters of the beta distribution by n and s. The distribution itself will be denoted by $\beta(n, s)$. The general form for the density function for a member of the beta family is

$$f(x) = k(n, s)x^{s-1}(1 - x)^{n-s-1} \qquad 0 \leq x \leq 1, \tag{12.1}$$

where the parameters n and s must satisfy $n > s > 0$. Remember that one property of continuous probability density functions is that the total area under the density function must be equal to 1 [see (2.9) in Section 2.3]. The quantity $k(n, s)$ in (12.1) is simply the constant needed to satisfy this condition. The parameters n and s need not be integers, but if they are, then

$$k(n, s) = \frac{(n-1)!}{(s-1)!\,(n-s-1)!}. \tag{12.2}$$

Furthermore, since

$$(s - 1) + (n - s - 1) = n - 2,$$

it follows that this constant can be expressed using such combinations as

$$k(n, s) = (n - 1)\binom{n-2}{s-1}. \tag{12.3}$$

In order to sketch a graph of x vs. $f(x)$, we would evaluate (12.1) for several values of x between 0 and 1 and connect the points with a smooth curve. For example, if $n = 11$ and $s = 2$ and we use (12.3), we would find, for $x = .1$,

$$(12.4) \qquad f(.1) = k(11, 2)(.1)^{2-1}(1 - .1)^{11-2-1}$$

$$= (11 - 1)\binom{11 - 2}{2 - 1}(.1)^{2-1}(1 - .1)^{11-2-1}$$

$$= 10\left[\binom{9}{1}(.1)^1(.9)^8\right].$$

Although it is not very hard to evaluate (12.4) directly, it becomes even easier if you note that thc factor in brackets is really just a binomial probability corresponding to the probability of 1 "success" in 9 trials with a probability .1 of a "success." From Table A1 we find the bracketed term to be .3874, so

$$f(.1) = 10[.3874] = 3.874.$$

From this illustration you can see that the binomial tables can be used to good advantage in evaluating the beta density function. Also, in our particular application of using the beta family as prior densities for the parameter p, we can reasonably replace the argument x in (12.1) with a p. With this notation we would write

$$(12.5) \qquad f(p) = (n - 1)\binom{n - 2}{s - 1}p^{s-1}(1 - p)^{n-s-1} \qquad 0 \leq p \leq 1$$

when n and s are integers.

Example 12.5 Sketch the density function for a beta random variable having parameter values $n = 11$ and $s = 2$.

Using (12.5) and the binomial tables A1, we can evaluate $f(p)$ for $p = 0, .1, .2, \ldots, 1$. These results are given in Table 12.5. The corresponding graph is shown in Figure 12.5.

In making a sketch of a density function such as this, it would be helpful to know the value of p at which the mode occurs. By methods of calculus it is not hard

TABLE 12.5

p:	0	.1	.2	.3	.4	.5	.6	.7	.8, .9, 1.0
$f(p)$:	0	3.874	3.020	1.556	0.605	0.176	0.035	0.004	0.000

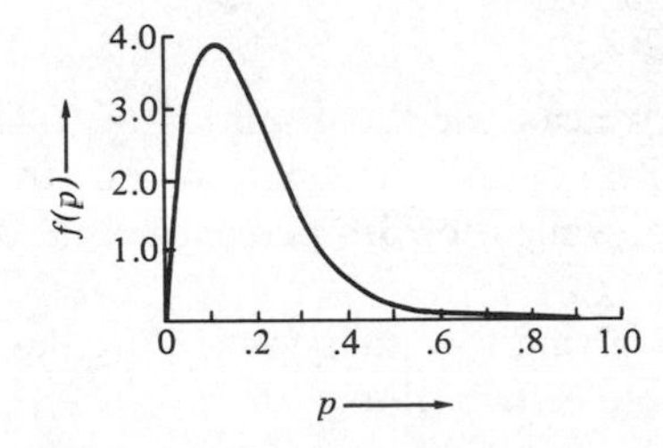

FIGURE 12.5
The Density Function for β(11, 2)

to show (Exercise 7) that for $n > 2$ and $s > 1$,

$$p_{\text{mode}} = \frac{s-1}{n-2}.$$

In Example 12.5, for example, the mode occurs at $(2-1)/(11-2) = \frac{1}{9}$. You can see that the density function sketched in Figure 12.5 is not symmetric. It will be symmetric only when $s = n/2$.

In addition to knowing the mode of the distribution, we will find it useful later to know the cumulative distribution function, the mean, and the variance of the distribution. It is an interesting fact that the cumulative distribution function of a random variable having a beta distribution can be related to the cumulative distribution function of a binomial random variable. Explicitly, if $X_\beta \sim \beta(n, s)$, then

$$F_{X_\beta}(p) = P[X_\beta \leq p] = P[X_B \geq s] = 1 - F_{X_B}(s-1), \tag{12.6}$$

where X_B is a binomial random variable having parameters "$n - 1$" (trials) and p (probability of a "success"). For example, if a random variable X_β has a $\beta(11, 2)$ distribution, what is the probability that X_β is less than or equal to .3 [i.e., what is $F_{X_\beta}(.3)$]? From Figure 12.5 you can get a rough guess as to the answer, since the area to the left of $p = .3$ is about .8 or .9. Using (12.6) and the cumulative binomial tables (Table A2), we find that

$$F_{X_\beta}(.3) = 1 - F_{X_B}(2-1) = 1 - .1493 = .8507.$$

In this case we took X_B to be a binomial random variable with $n - 1 = 10$ and $p = .3$.

Example 12.6 Sketch the cumulative distribution function for a beta random variable having parameter values $n = 11$ and $s = 2$.

Using (12.6) and the cumulative binomial tables, we can evaluate $F(p)$. The values of $F(p)$ for $p = 0, .15, .10, \ldots, .50$ are given in Table 12.6, and the corresponding graph is shown in Figure 12.6. In this case $F(p)$ is approximately 1 for $p > .5$.

TABLE 12.6

p:	0	.05	.10	.15	.20	.25	.30	.35	.40	.45	.50
$F(p)$:	0	.0861	.2639	.4557	.6242	.7560	.8507	.9140	.9536	.9767	.9893

As was done in Chapter 2, we can use the graph of the CDF (and/or the corresponding table) to find percentiles for this distribution. For this example the 10th percentile is slightly more than .05, while the 90th percentile is about .34. The median is about .16.

We will close this section by giving formulas for finding the mean and variance for a random variable having a beta distribution.

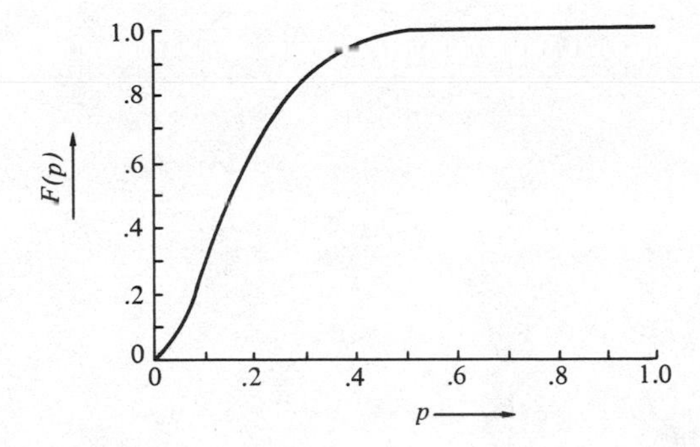

FIGURE 12.6
The Cumulative Distribution Function for $\beta(11, 2)$

Theorem 12.1 If X_β is a random variable having a beta distribution with parameters n and s, then

$$E(X_\beta) = \frac{s}{n} \quad \text{and} \quad V(X_\beta) = \frac{s(n-s)}{[n^2(n+1)]}.$$

Before proving this theorem, we can illustrate its application by an example.

Example 12.7 Find the mean, variance, and standard deviation for a beta random variable having parameters $n = 11$ and $s = 2$.

Using the results of Theorem 12.1, we find that

$$E(X_\beta) = \frac{s}{n} = \frac{2}{11} = .1818$$

$$V(X_\beta) = \frac{s(n-s)}{n^2(n+1)} = \frac{2(9)}{(11)^2(12)} = .0124$$

$$\sigma_{X_\beta} = \sqrt{.0124} = .1114.$$

● ● ● ● ●

Proof of Theorem 12.1: For any continuous random variable,

$$E(X) = \int_{-\infty}^{\infty} x \cdot f(x)\, dx.$$

Using the argument p as the variable of integration and remembering that $f(p)$ is positive only for values of p between 0 and 1, we get

$$(12.7) \qquad E(X_\beta) = \int_0^1 p f_{X_\beta}(p)\, dp = \int_0^1 p[k(n, s)(p)^{s-1}(1-p)^{n-s-1}]\, dp$$
$$= \int_0^1 k(n, s) p^s (1-p)^{n-s-1}\, dp.$$

If we now define $s^* = s + 1$ and $n^* = n + 1$, we can rewrite (12.7) as

$$\int_0^1 k(n, s) p^{s^*-1}(1-p)^{n^*-s^*-1}\, dp.$$

This integral would integrate to 1 if we had the appropriate constant, $k(n^*, s^*)$ instead of $k(n, s)$. Multiplying and dividing by $k(n^*, s^*)$, we find that

$$E(X_\beta) = \int_0^1 \frac{k(n^*, s^*)}{k(n^*, s^*)} k(n, s) p^{s^*-1}(1-p)^{n^*-s^*-1}\, dp$$

$$= \frac{k(n, s)}{k(n^*, s^*)} \int_0^1 k(n^*, s^*) p^{s^*-1}(1-p)^{n^*-s^*-1}\, dp$$

$$= \frac{k(n, s)}{k(n^*, s^*)}.$$

Now assuming that n and s are integers, we get [using (12.2)]

$$\frac{k(n, s)}{k(n^*, s^*)} = \frac{(n-1)!/(s-1)!\,(n-s-1)!}{(n^*-1)!/(s^*-1)!\,(n^*-s^*-1)!}$$

$$= \frac{(n-1)!/(s-1)!\,(n-s-1)!}{(n+1-1)!/(s+1-1)!\,((n+1)-(s+1)-1)!}$$

$$= \frac{(n-1)!/(s-1)!}{n!/s!} = \frac{s}{n}.$$

The proof that $V(X_\beta) = s(n-s)/[n^2(n+1)]$ is left to Exercise 11.

12.4 POSTERIOR DISTRIBUTION FOR THE BINOMIAL PARAMETER p

In the previous section we saw that a member of the beta family of distributions could be used for the prior distribution over the values of a binomial parameter p. Later in this section we will discuss how a *particular* member might be chosen, but for now let us simply assume that in some way Dee, the inspector at the Watzit company, has chosen a beta prior with parameters $n = 20$ and $s = 3$. How can this prior distribution be combined with sample information to get a posterior distribution over the possible values of p? More explicitly, let us assume that his sample information consisted of finding one defective item among 10 items sampled.

In Section 12.2 we saw that if p were treated like a discrete random variable, the posterior distribution would be found from

$$P[p = p_i \mid B] = \frac{P[p = p_i] \cdot P[B \mid p = p_i]}{\sum_j P[p = p_j] \cdot P[B \mid p = p_j]}. \tag{12.8}$$

We will use a continuous analog of equation (12.8) to find the posterior distribution when using a continuous prior. To distinguish between the prior and the posterior distributions, we will use primes and double primes. Hence we will let $f'(p)$ denote the prior density and $f''(p)$ denote the posterior density. Then

$f'(p)$ will be analogous to $P[p = p_i]$ and

$f''(p)$ will be analogous to $P[p = p_i \mid B]$.

The conditional probability will remain the same, except that p is not constrained to a discrete set of values $p_1, p_2, \ldots$. In the denominator of (12.8), the summation will be replaced with an integral. With these relationships in mind, we will have

$$f''(p) = \frac{f'(p) \cdot P[B \mid p]}{\int_0^1 f'(p)P[B \mid p]\, dp}. \tag{12.9}$$

Keep in mind that the integral in the denominator of (12.9) is simply the constant value needed to guarantee that $f''(p)$ will cover a total area of 1 and hence be a valid density function.

Remember that in the particular case of finding the posterior distribution for a binomial parameter p, the sample information is generally expressed by giving the number of observed successes, say X, in n trials. The event B then corresponds to the event $X = x$, where X is a binomial random variable having parameters n and p. To distinguish between this binomial parameter n and the parameters of the prior and posterior distributions, we will use n' and n'', respectively (as well as s' and s''), to denote the parameters of the prior and posterior distributions. The following theorem tells how prior and sample information can be combined to get the posterior probability distribution for p.

Theorem 12.2 If the prior probability density function for a random binomial parameter p is a beta distribution with parameters n' and s', and if a sample of n observations yields x successes, the posterior distribution for p will be a beta distribution with parameters

$$n'' = n + n' \qquad \text{and} \qquad s'' = x + s'.$$

Proof: In this case the prior density function is

$$f'(p) = k(n', s')p^{s'-1}(1-p)^{n'-s'-1} \qquad 0 \leq p \leq 1$$

and the conditional probability is given by

$$P[B \mid p] = P[X = x \mid n, p] = \binom{n}{x} p^x (1-p)^{n-x} \qquad x = 0, 1, \ldots, n.$$

Using these expressions in (12.9) gives

$$f''(p) = \frac{k(n', s')p^{s'-1}(1-p)^{n'-s'-1}\binom{n}{x}p^x(1-p)^{n-x}}{\int_0^1 k(n', s')p^{s'-1}(1-p)^{n'-s'-1}\binom{n}{x}p^x(1-p)^{n-x}\, dp}. \tag{12.10}$$

Remembering our earlier comment that the integral in the denominatior of (12.9) is simply a constant, we rewrite (12.10) replacing the denominator by C and combining exponents on p and $(1-p)$ in the numerator to get

$$f''(p) = \frac{k(n', s')\binom{n}{x}}{C} p^{(x+s'-1)}(1-p)^{(n+n')-(x+s')-1}. \tag{12.11}$$

If we define

$$n'' = n + n' \qquad \text{and} \qquad s'' = x + s'$$

and note that the constant $\left[k(n', s')\binom{n}{x}\Big/C\right]$ is simply the constant necessary to make

$f''(p)$ integrate to 1, it follows that (12.11) can be written as

$$f''(p) = k(n'', s'')p^{s''-1}(1 - p)^{n''-s''-1}.$$

Since this density function is exactly of the form of (12.1), it follows that the posterior distribution must be a beta distribution with parameters as defined in the statement of this theorem.

Example 12.8 The inspector at the Watzit company has assigned a beta prior distribution with parameters $n' = 20$ and $s' = 3$ to the probability p that an item produced by a watzit maker will be defective. In a subsequent sample he found one defective among 10 inspected items. What should his posterior distribution be? Also, graph the prior and posterior probability distributions.

Since 10 items were inspected with one of those items being found defective, we have $n = 10$ and $x = 1$. Hence the posterior distribution is a beta with parameters

$$n'' = n + n' = 10 + 20 = 30$$

and

$$s'' = x + s' = 1 + 3 = 4.$$

The graphs of these distributions are given in Figure 12.7.

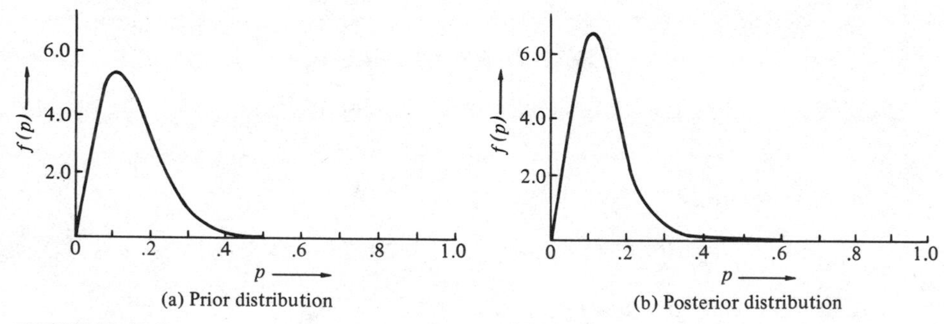

(a) Prior distribution

(b) Posterior distribution

FIGURE 12.7

It might be instructive to compare the means and standard deviations of the prior and posterior distributions for this example. The prior mean was $s'/n' = 3/20 = .150$, while the posterior mean is $s''/n'' = 4/30 = .133$. Notice that the sample proportion x/n was $1/10 = .100$, so that the posterior mean lies between the prior mean and the sample mean. The significance of this fact is that one's judgment about the mean of the distribution of p is always modified in the direction of the sample information. If you originally thought p was "large" (quantified by the value of the prior mean) but observed a sample proportion smaller than you would have expected, your posterior mean will be smaller than your prior mean. Likewise, if you observe a sample proportion larger than you would have expected, your posterior mean will be larger than your prior mean. The posterior mean lies between the prior mean and the sample mean because it actually is a weighted average of these two means. If

$n' > n$, the posterior mean will be closer to the prior mean, while if $n > n'$, the posterior mean will be closer to the sample mean.

What is the relationship between the prior and posterior standard deviations? Can you tell by comparing the graphs in Figure 12.7? The prior standard deviation is

$$\sigma' = \sqrt{\frac{s'(n' - s')}{n'^2(n' + 1)}} = \sqrt{\frac{3(17)}{(20)^2(21)}} = .078,$$

while the posterior standard deviation is

$$\sigma'' = \sqrt{\frac{s''(n'' - s'')}{n''^2(n'' + 1)}} = \sqrt{\frac{4(26)}{(30)^2(31)}} = .061.$$

In most (but not all) situations the posterior standard deviation will be smaller than the prior standard deviation, as is the case in this example. This is indicative of the fact that as more information becomes available about the value of p, the amount of uncertainty about its value, as reflected by the standard deviation, tends to get smaller. In fact, if n is very large relative to n', the posterior mean will be practically the same as the sample mean:

$$\frac{s''}{n''} = \frac{x + s'}{n + n'} \approx \frac{x}{n}.$$

At the same time the uncertainty, as measured by the posterior standard deviation, becomes essentially zero, since

$$\sigma'' = \frac{(x + s')[n + n' - (x + s')]}{(n + n')^2(n + n' + 1)} \approx 0$$

for large n. Note that this is consistent with the ordinary or classical situation where only sample information is used to get information about p: for large n we feel very confident that the sample proportion x/n will be quite close to p.

In Example 12.2 we considered a situation where additional sample information became available after the prior had already been revised in view of some initial sample information. In that case we used the first posterior distribution as a "new" prior distribution and repeated the process of revising the prior. Fortunately, it is very easy to perform similar repeated revisions in the case of using a continuous beta prior for the possible values of p. This is so, however, because of the fact that the posterior distribution is also a beta distribution if, in fact, we start with a beta prior distribution. It follows from Theorem 12.2 that the "old" posterior distribution can be used as a "new" prior distribution when additional sample information becomes available.

Example 12.9 Demetrius, an amateur silversmith, has made a medallion with an eagle on one side. He is interested in p, the probability that the medallion will land eagle side up when he flips it. He feels that his medallion is reasonably symmetric, so he chooses a beta prior distribution over the values of p having parameters $s' = 12$ and $n' = 24$. What should his posterior distribution for p be if he flips the medallion 20 times and observes it to land eagle side up 8 times? What should it be if he then flips it 10 additional times, observing 6 eagles?

After the first 20 trials the prior distribution should be revised to a beta posterior distribution having parameters

$$n'' = n + n' = 20 + 24 = 44$$

and

$$s'' = x + s' = 8 + 12 = 20.$$

When a second set of experimental data becomes available, this posterior distribution will be used as the current or updated prior. The new posterior distribution would then have parameters

$$n'' = n + n' = 10 + 44 = 54$$

and

$$s'' = x + s' = 6 + 20 = 26.$$

Note that this same result would have been obtained had the sample information been combined to 14 eagles observed in 30 trials and then used with the original prior.

Consider for a moment how you would combine sample information about a proportion from two different samples. We did this in the previous example: one sample showed 8 eagles in 20 flips, while a second showed 6 in 10 flips; these were combined to give 14 eagles in 30 flips. More generally, if you observe x_1 "successes" in n_1 trials and then observe an additional x_2 successes in n_2 trials, this information would be combined to give $(x_1 + x_2)$ successes in $(n_1 + n_2)$ trials. Classically, a point estimate of p would then be given by the ratio

$$\frac{x_1 + x_2}{n_1 + n_2}. \tag{12.12}$$

With this in mind, think about how prior and sample information are combined into the posterior distribution. In particular note that the posterior mean is given by

$$\frac{x + s'}{n + n'}. \tag{12.13}$$

Now compare (12.12) and (12.13) and notice the great similarity. The prior parameters n' and s' are in some sense like additional sample information. The prior parameters can be thought of as being "equivalent to" having made n' observations and having observed s' "successes." That is, your prior "experience" is quantified by translating it into equivalent sample information.

Specifically, how does one quantify past experience? One way this can be done is by asking two questions: (1) What does your experience tell you about the best guess about the value of p? (2) Within what range of values would you be quite certain that p will lie? The first question relates to a point estimate of p, while the second relates to an interval estimate of p. Demetrius, the amateur silversmith, answered these questions as follows. His best guess about the value of p was .5. He felt, however, that p would very likely lie between .3 and .7. Setting the prior mean equal to his best guess, he got

$$\frac{s'}{n'} = .5. \tag{12.14}$$

Using the rule of thumb that for continuous distributions which are not too skewed,

90% or more of the observations will usually fall within 2 standard deviations of the mean, he obtained a second equation:

$$(12.15) \qquad 4\sigma' = 4\sqrt{\frac{s'(n'-s')}{(n')^2(n'+1)}} = .7 - .3 = .4.$$

These two equations can be solved simultaneously for s' and n'. In particular, (12.15) can be rewritten as

$$4\sqrt{\frac{(s'/n')(1-s'/n')}{n'+1}} = .4.$$

Substituting $s'/n' = .5$ into this gives

$$4\sqrt{\frac{(.5)(1-.5)}{n'+1}} = .4.$$

$$\sqrt{n'+1} = 10(.5) = 5$$

$$n' = 5^2 - 1 = 24.$$

Substituting $n' = 24$ into (12.14) yields $s' = 12$. These are the values that Demetrius used for his prior distribution.

12.5 BAYESIAN ESTIMATION FOR THE BINOMIAL PARAMETER *p*

In Section 2.6 we discussed three measures of central tendency for the probability distribution of a random variable. That discussion is relevant in this section because here we treat the parameter p as a random variable having a certain probability distribution. Before reading on in this section you should review Section 2.6. In that section we set up a hypothetical situation where you were to guess at the value that a random variable would assume. We saw that the "best guess" would depend on the kind of penalty or loss incurred for wrong guesses. In an "all-or-nothing situation" (i.e., a situation where you get a certain reward only for a correct guess), the best choice is the mode of the distribution. If the penalty is related to the square of your error [i.e., related to $(x - g)^2$, where x is the observed value while g is your guess], the best guess is the mean of the distribution. If the penalty is related to the absolute error (i.e., to $|x - g|$), the best guess is the median of the distribution.

In the Bayesian situation we treat p as a random variable having a certain probability distribution. Our "guess" at the value that the random variable will assume is actually our point estimate of p. The best guess or best estimate of p will depend on the kind of loss incurred by a wrong guess. The best guess could be the mode, mean, or median of the distribution of p, depending on the type of loss structure. It could also be some other quantity if the loss structure is not one of the three types described earlier.

Example 12.10 Consider the primitive watzit maker of Examples 12.1 to 12.3. There it was assumed that p could only assume three values, .01, .05, or .10, and that the prior probabilities over these values were .60, .30, and .10. After a single item was inspected and found to be defective, the probabilities were revised to .1935, .4839, .3226. The inspector Dee has made a bet with a coworker. Dee claims that he can guess the dial setting (i.e., the value of p) and his coworker claims otherwise. The loser of the bet will take the other to dinner. What should Dee guess (estimate) p to be based only on his prior information? What should his guess be based on the posterior distribution?

This is certainly an all-or-nothing situation, so the best guess would be the mode. Based only on prior information, Dee's guess would be the mode of the prior distribution, namely, .01. Based on the posterior distribution, his guess would be the posterior mode, namely, .05.

In this example Dee's estimate of p was the mode of the distribution of p because of the all-or-nothing loss structure. Had the loss been proportional to $(x - g)^2$ (referred to as "squared-error loss"), his estimate would have been the mean of the appropriate distribution. Based only on prior information, the estimate would have been

$$E'(p) = \sum_{i=1}^{3} p_i \cdot P[p = p_i]$$
$$= (.01)(.60) + (.05)(.30) + (.10)(.10) = .031.$$

Based on the posterior distribution, his guess would have been

$$E''(p) = \sum_{i=1}^{3} p_i \cdot P[p = p_i \mid B]$$
$$= (.01)(.1935) + (.05)(.4839) + (.10)(.3226) = .05839.$$

Similarly, if the loss function had been proportional to the absolute error, Dee would have used the appropriate median as the estimate of p, either

$$\tilde{\mu}'_p = .01 \qquad \text{or} \qquad \tilde{\mu}''_p = .05.$$

Since the same kind of reasoning holds for continuous probability distributions, when p is treated like a continuous random variable, a point estimate of p can be given by the mean, median, or mode of the continuous prior or posterior probability distribution for p.

Example 12.11 In Example 12.8 we assumed that Dee chose a beta prior for the prior probability distribution for p having parameters $n' = 20$ and $s' = 3$. Having found one defective item among 10 inspected, he found the posterior probability distribution for p to be a beta distribution with parameters $n'' = 30$ and $s'' = 4$. Find estimates for p based on the prior and posterior distributions using (a) the mean and (b) the mode of the distributions.

Using the mean, the estimates would be

$$\mu'_p = \frac{s'}{n'} = .150 \qquad \text{or} \qquad \mu''_p = \frac{s''}{n''} = .133.$$

Using the mode, the estimates would be

$$\text{prior mode} = \frac{s' - 1}{n' - 2} = .105$$

or

$$\text{posterior mode} = \frac{s'' - 1}{n'' - 2} = .107.$$

Note that the median of a beta distribution is not as easy to find as the mean or mode, so we did not attempt to find it in this example. Recall (see Section 2.6, in particular Figure 2.17) that for skewed, unimodal distributions, the median will lie between the mean and the mode. While "between" does not mean exactly the midpoint between these values, that would not be a bad guess as to the location of the median.

If p is really a continuous random variable, an all-or-nothing loss function would not be reasonable, since it would be impossible to be *exactly* right when guessing at the value of p. This, of course, holds true for any continuous random variable X, since $P[X = g] = 0$, no matter what guess g you might make. A more reasonable situation might be to have a penalty (loss) of zero if your guess is within some small value of the observed value, say within .01, and some positive loss otherwise. The mode would be a good guess in a situation like this.

In a case where some tolerance is allowed on either side of the guess, the value of p is actually being estimated by an interval. Let us consider another situation where an interval can be used to guess at the value of a random variable. Say that your brother John is coming to visit you for a weekend and he asks you to arrange a date. You randomly select a date for him from among your acquaintances. John did not specify any height constraints for his date, but is now curious as to what that height might be. If he knows that the heights of young women follow a normal distribution with mean 65 inches and standard deviation 2.5 inches, between what two values can he be 90% sure that the height of his date will lie? That is, between what two values should he guess the height of his date to be if he wishes to have a 90% chance of being correct? In Figure 12.8 is sketched the distribution of the heights of young women. You can see that the 5th and 95th percentiles will be values satisfying the desired conditions. That is, John can be 90% sure that the height of his date will fall between 60.9 and 69.1 inches. The interval (60.9, 69.1) can be thought of as an interval estimate of the value of the height of John's date.

An approach similar to this one can be used to give interval estimates for the value of p. If the probability distribution of p is known, the probability that p will lie between its $100(\alpha/2)$ percentile and $100(1 - \alpha/2)$ percentile is $1 - \alpha$. For example, there is a 90% probability that p will lie between the 5th and 95th percentiles of the distribution of p. Conceptually, these percentiles are easy to find; however, in practice, finding them can be difficult if tables cannot be used. (Of course, if a computer is available, this difficulty is easily overcome.) To distinguish the Bayesian situation where p is treated like a random variable from the classical situation where p is treated like a constant, the interval found is usually referred to as a *credible interval* rather than a confidence interval.

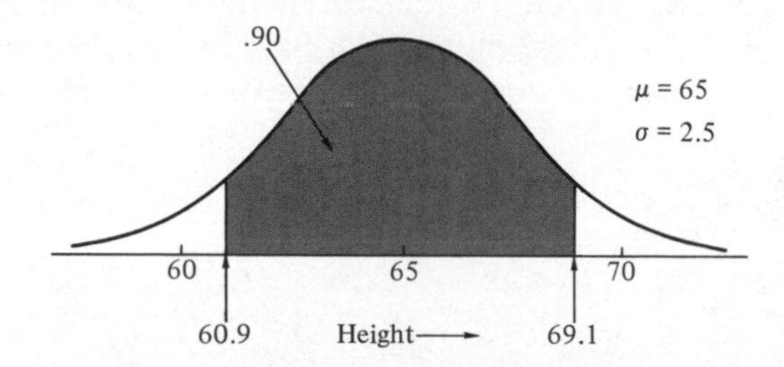

FIGURE 12.8

Example 12.12 An investigator has decided that the prior distribution for a proportion p should be a beta distribution with parameters $n' = 11$ and $s' = 2$. Using this distribution, find an 80% credible interval for p.

This is the beta distribution used in Examples 12.5 and 12.6. You might turn back to Figure 12.5 and see if you can make a rough guess as to two values between which there is an 80% chance that p will lie. More precisely, we would need to find the 10th and 90th percentiles of this distribution. Since from (12.6) we have

$$F_{X_\beta}(p) = 1 - F_{X_B}(s - 1),$$

we can use the cumulative binomial tables (as we did in Example 12.6) to find

$$F(.05) = .0861, \qquad F(.10) = .2639$$

so that the 10th percentile (by interpolation) is about .054. Similarly,

$$F(.30) = .8507, \qquad F(.35) = .9140,$$

so the 90th percentile is about .339. Hence an 80% credible interval would be given by (.054, .339).

It should be clear from Figure 12.5 and from our comments there that this distribution is not symmetric. If a distribution is symmetric (or at least approximately so), the shape of it might remind you of another familiar distribution. What are you reminded of when you look at the distribution in Figure 12.7(b)? This graph looks somewhat like a normal distribution and it would look more so if s/n were closer to 1/2. Because of the fact that the normal distribution can be used to approximate the beta distribution in certain cases, the percentiles needed to get an approximate credible interval can be found by using the normal tables.

Example 12.13 Find a 90% credible interval for the value of p, the probability that Demetrius' medallion will land eagle side up, based on the assumption that he used a beta prior for p having parameters $n' = 24$ and $s' = 12$ and given that he observed 14 eagles in 30 trials (see Example 12.9). Also find a 90% confidence interval for p based only on the sample data using classical methods.

The posterior distribution over the possible values of p is a beta distribution with parameters $n'' = 54$ and $s'' = 26$. The mean and standard deviation of this posterior distribution are

$$(12.16) \qquad \mu'' = \frac{s''}{n''} = .4815 \qquad \text{and} \qquad \sigma'' = \sqrt{\frac{s''(n'' - s'')}{(n'')^2(n'' + 1)}} = .0674.$$

To approximate the 5th and 95th percentiles of this distribution, we can use the same percentiles of a normal distribution having mean and standard deviation equal to the values given in (12.16). A sketch of this normal distribution is shown in Figure 12.9. The percentiles are shown on the graph. The percentiles were found from

$$P[X \leq c] = P\left[\frac{X - \mu''}{\sigma''} \leq \frac{c - \mu''}{\sigma''} = -1.645\right] = .05,$$

so

$$c = \mu'' - 1.645\sigma'' = .4815 - .1109 = .3706$$

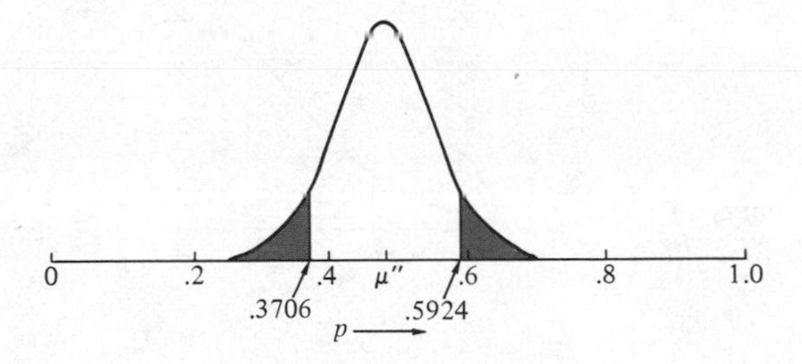

FIGURE 12.9

for the 5th percentile and

$$\mu'' + 1.645\sigma'' = .4815 + .1109 = .5924$$

for the 95th percentile. Consequently, a 90% credible interval is given approximately by (.3706, .5924).

A 90% confidence for p based only on the sample data using equation (7.36) would be

$$\frac{x}{n} \pm z_{.95}\sqrt{\frac{(x/n)(1 - x/n)}{n}} = \frac{14}{30} \pm 1.645\sqrt{\frac{(\frac{14}{30})(\frac{16}{30})}{30}},$$

which simplifies to $.4667 \pm .1499$ or (.3168, .6166).

We have now discussed how to find point estimates and interval estimates of the binomial parameter p based on the prior and/or posterior probability distribution over the values of p. Estimation is only one aspect of statistical inference. A second aspect of statistical inference is testing of hypotheses. Hypothesis testing about p in a Bayesian framework can be performed by using the posterior distribution. Actually, it can be viewed as a decision-making problem if appropriate costs are assigned to making a wrong decision (i.e., to type I and type II errors). We will not pursue this approach here, since decision theory will be the topic of the next chapter.

12.6 BAYES' STATISTICS APPLIED TO THE PARAMETER μ OF A NORMAL DISTRIBUTION

In the previous sections we have mentioned Demetrius, the amateur silversmith, in terms of estimating the probability p that one of his medallions will land eagle side up. Actually, there are other aspects of these medallions that might be of more interest. For example, if you were going to buy a medallion from Demetrius, you would be interested in knowing the amount of silver in it. A high price for the medallion might be justified if it contained a large amount of silver, but not otherwise. Let us assume that X, the amount of silver in a large medallion (8 cm in diameter), is actually a random variable due to the variability in the alloy composition and how completely the medallion mold is filled. Further, assume that X follows approximately a normal distribution having an unknown mean μ. From past experience it is estimated that $\sigma = 8$ grams (see Figure 12.10). If you felt that the mean was about 200 grams, you would be willing to buy a medallion. So far, Demetrius has made 9 eagle

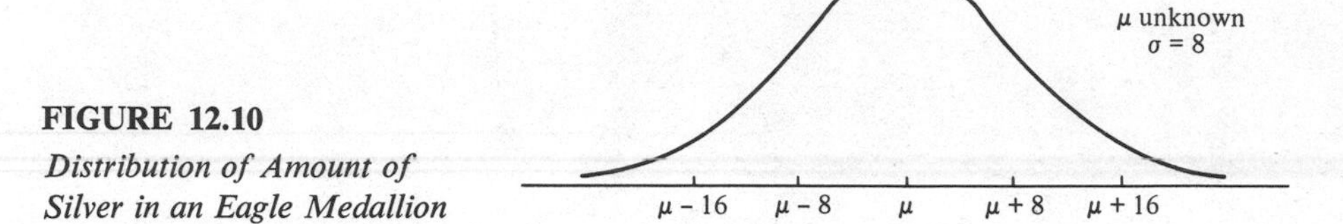

FIGURE 12.10
Distribution of Amount of Silver in an Eagle Medallion

medallions, but these are already sold. If you order one now, Demetrius will make another for you. How can you make a wise decision? Based on classical techniques you could use $\bar{X}$, the sample mean of the 9 medallions already struck, to give a point estimate for μ. (You might also compute an interval estimate of μ or test some hypothesis about μ.)

Let us assume that in the past Demetrius has designed and made many different styles of medallions of this size. For each of these different styles, the amount of silver in a specific style of medallion has had a different mean (e.g., for a sun medallion the mean was 205 grams, for a horse medallion the mean was 198 grams, etc.). In a sense, then, the mean μ can be considered to be a random variable. Let us assume that the parameter μ for a particular type of medallion is actually the observed value of a random variable having a normal distribution with mean $\mu' = 203$ grams and standard deviation $\sigma' = 6$ grams (see Figure 12.11). This distribution can then be considered to be the prior distribution over the possible values that μ can assume. Based *only* on this information, what single value would you guess the mean amount of silver in an eagle medallion to be? The best guess would be the mean of this distribution, namely, 203 grams. On this basis alone, you should be willing to order an eagle medallion from Demetrius, based on the 200-gram criterion mentioned earlier.

It would seem reasonable at this point to compare the information based only on prior information with information based on the 9 sample observations. If the sample mean were observed to be $\bar{x} = 205$ grams, say, you would probably feel quite comfortable with a decision to order a medallion, since the prior mean and sample mean both indicate that the true mean will be 200 grams or more. On the other hand, if the sample mean were 198 grams, you might feel it harder to made a decision, since the prior mean and sample mean are in "disagreement."

Sample information and the prior distribution can be combined to give a

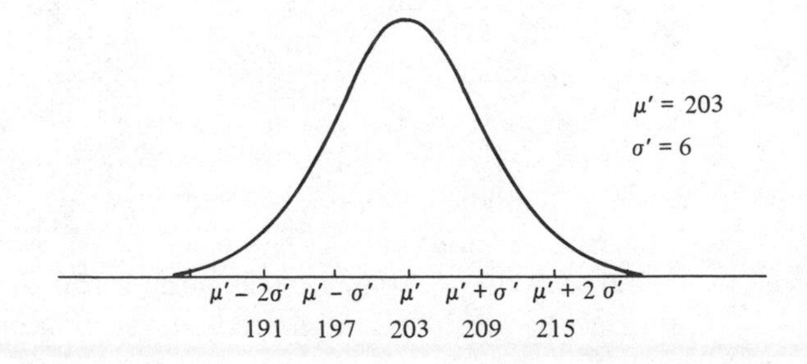

FIGURE 12.11
(Prior) Distribution of μ, the Mean Amount of Silver in a Demetrius Medallion

posterior distribution over the possible values of the parameter μ. The posterior distribution can be found by using a formula analogous to equation (12.9). The main difference is that the conditional probability $P[B \mid p]$, which was actually a binomial probability, has to be replaced with a continuous conditional density function. In particular, since the observed data is a sample mean, the conditional density function must be $f(\bar{x} \mid \mu)$. The analogous formula to (12.9) is

$$f''(\mu) = \frac{f'(\mu) \cdot f(\bar{x} \mid \mu)}{\int_{-\infty}^{\infty} f'(\mu) f(\bar{x} \mid \mu)\, d\mu}. \tag{12.17}$$

Here, too, keep in mind that the integral in the denominator of (12.17) is simply a constant, which we will denote by C, needed to guarantee that $f''(\mu)$ will be a valid density function [cf. discussion following (12.9)]. In this particular example, $\bar{X}$ is a random variable having a normal distribution with mean μ and variance σ^2/n and μ is a random variable having a normal distribution with mean μ' and variance σ'^2. Using the form of the normal density given in equation (4.25) of Section 4.8 in conjunction with (12.17), we can prove the following theorem.

Theorem 12.3 If the prior probability density function for a normal parameter μ is a normal distribution with mean μ' and variance σ'^2 and if the sample mean of n observations is $\bar{x}$, the posterior distribution for μ will be a normal distribution with parameters

$$\mu'' = \left(\frac{\sigma^2}{n}\mu' + \sigma'^2\bar{x}\right)\Big/\left(\frac{\sigma^2}{n} + \sigma'^2\right) \quad \text{and} \quad \sigma''^2 = 1\Big/\left(\frac{1}{\sigma'^2} + \frac{1}{\sigma^2/n}\right). \tag{12.18}$$

Proof: (This proof uses a large number of algebraic manipulations. You may wish to skip over this proof if it seems too difficult.) In this case the prior density is given by

$$f'(\mu) = \frac{1}{\sqrt{2\pi}\sigma'} \exp\left[-\frac{1}{2}\frac{(\mu - \mu')^2}{\sigma'^2}\right]$$

and the conditional density is given by

$$f(\bar{x} \mid \mu) = \frac{1}{\sqrt{2\pi}(\sigma/\sqrt{n})} \exp\left[-\frac{1}{2}\frac{(\bar{x} - \mu)^2}{\sigma^2/n}\right].$$

Using these expressions in (12.17) gives

$$f''(\mu) = \frac{\dfrac{1}{\sqrt{2\pi}\sigma'} \exp\left[-\dfrac{1}{2}\dfrac{(\mu - \mu')^2}{\sigma'^2}\right] \dfrac{1}{\sqrt{2\pi}(\sigma/\sqrt{n})} \exp\left[-\dfrac{1}{2}\dfrac{(\bar{x} - \mu)^2}{\sigma^2/n}\right]}{C}.$$

Remembering that when exponentials are multiplied together the exponents are simply added and combining constants, we get

$$f''(\mu) = \left[\frac{1}{2\pi\sigma'(\sigma/\sqrt{n})C}\right] \exp\left\{-\frac{1}{2}\left[\frac{(\mu - \mu')^2}{\sigma'^2} + \frac{(\bar{x} - \mu)^2}{\sigma^2/n}\right]\right\}. \tag{12.19}$$

Now consider the portion of the exponent in brackets. By expanding the square of each term, getting a common denominator and writing the result as a single squared term, we can simplify the exponent in (12.19) as follows:

$$(12.20) \qquad \frac{(\mu-\mu')^2}{\sigma'^2}+\frac{(\bar{x}-\mu)^2}{\sigma^2/n}=\frac{\mu^2-2\mu\mu'+\mu'^2}{\sigma'^2}+\frac{\bar{x}^2-2\mu\bar{x}+\mu^2}{\sigma^2/n}$$

$$=\frac{(\sigma^2/n)(\mu^2-2\mu\mu'+\mu'^2)+\sigma'^2(\bar{x}^2-2\mu\bar{x}+\mu^2)}{\sigma'^2\sigma^2/n}$$

$$=\left[\mu^2\left(\frac{\sigma^2}{n}+\sigma'^2\right)-2\mu\left(\frac{\sigma^2}{n}\mu'+\sigma'^2\bar{x}\right)+\frac{\sigma^2}{n}\mu'^2+\sigma'^2\bar{x}^2\right]\Big/\sigma'^2(\sigma^2/n).$$

If we now let

$$\mu''=\left(\frac{\sigma^2}{n}\mu'+\sigma'^2\bar{x}\right)\Big/\left(\frac{\sigma^2}{n}+\sigma'^2\right)$$

and

$$\sigma''^2=\frac{\sigma'^2\sigma^2/n}{\left(\frac{\sigma^2}{n}+\sigma'^2\right)}=\frac{1}{\frac{1}{\sigma^2/n}+\frac{1}{\sigma'_2}},$$

(12.20) simplifies to

$$(12.21) \qquad \frac{\mu^2-2\mu\mu''+\mu''^2}{\sigma''^2}-\frac{\mu''^2}{\sigma''^2}+\frac{(\sigma^2/n)\mu'^2+\sigma'^2\bar{x}^2}{\sigma'^2\sigma^2/n}=\frac{(\mu-\mu'')^2}{\sigma''^2}+K,$$

where K is a constant representing the last two terms in (12.21). With another algebraic manipulation after substituting this last expression in for the exponent in (12.19), we can write

$$f''(\mu)=\left[\frac{e^{-(1/2)K}}{2\pi\sigma'(\sigma/\sqrt{n})C}\right]\exp\left[-\frac{1}{2}\frac{(\mu-\mu'')^2}{\sigma''^2}\right].$$

Since the constant factor here simply assures that $f''(\mu)$ will be a density function, because of the form of this density it must be that the posterior distribution of μ is normal with parameters μ'' and σ''^2.

Unfortunately, these expressions are not in a very simple form. You should be able to see, however, that the posterior mean μ'' is just a linear combination (i.e., a weighted average) of the prior mean μ' and the sample mean $\bar{x}$.

Example 12.14 Assume that the prior distribution over the values of μ, the mean amount of silver in a Demetrius medallion, has a normal distribution having mean $\mu'=203$ and standard deviation $\sigma'=6$. If a sample of 9 observations of eagle medallions had a sample mean of 198 grams with the standard deviation of the amount of silver in a particular style of medallion being known to be $\sigma=8$, find the posterior distribution for μ.

Here we use Theorem 12.3 with $\mu'=203$, $\sigma'=6$, $n=9$, $\bar{x}=198$, and $\sigma=8$, to get

$$\mu''=\left[\frac{(8)^2}{9}(203)+(6)^2(198)\right]\Big/\left[\frac{(8)^2}{9}+6^2\right]$$

$$=\frac{8571.55}{43.11}=198.82$$

and

$$\sigma''^2=\frac{1}{\frac{9}{64}+\frac{1}{36}}=\frac{1}{.1684}=5.94.$$

So the posterior distribution for μ is normal with a mean of 198.82 and a variance of 5.94.

How does the posterior variance compare with the prior variance and how does the posterior mean compare with the prior mean and the sample mean? The posterior variance is smaller than the prior variance. This will be the case here, because the posterior distribution reflects more information than the prior. As we indicated earlier, since the posterior mean is a weighted average of the prior mean and sample mean, the posterior mean will fall between these two values. To which mean is the posterior mean closer? In this case it is closer to the sample mean. Let us see if we can see why this should be true. We saw earlier that μ could be estimated by using either the prior mean, μ', or the sample mean, $\bar{X}$. Each of these estimates has a variance associated with it, reflecting uncertainty about the estimate. The variance for the estimate μ' was $\sigma'^2 = 36$, while the variance for $\bar{X}$ was $\sigma^2/n = 64/9 = 7.11$. The weighted average estimate will always be closer to the quantity having the smaller variance. In this case σ^2/n is smaller than σ'^2, so the estimate (198.8) was closer to $\bar{X}$ (198) than to μ' (203).

The formulas for finding the parameters μ'' and σ''^2 of the posterior distribution can be simplified somewhat by introducing a new parameter for use with the prior distribution, n'. In some cases, although not with the medallion example, n' can be interpreted as an "equivalent sample size" in much the same way as it was with the beta distribution parameter, as we shall see. Define n' by

$$n' = \frac{\sigma^2}{\sigma'^2}. \tag{12.22}$$

With this definition for n', we have

$$\sigma'^2 = \frac{\sigma^2}{n'}.$$

Substituting this into (12.18), we find that

$$\mu'' = \frac{(\sigma^2/n)\mu' + (\sigma^2/n')\bar{x}}{\sigma^2/n + \sigma^2/n'} = \frac{\mu'/n + \bar{x}/n'}{1/n + 1/n'} = \frac{n'\mu' + n\bar{x}}{n' + n} \tag{12.23}$$

and

$$\sigma''^2 = 1\bigg/\left(\frac{1}{\sigma^2/n'} + \frac{1}{\sigma^2/n}\right) = 1\bigg/\left(\frac{n'}{\sigma^2} + \frac{n}{\sigma^2}\right) = \frac{\sigma^2}{n' + n}. \tag{12.24}$$

Thinking of n' as equivalent to a number of observations based on prior experience, we see that μ'' is a simple linear combination of μ' and $\bar{X}$, the weights being "sample size." Furthermore, the variance σ''^2 is like the classical expression for the variance of $\bar{X}$, with the "total sample size" being given by $n + n'$.

Example 12.15 Assuming a prior distribution with parameters $\mu' = 203$ and $\sigma' = 6$ and a sample mean of 198 based on 9 observations as in Example 12.14, find the posterior distribution for μ using (12.23) and (12.24).

We first need to find n'. We have

$$n' = \frac{\sigma^2}{\sigma'^2} = \frac{64}{36} = 1.778.$$

(The fact that n' is not an integer really does not affect the remaining computations. It does make it harder to interpret, however.) Using this value, we get

$$\mu'' = \frac{1.778(203) + 9(198)}{1.778 + 9}$$

$$= \frac{2142.934}{10.778} = 198.82$$

and

$$\sigma''^2 = \frac{64}{1.778 + 9} = 5.94.$$

Once the posterior distribution is found, it can be used for purposes of estimation. In Section 12.5 we showed how the posterior distribution for p could be used to estimate or guess at the value of p. In that situtation we saw that the best guess or estimate would depend on the type of loss function (i.e., the penalty incurred for wrong guesses). The best guess might be the mean, median, or mode of the posterior distribution. In the situation considered in this section the posterior distribution is a normal distribution; hence the mean, median, and mode are all equal. Consequently, to get a point estimate for μ, we simply use the posterior mean μ''. How will this point estimate compare with a point estimate in the classical case where the estimate is based only on sample information? Since from (12.23) we know that

$$\mu'' = \frac{n'\mu' + n\bar{x}}{n' + n},$$

we can see that for large n, $\mu'' \approx \bar{x}$, so the classical and Bayes' point estimates are practically equal.

To find an interval estimate, a credible interval, we use the appropriate percentiles of the posterior distribution. For example, a 90% credible interval for μ would be given by the 5th and 95th percentiles of the posterior distribution of μ.

Example 12.16 In Example 12.14 we found the posterior distribution for μ to be a normal distribution having mean $\mu'' = 198.82$ and variance $\sigma''^2 = 5.94$. Use this distribution to find a 90% credible interval for μ.

The posterior distribution for μ is shown in Figure 12.12. The standard deviation here is $\sigma'' = \sqrt{5.94} = 2.44$. The 5th percentile is found from

$$\mu'' - 1.645\sigma'' = 198.82 - (1.645)(2.44) = 194.81,$$

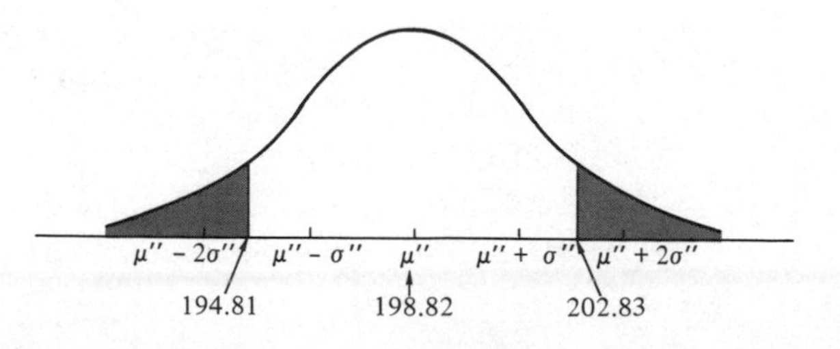

FIGURE 12.12
(Posterior) Distribution of μ, the Mean Amount of Silver in a Demetrius Medallion

while the 95th percentile is found from

$$\mu'' + 1.645\sigma'' = 198.82 + (1.645)(2.44) = 202.83.$$

These values are shown in Figure 12.12. The 90% credible interval is (194.81, 202.83).

In general, a $1 - \alpha$ credible interval for μ is given by

$$(\mu'' - z_{1-\alpha/2}\sigma'', \mu'' + z_{1-\alpha/2}\sigma''). \tag{12.25}$$

How does an interval like this compare with a confidence interval found by classical methods? We said earlier that for large n, μ'' will be approximately equal to $\bar{x}$. Likewise, if n is large,

$$\sigma''^2 = \frac{\sigma^2}{n' + n} \approx \frac{\sigma^2}{n}.$$

Hence if n is large, (12.25) will be approximately

$$\left(\bar{x} - z_{1-\alpha/2}\frac{\sigma}{\sqrt{n}}, \bar{x} + z_{1-\alpha/2}\frac{\sigma}{\sqrt{n}}\right),$$

which is just an ordinary $1 - \alpha$ confidence interval for μ. Without going into detail, we will say that although for large n Bayesian credible intervals and classical confidence intervals are approximately the same numerically, the interpretations given these numbers are rather different. The interested reader is encouraged to consult other textbooks for further discussion on these matters.

EXERCISES

1. Based solely on your own past experience, give a subjective point estimate of p in the following situations:

(a) The proportion of automobile drivers who exceed the speed limit on limited access highways.

(b) The proportion of truck drivers who exceed the speed limit on limited access highways.

(c) The proportion of families in your town who have two or more automobiles.

(d) The proportion of families in your town who have a telephone in their homes.

(e) The proportion of families in your town who have at least one dog as a pet.

(f) The proportion of automobiles in your town which are the same make as your car (or the car of a friend if you do not happen to own a car).

(g) The proportion of checks cashed at a local grocery store which are not covered by sufficient funds.

2. The following table is similar to Tables 12.1 and 12.2 and can be used to find the posterior distribution assigned to the quantity θ. Complete the table and find the posterior distribution.

θ	*Prior Probability*	$P[E \mid \theta]$	*Product*	*Posterior Probability,* $P(\theta \mid E)$
0	.2	.6		
1	.3	.5		
2		.4		
	___	___	___	___

3. A machine producing watzits produces either 10, 20, or 30% defectives. Past records indicate that these rates of defectives occur with probability .5, .3, and .2, respectively. (a) Find the posterior distribution given that an inspected item is found to be nondefective. (b) Using the posterior distribution found in part (a) as a new prior distribution, find the posterior distribution given that a second inspected item is found to be defective.

4. (a) Using the prior distribution given initially in Exercise 3, find the posterior distribution given that an inspected item is found to be defective. (b) Using the posterior distribution found in part (a) as a new prior distribution, find the posterior distribution given that a second inspected item is found to be nondefective.

5. Jane, an inspector at the Snagg and Wrun Hosiery Company, believes that the proportion of stockings that will fail the "spot run" test is either 5%, 10%, or 15% with probabilities .7, .2, and .1, respectively. Jane inspects stockings from cartons of 1000 by following one of two inspection schemes": (A) inspect 10 stockings and find the number that fail the test or (B) continue inspecting stockings until two are found that fail the test. How should Jane revise her prior probability distribution if under scheme A she finds 2 failures? How should she revise it if under scheme B she had to inspect 10 before finding the second failure: (*Hint:* The conditional probabilities in the former case can be found by using the binomial distribution and in the latter case can be found using the negative binomial distribution.)

6. In equations (12.2) and (12.3), two expressions are given for the constant $k(n, s)$ when n and s are integers. Show that these expressions are equivalent. That is, show that

$$\frac{(n-1)!}{(s-1)!\,(n-s-1)!} = (n-1)\binom{n-2}{s-1}.$$

• *7.* Show that if $n > 2$ and $s > 1$, the mode of a beta distribution occurs at $(s-1)/(n-2)$. [*Hint:* Find the first derivative of $f(p)$ as defined in (12.5). Set this derivative equal to zero and solve for p.]

8. (a) If $f_X(x)$ is a beta density function having parameters $n = 4$ and $s = 2$, graph this density function. Also find (b) $E(X)$; (c) $V(X)$; (d) the mode of this distribution.

9. (a) If $f_X(x)$ is a beta density function having parameters $n = 4$ and $s = 1$, graph this density function. Also find (b) $E(X)$; (c) $V(X)$; (d) the mode of this distribution.

10. (a) If $f_X(x)$ is a beta density function having parameters $n = 5$, $s = 4$, graph this density function. Also find (b) $E(X)$; (c) $V(X)$; (d) the mode of this distribution.

• *11.* Show that if $f_X(x)$ is a beta density function having parameters n and s, $E(X^2) = s(s+1)/n(n+1)$. Use this result and the fact that $E(X) = s/n$ to show that $V(X) = s(n-s)/n^2(n+1)$. (*Hint:* Follow the proof of Theorem 12.1 of Section 12.3.)

12. (a) If $f_X(x)$ is a beta density function having parameters $n = 4$ and $s = 2$, what is the median of this distribution? (b) Graph the cumulative distribution function of X (cf. Exercise 8).

13. (a) If $f_X(x)$ is a beta density function having parameters $n = 4$ and $s = 1$, graph the cumulative distribution function of X. Using this graph, find the (b) 10th, (c) 50th, (d) 90th percentile of the distribution (cf. Exercise 9).

14. (a) If $f_X(x)$ is a beta density function having parameters $n = 5$ and $s = 4$, graph the cumulative distribution function of X. Using this graph, find the (b) 25th, (c) 50th, (d) 75th percentile of the distribution (cf. Exercise 10).

15. Using the results of Exercise 2, find the mode, mean, and variance of (a) the prior distribution; (b) the posterior distribution.

16. Jane feels that a beta prior distribution with parameters $n' = 50$ and $s' = 4$ adequately describes the distribution for p, the proportion of stockings which will fail the "spot run" test (see Exercise 5). What should her posterior distribution for p be if she inspects 10 stockings and finds that (a) none, (b) 2 fail the test?

17. An experimenter is concerned with estimating the value of a binomial parameter p. (a) If the experimenter chooses a beta prior with parameters $n' = 4$ and $s' = 2$ and observes 4 successes in 6 trials, what posterior distribution should he use? (b) Graph the prior and posterior distributions on the same axis (see Exercise 8). Find the mode, mean, and standard deviation of (c) the prior distribution; (d) the posterior distribution. (*To think about:* Does the experimenter's choice of prior indicate a lot of confidence in his knowledge about the value of p? Why?)

18. Mr. Dee Syder, the inspector at the Watzit Manufacturing Company, feels that the watzit maker is probably producing about 15% defectives, but he would not be too surprised if the proportion of defectives was as low as 5% or as high as 40%. What parameters would best describe the prior distribution for p that Dee should use: (a) $n' = 10$, $s' = 2$; (b) $n' = 100$, $s' = 15$; or (c) $n' = 10$, $s' = 5$? Briefly explain the reasons for your choice.

19. If Dee inspected 10 items and found 1 to be defective, what would his posterior distribution for p be if he used as a prior distribution for p a beta with parameters (a) $n' = 10$, $s' = 2$; (b) $n' = 100$, $s' = 15$; (c) $n' = 10$, $s' = 5$?

20. In Exercise 1 you gave a subjective point estimate for p for seven different situations. (a) For which of those situations did you feel the *most confident* that your estimate was very close to correct? For this particular situation, find parameters n' and s' which would adequately describe your prior probability distribution over the possible values of p. (b) Repeat this process for the situation for which you feel *least confident* that your estimate was very close to correct. (*Hint:* Review the discussion at the end of Section 12.4.)

21. Using the prior distribution you found in Exercise 20(a), what would your posterior distribution be if you observed 7 "successes" in 20 trials? 13 "successes" in 20 trials? Repeat this process using the prior distribution found in Exercise 20(b).

22. Using each of the posterior distributions for p found in Exercise 19, find a point estimate for p if (a) the mean of the posterior distribution is used as the estimator; (b) the mode of the posterior distribution is used as the estimator.

23. (a) Using the posterior distribution for p found in Exercise 19(a), find a 90% credible interval for p. (b) Do the same using the posterior distribution found in Exercise 19(c).

24. If a posterior distribution for p is found to be a beta distribution with parameters $n'' = 4$ and $s'' = 2$, find an 80% credible interval for p. (*Hint:* Use the results of Exercise 12.)

25. If a posterior distribution for p is found to be a beta distribution with parameters

$n'' = 5$ and $s'' = 4$, find (a) a 90% credible interval for p; (b) an 80% credible interval for p. (*Hint:* Use the graph of the CDF found in Exercise 14.)

26. If a posterior distribution for p is found to be a beta distribution with parameters $n'' = 40$ and $s'' = 24$, find a 95% credible interval for p. (*Hint:* Use a normal approximation.)

27. Let p denote the proportion of automobiles in your town that are the same make as your car (or the car of a friend if you do not happen to own a car) [see Exercise 1(f)]. (a) What is the make? (b) What is your subjective point estimate of p? (c) Treating p as a random variable, what would be your prior probability distribution over the possible values of p? (That is, what would be the values of n' and s'?) (d) Perform the experiment of observing 50 cars and finding the number of these which are of the same make as yours. How many of the same make were there? (e) What is the posterior probability distribution for p? (f) Find a 90% credible interval for p.

28. The prior distribution for the parameter μ is normal with a mean $\mu' = 100$ and $\sigma'^2 = (40)^2$. A sample of 25 observations is taken from a $N(\mu, \sigma^2 = (10)^2)$ distribution. The sample mean was found to be 90. (a) Find the posterior distribution for μ. (b) Find a point estimate for μ. (c) Find a 95% credible interval for μ.

13

DECISION THEORY

13.1 INTRODUCTION

Every day each of us is faced with making decisions about which of several alternative actions to follow. The great majority of these decisions would be considered minor. Should I wear red socks or green socks today? Should I eat a good breakfast, have coffee and a doughnut, or skip breakfast altogether? Should I study psychology or go to the ballgame? These decisions might be considered minor in the sense that the consequences of choosing one action when another might have been better are mild. Wearing red socks when the green ones go better with your slacks may cause you to receive some teasing from friends but will certainly not cause you to be an outcast from society!

Once in a while we are faced with more serious decisions. Should I continue in school or drop out? Should I marry now or wait until after graduation? Should I buy a small compact car or a larger station wagon? Should I buy a new car or a used car? In situations such as these, the consequences of making wrong decisions can have effects for months and years in the future. Making the correct decision in such situations is very important, and you can sometimes get some advice from others. For personal decisions relating to marriage or family you may consult a clergyman, psychiatrist, or family counselor. For decisions involving finances you may consult a banker or financial advisor. One of the first things you may be advised to do is to make a list of possible alternative actions. Then for each action you will have to consider the pros and cons. However, since life is complex, there is generally no simple formula telling you how to put all of this information into an equation to get the correct course of action.

One of the things that makes decision making difficult is uncertainty related to future circumstances. Purchase of a larger station wagon instead of a small compact might be the correct decision for you if the price of gasoline remains the same as it currently is, but the compact might be a better choice if the price of gasoline doubles within the next year or two. The purchase of a new car may be financially feasible for you if your family remains healthy, but a used car may be a better choice if illness in the family leads to high medical bills. Excluding illegal or immoral actions, under various conditions almost any decision might turn out to be a good one. The difficulty which arises is that we cannot accurately predict future circumstances.

In this chapter we will consider the problem of decision making under some simplifying assumptions. We will assume that the number of alternative actions is rather small, that the number of different "future circumstances" or possible "states of nature" is rather small, and that a dollar value can be attached to each action under each "state of nature." Although these conditions may not be satisfied very often in practice, the general ideas involved can be helpful in many decision-making problems. Furthermore, when viewed properly, problems that at first do not appear to be problems in decision making can be couched in those terms. A problem in hypothesis testing, for example, can be viewed as a decision problem in that one must decide whether to accept H_0 or H_a as the true state of nature. (For instance, in Example 8.3 of Section 8.2, the picture-happy tourist had to make a decision. Although the problem was expressed in terms of hypothesis testing, it could also be couched in terms of decision making.)

13.2 PAYOFF TABLES AND LOSS TABLES

In a decision-making problem we will assume that a decision maker is faced with a set of possible actions, $A_1, A_2, \ldots, A_m$ and a set of possible states of nature, $\theta_1, \theta_2, \ldots, \theta_n$. (We are assuming that each of these sets is finite, although they could also correspond to countably infinite or continuous sets of possibilities. Also, m and n need not be equal.) For each action A_i and state of nature θ_j there is some corresponding reward or payoff to the decision maker. To see how these payoffs may be calculated, let us consider a simple illustration.

Tim is the owner of a small bakery, the Bread Shed, specializing in various kinds of breads. Early in the morning Tim must decide on how many loaves of each kind of bread to make. Let us assume that his primary concern is with a raisin-nut bread, which costs \$.40 a loaf to make and sells for \$1.00 a loaf. For simplicity, assume that bread not sold by the end of the day must be discarded at a total loss. Because of the size of his ovens, baking pans, mixing bowls, and so on, Tim has found it to be most efficient to bake loaves in multiples of 12 up to the capacity of his oven. This means that his set of possible actions is to bake 0, 12, 24, 36, or 48 loaves. Which action should he follow? (See Figure 13.1.) The best course of action will depend on the demand (i.e., the number of customers who would want to buy

FIGURE 13.1
Tim Trying to Decide What to Do

this bread today). Unfortunately, Tim does not know beforehand how many of his customers will want bread. As a simplifying assumption, let us assume that there are always 200 customers coming into the bakery each day but that the proportion of these wanting to buy raisin-nut bread is either 0, 5, 10, 15, or 20%. These values constitute the possible states of nature. With this information available, we would like to find the "reward" or "payoff" to Tim when he chooses a particular action A_i when the actual state of nature turns out to be θ_j. Specifically, let us say he chooses action A_2, bake 12 loaves, when the true state of nature is $\theta_3 = 10\%$ (i.e., 10% of 200 or 20 customers want to buy bread). His cost of baking is

$$(\$.40/\text{loaf}) \times (12 \text{ loaves}) = \$4.80,$$

while his sales are

$$(\$1.00/\text{loaf}) \times (12 \text{ loaves}) = \$12.00,$$

so his net payoff is 12 − 4.80 = \$7.20. We could write this as

$$p(A_2 \mid \theta_3),$$

which is the payoff (*not the probability*) corresponding to action 2 when the state of nature is θ_3. What would the payoff be if action 2 were followed when the state of nature is θ_2 (i.e., 5% or 10 of the customers want to buy bread)? It would be

$$\begin{aligned} p(A_2 \mid \theta_2) &= \text{sales} - \text{cost} \\ &= (\$1.00/\text{loaf}) \times (10 \text{ loaves}) - (\$.40/\text{loaf}) \times (12 \text{ loaves}) \\ &= \$10.00 - 4.80 = \$5.20. \end{aligned}$$

Note that the same result could be obtained by considering profit/loaf on loaves sold minus loss/loaf on loaves unsold. In this case 10 loaves would have been sold and 2 would have been unsold, to give

$$\begin{aligned} p(A_2 \mid \theta_2) &= (\$.60/\text{loaf}) \times (10 \text{ loaves sold}) - (\$.40/\text{loaf}) \times (2 \text{ loaves unsold}) \\ &= \$6.00 - .80 = \$5.20. \end{aligned}$$

The easiest way to summarize the payoffs for each possible state of nature is to do so in a table called, unsurprisingly, a payoff table. In our payoff tables we will put the possible states of nature across the top and the different actions down the left-hand margin. (We should point out that some other authors reverse this labeling.) The payoff table for Tim's problem is given in Table 13.1. (You should try to get a few of these entries yourself so that you can be sure you can calculate payoffs.)

TABLE 13.1

Payoff Table for Tim's Raisin-Nut Bread

		State of Nature (Demand) (Proportion of Customers Wanting Raisin-Nut Bread)				
		θ_1: 0	θ_2: .05	θ_3: .10	θ_4: .15	θ_5: .20
Possible Action	A_1: Bake 0	0	0	0	0	0
	A_2: Bake 12	−4.80	5.20	7.20	7.20	7.20
	A_3: Bake 24	−9.60	0.40	10.40	14.40	14.40
	A_4: Bake 36	−14.40	−4.40	5.60	15.60	21.60
	A_5: Bake 48	−19.20	−9.20	0.80	10.80	20.80

Example 13.1 One of Tim's specialties is a Norwegian fruited sweetbread, Jule Kaga. This bread is more expensive to make, costing \$.90/loaf, and sells for more, \$2.00/loaf, so the sales are usually lower than for some of his other breads. Tim has decided that it is only practical to make 0, 5, 10, or 15 loaves of Jule Kaga. The proportion of his customers who want to buy this bread is either 0, 3, 6, or 9%. Find the payoff table.

If we express the payoff as sales minus cost, the payoff would be

$$p(A_i \mid \theta_j) = \text{sales} - \text{cost}$$
$$= (\$2.00/\text{loaf}) \times (\text{number sold}) - (\$.90/\text{loaf}) \times (\text{number made}).$$

Using this expression, we find that if $A_3 = \{10 \text{ loaves made}\}$ and $\theta_2 = \{3\% \text{ or } 6 \text{ customers want this bread}\}$,

$$p(A_3 \mid \theta_2) = (\$2.00/\text{loaf}) \times (6 \text{ loaves sold}) - (\$.90/\text{loaf}) \times (10 \text{ loaves made})$$
$$= \$12.00 - \$9.00 = \$3.00.$$

(Note that we are again assuming that Tim makes no profit at all from the unsold loaves.) Similarly, we find that for A_3 and $\theta_3 = \{6\%$ or 12 customers want this bread$\}$,

$$p(A_3 \mid \theta_3) = (\$2.00/\text{loaf}) \times (10 \text{ loaves sold}) - (\$.90/\text{loaf}) \times (10 \text{ loaves made})$$
$$= \$20.00 - \$9.00 = \$11.00.$$

Continuing in this way, we find the payoff table to be as given in Table 13.2.

TABLE 13.2

Payoff Table for Tim's Norwegian Fruited Sweetbread

		State of Nature (Demand) *(Proportion of Customers Wanting Jule Kaga)*			
		θ_1: *0*	θ_2: *.03*	θ_3: *.06*	θ_4: *.09*
	A_1: *Bake 0*	0	0	0	0
	A_2: *Bake 5*	−4.50	5.50	5.50	5.50
Possible Action	A_3: *Bake 10*	−9.00	3.00	11.00	11.00
	A_4: *Bake 15*	−13.50	−1.50	10.50	16.50

If Tim knew for sure what the state of nature would be on a given day for a particular bread, it would be rather easy for him to make the best decision, that is, to choose the action leading to the highest payoff. For example, from Table 13.2 you can see that if $\theta = \theta_1 = 0$, the best action is A_1: bake 0. Likewise, for $\theta = \theta_2 = .03$, the best action is A_2: bake 5; and so on. (You must be somewhat careful here, however, to consider only those courses of action that are available. In some sense, if $\theta = \theta_2 = .03$, so that 6 customers will want Jule Kaga, the best course of action (leading to a profit of \$6.60) would be to bake 6 loaves. However, because of size and space considerations, baking 6 loaves is *not* one of the possible actions. Among the courses of action that are actually possible alternatives, the best is to bake 5.) Notice also in this table that in comparing actions pairwise, none is uniformly preferred to any other. In comparing A_2 with A_3 for example, A_2 would be a better course of action if $\theta = \theta_1$ or θ_2, while A_3 would be a better course of action if $\theta = \theta_3$ or θ_4.

Now consider the payoff table for raisin-nut bread given in Table 13.1. What is the best course of action if $\theta = \theta_1 = 0$? A_1 is best in this case. What if $\theta = \theta_3 = .10$? In this case A_3 is the best action, since the payoff under this action is \$10.40, which is larger than under any other action. If you were to compare the actions pairwise, you would find that something different happens here than with comparisons for Table 13.2. In particular, in comparing actions A_4 and A_5, you would find that under no state of nature would action A_5 be preferred. Since it is never better for Tim to bake 48 loaves than to bake 36 loaves, it would be foolish for him to ever consider baking 48. For all practical purposes, this action can be eliminated from further consideration.

Definition 13.1 If for two actions A_i and A_j, action A_i is never preferred to action A_j for any possible state of nature, that is, if

$$p(A_i \mid \theta_k) \leq p(A_j \mid \theta_k) \text{ for all } k,$$

then action A_i is said to be *inadmissible*.

An inadmissible action can be eliminated from further consideration in the decision process, thus somewhat simplifying the decision problem.

Example 13.2 Using the payoff table given in Table 13.3, find the best action when $\theta = \theta_1, \theta_2, \theta_3$, and θ_4. Also determine which actions, if any, are inadmissible.

TABLE 13.3
Payoff Table for Example 13.2

		State of Nature			
		θ_1	θ_2	θ_3	θ_4
	A_1	−1	2	3	8
Possible	A_2	−2	3	4	4
Action	A_3	−4	4	4	6
	A_4	−3	5	5	7

If $\theta = \theta_1$, $p(A_1 \mid \theta_1) = -1 > p(A_i \mid \theta_1)$ for $i = 2, 3, 4$, so A_1 would be the best action in this case. If $\theta = \theta_2$ or θ_3, the maximum payoff occurs when action A_4 is taken, so it is best. Action A_1 is best when $\theta = \theta_4$, since the payoff of 8 is the largest. In trying to find inadmissible actions, we need only worry about actions A_2 and A_3 being inadmissible. Since A_1 and A_4 are "best" under some state of nature, they must be admissible. In comparing A_3 and A_4, we see that

$$p(A_3 \mid \theta_k) < p(A_4 \mid \theta_k) \qquad \text{for } k = 1, 2, 3, 4,$$

so A_3 is inadmissible. It now remains to compare A_2 with A_1 and with A_4. Comparing A_2 with A_1, we see that A_2 is better if $\theta = \theta_3$ but not otherwise, and comparing A_2 with A_4, we see that A_2 is better if $\theta = \theta_1$ but not otherwise. Consequently, A_2 is not inadmissible. Only A_3 is inadmissible.

Consider once again the payoff table in Table 13.1 corresponding to the payoffs for making raisin-nut bread. What might Tim's feeling be at the end of the day if he followed action A_2 (baking 12 loaves) in the morning but found that the state of nature was actually $\theta = \theta_3 = .10$ (i.e., if 20 customers would have bought raisin-nut bread had it been available)? Although he made a profit of \$7.20 for selling all 12 loaves he baked, he could have made a profit of \$10.40 had he chosen to bake 24 loaves (action A_3). He has a certain feeling of regret for having lost the opportunity to make an additional \$3.20. What might his feelings be on this day had he started by following action A_4 (baking 36 loaves)? He would also feel some regret for having 12 additional leftover loaves, costing him an extra \$4.80 worth of ingredients. In the one case if he followed A_2 he would bemoan the loss of an opportunity to earn more than he did, while in the case of baking too much (as in following action A_4) he would bemoan the actual loss of the cost of ingredients. The loss that he would associate with an action A_i when he "should have" followed a different action can be given by

$$l(A_i | \theta_j) = \max_k p(A_k | \theta_j) - p(A_i | \theta_j). \tag{13.1}$$

For example, if $j = 3$, the maximum payoff would correspond to action A_3 and would be \$10.40. Tim's loss from following action A_2 would be

$$\begin{aligned} l(A_2 | \theta_3) &= \max_k p(A_k | \theta_3) - p(A_2 | \theta_3) \\ &= 10.40 - 7.20 = 3.20. \end{aligned}$$

Similarly,

$$\begin{aligned} l(A_4 | \theta_3) &= \max_k p(A_k | \theta_3) - p(A_4 | \theta_3) \\ &= 10.40 - 5.60 = 4.80. \end{aligned}$$

This information can be summarized in a loss table, that is, a table whose entries are $l(A_i | \theta_j)$. The loss table for the raisin-nut bread is given in Table 13.4. You can see from this table that $l(A_3 | \theta_3) = 0$, since when $\theta = \theta_3 = .10$, the best action among the five available options was to bake 24 loaves (A_3). Shouldn't he have some regrets about having 4 leftover loaves? In one sense, yes. But since baking just 20 loaves

TABLE 13.4
Loss Table for Tim's Raisin-Nut Bread

		State of Nature (Demand) (Proportion of Customers Wanting Raisin-Nut Bread)				
		θ_1: 0	θ_2: .05	θ_3: .10	θ_4: .15	θ_5: .20
Possible Action	A_1: Bake 0	0	5.20	10.40	15.60	21.60
	A_2: Bake 12	4.80	0	3.20	8.40	14.40
	A_3: Bake 24	9.60	4.80	0	1.20	7.20
	A_4: Bake 36	14.40	9.60	4.80	0	0
	A_5: Bake 48	19.20	14.40	9.60	4.80	0.80

was not one of his options, he must not feel regret for not doing something that he could not possibly do! This would change, of course, if his set of possible actions were different. However, given the constraints of his set of possible actions, he did the best he could. About the 4 leftover loaves, he can only say "C'est la vie" and go on from there.

Example 13.3 Find the loss table that Tim would have for his decision problem concerning the amount of Norwegian fruited sweetbread (Jule Kaga) to bake on a given day.

Using the payoff table given in Table 13.2, we find the loss table given in Table 13.5. Some specific calculations are the following: If $\theta = \theta_2 = .03$, the maximum payoff occurs under action A_2, namely, 5.50. Consequently,

$$l(A_4 | \theta_2) = \max_k p(A_k | \theta_2) - p(A_4 | \theta_2)$$
$$= 5.50 - (-1.50) = 7.00.$$

Similarly, if $\theta = \theta_3 = .06$, we find that

$$l(A_2 | \theta_3) = \max_k p(A_2 | \theta_3) - p(A_2 | \theta_3)$$
$$= 11.00 - 5.50 = 5.50.$$

TABLE 13.5

Loss Table for Tim's Norwegian Fruited Sweetbread

		State of Nature (Demand) (Proportion of Customers Wanting Jule Kaga)			
		θ_1: *0*	θ_2: *.03*	θ_3: *.06*	θ_4: *.09*
Possible Action	A_1: *Bake 0*	0	5.50	11.00	16.50
	A_2: *Bake 5*	4.50	0	5.50	11.00
	A_3: *Bake 10*	9.00	2.50	0	5.50
	A_4: *Bake 15*	13.50	7.00	0.50	0

13.3 DECISION MAKING WITHOUT PROBABILITY

The purpose of making a payoff table or loss table is to be able to see at a glance the possible rewards for each action. Once these tables are constructed they can be used to decide which course of action to follow. Of course, since the state of nature is unknown at the time an action must be chosen, a certain amount of risk is made in choosing any course of action in most real problems. One decision rule or strategy

TABLE 13.6
A Simple Payoff Table

		State of Nature		
		θ_1	θ_2	θ_3
Possible Action	A_1	5	6	7
	A_2	4	12	16
	A_3	−2	−3	17

that can be used is a very conservative one. To illustrate this, consider the simple payoff table given in Table 13.6. For each action consider the worst that can happen under all possible states of nature, that is, the minimum payoff under all states of nature: $\min p(A_i \mid \theta_k)$. For the first action the minimum occurs for $\theta = \theta_1$:

$$\min_k p(A_1 \mid \theta_k) = p(A_1 \mid \theta_1) = 5.$$

For the second action the minimum is $p(A_2 \mid \theta_1) = 4$, while for the third it is $p(A_3 \mid \theta_2) = -3$. By choosing the action corresponding to the largest of these minima (i.e., the *maxi*mum of the *mini*ma), we obtain the highest guaranteed return. This strategy is called the *maximin payoff* strategy. By choosing action A_1 in this illustration, one would be guaranteed a return of 5 (dollars). Under actions A_2 and A_3, the guaranteed amounts are 4 and -3, respectively.

Definition 13.2 The *maximin payoff strategy* for making a decision is to choose the action A_j for which

$$\min_k p(A_j \mid \theta_k)$$

is at least as large as $\min_k p(A_i \mid \theta_k)$ for $i \neq j$.

This decision strategy is sometimes considered to be conservative. Do you agree with this assessment? Compare the payoffs for actions A_1 and A_2. Which action would you prefer? Given no other information, most people would choose A_2, preferring to risk a slightly smaller payoff should θ_1 be the state of nature for the larger return should the state be θ_2 or θ_3.

Example 13.4 What action should Tim take in terms of the number of loaves of raisin-nut bread to bake if he wishes to follow the maximin payoff strategy?

By examining the payoff table given in Table 13.1 of Section 13.2, it is easy to see that the minimum payoff for action A_1 is zero. For actions A_2 through A_4, the minimum payoffs are -4.80, -9.60, and -14.40, respectively. (Note that since action A_5 is inadmissible, we need not consider it.) The maximum of these minima is zero, so action A_1 should be taken.

You see in this case that the maximin strategy leads to a decision to bake no raisin-nut bread. To avoid possible losses, this strategy dictates that none of this type of bread be baked. Obviously, if this strategy is followed for all types of bread, Tim will have to close up shop! He will have to consider some other strategy or go into some other business.

Instead of being very conservative or pessimistic, one might take the other extreme and be optimistic. For each action, find the best that can happen under various states of nature, that is, find for each A_i

$$\max_k p(A_i \mid \theta_k).$$

Then, hoping for the "best of all possible worlds," choose the action corresponding to the largest or maximum of these maxima. Referring to the payoff table in Table 13.6, the maximum payoff for action A_1 is

$$\max_k p(A_1 \mid \theta_k) = p(A_1 \mid \theta_3) = 7,$$

while for actions A_2 and A_3 the maxima are 16 and 17, respectively. The action corresponding to a payoff of 17, the largest of these maxima, is A_3.

Definition 13.3 The *maximax payoff strategy* for making a decision is to choose the action A_j for which

$$\max_k p(A_j \mid \theta_k)$$

is at least as large as $\max_k p(A_i \mid \theta_k)$ for $i \neq j$.

Would you agree that this strategy could be considered too optimistic? Compare the payoffs in Table 13.6 corresponding to actions A_2 and A_3. Choosing action A_3 results in a slightly higher payoff if state of nature θ_3 occurs, but this may not justify the possible small (negative, in fact) payoffs that will accrue under this action if θ_1 or θ_2 occur.

Example 13.5 Consider the payoff table given in Table 13.7(a). Find the actions that should be taken in following a maximin payoff strategy and a maximax payoff strategy.

In Table 13.7(b) we have put two extra columns, one corresponding to the minimum possible payoff for each action, the other corresponding to the maximum possible payoff for each action. The largest of the minima (denoted by an asterisk) corresponds to action A_1, so this would be the action under the maximin payoff strategy. The largest of the maxima (denoted by a double asterisk) corresponds to action A_3, so this would be the action under the maximax payoff strategy.

TABLE 13.7

		State of Nature				
		θ_1	θ_2	θ_3	$\min_k p(A_i \mid \theta_k)$	$\max_k p(A_i \mid \theta_k)$
Possible Action	A_1	12	24	36	12*	36
	A_2	32	17	10	10	32
	A_3	−2	38	18	−2	38**
		(a)			(b)	

TABLE 13.8
Two Simple Payoff Tables

(a)

		State of Nature θ_1	θ_2
Possible Action	A_1	99	100
	A_2	2	101

(b)

		State of Nature θ_1	θ_2
Possible Action	A_1	99	100
	A_2	999	98

It is not hard to give examples for which blindly following either of these two strategies would not make sense. We will leave it to you to decide which of these strategies would be reasonable to follow and which would be unreasonable for the simple payoff tables given in Table 13.8. In view of the fact that one of these strategies may be too pessimistic while the other may be too optimistic, we would like to find a strategy that might be more "center of the road." One such strategy involves consideration of the loss table rather than the payoff table.

When considering a loss table instead of a payoff table, we must remember that losses are undesirable, so we would like to minimize losses. As an example of a decision strategy involving loss tables, let us again consider the loss table for Tim's Jule Kaga bread (Table 13.5). For convenience, this table has been repeated in Table 13.9 but with some minor modifications in the labeling. Now consider the worst that could happen if Tim chooses to bake 0. By examining the first row of the loss table you can see that the largest loss is \$16.50. This value has been circled, as have the values corresponding to the largest (maximum) losses he could have if he chose to bake 5, 10, or 15. For easy reference these values are listed to the right of the loss table. Since these losses or regrets are undesirable, Tim would like to minimize them. Consequently, it would be reasonable for him to follow the action that *min*imizes the

TABLE 13.9
Modified Loss Table for Tim's Norwegian Fruited Sweetbread

		State of Nature θ_1: 0	θ_2: .03	θ_3: .06	θ_4: .09	Maximum Loss
Possible Action	A_1: Bake 0	0	5.50	11.00	16.50	16.50
	A_2: Bake 5	4.50	0	5.50	11.00	11.00
	A_3: Bake 10	9.00	2.50	0	5.50	9.00
	A_4: Bake 15	13.50	7.00	0.50	0	13.50

maximum regret. In this case the action "bake 10" could lead to a maximum loss of \$9.00. Since this is smaller than the other maxima, the action "bake 10" would be a reasonable one to follow.

Definition 13.4 The *minimax loss strategy* for making a decision is to choose the action A_j for which

$$\max_k l(A_j \mid \theta_k)$$

is a minimum, that is, is at least as small as $\max_k l(A_i \mid \theta_k)$ for $i \neq j$.

By studying the payoff table (Table 13.2) for Tim's Jule Kaga decision problem, you can see that the conservative maximin payoff strategy would lead to a decision to bake 0 loaves, while the optimistic maximax payoff strategy would dictate baking 15 loaves. You can see from this example that the minimax loss strategy may lead to actions that are more middle-of-the-road than these other two strategies.

Example 13.6 Craig runs a gas station which has a pump that can blend grades of gasoline. A valve in this pump is used to blend regular and premium gasoline. Properly adjusted, the valve allows a mix of 75% regular and 25% premium. If slightly out of adjustment, the mix is 60% regular and 40% premium; if badly out of adjustment, the mix is 40% regular and 60% premium. The current price differential between regular and premium gasoline is 8 cents/gallon. The valve can be checked and adjusted only at the beginning of the day. The cost of checking the valve is \$20 and the cost of adjusting the valve is another \$20. Craig has three options: to not check the valve, to check the valve and adjust it only if it is badly out of adjustment, or to check the valve and adjust it if it is either slightly or badly out of adjustment. Assuming that this pump dispenses 3000 gallons of gasoline per day, construct a payoff table, a loss table, and determine what action Craig should follow under the maximin payoff, maximax payoff, and minimax loss strategies.

Let us denote the three options by A_1, A_2, and A_3, respectively (where A_1 = do not check valve, etc.) and the three possible states of nature by θ_1 (valve properly adjusted), θ_2 (valve slightly out of adjustment), and θ_3 (valve badly out of adjustment). It is easy to see that if the valve is properly adjusted and he does not check it, his payoff is

$$p(A_1 \mid \theta_1) = 0,$$

while if he does check it he has lost the \$20 cost of checking, so

$$p(A_2 \mid \theta_1) = p(A_3 \mid \theta_1) = -20.$$

If the state of nature is θ_2 (valve slightly out of adjustment), he will have sold a number of gallons of premium gasoline at the price of regular. In particular, he will dispense

$$.40(3000) = 1200$$

gallons of premium instead of

$$.25(3000) = 750$$

gallons of premium, losing a total of

$$(450 \text{ gal}) \times (\$.08/\text{gal}) = \$36.00.$$

Consequently, if he choose action A_1, not to check the valve, when θ_2 is the state of nature he will lose $36:

$$p(A_1 | \theta_2) = -36.$$

If he follows strategy A_2, checking the valve but only making the adjustment if it is badly out of adjustment, his payoff will be

$$\begin{aligned} p(A_2 | \theta_2) &= \text{cost of checking} + \text{loss of revenue} \\ &= -20 + -36 = -56. \end{aligned}$$

If he checks and adjusts the valve (strategy A_3), he will not lose the revenue for selling a lot of premium gas at regular price, so his payoff will be

$$\begin{aligned} p(A_3 | \theta_2) &= \text{cost of checking} + \text{cost of adjustment} \\ &= -20 + -20 = -40. \end{aligned}$$

Finally, if the state of nature is θ_3 (valve badly out of adjustment), his loss of revenue for selling premium gasoline at the price of regular would be

$$[(.60)(3000) - (.25)(3000)] \text{ gal} \times (\$.08/\text{gal}) = \$84.$$

Therefore, $p(A_1 | \theta_3) = -84$. You can verify for yourself that

$$p(A_2 | \theta_3) = p(A_3 | \theta_3) = -40.$$

All of this information is summarized in the payoff table in Table 13.10(a).

TABLE 13.10

Payoff and Loss Tables for Craig's Blending Pump Problem

(a) Payoff table

Possible Action	State of Nature θ_1	θ_2	θ_3
A_1	0	−36	(−84)
A_2	−20	(−56)	−40
A_3	−20	(−40)	(−40)

(b) Loss table

Possible Action	State of Nature θ_1	θ_2	θ_3
A_1	0	0	44
A_2	(20)	(20)	0
A_3	(20)	4	0

The loss table is given in Table 13.10(b). Since, for example, when $\theta = \theta_2$, the largest payoff is −36, the corresponding loss would be zero. With this in mind we find that

$$\begin{aligned} l(A_2 | \theta_2) &= \max_k p(A_k | \theta_2) - p(A_2 | \theta_2) \\ &= -36 - (-56) = 20. \end{aligned}$$

You can verify that the other entries in the loss table are correct.

To help in finding the maximin payoff strategy, the minimum payoff for each action has been circled. (Two values are tied under action A_3.) These minima are −84, −56, and −40. Since the largest of these is −40, the maximin payoff strategy would call for action A_3: check the valve and make adjustments if necessary.

Since the largest payoff in the table is zero, the corresponding action A_1 would be followed as the maximax payoff strategy. This is clearly an optimistic strategy, since not checking the pump is tantamount to assuming that all is well.

To help in finding the minimax loss strategy, the maximum loss for each action has been circled. These maxima are 44, 20, and 20. In this case both actions A_2 and A_3 lead to a maximum loss of 20. Closer study of the loss table in this case would lead to the conclusion that A_3 is a better action than A_2.

You will notice from this example that while entries in a payoff table can be positive, negative, or zero, entries in a loss table must always be nonnegative. This of course follows from the way losses were defined in equation (13.1).

In this section we have discussed different strategies that might be employed by a decision maker faced with choosing one of several alternative actions. In many situations it would be foolish to simply decide to follow one particular strategy no matter what the surrounding circumstances might be. However, by examining the actions suggested by these strategies, you can get some indication as to reasonable courses of action.

13.4 DECISION MAKING WITH PROBABILITY

One of the main deficiencies of the decision strategies defined in the last section is that they all assume that you have no idea whatsoever which state of nature will occur. We pointed out that if you knew for sure which state of nature would occur, say θ_j, then choosing the appropriate action is easy: choose A_i such that $p(A_i | \theta_j)$ is a maximum. (In fact, we needed to find this maximum value in order to calculate losses.) Fortunately, it is often the case that a decision maker does have some idea about which states of nature are likely to occur. For instance, if Tim has had his bakery shop open for even a short time, he will have an idea whether a relatively large or relatively small proportion of his customers will want a particular type of bread. Similarly, if Craig has checked the mixing valve in his blending pump several times, he will have some idea as to the probability that the valve will be in adjustment, slightly out of adjustment, or badly out of adjustment. Often, a decision maker can assign probabilities to the various possible states of nature. In many cases the probabilities that a decision maker will assign will be subjective probabilities. In other cases these probabilities may be based on relative frequencies. No matter how they are obtained, these probabilities can be useful in choosing a decision strategy.

Consider for a moment Craig's decision problem relative to the blending pump. What action would you advise him to take if you felt that there was a 95% chance that the true state of nature was θ_1: the valve is properly adjusted? Certainly, you would advise him not to check the machine (i.e., to follow action A_1). The proper advice would be harder to give if the probability assigned to any one state were not so large. Assume, for example, that Craig has found that if he adjusts the valve on Monday mornings, then throughout the week the probability assigned to each state of nature is as follows:

	State of Nature		
	θ_1: *in adjustment*	θ_2: *slightly out*	θ_3: *badly out*
Probability, $P[\Theta = \theta_i]$	.60	.30	.10

Based on this set of probabilities, what would be the expected payoff for following action A_1? From Table 13.10(a) we find the payoffs under the three states of nature to be:

$$\text{Payoff under } A_1 \text{ given } \theta: \quad 0 \quad -36 \quad -84.$$

The expected payoff would be

$$E(p(A_1)) = \sum p(A_1 | \theta_i) \, P[\Theta = \theta_i]$$
$$= (0)(.60) + (-36)(.30) + (-84)(.10) = \$-19.20.$$

Similarly, we can find the expected payoff under actions A_2 and A_3 to be

$$E(p(A_2)) = \sum p(A_2 | \theta_i) \cdot P[\Theta = \theta_i]$$
$$= (-20)(.60) + (-56)(.30) + (-40)(.10) = \$-32.80$$

and

$$E(p(A_3)) = (-20)(.60) + (-40)(.30) + (-40)(.10) = \$-28.00.$$

Given this information, what course of action would you suggest that Craig follow? Since the expected payoff is largest when A_1 (no checking) is followed, this would seem to be the best course of action. (Actually, you might recommend that he replace the entire valve or pump, but that would be an entirely different decision problem!) We can define this strategy as follows.

Definition 13.5 The expected payoff strategy for making a decision when probabilities can be assigned to each state of nature is to choose the action A_j for which the expected payoff is as large as possible:

$$E(p(A_j)) \geq E(p(A_i)) \quad \text{for} \quad i \neq j.$$

Example 13.7 After several months of experience, Tim believes that the probabilities associated with the various proportions of customers who will want to buy raisin-nut bread is as follows:

State of nature:	$\theta_1 = 0$	$\theta_2 = .05$	$\theta_3 = .10$	$\theta_4 = .15$	$\theta_5 = .20$
Probability:	.10	.15	.20	.35	.20.

Use this information along with the payoff table given in Table 13.1 to find the best action for Tim to follow if he uses the expected payoff criterion.

We have reproduced Table 13.1 in Table 13.11 with some slight modifications. We have deleted the inadmissible action A_5 (bake 48 loaves) and have added the probabilities corresponding to each state of nature at the bottom of this table. It should be clear that the expected payoff for the action bake 0 will be zero, since the

TABLE 13.11

Modified Payoff Table for Tim's Raisin-Nut Bread Problem

		State of Nature					
		θ_1: 0	θ_2: .05	θ_3: .10	θ_4: .15	θ_5: .20	$E(p(A_i))$
	A_1: Bake 0	0	0	0	0	0	0
Possible Action	A_2: Bake 12	−4.80	5.20	7.20	7.20	7.20	5.70
	A_3: Bake 24	−9.60	0.40	10.40	14.40	14.40	9.10
	A_4: Bake 36	−14.40	−4.40	5.60	15.60	21.60	8.80
	$P[\Theta = \theta_i]$	.10	.15	.20	.35	.20	

payoff for each state of nature is zero. The expected payoff for action A_2, bake 12 loaves, is found from

$$\begin{aligned} E(p(A_2)) &= \sum p(A_2 \mid \theta_i) \cdot P[\Theta = \theta_i] \\ &= (-4.80)(.10) + 5.20(.15) + 7.20(.20) + 7.20(.35) + 7.20(.20) \\ &= 5.70. \end{aligned}$$

Similarly, $E(p(A_3))$ and $E(p(A_4))$ can be found to be 9.10 and 8.80, respectively. These values are shown in the right-hand column of Table 13.11. Since the largest of these expected values, \$9.10, corresponds to action A_3, bake 24 loaves, this is the action that should be followed.

Instead of looking at the expected payoff to determine an action, we might consider using loss tables and calculating expected losses for each possible action. Since losses are to be avoided, it would seem reasonable to choose the action leading to the smallest expected loss.

Definition 13.6 The expected loss strategy for making a decision when probabilities can be assigned to each state of nature is to choose the action A_j for which the expected loss is as small as possible:

$$E(l(A_j)) \leq E(l(A_i)) \qquad \text{for} \qquad i \neq j.$$

Example 13.8 Using the same probabilities for each state of nature as Tim gave in Example 13.7 along with the loss table given in Table 13.4, find the best action for Tim to follow in terms of the number of loaves of raisin-nut bread to bake if he uses the expected loss criterion.

We have reproduced a modified version of the loss table for this problem in Table 13.12. Here, too, the inadmissible action A_5 has been deleted and the probabil-

TABLE 13.12
Modified Loss Table for Tim's Raisin-Nut Bread Problem

		State of Nature					
		$\theta_1: 0$	$\theta_2: .05$	$\theta_3: .10$	$\theta_4: .15$	$\theta_5: .20$	$E(l(A_i))$
Possible Action	A_1: Bake 0	0	5.20	10.40	15.60	21.60	12.64
	A_2: Bake 12	4.80	0	3.20	8.40	14.40	6.94
	A_3: Bake 24	9.60	4.80	0	1.20	7.20	3.54
	A_4: Bake 36	14.40	9.60	4.80	0	0	3.84
	$P[\Theta = \theta_i]$	.10	.15	.20	.35	.20	

ities corresponding to each state of nature have been added to the bottom of the table. The expected losses are found in much the same way as expected payoffs: e.g.,

$$\begin{aligned} E(l(A_1)) &= \sum l(A_1 \mid \theta_i) \cdot P[\Theta = \theta_i] \\ &= 0(.10) + 5.20(.15) + 10.40(.20) + 15.60(.35) + 21.60(.20) \\ &= 12.64. \end{aligned}$$

This value, along with the expected losses for each other action, is given in the right-hand column of Table 13.12. Since the smallest of these values, \$3.54, corresponds to action A_3, this would be the action Tim should take under the expected loss strategy.

In comparing the results of Example 13.7 and 13.8, we see that both the expected payoff strategy and the expected loss strategy lead to following the same action, action A_3: bake 24 loaves. As we shall see shortly, it is always the case that *these two strategies will always lead to the same action.* Before doing this, examine Table 13.13, where we have presented along with each possible action the corresponding expected payoff and expected loss. What kind of relationships do you see? You may notice that when the expected payoff is large for a given action, the expected loss is small. Furthermore, you may see that the sum of the expected payoff and the expected loss is equal to a constant, \$12.64 in this case. If it is true that for all i,

$$E(p(A_i)) + E(l(A_i)) = \text{constant}, \tag{13.2}$$

TABLE 13.13
Comparison of Expected Payoffs and Expected Losses

Action:	A_1	A_2	A_3	A_4
$E(p(A_i))$:	0.00	5.70	9.10	8.80
$E(l(A_i))$:	12.64	6.94	3.54	3.84

it must be true that when $E(p(A_i))$ is as large as possible, $E(l(A_i))$ will be as small as possible. That equation (13.2) is actually true can be seen from the following theorem.

Theorem 13.1 Consider a decision problem with possible actions A_i, states of nature θ_j, and payoffs $p(A_i|\theta_j)$. If probabilities can be assigned to the occurrence of each state of nature, $P[\Theta = \theta_j]$, then for all i

$$E(p(A_i)) + E(l(A_i)) = C,$$

where C is a constant equal to $\sum_j (\max_k p(A_k|\theta_j)) \cdot P[\Theta = \theta_j]$.

Proof: We can calculate $E(L(A_i))$ by using the definition of loss given in equation (13.1). The rest of the proof follows by properties of summations and ordinary algebra. We have

$$\begin{aligned} E(l(A_i)) &= \sum_j l(A_i|\theta_j) \cdot P[\Theta = \theta_j] \\ &= \sum_j [\max_k p(A_k|\theta_j) - p(A_i|\theta_j)] \cdot P[\Theta = \theta_j] \\ &= \sum_j [\max_k p(A_k|\theta_j)] \cdot P[\Theta = \theta_j] - \sum_j p(A_i|\theta_j) \cdot P[\Theta = \theta_j] \\ &= C - E(p(A_i)). \end{aligned}$$

From this last equality we conclude that

$$E(p(A_i)) + E(l(A_i)) = C,$$

as stated in the theorem.

Before moving on to the next section, let us try to understand just what C represents. Recall from Section 13.2 that

$$\max_k p(A_k|\theta_j)$$

gives the maximum payoff possible when θ_j is the true state of nature. For instance, in examining Tim's payoff table for raisin-nut bread (either Table 13.1 or 13.11), we can see that when $\theta = \theta_1$, the maximum payoff, which happens to correspond to action A_1, is 0. When $\theta = \theta_2$, the maximum payoff is 5.20; when $\theta = \theta_3$, the maximum is 10.40; and so on. These maximum values represent the very best that Tim can do with raisin-nut bread. If he always knew, at the beginning of the day, what the value of θ would be on that day, he would choose the best action and, when $\theta = \theta_j$, he would get a payoff in the amount of $\max p(A_k|\theta_j)$. However, since θ will be equal to θ_j with probability $P[\Theta = \theta_j]$, Tim will earn, on the average, the amount

$$\sum_j [\max_k p(A_k|\theta_j)] \cdot P[\Theta = \theta_j]. \tag{13.3}$$

This is the constant C defined earlier, so C represents the very best the decision maker can hope to do when the states of nature are known beforehand. For the raisin-nut bread we find that

$$\begin{aligned} C &= (\text{best payoff when } \theta = \theta_1) \cdot P[\Theta = \theta_1] \\ &\quad + \ldots + (\text{best payoff when } \theta = \theta_5) \cdot P[\Theta = \theta_5] \\ &= 0(.10) + 5.20(.15) + 10.40(.20) + 15.60(.35) + 21.60(.20) \\ &= 12.64. \end{aligned}$$

If Tim knew, at the start of the day, what the value of θ would be that day, he would take the appropriate action for that day (the action might be A_1, A_2, A_3, or A_4, depending on the value of θ). In so doing he would get an average payoff per day of \$12.64 for making raisin-nut bread. We must emphasize, however, that since he *does not know* the value of θ, he can do best by following the expected payoff strategy, to take action A_3, with a resulting average payoff per day of \$9.10. You can see from this that Tim's ignorance of the true state of nature will cost him, on the average, \$3.54/day. Note that this is exactly his expected loss for following action A_3 [i.e., $E(l(A_3)) = \$3.54$].

13.5 DECISION MAKING WITH SAMPLE INFORMATION

We have seen that Tim's ignorance (i.e., lack of information about the state of nature regarding the demand for raisin-nut bread) will cost him an average of \$3.54/day. Tim is not unique in this regard; all of us share this problem of ignorance to some degree. How can Tim, or any other decision maker, improve his situation? He may be able to do better by gaining some information about the state of nature for the coming day. In this case the states of nature are the proportions of customers who will want to buy raisin-nut bread on a given day. Tim might be able to get information about this proportion by calling a randomly selected customer and asking whether or not that customer intends to buy raisin-nut bread. (Of course, Tim had better be careful about when he calls the customers, since a call at 5: 30 A.M. may do more harm than good!)

Let us say that after a random selection, Tim calls Sally and finds that she intends to buy a loaf of raisin-nut bread. How can he use this information to make a decision as to how many loaves to bake? One bit of information he has gained is that he is now absolutely certain that θ will not be zero. Actually, he can use her answer to say more than this. He can modify the probability distribution assigned to the different states of nature in view of this sample information by using Bayes' theorem. (This would be a good time to go back and review Sections 1.11, 12.1, and 12.2.) He can then use this revised or posterior probability distribution to determine the appropriate action based on the expected payoff criterion. As an illustration of this, let's use Bayes' theorem to revise the initial probability distribution in view of Sally's "yes" response to Tim's call. The easiest way to do this is to make a table similar to Table 1.1 or 12.1. We have done this in Table 13.14. In this case we use the event B to represent "a 'yes' answer on a phone call." We could also define a random variable X to be the number of "yes" answers on a single phone call. In either case the conditional probabilities are relatively easy to find. For example,

$$P[B \mid \theta = .05] = P[X = 1 \mid \theta = .05] = .05,$$

and so on. Remember that Bayes' theorem is simply an indirect way of calculating a conditional probability. In this case, for example,

$$P[\theta = .05 \mid B] = \frac{P[(\theta = .05) \cap B]}{P(B)}$$

$$= \frac{.0075}{.1200} = .0625.$$

TABLE 13.14

Posterior Distribution Given a "Yes" Answer on One Call

Value of θ, θ_i	Prior Probability, $P[\Theta = \theta_i]$	Conditional Probability, $P[B \mid \theta_i]$	Product, $P[\Theta = \theta_i] \cdot P[B \mid \theta_i]$	Posterior Probability, $P[\Theta = \theta_i] \cdot P[B \mid \theta_i]/\sum$
0.00	.10	.00	0	0
.05	.15	.05	.0075	.0625
.10	.20	.10	.0200	.1667
.15	.35	.15	.0525	.4375
.20	.20	.20	.0400	.3333
			$\sum = .1200$	

The posterior probabilities given in the last column of Table 13.14 can now be used to find the expected payoff under each possible action. For example,

$$\begin{aligned} E(p(A_2 \mid B)) &= \sum p(A_2 \mid \theta_i) P[\theta = \theta_i \mid B] \\ &= (-4.80)(0) + 5.20(.0625) + 7.20(.1667) + 7.20(.4375) + 7.20(.3333) \\ &= 7.08. \end{aligned}$$

You can verify for yourself that the expected payoffs for the other actions, all rounded to the nearest penny, are

$$E(p(A_1 \mid B)) = 0, \qquad E(p(A_3 \mid B)) = \$12.86, \qquad E(p(A_4 \mid B)) = \$14.68.$$

Since the largest of these expected payoffs is \$14.68, corresponding to action A_4, if Tim receives a "yes" answer to his call, his action should be to bake 36 loaves. Consequently, on days when he receives a "yes" answer, he can expect a return of \$14.68, on the average.

Example 13.9 Find the best action for Tim to follow if he makes a single phone call to a customer and receives the answer "no" to his question about the customer's intention to buy raisin-nut bread. Also determine his expected payoff for following this action.

Let us define the event C to be "a 'no' answer on a phone call." Using event C or the random variable X defined earlier to be the number of "yes" answers on a single phone call, we would have, for example,

$$P[C \mid \theta = .05] = P[X = 0 \mid \theta = .05] = .95.$$

Using results like this in the "conditional probability" column of Table 13.15, we can calculate the posterior probabilities over values of θ given the event C. The posterior probabilities are given in the last column. (Note that due to roundoff, these values sum to .9999 rather than 1.) Using these probabilities, we find the expected payoff under action A_1 to be 0 [i.e., $E(p(A_1 \mid C)) = 0$], while

$$\begin{aligned} E(p(A_2 \mid C)) &= \sum p(A_2 \mid \theta_i) \cdot P[\Theta = \theta_i \mid C] \\ &= (-4.80)(.1136) + 5.20(.1619) + 7.20(.2045) + 7.20(.3381) + 7.20(.1818) \\ &= 5.51. \end{aligned}$$

TABLE 13.15
Posterior Distribution Given a "No" Answer on One Call

Value of θ, θ_i	Prior Probability, $P[\Theta = \theta_i]$	Conditional Probability, $P[C\mid\theta_i]$	Product, $P[C\mid\theta_i] \cdot P[\Theta = \theta_i]$	Posterior Probability, $P[C\mid\theta_i] \cdot P[\Theta = \theta_i]/\Sigma$
0	.10	1	.1000	.1136
.05	.15	.95	.1425	.1619
.10	.20	.90	.1800	.2045
.15	.35	.85	.2975	.3381
.20	.20	.80	.1600	.1818
			$\Sigma = .8800$	

You can verify that $E(p(A_3 \mid C)) = 8.59$, while $E(P(A_4 \mid C)) = 8.00$. Since the largest of these expected payoffs is \$8.59, corresponding to action A_3, if Tim receives a "no" answer to his call, his action should be to bake 24 loaves.

In summary, we can see that Tim's actions will be somewhat different if he makes a phone call to get additional information than if he does not. With no phone call, his action is always A_3, to bake 24 loaves. If he follows this strategy, he will make an average of \$9.10/day for selling raisin-nut bread. If he does make a phone call and receives a "yes" answer, his action will be A_4, to bake 36 loaves. On such days he can expect to make an average of \$14.68/day. If he receives a "no" answer, his action will be A_3, to bake 24 loaves, in which case he will expect to make an average of 8.59/day. To find his overall average profit from raisin-nut bread on a day when he makes one call, Tim could use

$$\sum (\text{expected payoff} \mid X = x) \cdot P[X = x],$$

where $X =$ number of yes answers. We have seen that the expected payoffs are either \$8.59 or \$14.68, but we need to find

$$P[X = 0] = P[\text{zero yes answers}] = P[C].$$

This value is simply $\Sigma = .8800$, the sum of the fourth column in Table 13.15. Similarly, from Table 13.14 we find that

$$P[X = 1] = P[\text{1 yes answer}] = P[B] = .1200.$$

Using these numbers, we find the average profit on a day when he makes one phone call to be

$$(8.59)(.88) + (14.68)(.12) = 9.32.$$

From this we see that the additional information gained from making a single phone call allows Tim to increase his average profit per day for selling raisin-nut bread from \$9.10 to \$9.32. The value of this sample information is worth, on the average, 22 cents/day. Would you suggest that Tim make such a phone call? If the cost, in time and effort, is less than 22 cents, then from a profit viewpoint it would be worthwhile to make such a call.

You might expect that the more information Tim can get, the higher average payoff he can expect to make. To some degree this is true. Although it would be cumbersome to show all the details of the calculations, we can show you the results that Tim would obtain if he were to make three phone calls. If we assume that it is more realistic to expect Tim to sample three of his customers without replacement rather than with replacement, then X, the number of "yes" answers he will receive from three randomly chosen customers, will follow a hypergeometric distribution. Using this distribution, we can find the conditional probabilities $P[X = x \mid \theta]$, and these in turn can be used to find the posterior distribution of θ given x. These results are summarized in Table 13.16. (Because of the size of this table it is more convenient to change the format to using rows instead of columns.) The last column, giving the sums of the products, actually gives the probabilities that Tim will get 0, 1, 2, or 3 "yes" answers, respectively. By using each posterior distribution, the action leading to the highest expected payoff, given the value of X, can be found. For example, if $X = 0$, then the expected profit is largest under action A_3, this value being \$7.48. In all other situations the optimal action is A_4: bake 36 loaves. For example, on a day when Tim receives three "yes" answers and follows action A_4, he can expect a payoff of \$18.08. (Unfortunately, this does not happen often, the probability of three "yes" answers being only .0028.) From the results summarized in Table 13.17, we can find the expected payoff when three calls are made. We have

$$\$7.48(.6905) + 13.94(.2619) + 16.55(.0449) + 18.08(.0028) = \$9.61$$

(to the nearest penny). Recall that Tim's expected payoff based on prior information only with no sample information is \$9.10/day. By making three calls the expected

TABLE 13.16

Posterior Distributions When Three Calls Are Made

θ:	0	.05	.10	.15	.20	
Prior:	.10	.15	.20	.35	.20	
Conditional $P[X = 0 \mid \theta]$:	1	.8567	.7278	.6125	.5101	
Product:	.1000	.1285	.1456	.2144	.1020	$\Sigma = .6905$
Posterior:	.1448	.1861	.2109	.3105	.1477	
Conditional $P[X = 1 \mid \theta]$:	0	.1367	.2453	.3281	.3874	
Product:	0	.0205	.0491	.1148	.0775	$\Sigma = .2619$
Posterior:	0	.0783	.1875	.4384	.2959	
Conditional $P[X = 2 \mid \theta]$:	0	.0065	.0260	.0563	.0950	
Product:	0	.0010	.0052	.0197	.0190	$\Sigma = .0449$
Posterior:	0	.0217	.1159	.4389	.4234	
Conditional $P[X = 3 \mid \theta]$:	0	.0001	.0009	.0031	.0075	
Product:	0	.0000+	.0002	.0011	.0015	$\Sigma = .0028$
Posterior:	0	.0054	.0647	.3903	.5396	

TABLE 13.17
Summary of Payoffs and Probabilities When Three Calls Are Made

X (number of yes answers):	0	1	2	3
Optimal action given x:	A_3	A_4	A_4	A_4
Average payoff for optimal action:	$7.48	13.94	16.55	18.08
$P[X = x]$:	.6905	.2619	.0449	.0028

value goes up to $9.61/day, so the expected value of this sample information is $.51/day.

Can Tim keep increasing his expected payoff per day for raisin-nut bread by getting more and more information? No! There is an upper limit to his profit. You may recall from our remarks following the proof of Theorem 13.1 that if Tim knew exactly what the state of nature would be at the start of the day, his expected payoff would be

$$\sum_j [\max_k p(A_k \mid \theta_j)] \cdot P[\Theta = \theta_j] = \$12.64.$$

[See equation (13.3).] If, instead of making three phone calls to his customers, Tim could make a single phone call to Mother Nature, who would tell him exactly what the state of nature would be, what would such a call be worth to him? What would it be worth to Tim to be able to obtain perfect information? Since he can raise his expected payoff from $9.10 to $12.64, this would be worth

$$\$12.64 - 9.10 = \$3.54.$$

From a profit point of view, if the cost of a long-distance phone call to Mother Nature was less than $3.54, it would pay him to make the call. In this situation the expected value of perfect information (EVPI) is $3.54. You will notice that this is exactly the same value as the minimum expected loss incurred when the expected loss strategy is used (see Example 13.8).

(It is important to note that $12.64/day is the very best Tim can do if, in fact, the initial distribution over the values of θ is really the correct distribution and assuming that costs and selling prices remain fixed at the levels specified earlier. If Tim invested in an advertizing campaign or changed his selling price or costs, he would be able to change the upper limit of $12.64/day.)

13.6 ESTIMATION AS A DECISION PROBLEM

In the original statement of the problem related to Tim's decision as to the amount of bread to bake, we assumed that his possible actions were restricted by conditions such as oven size, baking pans, mixing bowls, and so on. Consequently, his possible actions were assumed to be to bake 0, 12, 24, 36 or 48 loaves. Since the states of nature were assumed to be 0, 5%, . . . , 20% of his customers (or, equivalently, 0, 10, . . . , 40 customers), it would usually be impossible for him to bake exactly the right number of loaves. That is, unless he baked none, he would have baked too much or too little bread, depending on the state of nature. He might be able to save himself some frustration (and perhaps increase his expected payoff in the bargain) if he would let his possible actions conform to the possible states of nature. This would mean that he should decide to bake either 0, 10, 20, 30, or 40 loaves of raisin-nut bread. Clearly, if the state of nature is a demand of 20, for example, then Tim's best course of action would be to bake 20 loaves, to bake 10 if the demand is 10, and so on. Of course Tim still will not know at the beginning of the day just what the state of nature will be. He can think of his problem, however, as a problem in estimation. If he estimates θ to be .10, he is estimating the demand to be 20 loaves, so he should bake 20 loaves.

To see what action he should follow from a decision theory point of view, let us first find the payoff table for this problem. Remember that this bread costs \$.40/loaf to make and sells for \$1.00/loaf (for a profit of \$.60/loaf). For instance, if he bakes 20 but $\theta = .05$, then 10 will be sold, so

$$p(20 \mid \theta = .05) = (.60 \times \text{number sold}) - (.40 \times \text{number leftover})$$
$$= (.60)(10) - (.40)(10) = +2.00.$$

If he bakes 20 and $\theta = .10$, .15, or .20, all loaves will be sold, yielding a payoff of \$12.00. All of this information is summarized in Table 13.18.

TABLE 13.18
Payoff Table

		State of Nature				
		θ_1: 0	θ_2: .05	θ_3: .10	θ_4: .15	θ_5: .20
	A_1: 0	0	0	0	0	0
Possible	A_2: 10	−4	6	6	6	6
Action	A_3: 20	−8	2	12	12	12
	A_4: 30	−12	−2	8	18	18
	A_5: 40	−16	−6	4	14	24

The loss table can be found from this payoff table. For instance, when $\theta = .10$, the best action is to bake 20, so

$$l(20 \mid \theta = .10) = 0.$$

Also,

$$l(30 \mid \theta = .10) = \max_k p(A_k \mid \theta = .10) - p(30 \mid \theta = .10)$$
$$= 12 - 8 = 4.$$

This information is summarized in the loss table given in Table 13.19. An extra row has been added to the bottom of this table, showing the probabilities for each state

TABLE 13.19
Loss Table

		State of Nature				
		θ_1: 0	θ_2: .05	θ_3: .10	θ_4: .15	θ_5: .20
	A_1: 0	0	6	12	18	24
Possible	A_2: 10	4	0	6	12	18
Action	A_3: 20	8	4	0	6	12
	A_4: 30	12	8	4	0	6
	A_5: 40	16	12	8	4	0
	$P[\Theta = \theta]$	.10	.15	.20	.35	.20

of nature. Using these probabilities, we can easily find the expected losses for each action. We have

$$E(l(A_1)) = 0(.10) + 6(.15) + 12(.20) + 18(.35) + 24(.20)$$
$$= 14.40.$$

Similarly, you can verify that the expected losses for actions A_2 to A_5 are 9.40, 5.90, 4.40, and 6.40, respectively. Using the expected loss criterion, Tim's best action would be A_4: bake 30 loaves.

Now examine the loss table (Table 13.19) and see if you can detect any patterns. One characteristic that stands out is the fact that there are zeros down the diagonal. Furthermore, for each position above the diagonal there is an increase of 6 and for each position below the diagonal there is an increase of 4. There are good reasons for observing these particular numbers. Consider the particular state of nature $\theta = \theta_3 = .10$. It should be true that a correct decision leads to a zero loss. In terms of estimation, this would mean that θ was correctly estimated to be θ_3, action A_3 would be taken and no loss would be incurred. If θ was wrongly estimated to be θ_4, action A_4 would have been followed. Forty loaves would be baked instead of 30, leading to a \$4 loss of ingredients. An additional \$4 would be lost for each additional extra 10 loaves baked. On the other hand, if θ actually equaled θ_3 but was wrongly estimated to be θ_2, action A_2 would have been followed. Twenty loaves would have been baked instead of 30, leading to a loss of potential profit for selling those extra loaves of \$6. An additional \$6 of potential profit would be lost for each additional 10 loaves left unbaked. From an estimation point of view, the action taken corresponds to an estimate of θ: if θ is estimated to be θ_i, follow action A_i. Instead of writing the possible actions in terms of the number of loaves of bread to bake, we could write the possible actions as estimates of θ. Denoting the estimate of θ by $\hat{\theta}$, we could express the losses incurred for errors in estimation as follows:

$$l(\hat{\theta} \mid \theta) = \begin{cases} 0 & \text{if } \hat{\theta} = \theta \\ (.40)(200)(\hat{\theta} - \theta) & \text{if } \hat{\theta} > \theta \\ (.60)(200)(\theta - \hat{\theta}) & \text{if } \hat{\theta} < \theta. \end{cases}$$

(Since θ represents a proportion, we need to multiply by 200, the number of customers, to get the loss in terms of the number of loaves times profit or loss per loaf.) Since the ingredients cost \$.40/loaf, we can define

(13.4) $$C_o = \text{cost of overestimation} = (\$.40/\text{loaf})(200).$$

Since a profit of \$.60/loaf is made, we can also define

(13.5) $$C_u = \text{cost of underestimation} = (\$.60/\text{loaf})(200).$$

The "best" estimate of θ, in terms of minimizing the expected loss, can be expressed in terms of C_o, C_u, and the probability distribution of θ. The method for finding the estimate is given in the following theorem, which we state without proof.

Theorem 13.2 Let Θ represent a parameter whose observed value will be determined by a probability distribution $P[\Theta = \theta]$. Assume that the loss in estimating the value that Θ will assume by $\hat{\theta}$ when the value is actually θ is

$$l(\hat{\theta}\,|\,\theta) = \begin{cases} 0 & \text{if } \hat{\theta} = \theta \\ C_o(\hat{\theta} - \theta) & \text{if } \hat{\theta} > \theta \\ C_u(\theta - \hat{\theta}) & \text{if } \hat{\theta} < \theta, \end{cases}$$

where C_o and C_u denote the costs of overestimation and underestimation, respectively. Then the estimate of the value that Θ will assume that will minimize the expected loss is the

$$\frac{C_u}{C_o + C_u}$$

fractile of the probability distribution of Θ.

[*Note:* If in Theorem 13.2 the costs of over- and underestimation are equal, the loss function can be expressed by

$$l(\hat{\theta}\,|\,\theta) = C\,|\hat{\theta} - \theta|$$

and the best estimate of θ will be the median of the probability distribution of Θ (cf. Section 12.6).]

Example 13.10 Using the probability distribution over the values of θ,

θ:	0	.05	.10	.15	.20
$P[\Theta = \theta]$:	.10	.15	.20	.35	.20,

find the estimate of θ and corresponding action Tim should take if the cost of ingredients for raisin-nut bread is \$.40/loaf and the profit is \$.60/loaf.

Since there are 200 customers, we have, from (13.4) and (13.5),

$$C_o = (.40)(200) \quad \text{and} \quad C_u = (.60)(200),$$

so we need to find the

$$\frac{C_u}{C_o + C_u} = .60$$

fractile (i.e., the 60th percentile of this probability distribution). The definition of fractiles was given in Definition 2.13 of Section 2.6. Using either this definition or the graphical technique as illustrated in Figure 2.15, we see that the 60th percentile is $\theta = .15$. This would mean that $\hat{\theta} = .15$, so the action would be to bake 30 loaves. (Note that this result is consistent with the one found using the loss table in Table 13.19. Also note that the factor 200, the number of customers, cancels so that the same percentile is appropriate for any number of customers.)

The estimation procedure described by Theorem 13.2 will be valid whether Θ is a discrete or continuous random variable. If, for example, Tim had considered θ to be a continuous quantity assuming values over the interval [0, 1], he might have used as a probability distribution over the values of θ a beta distribution with parameters $n' = 20$ and $s' = 3$. The estimate of θ that would minimize the expected loss would be the 60th percentile of this distribution.

Example 13.11 Shelly operates a small concession stand called Shelly's Snack Shack. Among other items Shelly sells hot chocolate. The sales depend on the weather: the cooler the weather the more sales. In view of the weather report, she has determined that the proportion θ of her customers who will want hot chocolate today is a random quantity with probability distribution as follows:

θ:	.10	.20	.30	.40	.50
$P[\Theta = \theta]$:	.10	.35	.30	.20	.05.

She usually has about 300 customers per day. Hot chocolate costs her 12 cents/cup and she sells it for 20 cents/cup. Leftover chocolate is not very good, so she must discard any leftover. How many cups should she make?

Shelly's cost of overestimation is 12 cents/cup and cost of underestimation (loss of potential profit) is 8 cents/cup. She would estimate θ to be the

$$\frac{8}{12 + 8} = .40$$

fractile (i.e., the 40th percentile of the distribution of Θ). This corresponds to the value $\theta = .20$. Consequently, Shelly should make $(300)(.20) = 60$ cups of hot chocolate to minimize her expected loss.

When estimating the value that Θ will assume, using a particular fractile will give an estimate that minimizes the expected loss, assuming that the loss function is of the form given in Theorem 13.2. If the loss function is of a different form, a different estimate may be required to minimize the expected loss. For example, if the loss function is quadratic, the mean of the distribution of Θ will minimize the expected loss.

13.7 UTILITY

In the last three sections we have discussed decision strategies and estimation based on either maximizing expected payoff or equivalently, minimizing expected loss. You remember that an expected value is a long-run average value of a random variable, not the value you actually "expect" to observe the random variable to be. For example, when Tim follows action A_3, bake 24 loaves of raisin-nut bread, his payoff on a particular day will be either

$$-9.60, \quad 0.40, \quad 10.40, \quad 14.40, \quad \text{or} \quad 14.40$$

depending on whether $\theta = 0, .05, \ldots,$ or .20 (see Example 13.7 and Table 13.11). After a large number of days of following this strategy he would get, *on the average*, about $E(p(A_3)) = \$9.10$/day. Since expected values are based on average values over a large number of trials, it may not be reasonable to use expectations as a decision criterion if you have only a single (or a few) opportunities to follow a particular strategy.

As an illustration, imagine yourself as a contestant on a TV game show. The emcee, Smilin' Sam, gives you a choice early in the show of taking \$100 cash or whatever is behind the door of your choice! Behind one door is a prize worth \$400 and behind the other is a booby prize—such as a jar of peanut butter worth \$1. Which would you choose? At the urgings of the members of the audience, you might decide to choose one of the doors. If you ignored the audience momentarily and set up a payoff table, you would have a table such as the one in Table 13.20. It is fairly easy to assign a probability to each state of nature in this case, namely, a probability of .5 to each. Under action A_1 you are sure to get \$100, so $E(p(A_1)) = \$100$. For action A_2 you would have

$$E(p(A_2)) = 400(.5) + 1(.5) = \$200.50.$$

If you base your reasoning on maximizing the expected payoff, you would follow the urgings of the audience and choose a door.

Now assume that you have been selected by Sam to get a chance at the super-colossal grand prize. You now have a choice of \$25,000 cash or whatever is behind the door of your choice. Behind one door is a prize worth \$100,000 and behind the other is another jar of peanut butter. What would you choose to do in this case? You will still receive urgings from the audience to go for the big money. Would you listen to them? Calculating expectations it is easy to see that

$$E(p(A_1)) = \$25{,}000$$

and that

$$E(p(A_2)) = 100{,}000(.5) + 1(.5) = \$50{,}000.50,$$

so using the criterion of maximizing expectation, you would have to choose A_2. Unfortunately, you will not have the chance to play this game every week, so long-run

TABLE 13.20

Payoff Table for Smilin' Sam's Show

		State of Nature	
		Right Door Chosen	*Wrong Door Chosen*
Possible Action	A_1: *Take \$100*	100	100
	A_2: *Choose a Door*	400	1

averages may not be very meaningful! Most people in this situation would adopt a "bird in the hand" philosophy and follow action A_1. What would you do? (Think for a moment of the response of your parents, spouse, or friends if you were to come home with a $1 jar of peanut butter!)

One reason for the difficulty here is that to most people the value of money is not "linear" for all values. This is not a question of purchasing power or taxes, but more a question of one's own personal attitude toward money. To you $2 might mean twice as much as $1, $20 twice as much as $10, . . . , $2000 twice as much as $1000. However, for most people $2,000,000 is not worth twice as much as $1,000,000, for example. There are few things that you could not do with $1 million that you could do with $2 million.

One method for describing an individual's attitude toward the value of money or toward risk is to use "utility." (The term "utility" as used here is different from the term as used in economics.) Basically, a numerical value is subjectively assigned by an individual to each possible payoff in a decision-making problem. These numerical values, called utilities, are then used instead of the payoffs themselves to assist in decision making. This assignment of numbers to payoffs can be thought of as determined by a mathematical function. To a payoff of m dollars is assigned a utility $U(m)$. It is assumed that the function is such that numerical assignments are made in a logical and consistent manner. For example, if A is greater than B, then $U(A)$ will be greater than $U(B)$.

In general, it would be difficult to obtain a concise mathematical function to describe an individual's utility function $U(m)$, so we will simplify the problem by determining the utility for a few values, say $m_1, m_2, \ldots, m_k$, plotting these values, and connecting them with a smooth curve. Rather than try to describe such a curve over all possible dollar values, we will restrict our attention to a certain range of interest. For example, since no one ever pays money to be on Smilin' Sam's show and since the largest prize is $100,000, the range of interest is 0 to 100,000. One way to start the determination of utility is to assign a value of zero to the smallest possible value and a value of 1 to the largest. In this case we would have

$$U(0) = 0 \qquad \text{and} \qquad U(100{,}000) = 1.$$

Now let us try to determine the utility to assign to the value $25,000, say. Assume that you are on Smilin' Sam's show. You may take $25,000 and go home or can take a chance on $100,000. Instead of choosing a door, however, you are given a chance to spin a wheel divided into 100 parts. A certain proportion of these parts have a W and the rest have an L. You win if the wheel stops on a W (see Figure 13.2). If the proportion of W's, call it p, is equal to .5, the situation is exactly the same as choosing one of two doors. Now let us say that Sam tried to entice you to spin the wheel by offering to change some of the L's to W's. You certainly would agree to spin the wheel if the proportion of W's was $p = 1.0$, since you would be sure to win! How about if p were .95? .90? .60? The smallest value of p that would entice you to spin the wheel instead of taking a sure $25,000 would correspond to your utility for this value. For example, if p were .75 for you, then

$$U(25{,}000) = .75.$$

FIGURE 13.2
Smilin' Sam's Spinning Wheel

(Actually, p should be the value that makes you feel indifferent between the two actions, but this is essentially the same as the smallest value that would entice you to spin the wheel.) To find your utility for another value, say \$10,000, you would repeat this process. You might determine that

$$U(10{,}000) = .40.$$

These points, together with $(0, U(0) = 0)$ and $(100{,}000, U(100{,}000) = 1)$, are shown in Figure 13.3. A smooth curve connects these points and can be used to find other utility values. For example, from the curve we could estimate the utility of \$60,000 to be about .95.

In Figure 13.4 are utility curves determined by two past contestants on Smilin' Sam's show. In Figure 13.4(a) is the curve for Fred, who will always take a fair bet. For instance, if $p = .25$, the expected payoff for spinning is

$$0(.75) + 100{,}000(.25) = \$25{,}000,$$

so Fred considers spinning the wheel with $p = .25$ as good as a sure \$25,000. Following this reasoning for all money values, Fred obtains a straight line. In Figure 13.4(b) is the utility curve for Diane, who is relatively rich and daring. She likes to have the crowd excited and will go for the big money even though the chances of getting it might be relatively small. For example, she would just as soon spin the wheel when

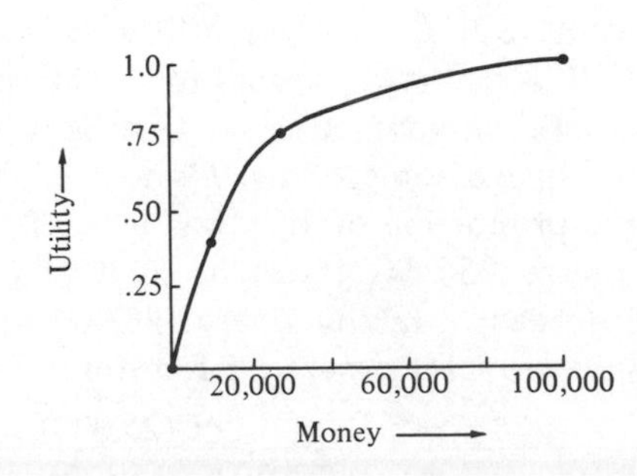

FIGURE 13.3
Utility Curve

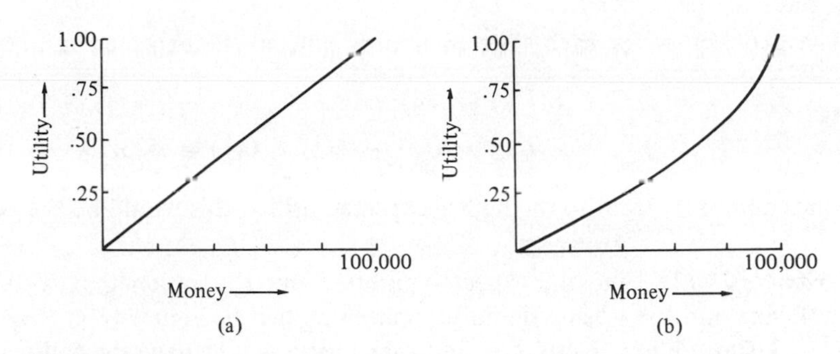

FIGURE 13.4

Utility Curves for Past Contestants

$p = .15$ as she would take a sure $25,000, even though her expected value would be only

$$0(.85) + 100{,}000(.15) = \$15{,}000.$$

In characterizing different attitudes toward risk, we might call Diane a risk seeker, Fred risk neutral, and a person with a curve as in Figure 13.3 a risk avoider. One must be careful in characterizing people as always having the same attitude. Diane may be a risk seeker on a TV game show, but she might be a risk avoider in a managerial role. How would you classify yourself? In the game show situation, would you rather take the sure money? Would you classify yourself the same way in other situations?

Now let us see how utility can be used in decision making. Say that Smilin' Sam offers you $40,000 cash or a chance to pick one of three doors. Behind one door is $100,000, behind a second is $20,000, but behind the last door is that jar of peanut butter (which for computational convenience we will take to be worth zero). To decide which to take, the sure cash (A_1) or a choice of doors (A_2), you would calculate the expected utility for each action and choose the action with the higher expected utility. Using the utility curve in Figure 13.3, we can find the utility associated with each payoff. These values are shown in the utility payoff table shown in Table 13.21.

TABLE 13.21

Utility Payoff Table

		State of Nature		
		θ_1: *Door Chosen Has $100,000*	θ_2: *Door Chosen Has $20,000*	θ_3: *Door Chosen Has $0*
Possible Action	A_1: *Sure Cash ($40,000)*	.88	.88	.88
	A_2: *Choose Door*	1	.60	0

The probability for each state of nature is $\frac{1}{3}$, so the expected utilities are

$$E(U_{A_1}) = .88$$
$$E(U_{A_2}) = 1(\tfrac{1}{3}) + .6(\tfrac{1}{3}) + 0(\tfrac{1}{3}) = .533.$$

Since action A_1 leads to the higher expected utility, this would be the action followed.

Example 13.12 Find the expected utilities and the action that would be followed by Diane and Fred using the utility curves shown in Figure 13.4.

Using Figure 13.4(b), we find that Diane would have the following utilities:

$$U(20{,}000) = .10 \qquad \text{and} \qquad U(40{,}000) = .25.$$

She would find that

$$E(U_{A_1}) = U(40{,}000) = .25$$

and

$$E(U_{A_2}) = U(100{,}000)(\tfrac{1}{3}) + U(20{,}000)(\tfrac{1}{3}) + U(0)(\tfrac{1}{3})$$
$$= 1(\tfrac{1}{3}) + .10(\tfrac{1}{3}) + 0(\tfrac{1}{3}) = .36\bar{6}.$$

She would follow action A_2, since this leads to the higher expected utility.

To find Fred's action we could read the graph in Figure 13.4(a) (although this could lead to some errors) or we could note that the mathematical equation for this straight line is simply $U(m) = m/(100{,}000)$. From this we have

$$U(20{,}000) = .2 \qquad \text{and} \qquad U(40{,}000) = .4,$$

so

$$E(U_{A_1}) = U(40{,}000) = .4$$

and

$$E(U_{A_2}) = U(100{,}000)(\tfrac{1}{3}) + U(20{,}000)(\tfrac{1}{3}) + U(0)(\tfrac{1}{3})$$
$$= 1(\tfrac{1}{3}) + .2(\tfrac{1}{3}) + 0(\tfrac{1}{3}) = .4.$$

Since the two expected utilities are the same, Fred would be indifferent to these two actions. Actually, assigning some positive value to a jar of peanut butter would make Fred lean to A_2, choosing a door. [Note that since Fred will accept a fair bet, he can use money rather than utility. He finds that

$$E(p(A_1)) = \$40{,}000$$

and

$$E(p(A_2)) = 100{,}000(\tfrac{1}{3}) + 20{,}000(\tfrac{1}{3}) + 0(\tfrac{1}{3})$$
$$= \$40{,}000.$$

Ignoring the value of a jar of peanut butter, he would be indifferent to these two actions. Counting the value of the peanut butter, he would tend to take action A_2.]

EXERCISES

1. A payoff table has been determined to be the following:

		State of Nature θ_1	θ_2	θ_3
Possible Action	A_1	−100	50	300
	A_2	200	0	500
	A_3	−300	400	50
	A_4	100	−100	400

(a) What action should be followed if the state of nature is known to be θ_1? θ_2? θ_3? (b) What actions, if any, are inadmissible?

2. A payoff table has been determined to be the following:

		State of Nature θ_1	θ_2	θ_3	θ_4
Possible Action	A_1	−100	200	300	100
	A_2	50	0	400	−100
	A_3	300	500	150	150

(a) What action should be followed if the state of nature is known to be θ_1? θ_2? θ_3? θ_4? (b) What actions, if any, are inadmissible?

3. Vicki has purchased a new dress for $125. She wishes to wear it to dinner tonight, but she knows there is a possibility of rain. If it rains, it may be a light shower or a heavy downpour. If unprotected, even a light shower will totally ruin the dress. If she carries an umbrella, a light shower will do no harm, but a downpour will cause $60 worth of damage. A good raincoat, however, will protect the dress even against a downpour. Vicki can purchase an umbrella for $10 or a good raincoat for $40. Construct a payoff table that Vicki could use for making a decision about possible purchases.

4. Abdul owns a large parcel of land in the desert. The El Gusho Oil Company has offered to buy drilling rights for a flat fee of $40,000, for 2% of the profits, or for $20,000 plus 1% of the profits. Based on results of drilling on his neighbors' land, Abdul knows that there may be no oil, a small pool worth $3 million in profits, or a large pool worth $6 million in profits. Construct a payoff table that Abdui could use for making a decision about which offer to take.

5. Construct a loss table corresponding to the payoff table given in Exercise 1.

6. Construct a loss table corresponding to the payoff table given in Exercise 2.

7. Construct a loss table that Vicki (Exercise 3) could use for making a decision about possible purchases.

8. Construct a loss table that Abdul (Exercise 4) could use for making a decision about which offer to take.

9. A payoff table has been determined to be the following:

		State of Nature		
		θ_1	θ_2	θ_3
	A_1	−500	0	500
Possible	A_2	−300	−100	700
Action	A_3	0	500	1000
	A_4	100	−200	500

What action should be taken if the (a) maximin payoff, (b) maximax payoff, (c) minimax loss strategy is followed?

10. A payoff table has been determined to be the following:

		State of Nature			
		θ_1	θ_2	θ_3	θ_4
Possible	A_1	−500	−300	0	100
Action	A_2	−100	200	0	−100
	A_3	−450	100	50	50

What action should be taken if the (a) maximin payoff, (b) maximax payoff, (c) minimax loss strategy is followed?

11. What action should Vicki take if she follows the (a) maximin payoff?; (b) maximax payoff?; (c) minimax loss strategy? Use the payoff and loss tables constructed in Exercises 3 and 7.

12. What action should Abdul take if he follows the (a) maximin payoff?; (b) maximax payoff?; (c) minimax loss strategy? Use the payoff and loss tables constructed in Exercises 4 and 8.

13. A payoff table has been determined to be the following:

		State of Nature			
		θ_1	θ_2	θ_3	θ_4
Possible	A_1	−1000	0	500	2000
Action	A_2	200	700	−300	800
	A_3	500	900	100	−200

If the probability distribution over the possible states of nature is

θ:	θ_1	θ_2	θ_3	θ_4
$P[\Theta = \theta]$:	.1	.2	.3	.4,

find (a) the expected payoff for each possible action; (b) the expected loss for each possible action. What action should be taken to (c) maximize expected payoff?; (d) minimize expected loss?

14. Repeat Exercise 13 if the probability distribution over the possible states of nature is

θ:	θ_1	θ_2	θ_3	θ_4
$P[\Theta = \theta]$:	.4	.3	.2	.1.

15. What decision should Vicki make concerning possible purchases if she follows the maximum expected payoff strategy and the probability of a light shower only is .2 while the probability of a heavy downpour is .1? Use the payoff table constructed in Exercise 3.

16. A geologist has told Abdul that the probability of finding a small pool of oil is .2 and the probability of finding a large pool is .3. Which offer should Abdul take if he follows the minimum expected loss strategy? (See Exercises 4 and 8.)

17. Assume that Tim the baker decides to call two of his customers to get information about the state of nature, the demand for raisin-nut bread on a given day. (For simplicity, assume that customers are chosen with replacement.) If the prior probability distribution is

θ:	0	.05	.10	.15	.20
$P[\Theta = \theta]$:	.10	.15	.20	.35	.20,

find the posterior distribution and the expected payoff when he uses the optimal action if he receives (a) no, (b) one, (c) two "yes" answers to the question "Do you plan to buy raisin-nut bread today?" (d) Find the overall expected payoff on days when he makes two phone calls. (e) What is the expected value of the sample information?

18. The inspector at Watzit Manufacturing, Dee Syder, knows that the watzit maker (Example 12.1 of Section 12.2) produces watzits with a defective rate of either 1, 5, or 10%. Dee has three actions from which to choose at the start of a shift: A_1: make no adjustment; A_2: make an external adjustment that will change a 10% rate to 5% or a 5% rate to 1%; A_3: make an internal adjustment that will change either a 10% or a 5% rate to 1%. The cost of an external adjustment is \$25 and the cost of an internal adjustment is \$40. One thousand watzits are produced per shift at a profit of \$1 each. Items can be inspected at a cost of \$2 each. For simplicity assume that the number of defective items produced is equal to 1000 times the rate (i.e., if the rate is 5%, 50 defectives are produced). Defective items are worthless.

(a) Construct a payoff table and a loss table for this problem.

(b) Find the action that Dee should take under the minimax, maximax, and maximin strategies.

(c) If the probability distribution over the possible values of defective rates is

p_i:	.01	.05	.10
$P[p = p_i]$:	.60	.30	.10,

find the action Dee should take under the maximum expected payoff strategy.

(d) What action should Dee take if he inspects one item and finds it to be nondefective?

(e) What action should Dee take if he inspects one item and finds it to be defective?

(f) What is the expected value of this sample information? Would it pay Dee to inspect a single item?

19. Referring to Exercise 18 and assuming the prior distribution as given in part (c) of that exercise, what action should Dee take if he inspects two items and finds (a) no defective items?; (b) one defective item?; (c) two defective items? (d) What is the expected value of this sample information? Would it pay Dee to inspect two items?

20. In Example 13.1 of Section 13.2 we saw that Norwegian fruited sweetbread, Jule Kaga, costs \$.90/loaf to make and sells for \$2.00/loaf. Assume that the proportion of Tim's customers who will want to buy this bread is either 0, 3, 6, or 9% and that his possible set of actions is to bake either 0, 6, 12, or 18 loaves. (Remember that we assume that Tim has 200 customers each day.) (a) Construct a payoff table and a loss table. (b) If the probability distribution over the states of nature is

θ:	0	.03	.06	.09
$P[\Theta = \theta]$:	.10	.40	.30	.20,

find Tim's best action using the expected loss criterion. (c) Find the values of C_o, the cost of overestimation, and C_u, the cost of underestimation. (d) Find the $C_u/(C_o + C_u)$ fractile of the distribution of Θ. Compare this fractile with the results obtained in part (b).

21. If in Exercise 20 the possible states of nature are all values of θ between 0 and 1 and if the probability distribution over these values is a beta distribution with parameters $n' = 20$ and $s' = 1$, find the $C_u/(C_o + C_u)$ fractile of this distribution and determine the number of loaves of Jule Kaga that Tim should bake.

22. If for raisin-nut bread the possible values for θ, the proportion of customers who will want to buy this bread, are between 0 and 1 and if the probability distribution over these values is a beta distribution with parameters $n' = 20$ and $s' = 3$, the estimate of θ that would minimize Tim's expected loss will be the 60th percentile of this distribution. Find this percentile. How many loaves of raisin-nut bread should Tim bake? (Cf. Example 13.10 of Section 13.6 and following.)

23. The demand for hot dogs at a baseball game follows (approximately) a normal distribution, with a mean of 700 and a standard deviation of 150. Chester runs the concession stand and needs to order hot dogs for the next game. Hot dogs cost 25 cents and sell for 75 cents. Assuming that leftover hot dogs have no value, how many hot dogs should Chester order to minimize his expected loss? How many should he order if he can sell leftover hot dogs to his neighbor Moose for 10 cents each?

24. Consider the two utility curves that have been constructed by two past contestants on Smilin' Sam's show (Section 13.7). One curve was made by timid Tessie and the other by bold Betty. Which curve would you say was made by Tessie? Why?

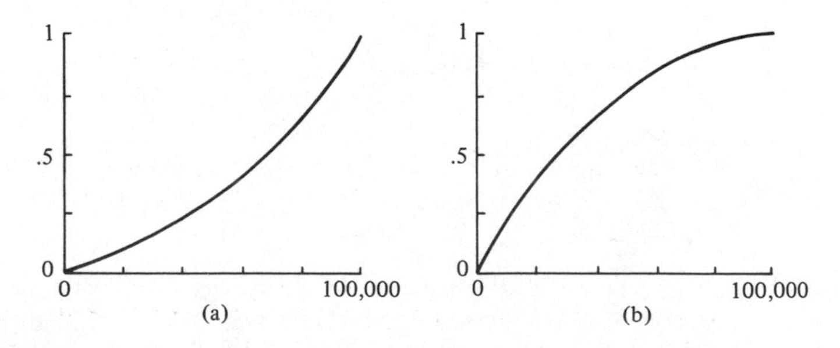

(a) (b)

25. Using the utility curve (a) from Exercise 24, find the utility associated with \$25,000, \$50,000, and \$75,000. Do the same using the utility curve in (b).

26. Construct your own utility curve over the range of values 0 to \$100,000. Follow the guidelines given in Section 13.7.

27. A contestant on Smilin' Sam's show is given a choice of either taking a sure \$25,000 or rolling a fair die. If the die is rolled and a 1 comes up, the contestant wins \$100,000; if a 2 or a 3 comes up, the prize is \$36,000; if a 4 or a 5 comes up, the prize is \$9000; but if a 6 comes up, the prize is \$0. (a) Find the expected value to a contestant who rolls a die. Based on expected dollar value, which choice is better? (b) Belinda is a contestant on the show who has a utility function given by

$$u = \sqrt{\frac{x}{100{,}000}},$$

where x is the amount of the prize. What utility would Belinda assign to \$25,000? What would be her expected utility if she were to roll the die? Based on her utility function, which choice is better? (c) Casper has a utility function given by

$$u = \left(\frac{x}{100{,}000}\right)^2.$$

What utility would Casper assign to \$25,000? What would be his expected utility if he were to roll the die? Based on his utility function, which choice is better? (*To think about:* Using your own utility curve as constructed in Exercise 26, what choice would you make?)

A1

SUMMATION NOTATION

If we have a sequence of n numbers to add together, say the numbers $x_1, x_2, \ldots, x_n$, we can express the sum by

$$x_1 + x_2 + \ldots + x_n.$$

A more-abbreviated notation is to use a capital sigma (the Greek letter for S, reminding us of the word "sum") to represent the operation of summation. We define

$$\text{(A.1)} \qquad \sum_{i=1}^{n} x_i = x_1 + x_2 + \ldots + x_n.$$

If it is clear from the context of the problem just what numbers x_i are to be added together, we might write any of the following:

$$\sum_{i=1}^{n} x_i, \qquad \sum_{i} x_i, \qquad \sum x_i, \qquad \sum_{\text{all } x} x.$$

If a represents a constant, that is, a quantity whose value does not depend on which x_i value is being considered, the following properties hold:

$$\sum_{i=1}^{n} (ax_i) = a\left(\sum_{i=1}^{n} x_i\right);$$

$$\sum_{i=1}^{n} a = na;$$

$$\sum_{i=1}^{n} (x_i + y_i) = \left(\sum_{i=1}^{n} x_i\right) + \left(\sum_{i=1}^{n} y_i\right).$$

To see that each of these equalities hold true, we simply need to use equation (A.1) and fundamental properties of addition. We have

$$\sum_{i=1}^{n} (ax_i) = ax_1 + ax_2 + \ldots + ax_n$$

$$= a(x_1 + x_2 + \ldots + x_n) = a\left(\sum_{i=1}^{n} x_i\right);$$

$$\sum_{i=1}^{n} a = \underbrace{a + a + \ldots + a}_{n \text{ terms}} = na;$$

$$\sum_{i=1}^{n} (x_i + y_i) = (x_1 + y_1) + (x_2 + y_2) + \ldots + (x_n + y_n)$$

$$= (x_1 + x_2 + \ldots + x_n) + (y_1 + y_2 + \ldots + y_n)$$

$$= \sum_{i=1}^{n} x_i + \sum_{i=1}^{n} y_i.$$

For example, if $n = 4$, $a = 5$, $x_1 = 2$, $x_2 = 4$, $x_3 = 7$, and $x_4 = 9$, then

$$\sum_{i=1}^{4} x_i = 22,$$

so

$$\sum_{i=1}^{n} ax_i = \sum_{i=1}^{4} 5(x_i) = 5 \sum_{i=1}^{4} x_i = 5(22) = 110$$

and

$$\sum_{i=1}^{n} a = \sum_{i=1}^{4} (5) = 4(5) = 20.$$

If, in addition, we have that $y_1 = 3$, $y_2 = -1$, $y_3 = 5$, and $y_4 = 12$, then

$$\sum_{i=1}^{4} y_i = 19$$

and

$$\sum_{i=1}^{4} (x_i + y_i) = \sum_{i=1}^{4} x_i + \sum_{i=1}^{4} y_i = 22 + 19 = 41.$$

If you feel uncertain about these results, you should write out the individual terms of $\sum_{i=1}^{4} ax_i$, $\sum_{i=1}^{4} a$, and $\sum_{i=1}^{4} (x_i + y_i)$ and add them together. For further detail on this subject, we refer you to a text on college algebra.

A2

TABLES AND HOW TO USE THEM

Tables are important time-savers for persons learning or applying statistics. They remain useful even to persons having hand calculators or even computers available to them, because in most cases answers can be found more quickly by referring to tables then by other means. In the following we will explain and illustrate how the tables can be used.

Table A1. Binomial Probabilities (Individual Terms) These tables give

$$P[X_B = x \mid n, p] = \binom{n}{x} p^x q^{n-x},$$

where X_B represents a binomial random variable having parameters n and p. We generally think of X_B as the number of "successes" in n independent trials when p is the probability of a success. Tabled values are for $x = 0, 1, \ldots, n$, with $n = 1, 2, \ldots, 20$, and $p = .05, .10, .15, \ldots, .50$.

If you wish to find $P[X_B = 2 \mid n = 3, p = .15]$, first locate $n = 3$ in the leftmost column. Next find the value $x = 2$ and read across to the column $p = .15$. The result is .0574, as you can see highlighted in Figure A1. All the key values have been circled.

These tables can also be used when the probability of a "success" is greater than .5. In such a case, however, it is either necessary to explicitly redefine what is meant by a success or to define the random variable X_B^*, say, to be the number of

TABLE A1

Binomial Probabilities (Individual Terms)

$$P[X_B = x \mid n, p] = \binom{n}{x} p^x q^{n-x}$$

		p									
n	*x*	.05	.10	.15	.20	.25	.30	.35	.40	.45	.50
1	0	.9500	.9000	.8500	.8000	.7500	.7000	.6500	.6000	.5500	.5000
	1	.0500	.1000	.1500	.2000	.2500	.3000	.3500	.4000	.4500	.5000
2	0	.9025	.8100	.7225	.6400	.5625	.4900	.4225	.3600	.3025	.2500
	1	.0950	.1800	.2550	.3200	.3750	.4200	.4550	.4800	.4950	.5000
	2	.0025	.0100	.0225	.0400	.0625	.0900	.1225	.1600	.2025	.2500
3	0	.8574	.7290	.6141	.5120	.4219	.3430	.2746	.2160	.1664	.1250
	1	.1354	.2430	.3251	.3840	.4219	.4410	.4436	.4320	.4084	.3750
	2	.0071	.0270	.0574	.0960	.1406	.1880	.3341	.3750	.3341	
	3	.0001	.0010	.0034	.0080	.0156	.0270	.0429	.0640	.0911	.1250
4	0	.8145	.6561	.5220	.4096	.3164	.2401	.1785	.1296	.0915	.0625
	1	.1715	.2916	.3685	.4096	.4219	.4116	.3845	.3456	.2995	.2500
	2	.0135	.0486	.0975	.1536	,2109	.2646	.3105	.3456	.3675	.3750
	3	.0005	.0036	.0115	.0256	.0469	.0756	.1115	.1536	.2005	.2500
	4	.0000	.0001	.0005	.0016	.0039	.0081	.0150	.0256	.0410	.0625

FIGURE A1
Portion of Table A1

"failures" in n trials with $p^* = 1 - p$ equal to the probability of a failure on a single trial. If there are x "successes," there must be $n - x$ "failures"; consequently,

$$P[X_B = x \mid n, p] = P[X_B^* = n - x \mid n, p^*].$$

For example, if you wish to find $P[X_B = 1 \mid n = 4, p = .55]$, instead find

$$P[X_B^* = 3 \mid n = 4, p^* = .45] = .2005.$$

The key values are circled and highlighted in Figure A1.

Table A2. Binomial Cumulative Distribution Function These tables give

$$F_{X_B}(x) = P[X_B \leq x \mid n, p] = \sum_{k=0}^{x} \binom{n}{k} p^k q^{n-k},$$

where X_B represents a binomial random variable having parameters n and p. Tabled values are for $x = 0, 1, \ldots, n - 1$, with $n = 2, 3, \ldots, 20$, $p = .05, .10, .15, \ldots, .50$.

If you wish to find $P[X \leq 2 \mid n = 4, p = .20]$, first locate $n = 4$ in the leftmost column. Next find the value $x = 2$ and read across to the column $p = .20$. The result is .9728, as you can see highlighted in Figure A2. All the key values have been circled.

The tables can also be used when the probability of a "success" is greater than .50. In this case we can express the probability in terms of the number of "failures" rather than the number of "successes." If we consider

$$0, 1, 2, \ldots, x \text{ "successes" in } n \text{ trials},$$

this is exactly the same as

$$n, n-1, n-2, \ldots, n-x \text{ "failures" in } n \text{ trials.}$$

Consequently,

$$P[X_B \leq x \mid n, p] = P[X_B^* \geq n - x \mid n, p^*],$$

where $p^* = 1 - p$ is the probability of a failure. Using the complement of the right-hand side, we have

$$P[X_B \leq x \mid n, p] = 1 - P[X_B^* \leq n - x - 1 \mid n, p^*].$$

For example, if you wish to find $P[X_B \leq 3 \mid n = 5, p = .6]$, instead find

$$1 - P[X_B^* \leq 1 \mid n = 5, p^* = .40] = 1 - .3370 = .6630.$$

The key values are circled and highlighted in Figure A2.

TABLE A2

Binomial Cumulative Distribution Function

$$F_{X_B}(x) = \sum_{k=0}^{x} \binom{n}{k} p^k (1-p)^{n-k}$$

n	x	.05	.10	.15	.20	.25	.30	.35	.40	.45	.50
2	0	.9025	.8100	.7225	.6400	.5625	.4900	.4225	.3600	.3025	.2500
	1	.9975	.9900	.9775	.9600	.9375	.9100	.8775	.8400	.7975	.7500
3	0	.8574	.7290	.6141	.5120	.4219	.3430	.2746	.2160	.1664	.1250
	1	.9928	.9720	.9392	.8060	.8438	.7840	.7182	.6480	.5748	.5000
	2	.9999	.9990	.9966	.9920	.0844	.9730	.9571	.9360	.9089	.8750
4	0	.8145	.6561	.5220	.4096	.3164	.2401	.1785	.1296	.0915	.0625
	1	.9860	.9477	.8905	.8192	.7383	.6517	.5630	.4752	.3910	.3125
	2	.9995	.9963	.9880	.9728	.9492	.9163	.8735	.8208	.7585	.6875
	3	.0000	.9999	.9995	.9984	.9961	.9919	.9850	.9744	.9590	.9375
5	0	.7738	.5905	.4437	.3277	.2373	.1681	.1160	.0776	.0503	.0342
	1	.9774	.9185	.8352	.7373	.6328	.5282	.4284	.3370	.2562	.1875
	2	.9988	.9914	.9734	.9421	.8965	.8369	.7648	.6826	.5931	.5000
	3	.0000	.9995	.9978	.9933	.9844	.9692	.9460	.9130	.8688	.8125
	4	.0000	.0000	.9999	.9997	.9990	.9976	.9947	.9898	.9815	.9688

FIGURE A2
Portion of Table A2

Table A3. Poisson Probabilities (Individual Terms) These tables give

$$P[X_P = x \mid \lambda t] = \frac{e^{-\lambda t}(\lambda t)^x}{x!},$$

where X_P represents a Poisson random variable having parameter λt. (If λ and t are given individually, their product will equal the parameter value.) Tabled values are for $\lambda t = 0.1, 0.2, 0.3, \ldots, 10.0$ and $11, 12, \ldots, 20$. The values of x tabled are 0, 1, 2, and so on, the last value corresponding to a probability of .0001 or less.

If you wish to find $P[X_P = 4 \mid \lambda t = 0.8]$, first locate the column headed 0.8, then read down that column to the row corresponding to $x = 4$. The result is .0077, as you can see highlighted in Figure A3. All the key values have been circled.

TABLE A3

Poisson Probabilities (Individual Terms)

$$P[X_p = x \mid \lambda t] = e^{-\lambda t}(\lambda t)^x / x!$$

x	.1	.2	.3	.4	.5	.6	.7	.8	.9	1.0
0	.9048	.8187	.7408	.6703	.6065	.5488	.4966	.4493	.4066	.3679
1	.0905	.1637	.2222	.2681	.3033	.3293	.3476	.3595	.3659	.3670
2	.0045	.0164	.0333	.0536	.0758	.0988	.1217	.1438	.1647	.1839
3	.0002	.0011	.0033	.0072	.0126	.0198	.0284	.0353	.0404	.0613
4	.0000	.0001	.0002	.0007	.0016	.0030	.0050	.0077	.0111	.0153
5	.0000	.0000	.0000	.0001	.0002	.0004	.0007	.0012	.0020	.0031
6	.0000	.0000	.0000	.0000	.0000	.0000	.0001	.0002	.0003	.0005
7	.0000	.0000	.0000	.0000	.0000	.0000	.0000	.0000	.0000	.0001

FIGURE A3
Portion of Table A3

Table A4. The Standard Normal Cumulative Distribution Function These tables give

$$F(z) = P[Z \leq z],$$

where Z represents a standard normal random variable. Tabled values are for $z =$ 0.00, 0.01, 0.02, . . . , 3.99.

If you wish to find $P[Z \leq .05]$, locate the column headed z and read down the column to .05. To the right of the number .05 read .5199, which is $F(.05)$. The starting column has been highlighted and the answer circled in Figure A4. To find

TABLE A4

The Standard Normal Cumulative Distribution Function

$$F(z) = P[Z \leq z]$$

z	F(z)	z	F(z)	z	F(z)	z	F(z)
.00	.5000						
.01	.5040	.51	.6950	1.01	.8438	1.51	.9345
.02	.5080	.52	.6985	1.02	.8461	1.52	.9357
.03	.5120	.53	.7019	1.03	.8485	1.53	.9370
.04	.5160	.54	.7054	1.04	.8508	1.54	.9382
.05	.5199	.55	.7088	1.05	.8531	1.55	.9394

FIGURE A4
Portion of Table A4

$F(z)$ for values $z < 0$, use the fact that the distribution is symmetric about the value zero. For example, $F(-1) = 1 - F(1) = .1587$.

The tables can also be used to find percentiles of the standard normal distribution. The 85th percentile, for example, is the value $z_{.85}$ for which $F(z_{.85}) = .85$. To find this value, first locate the column headed $F(z)$ and read down this column to the value .85. When you try this you will not find exactly the value .8500, but you will find two values close to this, .8485 and .8508. Since the latter value is closer, we could read the number to the left of this, 1.04, as an approximate answer. The starting column has been highlighted and the approximate answer circled in Figure A4. A slightly better answer can be found by interpolating between 1.03 and 1.04 to get 1.037 as an approximation to the 85th percentile. By the symmetry of the standard normal distribution, we can say that the 15th percentile is approximately -1.037. For convenience some of the commonly used percentiles are listed at the end of this table.

Table A5. Random Numbers This table consists of 2500 digits (50 rows by 50 columns) consisting of integers 0, 1, . . . , 9 in a "random" ordering. In a large sampling experiment you would probably use a larger table consisting of many pages of such digits or random digits as generated by a computer. In using such tables it is important to choose a random starting point, since always starting in the same place (e.g., the upper left corner) will always lead to the same set of digits. One way of doing this is to place a finger "randomly" on the page to find a column between 1 and 50, then repeat this to get a row between 1 and 50. For example, if we got column 21 and row 5, the starting point would be determined by this value. Numbers could then be read from left to right (as we will do), from right to left, from top to bottom, or in any systematic way that you choose (see Figure A5).

To choose a random sample from a population of size N, the elements of the population should be numbered from 1 to N. If $N = 100$ and a random sample of size 13 were desired, then using the random starting point we would take two-digit numbers to get 64, 03, 23, . . . , 17, 17. The elements corresponding to these numbers would constitute the sample. If the sample was to be taken with replacement, the

TABLE A5

Random Numbers

			column 21		
	10 48 01 50 11	01 53 60 20 11	81 64 79 16 46	69 17 91 41 94	62 59 03 62 07
	22 36 84 65 73	25 59 58 53 93	30 99 58 91 98	27 98 25 34 02	93 96 53 40 95
	24 13 04 83 60	22 52 79 72 65	76 39 36 48 09	15 17 92 48 30	49 34 03 20 81
	42 16 79 30 93	06 24 36 16 80	07 85 61 63 76	39 44 05 35 37	71 34 15 70 04
row 5	37 57 03 99 75	81 83 71 66 56	64 03 23 66 53	98 95 11 68 77	12 17 17 68 33
	77 92 10 69 07	11 00 84 27 51	27 75 65 34 98	18 60 27 06 59	90 65 51 50 53
	99 56 27 29 05	56 42 06 99 94	98 87 23 10 16	71 19 41 87 38	44 01 34 88 40

FIGURE A5
Portion of Table A5

repeated value of "17" would be allowed. If it was to be taken without replacement, repeated values would be ignored. In this example the first 12 numbers do not repeat, but the next two numbers in the list, 17 and 68, duplicate numbers chosen earlier, so these would be ignored and the number 33 would be the last one chosen. Note that we would take "00" to correspond to 100.

If $N = 1000$, three-digit numbers would be chosen. Using the random starting point, the first members of the sample would be 640, 323, 665, etc. If $N = 50$, we could choose two-digit numbers and ignore all those beyond 50. Hence we would ignore the numbers 64, 66, 53, and so on, but would include in the sample 03, 23, 11, and so on. A second alternative would be to divide each number by 50 and take the remainder to be the random number, since the remainders would all be equally likely. This would give a sample of 14, 03, 23, 16, 03, and so on. This avoids ignoring half of the digits and thus is more efficient. If $N < 100$, say, but does not divide evenly into 100, a similar procedure can be used. For example, if $N = 40$, we could consider all two-digit numbers between 1 and 80 (ignoring the others). For these numbers, if we divide by 40, all remainders are equally likely. (A remainder of zero would correspond to the number 40.) Using our starting point, the sample would be 24, 03, 23, 26, 13, and so on.

Table A6. Percentage Points of the Student's t Distribution This table gives percentiles (percentage points) corresponding to 60, 75, 90, 95, 97.5, 99, 99.5, 99.75, 99.9, and 99.95% for random variables having a Student's t distribution with $\nu = 1, 2, 3, \ldots, 30$ and 40, 60, 120, and ∞ degrees of freedom. If you wish to find the value $t_{4,.90}$, the 90th percentile of a Student's t distribution with 4 degrees of freedom, first locate the column headed .90 and read down the column to the row corresponding to $\nu = 4$. The value 1.533 is the desired value:

$$t_{4,.90} = 1.533.$$

The key values are circled and highlighted in Figure A6. Since as $\nu \to \infty$, the t distributions approach a standard normal, the last row of this table gives percentiles of the standard normal distribution. In each case, lower percentile values (40, 25, 10,

TABLE A6

Percentage Points of the Student's t Distribution

	$F(t)$									
ν	.60	.75	.90	.95	.975	.99	.995	.9975	.999	.9995
1	.325	1.000	3.078	6.314	12.706	31.821	63.657	127.32	318.31	636.62
2	.289	.816	1.886	2.920	4.303	6.965	9.925	14.089	22.327	31.598
3	.277	.765	1.638	2.353	3.182	4.541	5.841	7.453	10.214	12.024
4	.271	.741	1.533	2.132	2.776	3.747	4.604	5.598	7.173	8.610
5	.267	.727	1.476	2.015	2.571	3.365	4.032	4.773	5.893	6.869

FIGURE A6
Portion of Table A6

etc.) can be found by using the fact that the distributions are symmetric about zero. For example,

$$t_{4,.10} = -1.533.$$

Table A7. Percentage Points of the Chi-square Distribution This table gives percentiles (percentage points) corresponding to 0.5, 1, 2.5, 5, 10, 25, 50, 75, 90, 95, 97.5, 99, 99.5, and 99.9% for random variables having a chi-square distribution with $\nu = 1, 2, 3, \ldots, 30$ and $40, 50, \ldots, 100$ degrees of freedom. If you wish to find the 5th percentile corresponding to 4 degrees of freedom, $\chi^2_{4,.05}$, first locate the column headed .05 and read down the column to the row corresponding to $\nu = 4$. In the table you will find

$$\chi^2_{4,.05} = .710723.$$

The key values are circled and highlighted in Figure A7. Since the chi-square distribution is skewed, it is necessary to table both low and high percentiles. You can see, for example, that

$$\chi^2_{4,.95} = 9.48773.$$

TABLE A7

Percentage Points of the χ^2 Distribution

ν	.005	.01	.025	$F(\chi^2)$ .05	.10	.25	.50
1	392704.10^{-10}	157088.10^{-9}	982069.10^{-9}	393214.10^{-8}	.0157908	.1015308	.454936
2	.0100251	.0201007	.0506356	.102587	.210721	.575364	1.28629
3	.0717218	.114832	.215795	.351846	.584374	1.212534	2.36597
4	.206989	.297109	.484419	.710723	1.063623	1.92256	3.35669

TABLE A7 *(continued)*

ν	.75	.90	.95	.975	.99	.995	.999
1	1.32330	2.70554	3.84146	5.02389	6.63490	7.87944	10.828
2	2.77259	4.60517	5.99146	7.37776	9.21034	10.5966	13.816
3	4.10834	6.25139	7.81473	9.34840	11.34449	12.8382	16.266
4	5.38527	7.77944	9.48773	11.1433	13.2767	14.8603	18.467

FIGURE A7
Portion of Table A7

Table A8. Percentage Points of the F Distribution This table gives percentiles (percentage points) corresponding to 90, 95, 97.5, and 99.5% for random variables having F distributions. The degrees of freedom for the numerator are $\nu_1 = 1, 2, \ldots, 10$ and 12, 15, 20, 24, 30, 40, 60, 120, and ∞. The degrees of freedom for the denominator are $\nu_2 = 1, 2, \ldots 30$, and 40, 60, 120, and ∞. If you wish to find the value $f_{3,4,.90}$, the 90th percentile of an F distribution with 3 degrees of freedom for the numerator and 4 for the denominator, first locate the page corresponding to .90. Next, locate the column headed "3" and read down that column to the row corresponding to $\nu_2 = 4$. In the table you will find

$$f_{3,4,.90} = 4.19.$$

The key values are circled and highlighted in Figure A8.

These tables can also be used to find percentiles corresponding to 10, 5, 2.5, and .5% by using the relationship

$$f_{\nu_1,\nu_2,\alpha} = \frac{1}{f_{\nu_2,\nu_1,1-\alpha}}.$$

For example,

$$f_{6,5,.10} = \frac{1}{f_{5,6,.90}} = \frac{1}{3.11} = 0.322.$$

The key values here are also circled and highlighted.

TABLE A8

Percentage Points of the F Distribution

F(f)=.90

ν_2 \ ν_1	1	2	3	4	5	6	7	8	9	10	12	15	20
1	39.86	49.50	53.59	55.83	57.24	58.20	58.91	59.44	59.86	60.19	60.71	61.22	61.74
2	8.53	9.00	9.16	9.24	9.29	9.33	9.35	9.37	9.38	9.39	9.41	9.42	9.44
3	5.54	5.46	5.39	5.34	5.31	5.28	5.27	5.25	5.24	5.23	5.22	5.20	5.18
4	4.54	4.32	4.19	4.11	4.05	4.01	3.98	3.95	3.94	3.92	2.90	3.87	3.84
5	4.06	3.78	3.62	3.52	3.40	3.40	3.37	3.34	3.32	3.30	3.27	3.24	3.21
6	3.78	3.46	3.29	3.18	3.11	3.05	3.01	2.98	2.96	2.94	2.90	2.87	2.84

FIGURE A8
Portion of Table A8

Table A9. Fisher's Transformation of r to z' This table gives the value

$$z' = \tfrac{1}{2} \ln\left(\frac{1+r}{1-r}\right)$$

for values of $r = .000, .001, .002, \ldots, .999$. For example, if you wish to find the value of z' corresponding to $r = .152$, first locate the value .15 in the left-hand column labeled "r." Next read across this row to the column headed ".002," since the third decimal place is a "2." The corresponding value is .1532, the value of z'. The key values are circled and highlighted in Figure A9. Since the transformation is symmetric with respect to the origin, the value z' corresponding to $r = -.152$, for example, would be $-.1532$, the negative of the z' value corresponding to $r = +.152$.

These tables can also be used for the inverse transformation. That is, if a value of z' is given, the corresponding value of r can be found. For example, if $z' = .1755$, try to locate this value in the body of the table. If it cannot be found, you may locate

TABLE A9

Fisher's Transformation of r to $z' = \frac{1}{2} \ln [(1 + r)/(1 - r)]$

r	r (third decimal) .000	.001	.002	.003	.004	.005	.006	.007	.008	.009
.00	.0000	.0010	.0020	.0030	.0040	.0050	.0060	.0070	.0080	.0090
.01	.0100	.0110	.0120	.0130	.0140	.0150	.0160	.0170	.0180	.0190
.02	.0200	.0210	.0220	.0230	.0240	.0250	.0260	.0270	.0280	.0290
.03	.0300	.0310	.0320	.0330	.0340	.0350	.0360	.0370	.0380	.0390
.04	.0400	.0410	.0420	.0430	.0440	.0450	.0460	.0470	.0480	.0490
.15	.1511	.1522	.1532	.1542	.1552	.1563.	1573	.1583	.1593	.1604
.16	.1614	.1624	.1634	.1645	.1655	.1665	.1676	.1686	.1696	.1706
.17	.1717	.1727	.1737	.1748	.1758	.1768	.1779	.1789	.1799	.1809
.18	.1820	.1830	.1841	.1851	.1861	.1872	.1882	.1892	.1903	.1913
.19	.1923	.1934	.1944	.1955	.1965	.1976	.1986	.1996	.20007	.2017

FIGURE A9
Portion of Table A9

the value closest to it to get an approximate answer. In this case the closest value is .1758. The corresponding value of r is found by first reading left across the row to the "r" column to get the first two decimal places of r, namely, .17. Reading above the number .1758 to the first row, you can find the third decimal place to be .004. Consequently, the approximate inverse value would be $r = .174$ (see Figure A9). Using interpolation, the approximate answer would be $r = .1737$.

Table A10. Confidence Bands for Correlation ρ These charts allow confidence intervals for the correlation ρ for a bivariate normal distribution to be found for confidence levels of 90, 95, and 99%. The bands on the charts correspond to various sample sizes n, although for other sample sizes you can use interpolation. If you wish to find a 90% confidence interval for ρ when $n = 8$ and $r = .20$, proceed as follows. First find the chart corresponding to 90% (the first chart). Next locate the value $r = .20$ on the lower horizontal scale. Trace a line from this point upward until it intersects the first curve labeled "$n = 8$." Move from this point horizontally along a straight line until the line intersects the vertical scale. The value at the intersection, $-.45$, is the lower limit of the confidence interval. Now continue to trace the line from $r = .20$ upward until it intersects the second curve labeled "$n = 8$." Moving horizontally from this point to the vertical scale, find the value at the intersection. This value, approximately $+.69$, is the upper limit of the confidence interval. Consequently, a 90% confidence interval is about $(-.45, +.69)$ when $n = 8$ and $r = .20$. The key values are circled and shaded in Figure A10.

Table A11. Critical Values of the Wilcoxon Signed Rank Statistic This table gives critical values for the Wilcoxon signed rank test for sample sizes $n = 4, 5, \ldots, 15$. It can be used for either one-sided or two-sided alternatives. The tabled values are critical values corresponding to a particular sample size and level of significance. For

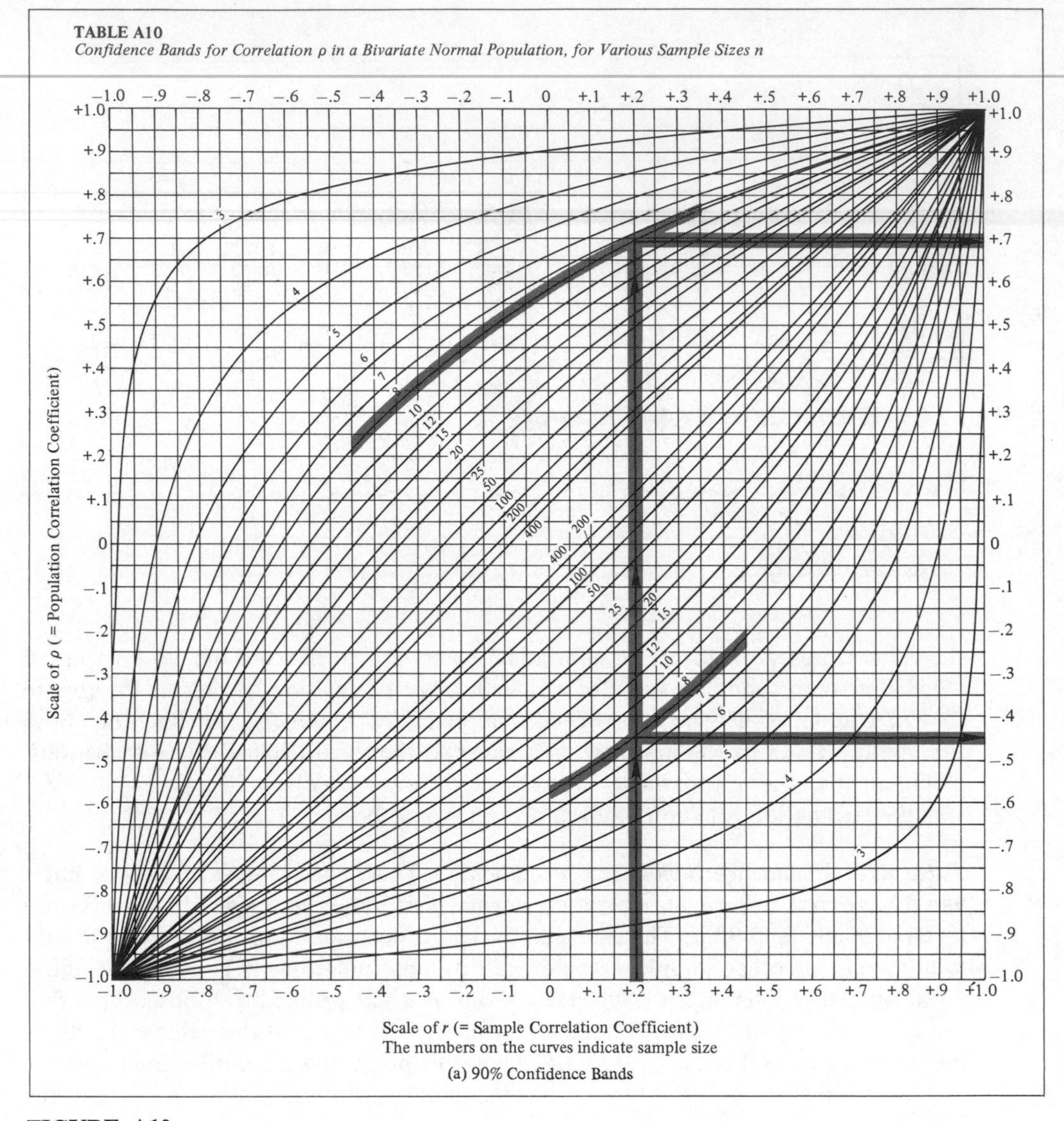

FIGURE A10

Table A10, Showing Key Values

a *two-sided* alternative when $n = 7$ and $\alpha = .05$, for example, find the probabilities $p = \alpha/2 = .025$ and $p = 1 - \alpha/2 = .975$ along the top of the table and read down these columns to the row $n = 7$. The critical values are 2 and 26. Hence reject H_0 if $T^+ \leq 2$ or if $T^+ \geq 26$ (see Figure A11). For a one-sided alternative where H_0 will be rejected if T^+ is "too small," the critical value corresponding to $n = 5$ and $\alpha = .10$ would be 2. That is, reject H_0 if T^+ is less than or equal to 2. Note that for small sample sizes and small α (e.g., $n = 5$, $\alpha = .01$), there is a dash in the body of the table. This means that when the null hypothesis is true, the most extreme values of

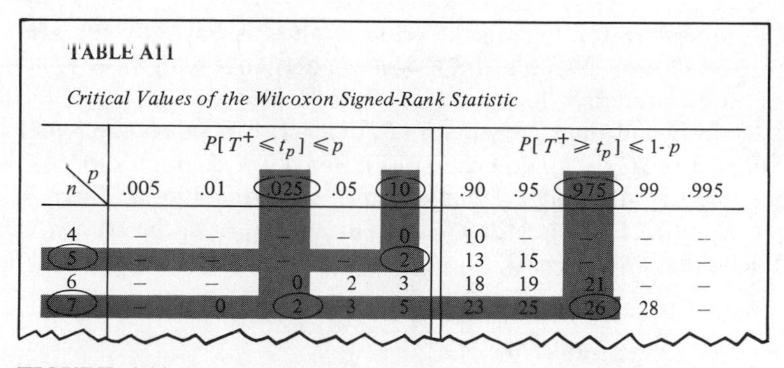

TABLE A11

Critical Values of the Wilcoxon Signed-Rank Statistic

	$P[T^+ \leqslant t_p] \leqslant p$					$P[T^+ \geqslant t_p] \leqslant 1-p$				
n \ p	.005	.01	.025	.05	.10	.90	.95	.975	.99	.995
4	–	–	–	–	0	10	–	–	–	–
5	–	–	–	–	2	13	15	–	–	–
6	–	–	0	2	3	18	19	21	–	–
7	–	0	2	3	5	23	25	26	28	–

FIGURE A11
Portion of Table A11

T^+ occur by chance with probability greater than α. Consequently, an ordinary critical value cannot be found for this level of significance.

Table A12. Critical Values of the Wilcoxon Rank Sum Statistic In this table n_1 corresponds to the smaller sample size. Critical values are given for sample sizes $n_1 \leq n_2$ for $n_2 = 3, 4, \ldots, 10$. In some cases, if n_1 is small, no critical region for $\alpha \leq .10$ can be found, so these values were omitted from the tables. For a *two-sided* alternative with $n_1 = 2$, $n_2 = 5$, and $\alpha = .10$, for example, find the probabilities $p = \alpha/2 = .05$ and $p = 1 - \alpha/2 = .95$ along the top of the table. Read down these columns to the values $n_1 = 2$ and $n_2 = 5$. The values 3 and 13 are the critical values. Reject H_0 if $T_1 \leq 3$ or $T_1 \geq 13$ (see Figure A12). For one-sided alternatives read down the column corresponding to either $p = \alpha$ or $p = 1 - \alpha$ as required.

Table A13. Critical Values of the Total Number of Runs Statistic The tables give critical values of the total number of runs statistic for selected values of n_1 and n_2.

TABLE A12

Critical Values of the Wilcoxon Rank Sum Statistic

		$P[T_1 \leqslant t_p] \leqslant p$					$P[T_1 \geqslant t_p] \leqslant 1-p$				
n_1	n_2 \ p	.005	.01	.025	.05	.10	.90	.95	.975	.99	.995
1	9	–	–	–	–	1	10	–	–	–	–
1	10	–	–	–	–	1	11	–	–	–	–
2	3	–	–	–	–	3	9	–	–	–	–
2	4	–	–	–	–	3	11	–	–	–	–
2	5	–	–	–	3	4	12	13	–	–	–
2	6	–	–	–	3	4	14	15	–	–	–

FIGURE A12
Portion of Table A12

The procedure for finding the critical values is basically the same as that used in Table A12. For example, if a *two-sided* alternative with $n_1 = 6$, $n_2 = 6$, and $\alpha = .05$ is being considered, locate the columns headed with $p = \alpha/2 = .025$ and $p = 1 - \alpha/2 = .975$ and the row with "6 6." The critical values are 3 and 11 (i.e., reject H_0 if $R \leq 3$ or $R \geq 11$; see Figure A13). For values of n_1 or n_2 *not in the table*, it may be possible to "interpolate." For example, the critical value for $p = .025$ and $n_1 = 6$, $n_2 = 6$ is "3." Likewise, for $p = .025$, $n_1 = 6$, $n_2 = 8$, the critical value is also "3." It follows that when $p = .025$, $n_1 = 6$, and $n_2 = 7$, the critical value will once again be "3."

TABLE A13

Critical values of the Total Number of Runs Statistic

	p	$P[R \leq r_p] \leq p$					$P[R \geq r_p] \leq 1-p$				
n_1	n_2	005	.01	.025	.05	.10	.90	.95	.975	.99	.995
4	17	3	3	4	4	5	–	–	–	–	–
4	20	3	3	4	4	5	–	–	–	–	–
6	6	2	2	3	3	4	10	11	11	12	12
6	8	3	3	3	4	5	11	12	12	13	13
6	10	3	3	4	5	5	12	12	13	–	–

FIGURE A13
Portion of Table A13

Table A14. Critical Values of the Kolmogorov–Smirnov Statistic This table gives critical values for the Kolmogorov–Smirnov statistic for samples of size $n = 1, 2, \ldots, 40$ and for values of $\alpha = 1 - p = .01, .02, .05, .10$, and .20. For example, if $n = 4$ and $\alpha = .10$, locate the value $p = 1 - \alpha = .90$ across the top and read down to the row corresponding to $n = 4$. The critical value is .565 and the null hypothesis would be rejected if the observed value of D is greater than or equal to .565 (see Figure A14).

TABLE A14

Critical Values of the Kolmogorov–Smirnov Statistic

$P[D \geq d_p] \leq 1-p$

	p				
n	.80	.90	.95	.98	.99
1	.900	.950	.975	.990	.995
2	.684	.776	.842	.900	.929
3	.565	.636	.708	.785	.829
4	.493	.565	.624	.689	.734
5	.447	.509	.563	.627	.669

FIGURE A14
Portion of Table A14

Table A15. Critical Values of the Kruskal–Wallis Statistic This table gives critical values of the Kruskal–Wallis statistic for certain values of $n_1 \leq n_2 \leq n_3$ and for values of $\alpha = 1 - p = .01, .05$, and .10. For example, if $n_1 = 1, n_2 = 3, n_3 = 3$, and $\alpha = .05$, first locate the triple "1 3 3" under the columns headed "n_1 n_2 n_3," then read across this row to the column headed $p = 1 - \alpha = .95$ to find the critical value 5.1429. The null hypothesis would be rejected if the observed values of H is greater than or equal to 5.1429 (see Figure A15).

TABLE A15

Critical Values of the Kruskal-Wallis Statistic

$P[H \geqslant h_p] \leqslant 1 - p$

n_1	n_2	n_3	.90	p .95	.99
1	2	3	4.2857	–	–
1	2	4	4.5000	–	–
1	2	5	4.2000	5.0000	–
1	3	3	4.5714	5.1429	–
1	3	4	4.0556	5.2083	–

FIGURE A15
Portion of Table A15

Table A16. Critical Values of the Friedman Statistic This table gives critical values of the Friedman statistic when comparing k populations ($k = 3$ or 4) by using b blocks and values of $\alpha = 1 - p = .10, .05, .025, .01$, and .005. For example, to find the critical value when $k = 3$ and $b = 4$ with $\alpha = .05$, first locate the pair "3 4" under the columns headed "k b," then read across this row to the column headed $p = 1 - \alpha = .95$ to find the critical value 6.500. The null hypothesis would be rejected if the observed value of Q is greater than or equal to 6.500 (see Figure A16).

TABLE A16

Critical Values of the Friedman Statistic

$P[Q \geqslant q_p] \leqslant 1 - p$

k	b	.90	.95	p .975	.99	.995
3	3	6.000	6.000	–	–	–
3	4	6.000	6.5000	8.000	8.000	8.000
3	5	5.200	6.400	7.600	8.400	10.000

FIGURE A16
Portion of Table A16

Table A17. Critical Values of Spearman's Rank Correlation Statistic This table gives critical values of Spearman's rank correlation statistic for sample sizes $n = 4, 5, \ldots, 30$ and values of $\alpha = .001, .005, .01, .025, .05$, and $.10$ (for one-sided alternatives). Since the distribution of R_S is symmetric about zero, percentiles corresponding to p and $1 - p$ are negatives of each other. Consequently, if $n = 6$ and $\alpha = .05$, for example, with a two-sided alternative, the critical values would be found by locating the column $p = 1 - \alpha/2 = .975$ and the row $n = 6$ to find the value .886 (see Figure A17). The null hypothesis would be rejected if the observed value of R_S is greater than or equal to $+.886$ or less than or equal to $-.886$.

TABLE A17

Critical Values of Spearman's Rank Correlation Statistic

$P[R_S \geq r_p] \leq 1 - p$

	p					
n	.90	.95	.975	.99	.995	.999
4	1.000	1.000	–	–	–	–
5	.800	.900	1.000	1.000	–	–
6	.657	.829	.886	.943	1.000	–
7	.571	.714	.786	.893	.929	1.000

FIGURE A17
Portion of Table A17

Table A18. Squares and Square Roots This table gives the square, the square root, and the square root of 10 times n for $n = 1.0, 1.1, \ldots, 9.9$. The squares and square roots should require no further explanation. To find the square root of 14, for example, note that 14 equals 10 times 1.4, so the $\sqrt{10n}$ column can be used to find $\sqrt{14} = 3.742$ (see Figure A18). To find the square root of 140, for example, write $140 = 1.4 \times 10^2$, so

$$\sqrt{140} = \sqrt{1.4 \times 10^2} = \sqrt{1.4}\sqrt{10^2} = (1.183)(10) = 11.83.$$

TABLE A18

Square and Square Roots

n	n^2	$\sqrt{n}$	$\sqrt{10n}$	n	n^2	$\sqrt{n}$	$\sqrt{10n}$
1.0	1.00	1.000	3.162	5.5	30.25	2.345	7.416
1.1	1.21	1.049	3.317	5.6	31.36	2.366	7.483
1.2	1.44	1.095	3.464	5.7	32.49	2.387	7.550
1.3	1.69	1.140	3.606	5.8	33.64	2.408	7.616
1.4	1.96	1.183	3.742	5.9	34.81	2.429	7.681

FIGURE A18
Portion of Table A18

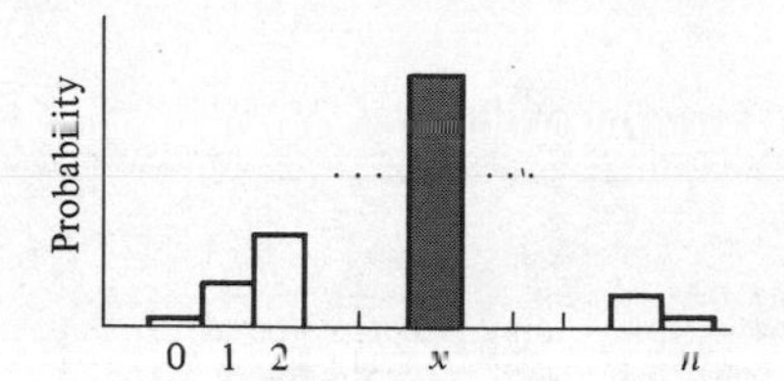

Tables give the probability corresponding to the shaded area.

TABLE A1 *Binomial Probabilities (Individual Terms)*

$$P[X_B = x \mid n, p] = \binom{n}{x} p^x q^{n-x}$$

n	x	p = .05	.10	.15	.20	.25	.30	.35	.40	.45	.50
1	0	.9500	.9000	.8500	.8000	.7500	.7000	.6500	.6000	.5500	.5000
	1	.0500	.1000	.1500	.2000	.2500	.3000	.3500	.4000	.4500	.5000
2	0	.9025	.8100	.7225	.6400	.5625	.4900	.4225	.3600	.3025	.2500
	1	.0950	.1800	.2550	.3200	.3750	.4200	.4550	.4800	.4950	.5000
	2	.0025	.0100	.0225	.0400	.0625	.0900	.1225	.1600	.2025	.2500
3	0	.8574	.7290	.6141	.5120	.4219	.3430	.2746	.2160	.1664	.1250
	1	.1354	.2430	.3251	.3840	.4219	.4410	.4436	.4320	.4084	.3750
	2	.0071	.0270	.0574	.0960	.1406	.1890	.2389	.2880	.3341	.3750
	3	.0001	.0010	.0034	.0080	.0156	.0270	.0429	.0640	.0911	.1250
4	0	.8145	.6561	.5220	.4096	.3164	.2401	.1785	.1296	.0915	.0625
	1	.1715	.2916	.3685	.4096	.4219	.4116	.3845	.3456	.2995	.2500
	2	.0135	.0486	.0975	.1536	.2109	.2646	.3105	.3456	.3675	.3750
	3	.0005	.0036	.0115	.0256	.0469	.0756	.1115	.1536	.2005	.2500
	4	.0000	.0001	.0005	.0016	.0039	.0081	.0150	.0256	.0410	.0625
5	0	.7738	.5905	.4437	.3277	.2373	.1681	.1160	.0778	.0503	.0312
	1	.2036	.3280	.3915	.4096	.3955	.3602	.3124	.2592	.2059	.1562
	2	.0214	.0729	.1382	.2048	.2637	.3087	.3364	.3456	.3369	.3125
	3	.0011	.0081	.0244	.0512	.0879	.1323	.1811	.2304	.2757	.3125
	4	.0000	.0004	.0022	.0064	.0146	.0284	.0488	.0768	.1128	.1562
	5	.0000	.0000	.0001	.0003	.0010	.0024	.0053	.0102	.0185	.0312
6	0	.7351	.5314	.3771	.2621	.1780	.1176	.0754	.0467	.0277	.0156
	1	.2321	.3543	.3993	.3932	.3560	.3025	.2437	.1866	.1359	.0938
	2	.0305	.0984	.1762	.2458	.2966	.3241	.3280	.3110	.2780	.2344
	3	.0021	.0146	.0415	.0819	.1318	.1852	.2355	.2765	.3032	.3125
	4	.0001	.0012	.0055	.0154	.0330	.0595	.0951	.1382	.1861	.2344
	5	.0000	.0001	.0004	.0015	.0044	.0102	.0205	.0369	.0609	.0938
	6	.0000	.0000	.0000	.0001	.0002	.0007	.0018	.0041	.0083	.0156
7	0	.6983	.4783	.3206	.2097	.1335	.0824	.0490	.0280	.0152	.0078
	1	.2573	.3720	.3960	.3670	.3115	.2471	.1848	.1306	.0872	.0547
	2	.0406	.1240	.2097	.2753	.3115	.3177	.2985	.2613	.2140	.1641
	3	.0036	.0230	.0617	.1147	.1730	.2269	.2679	.2903	.2918	.2734
	4	.0002	.0026	.0109	.0287	.0577	.0972	.1442	.1935	.2388	.2734
	5	.0000	.0002	.0012	.0043	.0115	.0250	.0466	.0774	.1172	.1641
	6	.0000	.0000	.0001	.0004	.0013	.0036	.0084	.0172	.0320	.0547
	7	.0000	.0000	.0000	.0000	.0001	.0002	.0006	.0016	.0037	.0078
8	0	.6634	.4305	.2725	.1678	.1001	.0576	.0319	.0168	.0084	.0039
	1	.2793	.3826	.3847	.3355	.2670	.1977	.1373	.0896	.0548	.0312
	2	.0515	.1488	.2376	.2936	.3115	.2965	.2587	.2000	.1569	.1094
	3	.0054	.0331	.0839	.1468	.2076	.2541	.2786	.2787	.2568	.2188
	4	.0004	.0046	.0185	.0459	.0865	.1361	.1875	.2322	.2627	.2734
	5	.0000	.0004	.0026	.0092	.0231	.0467	.0808	.1239	.1719	.2188
	6	.0000	.0000	.0002	.0011	.0038	.0100	.0217	.0413	.0703	.1094
	7	.0000	.0000	.0000	.0001	.0004	.0012	.0033	.0079	.0164	.0312
	8	.0000	.0000	.0000	.0000	.0000	.0001	.0002	.0007	.0017	.0039

TABLE A1 *Binomial Probabilities (Individual Terms)* (continued)

n	x	p = .05	.10	.15	.20	.25	.30	.35	.40	.45	.50
9	0	.6302	.3874	.2316	.1342	.0751	.0404	.0207	.0101	.0046	.0020
	1	.2985	.3874	.3679	.3020	.2253	.1556	.1004	.0005	.0339	.0176
	2	.0629	.1722	.2597	.3020	.3003	.2668	.2162	.1612	.1110	.0703
	3	.0077	.0446	.1069	.1762	.2336	.2668	.2716	.2508	.2119	.1641
	4	.0006	.0074	.0283	.0661	.1168	.1715	.2194	.2508	.2600	.2461
	5	.0000	.0008	.0050	.0165	.0389	.0735	.1181	.1672	.2128	.2461
	6	.0000	.0001	.0006	.0028	.0087	.0210	.0424	.0743	.1160	.1641
	7	.0000	.0000	.0000	.0003	.0012	.0039	.0098	.0212	.0407	.0703
	8	.0000	.0000	.0000	.0000	.0001	.0004	.0013	.0035	.0083	.0176
	9	.0000	.0000	.0000	.0000	.0000	.0000	.0001	.0003	.0008	.0020
10	0	.5987	.3487	.1969	.1074	.0563	.0282	.0135	.0060	.0025	.0010
	1	.3151	.3874	.3474	.2684	.1877	.1211	.0725	.0403	.0207	.0098
	2	.0746	.1937	.2759	.3020	.2816	.2335	.1757	.1209	.0763	.0439
	3	.0105	.0574	.1298	.2013	.2503	.2668	.2522	.2150	.1665	.1172
	4	.0010	.0112	.0401	.0881	.1460	.2001	.2377	.2508	.2384	.2051
	5	.0001	.0015	.0085	.0264	.0584	.1029	.1536	.2007	.2340	.2461
	6	.0000	.0001	.0012	.0055	.0162	.0368	.0689	.1115	.1596	.2051
	7	.0000	.0000	.0001	.0008	.0031	.0090	.0212	.0425	.0746	.1172
	8	.0000	.0000	.0000	.0001	.0004	.0014	.0043	.0106	.0229	.0439
	9	.0000	.0000	.0000	.0000	.0000	.0001	.0005	.0016	.0042	.0098
	10	.0000	.0000	.0000	.0000	.0000	.0000	.0000	.0001	.0003	.0010
11	0	.5688	.3138	.1673	.0859	.0422	.0198	.0088	.0036	.0014	.0004
	1	.3293	.3835	.3248	.2362	.1549	.0932	.0518	.0266	.0125	.0055
	2	.0867	.2131	.2866	.2953	.2581	.1998	.1395	.0887	.0513	.0269
	3	.0137	.0710	.1517	.2215	.2581	.2568	.2254	.1774	.1259	.0806
	4	.0014	.0158	.0536	.1107	.1721	.2201	.2428	.2365	.2060	.1611
	5	.0001	.0025	.0132	.0388	.0803	.1321	.1830	.2207	.2360	.2256
	6	.0000	.0003	.0023	.0097	.0268	.0566	.0985	.1471	.1931	.2256
	7	.0000	.0000	.0003	.0017	.0064	.0173	.0379	.0701	.1128	.1611
	8	.0000	.0000	.0000	.0002	.0011	.0037	.0102	.0234	.0462	.0806
	9	.0000	.0000	.0000	.0000	.0001	.0005	.0018	.0052	.0126	.0269
	10	.0000	.0000	.0000	.0000	.0000	.0000	.0002	.0007	.0021	.0054
	11	.0000	.0000	.0000	.0000	.0000	.0000	.0000	.0000	.0002	.0005
12	0	.5404	.2824	.1422	.0687	.0317	.0138	.0057	.0022	.0008	.0002
	1	.3413	.3766	.3012	.2062	.1267	.0712	.0368	.0174	.0075	.0029
	2	.0988	.2301	.2924	.2835	.2323	.1678	.1088	.0639	.0339	.0161
	3	.0173	.0852	.1720	.2362	.2581	.2397	.1954	.1419	.0923	.0537
	4	.0021	.0213	.0683	.1329	.1936	.2311	.2367	.2128	.1700	.1208
	5	.0002	.0038	.0193	.0532	.1032	.1585	.2039	.2270	.2225	.1934
	6	.0000	.0005	.0040	.0155	.0401	.0792	.1281	.1766	.2124	.2256
	7	.0000	.0000	.0006	.0033	.0115	.0291	.0591	.1009	.1489	.1934
	8	.0000	.0000	.0001	.0005	.0024	.0078	.0199	.0420	.0762	.1208
	9	.0000	.0000	.0000	.0001	.0004	.0015	.0048	.0125	.0277	.0537
	10	.0000	.0000	.0000	.0000	.0000	.0002	.0008	.0025	.0068	.0161
	11	.0000	.0000	.0000	.0000	.0000	.0000	.0001	.0003	.0010	.0029
	12	.0000	.0000	.0000	.0000	.0000	.0000	.0000	.0000	.0001	.0002

TABLE A1 *Binomial Probabilities (Individual Terms)* (continued)

n	x	p = .05	.10	.15	.20	.25	.30	.35	.40	.45	.50
13	0	.5133	.2542	.1209	.0550	.0238	.0097	.0037	.0013	.0004	.0001
	1	.3512	.3672	.2774	.1787	.1029	.0540	.0259	.0113	.0045	.0016
	2	.1109	.2448	.2937	.2680	.2059	.1388	.0836	.0453	.0220	.0095
	3	.0214	.0997	.1900	.2457	.2517	.2181	.1651	.1107	.0660	.0349
	4	.0028	.0277	.0838	.1535	.2097	.2337	.2222	.1845	.1350	.0873
	5	.0003	.0055	.0266	.0691	.1258	.1803	.2154	.2214	.1989	.1571
	6	.0000	.0008	.0063	.0230	.0559	.1030	.1546	.1968	.2169	.2095
	7	.0000	.0001	.0011	.0058	.0186	.0442	.0833	.1312	.1775	.2095
	8	.0000	.0000	.0001	.0011	.0047	.0142	.0336	.0656	.1089	.1571
	9	.0000	.0000	.0000	.0001	.0009	.0034	.0101	.0243	.0495	.0873
	10	.0000	.0000	.0000	.0000	.0001	.0006	.0022	.0065	.0162	.0349
	11	.0000	.0000	.0000	.0000	.0000	.0001	.0003	.0012	.0036	.0095
	12	.0000	.0000	.0000	.0000	.0000	.0000	.0000	.0001	.0005	.0016
	13	.0000	.0000	.0000	.0000	.0000	.0000	.0000	.0000	.0000	.0001
14	0	.4877	.2288	.1028	.0440	.0178	.0068	.0024	.0008	.0002	.0001
	1	.3593	.3559	.2539	.1539	.0832	.0407	.0181	.0073	.0027	.0009
	2	.1229	.2570	.2912	.2501	.1802	.1134	.0634	.0317	.0141	.0056
	3	.0259	.1142	.2056	.2501	.2402	.1943	.1366	.0845	.0462	.0222
	4	.0037	.0349	.0998	.1720	.2202	.2290	.2022	.1549	.1040	.0611
	5	.0004	.0078	.0352	.0860	.1468	.1963	.2178	.2066	.1701	.1222
	6	.0000	.0013	.0093	.0322	.0734	.1262	.1759	.2066	.2088	.1833
	7	.0000	.0002	.0019	.0092	.0280	.0618	.1082	.1574	.1952	.2095
	8	.0000	.0000	.0003	.0020	.0082	.0232	.0510	.0918	.1398	.1833
	9	.0000	.0000	.0000	.0003	.0018	.0066	.0183	.0408	.0762	.1222
	10	.0000	.0000	.0000	.0000	.0003	.0014	.0049	.0136	.0312	.0611
	11	.0000	.0000	.0000	.0000	.0000	.0002	.0010	.0033	.0093	.0222
	12	.0000	.0000	.0000	.0000	.0000	.0000	.0001	.0005	.0019	.0056
	13	.0000	.0000	.0000	.0000	.0000	.0000	.0000	.0001	.0002	.0009
	14	.0000	.0000	.0000	.0000	.0000	.0000	.0000	.0000	.0000	.0001
15	0	.4633	.2059	.0874	.0352	.0134	.0047	.0016	.0005	.0001	.0000
	1	.3658	.3432	.2312	.1319	.0668	.0305	.0126	.0047	.0016	.0005
	2	.1348	.2669	.2856	.2309	.1559	.0916	.0476	.0219	.0090	.0032
	3	.0307	.1285	.2184	.2501	.2252	.1700	.1110	.0634	.0318	.0139
	4	.0049	.0428	.1156	.1876	.2252	.2186	.1792	.1268	.0780	.0417
	5	.0006	.0105	.0449	.1032	.1651	.2061	.2123	.1859	.1404	.0916
	6	.0000	.0019	.0132	.0430	.0917	.1472	.1906	.2066	.1914	.1527
	7	.0000	.0003	.0030	.0138	.0393	.0811	.1319	.1771	.2013	.1964
	8	.0000	.0000	.0005	.0035	.0131	.0348	.0710	.1181	.1647	.1964
	9	.0000	.0000	.0001	.0007	.0034	.0116	.0298	.0612	.1048	.1527
	10	.0000	.0000	.0000	.0001	.0007	.0030	.0096	.0245	.0515	.0916
	11	.0000	.0000	.0000	.0000	.0001	.0006	.0024	.0074	.0191	.0417
	12	.0000	.0000	.0000	.0000	.0000	.0001	.0004	.0016	.0052	.0139
	13	.0000	.0000	.0000	.0000	.0000	.0000	.0001	.0003	.0010	.0032
	14	.0000	.0000	.0000	.0000	.0000	.0000	.0000	.0000	.0001	.0005
	15	.0000	.0000	.0000	.0000	.0000	.0000	.0000	.0000	.0000	.0000

TABLE A1 *Binomial Probabilities* (*Individual Terms*) (continued)

						p					
n	*x*	.05	.10	.15	.20	.25	.30	.35	.40	.45	.50
16	0	.4401	.1853	.0743	.0281	.0100	.0033	.0010	.0003	.0001	.0000
	1	.3706	.3294	.2097	.1126	.0535	.0228	.0087	.0030	.0009	.0002
	2	.1463	.2745	.2775	.2111	.1336	.0732	.0353	.0150	.0056	.0018
	3	.0359	.1423	.2285	.2463	.2079	.1465	.0888	.0468	.0215	.0085
	4	.0061	.0514	.1311	.2001	.2252	.2040	.1553	.1014	.0572	.0278
	5	.0008	.0137	.0555	.1201	.1802	.2099	.2008	.1623	.1123	.0667
	6	.0001	.0028	.0180	.0550	.1101	.1649	.1982	.1983	.1684	.1222
	7	.0000	.0004	.0045	.0197	.0524	.1010	.1524	.1889	.1969	.1746
	8	.0000	.0001	.0009	.0055	.0197	.0487	.0923	.1417	.1812	.1964
	9	.0000	.0000	.0001	.0012	.0058	.0185	.0442	.0840	.1318	.1746
	10	.0000	.0000	.0000	.0002	.0014	.0056	.0167	.0392	.0755	.1222
	11	.0000	.0000	.0000	.0000	.0002	.0013	.0049	.0142	.0337	.0667
	12	.0000	.0000	.0000	.0000	.0000	.0002	.0011	.0040	.0115	.0278
	13	.0000	.0000	.0000	.0000	.0000	.0000	.0002	.0008	.0029	.0085
	14	.0000	.0000	.0000	.0000	.0000	.0000	.0000	.0001	.0005	.0018
	15	.0000	.0000	.0000	.0000	.0000	.0000	.0000	.0000	.0001	.0002
	16	.0000	.0000	.0000	.0000	.0000	.0000	.0000	.0000	.0000	.0000
17	0	.4181	.1668	.0631	.0225	.0075	.0023	.0007	.0002	.0000	.0000
	1	.3741	.3150	.1893	.0957	.0426	.0169	.0060	.0019	.0005	.0001
	2	.1575	.2800	.2673	.1914	.1136	.0581	.0260	.0102	.0035	.0010
	3	.0415	.1556	.2359	.2393	.1893	.1245	.0701	.0341	.0144	.0052
	4	.9076	.0605	.1457	.2093	.2209	.1868	.1320	.0796	.0411	.0182
	5	.0010	.0175	.0668	.1361	.1914	.2081	.1849	.1379	.0875	.0472
	6	.0001	.0039	.0236	.0680	.1276	.1784	.1991	.1839	.1432	.0944
	7	.0000	.0007	.0065	.0267	.0668	.1201	.1685	.1927	.1841	.1484
	8	.0000	.0001	.0014	.0084	.0279	.0644	.1134	.1606	.1883	.1855
	9	.0000	.0000	.0003	.0021	.0093	.0276	.0611	.1070	.1540	.1855
	10	.0000	.0000	.0000	.0004	.0025	.0095	.0263	.0571	.1008	.1484
	11	.0000	.0000	.0000	.0001	.0005	.0026	.0090	.0242	.0525	.0944
	12	.0000	.0000	.0000	.0000	.0001	.0006	.0024	.0081	.0215	.0472
	13	.0000	.0000	.0000	.0000	.0000	.0001	.0005	.0021	.0068	.0182
	14	.0000	.0000	.0000	.0000	.0000	.0000	.0001	.0004	.0016	.0052
	15	.0000	.0000	.0000	.0000	.0000	.0000	.0000	.0001	.0003	.0010
	16	.0000	.0000	.0000	.0000	.0000	.0000	.0000	.0000	.0000	.0001
	17	.0000	.0000	.0000	.0000	.0000	.0000	.0000	.0000	.0000	.0000
18	0	.3972	.1501	.0536	.0180	.0056	.0016	.0004	.0001	.0000	.0000
	1	.3763	.3002	.1704	.0811	.0338	.0126	.0042	.0012	.0003	.0001
	2	.1683	.2835	.2556	.1723	.0958	.0458	.0190	.0069	.0022	.0006
	3	.0473	.1680	.2406	.2297	.1704	.1046	.0547	.0246	.0095	.0031
	4	.0093	.0700	.1592	.2153	.2130	.1681	.1104	.0614	.0291	.0117
	5	.0014	.0218	.0787	.1507	.1988	.2017	.1664	.1146	.0666	.0327
	6	.0002	.0052	.0301	.0816	.1436	.1873	.1941	.1655	.1181	.0708
	7	.0000	.0010	.0091	.0350	.0820	.1376	.1792	.1892	.1657	.1214
	8	.0000	.0002	.0022	.0120	.0376	.0811	.1327	.1734	.1864	.1669
	9	.0000	.0000	.0004	.0033	.0139	.0386	.0794	.1284	.1694	.1855
	10	.0000	.0000	.0001	.0008	.0042	.0149	.0385	.0771	.1248	.1669
	11	.0000	.0000	.0000	.0001	.0010	.0046	.0151	.0374	.0742	.1214

TABLE A1 *Binomial Probabilities (Individual Terms)* (continued)

n	x	p .05	.10	.15	.20	.25	.30	.35	.40	.45	.50
18	12	.0000	.0000	.0000	.0000	.0002	.0012	.0047	.0145	.0354	.0708
	13	.0000	.0000	.0000	.0000	.0000	.0002	.0012	.0045	.0134	.0327
	14	.0000	.0000	.0000	.0000	.0000	.0000	.0002	.0011	.0039	.0117
	15	.0000	.0000	.0000	.0000	.0000	.0000	.0000	.0002	.0009	.0031
	16	.0000	.0000	.0000	.0000	.0000	.0000	.0000	.0000	.0001	.0006
	17	.0000	.0000	.0000	.0000	.0000	.0000	.0000	.0000	.0000	.0001
	18	.0000	.0000	.0000	.0000	.0000	.0000	.0000	.0000	.0000	.0000
19	0	.3774	.1351	.0456	.0144	.0042	.0011	.0003	.0001	.0000	.0000
	1	.3774	.2852	.1529	.0685	.0268	.0093	.0029	.0008	.0002	.0000
	2	.1787	.2852	.2428	.1540	.0803	.0358	.0138	.0046	.0013	.0003
	3	.0533	.1796	.2428	.2182	.1517	.0869	.0422	.0175	.0062	.0018
	4	.0112	.0798	.1714	.2182	.2023	.1491	.0909	.0467	.0203	.0074
	5	.0018	.0266	.0907	.1636	.2023	.1916	.1468	.0933	.0497	.0222
	6	.0002	.0069	.0374	.0955	.1574	.1916	.1844	.1451	.0949	.0518
	7	.0000	.0014	.0122	.0443	.0974	.1525	.1844	.1797	.1443	.0961
	8	.0000	.0002	.0032	.0166	.0487	.0981	.1489	.1797	.1771	.1442
	9	.0000	.0000	.0007	.0051	.0198	.0514	.0980	.1464	.1771	.1762
	10	.0000	.0000	.0001	.0013	.0066	.0220	.0528	.0976	.1449	.1762
	11	.0000	.0000	.0000	.0003	.0018	.0077	.0233	.0532	.0970	.1442
	12	.0000	.0000	.0000	.0000	.0004	.0022	.0083	.0237	.0529	.0961
	13	.0000	.0000	.0000	.0000	.0001	.0005	.0024	.0085	.0233	.0518
	14	.0000	.0000	.0000	.0000	.0000	.0001	.0006	.0024	.0082	.0222
	15	.0000	.0000	.0000	.0000	.0000	.0000	.0001	.0005	.0022	.0074
	16	.0000	.0000	.0000	.0000	.0000	.0000	.0000	.0001	.0005	.0018
	17	.0000	.0000	.0000	.0000	.0000	.0000	.0000	.0000	.0001	.0003
	18	.0000	.0000	.0000	.0000	.0000	.0000	.0000	.0000	.0000	.0000
	19	.0000	.0000	.0000	.0000	.0000	.0000	.0000	.0000	.0000	.0000
20	0	.3585	.1216	.0388	.0115	.0032	.0008	.0002	.0000	.0000	.0000
	1	.3774	.2702	.1368	.0576	.0211	.0068	.0020	.0005	.0001	.0000
	2	.1887	.2852	.2293	.1369	.0669	.0278	.0100	.0031	.0008	.0002
	3	.0596	.1901	.2428	.2054	.1339	.0716	.0323	.0123	.0040	.0011
	4	.0133	.0898	.1821	.2182	.1897	.1304	.0738	.0350	.0139	.0046
	5	.0022	.0319	.1028	.1746	.2023	.1789	.1272	.0746	.0365	.0148
	6	.0003	.0089	.0454	.1091	.1686	.1916	.1712	.1244	.0746	.0370
	7	.0000	.0020	.0160	.0545	.1124	.1643	.1844	.1659	.1221	.0739
	8	.0000	.0004	.0046	.0222	.0609	.1144	.1614	.1797	.1623	.1201
	9	.0000	.0001	.0011	.0074	.0271	.0654	.1158	.1597	.1771	.1602
	10	.0000	.0000	.0002	.0020	.0099	.0308	.0686	.1171	.1593	.1762
	11	.0000	.0000	.0000	.0005	.0030	.0120	.0336	.0710	.1185	.1602
	12	.0000	.0000	.0000	.0001	.0008	.0039	.0136	.0355	.0727	.1201
	13	.0000	.0000	.0000	.0000	.0002	.0010	.0045	.0146	.0366	.0739
	14	.0000	.0000	.0000	.0000	.0000	.0002	.0012	.0049	.0150	.0370
	15	.0000	.0000	.0000	.0000	.0000	.0000	.0003	.0013	.0049	.0148
	16	.0000	.0000	.0000	.0000	.0000	.0000	.0000	.0003	.0013	.0046
	17	.0000	.0000	.0000	.0000	.0000	.0000	.0000	.0000	.0002	.0011
	18	.0000	.0000	.0000	.0000	.0000	.0000	.0000	.0000	.0000	.0002
	19	.0000	.0000	.0000	.0000	.0000	.0000	.0000	.0000	.0000	.0000
	20	.0000	.0000	.0000	.0000	.0000	.0000	.0000	.0000	.0000	.0000

Adapted from Table III.1—Individual Terms, Binomial Distribution, from *CRC Handbook of Tables for Probability and Statistics* (1966).

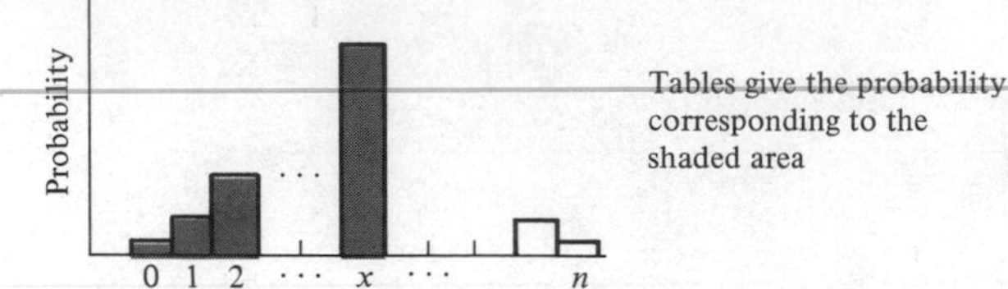

TABLE A2 *Binomial Cumulative Distribution Function*

$$F_{X_B}(x) = \sum_{k=0}^{x} \binom{n}{k} p^k (1-p)^{n-k}$$

n	x	p = 0.05	0.10	0.15	0.20	0.25	0.30	0.35	0.40	0.45	0.50
2	0	0.9025	0.8100	0.7225	0.6400	0.5625	0.4900	0.4225	0.3600	0.3025	0.2500
	1	0.9975	0.9900	0.9775	0.9600	0.9375	0.9100	0.8775	0.8400	0.7975	0.7500
3	0	0.8574	0.7290	0.6141	0.5120	0.4219	0.3430	0.2746	0.2160	0.1664	0.1250
	1	0.9928	0.9720	0.9392	0.8960	0.8438	0.7840	0.7182	0.6480	0.5748	0.5000
	2	0.9999	0.9990	0.9966	0.9920	0.9844	0.9730	0.9571	0.9360	0.9089	0.8750
4	0	0.8145	0.6561	0.5220	0.4096	0.3164	0.2401	0.1785	0.1296	0.0915	0.0625
	1	0.9860	0.9477	0.8905	0.8192	0.7383	0.6517	0.5630	0.4752	0.3910	0.3125
	2	0.9995	0.9963	0.9880	0.9728	0.9492	0.9163	0.8735	0.8208	0.7585	0.6875
	3	1.0000	0.9999	0.9995	0.9984	0.9961	0.9919	0.9850	0.9744	0.9590	0.9375
5	0	0.7738	0.5905	0.4437	0.3277	0.2373	0.1681	0.1160	0.0778	0.0503	0.0312
	1	0.9774	0.9185	0.8352	0.7373	0.6328	0.5282	0.4284	0.3370	0.2562	0.1875
	2	0.9988	0.9914	0.9734	0.9421	0.8965	0.8369	0.7648	0.6826	0.5931	0.5000
	3	1.0000	0.9995	0.9978	0.9933	0.9844	0.9692	0.9460	0.9130	0.8688	0.8125
	4	1.0000	1.0000	0.9999	0.9997	0.9990	0.9976	0.9947	0.9898	0.9815	0.9688
6	0	0.7351	0.5314	0.3771	0.2621	0.1780	0.1176	0.0754	0.0467	0.0277	0.0156
	1	0.9672	0.8857	0.7765	0.6554	0.5339	0.4202	0.3191	0.2333	0.1636	0.1094
	2	0.9978	0.9842	0.9527	0.9011	0.8306	0.7443	0.6471	0.5443	0.4415	0.3438
	3	0.9999	0.9987	0.9941	0.9830	0.9624	0.9295	0.8826	0.8208	0.7447	0.6562
	4	1.0000	0.9999	0.9996	0.9984	0.9954	0.9891	0.9777	0.9590	0.9308	0.8906
	5	1.0000	1.0000	1.0000	0.9999	0.9998	0.9993	0.9982	0.9959	0.9917	0.9844
7	0	0.6983	0.4783	0.3206	0.2097	0.1335	0.0824	0.0490	0.0280	0.0152	0.0078
	1	0.9556	0.8503	0.7166	0.5767	0.4449	0.3294	0.2338	0.1586	0.1024	0.0625
	2	0.9962	0.9743	0.9262	0.8520	0.7564	0.6471	0.5323	0.4199	0.3164	0.2266
	3	0.9998	0.9973	0.9879	0.9667	0.9294	0.8740	0.8002	0.7102	0.6083	0.5000
	4	1.0000	0.9998	0.9988	0.9953	0.9871	0.9712	0.9444	0.9037	0.8471	0.7734
	5	1.0000	1.0000	0.9999	0.9996	0.9987	0.9962	0.9910	0.9812	0.9643	0.9375
	6	1.0000	1.0000	1.0000	1.0000	0.9999	0.9998	0.9994	0.9984	0.9963	0.9922
8	0	0.6634	0.4305	0.2725	0.1678	0.1001	0.0576	0.0319	0.0168	0.0084	0.0039
	1	0.9428	0.8131	0.6572	0.5033	0.3671	0.2553	0.1691	0.1064	0.0632	0.0352
	2	0.9942	0.9619	0.8948	0.7969	0.6785	0.5518	0.4278	0.3154	0.2201	0.1445
	3	0.9996	0.9950	0.9786	0.9437	0.8862	0.8059	0.7064	0.5941	0.4770	0.3633
	4	1.0000	0.9996	0.9971	0.9896	0.9727	0.9420	0.8939	0.8263	0.7396	0.6367
	5	1.0000	1.0000	0.9998	0.9988	0.9958	0.9887	0.9747	0.9502	0.9115	0.8555
	6	1.0000	1.0000	1.0000	0.9999	0.9996	0.9987	0.9964	0.9915	0.9819	0.9648
	7	1.0000	1.0000	1.0000	1.0000	1.0000	0.9999	0.9998	0.9993	0.9983	0.9961
9	0	0.6302	0.3874	0.2316	0.1342	0.0751	0.0404	0.0207	0.0101	0.0046	0.0020
	1	0.9288	0.7748	0.5995	0.4362	0.3003	0.1960	0.1211	0.0705	0.0385	0.0195
	2	0.9916	0.9470	0.8591	0.7382	0.6007	0.4628	0.3373	0.2318	0.1495	0.0898
	3	0.9994	0.9917	0.9661	0.9144	0.8343	0.7297	0.6089	0.4826	0.3614	0.2539
	4	1.0000	0.9991	0.9944	0.9804	0.9511	0.9012	0.8283	0.7334	0.6214	0.5000
	5	1.0000	0.9999	0.9994	0.9969	0.9900	0.9747	0.9464	0.9006	0.8342	0.7461
	6	1.0000	1.0000	1.0000	0.9997	0.9987	0.9957	0.9888	0.9750	0.9502	0.9102
	7	1.0000	1.0000	1.0000	1.0000	0.9999	0.9996	0.9986	0.9962	0.9909	0.9805
	8	1.0000	1.0000	1.0000	1.0000	1.0000	1.0000	0.9999	0.9997	0.9992	0.9980

TABLE A2 *Binomial Distribution Function* (continued)

n	x	0.05	0.10	0.15	0.20	0.25	0.30	0.35	0.40	0.45	0.50
						p					
10	0	0.5987	0.3487	0.1969	0.1074	0.0563	0.0282	0.0135	0.0060	0.0025	0.0010
	1	0.9139	0.7361	0.5443	0.3758	0.2440	0.1493	0.0860	0.0464	0.0232	0.0107
	2	0.9885	0.9298	0.8202	0.6778	0.5256	0.3828	0.2616	0.1673	0.0996	0.0547
	3	0.9990	0.9872	0.9500	0.8791	0.7759	0.6496	0.5138	0.3823	0.2660	0.1719
	4	0.9999	0.9984	0.9901	0.9672	0.9219	0.8497	0.7515	0.6331	0.5044	0.3770
	5	1.0000	0.9999	0.9986	0.9936	0.9803	0.9527	0.9051	0.8338	0.7384	0.6230
	6	1.0000	1.0000	0.9999	0.9991	0.9965	0.9894	0.9740	0.9452	0.8980	0.8281
	7	1.0000	1.0000	1.0000	0.9999	0.9996	0.9984	0.9952	0.9877	0.9726	0.9453
	8	1.0000	1.0000	1.0000	1.0000	1.0000	0.9999	0.9995	0.9983	0.9955	0.9893
	9	1.0000	1.0000	1.0000	1.0000	1.0000	1.0000	1.0000	0.9999	0.9997	0.9990
11	0	0.5688	0.3138	0.1673	0.0859	0.0422	0.0198	0.0088	0.0036	0.0014	0.0005
	1	0.8981	0.6974	0.4922	0.3221	0.1971	0.1130	0.0606	0.0302	0.0139	0.0059
	2	0.9848	0.9104	0.7788	0.6174	0.4552	0.3127	0.2001	0.1189	0.0652	0.0327
	3	0.9984	0.9815	0.9306	0.8389	0.7133	0.5696	0.4256	0.2963	0.1911	0.1133
	4	0.9999	0.9972	0.9841	0.9496	0.8854	0.7897	0.6683	0.5328	0.3971	0.2744
	5	1.0000	0.9997	0.9973	0.9883	0.9657	0.9218	0.8513	0.7535	0.6331	0.5000
	6	1.0000	1.0000	0.9997	0.9980	0.9924	0.9784	0.9499	0.9006	0.8262	0.7256
	7	1.0000	1.0000	1.0000	0.9998	0.9988	0.9957	0.9878	0.9707	0.9390	0.8867
	8	1.0000	1.0000	1.0000	1.0000	0.9999	0.9994	0.9980	0.9941	0.9852	0.9673
	9	1.0000	1.0000	1.0000	1.0000	1.0000	1.0000	0.9998	0.9993	0.9978	0.9941
	10	1.0000	1.0000	1.0000	1.0000	1.0000	1.0000	1.0000	1.0000	0.9998	0.9995
12	0	0.5404	0.2824	0.1422	0.0687	0.0317	0.0138	0.0057	0.0022	0.0008	0.0002
	1	0.8816	0.6590	0.4435	0.2749	0.1584	0.0850	0.0424	0.0196	0.0083	0.0032
	2	0.9804	0.8891	0.7358	0.5583	0.3907	0.2528	0.1513	0.0834	0.0421	0.0193
	3	0.9978	0.9744	0.9078	0.7946	0.6488	0.4925	0.3467	0.2253	0.1345	0.0730
	4	0.9998	0.9957	0.9761	0.9274	0.8424	0.7237	0.5833	0.4382	0.3044	0.1938
	5	1.0000	0.9995	0.9954	0.9806	0.9456	0.8822	0.7873	0.6652	0.5269	0.3872
	6	1.0000	0.9999	0.9993	0.9961	0.9857	0.9614	0.9154	0.8418	0.7393	0.6128
	7	1.0000	1.0000	0.9999	0.9994	0.9972	0.9905	0.9745	0.9427	0.8883	0.8062
	8	1.0000	1.0000	1.0000	0.9999	0.9996	0.9983	0.9944	0.9847	0.9644	0.9270
	9	1.0000	1.0000	1.0000	1.0000	1.0000	0.9998	0.9992	0.9972	0.9921	0.0807
	10	1.0000	1.0000	1.0000	1.0000	1.0000	1.0000	0.9999	0.9997	0.9989	0.9968
	11	1.0000	1.0000	1.0000	1.0000	1.0000	1.0000	1.0000	1.0000	0.9999	0.9998
13	0	0.5133	0.2542	0.1209	0.0550	0.0238	0.0097	0.0037	0.0013	0.0004	0.0001
	1	0.8646	0.6213	0.3983	0.2336	0.1267	0.0637	0.0296	0.0126	0.0049	0.0017
	2	0.9755	0.8661	0.6920	0.5017	0.3326	0.2025	0.1132	0.0579	0.0269	0.0112
	3	0.9969	0.9658	0.8820	0.7473	0.5843	0.4206	0.2783	0.1686	0.0929	0.0461
	4	0.9997	0.9935	0.9658	0.9009	0.7940	0.6543	0.5005	0.3530	0.2279	0.1334
	5	1.0000	0.9991	0.9925	0.9700	0.9198	0.8346	0.7159	0.5744	0.4268	0.2905
	6	1.0000	0.9999	0.9987	0.9930	0.9757	0.9376	0.8705	0.7712	0.6437	0.5000
	7	1.0000	1.0000	0.9998	0.9988	0.9944	0.9818	0.9538	0.9023	0.8212	0.7095
	8	1.0000	1.0000	1.0000	0.9998	0.9990	0.9960	0.9874	0.9679	0.9302	0.8666
	9	1.0000	1.0000	1.0000	1.0000	0.9999	0.9993	0.9975	0.9922	0.9797	0.9539
	10	1.0000	1.0000	1.0000	1.0000	1.0000	0.9999	0.9997	0.9987	0.9959	0.9888
	11	1.0000	1.0000	1.0000	1.0000	1.0000	1.0000	1.0000	0.9999	0.9995	0.9983
	12	1.0000	1.0000	1.0000	1.0000	1.0000	1.0000	1.0000	1.0000	1.0000	0.9999
14	0	0.4877	0.2288	0.1028	0.0440	0.0178	0.0068	0.0024	0.0008	0.0002	0.0001
	1	0.8470	0.5846	0.3567	0.1979	0.1010	0.0475	0.0205	0.0081	0.0029	0.0009

TABLE A2 *Binomial Distribution Function* (continued)

n	x	0.05	0.10	0.15	0.20	0.25	0.30	0.35	0.40	0.45	0.50
14	2	0.9699	0.8416	0.6479	0.4481	0.2811	0.1608	0.0839	0.0398	0.0170	0.0065
	3	0.9958	0.9559	0.8535	0.6982	0.5213	0.3552	0.2205	0.1243	0.0632	0.0287
	4	0.9996	0.9908	0.9533	0.8702	0.7415	0.5842	0.4227	0.2793	0.1672	0.0898
	5	1.0000	0.9985	0.9885	0.9561	0.8883	0.7805	0.6405	0.4859	0.3373	0.2120
	6	1.0000	0.9998	0.9978	0.9884	0.9617	0.9067	0.8164	0.6925	0.5461	0.3953
	7	1.0000	1.0000	0.9997	0.9976	0.9897	0.9685	0.9247	0.8499	0.7414	0.6047
	8	1.0000	1.0000	1.0000	0.9996	0.9978	0.9917	0.9757	0.9417	0.8811	0.7880
	9	1.0000	1.0000	1.0000	1.0000	0.9997	0.9983	0.9940	0.9825	0.9574	0.9102
	10	1.0000	1.0000	1.0000	1.0000	1.0000	0.9998	0.9989	0.9961	0.9886	0.9713
	11	1.0000	1.0000	1.0000	1.0000	1.0000	1.0000	0.9999	0.9994	0.9978	0.9935
	12	1.0000	1.0000	1.0000	1.0000	1.0000	1.0000	1.0000	0.9999	0.9997	0.9991
	13	1.0000	1.0000	1.0000	1.0000	1.0000	1.0000	1.0000	1.0000	1.0000	0.9999
15	0	0.4633	0.2059	0.0874	0.0352	0.0134	0.0047	0.0016	0.0005	0.0001	0.0000
	1	0.8290	0.5490	0.3186	0.1671	0.0802	0.0353	0.0142	0.0052	0.0017	0.0005
	2	0.9638	0.8159	0.6042	0.3980	0.2361	0.1268	0.0617	0.0271	0.0107	0.0037
	3	0.9945	0.9444	0.8227	0.6482	0.4613	0.2969	0.1727	0.0905	0.0424	0.0176
	4	0.9994	0.9873	0.9383	0.8358	0.6865	0.5155	0.3519	0.2173	0.1204	0.0592
	5	0.9999	0.9978	0.9832	0.9389	0.8516	0.7216	0.5643	0.4032	0.2608	0.1509
	6	1.0000	0.9997	0.9964	0.9819	0.9434	0.8689	0.7548	0.6098	0.4522	0.3036
	7	1.0000	1.0000	0.9996	0.9958	0.9827	0.9500	0.8868	0.7869	0.6535	0.5000
	8	1.0000	1.0000	0.9999	0.9992	0.9958	0.9848	0.9578	0.9050	0.8182	0.6964
	9	1.0000	1.0000	1.0000	0.9999	0.9992	0.9963	0.9876	0.9662	0.9231	0.8491
	10	1.0000	1.0000	1.0000	1.0000	0.9999	0.9993	0.9972	0.9907	0.9745	0.9408
	11	1.0000	1.0000	1.0000	1.0000	1.0000	0.9999	0.9995	0.9981	0.9937	0.9824
	12	1.0000	1.0000	1.0000	1.0000	1.0000	1.0000	0.9999	0.9997	0.9989	0.9963
	13	1.0000	1.0000	1.0000	1.0000	1.0000	1.0000	1.0000	1.0000	0.9999	0.9995
	14	1.0000	1.0000	1.0000	1.0000	1.0000	1.0000	1.0000	1.0000	1.0000	1.0000
16	0	0.4401	0.1853	0.0743	0.0281	0.0100	0.0033	0.0010	0.0003	0.0001	0.0000
	1	0.8108	0.5147	0.2839	0.1407	0.0635	0.0261	0.0098	0.0033	0.0010	0.0003
	2	0.9571	0.7892	0.5614	0.3518	0.1971	0.0994	0.0451	0.0183	0.0066	0.0021
	3	0.9930	0.9316	0.7899	0.5981	0.4050	0.2459	0.1339	0.0651	0.0281	0.0106
	4	0.9991	0.9830	0.9209	0.7982	0.6302	0.4499	0.2892	0.1666	0.0853	0.0384
	5	0.9999	0.9967	0.9765	0.9183	0.8103	0.6598	0.4900	0.3288	0.1976	0.1051
	6	1.0000	0.9995	0.9944	0.9733	0.9204	0.8247	0.6881	0.5272	0.3660	0.2272
	7	1.0000	0.9999	0.9989	0.9930	0.9729	0.9256	0.8406	0.7161	0.5629	0.4018
	8	1.0000	1.0000	0.9998	0.9985	0.9925	0.9743	0.9329	0.8577	0.7441	0.5982
	9	1.0000	1.0000	1.0000	0.9998	0.9984	0.9929	0.9771	0.9417	0.8759	0.7728
	10	1.0000	1.0000	1.0000	1.0000	0.9997	0.9984	0.9938	0.9809	0.9514	0.8949
	11	1.0000	1.0000	1.0000	1.0000	1.0000	0.9997	0.9987	0.9951	0.9851	0.9616
	12	1.0000	1.0000	1.0000	1.0000	1.0000	1.0000	0.9998	0.9991	0.9965	0.9894
	13	1.0000	1.0000	1.0000	1.0000	1.0000	1.0000	1.0000	0.9999	0.9994	0.9979
	14	1.0000	1.0000	1.0000	1.0000	1.0000	1.0000	1.0000	1.0000	1.0000	0.9997
	15	1.0000	1.0000	1.0000	1.0000	1.0000	1.0000	1.0000	1.0000	1.0000	1.0000
17	0	0.4181	0.1668	0.0631	0.0225	0.0075	0.0023	0.0007	0.0002	0.0000	0.0000
	1	0.7922	0.4818	0.2525	0.1182	0.0501	0.0193	0.0067	0.0021	0.0006	0.0001
	2	0.9497	0.7618	0.5198	0.3096	0.1637	0.0774	0.0327	0.0123	0.0041	0.0012
	3	0.9912	0.9174	0.7556	0.5489	0.3530	0.2019	0.1028	0.0464	0.0184	0.0064
	4	0.9988	0.9779	0.9013	0.7582	0.5739	0.3887	0.2348	0.1260	0.0596	0.0245

TABLE A2 *Binomial Distribution Function* (continued)

n	x	p 0.05	0.10	0.15	0.20	0.25	0.30	0.35	0.40	0.45	0.50
17	5	0.9999	0.9953	0.9681	0.8943	0.7653	0.5968	0.4197	0.2639	0.1471	0.0717
	6	1.0000	0.9992	0.9917	0.9623	0.8929	0.7752	0.6188	0.4478	0.2902	0.1662
	7	1.0000	0.9999	0.9983	0.9891	0.9598	0.8954	0.7872	0.6405	0.4743	0.3145
	8	1.0000	1.0000	0.9997	0.9974	0.9876	0.9597	0.9006	0.8011	0.6626	0.5000
	9	1.0000	1.0000	1.0000	0.9995	0.9969	0.9873	0.9617	0.9081	0.8166	0.6855
	10	1.0000	1.0000	1.0000	0.9999	0.9994	0.9968	0.9880	0.9652	0.9174	0.8338
	11	1.0000	1.0000	1.0000	1.0000	0.9999	0.9993	0.9970	0.9894	0.9699	0.9283
	12	1.0000	1.0000	1.0000	1.0000	1.0000	0.9999	0.9994	0.9975	0.9914	0.9755
	13	1.0000	1.0000	1.0000	1.0000	1.0000	1.0000	0.9999	0.9995	0.9981	0.9936
	14	1.0000	1.0000	1.0000	1.0000	1.0000	1.0000	1.0000	0.9999	0.9997	0.9988
	15	1.0000	1.0000	1.0000	1.0000	1.0000	1.0000	1.0000	1.0000	1.0000	0.9999
	16	1.0000	1.0000	1.0000	1.0000	1.0000	1.0000	1.0000	1.0000	1.0000	1.0000
18	0	0.3972	0.1501	0.0536	0.0180	0.0056	0.0016	0.0004	0.0001	0.0000	0.0000
	1	0.7735	0.4503	0.2241	0.0991	0.0395	0.0142	0.0046	0.0013	0.0003	0.0001
	2	0.9419	0.7338	0.4797	0.2713	0.1353	0.0600	0.0236	0.0082	0.0025	0.0007
	3	0.9891	0.9018	0.7202	0.5010	0.3057	0.1646	0.0783	0.0328	0.0120	0.0038
	4	0.9985	0.9718	0.8794	0.7164	0.5187	0.3327	0.1886	0.0942	0.0411	0.0154
	5	0.9998	0.9936	0.9581	0.8671	0.7175	0.5344	0.3550	0.2088	0.1077	0.0481
	6	1.0000	0.9988	0.9882	0.9487	0.8610	0.7217	0.5491	0.3743	0.2258	0.1189
	7	1.0000	0.9998	0.9973	0.9837	0.9431	0.8593	0.7283	0.5634	0.3915	0.2403
	8	1.0000	1.0000	0.9995	0.9957	0.9807	0.9404	0.8609	0.7368	0.5778	0.4073
	9	1.0000	1.0000	0.9999	0.9991	0.9946	0.9790	0.9403	0.8653	0.7473	0.5927
	10	1.0000	1.0000	1.0000	0.9998	0.9988	0.9939	0.9788	0.9424	0.8720	0.7597
	11	1.0000	1.0000	1.0000	1.0000	0.9998	0.9986	0.9938	0.9797	0.9463	0.8811
	12	1.0000	1.0000	1.0000	1.0000	1.0000	0.9997	0.9986	0.9942	0.9817	0.9519
	13	1.0000	1.0000	1.0000	1.0000	1.0000	1.0000	0.9997	0.9987	0.9951	0.9846
	14	1.0000	1.0000	1.0000	1.0000	1.0000	1.0000	1.0000	0.9998	0.9990	0.9962
	15	1.0000	1.0000	1.0000	1.0000	1.0000	1.0000	1.0000	1.0000	0.9999	0.9993
	16	1.0000	1.0000	1.0000	1.0000	1.0000	1.0000	1.0000	1.0000	1.0000	0.9999
19	0	0.3774	0.1351	0.0456	0.0144	0.0042	0.0011	0.0003	0.0001	0.0000	0.0000
	1	0.7547	0.4203	0.1985	0.0829	0.0310	0.0104	0.0031	0.0008	0.0002	0.0000
	2	0.9335	0.7054	0.4413	0.2369	0.1113	0.0462	0.0170	0.0055	0.0015	0.0004
	3	0.9868	0.8850	0.6841	0.4551	0.2630	0.1332	0.0591	0.0230	0.0077	0.0022
	4	0.9980	0.9648	0.8556	0.6733	0.4654	0.2822	0.1500	0.0696	0.0280	0.0096
	5	0.9998	0.9914	0.9463	0.8369	0.6678	0.4739	0.2968	0.1629	0.0777	0.0318
	6	1.0000	0.9983	0.9837	0.9324	0.8251	0.6655	0.4812	0.3081	0.1727	0.0835
	7	1.0000	0.9997	0.9959	0.9767	0.9225	0.8180	0.6656	0.4878	0.3169	0.1796
	8	1.0000	1.0000	0.9992	0.9933	0.9713	0.9161	0.8145	0.6675	0.4940	0.3238
	9	1.0000	1.0000	0.9999	0.9984	0.9911	0.9674	0.9125	0.8139	0.6710	0.5000
	10	1.0000	1.0000	1.0000	0.9997	0.9977	0.9895	0.9653	0.9115	0.8159	0.6762
	11	1.0000	1.0000	1.0000	1.0000	0.9995	0.9972	0.9886	0.9648	0.9129	0.8204
	12	1.0000	1.0000	1.0000	1.0000	0.9999	0.9994	0.9969	0.9884	0.9658	0.9165
	13	1.0000	1.0000	1.0000	1.0000	1.0000	0.9999	0.9993	0.9969	0.9891	0.9682
	14	1.0000	1.0000	1.0000	1.0000	1.0000	1.0000	0.9999	0.9994	0.9972	0.9904
	15	1.0000	1.0000	1.0000	1.0000	1.0000	1.0000	1.0000	0.9999	0.9995	0.9978
	16	1.0000	1.0000	1.0000	1.0000	1.0000	1.0000	1.0000	1.0000	0.9999	0.9996
	17	1.0000	1.0000	1.0000	1.0000	1.0000	1.0000	1.0000	1.0000	1.0000	1.0000

TABLE A2 *Binomial Distribution Function* (continued)

n	x	0.05	0.10	0.15	0.20	0.25	0.30	0.35	0.40	0.45	0.50
						p					
20	0	0.3585	0.1216	0.0388	0.0115	0.0032	0.0008	0.0002	0.0000	0.0000	0.0000
	1	0.7358	0.3917	0.1756	0.0692	0.0243	0.0076	0.0021	0.0005	0.0001	0.0000
	2	0.9245	0.6769	0.4049	0.2061	0.0913	0.0355	0.0121	0.0036	0.0009	0.0002
	3	0.9841	0.8670	0.6477	0.4114	0.2252	0.1071	0.0444	0.0160	0.0049	0.0013
	4	0.9974	0.9568	0.8298	0.6296	0.4148	0.2375	0.1182	0.0510	0.0189	0.0059
	5	0.9997	0.9887	0.9327	0.8042	0.6172	0.4164	0.2454	0.1256	0.0553	0.0207
	6	1.0000	0.9976	0.9781	0.9133	0.7858	0.6080	0.4166	0.2500	0.1299	0.0577
	7	1.0000	0.9996	0.9941	0.9679	0.8982	0.7723	0.6010	0.4159	0.2520	0.1316
	8	1.0000	0.9999	0.9987	0.9900	0.9591	0.8867	0.7624	0.5956	0.4143	0.2517
	9	1.0000	1.0000	0.9998	0.9974	0.9861	0.9520	0.8782	0.7553	0.5914	0.4119
	10	1.0000	1.0000	1.0000	0.9994	0.9961	0.9829	0.9468	0.8725	0.7507	0.5881
	11	1.0000	1.0000	1.0000	0.9999	0.9991	0.9949	0.9804	0.9435	0.8692	0.7483
	12	1.0000	1.0000	1.0000	1.0000	0.9998	0.9987	0.9940	0.9790	0.9420	0.8684
	13	1.0000	1.0000	1.0000	1.0000	1.0000	0.9997	0.9985	0.9935	0.9786	0.9423
	14	1.0000	1.0000	1.0000	1.0000	1.0000	1.0000	0.9997	0.9984	0.9936	0.9793
	15	1.0000	1.0000	1.0000	1.0000	1.0000	1.0000	1.0000	0.9997	0.9985	0.9941
	16	1.0000	1.0000	1.0000	1.0000	1.0000	1.0000	1.0000	1.0000	0.9997	0.9987
	17	1.0000	1.0000	1.0000	1.0000	1.0000	1.0000	1.0000	1.0000	1.0000	0.9998
	18	1.0000	1.0000	1.0000	1.0000	1.0000	1.0000	1.0000	1.0000	1.0000	1.0000

Reproduced from Table I of *Probability and Statistics for Engineers* (1965), by Miller and Freund, Prentice-Hall.

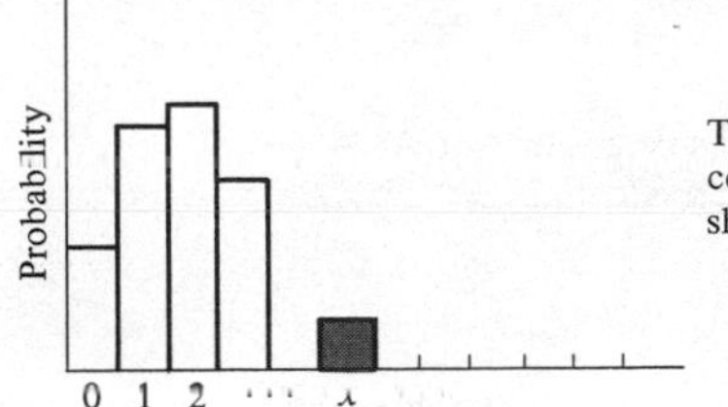

Tables give the probability corresponding to the shaded area.

TABLE A3 *Poisson Probabilities (Individual Terms)*

$$P[X_P = x \mid \lambda t] = e^{-\lambda t}(\lambda t)^x / x!$$

x	λt = 0.1	0.2	0.3	0.4	0.5	0.6	0.7	0.8	0.9	1.0
0	.9048	.8187	.7408	.6703	.6065	.5488	.4966	.4493	.4066	.3679
1	.0905	.1637	.2222	.2681	.3033	.3293	.3476	.3595	.3659	.3679
2	.0045	.0164	.0333	.0536	.0758	.0988	.1217	.1438	.1647	.1839
3	.0002	.0011	.0033	.0072	.0126	.0198	.0284	.0383	.0494	.0613
4	.0000	.0001	.0002	.0007	.0016	.0030	.0050	.0077	.0111	.0153
5	.0000	.0000	.0000	.0001	.0002	.0004	.0007	.0012	.0020	.0031
6	.0000	.0000	.0000	.0000	.0000	.0000	.0001	.0002	.0003	.0005
7	.0000	.0000	.0000	.0000	.0000	.0000	.0000	.0000	.0000	.0001

x	λt = 1.1	1.2	1.3	1.4	1.5	1.6	1.7	1.8	1.9	2.0
0	.3329	.3012	.2725	.2466	.2231	.2019	.1827	.1653	.1496	.1353
1	.3662	.3614	.3543	.3452	.3347	.3230	.3106	.2975	.2842	.2707
2	.2014	.2169	.2303	.2417	.2510	.2584	.2640	.2678	.2700	.2707
3	.0738	.0867	.0998	.1128	.1255	.1378	.1496	.1607	.1710	.1804
4	.0203	.0260	.0324	.0395	.0471	.0551	.0636	.0723	.0812	.0902
5	.0045	.0062	.0084	.0111	.0141	.0176	.0216	.0260	.0309	.0361
6	.0008	.0012	.0018	.0026	.0035	.0047	.0061	.0078	.0098	.0120
7	.0001	.0002	.0003	.0005	.0008	.0011	.0015	.0020	.0027	.0034
8	.0000	.0000	.0001	.0001	.0001	.0002	.0003	.0005	.0006	.0009
9	.0000	.0000	.0000	.0000	.0000	.0000	.0001	.0001	.0001	.0002

x	λt = 2.1	2.2	2.3	2.4	2.5	2.6	2.7	2.8	2.9	3.0
0	.1225	.1108	.1003	.0907	.0821	.0743	.0672	.0608	.0550	.0498
1	.2572	.2438	.2306	.2177	.2052	.1931	.1815	.1703	.1596	.1494
2	.2700	.2681	.2652	.2613	.2565	.2510	.2450	.2384	.2314	.2240
3	.1890	.1966	.2033	.2090	.2138	.2176	.2205	.2225	.2237	.2240
4	.0992	.1082	.1169	.1254	.1336	.1414	.1488	.1557	.1622	.1680
5	.0417	.0476	.0538	.0602	.0668	.0735	.0804	.0872	.0940	.1008
6	.0146	.0174	.0206	.0241	.0278	.0319	.0362	.0407	.0455	.0504
7	.0044	.0055	.0068	.0083	.0099	.0118	.0139	.0163	.0188	.0216
8	.0011	.0015	.0019	.0025	.0031	.0038	.0047	.0057	.0068	.0081
9	.0003	.0004	.0005	.0007	.0009	.0011	.0014	.0018	.0022	.0027
10	.0001	.0001	.0001	.0002	.0002	.0003	.0004	.0005	.0006	.0008
11	.0000	.0000	.0000	.0000	.0000	.0001	.0001	.0001	.0002	.0002
12	.0000	.0000	.0000	.0000	.0000	.0000	.0000	.0000	.0000	.0001

x	λt = 3.1	3.2	3.3	3.4	3.5	3.6	3.7	3.8	3.9	4.0
0	.0450	.0408	.0369	.0334	.0302	.0273	.0247	.0224	.0202	.0183
1	.1397	.1304	.1217	.1135	.1057	.0984	.0915	.0850	.0789	.0733
2	.2165	.2087	.2008	.1929	.1850	.1771	.1692	.1615	.1539	.1465
3	.2237	.2226	.2209	.2186	.2158	.2125	.2087	.2046	.2001	.1954
4	.1734	.1781	.1823	.1858	.1888	.1012	.1031	.1944	.1951	.1954
5	.1075	.1140	.1203	.1264	.1322	.1377	.1429	.1477	.1522	.1563
6	.0555	.0608	.0662	.0716	.0771	.0826	.0881	.0936	.0989	.1042
7	.0246	.0278	.0312	.0348	.0385	.0425	.0466	.0508	.0551	.0595
8	.0095	.0111	.0129	.0148	.0169	.0191	.0215	.0241	.0269	.0298
9	.0033	.0040	.0047	.0056	.0066	.0076	.0089	.0102	.0116	.0132

TABLE A3 *Poisson Probabilities (Individual Terms)* (continued)

x	λt 3.1	3.2	3.3	3.4	3.5	3.6	3.7	3.8	3.9	4.0
10	.0010	.0013	.0016	.0019	.0023	.0028	.0033	.0039	.0045	.0053
11	.0003	.0004	.0005	.0006	.0007	.0009	.0011	.0013	.0016	.0019
12	.0001	.0001	.0001	.0002	.0002	.0003	.0003	.0004	.0005	.0006
13	.0000	.0000	.0000	.0000	.0001	.0001	.0001	.0001	.0002	.0002
14	.0000	.0000	.0000	.0000	.0000	.0000	.0000	.0000	.0000	.0001

x	λt 4.1	4.2	4.3	4.4	4.5	4.6	4.7	4.8	4.9	5.0
0	.0166	.0150	.0136	.0123	.0111	.0101	.0091	.0082	.0074	.0067
1	.0679	.0630	.0583	.0540	.0500	.0462	.0427	.0395	.0365	.0337
2	.1393	.1323	.1254	.1188	.1125	.1063	.1005	.0948	.0894	.0842
3	.1904	.1852	.1798	.1743	.1687	.1631	.1574	.1517	.1460	.1404
4	.1951	.1944	.1933	.1917	.1898	.1875	.1849	.1820	.1789	.1755
5	.1600	.1633	.1662	.1687	.1708	.1725	.1738	.1747	.1753	.1755
6	.1093	.1143	.1191	.1237	.1281	.1323	.1362	.1398	.1432	.1462
7	.0640	.0686	.0732	.0778	.0824	.0869	.0914	.0959	.1002	.1044
8	.0328	.0360	.0393	.0428	.0463	.0500	.0537	.0575	.0614	.0653
9	.0150	.0168	.0188	.0209	.0232	.0255	.0280	.0307	.0334	.0363
10	.0061	.0071	.0081	.0092	.0104	.0118	.0132	.0147	.0164	.0181
11	.0023	.0027	.0032	.0037	.0043	.0049	.0056	.0064	.0073	.0082
12	.0008	.0009	.0011	.0014	.0016	.0019	.0022	.0026	.0030	.0034
13	.0002	.0003	.0004	.0005	.0006	.0007	.0008	.0009	.0011	.0013
14	.0001	.0001	.0001	.0001	.0002	.0002	.0003	.0003	.0004	.0005
15	.0000	.0000	.0000	.0000	.0001	.0001	.0001	.0001	.0001	.0002

x	λt 5.1	5.2	5.3	5.4	5.5	5.6	5.7	5.8	5.9	6.0
0	.0061	.0055	.0050	.0045	.0041	.0037	.0033	.0030	.0027	.0025
1	.0311	.0287	.0265	.0244	.0225	.0207	.0191	.0176	.0162	.0149
2	.0793	.0746	.0701	.0659	.0618	.0580	.0544	.0509	.0477	.0446
3	.1348	.1293	.1239	.1185	.1133	.1082	.1033	.0985	.0938	.0892
4	.1719	.1681	.1641	.1600	.1558	.1515	.1472	.1428	.1383	.1339
5	.1753	.1748	.1740	.1728	.1714	.1697	.1678	.1656	.1632	.1606
6	.1490	.1515	.1537	.1555	.1571	.1584	.1594	.1601	.1605	.1606
7	.1086	.1125	.1163	.1200	.1234	.1267	.1298	.1326	.1353	.1377
8	.0692	.0731	.0771	.0810	.0849	.0887	.0925	.0962	.0998	.1033
9	.0392	.0423	.0454	.0486	.0519	.0552	.0586	.0620	.0654	.0688
10	.0200	.0220	.0241	.0262	.0285	.0309	.0334	.0359	.0386	.0413
11	.0093	.0104	.0116	.0129	.0143	.0157	.0173	.0190	.0207	.0225
12	.0039	.0045	.0051	.0058	.0065	.0073	.0082	.0092	.0102	.0113
13	.0015	.0018	.0021	.0024	.0028	.0032	.0036	.0041	.0046	.0052
14	.0006	.0007	.0008	.0009	.0011	.0013	.0015	.0017	.0019	.0022
15	.0002	.0002	.0003	.0003	.0004	.0005	.0006	.0007	.0008	.0009
16	.0001	.0001	.0001	.0001	.0001	.0002	.0002	.0002	.0003	.0003
17	.0000	.0000	.0000	.0000	.0000	.0000	.0001	.0001	.0001	.0001

TABLE A3 *Poisson Probabilities* (*Individual Terms*) (continued)

x	λt 6.1	6.2	6.3	6.4	6.5	6.6	6.7	6.8	6.9	7.0
0	.0022	.0020	.0018	.0017	.0015	.0014	.0012	.0011	.0010	.0009
1	.0137	.0126	.0116	.0106	.0098	.0090	.0082	.0076	.0070	.0064
2	.0417	.0390	.0364	.0340	.0318	.0296	.0276	.0258	.0240	.0223
3	.0848	.0806	.0765	.0726	.0688	.0652	.0617	.0584	.0552	.0521
4	.1294	.1249	.1205	.1162	.1118	.1076	.1034	.0992	.0952	.0912
5	.1579	.1549	.1519	.1487	.1454	.1420	.1385	.1349	.1314	.1277
6	.1605	.1601	.1595	.1586	.1575	.1562	.1546	.1529	.1511	.1490
7	.1399	.1418	.1435	.1450	.1462	.1472	.1480	.1486	.1489	.1490
8	.1066	.1099	.1130	.1160	.1188	.1215	.1240	.1263	.1284	.1304
9	.0723	.0757	.0791	.0825	.0858	.0891	.0923	.0954	.0985	.1014
10	.0441	.0469	.0498	.0528	.0558	.0588	.0618	.0649	.0679	.0710
11	.0245	.0265	.0285	.0307	.0330	.0353	.0377	.0401	.0426	.0452
12	.0124	.0137	.0150	.0164	.0179	.0194	.0210	.0227	.0245	.0264
13	.0058	.0065	.0073	.0081	.0089	.0098	.0108	.0119	.0130	.0142
14	.0025	.0029	.0033	.0037	.0041	.0046	.0052	.0058	.0064	.0071
15	.0010	.0012	.0014	.0016	.0018	.0020	.0023	.0026	.0029	.0033
16	.0004	.0005	.0005	.0006	.0007	.0008	.0010	.0011	.0013	.0014
17	.0001	.0002	.0002	.0002	.0003	.0003	.0004	.0004	.0005	.0006
18	.0000	.0001	.0001	.0001	.0001	.0001	.0001	.0002	.0002	.0002
19	.0000	.0000	.0000	.0000	.0000	.0000	.0000	.0001	.0001	.0001

x	λt 7.1	7.2	7.3	7.4	7.5	7.6	7.7	7.8	7.9	8.0
0	.0008	.0007	.0007	.0006	.0006	.0005	.0005	.0004	.0004	.0003
1	.0059	.0054	.0049	.0045	.0041	.0038	.0035	.0032	.0029	.0027
2	.0208	.0194	.0180	.0167	.0156	.0145	.0134	.0125	.0116	.0107
3	.0492	.0464	.0438	.0413	.0389	.0366	.0345	.0324	.0305	.0286
4	.0874	.0836	.0799	.0764	.0729	.0696	.0663	.0632	.0602	.0573
5	.1241	.1204	.1167	.1130	.1094	.1057	.1021	.0986	.0951	.0916
6	.1468	.1445	.1420	.1394	.1367	.1339	.1311	.1282	.1252	.1221
7	.1489	.1486	.1481	.1474	.1465	.1454	.1442	.1428	.1413	.1396
8	.1321	.1337	.1351	.1363	.1373	.1382	.1388	.1392	.1395	.1396
9	.1042	.1070	.1096	.1121	.1144	.1167	.1187	.1207	.1224	.1241
10	.0740	.0770	.0800	.0829	.0858	.0887	.0914	.0941	.0967	.0993
11	.0478	.0504	.0531	.0558	.0585	.0613	.0640	.0667	.0695	.0722
12	.0283	.0303	.0323	.0344	.0366	.0388	.0411	.0434	.0457	.0481
13	.0154	.0168	.0181	.0196	.0211	.0227	.0243	.0260	.0278	.0296
14	.0078	.0086	.0095	.0104	.0113	.0123	.0134	.0145	.0157	.0169
15	.0037	.0041	.0046	.0051	.0057	.0062	.0069	.0075	.0083	.0090
16	.0016	.0019	.0021	.0024	.0026	.0030	.0033	.0037	.0041	.0045
17	.0007	.0008	.0009	.0010	.0012	.0013	.0015	.0017	.0019	.0021
18	.0003	.0003	.0004	.0004	.0005	.0006	.0006	.0007	.0008	.0009
19	.0001	.0001	.0001	.0002	.0002	.0002	.0003	.0003	.0003	.0004
20	.0000	.0000	.0001	.0001	.0001	.0001	.0001	.0001	.0001	.0002
21	.0000	.0000	.0000	.0000	.0000	.0000	.0000	.0000	.0001	.0001

x	λt 8.1	8.2	8.3	8.4	8.5	8.6	8.7	8.8	8.9	9.0
0	.0003	.0003	.0002	.0002	.0002	.0002	.0002	.0002	.0001	.0001
1	.0025	.0023	.0021	.0019	.0017	.0016	.0014	.0013	.0012	.0011
2	.0100	.0092	.0086	.0079	.0074	.0068	.0063	.0058	.0054	.0050
3	.0269	.0252	.0237	.0222	.0208	.0195	.0183	.0171	.0160	.0150
4	.0544	.0517	.0491	.0466	.0443	.0420	.0398	.0377	.0357	.0337
5	.0882	.0849	.0816	.0784	.0752	.0722	.0692	.0663	.0635	.0607
6	.1191	.1160	.1128	.1097	.1066	.1034	.1003	.0972	.0941	.0911
7	.1378	.1358	.1338	.1317	.1294	.1271	.1247	.1222	.1197	.1171
8	.1395	.1392	.1388	.1382	.1375	.1366	.1356	.1344	.1332	.1318
9	.1256	.1269	.1280	.1290	.1299	.1306	.1311	.1315	.1317	.1318
10	.1017	.1040	.1063	.1084	.1104	.1123	.1140	.1157	.1172	.1186
11	.0749	.0776	.0802	.0828	.0853	.0878	.0902	.0925	.0948	.0970
12	.0505	.0530	.0555	.0579	.0604	.0629	.0654	.0679	.0703	.0728
13	.0315	.0334	.0354	.0374	.0395	.0416	.0438	.0459	.0481	.0504
14	.0182	.0196	.0210	.0225	.0240	.0256	.0272	.0289	.0306	.0324
15	.0098	.0107	.0116	.0126	.0136	.0147	.0158	.0169	.0182	.0194
16	.0050	.0055	.0060	.0066	.0072	.0079	.0086	.0093	.0101	.0109
17	.0024	.0026	.0029	.0033	.0036	.0040	.0044	.0048	.0053	.0058
18	.0011	.0012	.0014	.0015	.0017	.0019	.0021	.0024	.0026	.0029
19	.0005	.0005	.0006	.0007	.0008	.0009	.0010	.0011	.0012	.0014
20	.0002	.0002	.0002	.0003	.0003	.0004	.0004	.0005	.0005	.0006
21	.0001	.0001	.0001	.0001	.0001	.0002	.0002	.0002	.0002	.0003
22	.0000	.0000	.0000	.0000	.0001	.0001	.0001	.0001	.0001	.0001

x	λt 9.1	9.2	9.3	9.4	9.5	9.6	9.7	9.8	9.9	10
0	.0001	.0001	.0001	.0001	.0001	.0001	.0001	.0001	.0001	.0000
1	.0010	.0009	.0009	.0008	.0007	.0007	.0006	.0005	.0005	.0005
2	.0046	.0043	.0040	.0037	.0034	.0031	.0029	.0027	.0025	.0023
3	.0140	.0131	.0123	.0115	.0107	.0100	.0093	.0087	.0081	.0076
4	.0319	.0302	.0285	.0269	.0254	.0240	.0226	.0213	.0201	.0189
5	.0581	.0555	.0530	.0506	.0483	.0460	.0439	.0418	.0398	.0378
6	.0881	.0851	.0822	.0793	.0764	.0736	.0709	.0682	.0656	.0631
7	.1145	.1118	.1091	.1064	.1037	.1010	.0982	.0955	.0928	.0901
8	.1302	.1286	.1269	.1251	.1232	.1212	.1191	.1170	.1148	.1126
9	.1317	.1315	.1311	.1306	.1300	.1293	.1284	.1274	.1263	.1251
10	.1198	.1210	.1219	.1228	.1235	.1241	.1245	.1249	.1250	.1251
11	.0991	.1012	.1031	.1049	.1067	.1083	.1098	.1112	.1125	.1137
12	.0752	.0776	.0799	.0822	.0844	.0866	.0888	.0908	.0928	.0948
13	.0526	.0549	.0572	.0594	.0617	.0640	.0662	.0685	.0707	.0729
14	.0342	.0361	.0380	.0399	.0419	.0439	.0459	.0479	.0500	.0521
15	.0208	.0221	.0235	.0250	.0265	.0281	.0297	.0313	.0330	.0347
16	.0118	.0127	.0137	.0147	.0157	.0168	.0180	.0192	.0204	.0217
17	.0063	.0069	.0075	.0081	.0088	.0095	.0103	.0111	.0119	.0128
18	.0032	.0035	.0039	.0042	.0046	.0051	.0055	.0060	.0065	.0071
19	.0015	.0017	.0019	.0021	.0023	.0026	.0028	.0031	.0034	.0037

TABLE A3 *Poisson Probabilities (Individual Terms)* (continued)

x	λt 9.1	9.2	9.3	9.4	9.5	9.6	9.7	9.8	9.9	10
20	.0007	.0008	.0009	.0010	.0011	.0012	.0014	.0015	.0017	.0019
21	.0003	.0003	.0004	.0004	.0005	.0006	.0006	.0007	.0008	.0009
22	.0001	.0001	.0002	.0002	.0002	.0002	.0003	.0003	.0004	.0004
23	.0000	.0001	.0001	.0001	.0001	.0001	.0001	.0001	.0002	.0002
24	.0000	.0000	.0000	.0000	.0000	.0000	.0000	.0001	.0001	.0001

x	λt 11	12	13	14	15	16	17	18	19	20
0	.0000	.0000	.0000	.0000	.0000	.0000	.0000	.0000	.0000	.0000
1	.0002	.0001	.0000	.0000	.0000	.0000	.0000	.0000	.0000	.0000
2	.0010	.0004	.0002	.0001	.0000	.0000	.0000	.0000	.0000	.0000
3	.0037	.0018	.0008	.0004	.0002	.0001	.0000	.0000	.0000	.0000
4	.0102	.0053	.0027	.0013	.0006	.0003	.0001	.0001	.0000	.0000
5	.0224	.0127	.0070	.0037	.0019	.0010	.0005	.0002	.0001	.0001
6	.0411	.0255	.0152	.0087	.0048	.0026	.0014	.0007	.0004	.0002
7	.0646	.0437	.0281	.0174	.0104	.0060	.0034	.0018	.0010	.0005
8	.0888	.0655	.0457	.0304	.0194	.0120	.0072	.0042	.0024	.0013
9	.1085	.0874	.0661	.0473	.0324	.0213	.0135	.0083	.0050	.0029
10	.1194	.1048	.0859	.0663	.0486	.0341	.0230	.0150	.0095	.0058
11	.1194	.1144	.1015	.0844	.0663	.0496	.0355	.0245	.0164	.0106
12	.1094	.1144	.1099	.0984	.0829	.0661	.0504	.0368	.0259	.0176
13	.0926	.1056	.1099	.1060	.0956	.0814	.0658	.0509	.0378	.0271
14	.0728	.0905	.1021	.1060	.1024	.0930	.0800	.0655	.0514	.0387
15	.0534	.0724	.0885	.0989	.1024	.0992	.0906	.0786	.0650	.0516
16	.0367	.0543	.0719	.0866	.0960	.0992	.0963	.0884	.0772	.0646
17	.0237	.0383	.0550	.0713	.0847	.0934	.0963	.0936	.0863	.0760
18	.0145	.0256	.0397	.0554	.0706	.0830	.0909	.0936	.0911	.0844
19	.0084	.0161	.0272	.0409	.0557	.0699	.0814	.0887	.0911	.0888
20	.0046	.0097	.0177	.0286	.0418	.0559	.0692	.0798	.0866	.0888
21	.0024	.0055	.0109	.0191	.0299	.0426	.0560	.0684	.0783	.0846
22	.0012	.0030	.0065	.0121	.0204	.0310	.0433	.0560	.0676	.0769
23	.0006	.0016	.0037	.0074	.0133	.0216	.0320	.0438	.0559	.0669
24	.0003	.0008	.0020	.0043	.0083	.0144	.0226	.0328	.0442	.0557
25	.0001	.0004	.0010	.0024	.0050	.0092	.0154	.0237	.0336	.0446
26	.0000	.0002	.0005	.0013	.0029	.0057	.0101	.0164	.0246	.0343
27	.0000	.0001	.0002	.0007	.0016	.0034	.0063	.0109	.0173	.0254
28	.0000	.0000	.0001	.0003	.0009	.0019	.0038	.0070	.0117	.0181
29	.0000	.0000	.0001	.0002	.0004	.0011	.0023	.0044	.0077	.0125
30	.0000	.0000	.0000	.0001	.0002	.0006	.0013	.0026	.0049	.0083
31	.0000	.0000	.0000	.0000	.0001	.0003	.0007	.0015	.0030	.0054
32	.0000	.0000	.0000	.0000	.0001	.0001	.0004	.0009	.0018	.0034
33	.0000	.0000	.0000	.0000	.0000	.0001	.0002	.0005	.0010	.0020
34	.0000	.0000	.0000	.0000	.0000	.0000	.0001	.0002	.0006	.0012
35	.0000	.0000	.0000	.0000	.0000	.0000	.0000	.0001	.0003	.0007
36	.0000	.0000	.0000	.0000	.0000	.0000	.0000	.0001	.0002	.0004
37	.0000	.0000	.0000	.0000	.0000	.0000	.0000	.0000	.0001	.0002
38	.0000	.0000	.0000	.0000	.0000	.0000	.0000	.0000	.0000	.0001
39	.0000	.0000	.0000	.0000	.0000	.0000	.0000	.0000	.0000	.0001

Adapted from Table III.3—Individual terms, Poisson Distribution, from *CRC Handbook of Tables for Probability and Statistics* (1966).

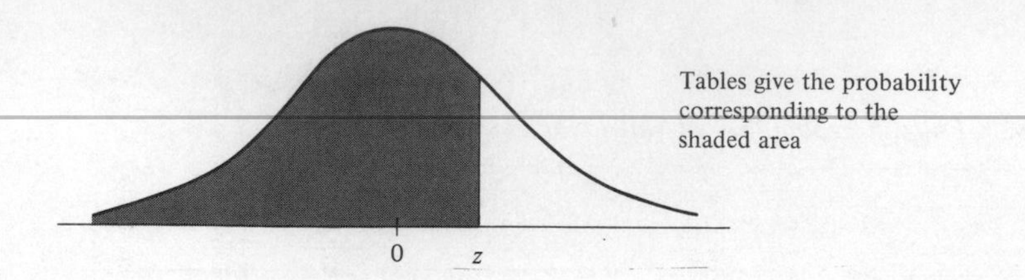

TABLE A4 *The Standard Normal Cumulative Distribution Function*

$$F(z) = P[Z \leq z]$$

z	$F(z)$	z	$F(z)$	z	$F(z)$	z	$F(z)$
.00	.5000						
.01	.5040	.51	.6950	1.01	.8438	1.51	.9345
.02	.5080	.52	.6985	1.02	.8461	1.52	.9357
.03	.5120	.53	.7019	1.03	.8485	1.53	.9370
.04	.5160	.54	.7054	1.04	.8508	1.54	.9382
.05	.5199	.55	.7088	1.05	.8531	1.55	.9394
.06	.5239	.56	.7123	1.06	.8554	1.56	.9406
.07	.5279	.57	.7157	1.07	.8577	1.57	.9418
.08	.5319	.58	.7190	1.08	.8599	1.58	.9429
.09	.5359	.59	.7224	1.09	.8621	1.59	.9441
.10	.5398	.60	.7257	1.10	.8643	1.60	.9452
.11	.5438	.61	.7291	1.11	.8665	1.61	.9463
.12	.5478	.62	.7324	1.12	.8686	1.62	.9474
.13	.5517	.63	.7357	1.13	.8708	1.63	.9484
.14	.5557	.64	.7389	1.14	.8729	1.64	.9495
.15	.5596	.65	.7422	1.15	.8749	1.65	.9505
.16	.5636	.66	.7454	1.16	.8770	1.66	.9515
.17	.5675	.67	.7486	1.17	.8790	1.67	.9525
.18	.5714	.68	.7517	1.18	.8810	1.68	.9535
.19	.5753	.69	.7549	1.19	.8830	1.69	.9545
.20	.5793	.70	.7580	1.20	.8849	1.70	.9554
.21	.5832	.71	.7611	1.21	.8869	1.71	.9564
.22	.5871	.72	.7642	1.22	.8888	1.72	.9573
.23	.5910	.73	.7673	1.23	.8907	1.73	.9582
.24	.5948	.74	.7704	1.24	.8925	1.74	.9591
.25	.5987	.75	.7734	1.25	.8944	1.75	.9599
.26	.6026	.76	.7764	1.26	.8962	1.76	.9608
.27	.6064	.77	.7794	1.27	.8980	1.77	.9616
.28	.6103	.78	.7823	1.28	.8997	1.78	.9625
.29	.6141	.79	.7852	1.29	.9015	1.79	.9633
.30	.6179	.80	.7881	1.30	.9032	1.80	.9641
.31	.6217	.81	.7910	1.31	.9049	1.81	.9649
.32	.6255	.82	.7939	1.32	.9066	1.82	.9656
.33	.6293	.83	.7967	1.33	.9082	1.83	.9664
.34	.6331	.84	.7995	1.34	.9099	1.84	.9671
.35	.6368	.85	.8023	1.35	.9115	1.85	.9678
.36	.6406	.86	.8051	1.36	.9131	1.86	.9686
.37	.6443	.87	.8078	1.37	.9147	1.87	.9693
.38	.6480	.88	.8106	1.38	.9162	1.88	.9699
.39	.6517	.89	.8133	1.39	.9177	1.89	.9706
.40	.6554	.90	.8159	1.40	.9192	1.90	.9713
.41	.6591	.91	.8186	1.41	.9207	1.91	.9719
.42	.6628	.92	.8212	1.42	.9222	1.92	.9726
.43	.6664	.93	.8238	1.43	.9236	1.93	.9732
.44	.6700	.94	.8264	1.44	.9251	1.94	.9738
.45	.6736	.95	.8289	1.45	.9265	1.95	.9744
.46	.6772	.96	.8315	1.46	.9279	1.96	.9750
.47	.6803	.97	.8340	1.47	.9292	1.97	.9756
.48	.6844	.98	.8365	1.48	.9306	1.98	.9761
.49	.6879	.99	.8389	1.49	.9319	1.99	.9767
.50	.6915	1.00	.8413	1.50	.9332	2.00	.9772

TABLE A4 *The Standard Normal Cumulative Distribution Function* (continued)

z	$F(z)$	z	$F(z)$	z	$F(z)$	z	$F(z)$
2.01	.9778	2.51	.9940	3.01	.9987	3.51	.9998
2.02	.9783	2.52	.9941	3.02	.9987	3.52	.9998
2.03	.9788	2.53	.9943	3.03	.9988	3.53	.9998
2.04	.9793	2.54	.9945	3.04	.9988	3.54	.9998
2.05	.9798	2.55	.9946	3.05	.9989	3.55	.9998
2.06	.9803	2.56	.9948	3.06	.9989	3.56	.9998
2.07	.9808	2.57	.9949	3.07	.9989	3.57	.9998
2.08	.9812	2.58	.9951	3.08	.9990	3.58	.9998
2.09	.9817	2.59	.9952	3.09	.9990	3.59	.9998
2.10	.9821	2.60	.9953	3.10	.9990	3.60	.9998
2.11	.9826	2.61	.9955	3.11	.9991	3.61	.9998
2.12	.9830	2.62	.9956	3.12	.9991	3.62	.9999
2.13	.9834	2.63	.9957	3.13	.9991	3.63	.9999
2.14	.9838	2.64	.9959	3.14	.9992	3.64	.9999
2.15	.9842	2.65	.9960	3.15	.9992	3.65	.9999
2.16	.9846	2.66	.9961	3.16	.9992	3.66	.9999
2.17	.9850	2.67	.9962	3.17	.9992	3.67	.9999
2.18	.9854	2.68	.9963	3.18	.9993	3.68	.9999
2.19	.9857	2.69	.9964	3.19	.9993	3.69	.9999
2.20	.9861	2.70	.9965	3.20	.9993	3.70	.9999
2.21	.9864	2.71	.9966	3.21	.9993	3.71	.9999
2.22	.9868	2.72	.9967	3.22	.9994	3.72	.9999
2.23	.9871	2.73	.9968	3.23	.9994	3.73	.9999
2.24	.9875	2.74	.9969	3.24	.9994	3.74	.9999
2.25	.9878	2.75	.9970	3.25	.9994	3.75	.9999
2.26	.9881	2.76	.9971	3.26	.9994	3.76	.9999
2.27	.9884	2.77	.9972	3.27	.9995	3.77	.9999
2.28	.9887	2.78	.9973	3.28	.9995	3.78	.9999
2.29	.9890	2.79	.9974	3.29	.9995	3.79	.9999
2.30	.9893	2.80	.9974	3.30	.9995	3.80	.9999
2.31	.9896	2.81	.9975	3.31	.9995	3.81	.9999
2.32	.9898	2.82	.9976	3.32	.9996	3.82	.9999
2.33	.9901	2.83	.9977	3.33	.9996	3.83	.9999
2.34	.9904	2.84	.9977	3.34	.9996	3.84	.9999
2.35	.9906	2.85	.9978	3.35	.9996	3.85	.9999
2.36	.9909	2.86	.9979	3.36	.9996	3.86	.9999
2.37	.9911	2.87	.9979	3.37	.9996	3.87	.9999
2.38	.9913	2.88	.9980	3.38	.9996	3.88	.9999
2.39	.9916	2.89	.9981	3.39	.9997	3.89	1.0000
2.40	.9918	2.90	.9981	3.40	.9997	3.90	1.0000
2.41	.9920	2.91	.9982	3.41	.9997	3.91	1.0000
2.42	.9922	2.92	.9982	3.42	.9997	3.92	1.0000
2.43	.9925	2.93	.9983	3.43	.9997	3.93	1.0000
2.44	.9927	2.94	.9984	3.44	.9997	3.94	1.0000
2.45	.9929	2.95	.9984	3.45	.9997	3.95	1.0000
2.46	.9931	2.96	.9985	3.46	.9997	3.96	1.0000
2.47	.9932	2.97	.9985	3.47	.9997	3.97	1.0000
2.48	.9934	2.98	.9986	3.48	.9997	3.98	1.0000
2.49	.9936	2.99	.9986	3.49	.9998	3.99	1.0000
2.50	.9938	3.00	.9986	3.50	.9998		

Some common percentiles:

1.282 .9000 1.645 .9500 1.960 .975 2.326 .990

Condensed from Table 1 of *Biometrika Tables for Statisticians,* Vol. 1 (1966), by Pearson and Hartley. Reproduced with permission of the trustees of Biometrika.

TABLE A5 *Random Numbers*

10 48 01 50 11	01 53 60 20 11	81 64 79 16 46	69 17 91 41 94	62 59 03 62 07
22 36 84 65 73	25 59 58 53 93	30 99 58 91 98	27 98 25 34 02	93 96 53 40 95
24 13 04 83 60	22 52 79 72 65	76 39 36 48 09	15 17 92 48 30	49 34 03 20 81
42 16 79 30 93	06 24 36 16 80	07 85 61 63 76	39 44 05 35 37	71 34 15 70 04
37 57 03 99 75	81 83 71 66 56	64 03 23 66 53	98 95 11 68 77	12 17 17 68 33
77 92 10 69 07	11 00 84 27 51	27 75 65 34 98	18 60 27 06 59	90 65 51 50 53
99 56 27 29 05	56 42 06 99 94	98 87 23 10 16	71 19 41 87 38	44 01 34 88 40
96 30 19 19 77	05 46 30 79 72	18 87 62 09 22	94 59 55 68 69	69 01 46 00 45
89 57 91 43 42	63 66 11 02 81	17 45 31 81 03	57 74 08 43 78	25 33 11 25 66
85 47 53 68 57	43 34 25 39 88	53 06 05 95 33	38 86 76 23 00	08 15 81 79 83
28 91 86 95 78	88 23 13 32 76	70 99 77 99 36	56 86 50 58 59	90 10 63 15 95
63 55 34 09 61	48 23 50 34 27	49 62 66 94 45	18 66 37 26 95	52 18 02 08 47
09 42 99 39 69	52 63 69 27 37	88 97 43 34 88	36 32 01 76 17	30 01 50 82 72
10 36 56 11 29	87 52 98 56 89	48 23 75 22 67	67 68 99 33 94	01 51 12 63 58
07 11 99 73 36	71 04 80 81 78	77 23 31 39 16	47 56 48 10 56	97 73 58 59 77
51 08 51 27 65	51 82 15 12 59	77 45 21 63 08	60 75 69 21 33	49 44 25 39 00
02 36 82 13 82	52 40 46 02 68	89 36 81 98 85	55 32 24 48 19	01 18 86 52 55
01 01 15 40 92	33 36 29 49 04	31 27 30 41 46	18 59 42 98 52	71 58 58 50 30
52 16 25 39 16	46 36 95 85 86	23 21 61 45 13	83 14 99 87 36	23 49 56 43 50
07 05 69 76 28	33 78 70 99 98	42 69 80 66 91	76 98 81 36 02	51 85 14 61 04
48 66 39 12 45	85 82 81 43 46	09 17 23 01 68	90 22 90 47 34	59 19 32 21 93
54 16 45 84 92	22 42 17 41 03	47 07 02 53 06	76 46 82 63 84	58 15 10 66 46
32 63 93 23 63	05 59 72 42 00	13 36 33 80 05	94 34 22 87 28	35 90 60 69 12
29 33 42 70 01	87 63 78 73 08	58 73 10 02 56	45 83 41 53 98	46 55 74 11 35
02 48 83 30 62	28 83 40 73 51	19 73 19 24 20	60 95 26 12 80	50 00 16 76 58
81 52 57 22 95	04 83 99 64 23	24 87 88 26 51	66 56 61 47 78	76 79 71 47 80
29 67 62 05 91	68 08 62 64 32	46 90 12 08 49	89 76 88 15 36	86 64 51 26 59
00 74 25 73 92	39 06 46 64 32	84 67 34 00 27	32 83 26 13 62	98 94 79 60 67
05 36 60 42 13	25 66 92 64 22	44 40 74 40 48	37 93 76 39 04	45 76 66 61 34
91 92 12 64 18	64 11 79 43 05	26 76 62 59 40	39 97 22 22 09	71 50 06 45 68
00 58 20 47 11	87 91 77 73 41	42 20 63 51 26	74 08 79 95 47	81 81 74 26 07
00 72 56 98 84	62 79 75 61 70	86 32 48 80 72	76 22 23 60 86	84 63 79 31 61
69 01 16 57 97	95 87 65 52 93	18 98 82 73 54	26 57 50 86 25	40 80 15 99 20
25 97 65 79 48	29 88 88 86 04	67 91 74 87 08	18 91 28 22 71	65 42 46 97 74
09 76 38 34 73	73 57 71 29 08	30 88 31 83 17	28 29 03 57 97	05 99 84 16 88
91 56 74 25 95	27 95 83 01 34	04 02 48 63 85	29 88 09 97 30	55 53 68 48 55
17 95 55 63 49	90 00 04 01 27	20 04 45 99 31	06 11 52 05 42	18 05 90 20 08
46 50 31 85 84	18 84 54 96 18	02 30 45 10 38	20 65 55 87 27	28 16 81 54 75
92 15 78 96 34	94 82 47 81 71	84 61 18 28 34	09 92 22 54 17	44 13 74 84 13
14 57 76 27 65	35 60 58 12 63	39 66 74 73 58	56 87 35 63 07	61 60 74 95 18
98 42 70 75 23	33 36 26 42 70	01 63 89 24 77	66 96 99 84 20	04 88 04 55 85
34 91 46 39 76	88 72 08 27 65	34 47 61 70 32	87 58 94 08 36	32 42 77 00 02
70 06 02 82 77	39 47 54 64 73	23 21 95 34 16	94 97 02 58 32	69 97 59 48 84
53 97 65 49 14	06 99 06 72 45	68 35 08 29 48	11 39 84 28 78	80 28 78 82 67
76 07 22 95 15	40 98 00 73 91	58 74 52 57 74	22 98 78 00 59	39 91 19 61 89
90 72 55 22 10	83 97 42 99 92	65 83 13 88 57	50 49 08 37 65	55 65 71 43 61
64 36 46 74 12	33 33 93 19 26	14 88 32 44 13	59 74 49 23 51	97 47 38 92 26
08 96 20 03 58	31 66 22 53 88	61 64 23 40 72	81 24 93 56 48	56 89 16 93 52
95 01 26 83 79	93 52 67 07 65	10 59 30 45 42	76 46 35 43 28	02 34 91 72 47
15 66 41 04 93	20 49 23 83 91	91 13 22 19 99	59 51 68 16 52	27 19 54 82 23

Adapted from Table XII.3—Random Units, from *CRC Handbook of Tables for Probability and Statistics* (1966).

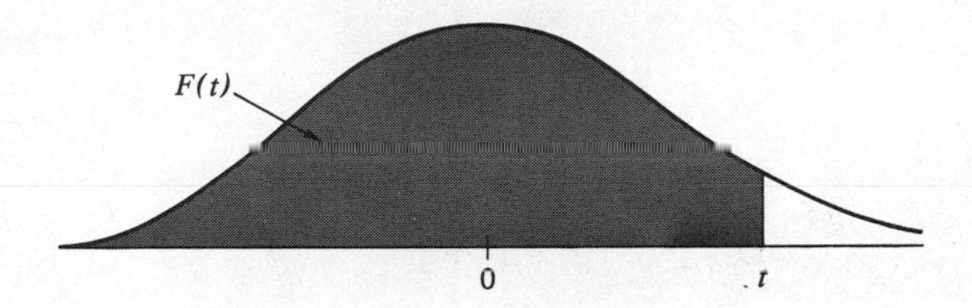

TABLE A6 *Percentage Points of the Student's t Distribution*

ν	$F(t)$ .60	.75	.90	.95	.975	.99	.995	.9975	.999	.9995
1	0·325	1·000	3·078	6·314	12·706	31·821	63·657	127·32	318·31	636·62
2	·289	0·816	1·886	2·920	4·303	6·965	9·925	14·089	22·327	31·598
3	·277	·765	1·638	2·353	3·182	4·541	5·841	7·453	10·214	12·924
4	·271	·741	1·533	2·132	2·776	3·747	4·604	5·598	7·173	8·610
5	0·267	0·727	1·476	2·015	2·571	3·365	4·032	4·773	5·893	6·869
6	·265	·718	1·440	1·943	2·447	3·143	3·707	4·317	5·208	5·959
7	·263	·711	1·415	1·895	2·365	2·998	3·499	4·029	4·785	5·408
8	·262	·706	1·397	1·860	2·306	2·896	3·355	3·833	4·501	5·041
9	·261	·703	1·383	1·833	2·262	2·821	3·250	3·690	4·297	4·781
10	0·260	0·700	1·372	1·812	2·228	2·764	3·169	3·581	4·144	4·587
11	·260	·697	1·363	1·796	2·201	2·718	3·106	3·497	4·025	4·437
12	·259	·695	1·356	1·782	2·179	2·681	3·055	3·428	3·930	4·318
13	·259	·694	1·350	1·771	2·160	2·650	3·012	3·372	3·852	4·221
14	·258	·692	1·345	1·761	2·145	2·624	2·977	3·326	3·787	4·140
15	0·258	0·691	1·341	1·753	2·131	2·602	2·947	3·286	3·733	4·073
16	·258	·690	1·337	1·746	2·120	2·583	2·921	3·252	3·686	4·015
17	·257	·689	1·333	1·740	2·110	2·567	2·898	3·222	3·646	3·965
18	·257	·688	1·330	1·734	2·101	2·552	2·878	3·197	3·610	3·922
19	·257	·688	1·328	1·729	2·093	2·539	2·861	3·174	3·579	3·883
20	0·257	0·687	1·325	1·725	2·086	2·528	2·845	3·153	3·552	3·850
21	·257	·686	1·323	1·721	2·080	2·518	2·831	3·135	3·527	3·819
22	·256	·686	1·321	1·717	2·074	2·508	2·819	3·119	3·505	3·792
23	·256	·685	1·319	1·714	2·069	2·500	2·807	3·104	3·485	3·767
24	·256	·685	1·318	1·711	2·064	2·492	2·797	3·091	3·467	3·745
25	0·256	0·684	1·316	1·708	2·060	2·485	2·787	3·078	3·450	3·725
26	·256	·684	1·315	1·706	2·056	2·479	2·779	3·067	3·435	3·707
27	·256	·684	1·314	1·703	2·052	2·473	2·771	3·057	3·421	3·690
28	·256	·683	1·313	1·701	2·048	2·467	2·763	3·047	3·408	3·674
29	·256	·683	1·311	1·699	2·045	2·462	2·756	3·038	3·396	3·659
30	0·256	0·683	1·310	1·697	2·042	2·457	2·750	3·030	3·385	3·646
40	·255	·681	1·303	1·684	2·021	2·423	2·704	2·971	3·307	3·551
60	·254	·679	1·296	1·671	2·000	2·390	2·660	2·915	3·232	3·460
120	·254	·677	1·289	1·658	1·980	2·358	2·617	2·860	3·160	3·373
∞	·253	·674	1·282	1·645	1·960	2·326	2·576	2·807	3·090	3·291

Condensed from Table 12 of *Biometrika Tables for Statisticians,* Vol. 1 (1966), by Pearson and Hartley. Reproduced with permission of the trustees of Biometrika.

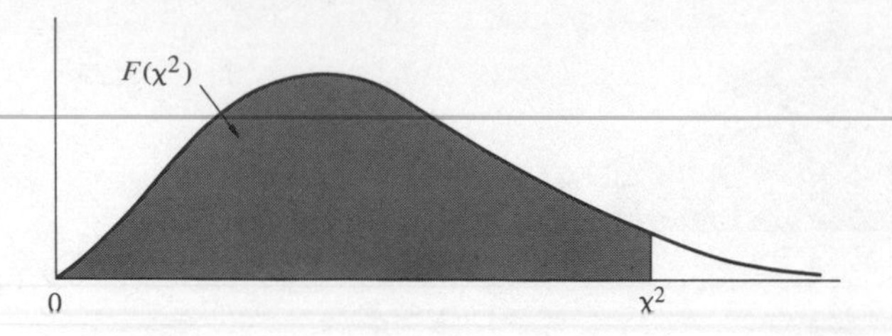

TABLE A7 *Percentage Points of the χ^2 Distribution*

ν	$F(\chi^2)$ .005	.01	.025	.05	.10	.25	.50
1	392704.10^{-10}	157088.10^{-9}	982069.10^{-9}	393214.10^{-8}	0·0157908	0·1015308	0·454936
2	0·0100251	0·0201007	0·0506356	0·102587	0·210721	0·575364	1·38629
3	0·0717218	0·114832	0·215795	0·351846	0·584374	1·212534	2·36597
4	0·206989	0·297109	0·484419	0·710723	1·063623	1·92256	3·35669
5	0·411742	0·554298	0·831212	1·145476	1·61031	2·67460	4·35146
6	0·675727	0·872090	1·23734	1·63538	2·20413	3·45460	5·34812
7	0·989256	1·239043	1·68987	2·16735	2·83311	4·25485	6·34581
8	1·34441	1·64650	2·17973	2·73264	3·48954	5·07064	7·34412
9	1·73493	2·08790	2·70039	3·32511	4·16816	5·89883	8·34283
10	2·15586	2·55821	3·24697	3·94030	4·86518	6·73720	9·34182
11	2·60322	3·05348	3·81575	4·57481	5·57778	7·58414	10·3410
12	3·07382	3·57057	4·40379	5·22603	6·30380	8·43842	11·3403
13	3·56503	4·10692	5·00875	5·89186	7·04150	9·29907	12·3398
14	4·07467	4·66043	5·62873	6·57063	7·78953	10·1653	13·3393
15	4·60092	5·22935	6·26214	7·26094	8·54676	11·0365	14·3389
16	5·14221	5·81221	6·90766	7·96165	9·31224	11·9122	15·3385
17	5·69722	6·40776	7·56419	8·67176	10·0852	12·7919	16·3382
18	6·26480	7·01491	8·23075	9·39046	10·8649	13·6753	17·3379
19	6·84397	7·63273	8·90652	10·1170	11·6509	14·5620	18·3377
20	7·43384	8·26040	9·59078	10·8508	12·4426	15·4518	19·3374
21	8·03365	8·89720	10·28293	11·5913	13·2396	16·3444	20·3372
22	8·64272	9·54249	10·9823	12·3380	14·0415	17·2396	21·3370
23	9·26043	10·19567	11·6886	13·0905	14·8480	18·1373	22·3369
24	9·88623	10·8564	12·4012	13·8484	15·6587	19·0373	23·3367
25	10·5197	11·5240	13·1197	14·6114	16·4734	19·9393	24·3366
26	11·1602	12·1981	13·8439	15·3792	17·2919	20·8434	25·3365
27	11·8076	12·8785	14·5734	16·1514	18·1139	21·7494	26·3363
28	12·4613	13·5647	15·3079	16·9279	18·9392	22·6572	27·3362
29	13·1211	14·2565	16·0471	17·7084	19·7677	23·5666	28·3361
30	13·7867	14·9535	16·7908	18·4927	20·5992	24·4776	29·3360
40	20·7065	22·1643	24·4330	26·5093	29·0505	33·6603	39·3353
50	27·9907	29·7067	32·3574	34·7643	37·6886	42·9421	49·3349
60	35·5345	37·4849	40·4817	43·1880	46·4589	52·2938	59·3347
70	43·2752	45·4417	48·7576	51·7393	55·3289	61·6983	69·3345
80	51·1719	53·5401	57·1532	60·3915	64·2778	71·1445	79·3343
90	59·1963	61·7541	65·6466	69·1260	73·2911	80·6247	89·3342
100	67·3276	70·0649	74·2219	77·9295	82·3581	90·1332	99·3341

TABLE A7 *Percentage Points of the χ^2 Distribution* (continued)

ν	$F(\chi^2)$						
	.75	.90	.95	.975	.99	.995	.999
1	1·32330	2·70554	3·84146	5·02389	6·63490	7·87944	10·828
2	2·77259	4·60517	5·99146	7·37776	9·21034	10·5966	13·816
3	4·10834	6·25139	7·81473	9·34840	11·3449	12·8382	16·266
4	5·38527	7·77944	9·48773	11·1433	13·2767	14·8603	18·467
5	6·62568	9·23636	11·0705	12·8325	15·0863	16·7496	20·515
6	7·84080	10·6446	12·5916	14·4494	16·8119	18·5476	22·458
7	9·03715	12·0170	14·0671	16·0128	18·4753	20·2777	24·322
8	10·2189	13·3616	15·5073	17·5345	20·0902	21·9550	26·125
9	11·3888	14·6837	16·9190	19·0228	21·6660	23·5894	27·877
10	12·5489	15·9872	18·3070	20·4832	23·2093	25·1882	29·588
11	13·7007	17·2750	19·6751	21·9200	24·7250	26·7568	31·264
12	14·8454	18·5493	21·0261	23·3367	26·2170	28·2995	32·909
13	15·9839	19·8119	22·3620	24·7356	27·6882	29·8195	34·528
14	17·1169	21·0641	23·6848	26·1189	29·1412	31·3194	36·123
15	18·2451	22·3071	24·9958	27·4884	30·5779	32·8013	37·697
16	19·3689	23·5418	26·2962	28·8454	31·9999	34·2672	39·252
17	20·4887	24·7690	27·5871	30·1910	33·4087	35·7185	40·790
18	21·6049	25·9894	28·8693	31·5264	34·8053	37·1565	42·312
19	22·7178	27·2036	30·1435	32·8523	36·1909	38·5823	43·820
20	23·8277	28·4120	31·4104	34·1696	37·5662	39·9968	45·315
21	24·9348	29·6151	32·6706	35·4789	38·9322	41·4011	46·797
22	26·0393	30·8133	33·9244	36·7807	40·2894	42·7957	48·268
23	27·1413	32·0069	35·1725	38·0756	41·6384	44·1813	49·728
24	28·2412	33·1962	36·4150	39·3641	42·9798	45·5585	51·179
25	29·3389	34·3816	37·6525	40·6465	44·3141	46·9279	52·618
26	30·4346	35·5632	38·8851	41·9232	45·6417	48·2899	54·052
27	31·5284	36·7412	40·1133	43·1945	46·9629	49·6449	55·476
28	32·6205	37·9159	41·3371	44·4608	48·2782	50·9934	56·892
29	33·7109	39·0875	42·5570	45·7223	49·5879	52·3356	58·301
30	34·7997	40·2560	43·7730	46·9792	50·8922	53·6720	59·703
40	45·6160	51·8051	55·7585	59·3417	63·6907	66·7660	73·402
50	56·3336	63·1671	67·5048	71·4202	76·1539	79·4900	86·661
60	66·9815	74·3970	79·0819	83·2977	88·3794	91·9517	99·607
70	77·5767	85·5270	90·5312	95·0232	100·425	104·215	112·317
80	88·1303	96·5782	101·879	106·629	112·329	116·321	124·839
90	98·6499	107·565	113·145	118·136	124·116	128·299	137·208
100	109·141	118·498	124·342	129·561	135·807	140·169	149·449

Condensed from Table 8 of *Biometrika Tables for Statisticians,* Vol. 1 (1966), by Pearson and Hartley. Reproduced with permission of the trustees of Biometrika.

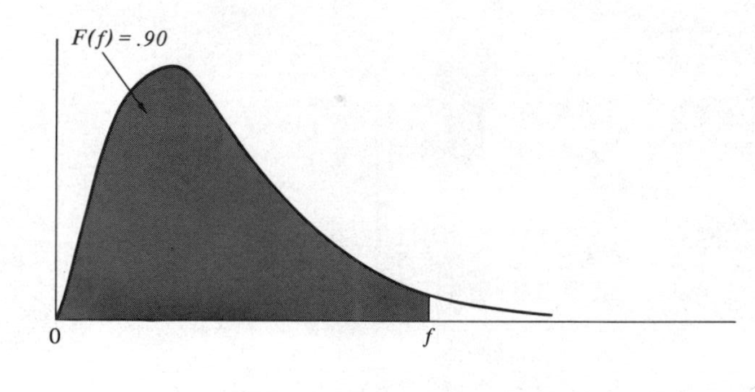

TABLE A8 *Percentage Points of the F Distribution*; $F(f) = .90$

ν_2 \ ν_1	1	2	3	4	5	6	7	8	9	10	12	15	20	24	30	40	60	120	∞
1	39·86	49·50	53·59	55·83	57·24	58·20	58·91	59·44	59·86	60·19	60·71	61·22	61·74	62·00	62·26	62·53	62·79	63·06	63·33
2	8·53	9·00	9·16	9·24	9·29	9·33	9·35	9·37	9·38	9·39	9·41	9·42	9·44	9·45	9·46	9·47	9·47	9·48	9·49
3	5·54	5·46	5·39	5·34	5·31	5·28	5·27	5·25	5·24	5·23	5·22	5·20	5·18	5·18	5·17	5·16	5·15	5·14	5·13
4	4·54	4·32	4·19	4·11	4·05	4·01	3·98	3·95	3·94	3·92	3·90	3·87	3·84	3·83	3·82	3·80	3·79	3·78	3·76
5	4·06	3·78	3·62	3·52	3·45	3·40	3·37	3·34	3·32	3·30	3·27	3·24	3·21	3·19	3·17	3·16	3·14	3·12	3·10
6	3·78	3·46	3·29	3·18	3·11	3·05	3·01	2·98	2·96	2·94	2·90	2·87	2·84	2·82	2·80	2·78	2·76	2·74	2·72
7	3·59	3·26	3·07	2·96	2·88	2·83	2·78	2·75	2·72	2·70	2·67	2·63	2·59	2·58	2·56	2·54	2·51	2·49	2·47
8	3·46	3·11	2·92	2·81	2·73	2·67	2·62	2·59	2·56	2·54	2·50	2·46	2·42	2·40	2·38	2·36	2·34	2·32	2·29
9	3·36	3·01	2·81	2·69	2·61	2·55	2·51	2·47	2·44	2·42	2·38	2·34	2·30	2·28	2·25	2·23	2·21	2·18	2·16
10	3·29	2·92	2·73	2·61	2·52	2·46	2·41	2·38	2·35	2·32	2·28	2·24	2·20	2·18	2·16	2·13	2·11	2·08	2·06
11	3·23	2·86	2·66	2·54	2·45	2·39	2·34	2·30	2·27	2·25	2·21	2·17	2·12	2·10	2·08	2·05	2·03	2·00	1·97
12	3·18	2·81	2·61	2·48	2·39	2·33	2·28	2·24	2·21	2·19	2·15	2·10	2·06	2·04	2·01	1·99	1·96	1·93	1·90
13	3·14	2·76	2·56	2·43	2·35	2·28	2·23	2·20	2·16	2·14	2·10	2·05	2·01	1·98	1·96	1·93	1·90	1·88	1·85
14	3·10	2·73	2·52	2·39	2·31	2·24	2·19	2·15	2·12	2·10	2·05	2·01	1·96	1·94	1·91	1·89	1·86	1·83	1·80
15	3·07	2·70	2·49	2·36	2·27	2·21	2·16	2·12	2·09	2·06	2·02	1·97	1·92	1·90	1·87	1·85	1·82	1·79	1·76
16	3·05	2·67	2·46	2·33	2·24	2·18	2·13	2·09	2·06	2·03	1·99	1·94	1·89	1·87	1·84	1·81	1·78	1·75	1·72
17	3·03	2·64	2·44	2·31	2·22	2·15	2·10	2·06	2·03	2·00	1·96	1·91	1·86	1·84	1·81	1·78	1·75	1·72	1·69
18	3·01	2·62	2·42	2·29	2·20	2·13	2·08	2·04	2·00	1·98	1·93	1·89	1·84	1·81	1·78	1·75	1·72	1·69	1·66
19	2·99	2·61	2·40	2·27	2·18	2·11	2·06	2·02	1·98	1·96	1·91	1·86	1·81	1·79	1·76	1·73	1·70	1·67	1·63
20	2·97	2·59	2·38	2·25	2·16	2·09	2·04	2·00	1·96	1·94	1·89	1·84	1·79	1·77	1·74	1·71	1·68	1·64	1·61
21	2·96	2·57	2·36	2·23	2·14	2·08	2·02	1·98	1·95	1·92	1·87	1·83	1·78	1·75	1·72	1·69	1·66	1·62	1·59
22	2·95	2·56	2·35	2·22	2·13	2·06	2·01	1·97	1·93	1·90	1·86	1·81	1·76	1·73	1·70	1·67	1·64	1·60	1·57
23	2·94	2·55	2·34	2·21	2·11	2·05	1·99	1·95	1·92	1·89	1·84	1·80	1·74	1·72	1·69	1·66	1·62	1·59	1·55
24	2·93	2·54	2·33	2·19	2·10	2·04	1·98	1·94	1·91	1·88	1·83	1·78	1·73	1·70	1·67	1·64	1·61	1·57	1·53
25	2·92	2·53	2·32	2·18	2·09	2·02	1·97	1·93	1·89	1·87	1·82	1·77	1·72	1·69	1·66	1·63	1·59	1·56	1·52
26	2·91	2·52	2·31	2·17	2·08	2·01	1·96	1·92	1·88	1·86	1·81	1·76	1·71	1·68	1·65	1·61	1·58	1·54	1·50
27	2·90	2·51	2·30	2·17	2·07	2·00	1·95	1·91	1·87	1·85	1·80	1·75	1·70	1·67	1·64	1·60	1·57	1·53	1·49
28	2·89	2·50	2·29	2·16	2·06	2·00	1·94	1·90	1·87	1·84	1·79	1·74	1·69	1·66	1·63	1·59	1·56	1·52	1·48
29	2·89	2·50	2·28	2·15	2·06	1·99	1·93	1·89	1·86	1·83	1·78	1·73	1·68	1·65	1·62	1·58	1·55	1·51	1·47
30	2·88	2·49	2·28	2·14	2·05	1·98	1·93	1·88	1·85	1·82	1·77	1·72	1·67	1·64	1·61	1·57	1·54	1·50	1·46
40	2·84	2·44	2·23	2·09	2·00	1·93	1·87	1·83	1·79	1·76	1·71	1·66	1·61	1·57	1·54	1·51	1·47	1·42	1·38
60	2·79	2·39	2·18	2·04	1·95	1·87	1·82	1·77	1·74	1·71	1·66	1·60	1·54	1·51	1·48	1·44	1·40	1·35	1·29
120	2·75	2·35	2·13	1·99	1·90	1·82	1·77	1·72	1·68	1·65	1·60	1·55	1·48	1·45	1·41	1·37	1·32	1·26	1·19
∞	2·71	2·30	2·08	1·94	1·85	1·77	1·72	1·67	1·63	1·60	1·55	1·49	1·42	1·38	1·34	1·30	1·24	1·17	1·00

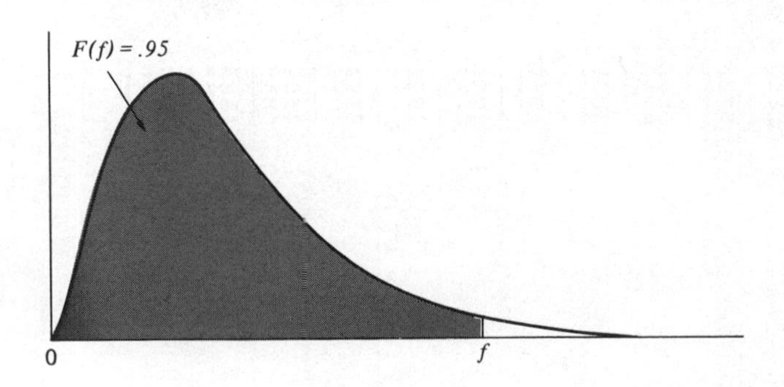

TABLE A8 *Percentage Points of the F Distribution*; $F(f) = .95$

ν_2 \ ν_1	1	2	3	4	5	6	7	8	9	10	12	15	20	24	30	40	60	120	∞
1	161·4	199·5	215·7	224·6	230·2	234·0	236·8	238·9	240·5	241·9	243·9	245·9	248·0	249·1	250·1	251·1	252·2	253·3	254·3
2	18·51	19·00	19·16	19·25	19·30	19·33	19·35	19·37	19·38	19·40	19·41	19·43	19·45	19·45	19·46	19·47	19·48	19·49	19·50
3	10·13	9·55	9·28	9·12	9·01	8·94	8·89	8·85	8·81	8·79	8·74	8·70	8·66	8·64	8·62	8·59	8·57	8·55	8·53
4	7·71	6·94	6·59	6·39	6·26	6·16	6·09	6·04	6·00	5·96	5·91	5·86	5·80	5·77	5·75	5·72	5·69	5·66	5·63
5	6·61	5·79	5·41	5·19	5·05	4·95	4·88	4·82	4·77	4·74	4·68	4·62	4·56	4·53	4·50	4·46	4·43	4·40	4·36
6	5·99	5·14	4·76	4·53	4·39	4·28	4·21	4·15	4·10	4·06	4·00	3·94	3·87	3·84	3·81	3·77	3·74	3·70	3·67
7	5·59	4·74	4·35	4·12	3·97	3·87	3·79	3·73	3·68	3·64	3·57	3·51	3·44	3·41	3·38	3·34	3·30	3·27	3·23
8	5·32	4·46	4·07	3·84	3·69	3·58	3·50	3·44	3·39	3·35	3·28	3·22	3·15	3·12	3·08	3·04	3·01	2·97	2·93
9	5·12	4·26	3·86	3·63	3·48	3·37	3·29	3·23	3·18	3·14	3·07	3·01	2·94	2·90	2·86	2·83	2·79	2·75	2·71
10	4·96	4·10	3·71	3·48	3·33	3·22	3·14	3·07	3·02	2·98	2·91	2·85	2·77	2·74	2·70	2·66	2·62	2·58	2·54
11	4·84	3·98	3·59	3·36	3·20	3·09	3·01	2·95	2·90	2·85	2·79	2·72	2·65	2·61	2·57	2·53	2·49	2·45	2·40
12	4·75	3·89	3·49	3·26	3·11	3·00	2·91	2·85	2·80	2·75	2·69	2·62	2·54	2·51	2·47	2·43	2·38	2·34	2·30
13	4·67	3·81	3·41	3·18	3·03	2·92	2·83	2·77	2·71	2·67	2·60	2·53	2·46	2·42	2·38	2·34	2·30	2·25	2·21
14	4·60	3·74	3·34	3·11	2·96	2·85	2·76	2·70	2·65	2·60	2·53	2·46	2·39	2·35	2·31	2·27	2·22	2·18	2·13
15	4·54	3·68	3·29	3·06	2·90	2·79	2·71	2·64	2·59	2·54	2·48	2·40	2·33	2·29	2·25	2·20	2·16	2·11	2·07
16	4·49	3·63	3·24	3·01	2·85	2·74	2·66	2·59	2·54	2·49	2·42	2·35	2·28	2·24	2·19	2·15	2·11	2·06	2·01
17	4·45	3·59	3·20	2·96	2·81	2·70	2·61	2·55	2·49	2·45	2·38	2·31	2·23	2·19	2·15	2·10	2·06	2·01	1·96
18	4·41	3·55	3·16	2·93	2·77	2·66	2·58	2·51	2·46	2·41	2·34	2·27	2·19	2·15	2·11	2·06	2·02	1·97	1·92
19	4·38	3·52	3·13	2·90	2·74	2·63	2·54	2·48	2·42	2·38	2·31	2·23	2·16	2·11	2·07	2·03	1·98	1·93	1·88
20	4·35	3·49	3·10	2·87	2·71	2·60	2·51	2·45	2·39	2·35	2·28	2·20	2·12	2·08	2·04	1·99	1·95	1·90	1·84
21	4·32	3·47	3·07	2·84	2·68	2·57	2·49	2·42	2·37	2·32	2·25	2·18	2·10	2·05	2·01	1·96	1·92	1·87	1·81
22	4·30	3·44	3·05	2·82	2·66	2·55	2·46	2·40	2·34	2·30	2·23	2·15	2·07	2·03	1·98	1·94	1·89	1·84	1·78
23	4·28	3·42	3·03	2·80	2·64	2·53	2·44	2·37	2·32	2·27	2·20	2·13	2·05	2·01	1·96	1·91	1·86	1·81	1·76
24	4·26	3·40	3·01	2·78	2·62	2·51	2·42	2·36	2·30	2·25	2·18	2·11	2·03	1·98	1·94	1·89	1·84	1·79	1·73
25	4·24	3·39	2·99	2·76	2·60	2·49	2·40	2·34	2·28	2·24	2·16	2·09	2·01	1·96	1·92	1·87	1·82	1·77	1·71
26	4·23	3·37	2·98	2·74	2·59	2·47	2·39	2·32	2·27	2·22	2·15	2·07	1·99	1·95	1·90	1·85	1·80	1·75	1·69
27	4·21	3·35	2·96	2·73	2·57	2·46	2·37	2·31	2·25	2·20	2·13	2·06	1·97	1·93	1·88	1·84	1·79	1·73	1·67
28	4·20	3·34	2·95	2·71	2·56	2·45	2·36	2·29	2·24	2·19	2·12	2·04	1·96	1·91	1·87	1·82	1·77	1·71	1·65
29	4·18	3·33	2·93	2·70	2·55	2·43	2·35	2·28	2·22	2·18	2·10	2·03	1·94	1·90	1·85	1·81	1·75	1·70	1·64
30	4·17	3·32	2·92	2·69	2·53	2·42	2·33	2·27	2·21	2·16	2·09	2·01	1·93	1·89	1·84	1·79	1·74	1·68	1·62
40	4·08	3·23	2·84	2·61	2·45	2·34	2·25	2·18	2·12	2·08	2·00	1·92	1·84	1·79	1·74	1·69	1·64	1·58	1·51
60	4·00	3·15	2·76	2·53	2·37	2·25	2·17	2·10	2·04	1·99	1·92	1·84	1·75	1·70	1·65	1·59	1·53	1·47	1·39
120	3·92	3·07	2·68	2·45	2·29	2·17	2·09	2·02	1·96	1·91	1·83	1·75	1·66	1·61	1·55	1·50	1·43	1·35	1·25
∞	3·84	3·00	2·60	2·37	2·21	2·10	2·01	1·94	1·88	1·83	1·75	1·67	1·57	1·52	1·46	1·39	1·32	1·22	1·00

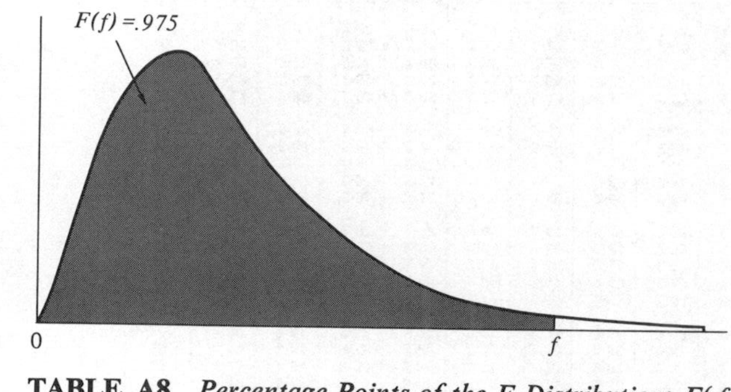

TABLE A8 *Percentage Points of the F Distribution*; $F(f) = .975$

ν_2 \ ν_1	1	2	3	4	5	6	7	8	9	10	12	15	20	24	30	40	60	120	∞
1	647·8	799·5	864·2	899·6	921·8	937·1	948·2	956·7	963·3	968·6	976·7	984·9	993·1	997·2	1001	1006	1010	1014	1018
2	38·51	39·00	39·17	39·25	39·30	39·33	39·36	39·37	39·39	39·40	39·41	39·43	39·45	39·46	39·46	39·47	39·48	39·49	39·50
3	17·44	16·04	15·44	15·10	14·88	14·73	14·62	14·54	14·47	14·42	14·34	14·25	14·17	14·12	14·08	14·04	13·99	13·95	13·90
4	12·22	10·65	9·98	9·60	9·36	9·20	9·07	8·98	8·90	8·84	8·75	8·66	8·56	8·51	8·46	8·41	8·36	8·31	8·26
5	10·01	8·43	7·76	7·39	7·15	6·98	6·85	6·76	6·68	6·62	6·52	6·43	6·33	6·28	6·23	6·18	6·12	6·07	6·02
6	8·81	7·26	6·60	6·23	5·99	5·82	5·70	5·60	5·52	5·46	5·37	5·27	5·17	5·12	5·07	5·01	4·96	4·90	4·85
7	8·07	6·54	5·89	5·52	5·29	5·12	4·99	4·90	4·82	4·76	4·67	4·57	4·47	4·42	4·36	4·31	4·25	4·20	4·14
8	7·57	6·06	5·42	5·05	4·82	4·65	4·53	4·43	4·36	4·30	4·20	4·10	4·00	3·95	3·89	3·84	3·78	3·73	3·67
9	7·21	5·71	5·08	4·72	4·48	4·32	4·20	4·10	4·03	3·96	3·87	3·77	3·67	3·61	3·56	3·51	3·45	3·39	3·33
10	6·94	5·46	4·83	4·47	4·24	4·07	3·95	3·85	3·78	3·72	3·62	3·52	3·42	3·37	3·31	3·26	3·20	3·14	3·08
11	6·72	5·26	4·63	4·28	4·04	3·88	3·76	3·66	3·59	3·53	3·43	3·33	3·23	3·17	3·12	3·06	3·00	2·94	2·88
12	6·55	5·10	4·47	4·12	3·89	3·73	3·61	3·51	3·44	3·37	3·28	3·18	3·07	3·02	2·96	2·91	2·85	2·79	2·72
13	6·41	4·97	4·35	4·00	3·77	3·60	3·48	3·39	3·31	3·25	3·15	3·05	2·95	2·89	2·84	2·78	2·72	2·66	2·60
14	6·30	4·86	4·24	3·89	3·66	3·50	3·38	3·29	3·21	3·15	3·05	2·95	2·84	2·79	2·73	2·67	2·61	2·55	2·49
15	6·20	4·77	4·15	3·80	3·58	3·41	3·29	3·20	3·12	3·06	2·96	2·86	2·76	2·70	2·64	2·59	2·52	2·46	2·40
16	6·12	4·69	4·08	3·73	3·50	3·34	3·22	3·12	3·05	2·99	2·89	2·79	2·68	2·63	2·57	2·51	2·45	2·38	2·32
17	6·04	4·62	4·01	3·66	3·44	3·28	3·16	3·06	2·98	2·92	2·82	2·72	2·62	2·56	2·50	2·44	2·38	2·32	2·25
18	5·98	4·56	3·95	3·61	3·38	3·22	3·10	3·01	2·93	2·87	2·77	2·67	2·56	2·50	2·44	2·38	2·32	2·26	2·19
19	5·92	4·51	3·90	3·56	3·33	3·17	3·05	2·96	2·88	2·82	2·72	2·62	2·51	2·45	2·39	2·33	2·27	2·20	2·13
20	5·87	4·46	3·86	3·51	3·29	3·13	3·01	2·91	2·84	2·77	2·68	2·57	2·46	2·41	2·35	2·29	2·22	2·16	2·09
21	5·83	4·42	3·82	3·48	3·25	3·09	2·97	2·87	2·80	2·73	2·64	2·53	2·42	2·37	2·31	2·25	2·18	2·11	2·04
22	5·79	4·38	3·78	3·44	3·22	3·05	2·93	2·84	2·76	2·70	2·60	2·50	2·39	2·33	2·27	2·21	2·14	2·08	2·00
23	5·75	4·35	3·75	3·41	3·18	3·02	2·90	2·81	2·73	2·67	2·57	2·47	2·36	2·30	2·24	2·18	2·11	2·04	1·97
24	5·72	4·32	3·72	3·38	3·15	2·99	2·87	2·78	2·70	2·64	2·54	2·44	2·33	2·27	2·21	2·15	2·08	2·01	1·94
25	5·69	4·29	3·69	3·35	3·13	2·97	2·85	2·75	2·68	2·61	2·51	2·41	2·30	2·24	2·18	2·12	2·05	1·98	1·91
26	5·66	4·27	3·67	3·33	3·10	2·94	2·82	2·73	2·65	2·59	2·49	2·39	2·28	2·22	2·16	2·09	2·03	1·95	1·88
27	5·63	4·24	3·65	3·31	3·08	2·92	2·80	2·71	2·63	2·57	2·47	2·36	2·25	2·19	2·13	2·07	2·00	1·93	1·85
28	5·61	4·22	3·63	3·29	3·06	2·90	2·78	2·69	2·61	2·55	2·45	2·34	2·23	2·17	2·11	2·05	1·98	1·91	1·83
29	5·59	4·20	3·61	3·27	3·04	2·88	2·76	2·67	2·59	2·53	2·43	2·32	2·21	2·15	2·09	2·03	1·96	1·89	1·81
30	5·57	4·18	3·59	3·25	3·03	2·87	2·75	2·65	2·57	2·51	2·41	2·31	2·20	2·14	2·07	2·01	1·94	1·87	1·79
40	5·42	4·05	3·46	3·13	2·90	2·74	2·62	2·53	2·45	2·39	2·29	2·18	2·07	2·01	1·94	1·88	1·80	1·72	1·64
60	5·29	3·93	3·34	3·01	2·79	2·63	2·51	2·41	2·33	2·27	2·17	2·06	1·94	1·88	1·82	1·74	1·67	1·58	1·48
120	5·15	3·80	3·23	2·89	2·67	2·52	2·39	2·30	2·22	2·16	2·05	1·94	1·82	1·76	1·69	1·61	1·53	1·43	1·31
∞	5·02	3·69	3·12	2·79	2·57	2·41	2·29	2·19	2·11	2·05	1·94	1·83	1·71	1·64	1·57	1·48	1·39	1·27	1·00

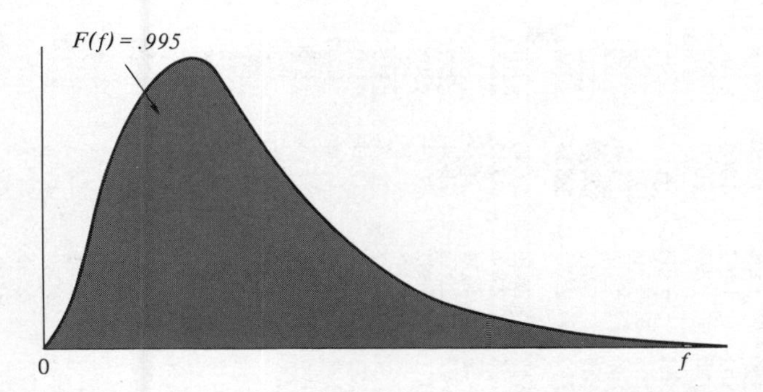

TABLE A8 *Percentage Points of the F Distribution*; $F(f) = .995$

ν_2 \ ν_1	1	2	3	4	5	6	7	8	9	10	12	15	20	24	30	40	60	120	∞
1	16211	20000	21615	22500	23056	23437	23715	23925	24091	24224	24426	24630	24836	24940	25044	25148	25253	25359	25465
2	198·5	199·0	199·2	199·2	199·3	199·3	199·4	199·4	199·4	199·4	199·4	199·4	199·4	199·5	199·5	199·5	199·5	199·5	199·5
3	55·55	49·80	47·47	46·19	45·39	44·84	44·43	44·13	43·88	43·69	43·39	43·08	42·78	42·62	42·47	42·31	42·15	41·99	41·83
4	31·33	26·28	24·26	23·15	22·46	21·97	21·62	21·35	21·14	20·97	20·70	20·44	20·17	20·03	19·89	19·75	19·61	19·47	19·32
5	22·78	18·31	16·53	15·56	14·94	14·51	14·20	13·96	13·77	13·62	13·38	13·15	12·90	12·78	12·66	12·53	12·40	12·27	12·14
6	18·63	14·54	12·92	12·03	11·46	11·07	10·79	10·57	10·39	10·25	10·03	9·81	9·59	9·47	9·36	9·24	9·12	9·00	8·88
7	16·24	12·40	10·88	10·05	9·52	9·16	8·89	8·68	8·51	8·38	8·18	7·97	7·75	7·65	7·53	7·42	7·31	7·19	7·08
8	14·69	11·04	9·60	8·81	8·30	7·95	7·69	7·50	7·34	7·21	7·01	6·81	6·61	6·50	6·40	6·29	6·18	6·06	5·95
9	13·61	10·11	8·72	7·96	7·47	7·13	6·88	6·69	6·54	6·42	6·23	6·03	5·83	5·73	5·62	5·52	5·41	5·30	5·19
10	12·83	9·43	8·08	7·34	6·87	6·54	6·30	6·12	5·97	5·85	5·66	5·47	5·27	5·17	5·07	4·97	4·86	4·75	4·64
11	12·23	8·91	7·60	6·88	6·42	6·10	5·86	5·68	5·54	5·42	5·24	5·05	4·86	4·76	4·65	4·55	4·44	4·34	4·23
12	11·75	8·51	7·23	6·52	6·07	5·76	5·52	5·35	5·20	5·09	4·91	4·72	4·53	4·43	4·33	4·23	4·12	4·01	3·90
13	11·37	8·19	6·93	6·23	5·79	5·48	5·25	5·08	4·94	4·82	4·64	4·46	4·27	4·17	4·07	3·97	3·87	3·76	3·65
14	11·06	7·92	6·68	6·00	5·56	5·26	5·03	4·86	4·72	4·60	4·43	4·25	4·06	3·96	3·86	3·76	3·66	3·55	3·44
15	10·80	7·70	6·48	5·80	5·37	5·07	4·85	4·67	4·54	4·42	4·25	4·07	3·88	3·79	3·69	3·58	3·48	3·37	3·26
16	10·58	7·51	6·30	5·64	5·21	4·91	4·69	4·52	4·38	4·27	4·10	3·92	3·73	3·64	3·54	3·44	3·33	3·22	3·11
17	10·38	7·35	6·16	5·50	5·07	4·78	4·56	4·39	4·25	4·14	3·97	3·79	3·61	3·51	3·41	3·31	3·21	3·10	2·98
18	10·22	7·21	6·03	5·37	4·96	4·66	4·44	4·28	4·14	4·03	3·86	3·68	3·50	3·40	3·30	3·20	3·10	2·99	2·87
19	10·07	7·09	5·92	5·27	4·85	4·56	4·34	4·18	4·04	3·93	3·76	3·59	3·40	3·31	3·21	3·11	3·00	2·89	2·78
20	9·94	6·99	5·82	5·17	4·76	4·47	4·26	4·09	3·96	3·85	3·68	3·50	3·32	3·22	3·12	3·02	2·92	2·81	2·69
21	9·83	6·89	5·73	5·09	4·68	4·39	4·18	4·01	3·88	3·77	3·60	3·43	3·24	3·15	3·05	2·95	2·84	2·73	2·61
22	9·73	6·81	5·65	5·02	4·61	4·32	4·11	3·94	3·81	3·70	3·54	3·36	3·18	3·08	2·98	2·88	2·77	2·66	2·55
23	9·63	6·73	5·58	4·95	4·54	4·26	4·05	3·88	3·75	3·64	3·47	3·30	3·12	3·02	2·92	2·82	2·71	2·60	2·48
24	9·55	6·66	5·52	4·89	4·49	4·20	3·99	3·83	3·69	3·59	3·42	3·25	3·06	2·97	2·87	2·77	2·66	2·55	2·43
25	9·48	6·60	5·46	4·84	4·43	4·15	3·94	3·78	3·64	3·54	3·37	3·20	3·01	2·92	2·82	2·72	2·61	2·50	2·38
26	9·41	6·54	5·41	4·79	4·38	4·10	3·89	3·73	3·60	3·49	3·33	3·15	2·97	2·87	2·77	2·67	2·56	2·45	2·33
27	9·34	6·49	5·36	4·74	4·34	4·06	3·85	3·69	3·56	3·45	3·28	3·11	2·93	2·83	2·73	2·63	2·52	2·41	2·29
28	9·28	6·44	5·32	4·70	4·30	4·02	3·81	3·65	3·52	3·41	3·25	3·07	2·89	2·79	2·69	2·59	2·48	2·37	2·25
29	9·23	6·40	5·28	4·66	4·26	3·98	3·77	3·61	3·48	3·38	3·21	3·04	2·86	2·76	2·66	2·56	2·45	2·33	2·21
30	9·18	6·35	5·24	4·62	4·23	3·95	3·74	3·58	3·45	3·34	3·18	3·01	2·82	2·73	2·63	2·52	2·42	2·30	2·18
40	8·83	6·07	4·98	4·37	3·99	3·71	3·51	3·35	3·22	3·12	2·95	2·78	2·60	2·50	2·40	2·30	2·18	2·06	1·93
60	8·49	5·79	4·73	4·14	3·76	3·49	3·29	3·13	3·01	2·90	2·74	2·57	2·39	2·29	2·19	2·08	1·96	1·83	1·69
120	8·18	5·54	4·50	3·92	3·55	3·28	3·09	2·93	2·81	2·71	2·54	2·37	2·19	2·09	1·98	1·87	1·75	1·61	1·43
∞	7·88	5·30	4·28	3·72	3·35	3·09	2·90	2·74	2·62	2·52	2·36	2·19	2·00	1·90	1·79	1·67	1·53	1·36	1·00

Condensed from Table 18 of *Biometrika Tables for Statisticians*, Vol. 1 (1966), by Pearson and Hartley. Reproduced with permission of the trustees of Biometrika.

TABLE A9 *Fisher's Transformation of r to* $z' = \frac{1}{2} \ln [(1 + r)/(1 - r)]$*

	r (*third decimal*)									
r	*.000*	*.001*	*.002*	*.003*	*.004*	*.005*	*.006*	*.007*	*.008*	*.009*
.00	.0000	.0010	.0020	.0030	.0040	.0050	.0060	.0070	.0080	.0090
.01	.0100	.0110	.0120	.0130	.0140	.0150	.0160	.0170	.0180	.0190
.02	.0200	.0210	.0220	.0230	.0240	.0250	.0260	.0270	.0280	.0290
.03	.0300	.0310	.0320	.0330	.0340	.0350	.0360	.0370	.0380	.0390
.04	.0400	.0410	.0420	.0430	.0440	.0450	.0460	.0470	.0480	.0490
.05	.0500	.0510	.0520	.0530	.0541	.0551	.0561	.0571	.0581	.0591
.06	.0601	.0611	.0621	.0631	.0641	.0651	.0661	.0671	.0681	.0691
.07	.0701	.0711	.0721	.0731	.0741	.0751	.0761	.0771	.0782	.0792
.08	.0802	.0812	.0822	.0832	.0842	.0852	.0862	.0872	.0882	.0892
.09	.0902	.0913	.0923	.0933	.0943	.0953	.0963	.0973	.0983	.0993
.10	.1003	.1013	.1024	.1034	.1044	.1054	.1064	.1074	.1084	.1094
.11	.1104	.1115	.1125	.1135	.1145	.1155	.1165	.1175	.1186	.1196
.12	.1206	.1216	.1226	.1236	.1246	.1257	.1267	.1277	.1287	.1297
.13	.1307	.1318	.1328	.1338	.1348	.1358	.1368	.1379	.1389	.1399
.14	.1409	.1419	.1430	.1440	.1450	.1460	.1471	.1481	.1491	.1501
.15	.1511	.1522	.1532	.1542	.1552	.1563	.1573	.1583	.1593	.1604
.16	.1614	.1624	.1634	.1645	.1655	.1665	.1676	.1686	.1696	.1706
.17	.1717	.1727	.1737	.1748	.1758	.1768	.1779	.1789	.1799	.1809
.18	.1820	.1830	.1841	.1851	.1861	.1872	.1882	.1892	.1903	.1913
.19	.1923	.1934	.1944	.1955	.1965	.1976	.1986	.1996	.2007	.2017
.20	.2027	.2038	.2048	.2059	.2069	.2079	.2090	.2100	.2111	.2121
.21	.2132	.2142	.2153	.2163	.2174	.2184	.2195	.2205	.2216	.2226
.22	.2237	.2247	.2258	.2268	.2279	.2289	.2300	.2310	.2321	.2331
.23	.2342	.2352	.2363	.2374	.2384	.2395	.2405	.2416	.2427	.2437
.24	.2448	.2458	.2469	.2480	.2490	.2501	.2512	.2522	.2533	.2543
.25	.2554	.2565	.2575	.2586	.2597	.2608	.2618	.2629	.2640	.2650
.26	.2661	.2672	.2683	.2693	.2704	.2715	.2726	.2736	.2747	.2758
.27	.2769	.2780	.2790	.2801	.2812	.2823	.2833	.2844	.2855	.2866
.28	.2877	.2888	.2899	.2909	.2920	.2931	.2942	.2953	.2964	.2975
.29	.2986	.2997	.3008	.3018	.3029	.3040	.3051	.3062	.3073	.3084
.30	.3095	.3106	.3117	.3128	.3139	.3150	.3161	.3172	.3183	.3194
.31	.3205	.3217	.3228	.3239	.3250	.3261	.3272	.3283	.3294	.3305
.32	.3316	.3328	.3339	.3350	.3361	.3372	.3383	.3395	.3406	.3417
.33	.3428	.3440	.3451	.3462	.3473	.3484	.3496	.3507	.3518	.3530
.34	.3541	.3552	.3564	.3575	.3586	.3598	.3609	.3620	.3632	.3643
.35	.3654	.3666	.3677	.3689	.3700	.3712	.3723	.3734	.3746	.3757
.36	.3769	.3780	.3792	.3803	.3815	.3826	.3838	.3850	.3861	.3873
.37	.3884	.3896	.3907	.3919	.3931	.3942	.3954	.3966	.3977	.3989
.38	.4001	.4012	.4024	.4036	.4047	.4059	.4071	.4083	.4094	.4106
.39	.4118	.4130	.4142	.4153	.4165	.4177	.4189	.4201	.4213	.4225
.40	.4236	.4248	.4260	.4272	.4284	.4296	.4308	.4320	.4332	.4344
.41	.4356	.4368	.4380	.4392	.4404	.4416	.4428	.4441	.4453	.4465
.42	.4477	.4489	.4501	.4513	.4526	.4538	.4550	.4562	.4574	.4587
.43	.4599	.4611	.4624	.4636	.4648	.4660	.4673	.4685	.4698	.4710
.44	.4722	.4735	.4747	.4760	.4772	.4784	.4797	.4809	.4822	.4834
.45	.4847	.4860	.4872	.4885	.4897	.4910	.4922	.4935	.4948	.4960
.46	.4973	.4986	.4999	.5011	.5024	.5037	.5049	.5062	.5075	.5088
.47	.5101	.5114	.5126	.5139	.5152	.5165	.5178	.5191	.5204	.5217
.48	.5230	.5243	.5256	.5269	.5282	.5295	.5308	.5321	.5334	.5347
.49	.5361	.5374	.5387	.5400	.5413	.5427	.5440	.5453	.5466	.5480

TABLE A9 *Fisher's Transformation of r to* $z' = \frac{1}{2} \ln [(1 + r)/(1 - r)]$* (continued)

	r (third decimal)									
r	*.000*	*.001*	*.002*	*.003*	*.004*	*.005*	*.006*	*.007*	*.008*	*.009*
.50	.5493	.5506	.5520	.5533	.5547	.5560	.5573	.5587	.5600	.5614
.51	.5627	.5641	.5654	.5668	.5682	.5695	.5709	.5722	.5736	.5750
.52	.5763	.5777	.5791	.5805	.5818	.5832	.5846	.5860	.5874	.5888
.53	.5901	.5915	.5929	.5943	.5957	.5971	.5985	.5999	.6013	.6027
.54	.6042	.6056	.6070	.6084	.6098	.6112	.6127	.6141	.6155	.6169
.55	.6184	.6198	.6213	.6227	.6241	.6256	.6270	.6285	.6299	.6314
.56	.6328	.6343	.6358	.6372	.6387	.6401	.6416	.6431	.6446	.6460
.57	.6475	.6490	.6505	.6520	.6535	.6550	.6565	.6580	.6595	.6610
.58	.6625	.6640	.6655	.6670	.6685	.6700	.6716	.6731	.6746	.6761
.59	.6777	.6792	.6807	.6823	.6838	.6854	.6869	.6885	.6900	.6916
.60	.6931	.6947	.6963	.6978	.6994	.7010	.7026	.7042	.7057	.7073
.61	.7089	.7105	.7121	.7137	.7153	.7169	.7185	.7201	.7218	.7234
.62	.7250	.7266	.7283	.7299	.7315	.7332	.7348	.7365	.7381	.7398
.63	.7414	.7431	.7447	.7464	.7481	.7498	.7514	.7531	.7548	.7565
.64	.7582	.7599	.7616	.7633	.7650	.7667	.7684	.7701	.7718	.7736
.65	.7753	.7770	.7788	.7805	.7823	.7840	.7858	.7875	.7893	.7910
.66	.7928	.7946	.7964	.7981	.7999	.8017	.8035	.8053	.8071	.8089
.67	.8107	.8126	.8144	.8162	.8180	.8199	.8217	.8236	.8254	.8272
.68	.8291	.8310	.8328	.8347	.8366	.8385	.8404	.8423	.8441	.8460
.69	.8480	.8499	.8518	.8537	.8556	.8576	.8595	.8614	.8634	.8653
.70	.8673	.8693	.8712	.8732	.8752	.8772	.8792	.8812	.8832	.8852
.71	.8872	.8892	.8912	.8933	.8953	.8973	.8994	.9014	.9035	.9056
.72	.9076	.9097	.9118	.9139	.9160	.9181	.9202	.9223	.9245	.9266
.73	.9287	.9309	.9330	.9352	.9373	.9395	.9417	.9439	.9461	.9483
.74	.9505	.9527	.9549	.9571	.9594	.9616	.9639	.9661	.9684	.9707
.75	.9730	.9752	.9775	.9798	.9822	.9845	.9868	.9892	.9915	.9939
.76	.9962	.9986	1.001	1.003	1.006	1.008	1.011	1.013	1.015	1.018
.77	1.020	1.023	1.025	1.028	1.030	1.033	1.035	1.038	1.040	1.043
.78	1.045	1.048	1.050	1.053	1.056	1.058	1.061	1.064	1.066	1.069
.79	1.071	1.074	1.077	1.079	1.082	1.085	1.088	1.090	1.093	1.096
.80	1.099	1.101	1.104	1.107	1.110	1.113	1.116	1.118	1.121	1.124
.81	1.127	1.130	1.133	1.136	1.139	1.142	1.145	1.148	1.151	1.154
.82	1.157	1.160	1.163	1.166	1.169	1.172	1.175	1.179	1.182	1.185
.83	1.188	1.191	1.195	1.198	1.201	1.204	1.208	1.211	1.214	1.218
.84	1.221	1.225	1.228	1.231	1.235	1.238	1.242	1.245	1.249	1.253
.85	1.256	1.260	1.263	1.267	1.271	1.274	1.278	1.282	1.286	1.290
.86	1.293	1.297	1.301	1.305	1.309	1.313	1.317	1.321	1.325	1.329
.87	1.333	1.337	1.341	1.346	1.350	1.354	1.358	1.363	1.367	1.371
.88	1.376	1.380	1.385	1.389	1.394	1.398	1.403	1.407	1.412	1.417
.89	1.422	1.427	1.432	1.437	1.442	1.447	1.452	1.457	1.462	1.467
.90	1.472	1.478	1.483	1.488	1.494	1.499	1.505	1.510	1.516	1.522
.91	1.528	1.533	1.539	1.545	1.551	1.558	1.564	1.570	1.576	1.583
.92	1.589	1.596	1.602	1.609	1.616	1.623	1.630	1.637	1.644	1.651
.93	1.658	1.666	1.673	1.681	1.689	1.697	1.705	1.713	1.721	1.730
.94	1.738	1.747	1.756	1.764	1.774	1.783	1.792	1.802	1.812	1.822
.95	1.832	1.842	1.853	1.863	1.874	1.886	1.897	1.909	1.921	1.933
.96	1.946	1.959	1.972	1.986	2.000	2.014	2.029	2.044	2.060	2.076
.97	2.092	2.110	2.127	2.146	2.165	2.185	2.205	2.227	2.249	2.273
.98	2.298	2.323	2.351	2.380	2.410	2.442	2.477	2.515	2.555	2.599
.99	2.647	2.700	2.759	2.826	2.903	2.994	3.106	3.250	3.453	3.800

*$z' = \frac{1}{2} \ln [(1 + r)/(1 - r)] = \tanh^{-1} r$.

TABLE A10 *Confidence Bands for Correlation ρ in a Bivariate Normal Population, for Various Sample Sizes n*

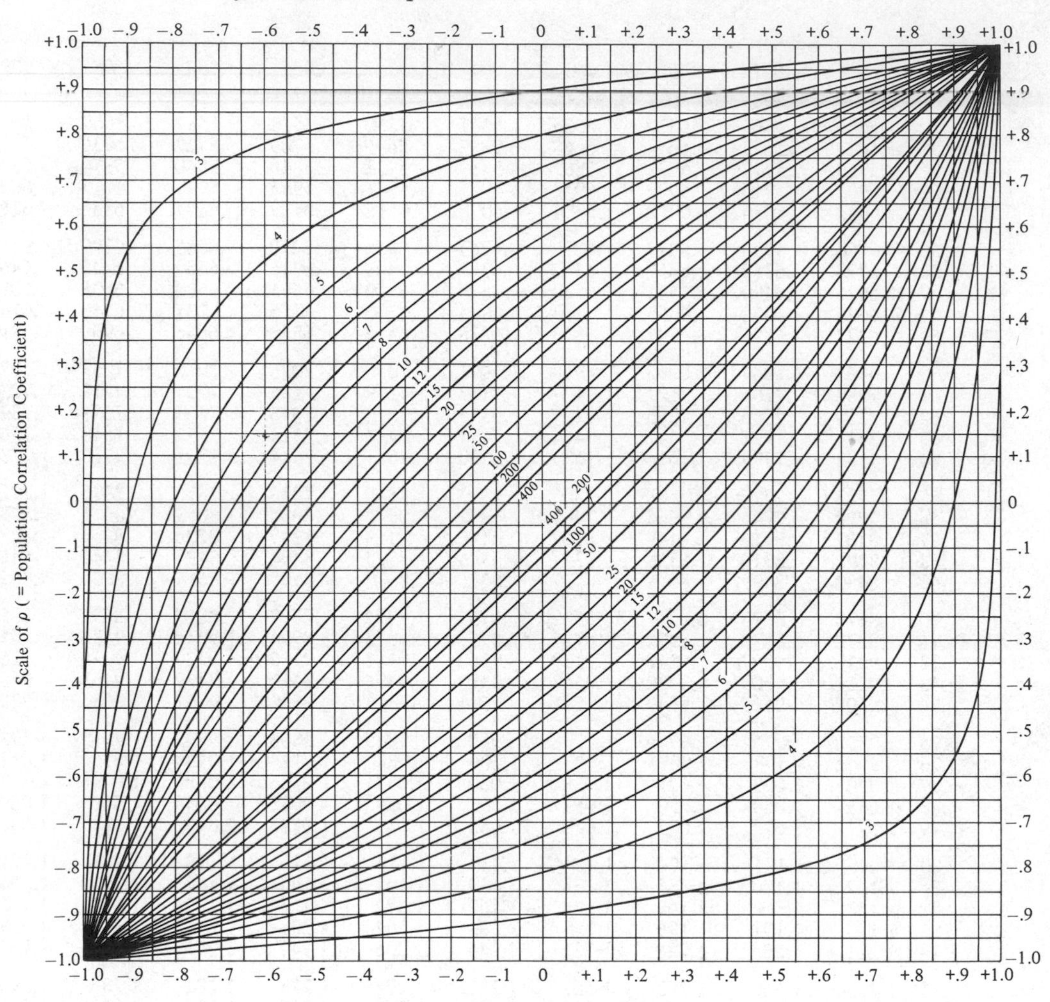

Scale of r (= Sample Correlation Coefficient)
The numbers on the curves indicate sample size

(a) 90% Confidence Bands

TABLE A10 *Confidence Bands for Correlation ρ in a Bivariate Normal Population, for Various Sample Sizes n* (continued)

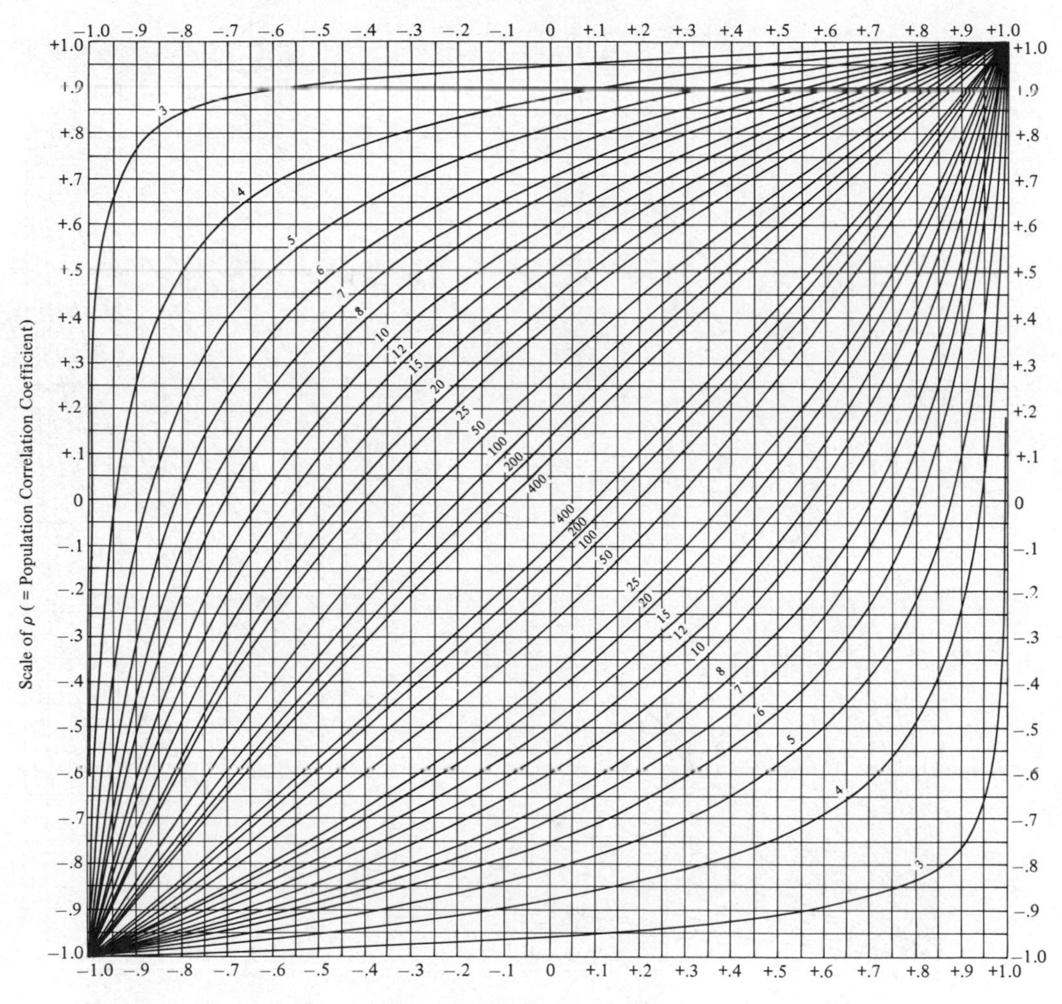

Scale of r (= Sample Correlation Coefficient)
The numbers on the curves indicate sample size

(b) 95% Confidence Bands

TABLE A10 *Confidence Bands for Correlation ρ in a Bivariate Normal Population, for Various Sample Sizes n* (continued)

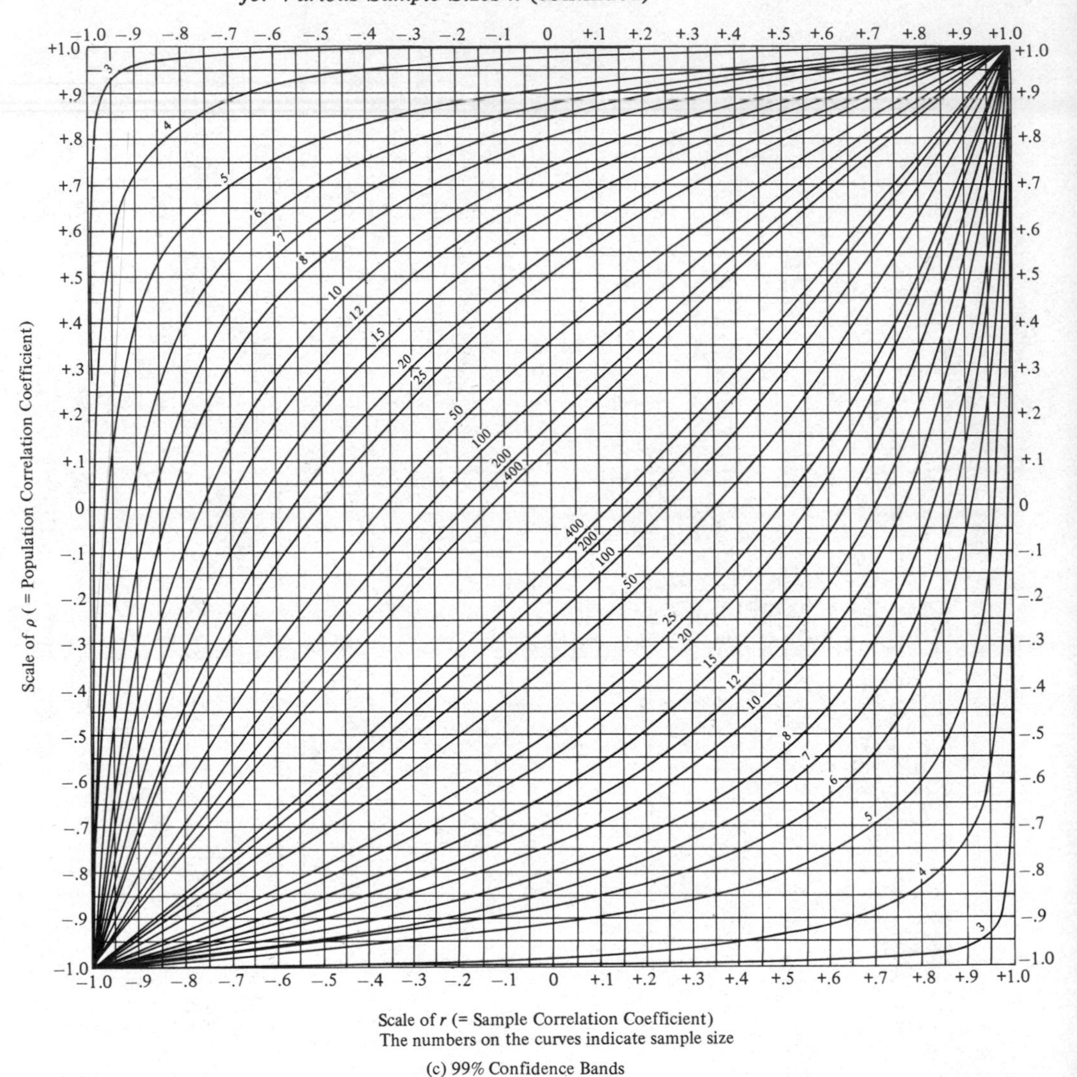

Scale of r (= Sample Correlation Coefficient)
The numbers on the curves indicate sample size

(c) 99% Confidence Bands

From F. N. David, *Tables of the Ordinates and Probability Integral of the Distribution of the Correlation Coefficient in Small Samples,* Cambridge University Press, for the Biometrika trustees (1938).

TABLE A11 *Critical Values of the Wilcoxon Signed Rank Statistic*

	$P[T^+ \leq t_p] \leq p$					$P[T^+ \geq t_p] \leq 1-p$				
$n \backslash p$	*.005*	*.01*	*.025*	*.05*	*.10*	*.90*	*.95*	*.975*	*.99*	*.995*
4	—	—	—	—	0	10	—	—	—	—
5	—	—	—	0	2	13	15	—	—	—
6	—	—	0	2	3	18	19	21	—	—
7	—	0	2	3	5	23	25	26	28	—
8	0	1	3	5	8	28	31	33	35	36
9	1	3	5	8	10	35	37	40	42	44
10	3	5	8	10	14	41	45	47	50	52
11	5	7	10	13	17	49	53	56	59	61
12	7	9	13	17	21	57	61	65	69	71
13	9	12	17	21	26	65	70	74	79	82
14	12	15	21	25	31	74	80	84	90	93
15	15	19	25	30	36	84	90	95	101	105

Adapted from Table C, Wilcoxon Signed-Rank Distributions, from C. H. Kraft and C. Van Eeden, *A Nonparametric Introduction to Statistics* (1969), Macmillan Publishing Co., New York.

TABLE A12 *Critical Values of the Wilcoxon Rank Sum Statistic*

			$P[T_1 \leq t_p] \leq p$					$P[T_1 \geq t_p] \leq 1 - p$				
n_1	n_2	p	.005	.01	.025	.05	.10	.90	.95	.975	.99	.995
1	9		—	—	—	—	1	10	—	—	—	—
1	10		—	—	—	—	1	11	—	—	—	—
2	3		—	—	—	—	3	9	—	—	—	—
2	4		—	—	—	—	3	11	—	—	—	—
2	5		—	—	—	3	4	12	13	—	—	—
2	6		—	—	—	3	4	14	15	—	—	—
2	7		—	—	—	3	4	16	17	—	—	—
2	8		—	—	3	4	5	17	18	19	—	—
2	9		—	—	3	4	5	19	20	21	—	—
2	10		—	—	3	4	6	20	22	23	—	—
3	3		—	—	—	6	7	14	15	—	—	—
3	4		—	—	—	6	7	17	18	—	—	—
3	5		—	—	6	7	8	19	20	21	—	—
3	6		—	—	7	8	9	21	22	23	—	—
3	7		—	6	7	8	10	23	25	26	27	—
3	8		—	6	8	9	11	25	27	28	30	—
3	9		6	7	8	10	11	28	29	31	32	33
3	10		6	7	9	10	12	30	32	33	35	36
4	4		—	—	10	11	13	23	25	26	—	—
4	5		—	10	11	12	14	26	28	29	30	—
4	6		10	11	12	13	15	29	31	32	33	34
4	7		10	11	13	14	16	32	34	35	37	38
4	8		11	12	14	15	17	35	37	38	40	41
4	9		11	13	14	16	19	37	40	42	43	45
4	10		12	13	15	17	20	40	43	45	47	48
5	5		15	16	17	19	20	35	36	38	39	40
5	6		16	17	18	20	22	38	40	42	43	44
5	7		16	18	20	21	23	42	44	45	47	49
5	8		17	19	21	23	25	45	47	49	51	53
5	9		18	20	22	24	27	48	51	53	55	57
5	10		19	21	23	26	28	52	54	57	59	61
6	6		23	24	26	28	30	48	50	52	54	55
6	7		24	25	27	29	32	52	55	57	59	60
6	8		25	27	29	31	34	56	59	61	63	65
6	9		26	28	31	33	36	60	63	65	68	70
6	10		27	29	32	35	38	64	67	70	73	75
7	7		32	34	36	39	41	64	66	69	71	73
7	8		34	35	38	41	44	68	71	74	77	78
7	9		35	37	40	43	46	73	76	79	82	84
7	10		37	39	42	45	49	77	81	84	87	89
8	8		43	45	49	51	55	81	85	87	91	93
8	9		45	47	51	54	58	86	90	93	97	99
8	10		47	49	53	56	60	92	96	99	103	105
9	9		56	59	62	66	70	101	105	109	112	115
9	10		58	61	65	69	73	107	111	115	119	122
10	10		71	74	78	82	87	123	128	132	136	139

Adapted from Table B, Wilcoxon, Mann-Whitney Distributions, from C. H. Kraft and C. Van Eeden, *A Nonparametric Introduction to Statistics* (1969), Macmillan Publishing Co., New York.

TABLE A13 *Critical Values of the Total Number of Runs Statistic*

		$P[R \leq r_p] \leq p$					$P[R \geq r_p] \leq 1 - p$				
n_1	n_2 \ p	.005	.01	.025	.05	.10	.90	.95	.975	.99	.995
2	6	—	—	—	—	2	—	—	—	—	—
2	8	—	—	—	2	2	—	—	—	—	—
2	10	—	—	—	2	2	—	—	—	—	—
2	12	—	—	2	2	2	—	—	—	—	—
2	14	—	—	2	2	2	—	—	—	—	—
2	17	—	—	2	2	2	—	—	—	—	—
2	20	—	2	2	2	3	—	—	—	—	—
4	6	—	2	2	3	3	9	9	9	—	—
4	8	2	2	3	3	3	9	—	—	—	—
4	10	2	2	3	3	4	—	—	—	—	—
4	12	2	3	3	4	4	—	—	—	—	—
4	14	2	3	3	4	4	—	—	—	—	—
4	17	3	3	4	4	5	—	—	—	—	—
4	20	3	3	4	4	5	—	—	—	—	—
6	6	2	2	3	3	4	10	11	11	12	12
6	8	3	3	3	4	5	11	12	12	13	13
6	10	3	3	4	5	5	12	12	13	—	—
6	12	3	4	4	5	6	12	13	13	—	—
6	14	4	4	5	5	6	13	13	—	—	—
6	17	4	5	5	6	6	13	—	—	—	—
6	20	4	5	6	6	7	—	—	—	—	—
8	8	3	4	4	5	5	13	13	14	14	15
8	10	4	4	5	6	6	13	14	15	15	16
8	12	4	5	6	6	7	14	15	16	16	17
8	14	5	5	6	7	7	15	16	16	17	17
8	17	5	6	7	7	8	16	16	17	—	—
8	20	6	6	7	8	9	16	17	17	—	—
11	11	5	6	7	7	8	16	17	17	18	19
11	14	6	7	8	8	9	17	18	19	20	20
11	17	7	8	9	9	10	18	19	20	21	22
11	20	8	8	9	10	11	19	20	21	22	22
14	14	7	8	9	10	11	19	20	21	22	23
14	17	8	9	10	11	12	21	22	23	24	24
14	20	9	10	11	12	13	22	23	24	25	25
17	17	10	10	11	12	13	23	24	25	26	26
17	20	11	11	13	13	15	24	25	26	27	28
20	20	12	13	14	15	16	26	27	28	29	30

From Swed and Eisenhart (1943), Tables for testing randomness of grouping in a sequence of alternatives, *Annals of Mathematical Statistics* 14:66–67.

TABLE A14 *Critical Values of the Kolmogorov–Smirnov Statistic*

$P[D \geq d_p] \leq 1 - p$

n	p .80	.90	.95	.98	.99
1	.900	.950	.975	.990	.995
2	.684	.776	.842	.900	.929
3	.565	.636	.708	.785	.829
4	.493	.565	.624	.689	.734
5	.447	.509	.563	.627	.669
6	.410	.468	.519	.577	.617
7	.381	.436	.483	.538	.576
8	.358	.410	.454	.507	.542
9	.339	.387	.430	.480	.513
10	.323	.369	.409	.457	.489
11	.308	.352	.391	.437	.468
12	.296	.338	.375	.419	.449
13	.285	.325	.361	.404	.432
14	.275	.314	.349	.390	.418
15	.266	.304	.338	.377	.404
16	.258	.295	.327	.366	.392
17	.250	.286	.318	.355	.381
18	.244	.279	.309	.346	.371
19	.237	.271	.301	.337	.361
20	.232	.265	.294	.329	.352
21	.226	.259	.287	.321	.344
22	.221	.253	.281	.314	.337
23	.216	.247	.275	.307	.330
24	.212	.242	.269	.301	.323
25	.208	.238	.264	.295	.317
26	.204	.233	.259	.290	.311
27	.200	.229	.254	.284	.305
28	.197	.225	.250	.279	.300
29	.193	.221	.246	.275	.295
30	.190	.218	.242	.270	.290
31	.187	.214	.238	.266	.285
32	.184	.211	.234	.262	.281
33	.182	.208	.231	.258	.277
34	.179	.205	.227	.254	.273
35	.177	.202	.224	.251	.269
36	.174	.199	.221	.247	.265
37	.172	.196	.218	.244	.262
38	.170	.194	.215	.241	.258
39	.168	.191	.213	.238	.255
40	.165	.189	.210	.235	.252

Adapted from Table 1 of Miller (1956), Table of percentage points of Kolmogorov Statistics, *Journal of American Statistical Association* 51:111–121.

TABLE A15 *Critical Values of the Kruskal–Wallis Statistic*

$P[H \geq h_p] \leq 1 - p$

n_1	n_2	n_3	p .90	.95	.99
1	2	3	4.2857	—	—
1	2	4	4.5000	—	—
1	2	5	4.2000	5.0000	—
1	3	3	4.5714	5.1429	—
1	3	4	4.0556	5.2083	—
1	3	5	4.0178	4.9600	—
1	4	4	4.1667	4.9667	6.6667
1	4	5	3.9873	4.9855	6.9545
1	5	5	4.1091	5.1273	7.3091
2	2	2	4.5714	—	—
2	2	3	4.5000	4.7143	—
2	2	4	4.4583	5.3333	—
2	2	5	4.3733	5.1600	6.5333
2	3	3	4.5556	5.3611	—
2	3	4	4.5111	5.4444	6.4444
2	3	5	4.6509	5.2509	6.9091
2	4	4	4.5545	5.4545	7.0364
2	4	5	4.5409	5.2727	7.2045
2	5	5	4.6231	5.3385	7.3385
3	3	3	5.0667	5.6889	7.2000
3	3	4	4.7091	5.7909	6.7455
3	3	5	4.5333	5.6485	7.0788
3	4	4	4.5455	5.5985	7.1439
3	4	5	4.5487	5.6564	7.4449
3	5	5	4.5451	5.7055	7.5780
4	4	4	4.6539	5.6923	7.6538
4	4	5	4.6187	5.6571	7.7604
4	5	5	4.5229	5.6657	7.8229
5	5	5	4.5600	5.7800	8.0000

Adapted from Table 6.1 of Kruskal and Wallis (1952), Use of ranks on one-criterion variance analysis, *Journal of American Statistical Association* 47:583–621.

TABLE A16 *Critical Values of the Friedman Statistic*

$P[Q \geq q_p] \leq 1 - p$

k	b	.90	.95	.975	.99	.995
3	3	6.000	6.000	—	—	—
3	4	6.000	6.500	8.000	8.000	8.000
3	5	5.200	6.400	7.600	8.400	10.000
3	6	5.333	7.000	8.333	9.000	10.333
3	7	5.429	7.143	7.714	8.857	10.286
3	8	5.250	6.250	7.750	9.000	9.750
3	9	5.556	6.222	8.000	8.667	10.667
4	2	6.000	6.000	—	—	—
4	3	6.600	7.400	8.200	9.000	9.000
4	4	6.300	7.800	8.400	9.600	10.200

TABLE A17 *Critical Values of Spearman's Rank Correlation Statistic*

$P[R_S \geq r_p] \leq 1 - p$

n	.90	.95	.975	.99	.995	.999
4	1.000	1.000	—	—	—	—
5	.800	.900	1.000	1.000	—	—
6	.657	.829	.886	.943	1.000	—
7	.571	.714	.786	.893	.929	1.000
8	.524	.643	.738	.833	.881	.952
9	.483	.600	.700	.783	.833	.917
10	.455	.564	.649	.745	.794	.879
11	.427	.536	.618	.709	.764	.855
12	.406	.503	.587	.678	.734	.825
13	.385	.483	.560	.648	.703	.797
14	.367	.464	.538	.626	.679	.771
15	.354	.446	.521	.604	.657	.750
16	.341	.429	.503	.585	.635	.729
17	.329	.414	.488	.566	.618	.711
18	.317	.401	.474	.550	.600	.692
19	.309	.391	.460	.535	.584	.675
20	.299	.380	.447	.522	.570	.660
21	.292	.370	.436	.509	.556	.647
22	.284	.361	.425	.497	.544	.633
23	.278	.353	.416	.486	.532	.620
24	.271	.344	.407	.476	.521	.608
25	.265	.337	.398	.466	.511	.597
26	.260	.331	.390	.457	.501	.586
27	.255	.324	.383	.449	.492	.576
28	.250	.318	.376	.441	.483	.567
29	.245	.312	.369	.433	.475	.557
30	.240	.306	.363	.426	.467	.548

Table A16 is adapted from Friedman (1937), The use of ranks to avoid the assumption of normality implicit in the analysis of variance, *Journal of American Statistical Association* 52:688–689. Table A17 is adapted from Glasser and Winter (1961), Critical values of the coefficient of rank correlation for testing the hypothesis of independence, *Biometrika* 48:444–448.

TABLE A18 *Squares and Square Roots*

n	n^2	$\sqrt{n}$	$\sqrt{10n}$	n	n^2	$\sqrt{n}$	$\sqrt{10n}$
1.0	1.00	1.000	3.162	5.5	30.25	2.345	7.416
1.1	1.21	1.049	3.317	5.6	31.36	2.366	7.483
1.2	1.44	1.095	3.464	5.7	32.49	2.387	7.550
1.3	1.69	1.140	3.606	5.8	33.64	2.408	7.616
1.4	1.96	1.183	3.742	5.9	34.81	2.429	7.681
1.5	2.25	1.225	3.873	6.0	36.00	2.449	7.746
1.6	2.56	1.265	4.000	6.1	37.21	2.470	7.810
1.7	2.89	1.304	4.123	6.2	38.44	2.490	7.874
1.8	3.24	1.342	4.243	6.3	39.69	2.510	7.937
1.9	3.61	1.378	4.359	6.4	40.96	2.530	8.000
2.0	4.00	1.414	4.472	6.5	42.25	2.550	8.062
2.1	4.41	1.449	4.583	6.6	43.56	2.569	8.124
2.2	4.84	1.483	4.690	6.7	44.89	2.588	8.185
2.3	5.29	1.517	4.796	6.8	46.24	2.608	8.246
2.4	5.76	1.549	4.899	6.9	47.61	2.627	8.307
2.5	6.25	1.581	5.000	7.0	49.00	2.646	8.367
2.6	6.76	1.612	5.099	7.1	50.41	2.665	8.426
2.7	7.29	1.643	5.196	7.2	51.84	2.683	8.485
2.8	7.84	1.673	5.292	7.3	53.29	2.702	8.544
2.9	8.41	1.703	5.385	7.4	54.76	2.720	8.602
3.0	9.00	1.732	5.477	7.5	56.25	2.739	8.660
3.1	9.61	1.761	5.568	7.6	57.76	2.757	8.718
3.2	10.24	1.789	5.657	7.7	59.29	2.775	8.775
3.3	10.89	1.817	5.745	7.8	60.84	2.793	8.832
3.4	11.56	1.844	5.831	7.9	62.41	2.811	8.888
3.5	12.25	1.871	5.916	8.0	64.00	2.828	8.944
3.6	12.96	1.897	6.000	8.1	65.61	2.846	9.000
3.7	13.69	1.924	6.083	8.2	67.24	2.864	9.055
3.8	14.44	1.949	6.164	8.3	68.89	2.881	9.110
3.9	15.21	1.975	6.245	8.4	70.56	2.898	9.165
4.0	16.00	2.000	6.325	8.5	72.25	2.915	9.220
4.1	16.81	2.025	6.403	8.6	73.96	2.933	9.274
4.2	17.64	2.049	6.481	8.7	75.69	2.950	9.327
4.3	18.49	2.074	6.557	8.8	77.44	2.966	9.381
4.4	19.36	2.098	6.633	8.9	79.21	2.983	9.434
4.5	20.25	2.121	6.708	9.0	81.00	3.000	9.487
4.6	21.16	2.145	6.782	9.1	82.81	3.017	9.539
4.7	22.09	2.168	6.856	9.2	84.64	3.033	9.592
4.8	23.04	2.191	6.928	9.3	86.49	3.050	9.644
4.9	24.01	2.214	7.000	9.4	88.36	3.066	9.695
5.0	25.00	2.236	7.071	9.5	90.25	3.082	9.747
5.1	26.01	2.258	7.141	9.6	92.16	3.098	9.798
5.2	27.04	2.280	7.211	9.7	94.09	3.114	9.849
5.3	28.09	2.302	7.280	9.8	96.04	3.130	9.899
5.4	29.16	2.324	7.348	9.9	98.01	3.146	9.950

A3

DATA ON STATISTICS STUDENTS

Student	Sex (0 male; 1, female)	Age	Height	Weight	Pulse Rate	GPA	Year	Student	Sex (0 male; 1, female)	Age	Height	Weight	Pulse Rate	GPA	Year
1	0	19	73.0	175	70	2.40	2	51	0	20	71.0	185	66	3.07	2
2	0	20	66.0	157	60	2.78	2	52	0	20	59.0	150	60	2.40	2
3	0	20	72.0	172	60	3.13	2	53	1	20	62.0	115	76	3.09	2
4	0	19	78.0	168	74	2.85	2	54	0	21	72.0	187	62	2.06	4
5	1	20	67.0	127	86	2.80	3	55	0	20	73.0	185	66	3.22	2
6	0	19	72.5	215	72	2.40	2	56	0	19	69.5	153	64	2.60	2
7	0	19	72.0	148	80	2.90	2	57	0	21	72.5	167	67	2.20	2
8	1	20	68.0	135	68	3.25	2	58	0	19	73.0	140	80	2.30	2
9	0	19	70.0	155	56	2.70	2	59	0	22	72.0	174	66	3.47	4
10	0	20	70.0	170	60	2.40	3	60	0	22	71.0	160	54	3.10	3
11	0	19	72.0	165	58	2.75	2	61	0	20	70.0	160	88	2.80	3
12	1	19	62.0	140	82	3.32	2	62	0	20	71.0	175	78	3.31	3
13	1	19	66.0	115	80	2.40	2	63	0	20	70.0	165	54	2.20	2
14	0	19	68.0	160	70	3.20	2	64	1	21	67.0	120	88	2.60	3
15	0	19	71.0	145	72	3.26	2	65	1	20	63.7	103	92	3.19	2
16	1	19	64.0	110	60	3.22	2	66	0	19	74.0	145	78	1.90	2
17	0	19	73.0	170	84	2.50	2	67	0	20	70.0	150	70	2.88	2
18	0	19	70.0	150	60	2.50	2	68	1	21	63.0	112	100	3.20	2
19	0	20	72.0	190	50	2.30	2	69	0	20	69.0	155	50	2.30	2
20	0	20	74.0	177	70	2.10	3	70	0	19	77.0	208	58	2.50	2
21	0	19	69.0	165	70	3.14	2	71	0	20	70.0	160	114	2.68	2
22	0	20	71.0	160	62	2.75	3	72	0	19	75.0	195	72	2.10	2
23	1	20	60.0	100	94	1.75	3	73	1	19	65.0	160	75	3.20	2
24	1	20	64.0	130	64	2.20	3	74	1	19	64.0	125	64	3.92	2
25	0	19	72.0	160	66	2.10	2	75	0	20	68.0	165	66	2.00	3
26	0	19	70.0	160	70	3.40	2	76	1	19	61.0	105	64	3.00	2
27	0	20	72.0	150	72	2.30	3	77	0	21	68.0	160	66	2.10	4
28	0	19	69.0	165	56	3.20	2	78	1	19	66.5	130	80	3.00	2
29	1	20	63.0	125	80	3.40	2	79	1	20	66.0	150	88	2.50	2
30	1	19	66.0	145	70	3.90	2	80	0	19	70.0	150	88	3.67	1
31	0	19	72.0	150	68	3.38	2	81	0	20	69.0	140	62	2.30	2
32	1	19	63.5	115	60	3.62	2	82	0	19	70.0	180	67	2.40	2
33	1	19	61.0	115	80	3.00	2	83	0	20	76.0	175	64	3.40	2
34	1	19	65.5	120	68	2.50	2	84	0	19	69.0	135	80	2.90	2
35	1	20	68.2	155	92	3.40	1	85	0	21	67.0	160	64	2.00	4
36	0	20	74.0	179	64	2.40	2	86	0	19	69.0	155	66	1.80	2
37	0	20	66.0	135	80	2.30	3	87	1	19	67.5	125	100	3.80	2
38	0	20	74.0	190	86	2.40	3	88	0	19	70.0	191	72	2.23	2
39	0	20	70.0	140	100	3.12	2	89	0	19	69.0	155	78	2.93	2
40	0	19	67.0	130	60	2.46	2	90	0	20	71.0	175	60	2.59	2
41	0	19	70.0	160	64	2.40	2	91	0	19	66.0	135	74	3.10	2
42	1	19	61.0	115	84	2.78	2	92	0	20	71.0	160	60	2.93	2
43	0	20	72.0	170	52	3.00	2	93	1	21	63.0	120	70	2.50	3
44	1	20	68.0	135	86	3.34	3	94	1	21	67.5	123	76	3.10	4
45	0	19	68.0	165	70	2.90	3	95	0	19	71.0	163	62	2.80	2
46	0	20	69.0	175	80	2.50	2	96	1	20	65.0	125	62	2.75	2
47	1	21	62.0	107	60	3.60	3	97	0	20	66.0	145	64	2.00	2
48	0	21	68.0	180	52	2.50	3	98	0	19	71.0	165	62	2.00	2
49	0	20	69.0	145	75	2.20	2	99	0	19	72.0	180	72	2.63	2
50	0	20	64.0	155	74	1.80	3	100	0	19	71.0	180	54	2.40	2

Student	Sex (0 male, 1, female)	Age	Height	Weight	Pulse Rate	GPA	Year	Student	Sex (0 male, 1, female)	Age	Height	Weight	Pulse Rate	GPA	Year
101	0	19	68.0	150	70	3.00	2	151	0	19	69.0	155	65	3.20	2
102	0	21	73.0	180	58	2.10	3	152	0	20	73.0	155	80	2.00	3
103	1	22	69.0	140	66	2.80	3	153	0	20	73.0	165	66	3.10	2
104	1	19	61.0	100	100	2.10	2	154	0	19	70.0	173	70	2.50	2
105	0	19	72.0	165	68	2.10	2	155	0	19	68.0	160	72	2.00	2
106	1	19	62.0	118	80	3.10	2	156	0	20	69.0	145	60	2.70	3
107	0	21	72.0	175	72	2.65	3	157	0	20	70.0	170	66	2.31	2
108	1	19	60.0	100	92	3.20	2	158	1	21	59.0	88	76	3.14	3
109	0	19	74.0	220	80	2.29	2	159	1	19	65.0	114	72	3.35	2
110	0	19	69.7	155	56	2.90	2	160	1	20	63.0	110	60	2.10	3
111	0	19	72.0	165	50	1.70	2	161	0	19	68.0	235	56	2.50	2
112	0	20	73.0	165	78	2.50	2	162	1	19	68.0	135	76	2.22	2
113	0	19	71.0	177	56	3.40	2	163	0	19	68.0	150	74	3.15	2
114	1	21	65.0	115	88	3.17	3	164	0	20	68.0	165	64	2.10	2
115	1	20	63.0	115	62	3.21	2	165	0	19	69.0	130	80	2.80	2
116	0	20	72.0	175	62	2.70	2	166	0	21	68.0	160	70	2.00	2
117	1	20	70.5	135	66	3.50	3	167	0	19	71.0	180	66	2.40	2
118	0	19	66.5	148	72	3.17	2	168	0	19	71.0	150	88	2.21	2
119	0	19	73.2	155	70	2.80	2	169	0	21	75.0	240	60	2.12	3
120	0	19	71.0	155	70	2.80	2	170	0	20	71.0	155	74	3.22	2
121	0	21	73.0	225	72	2.01	3	171	0	21	69.0	145	86	2.21	3
122	0	19	76.0	210	68	3.25	2	172	1	19	65.0	115	80	2.87	2
123	0	21	72.0	172	54	2.00	3	173	0	19	72.0	165	78	3.50	2
124	0	20	73.0	178	56	2.75	2	174	0	19	73.0	200	60	2.60	2
125	1	21	65.0	130	90	2.40	4	175	1	19	63.0	112	90	2.50	2
126	0	19	70.0	130	82	2.20	2	176	0	19	66.0	145	40	2.00	2
127	1	19	67.0	128	78	3.76	2	177	1	20	61.5	105	72	3.40	2
128	1	20	62.0	145	70	2.50	2	178	1	19	65.0	125	90	3.59	2
129	0	20	69.0	143	56	3.50	2	179	0	21	72.0	152	54	2.45	3
130	1	21	64.0	115	65	2.90	3	180	0	20	69.0	145	75	2.20	2
131	0	22	71.0	150	76	2.80	4	181	1	21	65.0	127	86	3.69	3
132	0	20	70.5	155	78	2.66	2	182	0	20	73.0	160	54	3.40	2
133	1	19	66.0	122	76	3.13	2	183	1	19	68.0	160	82	2.90	2
134	0	20	68.5	160	63	2.50	2	184	0	19	70.0	188	52	3.07	2
135	0	20	72.0	185	66	2.89	2	185	0	20	72.0	165	54	2.60	3
136	0	19	75.0	190	58	3.00	2	186	1	20	62.0	118	70	2.40	2
137	0	19	69.0	160	70	3.30	2	187	0	19	70.0	165	66	3.00	2
138	0	20	69.0	168	62	3.00	2	188	1	19	65.0	115	90	2.23	2
139	1	19	65.0	120	84	3.70	2	189	1	20	65.5	140	80	2.18	3
140	0	20	68.5	137	68	3.30	2	190	0	22	72.0	174	66	3.47	4
141	0	19	72.0	145	69	2.75	2	191	0	21	73.0	145	76	2.00	3
142	0	19	72.0	135	72	2.20	2	192	0	20	65.0	130	86	2.50	3
143	0	20	71.0	165	62	2.75	2	193	0	21	75.0	210	78	2.00	4
144	0	20	75.0	170	50	2.74	2	194	1	20	62.0	107	78	2.56	2
145	0	20	79.0	185	74	1.80	2	195	0	19	74.0	170	48	3.57	2
146	0	21	72.0	205	96	2.00	2	196	1	20	70.0	140	58	3.70	3
147	1	20	67.0	155	70	2.40	2	197	1	19	64.5	115	82	2.95	2
148	0	19	72.0	170	72	2.90	2	198	1	19	66.0	115	74	3.10	2
149	0	20	71.0	150	72	2.38	2	199	1	19	67.0	134	78	3.35	1
150	0	19	73.0	160	66	2.50	3	200	0	19	70.0	143	68	3.03	2

A4

SHORT ANSWERS TO ODD-NUMBERED EXERCISES

Chapter 1

3. $S = \{A\heartsuit, 2\heartsuit, \ldots, K\heartsuit, \text{etc.}\}$ (52 pts.)

5. $S = \{0, 1, 2, \ldots, 10\}$

7. (a) finite (b) finite

11. (a) Betty (b, d) Alan, Betty, Carl, Eric
(c) Alan, Betty, Carl, Diane, Eric, Frieda

13. (A, E), (A, F), (B, D), (C, D), (E, F)

15. 60

17. (a) 6,760,000 (b) 6,250,000 (c) 510,000

19. 9880

21. 165

23. (a) 720 (b) 480

27. 36, $\frac{1}{36}$

29. $A, \frac{1}{2}$; $B, \frac{1}{2}$; $C, \frac{1}{2}$; $D, \frac{2}{3}$

31. (a) .25 (b) .50 (c) .75 (d) .15 (e) .60

33. $\frac{1}{4}$

35. no

41. (a) $\frac{14}{36}$ (b) $\frac{15}{36}$ (c) $\frac{25}{36}$

43. $\frac{1}{9}$

45. (a) OK (b) violates axiom 1 (c) violates Theorem 1.4 (d) OK

47. equal, less than, greater than

49. $\frac{1}{3}$

51. (a) .0005 (b) .4234

53. .0316

57. .9744

59. (a) $\frac{7}{9}$ (b) .4705

61. (a) .2 (b) $\frac{1}{3}$

63. (a) $\frac{5}{8}$ (b) $\frac{3}{8}$

65. $\frac{2}{3}$

67. (a) .0714 (b) .28

Chapter 2

3. (a)

x:	0	1	2
$P[X = x]$:	.16	.48	.36

5. (a) OK (b) no; (2.3) is violated (c, d) no; (2.4) is violated

7. (a) $X \geq 1$ (b) $X \leq 1$ (c) $X \geq 2$ (d) .55, .70, .30

9. (b) $\frac{1}{4}, \frac{3}{4}, 1, 0$

11. (c) .5 (d) .15625, .84375, .6875

13.
$$F(x) = \begin{cases} 0 & x < 0 \\ .10 & 0 \leq x < 1 \\ .25 & 1 \leq x < 2 \\ .40 & 2 \leq x < 3 \\ .60 & 3 \leq x < 4 \\ 1.00 & 4 \leq x \end{cases}$$

15. (a) 0, .25, .25, 1

(b)

y:	1	2	4
$P[Y = y]$:	.25	.25	.50

17. (a)
$$F(y) = \begin{cases} 0 & y < 0 \\ \frac{y^2}{4} & 0 \leq y \leq 2 \\ 1 & y > 2 \end{cases}$$

(b) $F(1), 1 - F(1), F(3) - F(0), 1 - F(2)$

(c) $\frac{1}{4}, \frac{3}{4}, 1, 0$

19. (a)
$$F(x) = \begin{cases} 0 & x < 15 \\ 1012.5 - 191.25x + 12x^2 - .25x^3 & 15 \leq x \leq 17 \\ 1 & x > 17 \end{cases}$$

(b) .15625
.84375
.6875

21. (a) $1, \frac{1}{8}, \frac{63}{64}$ (b) $f(y) = \begin{cases} \frac{3}{8}(y-1)^2 & 1 \leq y \leq 3 \\ 0 & \text{otherwise} \end{cases}$

23. 2.65

25. (a) no (b) yes (c) .34375

27. 16, 8, 0, −20

29. (a) 1.05 (b) 12.6¢

31. $\frac{4}{3}$, 2

33. 16

35. .15, 0, −1

37. 1.85, 1, 1

39. 16, 16, 16

41. .5 (or any number between 0 and 1), 1.5 (or any number between 1 and 2), 3, 4

43. 3 (or any number between 2 and 4), 1, 4, 4

49. .9475, .9734

51. (a) 19.2 (b) .00192 (c) 19.2

53. $\frac{4}{3}, 2, \frac{2}{9}$

55. (a) 31 (b) 324 (c) 18

61. the cat (using standard scores); the owl (using raw scores)

63. (a) 1.5, .387 (b) .0476 (c) .25

Chapter 3

1. (a) .56 (b) .62

3. (a) .31 (b) .50 (c) .26

9. (a)

x:	0	1	2	3
$P[X = x]$:	.1	.4	.3	.2

(b)

y:	0	1	2	3	4
$P[Y = y]$:	.4	.2	.2	.1	.1

(c)

y:	0	1	2	3	4
$P[Y = y \mid x = 0]$:	1	0	0	0	0
$P[Y = y \mid x = 1]$:	$\frac{15}{40}$	$\frac{11}{40}$	$\frac{10}{40}$	$\frac{2}{40}$	$\frac{2}{40}$
$P[Y = y \mid x = 2]$:	$\frac{9}{30}$	$\frac{5}{30}$	$\frac{8}{30}$	$\frac{5}{30}$	$\frac{3}{30}$
$P[Y = y \mid x = 3]$:	$\frac{6}{20}$	$\frac{4}{20}$	$\frac{2}{20}$	$\frac{3}{20}$	$\frac{5}{20}$

(d) 0, 1.125, 1.6, 1.85

11. (a)
$$f(y|x=1) = \begin{cases} \dfrac{y+1}{4} & 0 \leq y \leq 2 \\ 0 & \text{otherwise} \end{cases}$$
(b) $\frac{7}{6}$

13. (a) dependent (b) independent (c) independent
15. (a) dependent (b) dependent
17. (a) 1.6 (b) 1.3 (c) 2.52 (d) \$.93
19. (a) .70 (b) 1.20 (c) .8333
21. (a) 2 (b) .5 (c) 21
23. 0, −.04
25. 2.8475, 1.6875
27. 0

Chapter 4

1. (a, c) yes (b, d) no; there are more than two possible outcomes
3. (a) .5905 (b) .8322 (c) .0368 (d) .5584
5. (a) .2508 (b) .1672
7. .001; assumption of independence is not reasonable
9. (a) 5, 2 (b) 20, 2 (c) 10, 3 (d) 90, 3
13. (a) .9568 (b) 279
15. (a) .1058 (b) .000245 (c) .0548 (d) 1, .9
17. .0005; might conclude that the assumption of intermingling is wrong
19. The joint distribution is not equal to the product of the marginals.
21. (a) .00486 (b) .999 (c) 30 (d) 270
23. about .07, about 36
25. 63.3
27. (a) .1821 (b) .6415 (c) .8784
29. (a) .2663 (b) .2637
31. (a) .0902 (b) .5940 (c) 2 (d) 2
33. (a) .75 (b) .5270 (c) .9592 (using interpolation in b and c)
35. .185
39. (b) $\frac{3}{5}$ (c) 7.5 (d) 2.083
41. $a = 1$, $b = 19$
43. (a) .8413 (b) .6915 (c) .2061 (d) .0668 (e) .9750
45. (a) .8413 (b) .747 (c) .091 (d) .370 (e) .747
47. (a) 2 (b) 2.75 (c) 4.025 (d) −2.935
49. (a) .524 (b) .4778
51. (a) .5792 (b) .036
53. (a) .1587 (b) .0625
55. (a) .0025 (b) .5 (c) .25
57. (a) 2 (b) .1353
59. (a) .8531 (b) .9394
61. (a) .0062 (b) likely unfair
63. (a) .9154 (b) .9154 (c) .0994
65. (a) .4060 (b) .3233
67. (a) .3033 (b) .0902

Chapter 5

3. (a) not random (b) should be representative
5. (a) should be representative (b) possibly biased
9. (a) 61 (b) 71.5 (c) 31.91
11. (a)

$\bar{x}$:	6	$6\frac{1}{3}$	$6\frac{2}{3}$	7	$7\frac{1}{3}$	$7\frac{2}{3}$	8
$P[\bar{X} = \bar{x}]$:	.027	.108	.225	.280	.225	.108	.027

(b)

$\tilde{x}$:	6	7	8
$P[\tilde{X} = \tilde{x}]$:	.216	.568	.216

13. (a) .20 (b) .432
15. (a) 123, 7.75 (b) 123, 3.1 (c) 123, 1.55
19. (a) $N(90, 225)$ (b) $N(0, 252)$ (c) .7475 (d) .1038
21. .9509
23. (a) .5762 (b) .9544 (c) .9986
25. 2.934
27. .0003

Chapter 6

9. (b) about 9000, about 22,000
11. 1959.54
13. (a) 2
17. 1,289,823.5, 1,135.70
19. 164, 170.68, 36.88
21. nominal, ratio, ordinal, ratio, nominal
23. 170.68, 36.88

Chapter 7

1. (a) 177.4, 172, 488.3 (b) 125.4, 127, 163.3
3. could use $\tilde{X}$ if the distribution is symmetric
5. M.L.E. of $\theta = \lambda t$ is $\bar{X}$, the sample mean
7. (a) 1 (b) .4

(c)

$\bar{x}$:	0	$\frac{1}{3}$	$\frac{2}{3}$	1	$\frac{4}{3}$	$\frac{5}{3}$	2
$P[\bar{X} = \bar{x}]$:	.008	.072	.240	.360	.240	.072	.008

(d)

$\tilde{x}$:	0	1	2
$P[\tilde{X} = \tilde{x}]$:	.104	.792	.104

9. $E(\tilde{X}) = 1$, $V(\tilde{X}) = .208$
11. $E(\bar{X}) = .5$, $V(\bar{X}) = .15$
15. $\sigma^2/3$, $\sigma^2/5$, $5\sigma^2$, $.8\sigma^2$; the estimator in (b) would be considered best
17. (a) both will be larger than $\bar{X}$ (b) $\mu + 2$, $(n/(n-2))\mu$
(c) σ^2/n, $n\sigma^2/(n-2)^2$
19. (a) $\mu/(n-1)$ (b) $(n\sigma^2 + \mu^2)/(n-1)^2$
21. (a) 1.282 (b) 1.645 (c) 2.326 (d) −1.282 (e) −1.645 (f) −2.326
23. (a) (11.71, 18.29) (b) (11.08, 18.92)
25. (207.35, 216.65)
27. (a) 1.397 (b) 1.753 (c) 2.326 (d) −1.440 (e) −.687 (f) −.255
29. (2.73, 3.23); assumed grades are approximately normal
31. (a) (69.19, 91.23) (b) (65.06, 93.54)
33. lengths are longer in 31; one reason is higher level of confidence
35. (−4.06, 6.06) (in thousands)
37. (18.42, 31.58)
41. (a) (−1.54, 1.58) (b) using $\nu = 8$, (−1.72, +1.76)
43. (a) .25 (b) .995 (c) 0 (d) .75 (e) .005 (f) .95
45. (a) .50 (b) .90
47. (a) (2.83, 13.97) (b) (2.72, 41.92)
49. (.303, .414); the claim seems to be too low
51. (a) 4.53 (b) .162 (c) 3.85 (d) .260 (e) 4.91 (f) .204
53. (a) .975 (b) .10
55. (.347, 8.49)
57. (.452, .648)
59. (.107, .293)
61. $f'(p) = 1 - 2p = 0 \Rightarrow p = \frac{1}{2}$
63. (−.244, .078)
65. (−.176, −.064)
67. 174
69. (a) 1692 (b) 6765
71. (a) 19.6 (b) 19.2 (c) 18.83
75. (a) 1.2 thousand (b) 1.14 thousand
77. (a) 50, 30, 20 (b) .6 thousand (c) .570 thousand
79. (a) 34, 42, 24 (or 34, 41, 25) (b) .513 thousand (c) .483 thousand

Chapter 8

1. (a) $\mu \neq 1000$ (b) $\mu < 32$ (c) $\sigma \geq 1$ mm
(d) X does not follow a Poisson distribution
3. (c) take H_0 to be "animal is rabid"
5. $\alpha = .20$, $\beta = .30$
7. (a) $\alpha = .159 = \beta$ (b) $\alpha = .081$, $\beta = .274$
9. (a) $c = 26.41$, $\beta = .236$ (b) $c = 28.225$, $\beta = .362$

11. (a) $c = 25$, $n = 43$ (b) $c = 25.62$, $n = 34$
13. assumption of normality not reasonable in either case
15. (b) $H_0: \mu \geq 16.83$, $H_a: \mu < 16.83$ (c) $\alpha = .036$
17. (a) test statistic $= t = 3.417$, so reject H_0 since $|t| > t_{3,.975} = 3.182$
(b) test statistic $= t = 3.036$, do not reject H_0
19. $t = 1.27$ does not exceed $t_{5,.95} = 2.015$, so do not reject H_0
21. (a) $z < 0$, do not reject H_0 since critical value is $+1.645$
(b) $z = 1.46$, do not reject H_0
23. $t = .74$ does not exceed $t_{4,.95} = 2.132$, so do not reject H_0
25. $t = -2.76$ is not less than $t_{8,.01} = -2.896$, so do not reject H_0
27. $|t| = |-.488|$ is not greater than $t_{10,.975} = 2.228$, so do not reject H_0
29. critical value is $\chi^2_{20,.05} = 10.8508$ (a) test statistic $= 10$, so reject H_0
(b) test statistic $= 32$, do not reject H_0
31. test statistic $= 91.97 > 83.30 = \chi^2_{60,.975}$, so reject H_0
33. critical value is $f_{15,24,.01} = .304$ (a) $F = 1.006$ not in critical region, do not reject H_0
(b) $F = .075$ is in critical region, so reject H_0
35. $F = 1.69$ does not exceed $f_{4,4,.95} = 6.39$, so do not reject H_0
37. critical region is values $\leq f_{13,5,.025} = .265$ or $\geq f_{13,5,.975} = 6.49$
(a) $F = .77$, do not reject H_0 (b) $F = 2.25$, do not reject H_0
39. critical value is $z_{.99} = 2.326$ (a) $z = -.50$, do not reject H_0
(b) $z = 2.5$, reject H_0
41. critical value is either $z_{.95} = 1.645$ or $z_{.99} = 2.326$, depending on choice of α; here $z = 1.5$, so do not reject H_0
43. (a) $|z| = 1.09$ (b) $|z| = 1.75$ neither exceed $|z_{.995}| = 2.576$, so do not reject H_0 in either case
45. $z = .494$ does not exceed $z_{.95} = 1.645$, do not reject H_0
47. (a) .091 (b) .091 (c) if $\alpha = .05$, do not reject H_0 (d) if $\alpha = .10$, reject H_0
49. (a) .5424 (b) .0026
53. (a) $c_1 = 0.4$, $c_2 = 39.6$ (b) $c_1 = 8.24$, $c_2 = 31.76$

Chapter 9

1. (a) stochastic (b) deterministic
3.

x:	1	2	3	4
$E(Y\|x)$:	.5833	.5555	.5417	.5333

not linear; all means do not fall on a straight line
5. both appear to be approximately linear
9. (a) $\sum e_i^2 = 1.60$ (b) 1.65 (c) 2.15
11. (a) 35.6 (b) 29.1
13. (a) yes (b) $\hat{y} = 78.3 + .298x$
17. (a) $\hat{y} = 11.75 - 2.25x$ (b) 4.05 (c) 1.88 (d) .1446
19. (a) 31.68 (b) 428.2 (c) .0136
21. (a) (8.23, 15.27) (b) $T = -.66$ is greater than $t_{5,.05} = -2.015$, so do not reject H_0
23. $T = -1.855$ is less than $t_{9,.05} = -1.833$, so reject H_0, conclude $\beta_1 < 0$
27. Residuals: −1.88 −3.63 4.72 3.33 −7.89 9.85 2.28 3.54 −0.32 −3.02 −6.98; the assumption of a common variance is reasonable
29. (a) Residuals: −18.3 11.7 2.7 3.7 −0.3 10.8 1.8 −1.2 −15.2 4.8 5.9 −1.1 −5.1 −14.1 13.9 (b) the assumption of a common variance is reasonable
31. (c) Residuals: −3 4 0 −3 2 13.2 2.2 −9.8 −9.8 4.2 −4.6 15.4 −8.6 −1.6 −0.6 SSE = 762 (e) the assumption of a common variance is not reasonable
(f) no

33. $\rho = 0$

		x = −1	x = 0	x = 1
y	0	0	.4	0
	1	.3	0	.3

35. $r^2 = .875$

39. .8758

41. (a) not normal, both skewed (b) could be approximately normal
(c) not normal since Y could have a lot of zero values

43. $r = .8023$, CI = (.23, .90)

45. (a) (−.462, .422) (b) (−.381, .690)

Chapter 10

1. (a) qualitative (b) quantitative (c) qualitative (d) quantitative
(e) qualitative

3. (a) $s_1^2 = 1.4539$, $s_2^2 = .9487$, $s_3^2 = 1.2934$ (b) $s_p^2 = 1.2142$

5. 6.054

7. $F_{2,24} = 4.25$

9. (a) $\hat{\mu} = 67.15$, $\hat{\mu}_1 = 68.56$, $\hat{\mu}_2 = 71.00$, $\hat{\mu}_3 = 61.89$
(c) 11.444, 9.444, −5.556, −3.556, . . . , −11.889, 6.111, 1.111 (d) yes

13.

Source	*df*	*SS*	*MS*	*E(MS)*	*F*
Among treatments	2	400.296	200.148	$\sigma^2 + \frac{9}{2}\sum \tau_i^2$	4.25
Within treatments	24	1129.111	47.046	σ^2	
Total	26	1529.407			

15. (a)

Source	*df*	*SS*	*MS*	*E(MS)*	*F*
Among treatments	2	2220.40	1110.20	$\sigma^2 + \frac{5}{2}\sum \tau_i^2$	10.08
Within treatments	12	1322.00	110.67	σ^2	
Total	14	3542.40			

(b)

Source	*df*	*SS*	*MS*	*E(MS)*	*F*
Among treatments	2	2083.73	1041.87	$\sigma^2 + \frac{5}{2}\sum \tau_i^2$	16.41
Within treatments	12	762.00	63.50	σ^2	
Total	14	2845.73			

17. 61.89 68.56 71.00 (all three underlined together)

19. (a)

Source	*df*	*SS*	*MS*	*E(MS)*	*F*
Factor A (frequencies)	2	39,251.06	19,625.53	$\sigma^2 + 6\sum \alpha_i^2$	5.74
Factor B (Intensity)	1	20,976.69	20,976.69	$\sigma^2 + 18\sum \beta_j^2$	6.13
Error	32	109,501.22	3,421.91	σ^2	
Total	35	169,728.97			

21. (a)

Source	*df*	*SS*	*MS*	*E(MS)*	*F*
Factor A (types of ownership)	2	882.849	441.424	$\sigma^2 + 9\sum \alpha_i^2$	11.08
Factor B (complexity of facilities)	2	287.791	143.896	$\sigma^2 + 9\sum \beta_j^2$	3.61
Interaction	4	276.142	69.036	$\sigma^2 + \frac{3}{2}\sum (\alpha\beta)_{ij}^2$	1.73
Error	45	1792.923	39.843	σ^2	
Total	53	3239.706			

23. (a)

Source	*df*	*SS*	*MS*	*E(MS)*	*F*
Treatments	2	400.296	200.148	$\sigma^2 + \frac{9}{2}\sum \tau_i^2$	10.30
Blocks	2	701.630	350.815	$\sigma^2 + \frac{9}{2}\sum \beta_j^2$	18.05
Error	22	427.481	19.431	σ^2	
Total	26	1529.407			

Chapter 11

1. reject H_0 if number of + signs ≤ 4 or ≥ 15; since $X = 13$, do not reject H_0
3. reject H_0 since $X = 11$ exceeds the critical value 10
5. Observed $z = 1.81$, which is less than 1.96, so do not reject H_0
7. $T^+ = 84.5$ exceeds the critical value 74, so reject H_0
9. $T^+ = 49$ exceeds the upper critical value 47, so reject H_0
11. observed $z = .85$, which does not exceed 1.645, so do not reject H_0
15. observed $z = -2.50$, which is less than -1.645, so reject H_0
17. observed $z = -1.51$, which is greater than -1.96, so do not reject H_0
19. $R = 8$, which is equal to the critical value, so reject H_0
21. $R = 3$ is less than the lower critical value of 4, so reject H_0
23. observed $X^2 = 5$, which does not exceed 15.1, so do not reject H_0
25. manager's claim is reasonable
27. observed $X^2 = 2.8$ exceeds 2.71, so reject H_0
29. observed $H = 3.86$ does not exceed 9.21, so do not reject H_0
31. observed $H = 10.17$ exceeds 6.25, so reject H_0
33. observed $X^2 = 5.60$ exceeds 3.84, so reject H_0
35. observed $X^2 = 18.81$, so reject H_0 even with $\alpha = .01$
37. observed $Q = 7.67$ does not exceed 11.3, so do not reject H_0 with $\alpha = .01$
39. observed $Q = 16.25$ exceeds 7.81, so reject H_0
41. $r_S = .123$
43. observed $r_S = .174$ does not exceed .657, so do not reject H_0
45.

r_S:	−1.	−.8	−.6	−.4	−.2	0	.2	.4	.6	.8	1.0
$P[R_S = r_S]$:	$\frac{1}{24}$	$\frac{3}{24}$	$\frac{1}{24}$	$\frac{4}{24}$	$\frac{2}{24}$	$\frac{2}{24}$	$\frac{2}{24}$	$\frac{4}{24}$	$\frac{1}{24}$	$\frac{3}{24}$	$\frac{1}{24}$

47. observed $X^2 = 20.96$ exceeds 9.49, so reject H_0

Chapter 12

3. (a)

p:	.10	.20	.30
Prob.:	.542	.289	.169

(b)

p:	.10	.20	.30
Prob.:	.3331	.3553	.3116

5. (a)

p:	.05	.10	.15
Prob:	.44049	.32678	.23273

(b) same as (a), except for possible rounding error

9. (b) .25 (c) .0375 (d) 0

13. (b) $\approx .04$ (c) $\approx .21$ (d) $\approx .55$

15. (a) 2, 1.3, .61 (b) 2, 1.171, .652

17. (a) $\beta(n'' = 10, s'' = 6)$ (c) .5, .5, .224 (d) .625, .6, .148

19. (a) $\beta(n'' = 20, s'' = 3)$ (b) $\beta(n'' = 110, s'' = 16)$ (c) $\beta(n'' = 20, s'' = 6)$

23. (a) $\approx (.05, .30)$ (b) $\approx (.15, .48)$

25. (a) $\approx (.46, .98)$ (b) $\approx (.56, .97)$

Chapter 13

1. (a) A_2, A_3, A_2 (b) A_4 is inadmissible

3.

		State of Nature		
		θ_1: *No Rain*	θ_2: *Light Rain*	θ_3: *Heavy Rain*
Possible Action	A_1: *No Purchase*	0	−125	−125
	A_2: *Buy Umbrella*	−10	−10	−70
	A_3: *Buy Raincoat*	−40	−40	−40

5.

	θ_1	θ_2	θ_3
A_1	300	350	200
A_2	0	400	0
A_3	500	0	450
A_4	100	500	100

7.

	θ_1	θ_2	θ_3
A_1	0	115	85
A_2	10	0	30
A_3	40	30	0

9. (a) A_3 (b) A_3 (c) A_3

11. (a) A_3 (b) A_1 (c) A_2

13. (a) 850, 390, 180 (b) 330, 790, 1000 (c) A_1 (d) A_1

15. A_2: buy umbrella

17. (a) A_3 (b) A_4 (c) A_4 (d) $9.48 (e) $.38

19. (a) A_1 (b) A_2 (c) A_3 (d) $.65; no, it would not pay

21. bake 9 loaves

23. (a) 765 (b) 810

25. (a) .15, .32, .58 (b) .48, .75, .94

27. (a) $31,667; die is better (b) .500, .467; do not roll die
(c) .0625, .212; die is better

INDEX